Bacterial Metabolism

Second Edition

Bacterial Metabolism

Second Edition

H. W. Doelle

Department of Microbiology
University of Queensland
St. Lucia, Brisbane, Australia

ACADEMIC PRESS New York San Francisco London 1975

A Subsidiary of Harcourt Brace Jovanovich, Publishers

ACADEMIC PRESS, INC.
111 Fifth Avenue, New York, New York 10003

United Kingdom Edition published by
ACADEMIC PRESS, INC. (LONDON) LTD.
24/28 Oval Road, London NW1

Library of Congress Cataloging in Publication Data

Doelle, H W
Bacterial metabolism.

Includes bibliographies and index.
1. Microbial metabolism. I. Title. [DNLM: 1. Bacteria–Metabolism. QW52 D651b]
QR88.D62 1975 589.9'01'33 74-10213
ISBN 0-12-219352-0

PRINTED IN THE UNITED STATES OF AMERICA

Contents

Preface

The tremendous progress made in some areas of bacterial metabolism has created a need for a second edition of this work. New sections have been added and the material presented in the first edition amended to include data on the regulatory mechanisms, bacterial growth kinetics as compared to enzyme kinetics, the glucose transport mechanism, and aerobic amino acid catabolism. The information presented has been rearranged to make it easier for the reader to compare the catabolic events of various bacterial groups.

Chapter 1 of the first edition appears as two in this one, the first dealing with the thermodynamics of biological reactions and the second with enzymes, coenzymes, and bacterial growth kinetics. The major change in Chapter 3 (Photosynthesis and Photometabolism) can be found in the section on photosynthetic phosphorylation in which greater emphasis has been given to its development in the different bacterial families. Data presented in Chapters 5–9 have been considerably rearranged. The role of the individual families (e.g., acetic acid bacteria, lactic acid bacteria, and Pseudomonadaceae) has been discussed in the catabolic events of anaerobic respiration (Chapter 4), aerobic respiration (Chapters 6–8), and fermentation (Chapter 9), as well as in the pathways of carbohydrate metabolism (Chapter 5). The inclusion of the regulatory mechanisms made it necessary to divide aerobic respiration into three separate chapters, one dealing with the chemolithotrophic bacteria (Chapter 6), another with chemoorganotrophic bacteria (Chapter 7), and the third with hydrocarbon metabolism (Chapter 8). Similarly, the chapter on fermentation has been subdivided so that the activities of the different bacterial families could be treated individually (e.g., propionic acid bacteria, saccharolytic clostridia, Enterobacteriaceae, lactic acid bacteria, and proteolytic clostridia).

I would like to point out that this second edition, as the first, deals only with the catabolic events of bacterial metabolism, the most important aspect in collating studies on microbial systematics and microbial biochemistry. Thus, the reader will find that all regulatory mechanisms discussed deal only with the appropriate catabolic regulation and not with aspects of biosynthetic pathways and their regulatory mechanisms. It is hoped that this edition will lead to a greater understanding of the increasing connection between microbial systematics and microbial biochemistry.

I am well aware that some publications or information not at my disposal may have been overlooked in compiling this work. As stated in the first edition, criticism, suggestions, and additional information will be greatly appreciated.

I am deeply indebted to all those who helped me in the preparation of this edition. The availability of the MEDLAR Search System at the National Library of Australia at Canberra enabled me to obtain information which would otherwise not have been available. Mr. P. Hodgson was extremely helpful in the computer search program. The critical evaluation of the organization of the text by my graduate and postgraduate students, particularly by Mr. A. Westwood, Mr. G. Gordon, and Dr. G. J. Manderson, and by Dr. I. C. MacRae and Dr. R. Boskot is very much appreciated. Thanks are also due to all those with whom I had the pleasure of discussing various aspects of the first edition during my study leave in 1970. I am grateful to publishers and authors for permission to reproduce material.

Finally, I wish to express my sincere gratitude to the staff of Academic Press for their cooperation in making publication of this edition possible.

H. W. Doelle

Preface to the First Edition

Microbial chemistry, which deals with the aspects of metabolic events occurring in microorganisms, has developed at a tremendous pace, from a mere diagnostic section of bacteriology to an independent branch that links the fields of microbiology and biochemistry. The work in this field is reflected by the large number of publications appearing in journals of biochemistry, microbiology, biological chemistry, and others.

There are a number of textbooks available on biochemistry, which deal with certain aspects of bacterial metabolism. However, since they are biochemically oriented, the microbial approach is very much neglected. In contrast to the biochemist whose main interest is directed toward the actual mechanism of reactions, and who therefore uses the microorganism more as a tool of convenience, the microbial chemist prefers to study the occurrence of biological reactions in the light of the particular microorganism and its environment, and to relate this information to biology as a whole. There are many problems which cannot be solved by the biochemist without the help of the microbiologist, and vice versa. Therefore, the aim of this book is to bring together the studies of microbial systematics and microbial chemistry with particular emphasis on catabolic events.

The importance of and the need for a work dealing specifically with bacterial metabolism are obvious, but its preparation has been a difficult task. I am well aware that certain sections may become outdated by the time the text reaches the bookstalls.

With perhaps questionable validity I have assumed that the person who will reach for this book will be a student who has had little or no training in the field of microbial chemistry, an advanced student or a teacher who would like additional information to enhance his present knowledge, or a research worker who will find himself confronted with problems in this field of study. This book should prove helpful to all of them.

The work begins with a chapter on thermodynamics and enzymes. Both of these topics are frequently neglected in metabolic studies, and yet they are particularly important, for example, in the evaluation of chemosynthetic pathways. With regard to enzymes, emphasis is given to the Commission Report on Enzyme Nomenclature of the International Union of Biochemists. I am very much in favor of a standardized system for the presentation of results. Where possible, the corresponding enzyme numbers are given; where this is not possible, suggestions are made as to which one of the known enzymes in the report could be responsible for the particular reaction.

The lack of uniform terminology is a problem in microbial chemistry. The second chapter of this book begins, therefore, with definitions of a number of terms used by the majority of research workers and which will be used throughout this work. In order to keep this list short, only the important terms are explained and defined. The second chapter of the book deals with the development of photosynthesis. As in all other chapters, the material is presented very generally at first and is then elaborated upon to include details and problems related to the various disciplines. A comprehensive general scheme will always be found at the beginning of a chapter, followed by a detailed treatment of the scheme. Metabolic events have been considered in the particular bacterial groups. It was the aim to stress not only the biochemical but also the biological problems facing the microbial chemist today.

Bacterial metabolism research includes studies of sources and storage of substances used for synthesis. These processes are intimately related to each other in the completion of overall metabolic events, but, for purposes of convenience, the main events leading to the production and storage of energy and the production of reducing substances have been dealt with under the titles of anaerobic respiration, aerobic respiration, and fermentation in Chapters 3, 5, and 6.

The metabolism of carbohydrates occurs through four major pathways all of which lead to pyruvate, and the principal energy-yielding reactions lead to the subsequent attack of this substrate. For this reason the pathways of carbohydrate catabolism have been treated separately (in Chapter 4) from that of pyruvate, which is dealt with under fermentation and respiration.

In the microbial world, a few organisms exist which may, under specific circumstances, behave quite differently from the general schemes discussed in Chapters 2–6. These organisms are the acetic acid bacteria, the lactic acid bacteria, and the pseudomonads, which are discussed in separate chapters in which only material supplementary to the general schemes has been added.

Bacterial metabolism, however, does not only comprise the carbon compounds but also, in some instances, nitrogenous compounds as energy sources. The last chapter, therefore, is devoted to the main events in this field.

In compiling this work, I am well aware that I may have overlooked publications or information not at my disposal. Criticism, information, or suggestions will, therefore, be very much appreciated.

I am deeply indebted to all those who helped me in the preparation of this book. The critical evaluation of the organization by my graduate and postgraduate students, in particular by Mr. G. J. Manderson, and of the text by Professor V. B. D. Skerman and Dr. I. C. MacRae is very much appreciated. Professor V. B. D. Skerman also gave valuable assistance not only in editorial but also in taxonomic problems. I would like to thank publishers and authors for permission to reproduce material, and all those who gave additional prepublished information. The valuable help of Mr. G. Heys in collecting the material, of Miss Barbara Heron in typing and correcting the manuscript, and of the photography section of the university is gratefully acknowledged.

Finally, I wish to express my sincere gratitude to Academic Press, who made this publication possible.

H. W. DOELLE

October, 1968

1

Thermodynamics of Biological Reactions

Concepts of Thermodynamics

To synthesize the major chemical constituents of a living cell, microorganisms require a source of energy. Under normal culture conditions, namely, an aqueous environment with an almost neutral pH and relatively low temperatures, only energy-yielding (exergonic) reactions ($-\Delta F$) can take place spontaneously. Synthetic reactions are endergonic. To synthesize new cell material, the individual chemical reactions involved must be linked or coupled with energy-yielding reactions. The study of these energy transactions and transfer mechanisms is based on thermodynamics, and it is possible to explain and understand these transactions only in terms of thermodynamics.

The field of thermodynamics often appears rather abstract to many students because its principles are usually developed within idealized systems. However, the working philosophy and the approach taken by thermodynamicists are really simple and logical and do not require special knowledge in chemistry or molecular theory. Apart from this, it is only necessary to deal with a small part of thermodynamic formalism in order to examine the nature of biological energy transfer. Thermodynamics is the most basic and exact section of physics that deals with masses of matter. It does not deal with the characteristics of atoms or molecules, but with the macroscopic characters of matter, such as volume, pressure, temperature, and density.

In thermodynamics, as in all other disciplines, there exist a few terms that are of importance for its understanding. As is common in physical sciences, some region is selected for special attention. These regions represent a collection of matter in which one wants to study energy changes during some physical or chemical process. This specifically selected region of study is called a "system." All other matter or regions apart from the one studied is called the "surrounding." Anything may be chosen as a thermodynamical system, and the partition from the surrounding may be purely conceptional or even a physical barrier.

A system in general is described in terms of a number of characteristics. These may be a certain pressure, volume, temperature, density, chemical composition, etc. For practical purposes, the thermodynamicist must limit the number of parameters for the description of a system in such manner that a simple mathematical and experimental treatment is possible. After the system has been described, it is necessary to specify the total energy content of the system before the process takes place, which is called the "initial state," and then again after the process has taken place, which is called its "final state." As the system proceeds from its initial to its final state it may either absorb energy from or deliver energy to its surroundings. The difference in the energy content of the system during its process from initial to final state must, of course, be counterbalanced by a corresponding and inverse change in the energy content of the surroundings. In order to describe the total energy content of any isolated system, in either the initial or the final state, the relevant macroscopic properties of the system, such as pressure, volume, temperature, composition, and heat content, must be specified by means of an equation called the "equation of state," which requires that the system be in equilibrium. In thermodynamics only the initial and final states are important in obtaining the energy differences between both states; the pathway taken by the process is irrelevant.

When the thermodynamicist studies energy changes during a process, he tries to reduce the number of variables to as few as possible. If, for example, the volume, pressure, or temperature of a system is held constant, it becomes easier to specify exactly the changes in energy content in going from the initial to the final state. This is the case in biological reactions, which normally take place in dilute aqueous solution with not just one but three of the most important variables—temperature, volume, and pressure—all constant throughout the system. We can therefore leave aside the large part of thermodynamics that deals with gases, solids, and pressure–volume changes.

On a number of occasions above, the terms "energy content" and "energy change" have been used. What is energy? Energy could be regarded as the capacity for doing work. The term "work," again, is usually used with respect to the manifestation of energy in its physical or chemical form. Energy can also exist in various other forms, such as electrical or thermal energy. This type of energy may be measured by conversion into work or into heat energy, which can be measured calorimetrically.

First Law of Thermodynamics

All events in the physical world conform to and are determined by the laws of thermodynamics (6). The first law states that "the total amount of energy in nature is constant" and is essentially the law of conservation of energy. The most familiar form of energy is heat. Virtually every physical or chemical process either absorbs heat from or delivers heat to its surroundings. When a process occurs with loss of heat, it is called "exothermic," and when it occurs with absorption of heat, it is called "endothermic." If the process is a physiological one, e.g., connected with an energy-producing or energy-requiring reaction, the terms "exergonic" and "endergonic" are used, respectively. Such transfer of heat or energy can take place only when there is a difference in temperature between the systems and the surroundings. Heat can only flow from a warmer to a cooler body.

If we take now an isolated system of a given energy content and give the symbol q to heat, the principle of conservation of energy says that the heat added to the system must appear as a "change in the internal energy" (ΔE) of the system, or in the amount of total work w done by the system on the surroundings. E or w may be zero.

$$q = \Delta E + w \qquad \text{or} \qquad \Delta E = q - w$$

The absolute value of E, which is called the "internal" or "intrinsic" energy, cannot be determined. However, we are concerned only with changes in E as they occur when a chemical reaction takes place. If the change had given rise to a decrease in the value considered, the delta sign would have been preceded by a minus. It is also important to realize that ΔE is independent of the pathway by which the change is carried out, whereas q and w are dependent on this pathway.

In many instances, the addition of energy as heat to the system results in a change of volume, the pressure remaining constant, which automatically implies the performance of work against the surroundings. Therefore

$$\Delta E = q - P\,\Delta V - w'$$

where P is the constant pressure, ΔV is the change in volume, and w' is useful work. As $P\,\Delta V$ is rarely a useful form of work, it has been found convenient to combine it with ΔE, so that

$$\Delta E + P\,\Delta V = q - w'$$

As biochemical reactions usually occur at constant pressure, that of the atmosphere, and not at constant volume, the heat change ($\Delta E + P\,\Delta V$) has been conveniently supplemented with another quantity, ΔH, the change of the heat content or "enthalpy change."

$$\begin{aligned}\Delta H &= q - w' \\ &= \Delta E + P\,\Delta V\end{aligned}$$

At constant pressure, then, ΔH is the heat absorbed in the reaction. If the reaction occurs not only under constant pressure but also at a constant volume, which means no work is done, then

$$\Delta H = \Delta E$$

The enthalpy change ΔH of a reaction is easily measured by direct calorimetry. However, because the bomb calorimeter is a constant-volume device, the heats of reactions so measured give changes in internal energy. These values of ΔE have then to be converted to ΔH, which can readily be done by the above formula, considering that at any temperature T,

$$PV = nRT$$

where n represents the number of moles and R is the gas constant (1.987 cal/mole degree). The equation to be used is then

$$\Delta H = \Delta E + nRT$$

Each chemical reaction proceeds to completion with a definite heat of reaction that is quantitatively related to the number of molecules reacting.

The heat given off in the combustion of an organic compound, such as glucose, will raise the temperature of the surrounding water in the jacket of the calorimeter to a degree that depends on the amount of water in the jacket and the number of molecules of glucose combusted. From such measurements, one can determine the molar enthalpy as

$$\Delta H = -673{,}000 \text{ cal/mole}$$

A calorie is the amount of energy required to raise the temperature of 1 gm of water from 14.5°C to 15.5°C and the negative sign indicates an exergonic reaction. This heat is produced because complex organic molecules have a large potential energy of configuration and are degraded to simple, stable products, such as CO_2 and water, which have a much lower energy content. From elementary physics one knows that there exists an equivalence between heat and work, the so-called mechanical equivalence of heat. This equivalence suggests that heat and mechanical work are easily interconvertible, which, of course, is not the case. Mechanical work can be completely converted into heat, but the reverse process in practice is never complete. In biological systems, heat is not a useful way of transferring energy because the different components of living organisms are essentially isothermal. If there is no temperature differential at all, heat cannot be converted into work under any circumstances.

Spontaneous physical or chemical changes have a direction that cannot be explained by the first law of thermodynamics. This law also does not provide a criterion as to whether and when such transformation occurs spontaneously. All systems tend toward the equilibrium state, in which all measurable parameters of the system become uniform throughout. Once this equilibrium state is approached, they no longer change back spontaneously to the nonuniform state. There exist, however, spontaneous transformations that occur despite the fact that the internal energy of the system at the final state is essentially the same as in the initial state. If a metal ion and a protein molecule are permitted to come in contact with each other, a complex will be formed spontaneously, although ΔE is +3 kcal (9). It is therefore possible that reactions for which ΔE is negative, zero, or even positive can occur spontaneously. Therefore, ΔE is no clear criterion for a transformation to occur spontaneously.

Second Law of Thermodynamics

As the first law of thermodynamics cannot explain the spontaneous transformation fully, we shall consult the second law, which states that "the total amount of entropy in nature is increasing." What is entropy? In order to understand this term (9) one must realize that total work is determined by two factors, an intensity factor and a capacity factor. In the case of chemical energy, the intensity factor would be mass and the capacity factor the chemical potential. A convenient unit for chemical energy has been found in the gram molecule, or mole, which naturally varies with the chemical compound. In general, chemical reactions continue until a minimal chemical potential has been reached. Therefore, the units of quantity are transferred spontaneously from a high to a low chemical potential, which is the equilibrium state. In this case, the capacity factor determines the direction of the chemical change.

In most chemical reactions, chemical energy is converted to thermal energy, because such factors as friction make the transformation inefficient.

This thermal energy also consists of an intensity factor, which is temperature, and a capacity factor, which is the entropy. This capacity factor can be described as being q/T. If the suggestion is accepted, that heat flows spontaneously from a hotter object to a cooler one the following equation should be characteristic for any transformation that can occur spontaneously.

$$(q/T)_{\text{system}} + (q/T)_{\text{surrounding}} = \text{a positive number}$$

or

$$\sum (q/T) > 0$$

If $\sum (q/T) = 0$, the system is at equilibrium.

Entropy is expressed in calories per mole degree and is represented by the symbol S. At any given temperature, solids have relatively low entropy, liquids an intermediate amount, and gases the highest entropy. The entropy S is that fraction of the enthalpy H that may not be utilized for the performance of useful work, because in most cases it has increased the random motions of the molecules in the system. The product $T \times S$, in which T is the absolute temperature, represents energy that is wasted in the form of random molecular motions. In terms of S, the second law states that, given the opportunity, any system will undergo spontaneous change in the direction that results in an increase in entropy. Equilibrium is attained when entropy has reached a maximum and no further change can occur spontaneously unless additional energy is supplied from outside the system.

The second law of thermodynamics therefore (*1*) defines entropy as a randomized state of energy that is unavailable to do work; (*2*) states that all physical and chemical processes proceed in such a way that the entropy of the system becomes the maximum possible, at which point there is an equilibrium; and (*3*) states that entropy or randomness in the universe always increases in irreversible processes. Under no circumstances does the entropy of the universe ever decrease.

Third Law of Thermodynamics

The third law of thermodynamics could simply be viewed as a defining statement for the arbitrary standard state of entropy (3). It states that "the entropy of all pure substances at absolute zero is zero." In fact, T in the earlier equations represents the absolute temperature. The definition of this absolute temperature arises from thermodynamic arguments. An ideal gas obeys the equation of state $PV = RT$, where T is temperature on a scale the zero of which coincides with -273.16°C, most commonly used as $-273°C$. It is said that at this temperature an ideal gas, by extrapolation of its behavior at higher temperature, would occupy zero volume.

Theoretically, it is possible that the entropy of a system may remain constant during a process; if so, the process is reversible. Real processes are nearly always irreversible because they are usually accompanied by friction, which always leads to an increase in entropy.

Because heat represents the kinetic energy of random molecular motion, the addition of heat increases the entropy. If the system is at equilibrium

$$q = T \Delta S$$

If the system is not at equilibrium, however, changes in the system may spontaneously increase the entropy even without addition of heat

$$T \Delta S > q$$

If the equations for the first and the second law for a system at equilibrium are combined, it is found that

$$\Delta H = T \Delta S - w'$$

The main interests in biological reactions, however, are in reactions proceeding in the direction that approaches equilibrium and at a single temperature. For such systems

$$\Delta H < T \Delta S - w'$$

The tendency to seek the position of maximum entropy is the driving force of all processes, and heat is either given up or absorbed by the system from the surroundings to allow the system plus surroundings to reach the state of maximum entropy. These changes of heat and entropy are related by a third dimension of energy: free energy.

Free Energy

The free energy function was created by Gibbs and Helmholtz, who fused the first and second law of thermodynamics. This free energy of a system is denoted by the symbol F, although most textbooks of physical chemistry use the symbol G, after Williard Gibbs (6, 14). It represents the component of the total energy of a system that is able to do work under isothermal conditions. It is therefore also a property of the system, which means it depends only on the state of the system and not on the path by which that state was reached. When the system undergoes a change of state, which takes place in most chemical reactions, a change in free energy occurs. It is in these changes in free energy rather than the absolute free energies of reactants and products that we are interested. Such a change is denoted by ΔF. Free energy is thus "useful" energy, whereas entropy is "degraded" energy. In general, the change in free energy, ΔF, is the energy that becomes available to be utilized for the accomplishment of work as a system proceeds toward equilibrium. Thus, if a system is changed from one state to another at the same temperature, the change in free energy associated with this change is

$$\Delta F = \Delta H - T\,\Delta S$$

When this reversible system is at equilibrium, the free energy change is zero, $\Delta F = 0$, and therefore

$$\Delta H = T\,\Delta S$$

The entropy of the system and its surrounding is at a maximum and the system is in its most probable state. These are the forms in which the laws of thermodynamics are most readily expressed for the description of biochemical systems. For such a system that is not at equilibrium

$$\Delta F = -w', \qquad \text{since} \qquad \Delta H = T\,\Delta S - w'$$

Therefore, systems not at equilibrium proceed spontaneously only in the direction of negative free energy changes and nearly to completion. Those systems characterized by a positive ΔF occur to only a very limited extent or may proceed backwards in the presence of products, provided no external energy driving force is available. The magnitude of ΔF for a chemical reaction therefore indicates only the extent to which it proceeds and the amount of energy available from it. It does not necessarily provide information concerning the rate of the reaction. As ΔF is a measure of a system's capacity for doing work, it is what determines whether or not a reaction may occur spontaneously. Spontaneous reactions can occur only if there is a decrease in free energy, in which case the reactions are termed "exergonic." If there is an increase in free energy, then work or energy must be put into the system to bring the change about. This type of reaction is termed "endergonic." As the increase in free energy is denoted by a negative sign, one would expect that if the free energy change of a reaction is -6000 calories, the reaction may occur spontaneously with a decrease in ΔF of 6000 calories. It does not follow that because a given reaction has a high negative value for ΔF the reaction takes place at a measurable rate. ΔF measures only the difference in free energy between the initial and final states of the reaction. To bring about an appreciable reaction, a catalyst may be necessary.

Thermodynamics of Chemical Reactions

Biological systems are mostly systems that need such catalysis in order to make reactions proceed at a measurable rate (see Chapter 2). Indeed, a reaction involving an increase of free energy can only proceed if it is coupled to a suitable source of free energy. Chemical reactions are most commonly exothermic. The heat evolved during an exothermic process, however, does not correspond to the change in free energy. In an ordinary chemical reaction only ΔH can be measured directly. The determination of free energy, ΔF, depends on accurate measurements of the electromotive forces (emf; see p. 14) or the equilibrium constant K of a reversible reaction.

The equilibrium constant K of a reversible reaction is the product of the active masses of the reaction products divided by the products of the active masses of reactants at equilibrium. For the reaction

$$\mathrm{A} + \mathrm{B} \rightleftharpoons \mathrm{C} + \mathrm{D}$$

the constant K can be calculated as follows

$$K = \frac{[\mathrm{C}] \times [\mathrm{D}]}{[\mathrm{A}] \times [\mathrm{B}]}$$

The more vigorously the reaction between substances A and B proceeds, the greater the proportion of reaction products in the equilibrium mixture and the greater the equilibrium constant K. It can then be said that the reaction mixture possesses a high potential energy. During the course of the reaction, this potential energy is reduced. Because this energy change is related quantitatively to the equilibrium constant, the constant K must be a mathematical function of the free energy change of the components of the reaction. This function can be expressed as

$$\Delta F^\circ = -RT \ln K$$

where R is the gas constant (1.987 cal/mole degree), T is the absolute temperature, and $\ln K$ is the natural logarithm of the equilibrium constant. ΔF° represents the change of free energy under standard conditions, e.g., the gain or loss on free energy in calories as 1 mole of the reactant is converted to 1 mole of product. The reactant must therefore be in a 1-molar concentration. This standard state is a convenient reference condition in which the activities are arbitrarily defined as unity for pure liquids or solids, gases at 1 atm, and compounds in solution at 1 M concentration at a given temperature, usually 25°C. Therefore, ΔF° is a constant for any given reaction and its values are additive. The standard free energy, ΔF°, however, must not be confused with the free energy, ΔF. Both are related for the considered reaction by the equation

$$\Delta F = \Delta F^\circ + RT \ln \frac{[C] \times [D]}{[A] \times [B]}$$

It is always ΔF and not ΔF° that determines whether or not the reaction may occur spontaneously. ΔF° is the value always tabulated for a given reaction because it is a defined quantity, whereas ΔF can have any value, depending on the conditions implied. Because most biological reactions occur at or near pH 7.0, the symbol $\Delta F'$ is used to indicate the standard free energy change at pH 7.0. This becomes important only, if H^+ is a reactant in the system. A chemical reaction, then, is thermodynamically possible if it is exothermic, i.e., is attended by a decrease of free energy.

Whether or not it actually takes place, however, depends on other factors. A body will not slide down a rough plane if the "frictional forces" are greater than the forces exerted by its free gravitational potential energy. Similarly, a chemical reaction can take place if it entails a fall of free energy but will not actually do so if the frictional forces tending to oppose it are too large. A chemical reaction requires for its accomplishment that the molecules shall be in a reactive state. Only those molecules with a high energy content are likely to react to form the product. In order for the reaction to proceed, the energy content of the entire population of molecules must be raised so that the activation energy barrier is overcome. Then the reaction proceeds quickly to its equilibrium. One way of doing this is to heat the mixture. In biological systems, however, enzymes, in their function as catalysts, take care of this requirement. The catalyst or enzyme lowers the activation energy of the reaction by allowing a much larger fraction of the molecular population to react at any one time. The catalyst can do this because it can form an unstable intermediate complex or compound with the substrate, which quickly decomposes to the product and thus provides a channel through the barrier that represents the activation energy. Once this channel of low activation energy is found, the reaction proceeds rapidly. Enzymes merely accelerate the approach to equilibrium but do not influence the equilibrium point attained. The free energy change is the same whether the reaction is catalyzed by the enzyme or occurs slowly by itself. Time and rate do not enter into the calculation, only the final state does. It follows from this that catalysts can only initiate and/or accelerate a reaction that is thermodynamically possible.

As an example of a practical calculation of the equilibrium constant and the free energy change of a chemical reaction (14), consider the following.

$$\text{Glucose 1-phosphate} \rightleftharpoons \text{glucose 6-phosphate}$$

is catalyzed by the enzyme phosphoglucomutase (α-D-glucose 1,6-diphosphate:α-D-glucose-1-phosphate phosphotransferase, EC 2.7.5.1). If the reaction is started by adding the enzyme to a 0.02 *M* glucose 1-phosphate solution at 25°C and at pH 7.0, it is found by chemical analysis that the reaction proceeds to an equilibrium at which the final concentration of glucose 6-phosphate has risen from zero to 0.019 *M*. If it is assumed that these measured concentrations are equal to the thermodynamically active masses, then

$$K = \frac{[\text{glucose 6-phosphate}]}{[\text{glucose 1-phosphate}]} = \frac{0.019}{0.001} = 19$$

Having the K value, one can calculate the standard free energy change of the above reaction.

$$\begin{aligned} \Delta F' &= -RT \ln K \\ &= -1.987 \times 298 \times \ln 19 \\ &= -1.987 \times 298 \times 2.303 \log_{10} 19 \\ &= -1745 \text{ cal/mole} \end{aligned}$$

In other words, there is a decline in free energy of 1745 cal when 1.0 mole of glucose 1-phosphate is converted to 1.0 mole of glucose 6-phosphate at 25°C.

Chemical reactions in living organisms, however, are in characteristically organized sequences called "metabolic pathways." These sequences must be dealt with as a whole, i.e., as the summation of the individual free energy steps

$$\text{A} + \text{B} \overset{-6000}{\rightleftharpoons} \text{C} + \text{D}$$

$$\text{D} + \text{E} \overset{-4000}{\rightleftharpoons} \text{G} + \text{H}$$

$$\text{H} + \text{I} \overset{+3000}{\rightleftharpoons} \text{J} + \text{K}$$

The theoretical feasibility of the overall reaction, from A + B to J + K, can be assessed by algebraic summation of the individual free energy (ΔF) changes (e.g., $-6000 - 4000 + 3000 = -7000$ cal/mole of A). The whole sequence will therefore tend to move spontaneously from left to right. This example shows quite clearly why the microbial chemist should be familiar with the free energy changes of reactions. They can provide him with a rough check on the accuracy of his ideas about metabolic pathways. No system of coupled reactions will proceed spontaneously unless $\Delta F'$ is negative for the overall system. What happens, now, to the free energy, 7000 cal/mole of A, that is left? In a normal chemical reaction it is evolved as heat and is lost from the system to its surroundings. In biological systems, however, this need not be the case, as exergonic processes may be coupled or connected to endergonic processes so that the former delivers energy to the latter. In such coupled systems, the endergonic process will take place only if the decline in free energy of the exergonic process to which it is coupled is larger than the gain in free energy of the endergonic process. The algebraic sum of these processes must always be negative if the reaction sequences are to proceed.

The sequential steps in the oxidation of an organic substrate are aimed at the gradual liberation of energy that is either used directly in an endergonic reaction or stored for subsequent release at a later stage of the pathway.

The most important mechanism of storage is the formation of an energy-rich intermediate. This compound has the role of conserving the free energy not as heat but as chemical energy. Several types of such compounds that occur in microorganisms are (*a*) derivatives of phosphoric acid, adenosine triphosphate (ATP), uridine triphosphate (UTP), acyl phosphates, and inorganic polyphosphates, and (*b*) derivatives of carboxylic acids, such as acetyl coenzyme A. Of these energy carriers, the most important is ATP. This is the carrier of chemical energy from the energy-yielding oxidation to those processes or reactions of the cell that cannot occur spontaneously and can proceed only if chemical energy is supplied. Adenosine triphosphate (ATP) is formed from adenosine diphosphate (ADP) in coupled reactions. The many chemical steps in the charging and discharging of the ATP system in the cell are catalyzed by enzyme systems. The few examples given indicate that the concept of free energy plays a most important role in biological systems.

Oxidation–Reduction Reactions

Energy-yielding reactions within organisms are of the nature of oxidations. "Oxidation" may be defined, in general, as the loss of electrons and "reduction" as the gain of electrons. The oxidation of molecular hydrogen can therefore be expressed as

$$H_2 - 2\,e^- \rightleftharpoons 2\,H^+$$

The electrons from this oxidation must be accepted by an oxidizing agent. For example, if a ferric salt is used the equation becomes

$$\begin{array}{r}H_2 - 2\,e^- \rightleftharpoons 2\,H^+ \\ 2\,Fe^{2+} + 2\,e^- \rightleftharpoons 2\,Fe^{2+} \\ \hline H_2 + 2\,Fe^{3+} \rightleftharpoons 2\,H^+ + 2\,Fe^{2+}\end{array}$$

Molecular oxygen can act as an oxidizing agent, in a similar manner, picking up either two or four electrons

$$O_2 + 2\,e^- \rightleftharpoons O_2^{2-} \rightleftharpoons H_2O_2$$

$$O_2 + 4\,e^- \rightleftharpoons 2\,O^{2-} \rightleftharpoons 2\,H_2O$$

$$\Delta F = -57 \text{ kcal/mole } H_2O$$

According to modern theory, an electric current is essentially a transfer of electrons (17), as the electron donor possesses a characteristic electron pressure and the corresponding electron acceptor, a characteristic electron affinity. It should therefore be possible to obtain direct proof of the transfer of electricity in oxidation–reduction reactions under suitable experimental conditions. This transfer of electricity could be a quantitative measure of the tendency of substances to donate or accept electrons and thus a means for calculating free energy changes for oxidation–reduction reactions. This quantitative measure is termed an "oxidation–reduction potential."

When a pure zinc rod is immersed in distilled water, some of the zinc ionizes.

$$Zn \rightleftharpoons Zn^{2+} + 2\,e^-$$

whereby Zn^{2+} passes into the solution and the freed electrons accumulate on the metal. The electrostatic forces, however, are responsible for the attraction of Zn^{2+} to the negatively charged electrons, which arranges the ions around the electrode in the form of a boundary layer. Between the electrode and this boundary layer of ions exists, therefore, a certain difference of potential. If the distilled water is now replaced by a normal solution of a soluble zinc salt, such as zinc sulfate, the reaction is driven to the left and a new equilibrium established between the electrode and

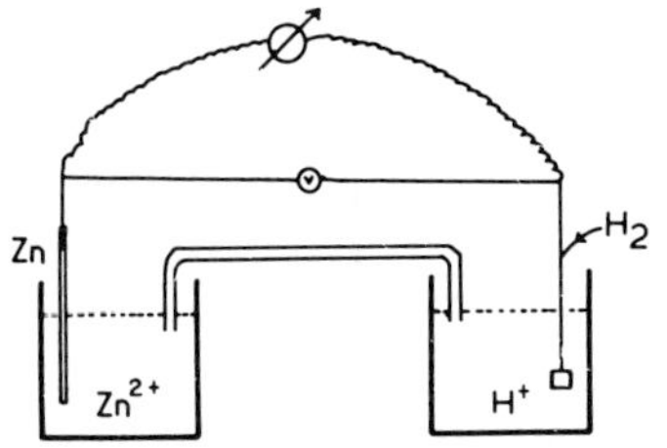
Zn
Zn^{2+}
H_2
H^+

Fig. 1.1. An electrolytic cell consisting of a Zn and a H_2 half-cell.

the solution. The potential difference that now exists between the zinc electrode and the normal solution of zinc ions is termed the "standard electrode potential" of zinc.

In a similar manner, if hydrogen gas at normal atmospheric pressure is adsorbed onto the surface of finely divided platinum and immersed in a normal solution of hydrochloric acid, a standard hydrogen electrode is produced.

It is not possible to determine the absolute potential difference of either of these electrodes because they represent only half-cell reactions. When two half-cells are coupled, the electromotive force (emf) of the cell is the algebraic difference of the potentials of the two half-cells, with the sign removed. By convention (2), the potential of the standard hydrogen electrode is equal to zero at all temperatures. Consequently, the potential of any other electrode system can be determined with reference to the normal hydrogen electrode.

If hydrogen and zinc electrodes are connected as in Fig. 1.1, the voltmeter will register the emf in volts, and the null point ammeter will indicate the direction of current flow (the reverse of the direction of electron flow). As the hydrogen potential is arbitrarily set at zero, the voltmeter measures, equally arbitrarily, the potential difference between the zinc and its ions. The sign that is placed before this potential difference is determined by the direction of the electron flow. Electrons will only flow from a place of low potential to one of high potential (from one potential to a more positive potential). Because the current flows from the hydrogen electrode to the zinc electrode, the electron flow is from the zinc to the hydrogen electrode. The potential difference between the zinc and its ions is more negative than that of the hydrogen and its ions and therefore carries a negative sign. In this case, therefore

$$E_h(\mathrm{Zn} \rightleftharpoons \mathrm{Zn}^{2+}) = -0.77\ \mathrm{V}$$

where E_h is the electrode potential with reference to the hydrogen electrode.

Although it is correct to refer to the potential difference between a metal and its ions, the convention assigns the potential to the metal. One therefore refers to the "potential of the zinc electrode." This potential, however, varies with the concentration of the ions in solution. When the activity of the ion Zn^{2+} is equal to unity (approximately true for 1 *M* solutions), the electrode potential, E_h, is equal to the standard potential, E_0.

In E_h the subscript *h* indicates a comparison with the standard hydrogen electrode. A standard hydrogen electrode, however, is very difficult to manipulate. It is therefore common to determine electrode potentials on the hydrogen scale indirectly by measuring the emf of a cell formed from the electrode in question and a convenient reference electrode the potential of which with respect to the hydrogen electrode is accurately known. The reference electrodes generally used are the calomel electrode and the silver–silver chloride electrode. If the saturated calomel electrode (sce) with an $E_h = +0.246$ were substituted as a reference electrode, the emf between the zinc electrode and this reference electrode would be

$$-0.77 - (+0.246) = 1.046 \text{ V}$$

and would be expressed by

$$E_{\text{sce}}\ (\text{Zn} \rightleftharpoons \text{Zn}^{2+}) = -1.046 \text{ V}$$

Because the ionization of zinc results in a loss of electrons, Zn^{2+} may thus be regarded as the oxidized form of zinc and the metal as the reduced form of zinc. The potential difference *E* between the zinc and its ions is expressed by the Nernst equation

$$E_h = E_0 + \frac{RT}{nF} \ln \frac{(A_{\text{ox}})}{(A_{\text{red}})}$$

where E_0 is the standard electrode potential, *R* is the gas constant equal to 8.314 J/degree/mole, *T* is the absolute temperature, *n* represents the valency of the ions involved in the reaction, *F* is the Faraday constant (equal to 96,494 C), which is necessary to convert one equivalent of an element to an equivalent of ions, and (A_{ox}) and (A_{red}) are the activities of the oxidized and reduced forms of the oxidation–reduction system.

Where a metal ion exists in two oxidation states (e.g., ferric and ferrous ions) in solution, a potential difference exists between the two. If an inert electrode, such as platinum, is immersed in an oxygen-free solution of ferric and ferrous ions, electrons will accumulate on the metal. If the platinum electrode is coupled to a hydrogen electrode, electrons will flow from the hydrogen half-cell to the iron half-cell. This direction of electron flow indicates that the potential of the former is lower than that of the latter. Similarly, electrons would flow from a Zn/Zn^{2+} electrode to an Fe^{2+}/Fe^{3+} electrode.

In biological oxidation–reduction systems the interactions are governed by the same laws.

At 30°C, which is the temperature frequently employed for electrode measurements, the factor 2.303 RT/nF (converting natural logarithm to the base 10) has a value of 0.06 for $n = 1$ and 0.03 for $n = 2$. Thus, for $n = 1$

$$E_h = E_0 + 0.06 \log \frac{(A_{\text{ox}})}{(A_{\text{red}})}$$

When the activity of the oxidant and reductant are the same, the expression

$$+0.06 \log \frac{(A_{\text{ox}})}{(A_{\text{red}})} = 0$$

and

$$E_h = E_0$$

The standard electrode potential is therefore the potential of an electrode in equilibrium with a unity activity of its ions. In the case of an oxidation–reduction system involving a solution of two ions at different oxidation states, E_0 will be the potential at which the ratio of the activities of the two is one. This value is characteristic for each oxidation–reduction system and gives a measure of the relative ability of that system to accept or donate electrons in oxidation–reduction reactions; i.e., it has nothing to do with their concentration except insofar as concentration affects the activity. The redox potential is therefore a measure of oxidizing (or reducing) intensity and not capacity, in the same way that pH is a measure of the acidity or alkalinity of a system but not its buffering power. The quantity of electrons that can be transferred depends on the concentration of the components of the redox system.

If the variations of E_0 for Fe^{2+}/Fe^{3+} systems are examined as a function of pH of the solution, it will be found that E_0 is the same over a wide range. Such constancy, however, is not found in systems in which hydrogen ions enter into the overall chemical reaction where the value varies with the pH.

Direct measurements of standard potentials are difficult to obtain because of the uncertainty of the exact concentration of the substances involved or the slowness of the establishment of equilibrium with the inert metal of the electrode (5).

MEASUREMENTS OF THE STANDARD POTENTIALS WHEN THE EXACT CONCENTRATIONS OF THE SUBSTANCES INVOLVED ARE UNCERTAIN. The pure oxidized form (100% oxidized and 0% reduced) of the system (e.g., quinone) is first dissolved in a solution of definite hydrogen ion concentration, such as a buffer solution. Known amounts of a reducing solution are then added in the absence of oxygen, and the solution is kept agitated by means of a nitrogen stream. The potential of an inert electrode (e.g., platinum or gold) immersed in the reacting solution is measured after each addition of the titrant. The point at which the potential undergoes a rapid change

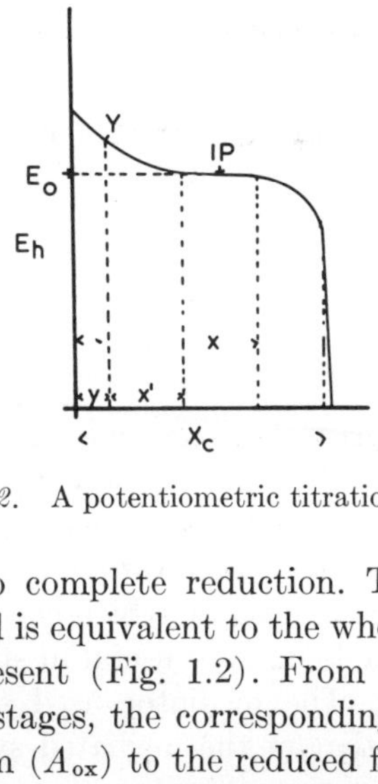
Y
IP
E_0
E_h
x
x'
X_c

Fig. 1.2. A potentiometric titration curve.

is that corresponding to complete reduction. The quantity of reducing solution (X_c) then added is equivalent to the whole of the oxidized organic compound originally present (Fig. 1.2). From the amounts of reducing agent added at various stages, the corresponding ratios of the concentrations of the oxidized form (A_{ox}) to the reduced form (A_{red}) may be calculated. From these, the standard potential is derived for several values:

$$E_h = E_0 - \frac{RT}{nF} \ln \frac{(A_{ox})}{(A_{red})}$$

$$= E_0 - \frac{RT}{nF} \ln \frac{X_c - X}{X}$$

When a system is partially reduced, the extent of reduction can be determined without any knowledge of the concentration of the reducing agent employed for titration, provided the E_0 of the system is known.

The initial point in the titration (Fig. 1.2) may be taken as point Y on the oxidation–reduction curve, equivalent to y milliliters of reductant. The curve is plotted commencing at Y with progressive additions of reductant. After the addition of X' milliliters of reductant

$$E_h = E_0 - \frac{RT}{nF} \ln \frac{(A_{\text{ox}})}{(A_{\text{red}})}$$

$$= E_0 - \frac{RT}{nF} \ln \frac{[X_c - (X' + y)]}{[X' + y]}$$

where the values of E_h, E_0, $(X_c - y)$, and X' are known and y can be derived.

When the value of E_0 is not known for the system under investigation, and when the initial state of reduction is not too great, the value of E_0 may be read from the inflection point (IP). From the value of X' at IP the value $M[= X_c - (X' + y)]$ represents 50% reduction and

$$\frac{M - X'}{2M} \times \frac{100}{1}$$

gives the percentage initial reduction.

DIRECT MEASUREMENTS OF THE STANDARD POTENTIALS IN RELATION TO THE SLOWNESS OF THE ESTABLISHMENT OF EQUILIBRIUM. The slower the equilibrium is obtainable, the greater may be the error in measuring potential. This sluggishness of a system can be altered by accelerating the process with an electromotively active system, which is called a "potential mediator." These mediators have been very useful in studying systems that are quite inactive themselves. The mediator used should have a midpoint potential approximate to that of the mediated system. When active and inactive systems come to an equilibrium, the two systems should have the same potentials for the conditions used. It is therefore essential that the mediator system be well chosen and be not converted by the system too far toward 100% oxidation or reduction. The mediator should itself not be of sufficient concentration to materially alter the state of oxidation of the system being measured. Oxidation–reduction indicators have been used frequently as mediators because their color changes are helpful in rapidly selecting the mediator and their potentials are fairly well established.

A number of curves obtained by potentiometric titration are presented in Fig. 1.3. The position of the curve on the oxidation–reduction scale depends on the standard potential of the system, which corresponds to

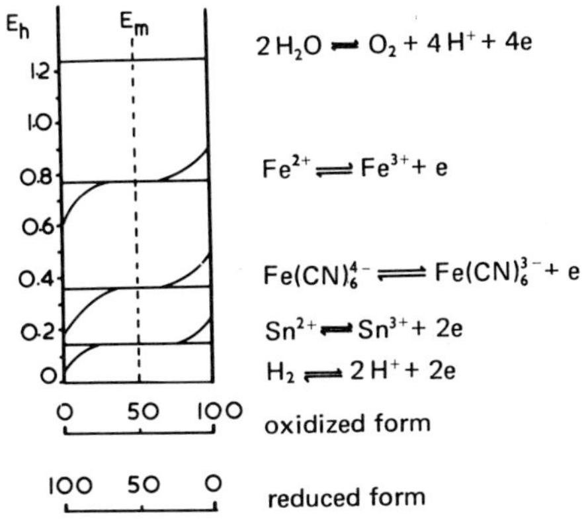
Eh
Em
1.2
1.0
0.8
0.6
0.4
0.2
0
2H2O ⇌ O2 + 4H+ + 4e
Fe2+ ⇌ Fe3+ + e
Fe(CN)6 4− ⇌ Fe(CN)6 3− + e
Sn2+ ⇌ Sn3+ + 2e
H2 ⇌ 2H+ + 2e
0 50 100
oxidized form
100 50 0
reduced form

Fig. 1.3. The relation between oxidation–reduction potential and relative concentration of oxidized and reduced forms (14) (reprinted with permission of the authors and Cambridge Univ. Press).

50% reduction. Its slope is determined by the number of electrons by which the oxidized and reduced states differ. The majority of biological systems involve either one- or two-electron changes and the potentials vary by 0.06 and 0.03 V, respectively, for each tenfold change in ratio of oxidant to reductant.

The midpoint at 50% reduction is designated by E_m, the subscript m being a reminder of midpoint. A number of authors also use the designation E_0'. As the values for E_m vary in systems that vary with the hydrogen ion concentration (H^+), the particular pH is signified by the addition of a corresponding number; e.g., E_{m7} signifies the potential at 50% reduction at pH 7.0. The Nernst equation in oxidation–reduction potential measurements would therefore be

$$E_h = E_m \frac{RT}{nF} \ln \frac{[A_{\text{ox}}]}{[A_{\text{red}}]}$$

When the oxidized form of one reversible system is reduced with the reduced form of another (or *vice versa*), the accuracy of the end point will be determined by the relative values of the E_m of each system. The potential before the equivalence point is determined by the titrated system because this is present in excess, and the potential after the equivalence point is determined by the titrant system. Where the respective standard potentials (midpoints) are reasonably far apart, a high accuracy can be obtained. The standard potentials should differ by at least 0.35 V if n is unity for both systems, 0.26 V if n is unity for one and two for the other, or 0.18 V if n is two for both.

TABLE 1.1
Electrode Potentials of Some O–R Systems[a]

	E_0' (V)	pH
H_2O/O_2	0.82	7.0
NO_2^-/NO_3^-	0.42	7.0
$H_2O_2/O_2 + H_2O$	0.30	7.0
Cytochrome a Fe^{2+}/Fe^{3+}	0.29	7.0
Cytochrome c Fe^{2+}/Fe^{3+}	0.22	7.0
Butyryl-CoA/crotonyl-CoA	0.19	7.0
Cytochrome b_2 Fe^{2+}/Fe^{3+}	0.12	7.4
Ubiquinone (red/ox)	0.10	7.4
Ascorbic acid/dehydroascorbic acid	0.08	6.4
Cytochrome b Fe^{2+}/Fe^{3+}	0.07	7.4
Succinic/fumaric acid	0.03	7.0
Methylene Blue (red/ox)	0.01	7.0
Malic/oxalacetic acid	−0.17	7.0
Lactic/pyruvic acid	−0.19	7.0
$NADH + H^+/NAD^+$	−0.32	7.0
Acetaldehyde + CoA/acetyl-CoA	−0.41	7.0
H_2/H^+	−0.42	7.0
α-Oxoglutaric acid/succinic acid + CO_2	−0.67	7.0
Pyruvic acid/acetic acid + CO_2	−0.70	7.0

[a] From White *et al.* (18). [Reprinted with permission of the authors and McGraw-Hill Book Co.]

Because $E_m(E_0')$ is a measure of the oxidation or reduction intensity of a system, it is possible to draw up a list of oxidation–reduction (O–R) systems in order of their standard electrode potentials. Any given system will be theoretically capable of being oxidized by a system more positive and will in turn be reduced by any system more negative than itself (Table 1.1).

Energy Relations in Oxidative Reactions

Because oxidative reactions yield energy, some quantitative aspects of oxidative changes in relation to energy production will be considered. It was stated above that the standard free energy change, in calories per mole, may be calculated from the equilibrium state in determining the equilibrium constant K. The determination of K depends on the availability of adequate analytical methods for the various components. When the difference in potential E_m between the two systems is large, equilibrium may lie so far in one direction that an accurate determination of the final concentration of the compound to be oxidized may be impossible. The free energy change associated with this reaction, however, may be calculated from the potentials of the two reacting systems

$$\Delta E_m = (RT/nF) \ln K \qquad \text{or} \qquad nF\,\Delta E_m = RT \ln K$$

Because

$$\begin{aligned} -\Delta F^\circ &= RT \ln K \\ &= nF\,\Delta E_m \end{aligned}$$

where ΔF° is the standard free energy of the reaction, n is the number of electrons (or hydrogen ions) involved, F is the Faraday constant (96,500 C), and ΔE_m is the difference between the E_m values of the two systems. The units of $F\,\Delta E_m$ are coulomb volts or joules, which can be converted to the usual units free energy, because 4.18 J = 1 gm cal. The value obtained for ΔF° is that for the oxidation of 1 mole of reductant.

EXAMPLE 1. Malate is oxidized to oxalacetate by cytochrome c under circumstances such that equimolar concentrations of each of the reactants initially exist. Because the E_m value of the malate–oxalacetate system is -0.17 V and of the cytochrome c–reduced cytochrome c is 0.22 V

$$\begin{aligned} \Delta F^\circ &= -nF\,\Delta E_m \\ &= \frac{-2 \times 96{,}500 \times [0.22 - (-0.17)]}{4.18} \\ &= -18{,}007 \text{ cal} \end{aligned}$$

Theoretically, if molecular oxygen were substituted for cytochrome c, $-45{,}715$ cal would be released because the E_m for the reduction of oxygen is $+0.82$ V. This amount of energy could then be available under physiological circumstances for doing useful work.

EXAMPLE 2.

$$\text{Lactate} + \text{acetaldehyde} \rightleftharpoons \text{pyruvate} + \text{ethanol}$$

At pH 7.0, E_m for the lactate–pyruvate system is -0.19 V and for the ethanol–acetaldehyde system is -0.20 V

$$\Delta F^\circ = -nF\,\Delta E_m$$

For this reaction $n = 2$, and

$$\Delta E_m = -0.20 - (-0.19)$$
$$= -0.01 \text{ V}$$

Therefore

$$\Delta F^\circ = \frac{-2 \times 96{,}500 \times (-0.01)}{4.18}$$
$$= 461.72 \text{ cal}$$

ENERGY STORAGE AND RELEASE. Within cells, a mechanism exists by which the free energy available from oxidation reactions may be utilized to drive endergonic processes (13). This is done largely by trapping this energy through the formation of a special class of phosphate compounds. The terminal phosphate group of ATP may be transferred to a second molecule, a so-called "phosphate acceptor," leaving behind ADP. The function of this reversible reaction will now be considered.

When ATP was first isolated from muscle, it was suspected to have something to do with the energy of muscle contraction. It was later found that when ATP is incubated with muscle fibers it undergoes enzymatic hydrolysis with the formation of ADP and inorganic phosphate

$$ATP^{4-} + H_2O \rightarrow ADP^{3+} + HPO_4^{2-} + H^+$$

During this hydrolysis, a considerable amount of heat was liberated. Because the free energy change of a chemical reaction can be determined from its equilibrium constant, by measuring this constant for the enzymatic reaction in which the terminal phosphate group of ATP is transferred, an approximate value for the free energy change of ATP hydrolysis has been obtained. If only one phosphate group of ATP is hydrolyzed, the free energy of this hydrolysis at pH 7.0 and 25°C is

$$\Delta F^\circ = -7000 \text{ cal/mole}$$

under standard conditions. It is likely, however, that this value of 7000 cal/mole does not represent the actual free energy released during the hydrolysis of ATP within the intact cell, because ADP and ATP are not present in the cell in equimolar concentrations. Furthermore, ATP, ADP, and inorganic phosphates are able to form complex compounds with Mg^{2+} ions, which can result in a shift of the equilibrium of ATP hydrolysis. With appropriate corrections (11), a free energy value of 12,000 cal/mole is

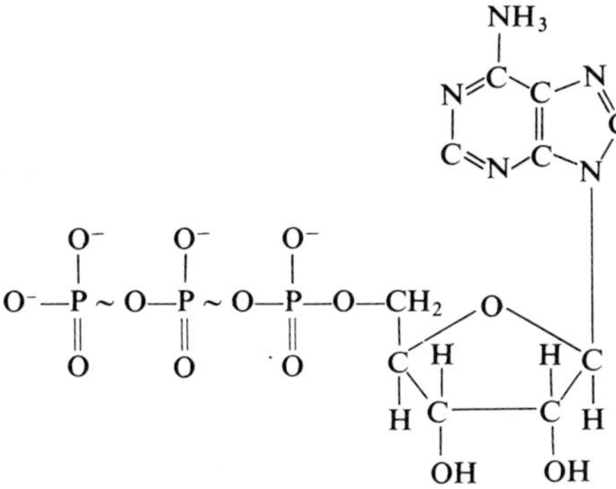
adenine
D-ribose

Fig. 1.4. Adenosine triphosphate (ATP).

more likely. The free energy of hydrolysis is significantly higher than that of simple esters or glycosides. This is one of the reasons ATP has been called a high-energy phosphate compound. Why, then, does the equilibrium of ATP hydrolysis lie farther in the direction of completion than do those of so-called "low-energy" compounds? Briefly, the larger the equilibrium constant, the greater the decrease in the quantity of free energy. There are two basic features of the ATP molecule that endow it with a relatively high free energy of hydrolysis; both are properties of the highly charged polyphosphate structure (Fig. 1.4). The chemical structure reveals three kinds of building blocks. First, there is a heterocyclic aromatic ring structure, called adenine, which is a derivative of 6-aminopurine. Attached to this base through a glycosidic linkage is a molecule of the five-carbon sugar D-ribose, to which is attached a phosphate group in ester linkage at the 5′ position, thus constituting adenosine monophosphate (AMP). This compound can contain a second (ADP) and a third (ATP) phosphate group in anhydric linkage with the 5′-phosphate. The third phosphate group is in linear anhydric linkage to the second group. All these compounds fall into the category of mononucleotides, which occur in each microbial cell. At pH 7.0, the linear polyphosphate structure of ATP is completely ionized, giving the ATP molecule four negative charges. These negative charges are very close to each other and repel each other very strongly. If the terminal phosphate group is hydrolyzed, this electrostatic stress between the groups is relieved, for the charges are now separated and partly distributed on the ADP^{3-} ion and the $phosphate^{2-}$ ion. Once separated, the ions will have little tendency to approach each other again because of the electrostatic repulsion. Low-energy phosphate bonds, in contrast, have no such repulsive force at pH 7.0 between the products of hydrolysis, because one of these normally has no charge at all.

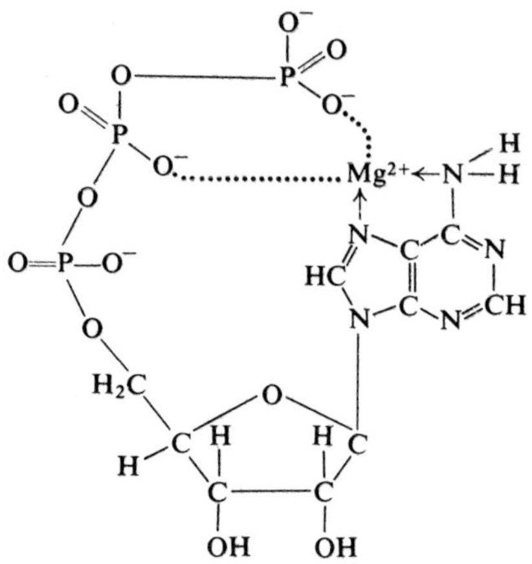
O⁻
O
O
P
O
O⁻
P
O⁻
Mg²⁺
N
H
H
O
N
C
C
N
O=P—O⁻
HC
N
C
N
CH
O
H₂C
O
C
H
H
C
H
C
C
OH
OH

Fig. 1.5. The structure of the Mg^{2+}–ATP complex in the cell.

A further reason for the high free energy of ATP hydrolysis is the fact that both reaction products, ADP and phosphoric acid, stabilize immediately as resonance hybrids. The new arrangement of the product's (ADP) electrons is made as soon as the bond is broken in a way that obtains a lower energy content. This resonance stabilization of the hydrolysis is a major reason for the relatively high free energy of hydrolysis of this phosphate ester class. It is quite common to use the term "high-energy phosphate bonds" for such bonds, which are universally designated by the symbol ~P. Lehninger (11) warns, however, that this term may be taken wrongly and may be misleading. It implies that the energy spoken of is in the bond and that when the bond is split, energy is set free. In physical chemistry the term "bond energy" means the energy required to break a given bond between two atoms. In other words, the free energy of hydrolysis is not localized in the actual chemical bond itself. In using the term "high-energy bond" as stated above, one should be aware that it means only that the difference in energy content between the reactants and products of hydrolysis is relatively high.

As the ATP molecule in the intact cell is highly charged at pH 7.0 and if each of the three phosphate groups is completely ionized, stable, soluble complexes with certain divalent cations, such as Mg^{2+} and Ca^{2+}, are formed (Fig. 1.5). This feature of ATP can also be related to its ability to act as an energy barrier. Because of this complex formation, very little free ATP anion exists as such in the cell. This geometrical structure is important insofar as the enzymes that make and use ATP have active sites to which the ATP structure must fit exactly in order to be functional.

TABLE 1.2
Free Energy of Hydrolysis of Phosphate Compounds[a]

	$\Delta F'$ (cal/mole)	Phosphate transfer potential	Direction of P group transfer
Phosphoenolpyruvate	−12,800	12.8	↓
1,3-Diphosphoglycerate	−11,800	11.8	
Acetyl phosphate	−10,100	10.1	
ATP	−7,000	7.0	
Glucose 1-phosphate	−5,000	5.0	
Fructose 6-phosphate	−3,800	3.8	
Glucose 6-phosphate	−3,300	3.3	
3-Phosphoglycerate	−3,100	3.1	
Glycerol 1-phosphate	−2,300	2.3	

[a] From Lehninger (11). [Reprinted with permission of the author and W. A. Benjamin, Inc.]

Reference to Table 1.2 may make one wonder why ATP is so unique if it is only one of many high-energy compounds in the cell, some of which have even greater free energy of hydrolysis than ATP. Generally speaking, it is the whole function of the ATP–ADP system to act as an intermediate bridge or linking system between phosphate compounds having a high transfer capacity and other compounds having a low transfer capacity. Adenosine diphosphate serves, in other words, as the specific enzymatic acceptor of phosphate groups from cellular phosphate compounds of very high potential. The ATP so formed can then donate its terminal phosphate group enzymatically to certain specific phosphate acceptors, such as glucose, with the formation of the corresponding phosphate derivatives. Moreover, all the reactions in the cell that cause phosphorylation of ADP to ATP at the expense of phosphate compounds of very high potential, as well as the terminal phosphate transfer reactions, are catalyzed by enzymes. Nearly all these enzymes are specific for ATP and ADP. These two compounds constitute virtually a shuttle service for phosphate groups from high-energy to low-energy compounds.

Conservation of Energy of Oxidation as ATP Energy

The conservation of energy from oxidoreduction reactions occurs in two ways (7).

1. The oxidation of a catabolite is followed by the reduction of a biosynthetic intermediate. The transfer of reducing equivalents is carried out by an intermediate oxidoreduction carrier, e.g., NAD^+ or $NADP^+$

(see p. 46). The general scheme of such a system is

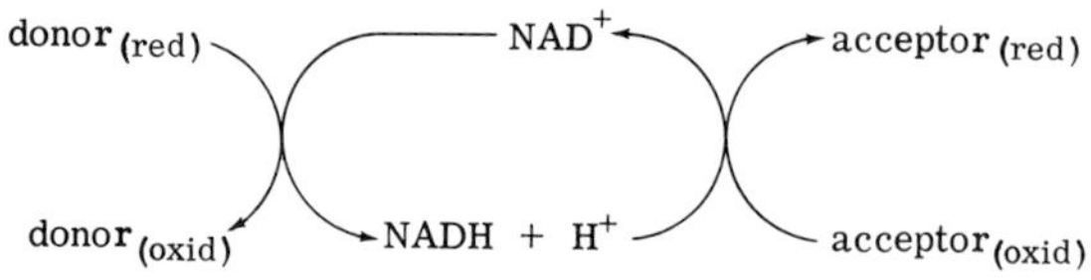
donor (red)
NAD^+
acceptor (red)
donor (oxid)
NADH + H^+
acceptor (oxid)

2. The oxidation of a catabolite is followed by an energy transfer to an endergonic biosynthetic reaction, whereby ATP is synthesized and again hydrolyzed

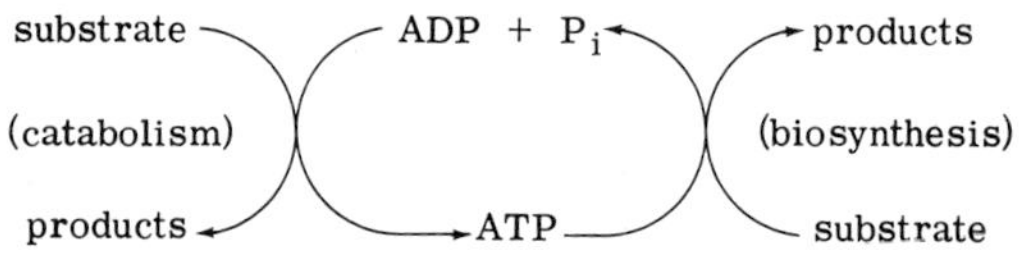
substrate
ADP + P_i
products
(catabolism)
(biosynthesis)
products
ATP
substrate

In both cases catabolic and biosynthetic reactions are coupled via oxidoreduction mechanisms and both result in the production, conservation, and release of energy in the form of the shuttle service for phosphate groups, as described in the previous section. We therefore divide all coupled energy-forming oxidoreduction reactions into two categories: (*1*) oxidative phosphorylation and (*2*) substrate level phosphorylation.

OXIDATIVE PHOSPHORYLATION. During the oxidation of a catabolite, electrons are set free that must be accepted by an oxidizing agent. If the catabolite is assumed to be molecular hydrogen and the oxidizing agent molecular oxygen, the potential difference between both would be $0.81 - (-0.42) = 1.23$ V, which is equivalent to a $\Delta F° = 57$ kcal (see p. 21). In the biological cell, however, molecular hydrogen is rarely a catabolite, as the catabolite transfers its hydrogen first to a hydrogen carrier, e.g., NAD^+ or $NADP^+$, and reduces this carrier; the cell then oxidizes $NADH + H^+$ or $NADPH + H^+$. The potential difference is thus reduced to 1.12 V or a $\Delta F° = +52$ kcal. This amount of energy would certainly damage or destroy the cell if released as heat. The cell therefore has an arrangement by which this biochemical reaction is subdivided in a number of small individual energy steps. In other words, $NADH + H^+$ and $NADPH + H^+$ are unable to react directly with molecular oxygen but do react with a number of individual intermediates. This stepwise reduction of the total potential energy through a chain of redox systems is made possible by a number of coupled reactions. The cell is thus able to conserve part of the energy as chemical energy (ATP). This stepwise electron carrier system is called the "respiratory chain" (Fig. 1.6).

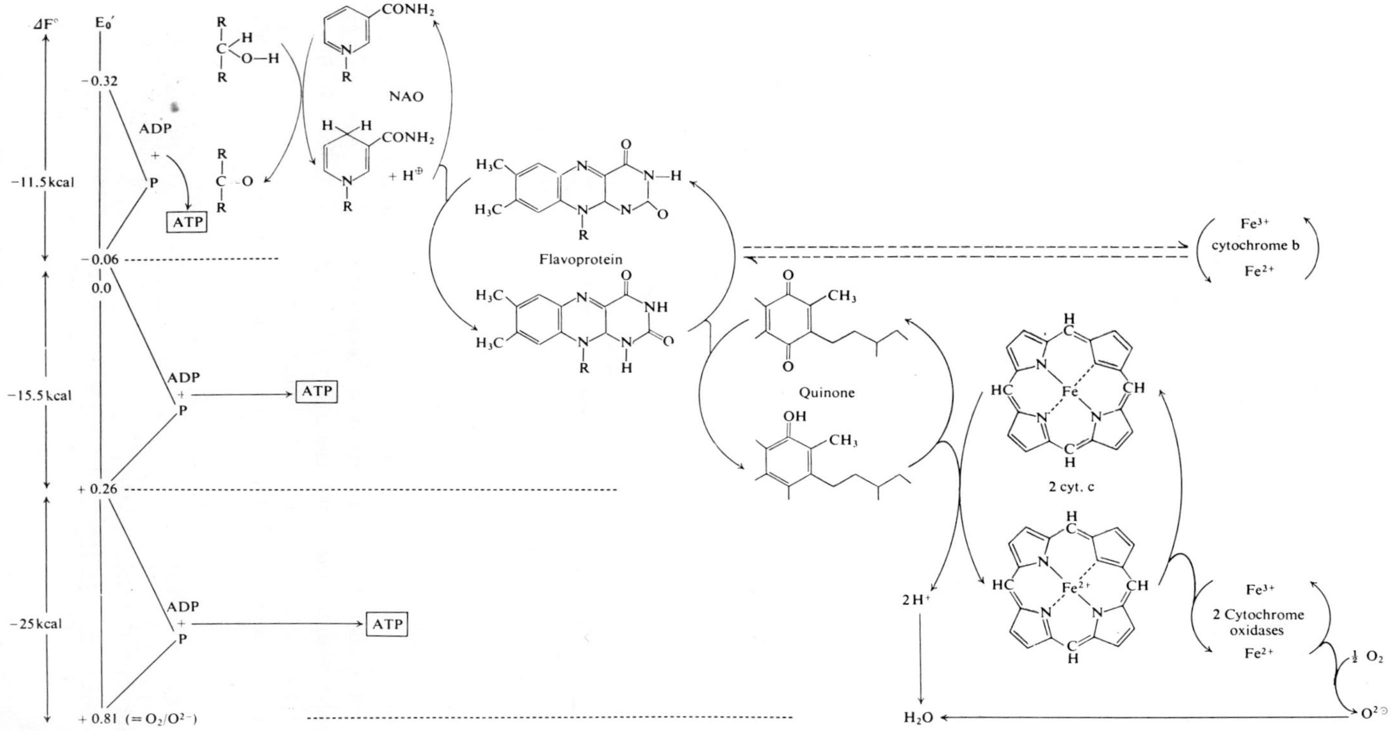
ΔF°
E₀′
−0.32
ADP
+
P
ATP
−11.5 kcal
−0.06
0.0
−15.5 kcal
ADP
+
P
ATP
+0.26
−25 kcal
ADP
+
P
ATP
+0.81 (= O₂/O²⁻)
NAO
+ H⊕
Flavoprotein
Quinone
2 cyt. c
Fe³⁺
cytochrome b
Fe²⁺
Fe³⁺
2 Cytochrome
oxidases
Fe²⁺
½ O₂
2H⁺
H₂O

Fig. 1.6. Sequence of redox systems in the respiratory chain [from Karlson (8)].

Although the electron transport mechanisms in mammalian systems are well studied, the knowledge of bacterial electron transport mechanisms has lagged well behind. The main reason for this is the great diversity of metabolic types among bacteria, which provides an even greater diversity in composition of the electron transport chains. A generalized scheme was well summarized by Dolin (4) (Fig. 1.7). The donated hy-

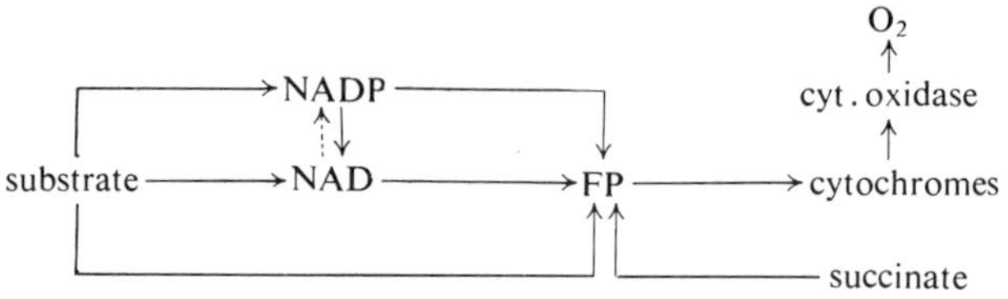
O2
cyt. oxidase
NADP
substrate
NAD
FP
cytochromes
succinate

Fig. 1.7. Generalized electron transport system.

drogen is transferred from the substrate to NAD^+, which itself then donates electrons to the cytochrome system via a flavoprotein (FP). The identity and number of cytochrome components vary from species to species and also with growth conditions. Bacteria may contain such pigments as the cytochromes a, a_3, b, c, and c_1 and others in a number of combinations. Some of the cytochromes have only been observed in bacteria. The main difference between the mammalian and the bacterial cytochrome systems, however, seems to be the presence of several oxidases in bacteria.

The possible pathway among the different bacteria could be summarized as shown in Fig. 1.8 (16). The cytochromes of the bacterial systems, which are attached to insoluble particulate matter within the cell, can have very high turnover rates. The numerous combinations of cytochrome components in different bacteria has led to the conclusion that the only requirement is for a mixture of several cytochromes with appropriately separated redox potentials. The most common ones are presented in Table 1.3, together with some electron donors, acceptors, and carriers. After the enzyme-catalyzed dehydrogenation of a substrate, two electrons pass through the components of the electron transport chain. This chain forms a transport sequence in which there occurs a stepwise increase in potential from that of $NADH + H^+/NAD^+$ to that of the oxygen electrode. Most

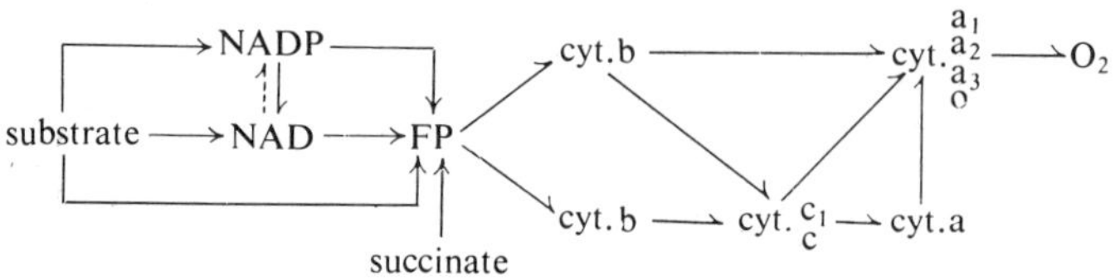
NADP
substrate
NAD
FP
succinate
cyt.b
cyt.
a1
a2
a3
o
O2
cyt.b
cyt.
c1
c
cyt.a

Fig. 1.8. Possible variations in the electron transport system of microorganisms.

TABLE 1.3

Some Electrode Potentials of Biological Interest[a]

Couple	E_0' (V at pH 7.0)
$2\,H_2O \rightleftharpoons O_2 + 4\,H^+ + 4\,e$	+0.816
$NO_2^- + H_2O \rightleftharpoons NO_3^- + 2\,H^+ + 2\,e$	+0.421
$H_2O_2 \rightleftharpoons O_2 + 2\,H^+ + 2\,e$	+0.295
Cyt. $a_3^{2+} \rightleftharpoons$ cyt. $a_3^{3+} + 1\,e$	+0.285
Cyt. $a^{2+} \rightleftharpoons$ cyt. $a^{3+} + 1\,e$	+0.290
Cyt. $c^{2+} \rightleftharpoons$ cyt. $c^{3+} + 1\,e$	+0.250
Succinate $\rightleftharpoons$ fumarate $+ 2\,H^+ + 2\,e$	+0.031
$H_2 \rightleftharpoons 2\,H^+ + 2\,e$ (pH 0)	0.0
Cyt. $b^{2+} \rightleftharpoons$ cyt. $b^{3+} + 1\,e$ (pH 7.4)	−0.040
Lactate $\rightleftharpoons$ pyruvate $+ 2\,H^+ + 2\,e$	−0.19
$FADH + H^+ \rightleftharpoons FAD^+ + 2\,H^+ + 2\,e$	−0.22
$NADH + H^+ \rightleftharpoons NAD^+ + 2\,H^+ + 2\,e$	−0.32
$NADPH + H^+ \rightleftharpoons NADP^+ + 2\,H^+ + 2\,e$	−0.324
$H_2 \rightleftharpoons 2\,H^+ + 2\,e$	−0.414
Glyceraldehyde 3-P $+ H_2O \rightleftharpoons$ 3-phosphoglycerate $+ 3\,H^+ + 2\,e$	−0.57
α-Ketoglutarate $+ H_2O \rightleftharpoons$ succinate $+ CO_2 + 2\,H^+ + 2\,e$	−0.673
Pyruvate $+ H_2O \rightleftharpoons$ acetate $+ CO_2 + 2\,H^+ + 2\,e$	−0.699

[a] From Dolin (4).

substrates (e.g., lactate) are electromotively passive. In the presence of the appropriate dehydrogenase, the oxidation of the substrate is catalyzed and the resulting two-component system (e.g., lactate–pyruvate) establishes the potential of the lower end of the electron transport chain. The oxidized substrate (e.g., pyruvate) may enter a metabolic system while the released electrons proceed independently along the transport system.

From Table 1.3 and Fig. 1.9, it can be seen that glyceraldehyde, α-ketoglutaric acid, and pyruvic acid—carbonyl and carboxyl compounds—are the most potent electron donors. Electrons from substrates usually enter the electron transport system through a carrier the potential of which lies in the vicinity of, or higher than, the potential for the substrate dehydrogenation. In the presence of the appropriate catalyst, electron flow will take place from the system of lower potential to the system of higher potential (more positive). The greater the difference in voltage, the farther the reactions will go toward completion.

The importance of the respiratory chain, with its redox cascades, lies in the possibilities of transforming the free energy obtained in every step into chemical energy by forming and storing ATP. The yield of $\sim$P thus depends on the ΔF of the reaction and on the number of steps available for energy conservation. If, for example, 1 mole $NADH + H^+$ reacts with

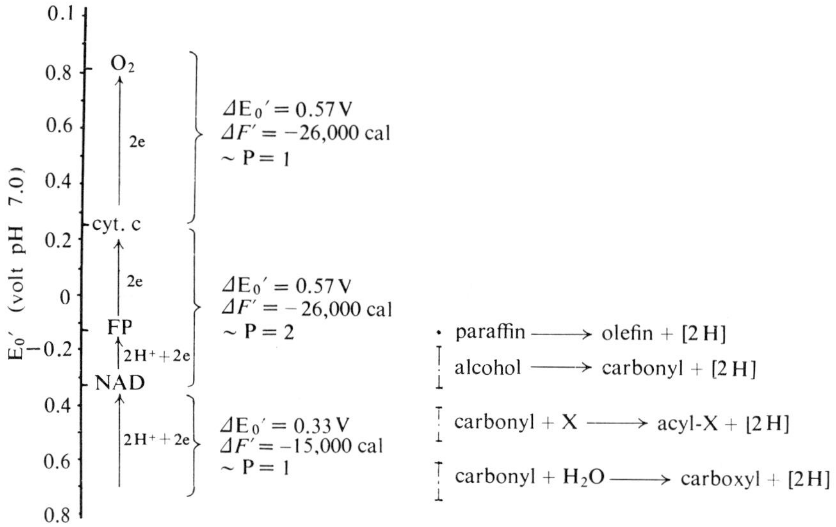

0.1
0.8
0.6
0.4
0.2
0
−0.2
0.4
0.6
0.8
E_0' (volt pH 7.0)
O_2
2e
cyt. c
2e
FP
$2H^+ + 2e$
NAD
$2H^+ + 2e$
$\Delta E_0' = 0.57$ V
$\Delta F' = -26{,}000$ cal
$\sim P = 1$
$\Delta E_0' = 0.57$ V
$\Delta F' = -26{,}000$ cal
$\sim P = 2$
$\Delta E_0' = 0.33$ V
$\Delta F' = -15{,}000$ cal
$\sim P = 1$
paraffin ⟶ olefin + [2 H]
alcohol ⟶ carbonyl + [2 H]
carbonyl + X ⟶ acyl-X + [2 H]
carbonyl + H_2O ⟶ carboxyl + [2 H]

Fig. 1.9. Comparison of the electrode potentials of coenzymes and substrate systems (4).

0.5 mole O_2, 52 kcal will be set free. It has been demonstrated above that for the formation of 3 moles ATP from ADP and inorganic phosphates only about 21 kcal are necessary; thus, a 40% energy conservation would be obtained. Figure 1.8 also indicates that substrates that are dehydrated not by NAD^+ but through flavoproteins (e.g., succinate), or even cytochrome c, must obtain less than 3 moles ATP. Wherever oxygen is the final electron acceptor, the results of such energetical calculations are usually expressed as P/O ratios. This ratio represents the equivalents of phosphate esterified per atom of oxygen taken up.

The key question in oxidative phosphorylation is still unsolved: what is the mechanism that leads from the release of free energy to the readily formed storage product ATP? At the moment there exist two main hypotheses.

1. The Slater theory of a chemical coupling via energy-rich intermediates, which should go parallel to the substrate phosphorylation (see p. 32)

2. The Mitchell theory of chemiosmosis, whereby it is necessary first to establish a difference in electrochemical concentration

The Slater theory states that the redox reaction is coupled with the formation of an energy-rich intermediate $C \sim A$. These intermediates can be hydrolyzed by dinitrophenol, which explains the function of this chemical compound as respiratory chain uncoupler. The intermediate $C \sim A$ now transfers the energy-rich $\sim A$ to a second compound, X, and

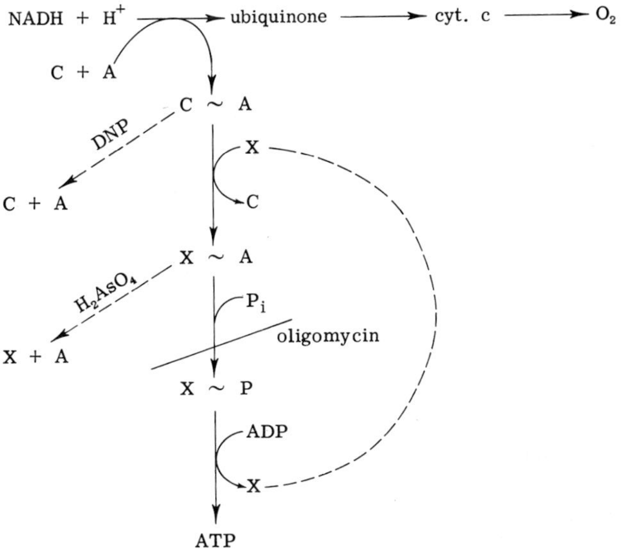

NADH + H⁺
ubiquinone
cyt. c
O₂
C + A
C ~ A
DNP
X
C + A
C
X ~ A
H₂AsO₄
Pᵢ
oligomycin
X + A
X ~ P
ADP
X
ATP

forms X ~ A. This energy-rich intermediate could, if so required, transfer its energy immediately to an endergonic process, such as reversal of the respiratory chain (see aerobic chemolithotrophs) or osmotic work. It can also react further with inorganic phosphate, however, to form X ~ P. This step can be intercepted by oligomycin, which explains the function of this compound as an uncoupling agent. The inorganic phosphate can now be transferred to ADP by X and ATP is formed.

The disadvantage of this hypothesis is that so far none of the proposed energy-rich intermediates have been found or isolated.

The Mitchell theory is based on the assumption that during the redox reactions of the membrane-bound enzymes, the H^+ ions can only be formed outside and the OH^- ions only inside the membrane. As the H^+ ions can be exchanged with the K^+, OH^-, and Cl^- ions, an electrochemical gradient develops that can do work. This gradient is thought to equilibrate at certain "coupling places" with the formation of energy-rich carrier. The energy of this carrier should be sufficient to attach the inorganic phosphate to ADP with the formation of ATP. The generation of a potential difference across the membrane, in other words, is the result of proton translocation (12, 15).

This theory depends solely on the assumption that there is an intact membrane structure, as is the case in higher organisms with mitochondria. In principle, it is the reverse of the process of active transport, the mechanism of which has still to be discovered. It is hoped that future research will clarify the problems of oxidative phosphorylation.

SUBSTRATE LEVEL PHOSPHORYLATION. An oxidoreduction reaction with a large and negative free energy change can be coupled to ATP synthesis and therefore does not need a further hydrogen transfer from NADH + H^+. Lehninger (11) demonstrated the foregoing by an actual reaction occurring in the cell as follows: The oxidation of an aldehyde to a carboxylic acid in aqueous solution is known to proceed with a large decline in free energy

$$R{-}\underset{\underset{O}{\|}}{C}{-}H + H_2O \rightarrow 2H + R{-}\underset{\underset{O}{\|}}{C}{-}O^- + H^+$$

$$\Delta F' \cong -7000 \text{ cal/mole}$$

In the cell, the oxidation of certain aldehydes takes place enzymatically in such a way that this energy is not simply lost as heat but is largely conserved. For example, 3-phosphoglyceraldehyde is oxidized to the acid 3-phosphoglycerate during glucose oxidation

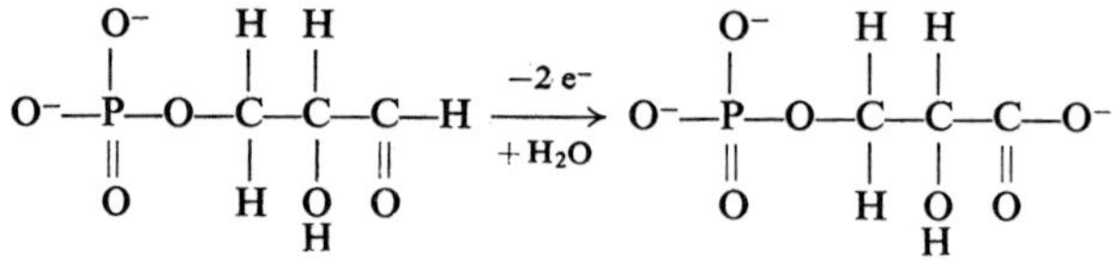

$$RCHO + HPO_4^{2-} + ADP^{3-} \rightarrow 2H + RCOO^- + ATP^{4-}$$
$$\Delta F' = 0 \text{ cal/mole}$$

This reaction does not take place in exactly this way but is coupled with the combination of one molecule of phosphate to one of ADP to form ATP.

TABLE 1.4
Standard Free Energy of Hydrolysis of Some Energy-Rich Compounds[a]

Compound	$\Delta F°$ (cal/mole)
ATP (−ADP + orthophosphate)	7,000
ATP (−AMP + pyrophosphate)	8,000
Pyrophosphate (−2 orthophosphate)	6,000
Creatine phosphate	8,000
Phosphoenolpyruvate	12,000
Phosphoglyceryl phosphate	11,000
Acetyl-coenzyme A	8,000
Aminoacyl AMP	7,000

[a] From Karlson (8).

The potential energy of the aldehyde group is transformed by the binding of an additional phosphate on to ADP, forming ATP. This conclusion is drawn from the free energy change in the reaction, which is approximately zero. The free energy decline of ~7000 cal/mole from the aldehyde oxidation is absorbed in the formation of ATP from ADP and phosphate, which requires an input of 7000 cal/mole (see Table 1.4). This phosphate transfer occurs in two separate steps, each catalyzed by a separate enzyme. When the aldehyde R—CHO was oxidized by the first enzyme, a large

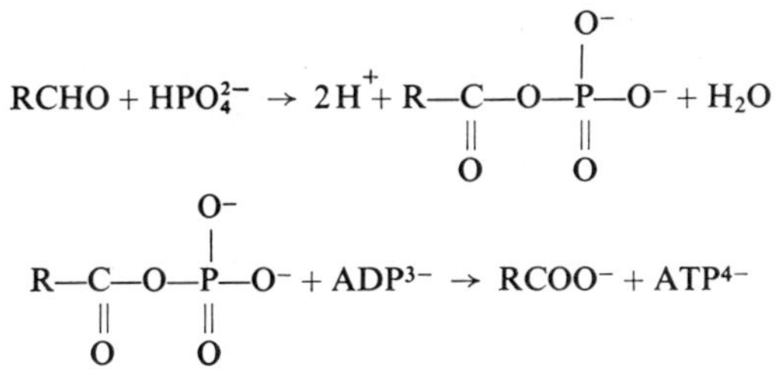

part of the free energy decline normally occurring when aldehydes are oxidized was conserved in the form of the phosphate derivative of the carboxylic acid. In the second reaction, the carboxyl phosphate group of the 1,3-diphosphoglycerate, the free energy hydrolysis of which is substantially higher than that of ATP, is enzymatically transferred to ADP and ATP is formed. The energy of oxidation of the aldehyde has thus been conserved in the form of ATP by two sequential reactions, in which the high-energy phosphate derivative of a carboxylic acid is the common intermediate.

Uphill reactions, whereby ATP energy is utilized to do chemical work, are very similar and can be demonstrated in the reaction

$$\text{glucose} + \text{fructose} \rightarrow \text{sucrose} + \text{water} \qquad \Delta F' = +5500 \text{ cal/mole}$$

If the sequence of reactions required are now dissected to analyze the energy changes, it is found that the energy-yielding process is the hydrolysis of ATP and the energy-requiring process is the formation of sucrose. Adenosine triphosphate is the common intermediate, linking the energy-yielding reaction and the energy-requiring synthesis of sucrose.

$$\text{RCHO} + \text{HPO}_4^{2-} \rightarrow 2\,\text{H}^+ + \text{RCOOPO}_3^{2-}$$

$$\text{RCOOPO}_3^{2-} + \text{ADP} \rightarrow \text{RCOO}^- + \text{ATP}$$

$$\text{ATP} + \text{glucose} \rightarrow\rightarrow \text{ADP} + \text{glucose 1-phosphate}$$

$$\text{glucose 1-phosphate} + \text{fructose} \rightarrow \text{sucrose} + \text{phosphate}$$

$$\text{sum: RCHO} + \text{glucose} + \text{fructose} \rightarrow 2\text{H}^+ + \text{RCOO}^- + \text{sucrose}$$

This overall equation shows that the energy yielded by oxidation of aldehyde to an acid was used to form sucrose from glucose and fructose.

A number of bacteria are able to form high molecular weight polyphosphates or metaphosphates. This process appears to be a form of "high-energy" phosphate storage. This polyphosphate is formed solely from the terminal phosphate group of ATP.

The division of the bacteria into aerobes and anaerobes cannot be rigidly maintained from the point of view of comparative biochemistry, although it still plays an important role in bacterial nutrition and classification. Among the chemosynthetic bacteria in general, those which reduce inorganic sulfur compounds in various oxidation states are anaerobes, those which reduce nitrogen compounds in various oxidation states are facultative anaerobes, and those which oxidize reduced sulfur and nitrogen compounds are obligate aerobes.

It is possible to estimate the maximum energy available for biochemical use in the electron systems that may be encountered in anaerobes. However, information on phosphorylations coupled to the electron transport systems in anaerobic bacteria is very scattered and will be considered with the individual bacterial groups. In general, there is no justification for assuming that the electron transport systems of anaerobic bacteria differ in principle or even in any major way from that of aerobes. There is almost no difference in the cytochrome-dependent electron transport systems of facultative anaerobes when grown either aerobically or anaerobically, for both terminate in cytochromes. There are intermediate systems, in which the bacteria have become less dependent on cytochromes, that, as in aerobically growing Lactobacillaceae, are completely independent of

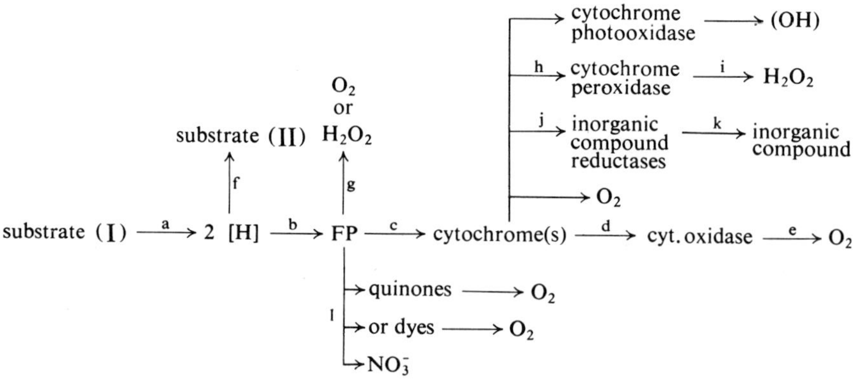
cytochrome photooxidase
(OH)
h
cytochrome peroxidase
i
H_2O_2
O_2 or H_2O_2
j
inorganic compound reductases
k
inorganic compound
substrate (II)
f
g
O_2
substrate (I)
a
2 [H]
b
FP
c
cytochrome(s)
d
cyt. oxidase
e
O_2
quinones
O_2
l
or dyes
O_2
NO_3^-

Fig. 1.10. General diagram of electron transport systems in aerobic, facultative anaerobic, and anaerobic bacteria.

cytochromes. Systems in obligate anaerobes will be discussed elsewhere (see Chapters 4 and 8). With all the presently available information, the tentative scheme for electron systems used by various microorganisms has been postulated by Dolin (4) and may function as a guide throughout bacterial metabolic studies (Fig. 1.10).

Depending on the pathway under consideration in Fig. 1.10, 2[H] may represent (*a*) reducing equivalents, (*b*) reduced pyridine nucleotides, or (*c*) reduced flavoprotein. Cytochrome photooxidase is used by photosynthetic bacteria. Dolin provides the following definitions in relation to Fig. 1.10.

Obligate aerobes: Cannot grow in the absence of O_2. They presumably have no functional fermentative metabolism. Pathway of electron transport—a,b,c,d,e.

Facultative anaerobes: Can grow in the presence or the absence of O_2 (in absence of O_2 they use fermentative pathways). There are two groups:

1. Cytochrome independent (e.g., lactic acid bacteria). Pathway of electron transport: aerobic—a,b,g; anaerobic—a,f (may be sole pathway of some representatives of this class).
2. Cytochrome dependent (e.g., coliforms). Pathway of electron transport: aerobic—a,b,c,d,e; a,b,g; anaerobic—a,f; a,b,c,j (nitrate reduction); a,b,l.

Obligate anaerobes: Cannot grow aerobically. There are two groups:

1. Cytochrome independent (e.g., *Clostridium*). Pathway of electron transport—a,f; a,b,g (not used under physiological conditions).
2. Cytochrome dependent (e.g., *Desulfovibrio*). Pathway of electron transport—a,b,c,j (reduction of sulfur ions; step c obscure).

Such quinones as menaquinone seem to play an important role in electron transfer in *Mycobacterium* (1) and *Corynebacterium* (10).

These examples illustrate the general working principle by which ATP is the energy carrier in sequential reactions involving flow of phosphate groups in coupled reactions.

References

1. Brodie, A. F., and Adelson, J. (1965). Respiratory chains and sites of coupled phosphorylation. Studies in a bacterial system give further evidence of a basic biochemical unity between forms. *Science* **149,** 265.
2. Clark, W. M. (1960). "Oxidation–Reduction Potentials of Organic Systems." Williams & Wilkins, Baltimore, Maryland.
3. Dawes, E. A. (1967). "Quantitative Problems in Biochemistry," 4th ed. Livingstone, Edinburgh.

4. Dolin, M. I. (1961). Survey of microbial electron transport mechanisms. *In* "The Bacteria" (I. C. Gunsalus and R. Y. Stanier, eds.), Vol. 2, p. 341ff. Academic Press, New York.
5. Glasstone, S. (1954). "Introduction to Electrochemistry," 6th printing. Van Nostrand-Reinhold, Princeton, New Jersey.
6. Glasstone, S. (1962). "Textbook of Physical Chemistry," 2nd ed. Macmillan, New York.
7. Hawker, L. E., and Linton, A. H. (1971). "Microorganisms. Function, Form, and Environment." Arnold, London.
8. Karlson, P. (1970). "Kurzes Lehrbuch der Biochemie," 7th ed. Thieme, Stuttgart.
9. Klotz, I. M. (1967). "Energy Changes in Biochemical Reactions." Academic Press, New York.
10. Krogstadt, D. J., and Howland, J. L. (1966). Role of menaquinone in *Corynebacterium diphtheriae* electron transport. *Biochim. Biophys. Acta* **118,** 189.
11. Lehninger, A. L. (1965). "Bioenergetics." Benjamin, New York.
12. Lieberman, E. A., and Skulacher, V. P. (1970). Conversion of biomembrane-produced energy into electric form. IV. General discussion. *Biochim. Biophys. Acta* **216,** 30.
13. Racker, E. (1965). "Mechanisms in Bioenergetics." Academic Press, New York.
14. Ramsay, J. A. (1965). "The Experimental Basis of Modern Biology." Cambridge Univ. Press, London and New York.
15. Robertson, R. N. (1968). The separation of protons and electrons as fundamental biological process. *Endeavour* **26,** 134.
16. Smith, L. (1961). Cytochrome systems in aerobic electron transport. *In* "The Bacteria" (I. C. Gunsalus and R. Y. Stanier, eds.), Vol. 2, p. 365. Academic Press, New York.
17. Vogel, A. I. (1961). "A Textbook of Quantitative Inorganic Analysis Including Elementary Instrumental Analysis," 3rd ed. Longmans, Green, New York.
18. White, A., Handler, P., and Smith, E. L. (1964). "Principles of Biochemistry," 3rd ed. McGraw-Hill, New York.

Supplementary Readings

Jacob, H. E. (1971). Das Redoxpotential in Bakterienkulturen. *Ztschr. f. allgem. Mikrobiol.* **11,** 691.

Minkoff, L., and Damadian, R. (1973). Caloric catastrophe. *Biophys. J.* **13,** 167.

San Pietro, A., and Gest H., eds. (1972). "Horizons of Bioenergetics." Academic Press, New York.

Questions

1. What is energy?
2. What is the first law of thermodynamics?
3. What is enthalpy?
4. What is the second law of thermodynamics?
5. What is entropy?
6. What is the third law of thermodynamics?
7. Explain the term "free energy."

8. What is the importance of the equilibrium constant K in chemical reactions?
9. Explain the formula $\Delta F = \Delta F° \times RT \ln K$.
10. A reaction is started by adding an enzyme to 0.05 moles of compound A at 25°C and pH 7.0. Chemical analysis reveals that the reaction proceeds to an equilibrium at which the final concentration of compound B has risen from zero to 0.038 M. Calculate the standard free energy change of this reaction.
11. How do you calculate the overall feasibility of reactions involved in a metabolic pathway?
12. What are the major differences between the thermodynamics of a mechanical and a biological system, expressed in general terms.
13. What is an oxidation–reduction potential?
14. What do we understand as being the electromotive force of a cell?
15. Explain the difference between the electrode potential and the standard potential of an electrode?
16. What are reference electrodes and what is their importance?
17. What does the Nernst equation express?
18. What is the standard electrode potential and its importance?
19. How do pH and temperature influence the oxidation–reduction potential of a system?
20. Discuss the different ways of making oxidation–reduction measurements.
21. Explain the general principle of energy storage in biological systems.
22. What are the reasons for calling such compounds as ATP "high-energy" compounds?
23. What is the difference between a high-energy bond and a high-energy compound?
24. What is the role of oxidation–reduction reactions in the conservation of energy?
25. Explain the differences between the Slater and the Mitchell theories of oxidative phosphorylation.

2

Enzymes, Coenzymes, and Bacterial Growth Kinetics

Enzymes

Catalytic Function

A high, negative value of ΔF indicates that a chemical reaction is likely to proceed spontaneously and that the products will greatly exceed the reactants at equilibrium (see Table 2.1). However, it does not guarantee that the reaction will proceed with measurable speed. There exists a kind of energy barrier that must be overcome before the reaction can proceed. The important quantity is the free energy of activation.

Reactions that fail to proceed notwithstanding a high negative value of ΔF can often be persuaded to do so in the presence of a catalyst. From the point of view of thermodynamics, a catalyst is something that lowers the free energy of activation. From the physical point of view, what probably happens is that the reactant combines temporarily with the catalyst. As a result, the energy of the reactant molecule is redistributed so that certain bonds become more liable to rupture by thermal agitation.

Enzymes are true catalysts because they do not influence the point of equilibrium of the reaction they catalyze, nor are they used up during catalysis. Like other catalysts, enzymes lower the activation energy of the reaction they catalyze in order to obtain an equilibrium state of the reaction. It is impossible for the catalysis to overshoot this equilibrium state. Every enzymatically catalyzed reaction keeps on reacting until the equilib-

TABLE 2.1
Redox Potentials at pH 7.0 (E_0') for Some Biochemical Redox Systems

Coenzymes			Substrates	
$\Delta F°$ (kcal)	E_0' (V)	Substance	E_0' (V)	Substance
			−0.47	Acetaldehyde/acetate
			−0.42	$H_2/2\ H^+$
	−0.32	NADH + H^+/NAD^+		
			−0.20	Ethanol/acetaldehyde
	−0.185	Riboflavin–P·H_2/riboflavin –P		
−11.5			−0.18	Lactate/pyruvate
	−0.06	Flavoproteins		
	−0.05	Phyllohydroquinone/ phylloquinone		
	−0.04	Cytochrome b		
			0.0	Succinate/fumarate
−15.5			+0.01	Methylene Blue/leukodye
			+0.20	Ascorbate/dehydro- ascorbate (pH 3.3)
	+0.26	Cytochrome c		
	+0.29	Cytochrome a		
−25.0			+0.81	$\frac{1}{2}O_2/O^{2-}$

rium state is obtained. This is one of the basic laws of enzymology. The only possibility for a reaction to continue beyond equilibrium occurs in coupled reactions, where the product of the first reaction is immediately catalyzed to a further product with the help of a different enzyme. Therefore, an organism will never be close to a chemical equilibrium stage, as a system in equilibrium cannot perform any work. This continuous aim toward an equilibrium stage is quite often referred to as the "steady state," wherein substrates must be continuously fed into a stationary system and the products of the reaction be taken out. As the organisms represent an open system, the steady state forms stationary concentrations that are different from the thermodynamically ruled chemical equilibriums. This is the main reason the reactions keep working toward the equilibrium state. The organism always gets its energy from these reactions.

All the enzymes the chemical compositions of which have been investigated are proteins. The methods that are used to separate and purify enzymes are the same as those used to separate and purify proteins. Enzymes are susceptible to influences and agents that are known to affect proteins. The molecules have a limited life within a cell, new enzymes being continually produced to replace the old. With exceptions, they are rapidly denatured, and their catalytic properties destroyed, at temperatures of 50°C and over and by the ions of heavy metals. Their catalytic activities are notably affected by pH. Enzymes differ in several respects from inorganic catalysts. Enzymes are more efficient than such inorganic catalysts as platinum, show greater specificity, and are less stable.

Enzymes consist of a protein component and a "prosthetic group." The latter may be removed reversibly, and in such cases the protein part is called "apoenzyme" and the prosthetic group "coenzyme" (43). The protein component determines the substrate specificity, i.e., it decides which one of the substrates is to be converted. On many occasions it also determines the direction of the reaction, i.e., it determines to which of the many possible reactions the substrate molecule will go. This role of the apoenzyme is particularly obvious whenever the same coenzyme is attached to different apoenzymes, with different reaction products as results. One and the same coenzyme can therefore catalyze different reactions, depending on the apoenzyme (43).

Much effort has been devoted to the study of enzymatic action (34). When an enzyme catalyzes a specific reaction, it first combines transiently with the substrate, the name given to the substance on which the enzyme acts, to form the enzyme–substrate complex. In this complex, there is a "lock-and-key" fit of the substrate molecule to a "patch" on the surface of the very large enzyme molecule (51). This patch is called the "active site," and because of the specific geometrical relationship of the chemical groups that combine with the substrate, it can only accept molecules having a complementary fit. During the formation of the enzyme–substrate complex, the enzyme molecule is twisted somewhat, which places some strain on the geometry of the substrate molecule. This renders it susceptible to attack by H^+ or OH^- ions or by specific functional groups of the enzyme. In this manner the substrate molecule is converted to its products, which now diffuse away from the active site. The enzyme molecule returns to its native shape, combines with a second substrate molecule, and repeats the cycle. Most enzymes can be inhibited by specific poisons, which may be structurally related to their normal substrate. Such inhibitors are very useful in analyzing enzyme-catalyzed reactions in cells and tissues. When enzymes act in a sequence, so that the product of one enzyme becomes the substrate for the next, and so on, we have a multienzyme system and the chains of reactions are known as "metabolic pathways."

Enzymes are not only chemically specific, they are also sterically specific when they act on substances containing asymmetric centers. The substrate may contain an asymmetric carbon atom, in which case it is usually found that the enzyme acts on only one of the optical isomers. Specificity of this

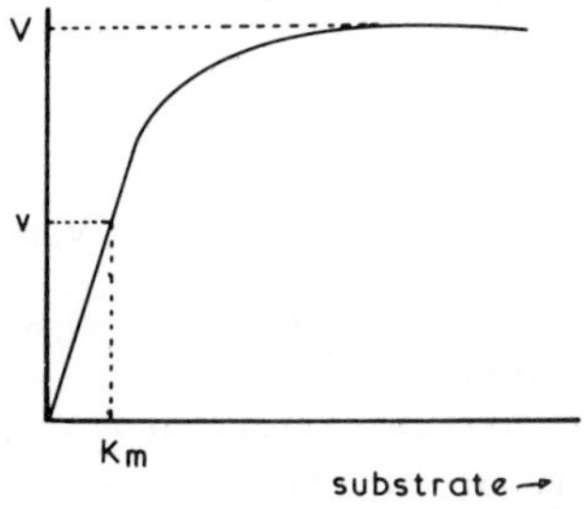
V
v
Km
substrate →

Fig. 2.1. Hyperbolic form of a typical substrate concentration curve.

type appears usually to be absolute. A good example is glyceraldehyde-3-phosphate dehydrogenase, which reacts only with the D isomer of DL-glyceraldehyde 3-phosphate.

Substrate concentration is one of the most important of the factors that determine the velocity of enzymatic reactions. The enzyme first forms a complex with its substrate; this subsequently breaks down, giving the free enzyme and the products of the reaction. One can therefore write

$$\text{Enzyme} + \text{substrate} \rightleftharpoons \text{enzyme–substrate}$$
$$E + S \rightleftharpoons ES$$
$$ES \rightarrow \underbrace{A + B}_{\text{products}} + E$$

If one starts with a given amount of enzyme and raises the substrate concentration gradually, more and more enzyme will be converted into the complex ES. The rate of reaction will increase until, finally, virtually all the enzyme is in the form of ES. The equilibrium constant could thus be calculated from

$$K_{eq} = \frac{[E] \times [S]}{[ES]}$$

The enzyme is then saturated and the reaction rate is maximal. When velocity is plotted against substrate concentration a section of a rectangular hyperbola is obtained (Fig. 2.1), which may be owing to many factors, such as inactivation at the temperature or the pH of the reaction. The saturation concentration changes from enzyme to enzyme and is different for each enzyme from substrate to substrate. It is very difficult to read these differences from the curve. A different approach, which is dependent on the measurement of initial velocity or half-maximal velocity, is therefore adopted in the study of enzymatic actions. At the point of half-maximal velocity, half of the entire enzyme is in the form of the ES complex, and the other half is free enzyme. Because the reaction rate should be proportional to the ES concentration,

$$K_m = [S] \text{ at half-maximal velocity}$$

In other words, that substrate concentration at which half-maximal reaction velocity is reached equals the dissociation constant of the enzyme–substrate complex. This constant is named the "Michaelis-Menten constant," or K_m. A large constant means that a high substrate concentration is necessary or that the enzyme possesses a low affinity for the substrate. These constants normally range between 10^{-2} and 10^{-5} mole/liter.

It is relatively easy to determine the constant K_m experimentally. In Fig. 2.1, it is shown that the maximal velocity V is reached when the enzyme is saturated by the substrate, i.e., when all the enzyme is converted to the complex ES. The mathematical expression for the enzyme velocity is

$$v = V\left(\frac{S}{K_m + S}\right)$$

where v is the reaction velocity, S is the substrate concentration, K_m is the Michaelis-Menten constant, and V is the maximal velocity of the reaction. If the variables v and S are then taken as reciprocals and drawn graphically, V and K_m can easily be determined on their corresponding coordinates (Fig. 2.2). A more detailed description of enzyme kinetics is given by Dixon and Webb (22), Karlson (43), and Gutfreund (34).

The reaction rate of an enzyme-catalyzed reaction is also used to define "enzyme units." One unit (U) of any enzyme is defined as that amount that will catalyze the transformation of 1 μmole of substrate per minute or, where more than one bond of each substrate molecule is attacked, 1 μEq

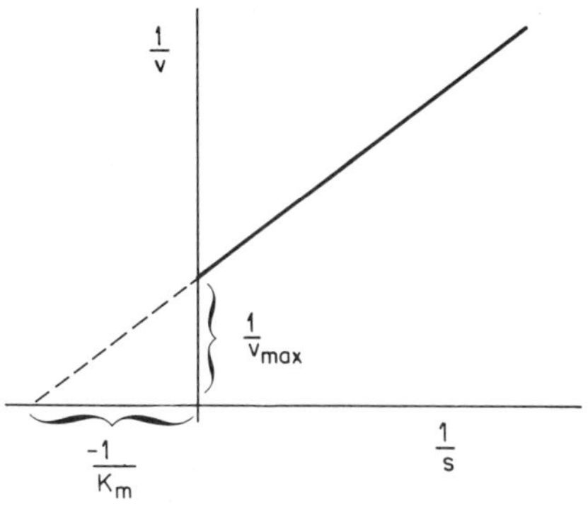
1/v
1/vmax
-1/Km
1/s

Fig. 2.2. Determination of V_{max} and K_m from the Lineweaver-Burk plot.

of the group concerned per minute under defined conditions. The temperature and all other conditions, including pH and substrate concentration, should be clearly stated. The International Union of Biochemists (IUB) recommends that, where practicable, the temperature be 30°C and that enzyme assays be based wherever possible on measurements of initial rates of reactions in order to avoid complications owing, for instance, to reversibility of reactions or to functions of inhibitory products. The substrate concentration, wherever possible, should be sufficient to saturate the enzyme, so that the kinetics in the standard assay approach zero order. Where a distinct suboptimal concentration of substrate must be used, it is recommended that, where feasible, the Michaelis constant be determined so that the observed rate may be converted into that which would be obtained on saturation with substrate.

Specific activity is expressed as units of enzyme per milligram of protein. "Molecular activity" is defined as units per micromole of enzyme at optimal substrate concentration, that is, as the number of molecules of substrate transformed per minute per molecule of enzyme. When the enzyme has a prosthetic group or catalytic center the concentration of which can be measured, the catalytic power can be expressed as catalytic center activity, i.e., the number of molecules transformed per minute per catalytic center.

Concentration of an enzyme in solution should be expressed as units per milliliter.

These are the recommendations of the definitions of the IUB, which every microbial chemist should follow strictly in his biochemical work.

Classification

It is also very important for a microbial chemist to know the principles on which the IUB has based its classification scheme. Enzymes that have been known for a long time still have their trivial names, e.g., trypsin and pepsin. The more recent names, however, have been devised with the suffix "-ase." This suffix is attached to the name of the reaction catalyzed. Enzymes that transfer groups are called "transferases"; dehydrogenating enzymes are called "dehydrogenases," etc. The IUB commission, however, recommended the use of this suffix only for single enzymes. A suffix should not be applied to systems containing more than one enzyme. When it is desired to name such a system on the basis of the overall reaction catalyzed by it, the word "system" should be included in the name, e.g., the "succinate oxidase system." Thus, only actions of single enzyme entities, and not of composite enzyme systems, are considered for classification.

The enzymes are divided into groups on the basis of the type of reaction catalyzed and this, together with the name(s) of the substrate(s), provides a basis for naming individual enzymes. The IUB decided further to recommend that the overall reaction, as expressed by the formal equation, should be taken as the basis, which means only the observed chemical change by the complete enzymatic reaction. In the case where a prosthetic group serves to catalyze the transfer from a donor to an acceptor (biotin and pyridoxine enzymes) the name of the prosthetic group is not included in the name of the enzyme. In exceptional cases, where alternative names are possible, the mechanism may be taken into account in distinguishing between the enzymes involved.

As a part of its classification, each enzyme has a number. Each enzyme number contains four elements, separated by points. The first figure indicates to which of the six main divisions the particular enzyme belongs.

1. Oxidoreductases
2. Transferases
3. Hydrolases
4. Lyases
5. Isomerases
6. Ligases (synthetases)

The second figure indicates the subclass and characterizes the type of group, bond, or link involved in the reaction. The third figure indicates the subsubclass and specifies in more detail the type of donor or group involved in the reaction. The fourth figure is the serial number of the enzyme in its subsubclass.

With this system it is possible to insert a new enzyme at the end of its subsubclass without disturbing any other numbers. Should it become necessary to create new classes, subclasses, or subsubclasses, they can be added without disturbing those already defined. It is the aim of the IUB that even at revisions the existing enzymes should not be renumbered to take account of any newly discovered enzyme. Such enzymes will be placed at the end of their respective sections and should be given new numbers. This should be done by some suitable authority and not by individuals.

In order to avoid the difficulties encountered by lengthy systematic names, the IUB adopted two kinds of names: (a) the systematic name of an enzyme, which will be formed in accordance with the above-mentioned definite rules, and (b) the trivial name, which will be sufficiently short for general use, but not necessarily very exact or systematic. The detailed rules for systematic and trivial nomenclature are outlined in Chapter 6 of the report of the IUB (40) and by Dixon and Webb (22) and Karlson (43). In order that the nomenclature system be understood, an example will be given with fructose-1,6-diphosphate aldolase, as it is known in the literature. This enzyme has the systematic name "fructose 1,6-diphosphate D-glyceraldehyde-3-phosphate-lyase," with the code number EC 4.1.2.13. This means:

4. The enzyme is a lyase, a group of enzymes that remove groups from their substrates (not by hydrolysis), leaving double bonds or, conversely, that add groups to double bonds
4.1. The enzyme is a carbon–carbon lyase
4.1.2. The enzyme is an aldehyde lyase with a serial number 13. The reaction catalyzed by this enzyme is

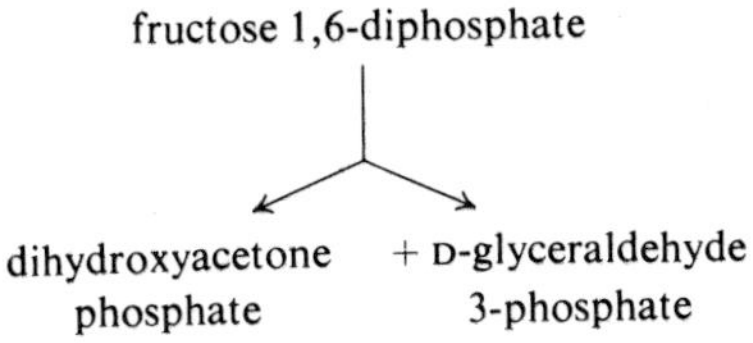
fructose 1,6-diphosphate
dihydroxyacetone phosphate
+ D-glyceraldehyde 3-phosphate

The recommended trivial name is "fructose diphosphate aldolase." The names "zymohexase" and "aldolase" are no longer recommended.

Coenzymes

In order to function, many enzymes require certain organic substances as cofactors. The cofactors, or coenzymes, generally act as acceptors or donors of groups of atoms that are removed from or contributed to the substrate. The terms "coenzyme" and "prosthetic group" are frequently used synonymously. Lately, the tendency has been to call the tightly bound groups that cannot be removed (e.g., by dialysis) "prosthetic groups," and those that dissociate easily "coenzymes" (43, 101). A third group of cofactors is the activators, frequently of a very simple nature, e.g., inorganic ions, which bring the enzyme itself into a catalytically active state.

Because there is no definite classification scheme available as yet for the coenzymes, this particular field will be treated in a way similar to that used to classify enzymes, namely, according to the reactions in the catalysis of which they are instrumental (see Table 2.2).

The most striking feature of these coenzymes is that the majority are either actual nucleotides or have some structural analogy with nucleotides. Many have a nitrogenous base at one end of the molecule and a phosphate group at the other, with or without a carbohydrate moiety between, or have two such structures joined in the form of a dinucleotide. The close relationship to vitamins is very interesting. These vitamins are necessary for the

TABLE 2.2
Coenzymes[a]

Coenzymes	Usual abbreviation	Group transferred	Corresponding vitamins
1. Hydrogen-transferring coenzymes			
Nicotinamide–adenine dinucleotide	NAD^+	Hydrogen	Nicotinamide
Nicotinamide–adenine dinucleotide phosphate	$NADP^+$	Hydrogen	Nicotinamide
Nicotinamide mononucleotide	—	Hydrogen	Nicotinamide
Flavin mononucleotide (riboflavin phosphate)	FMN	Hydrogen	Riboflavin
Flavin–adenine dinucleotide	FAD	Hydrogen	Riboflavin
Lipoic acid	Lip (S_2)	Hydrogen and acyl	—
Glutathione	GSH	Hydrogen	—
Ascorbate	—	—	—
Coenzyme Q	Q	Hydrogen	—
Cytochromes	cyt.	Electrons	
2. Group-transferring enzymes			
Adenosine triphosphate	ATP	Phosphate	
Phosphoadenyl sulfate	PAPS	Sulfate	—
Uridine diphosphate	UDP	Sugar, uronic acid	—
Cytidine diphosphate	CDP	Phosphoryl choline	—
Pyridoxal phosphate	PALP	Amino	Pyridoxine
Adenosyl methionine	—	Methyl	(Methionine)
Tetrahydrofolic acid	THF	Formyl	Pantothenic acid
Biotin	—	Carboxyl (CO_2)	Biotin
Coenzyme A	CoA	Acetyl	Pantothenic acid
Thiamine pyrophosphate	TPP	C_2-Aldehyde	Thiamine
3. Coenzymes of isomerases and lyases			
Uridine diphosphate	UDP	Sugar isomerization	—
Pyridoxal phosphate	PALP	Decarboxylation	Pyridoxine
Thiamine pyrophosphate	TPP	Decarboxylation	Thiamine
B_{12} coenzyme	—	Carboxyl displacement	Cobalamin

[a] From Karlson (43).

proper functioning of life processes and cannot be replaced by other substances. A biocatalytic function is known for many vitamins and generally a vitamin is the main or sole component of a coenzyme. A full understanding of coenzymes necessarily involves a detailed knowledge of the reactions in which they participate. However, a few introductory notes on their biochemistry will help in understanding their role in metabolism.

Hydrogen-Transferring Coenzymes

NICOTINAMIDE NUCLEOTIDE COENZYMES. The hydrogen-transferring enzymes of fermentation, respiration, and many other reactions utilize as their coenzymes dinucleotides, one of the bases of which is the pyridine derivative nicotinamide (Fig. 2.3; 14). These dinucleotides are therefore pyridine nucleotides. The nomenclature of the nicotinamide nucleotide coenzyme was, and partly still is, very much disputed. It is quite common to find in literature the names "diphosphopyridine nucleotide (DPN)" and "triphosphopyridine nucleotide (TPN)." Older literature contains a further number of names, such as "cozymase," "phosphocozymase," "coenzyme I (Co I)," "coenzyme II (Co II)," "codehydrogenase I," and "codehydrogenase II." In order to eliminate this confusion in the terminology of pyridine nucleotides the IUB made the following recommendations: The nicotinamide nucleotide coenzymes should in future be known by their chemical names "nicotinamide–adenine dinucleotide (NAD)" and "nicotinamide–adenine dinucleotide phosphate (NADP)," respectively. All other names should no longer be used. The mononucleotide should continue to be known as "nicotinamide mononucleotide (NMN)."

For the reduced form of NAD, the IUB allows two alternative abbreviations, namely, "reduced NAD" or, where it is desired to show the release of H^+ ion in the reduction, "NADH + H^+."

If the latter form is used, the oxidized form should then be written "NAD^+." The reaction should never be written $NAD \rightarrow NADH_2$. The same rules apply for NADP.

The pyridine ring in the coenzyme is attached in *N*-glycosidic linkage to ribose. This linkage is possible only with the pyridinium cation, which bears one hydrogen atom on the nitrogen (Fig. 2.4). Pyrophosphoric acid provides the linkage between nicotinamide riboside and adenosine.

In nicotinamide–adenine dinucleotide phosphate the adenosine moiety carries the additional phosphate group in the 2′ position (see Fig. 2.5). Because of the positive charge in the pyridine ring these coenzymes are also abbreviated NAD^+ and $NADP^+$. Their function is the reversible uptake of hydrogen. The pyridine ring becomes reduced, retaining only two double bonds, whereas the nitrogen loses its positive charge. Details of these reactions will be found in Karlson (43) and Dixon and Webb (22).

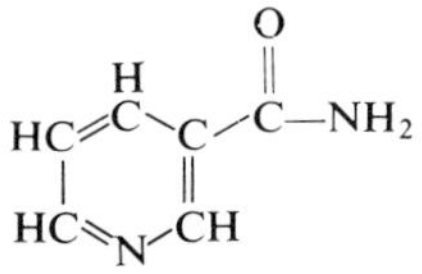
O
H
HC
C
C
C—NH2
HC
N
CH

Fig. 2.3. Nicotinamide.

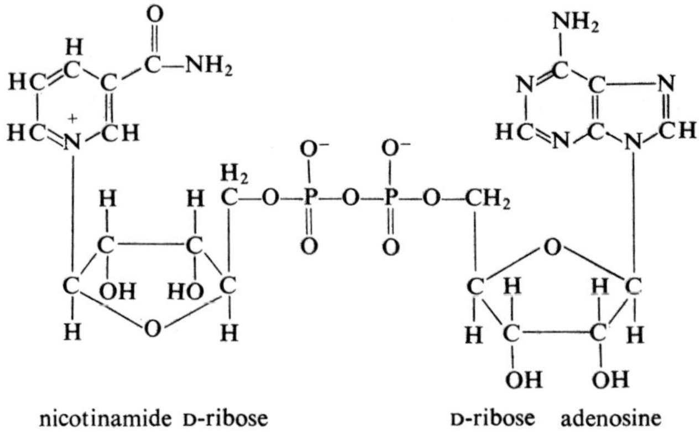
nicotinamide D-ribose
D-ribose adenosine

Fig. 2.4. Nicotinamide–adenine dinucleotide (NAD).

NAD^+ can form a complex with an enzyme, E I. In this catalytic system, the NAD can oxidize a substrate and in so doing is reduced to NADH + H^+. This reduced NAD can no longer form a complex with enzyme I and is released. The NADH + H^+ can then complex with an enzyme II at some other stage in a metabolic pathway and in this complex can reduce another substrate, itself being reoxidized to NAD^+ and again released, thus returning to the cycle (Fig. 2.6). NAD^+ therefore forms a link between two enzymes.

The production of NADH + H^+ during a reaction causes a rise in the spectrophotometer absorption band at 340 nm. Because this rise in absorp-

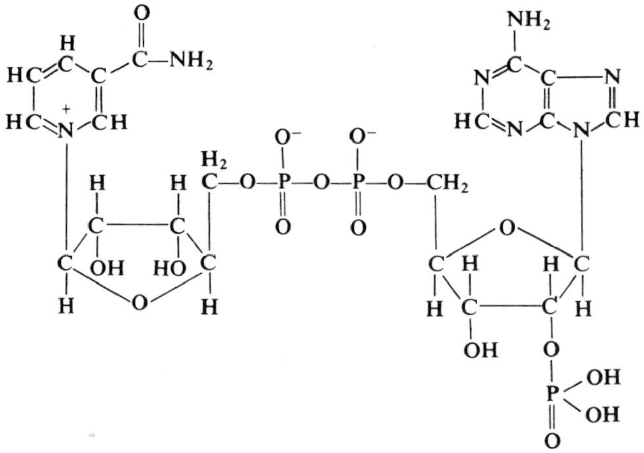

Fig. 2.5. Nicotinamide–adenine dinucleotide phosphate (NADP).

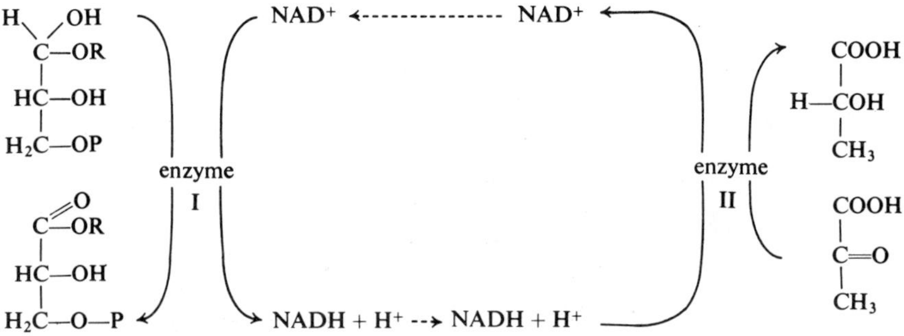
NAD+
NAD+
enzyme I
enzyme II
NADH + H+
NADH + H+
COOH
H—COH
CH3
C=O

Fig. 2.6. The role of $NAD^+/NADH + H^+$ in enzyme catalysis.

tion can be followed quite easily the transition can be followed with any spectrophotometer. The oxidation of reduced NAD, of course, can also be followed optically by way of the disappearance of the 340-nm peak. The increase or decrease of absorption per unit time is proportional to the enzyme concentration.

The reduction of NAD^+ occurs very frequently in biochemical systems; the reaction is symbolized in the following manner:

$$NAD^+ + CH_3{-}CH_2OH \rightleftharpoons CH_3CHO + NADH + H^+$$

This coenzyme functions particularly in dehydrogenation processes of primary and secondary alcohol groups and is associated mainly with a large number of dehydrogenases. As these dehydrogenation reactions are generally reversible, the importance of this coenzyme rests on its role of reversibly transferring electrons (with associated protons). The coenzyme does not react catalytically by itself but, in association with the apoenzyme, reacts with the substrate strictly stoichiometrically (Fig. 2.6). The reoxidation of reduced NAD also functions strictly stoichiometrically. Because such coenzymes as NAD can couple to different enzymes, they are links by means of which the exchange of material becomes possible; NAD occurs in all cells, and its role may be compared with that of ATP. Adenosine triphosphate is a universal phosphate carrier, and NAD is found universally as an electron carrier in cells.

FLAVINE NUCLEOTIDES. Flavoproteins are a class of oxidizing enzymes containing as electron acceptor flavin–adenine dinucleotide (FAD), which is an electron carrier similar to NAD in its action. This molecule is remarkable for having as a building block the vitamin riboflavin or vitamin B_2 (Fig. 2.7). In contrast to nicotinamide, riboflavin is an isoalloxazine derivative, which means a pteridine ring with a benzene ring fused onto

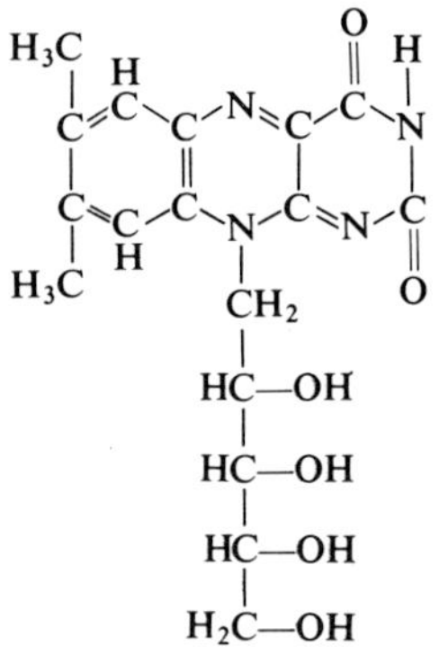

Fig. 2.7. Riboflavin (6,7-dimethyl-9-ribityl-isoalloxazine).

it. The side chain is a five-carbon polyhydroxy group. It is not an *N*-glycoside of ribose but is a derivative of ribitol.

Most flavoproteins contain the dinucleotide, rather than the mononucleotide. The linkage between the two mononucleotides is the same as in the case of NAD, a pyrophosphate bond (Fig. 2.8). The isoalloxazine ring acts here as a reversible redox system with the enzyme. Hydrogen is added to N-1 and N-10:

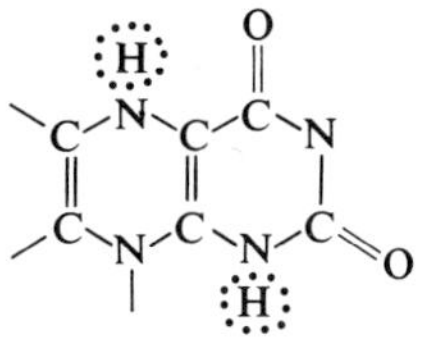

For the enzyme to retain its catalytic power the flavin system must be reoxidized. This is usually accomplished by another enzyme system. A number of flavoproteins react directly with molecular oxygen, reducing the oxygen molecule to H_2O_2, i.e., lactate oxidase (L-lactate:oxygen oxidoreductase, EC 1.1.3.2), pyruvate oxidase [pyruvate:oxygen oxidoreductase (phosphorylating), EC 1.2.3.3], and xanthine oxidase (xanthine:oxygen oxidoreductase, EC 1.2.3.2), whereas others use NAD or NADP as acceptor. There is some evidence that in the reduction of several flavoproteins only one electron is taken up (22). The result is a semiquinone with the properties of radicals (unpaired electrons). The semiquinone is very active and can easily donate the accepted electron, for example, to Fe^{3+} ions or to oxidized cytochromes. Some flavoproteins also contain tightly bound metal ions that probably participate in catalysis.

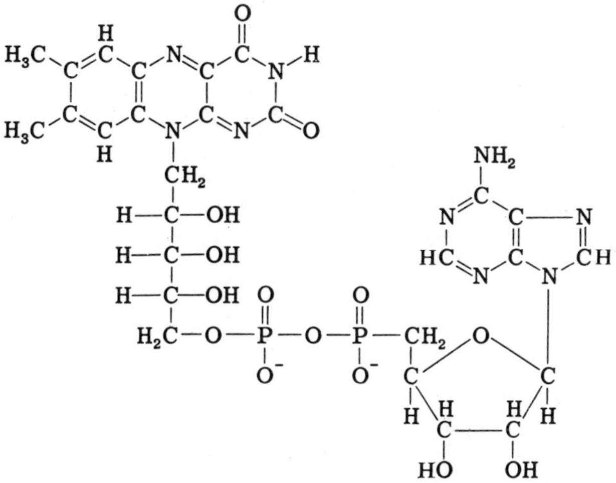

riboflavin D-ribose adenosine

Fig. 2.8. Flavin–adenine dinucleotide (FAD).

LIPOIC ACID. The name "lipoic acid" was first used by Reed *et al.* (76) for the crystalline organic acid isolated from an extract of beef liver. It was highly active in assays for a substance that replaces acetate in lactic acid bacteria (acetate-replacing factor) (31, 32). Bullock *et al.* (12) proposed the name "6-thioctic acid." However, "lipoic acid" has been adopted as the trivial name of 1,2-dithiolane-3-valeric acid (Fig. 2.9). Its biological activity depends on the great reactivity of the disulfide bond of the dithiolane ring (79). The polymerization to linear disulfides is a characteristic of 1,2-dithiolane. Lipoic acid exhibits a spectrophotometric absorption maximum at 330 nm and is widely distributed among microorganisms, plants, and animals. The only well-defined role of lipoic acid is as a prosthetic group in multienzyme complexes (28) that catalyze an oxidative decarboxylation of pyruvate and α-ketoglutarate to yield acetyl-CoA and succinyl-CoA, respectively. This acyl transfer mechanism is carried out together with NAD^+, thiamine pyrophosphate, and also with FAD^+ (see p. 384).

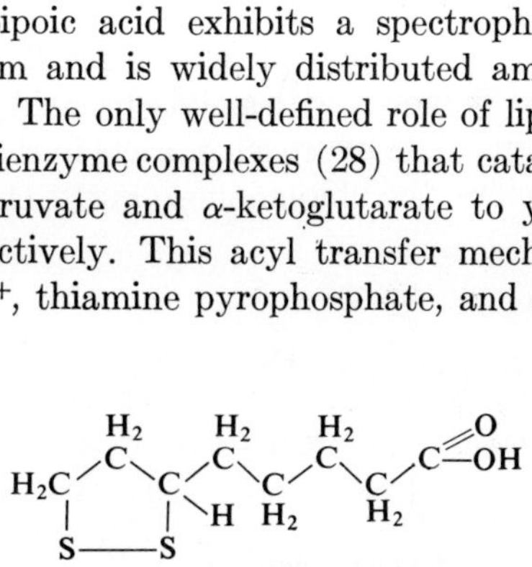

Fig. 2.9. Lipoic acid.

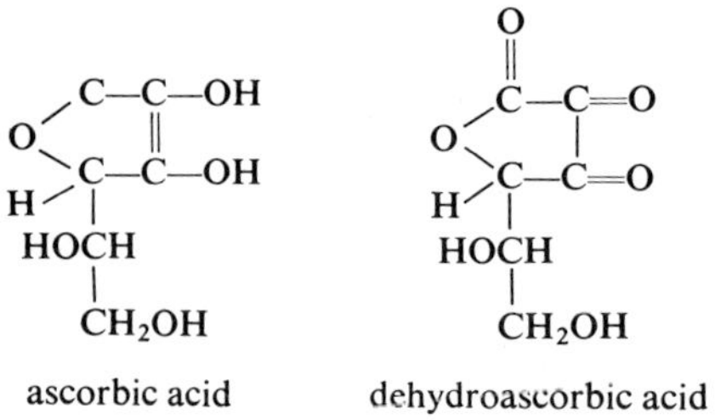
ascorbic acid
dehydroascorbic acid

Fig. 2.10. Ascorbic and dehydroascorbic acid.

GLUTATHIONE. Glutathione is a widely used and distributed sulfur-containing peptide, the functional group of which is the thiol group. It is oxidized by molecular oxygen under suitable conditions in the presence of traces of catalytic metals, and also by cytochrome c. In order to reduce the oxidized form back to the thiol form, either powerful reducing agents or the enzyme glutathione reductase [reduced NAD(P):oxidized glutathione oxidoreductase, EC 1.6.4.2] and NAD^+ or $NADP^+$ are needed. Because glutathione undergoes enzymatic oxidation and reduction, it can act as a biological hydrogen carrier, particularly in the chemoautotrophs (see p. 341). Its function, however, is still obscure (16, 22).

ASCORBIC ACID. L-Ascorbic acid or vitamin C is an effective reducing agent useful in culturing anaerobic bacteria. Although the substance exists in an oxidized and a reduced form, the name "ascorbic acid" is given to the reduced form only and the oxidized form is called "dehydroascorbic acid" (Fig. 2.10). Being a good reducing agent, ascorbate, like glutathione, may play a part in maintaining the activity of the —SH enzymes.

QUINONES. Quinones are widespread in living cells and some of them, mainly the methylated quinones with polyisoprenoid side chains, play an important part as intermediate hydrogen carriers in the respiratory system. These include two families, the ubiquinones (Fig. 2.11) and the vitamin K

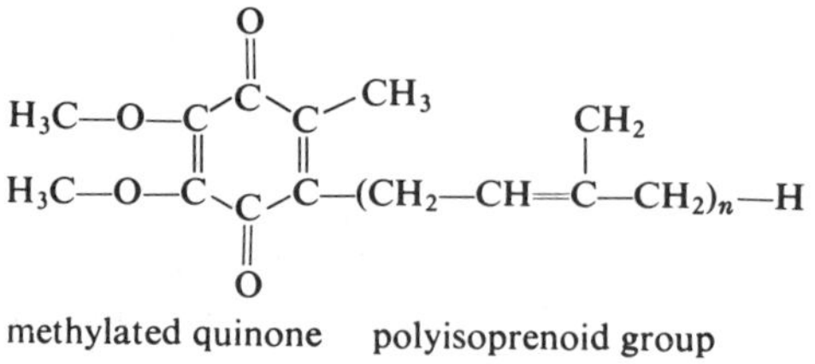
methylated quinone
polyisoprenoid group

Fig. 2.11. Ubiquinone.

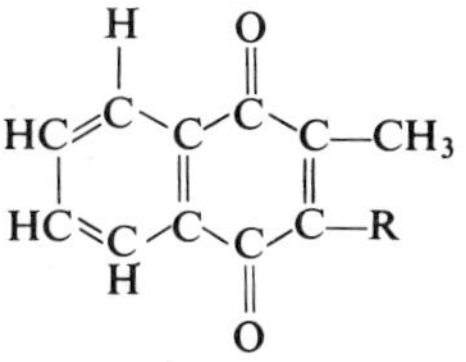

Fig. 2.12. Vitamin K.

group (Fig. 2.12). The vitamin K group consists of vitamin K, vitamin K_2, and menadione. Ubiquinone was first discovered by Morton (64). It has been variously known as "272 mμ substance," "SA," "coenzyme Q," and "Q_{275}," but the IUB recommended the term "ubiquinone," although "coenzyme Q" is still widely used. It is still difficult to say how many substrates depend for their oxidation on ubiquinone. The reduced form of ubiquinone (ubiquinol) is oxidized through the cytochrome system (35). The oxidation depends on cytochrome c and cytochrome oxidase and is quite often coupled to phosphorylation.

Vitamin K has been reported in the electron transport processes of photosynthetic phosphorylation in bacteria, although ferredoxin may replace it in this role.

Electron-Transferring Coenzymes

HEME-IRON ELECTRON TRANSFER PROTEINS. *Cytochromes.* At present the name "cytochromes" appears to include all intracellular hemeproteins with the exception of hemoglobin, myoglobin, peroxidase, and catalase (39, 42). The group includes substances with many different functions, but they all appear to act through oxidation and reduction of the iron. The nomenclature of individual cytochromes has become difficult owing to the isolation from different sources of cytochromes with similar properties. The recommendations of the IUB in regard to the classification and nomenclature can be found in Chapter 5 of the Commission's report (40). Four major groups of cytochromes have been established as follows:

1. Cytochrome a: cytochromes in which the heme prosthetic group contains a formyl side chain, e.g., heme a. The heme a is linked with the apoprotein noncovalently (42) and is therefore removable from the apoprotein by extraction with acid acetone. The pyridine ferrohemechrome (reduced cytochrome) shows an α-absorption band of 580–590 nm.

2. Cytochrome b: cytochromes with noncovalently bound protoheme as prosthetic group come into this category. The pyridine ferrohemechrome shows an α-absorption band of 556–558 nm (Fig. 2.13).

3. Cytochrome c: cytochromes c are defined as those cytochromes with

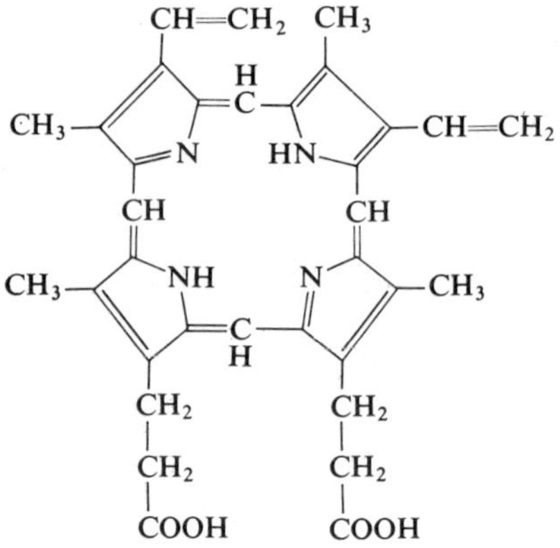

Fig. 2.13. Cytochrome b.

a covalently linked heme prosthetic group. To date, only thioether linkages are known. The pyridine ferrohemechrome shows an α-absorption band of 549–551 nm.

4. Cytochrome d: cytochromes that exhibit the same characteristics as cytochrome a, but with the α-absorption band of the pyridine ferrohemechrome at 600–620 nm, are classified as cytochrome d (42). These cytochromes were originally called "cytochrome a_2." They are autoxidizable and combine with CO and CN^- (25, 68, 88, 99, 100).

Bacterial cytochrome a: bacterial cytochrome a can be classified into two subgroups (42):

a. cytochrome a, which is similar to mammalian cytochrome a and shows α and γ bands at 600–605 nm and 440–445 nm, respectively. This cytochrome is termed "cytochrome a and a_3," if the spectrochemical responses of the α and γ bands toward CO and CN^- are similar to those of the mammalian cytochrome (aa_3) (93). In the presence of CO, the α band, 606 nm, and the γ band, 445 nm, of reduced cytochrome a_3 shift to approximately 590 and 430 nm, respectively. As the spectra of cytochrome a are not affected by CO, a separation between the two is possible (94).

b. cytochrome a_1, which has an α-absorption band between 585 and 595 nm and a γ band around 435–445 nm.

Bacterial cytochrome b: the range of α-absorption bands of all bacterial cytochromes b isolated so far was found to be from 557 to 562 nm. Very frequently a subdivision can be found into cytochrome b and cytochrome b_1. This differentiation was done in the past, and the redox pigment with the α-reduced band at 562–565 nm was named "cytochrome b" and all those with a reduced band at 557–560 nm were named "cytochromes b_1."

Two heme proteins also exist, which do not exhibit the characteristic spectra of cytochrome b but have protoheme bound noncovalently. Hemeprotein 558 was isolated from *Acetobacter suboxydans* (41) and hemeprotein P-450 from *Rhizobium japonicum.* The latter hemeprotein appears to function mainly as a terminal oxidase particularly, in mixed function oxidations (17, 33, 44, 80, 91).

Another variant of the b-type cytochrome appears to be cytochrome o, which obtained its name from oxidase (13). Besides cytochromes d, a_3, and a, it is a fourth unique type of CO-binding hemeprotein. The cytochrome o-CO complex exhibits α, β, and γ bands at 557–567, 532–537, and 415–420 nm, respectively. The absolute absorption spectrum of cytochrome o is still unknown, however (42).

Bacterial cytochrome c: although no purified bacterial c-type cytochrome has appeared to be of the mitochondrial type, comparative biochemistry of these cytochromes has revealed at least six different classes. The criteria used in these studies include absorption bands, ability to form a complex with such ligands as CO, molecular size, isoelectric pH, standard redox potential, and amino acid sequence of the peptide chains:

1. Cytochrome c_2: the first bacterial cytochrome to be purified came from *Rhodospirillum rubrum* (98). It is a readily soluble cytochrome c with a molecular weight of 12,000–14,000 and has a redox potential greater than +0.28 V. Cytochrome c_2 contains a single heme with a reduced α band at 550–552 nm. Since its discovery in *Rhodospirillum rubrum,* it has been found in a number of other photosynthetic bacteria (20).

2. Cytochrome c_3: this cytochrome was the first cytochrome obtained from a strictly anaerobic bacterium, *Desulfovibrio desulfuricans* (73). It has an extremely high reducing standard potential, −0.25 V, and is consequently readily autoxidizable. It contains three heme groups, but an analog with two hemes has been isolated from *D. gigas* (50). The molecular weight of cytochrome c_3 is approximately 13,000. As will be seen below, cytochrome c_3 is an integrated part of the bacterial electron transfer system, which is involved in the phosphoroclastic reaction (see p. 596), the reduction of sulfite (see p. 163), the reduction of sulfate (see p. 159), and the evolution of hydrogen gas by hydrogenase (see p. 117). A cytochrome c obtained from *Escherichia coli,* which has been designated "cytochrome c-552," appears to be very similar to cytochrome c_3, mainly because of its low redox potential, −0.2 V (15, 26, 27, 29, 30, 102).

3. Cytochrome c_4 and c_5: both cytochromes were isolated from *Azotobacter vinelandii.* Cytochrome c_4 exhibits a reduced α band at 551 nm and c_5 one at 554 nm. Both cytochromes have a molecular weight of 12,000. Their standard redox potentials are +0.3 and +0.32 V, respectively.

4. Cytochromes c′ and cc′: a further group of cytochromes exists having one or two covalently bound heme groups but distinctly different spectroscopic and chemical properties (6). The monoheme cytochrome c′ has a molecular weight range of between 13,000 and 29,000, and the diheme cytochrome cc′ one of between 27,000 and 30,000. Although neither cytochrome has a hemechrome peak in the visible region, their γ peaks are characteristic. Whereas the ferricytochrome shows absorption bands at 620–640, 520, and 390–400 nm, the ferrocytochrome exhibits a broad band near 550–560 nm and γ peaks near 420 and 435 nm. The standard redox potentials are identical and range from 0 to 0.1 V. Cytochrome c′ has been found in *Rhodopseudomonas palustris* (20, 36) and *Rhodospirillum molischianum,* whereas cytochrome cc′ is present in *Rhodospirillum rubrum* (37, 38), *Chromatium* (5), and one strain of *Pseudomonas denitrificans* (84, 85). Cytochrome cc′ and cytochrome c′ substitute for those hemeproteins, which were earlier designated for RHP (*Rhodospirillum* heme protein), cytochromoid c and cryptocytochrome c.

NONHEME-IRON ELECTRON TRANSFER PROTEINS. *Ferredoxin.* During the last two decades, a new electron carrier has been discovered that functions on the hydrogen side of NAD^+ and $NADP^+$, in contrast to the cytochromes, which function as electron acceptors of reduced NAD and reduced NADP. As these ferredoxins transfer hydrogen to NAD^+ and $NADP^+$, the oxidation–reduction potential is approximately -0.40 V (pH 7.0), which lies between molecular hydrogen and the $NAD^+/NADH + H^+$ system. As electron carriers, ferredoxins have only one feature in common. The electron donor is either hydrogen gas or a photochemical reaction that liberates electrons with a reducing power at least equal to that of molecular hydrogen (42). Hydrogen gas can also be substituted by a compound that can give rise to molecular hydrogen or an equivalent reducing power.

Ferredoxins are nonheme-iron-containing proteins and are mainly found in anaerobic bacteria and in chloroplasts (11). The first isolation was from *Clostridium pasteurianum* and the actual name was introduced about 1962 (63). It was left to Tagawa and Arnon (86) to establish the relationship between the *C. pasteurianum* ferredoxin and some proteins isolated from chloroplasts earlier. Ferredoxins play a vital role in photosynthesis, fermentation, and aerobic nitrogen fixation. In each of these processes, ferredoxin is an electron carrier, but its chemical constitution is different. To date, there are four main ferredoxins, which are all colored proteins with distinct absorption spectra.

1. Ferredoxin a is the *Clostridium pasteurianum* ferredoxin and generally occurs in all fermentative and nonphotosynthetic green anaerobic bacteria. Its absorption spectrum shows a single peak at 390 or 385 nm and a peak

in the UV region at about 280 nm, with a shoulder at 300 nm. The molecular weight of 6000 has been confirmed in a number of clostridial species (21, 52) and other bacteria (9, 10, 92). The molecule contains more than two iron and sulfide groups.

2. Ferredoxin a_1 is the *Chromatium* ferredoxin and occurs in all photosynthetic purple sulfur bacteria. The molecular weight was found to be 10,000 (65, 78). Ferredoxin a_1 appears to be much more electronegative (approximately −0.49 V at pH 7.0) than the other three ferredoxins. Its absorption maxima are the same as the *C. pasteurianum* ferredoxin, but it contains seven to eight iron and sulfide groups to the molecule.

3. Ferredoxin b, is the spinach chloroplast ferredoxin. Its absorption maxima are at 465, 425, 325, and 280 nm and the molecular weight appears to be 12,000, which is characteristic for all plant ferredoxins (7, 45, 57, 75, 83, 87), including algae. It contains two iron and 2 labile sulfide groups.

4. Ferredoxin c occurs in the aerobic nitrogen-fixing bacteria and was first characterized from *Azotobacter vinelandii*. With a molecular weight of 20,000, it is by far the largest ferredoxin molecule yet known (103). It has a distinctive absorption maximum at 400 nm. The molecule contains six iron and sulfide groups.

It is now the opinion of many workers that the *C. pasteurianum* ferredoxin possesses two independent one-electron sites, which means that all ferredoxins basically mediate single-electron transfers. Therefore, the bacterial ferredoxin can accommodate two transfers per molecule, whereas plant ferredoxins can accommodate only one.

Rubredoxin. In contrast to ferredoxin, rubredoxin contains only one iron atom per molecule and no inorganic sulfide (54, 58). It functions as a one-electron carrier in oxidation–reduction reactions and has a much higher oxidation–reduction potential than bacterial ferredoxin (−0.57 V). Rubredoxin shows absorption maxima at 490, 380, and 280 nm (53, 57) and has a molecular weight of approximately 6000. Rubredoxin has been found to participate in a number of biological reactions in which ferredoxin is also active (56). As was ferredoxin, rubredoxin was isolated first from *Clostridium pasteurianum* (53, 54).

PTERIDINE ELECTRON TRANSFER PROTEINS. *Flavodoxin.* If organisms are grown on media containing insufficient iron to support the synthesis of ferredoxin, flavodoxin can be substituted for ferredoxin (69). Bacterial flavodoxins are devoid of metal and labile sulfide (48, 59), possess a molecular weight of around 15,000, and contain 1 mole-equivalent of FMN (49, 59). For the couples oxidized flavodoxin–flavodoxin semiquinone and flavodoxin semiquinone–reduced flavodoxin, the standard redox potentials are −0.115 V and −0.372 V, respectively. As they substitute for ferredoxin, it is probable that the second redox couple is the more important reaction. So far, bacterial flavodoxins have been isolated from *C. pasteurianum* (47–49), *Desulfovibrio gigas* (23), *Peptostreptococcus elsdenii* (59, 60), and *Azotobacter vinelandii* (95). Differences in the amino acid composition of purified flavodoxins from *C. pasteurianum*, *C.* MP (61), and *Peptostreptococcus elsdenii* (89) may indicate further subdivisions.

Diaphorases. Diaphorases are flavoproteins capable of catalyzing the oxidation of a reduced pyridine nucleotide by organic dyes, such as Methylene Blue or Indophenol, or by such inorganic compounds as ferricyanide. However, they are not capable of reacting directly with molecular oxygen. Most of these enzymes have FMN or FAD as prosthetic group and a molecular weight between 50,000 and 70,000 (see Chapter 8).

Group-Transferring Coenzymes

PHOSPHATE CARRIERS. The two most important types of transport reaction are electron transport and phosphate transport. The first type is concerned with the production of energy and the second with its transfer from one process to another. The biological carriers of phosphates are the nucleoside phosphates ADP and ATP, which act as cofactors in transphosphorylation processes. The enzymes that catalyze the transfer reactions to and from these phosphate carriers are the kinases (EC 2.7.1–4). It is not clear how far other nucleoside phosphates, such as inosine diphosphate (IDP), guanosine diphosphate (GDP), cytidine diphosphate (CDP), and uridine diphosphate (UDP), can act as phosphate carriers. The mechanism of formation of ATP as an energy-rich compound has been dealt with above. It contains two "energy-rich bonds" and has a high capacity for group transfers. ATP action can be differentiated into (*a*) transfer of the terminal phosphate group with release of ADP; (*b*) transfer of the pyrophosphate group with a release of AMP; (*c*) transfer of the adenosyl monophosphate group with release of pyrophosphate; and (*d*) transfer of the adenosyl group with release of both orthophosphate and pyrophosphate.

The transfer of a terminal phosphate group of ATP was explained above. In the case of pyrophosphate, however, this type of transfer cannot occur, nor can it be used by the cell to participate in the phosphorylation of ADP during oxidative phosphorylation or fermentation. It must first be enzymatically hydrolyzed to orthophosphate by an enzyme called "inorganic pyrophosphatase" (pyrophosphate phosphohydrolase, EC 3.6.1.1). This enzymatic hydrolysis of pyrophosphate proceeds with a large negative free energy change.

It is thought that ATP is the main line of transfer of bond energy, whereas all the other nucleoside 5′-triphosphates and the deoxynucleoside 5′-triphosphates function to channel ATP energy into different biosynthetic pathways (15). Such a channeling process is made possible because the terminal phosphate group of ATP can be enzymatically transferred to the 5′-diphosphates of the other nucleosides by the action of enzymes called "nucleosidediphosphate kinases" (ATP:nucleosidediphosphate phosphotransferase, EC 2.7.4.6).

$$ATP + GDP \rightleftharpoons ADP + GTP$$

$$ATP + UDP \rightleftharpoons ADP + UTP$$

$$ATP + CDP \rightleftharpoons ADP + CTP$$

$$ATP + dADP \rightleftharpoons ADP + dATP$$

where dADP is deoxyadenine diphosphate.

All these phosphate carriers have about the same free energy of hydrolysis of the terminal phosphate group as ATP. However, only ADP can accept phosphate groups during oxidative phosphorylation. The ADP–ATP system is therefore necessary to phosphorylate GDP, UDP, etc., and to fill each channel with high-energy phosphate groups. Lehninger (51) illustrated this as shown in Fig. 2.14.

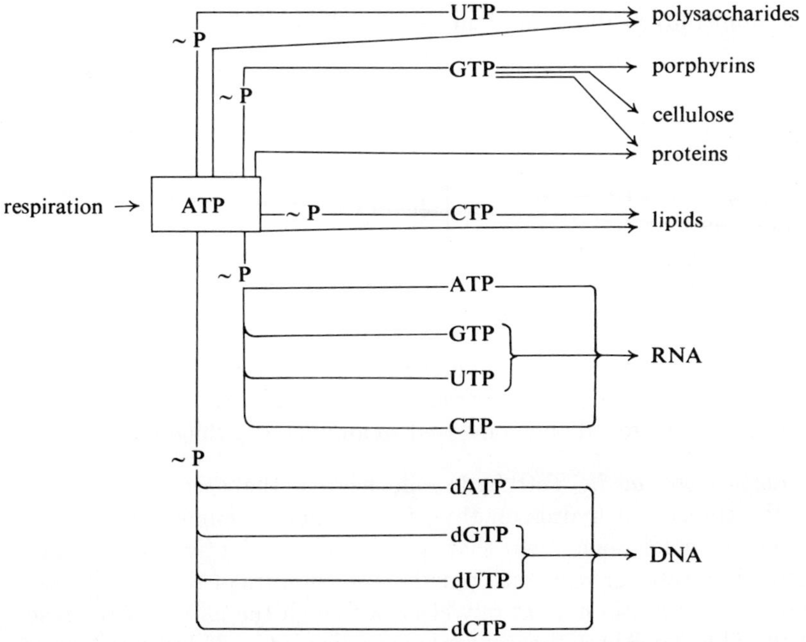
UTP
polysaccharides
~P
GTP
porphyrins
~P
cellulose
proteins
respiration →
ATP
~P
CTP
lipids
~P
ATP
GTP
UTP
CTP
RNA
~P
dATP
dGTP
dUTP
dCTP
DNA

Fig. 2.14. Distribution of energy from ATP for biosynthesis (reprinted with permission of the author and W. A. Benjamin, Inc.).

It appears probable that ATP occupies the central role in energy-transfer reactions because it may have been the first nucleoside 5′-triphosphate formed in the prebiotic era. It was therefore selected for the role of energy carrier in the first primitive cell. Certain thiol esters, e.g., acetyl-CoA, can be formed in cells without the intervention of ATP; in anaerobic glycolysis, probably a very primitive process of energy production, it serves as a precursor of ATP. On these grounds, the CoA derivatives could possibly represent the evolutionary ancestors of ATP.

ONE-CARBON GROUP CARRIER. Just as ATP can carry and transfer phosphate groups, coenzymes also exist that can transfer one-carbon groups, such as (*a*) a methyl group (CH_3) from methanol $HOCH_3$, (*b*) a hydroxymethyl group (CH_2OH) from serine ($HOCH_2 \cdot NH_2 \cdot COOH$), (*c*) a formyl group (CHO) from formic acid (HCOOH), or (*d*) a carboxyl group (COOH) from carbonic acid (HO—COOH). In the transfer of the methyl group, adenosylmethionine or "active methyl" appears to be the most important donor of such groups, wherein the methionine component plays the donating role. The transfer of these methyl groups is mostly carried out to amino groups. As the thioether-linked methyl group does not possess a high potential, ATP is necessary for its activation, in order to form first a sulfonium (43) in the form of adenosylmethionine:

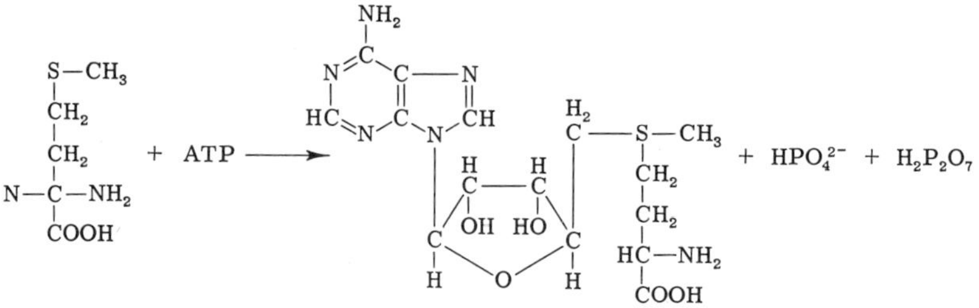
S—CH3
CH2
CH2
N—C—NH2
COOH
+ ATP
NH2
N
C
N
HC
CH
H2
C
S—CH3
+ HPO4 2−
+ H2P2O7
H
H
OH HO
CH2
CH2
HC—NH2
COOH
O

The methyl group, which is bound to the sulfur, can now be transferred to an atom that has a free electron pair. With this reaction, the sulfonium structure is destroyed and converted to an adenosyl thioether.

Tetrahydrofolic acid. Tetrahydrofolic acid is the coenzyme responsible for the transfer of hydroxymethyl, formyl, and formimino groups. It has been known for some time that a folic acid derivative is important in one-carbon transfer reactions, but the exact nature of this factor was long in doubt and is still not entirely clear, although the nature of the reacting group is known. The active carrier is not folic acid itself, but tetrahydrofolic acid (THFA). In bacterial metabolism it is the polyglutamate derivative of folic acid that stimulates growth of a number of strains. The activated

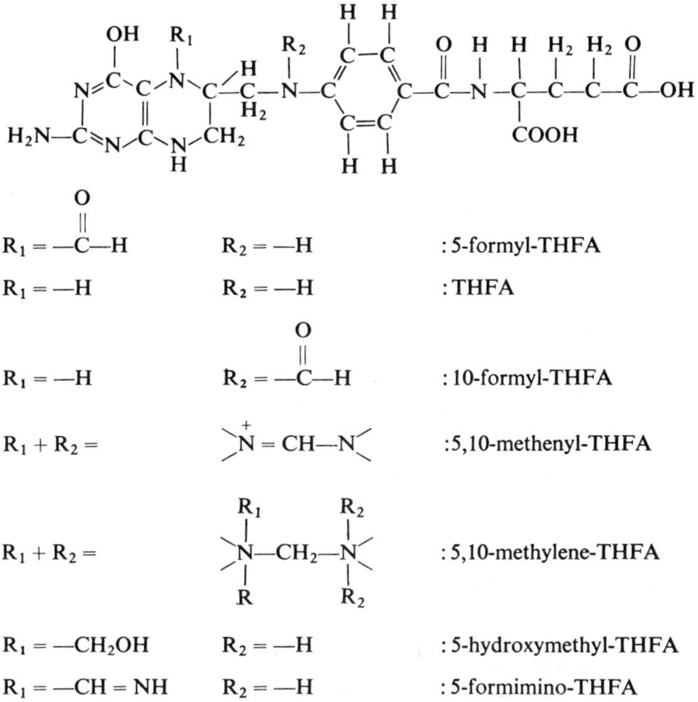
$R_1 = —CHO$ $R_2 = —H$: 5-formyl-THFA
$R_1 = —H$ $R_2 = —H$: THFA
$R_1 = —H$ $R_2 = —CHO$: 10-formyl-THFA
$R_1 + R_2 =$ $>\overset{+}{N} = CH—N<$: 5,10-methenyl-THFA
$R_1 + R_2 =$ $>N—CH_2—N<$: 5,10-methylene-THFA
$R_1 = —CH_2OH$ $R_2 = —H$: 5-hydroxymethyl-THFA
$R_1 = —CH = NH$ $R_2 = —H$: 5-formimino-THFA

Fig. 2.15. Tetrahydrofolic acid (THFA) and its derivatives.

formaldehyde is probably bound to N-10 as a hydroxymethyl group, but easily joins on to N-5, thus forming a ring system, and the new derivative is the 5,10-methylene-THFA (Fig. 2.15). An important donor for the hydroxymethyl group is serine, which is able to transfer its α-carbon atom to THFA.

The activated formic acid was identified in form of the 10-formyl-THFA. The high-energy intermediate, 5,10-methenyl-THFA, which converts easily to 10-formyl-THFA in weak basic solution, ought also to be mentioned. 10-formyl-THFA can also be formed from formic acid and THFA, a reaction that does require ATP. However, the C_1 fragment is normally formed during metabolism and ATP is therefore not necessary. Activated formic acid is mainly needed in purine synthesis.

Tetrahydrofolate is highly autooxidizable, is rapidly converted into dihydrofolate by molecular oxygen, and is also oxidized enzymatically by NADP. Folate is reduced to di- and tetrahydrofolate by reduced NADP and the enzyme dihydrofolate dehydrogenase (7,8-dihydrofolate:NADP oxidoreductase, EC 1.5.1.4). Tetrahydrofolate differs from other carriers

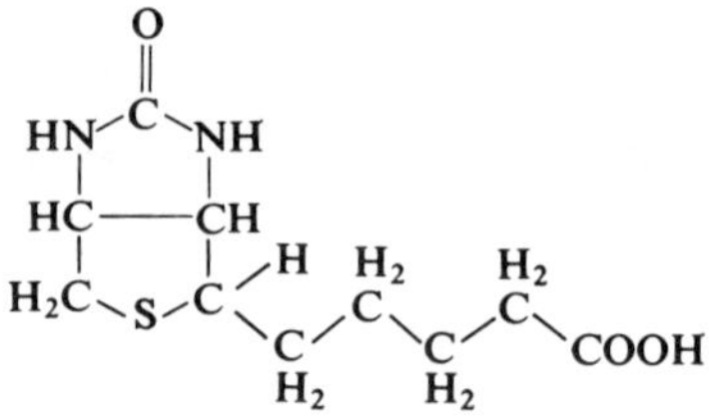
O
HN
C
NH
HC
CH
H
H2
H2
H2C
S
C
C
C
C
C
C
COOH
H2
H2

Fig. 2.16. Biotin.

in that the group transported may undergo a change while it is attached to the carrier, so that the group given up to the acceptor may not be identical with that which was taken up by it.

Biotin. Biotin is known to be an vitamin (46). It is a cyclic derivative of urea with a thiophan ring attached to it (Fig. 2.16). Its function as a cofactor of enzymatic reactions involving incorporation or transfer of CO_2 was discovered not so long ago. Biotin is linked to the enzyme protein by a peptide bond to the ϵ-amino group of a lysyl residue. The charging of biotin with carbon dioxide is an endergonic process that requires ATP (Fig. 2.17). The carbon dioxide attached to the nitrogen of biotin is the "active form of carbon dioxide" that participates in numerous carboxylation reactions, such as the formation of malonyl-CoA from acetyl-CoA. The very labile carboxybiotin has been isolated as a methyl ester.

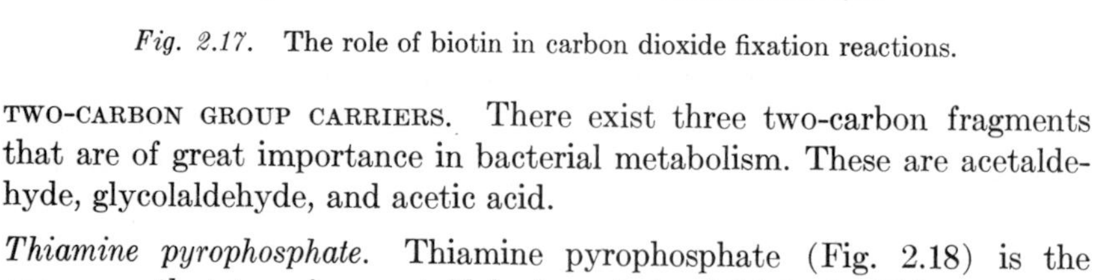
HCO3⁻
ATP
NH [enzyme]
NH[enzyme]
-OOC
biotin-enzyme
carboxybiotin-enzyme

Fig. 2.17. The role of biotin in carbon dioxide fixation reactions.

TWO-CARBON GROUP CARRIERS. There exist three two-carbon fragments that are of great importance in bacterial metabolism. These are acetaldehyde, glycolaldehyde, and acetic acid.

Thiamine pyrophosphate. Thiamine pyrophosphate (Fig. 2.18) is the coenzyme that transfers acetaldehyde and glycolaldehyde. This coenzyme has also been called "cocarboxylase," "diphosphothiamine (DPT)," and "vitamin B_1 pyrophosphate." The characteristic component is thiamine (vitamin B_1), the name referring to its sulfur content. The structural formula has two uncondensed heterocyclic rings: one pyrimidine and one thiazol. It therefore always carries a charge. It behaves as a prosthetic

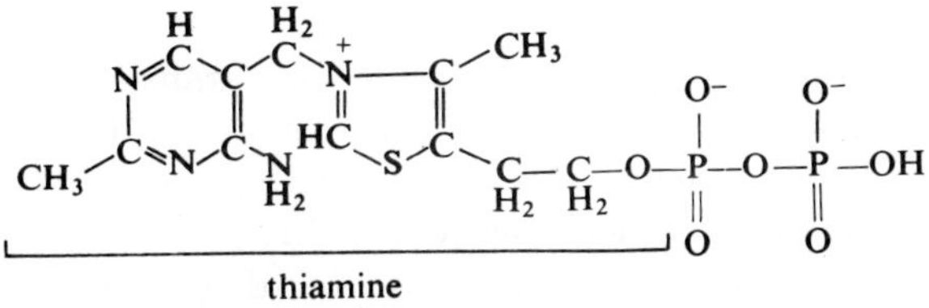
thiamine

Fig. 2.18. Thiamine pyrophosphate (TPP).

group because it remains bonded to the enzyme protein. In all reactions involving such α-keto acids as pyruvate, the C-2 of the thiazol ring binds the α-keto group in its polarized form, thus catalyzing the decarboxylation. The so-called active acetaldehyde on the C-2 of the thiazol ring is then transferred to the adjacent amino group and can now be set free or transferred to an acceptor (43). This intermediate is called "α-hydroxyethyl-thiamine pyrophosphate" and its fate depends on which enzyme is involved in the following reaction:

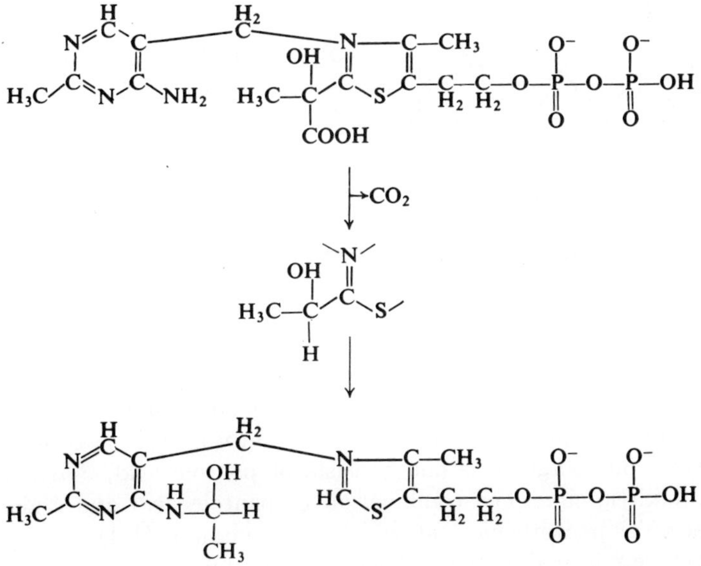

The most important reactions involving TPP are therefore the oxidative decarboxylations of α-keto acids, whereby the acetaldehyde residue can be transferred to lipoic acid, which functions as an oxidizing agent.

In the transketolase reaction TPP functions simply as a group-transferring coenzyme in transferring glycolaldehyde.

Coenzyme A. Recent studies of intermediary metabolism of fats and carbohydrates drew attention to the acyl-group transfer, particularly by carriers containing thiol groups, with which they form thiol esters. These energy-rich compounds share with energy-rich phosphate compounds the important function of biological energy transfer.

Coenzyme A (CoA) was originally discovered by Lipman in 1947 and transfers carboxylic acids. The letter "A" stands for acylation. Acids bound to CoA have a high capacity for group transfer. The two-carbon fragment or "active acetyl" known to be involved in carbohydrate metabolism was identified as acetyl-CoA. It was later found that not only acetyl-CoA but also acyl-CoA compounds generally are very important in metabolism, which is shown in the fact that over 60 enzymes act on acyl-CoA compounds. Hydrolysis of an acyl-CoA compound is exergonic to the extent of about 8000 cal/mole.

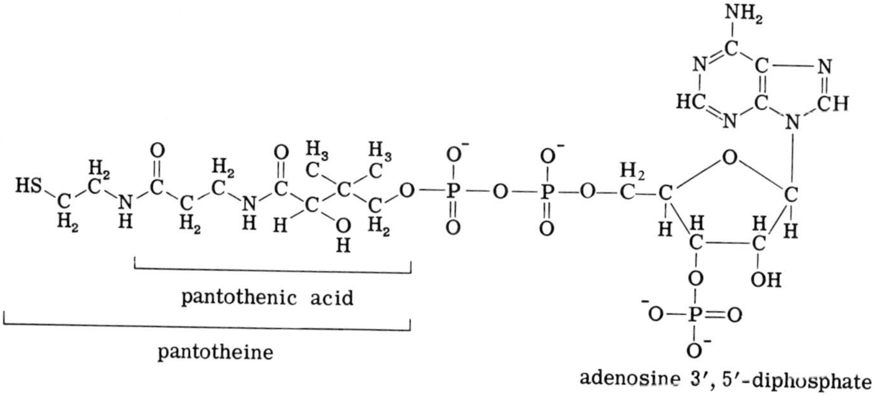
pantothenic acid
pantotheine
adenosine 3′, 5′-diphosphate

Fig. 2.19. Coenzyme A.

The chemical structure of coenzyme A is considerably more complex than that of cytochrome or NAD. It consists of adenosine 3′,5′-diphosphate and pantotheine phosphate (Fig. 2.19). Pantotheine is a growth factor for several microorganisms and consists of pantoic acid, β-alanine, and mercaptoethylamine. The combination of pantoic acid and β-alanine is also known as pantothenic acid, which is a B vitamin (1-4).

Acetyl-CoA is undoubtedly the most important CoA compound. The acetyl residue, CH_3CO—, is bound to the free SH group, which constitutes a very reactive thioester. In order to bring acetate, or any other carboxylic acid, into this compound, energy is required. This energy is most often derived from a strongly exergonic reaction, such as oxidative decarboxylation. Another possibility, of course, is from the cleavage of ATP, where an intermediate arises in the form of an anhydride with adenylic acid.

Coenzyme A is a colorless substance having a spectrophotometric absorption band in the region of 257 nm owing to the adenine residue. All the enzymes of subgroup EC 6.2.1 (acid–thiol ligases) bring about the formation of CoA thiol esters from free acids, making use of the energy of ATP or GTP. All acyltransferases (EC 2.3.1) transfer acyl groups to or from CoA. The transfer may be from a combination with another S, N, O, or C atom or from phosphate. Acetyl-CoA is also involved in reactions catalyzed by lyases. The acetyl group is added to a double bond in the acceptor molecule, either with or without hydrolysis of the thioester bond. The three synthase reactions, for example, form the important metabolites citrate, malate, and hydroxymethylglutaryl-CoA.

Coenzyme A is essential for the initiation of the tricarboxylic acid cycle (TCA), for the oxidative breakdown of fatty acids, and for various biosynthetic processes.

OTHER GROUP-TRANSFERRING COENZYMES. *Pyridoxal phosphate.* Pyridoxal phosphate is the coenzyme of amino acid metabolism (24). In some of its reactions, which are transaminations, it acts as an amino carrier. Pyridoxal phosphate is an excellent example of a single coenzyme capable of catalyzing completely different reactions. It is the active group not only of the aminotransferases but also for the decarboxylases and various lyases and synthetases. The mechanism of action is believed to be in all cases the formation of an azomethine (Schiff's base) (Fig. 2.20) by combination of its aldehyde group with the amino group of the substrate. The fate of this substance depends on the nature of the enzyme protein to which the pyridoxal phosphate is attached and on the group R.

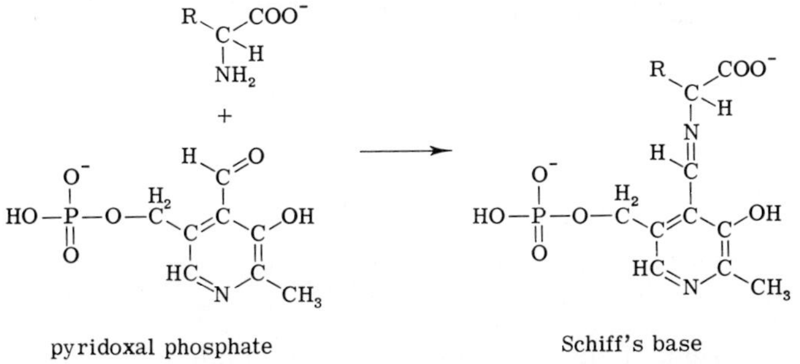
pyridoxal phosphate
Schiff's base

Fig. 2.20. Schiff's base formation of substrate and pyridoxal phosphate.

Vitamin B_{12}. Cobalamine, or vitamin B_{12}, is a coenzyme of a rather complex structure that shows certain relationships to the hemin system. Cobalamine acts also in isomerizations involving intramolecular transfer at C—C bonds. It can also act as a carrier of methyl groups.

SUMMARY. It is certainly not easy, at present, to make a sharp distinction between a prosthetic group forming a part of the enzyme, a coenzyme distinct from the enzyme but forming a part of the catalytic mechanism, and a substrate acting purely as a reactant in the enzyme-catalyzed reaction. It has been suggested (22), however, that firmness of combination with the enzyme protein should be used as a criterion for deciding whether a given substance is or is not a prosthetic group. This seems almost impracticable. There is, however, one definite difference between typical prosthetic groups and such carriers as NAD. True prosthetic groups undergo the whole catalytic process while attached to the same enzyme protein molecule. A carrier such as NAD must migrate from one enzyme protein to another in order to fulfill its catalytic function. If this is adopted, hemes, flavins, biotin, and pyridoxal phosphate would be considered as prosthetic groups, whereas NAD, NADP, and coenzyme A would be considered as carrier substrates.

Bacterial Growth Kinetics

Batch Cultures

When microbial cells are inoculated into a nutrient broth and incubated at a suitable temperature, a sequence of changes occurs that can be followed by various methods (74, 81). After a certain lag period, the organisms increase in mass and divide. Growth, in other words, represents an increase in the amount of living matter and leads to an increase in self-duplicating material. It is therefore an autocatalytic process in which the catalyzed reaction (growth) results in the production of more and more catalysts (living matter). Consequently, under ideal conditions for growth and reproduction, the amount of living matter increases, not in direct proportion to time, but according to a geometric progression with time (72). In other words, it multiplies itself by a constant factor in each successive unit of time. This geometric progression can be seen as being

$$2^0 \rightarrow 2^1 \rightarrow 2^2 \rightarrow 2^3 \cdots 2^n$$

for binary fission. If the number of cells at inoculum is N_0, the number of cells N after n divisions would thus be

$$N = N_0 \times 2^n \tag{2.1}$$

or, taking the logarithm

$$\log N = \log N_0 + n \log 2 \tag{2.2}$$

By rearranging this equation, it is possible to determine the number of generations or cell divisions having occurred between N_0 and N

$$n = \frac{\log N - \log N_0}{\log 2} \tag{2.3}$$

As log 2 equals 0.301, this equation becomes

$$n = 3.32(\log N - \log N_0) \tag{2.4}$$

If, under these ideal conditions, the time factor is included, then we arrive at the multiplication rate, r, of a culture

$$r = \frac{n}{t_1 - t_0} \tag{2.5}$$

or, substituting n from Eq. (2.4)

$$r = \frac{3.32(\log N - \log N_0)}{t_1 - t_0} \tag{2.6}$$

As cell growth proceeds in this logarithmic or exponential growth phase, nutrients are taken up from the nutrient broth and end products are excreted into it. This metabolic activity usually causes the pH value to change and, in the case of aerobic organisms, also decreases the availability of oxygen in the broth. Therefore, the process of growth causes the environment to change in such way that it is finally unable to support further growth. The culture then enters the stationary phase. The sequence of changes are referred to as the "growth cycle" and certainly depend on the sequence of interactions with the environment. Therefore, it must be assumed that no ideal condition exists in most of the presently used culture conditions. As ideal conditions are not normally present in a batch culture, a more frequent term used instead of the number of generations is g, the "generation time" or "doubling time," i.e., the time necessary for one dividing cycle

$$g = t/n \quad \text{or} \quad g = 1/r \tag{2.7}$$

or

$$n = t/g \tag{2.8}$$

It is not easy to determine the doubling time in a microbial culture, however, because of the difficulties encountered in determining the number of cells at N_0 and N_1 as the microbial population multiplies by a constant factor.

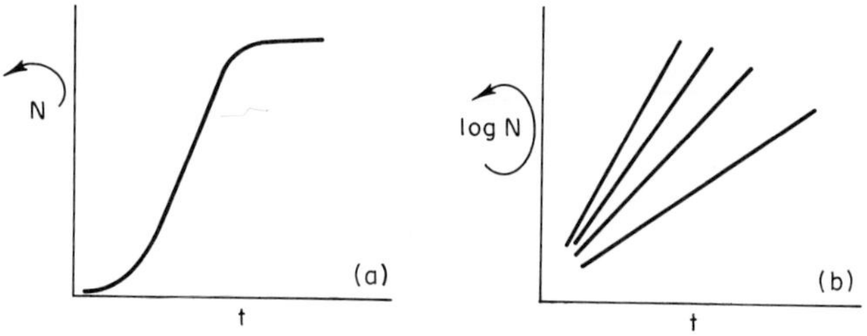
N
(a)
t
log N
(b)
t

Fig. 2.21. Graphical expressions of bacterial growth using actual cell numbers (a) or the logarithm of the number of cells (b) against time.

A consequence of this exponential nature of the growth of the unicellular organism is that the growth rate of cultures cannot be expressed simply in terms of the number of cells per unit time. This would give us the normal growth curve (see Fig. 2.21a). The growth of the culture must instead be expressed in terms of the logarithm of the number of cells per unit time. In plotting the logarithm of the cell number in the exponential phase against time, a straight line is obtained (see Fig. 2.21b). The steeper the slope, the greater the rate of growth and the shorter the doubling time. If generation time is determined according to this method, an average generation time can be obtained. This method, however, does not consider that a bacterial population always contains mutated cells or such that have lost their ability to divide.

Because of these difficulties in the determination of viable cells, a new method is used, whereby a growing bacterial population is regarded as an autocatalytic system. The bacterial mass X, instead of cell numbers, is the important parameter. The rate of change in the bacterial density would then be proportional at all times and would follow the kinetics of a first-order reaction. In order to establish this proportional relationship, let us start with Eq. (2.1)

$$N = N_0 \times 2^n$$

If n is substituted from Eq. (2.8)

$$N = N_0 \times 2^{t/g}$$

As the cell number is being replaced by bacterial density, the equation changes to

$$X = X_0 \times 2^{t/g}$$

or, in logarithmic form

$$\ln X = \ln X_0 + t/g \ln 2 \tag{2.9}$$

This equation is now differentiated with respect to t, assuming that $X_0 = 1$,

$$\frac{d(\ln X)}{dt} = 0 + \frac{\ln 2}{g} \tag{2.10}$$

or

$$\frac{d(\ln X)}{dt} = \frac{\ln 2}{g} \tag{2.11}$$

From Eqs. (2.9) and (2.11) then follows

$$\frac{dX}{dt} = X_0 \frac{\ln 2}{g} e^{\ln 2t/g}$$

$$= \frac{\ln 2}{g} X$$

or

$$\frac{1}{X}\frac{dX}{dt} = \frac{\ln 2}{g} \tag{2.12}$$

The exponential growth equation develops from Eqs. (2.11) and (2.12) as

$$\frac{1}{X}\frac{dX}{dt} = \frac{d(\ln X)}{dt} = \frac{\ln 2}{g} = \mu \tag{2.13}$$

In other words, μ represents the specific growth rate expressed in terms of the concentration of organisms present at a given time (55).

In order to explain the difference between the growth rate and the specific growth rate one can define as follows: (a) the actual rate of increase of concentration of organisms, dX/dt, is called the "growth rate"; (b) the rate of increase per unit of organism concentration, $(1/X)(dX/dt)$, is called the "specific growth rate." If the logarithmic graph expression is used, both growth rates can be determined. For this purpose, the expression $\ln 2/g$ is substituted in Eq. (2.9) by μ

$$\ln X = \mu t + \ln X_0 \tag{2.14}$$

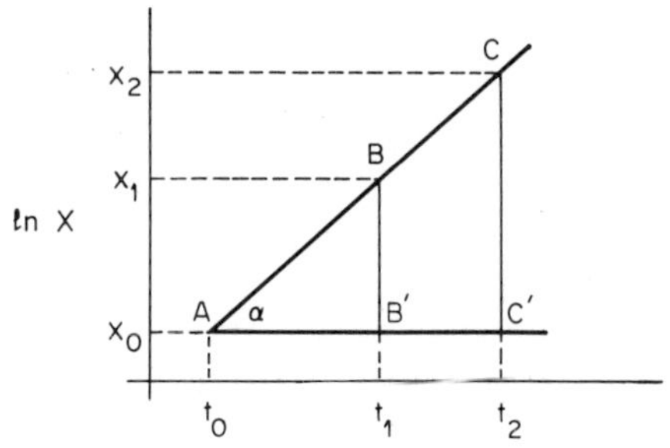
ln X
X2
X1
X0
A
α
B
B′
C
C′
t0
t1
t2

Fig. 2.22. Graphical determination of the specific growth rate. The bacterial population density (X) was measured at three different time intervals (t_0, t_1, t_2) during the logarithmic growth phase. The logarithm of these readings are plotted (A, B, C) and resulted in a straight line. The rate of increase of concentration of organisms can be obtained in using the tan of the angle at the logarithmic growth curve. B' and C' are extrapolated points to the chosen base line to obtain the gradient of the growth curve, that represents μ.

If t is plotted on the abscissa and $\ln X$ on the ordinate, a straight line is obtained with a slope depending on the rate of growth. The gradient of this line is then μ (see Fig. 2.22). Its value can be determined by the relationship

$$\tan \alpha = \mu = \frac{BB'}{AB'} = \frac{CC'}{AC'}$$

or directly from Eq. (2.14),

$$\mu = \frac{\ln X_1 - \ln X_0}{t}$$

In the case of the growth rate, however, the calculation is slightly different, as $1/X$ does not exist. The growth rate can therefore be calculated from any time interval, whereas the specific growth rate is always related to $\tan \alpha$. During the log phase, the growth rates should be reasonably constant.

If one takes the logarithmic expression into consideration, Eq. (2.7) then becomes the basic equation for the determination of generation time

$$g = \frac{t \ln 2}{\ln X - \ln X_0} = \frac{0.69t}{\ln X - \ln X_0} \tag{2.15}$$

From all these expressions, which firmly characterize the growth of a bacterial culture, it can be visualized that such cultures are markedly

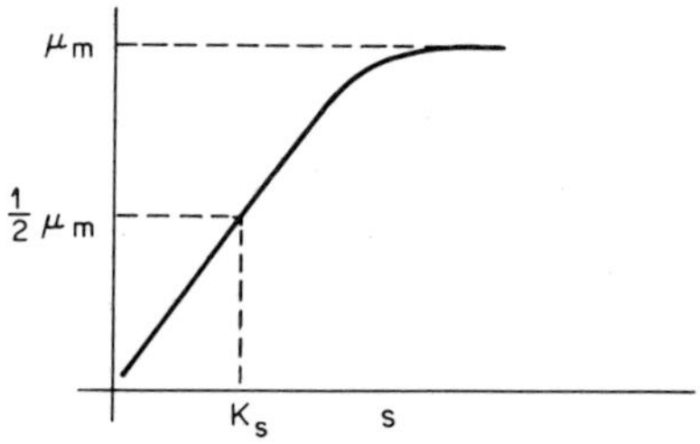
μm
½ μm
Ks
s

Fig. 2.23. The dependence of the maximal specific growth rate (μ_m) of bacteria on substrate concentration (s) and the establishment of the substrate saturation constant K_s.

influenced by the environment, particularly by the concentration of various essential nutrients. If μ is plotted on the ordinate and the substrate concentration, s, on the abscissa, a hyperbola similar to the one described for enzyme velocity measurements (see p. 41) is obtained (Fig. 2.23). Monod (62) showed this dependence in the mathematical expression

$$\mu = \mu_m \left(\frac{s}{K_s + s}\right) \tag{2.16}$$

where μ_m is the maximum value of μ (i.e., when s is no longer growth limiting) and K_s is a saturation constant (equal to the growth-limiting substrate concentration at $\frac{1}{2}\,\mu_m$). As in batch culture all nutrients are present initially in concentrations sufficient not to be limiting for the growth rate μ, the latter (μ) is generally equal to μ_m during the exponential growth phase. As K_s is always greater than 0, the values of the fraction is smaller than 1. The resemblance of Eq. (2.16) to the Michaelis-Menten equation describing the kinetics of enzyme systems should be noted. This resemblance is more than superficial and illustrates that both equations deal with the kinetics of systems the rates of which are controlled by a limiting saturation value of one component in the system. This is why it is possible to use the method of Lineweaver and Burk (Fig. 2.2) for the more exact determination of the value of K_s and μ_m. Equation (2.16) is therefore adjusted for this purpose.

$$\frac{1}{\mu} = \frac{K_s + s}{\mu_m s} = \frac{1}{s}\frac{K_s}{\mu_m} + \frac{1}{\mu_m} \tag{2.17}$$

If $1/\mu$ is plotted on the ordinate and $1/s$ on the abscissa, the straight line obtained defines the value $1/\mu_m$ on the ordinate and $-1/K_s$ on the abscissa (Fig. 2.24).

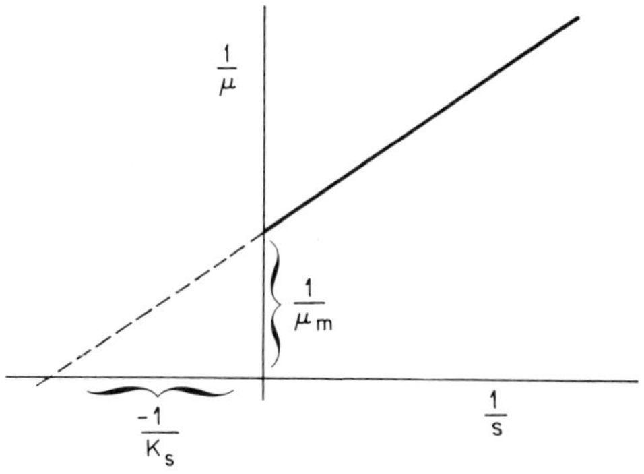
1/μ
1/μm
-1/Ks
1/s

Fig. 2.24. The determination of μ_m and K_s using the Lineweaver–Burk plot (see also Fig. 2.2).

Monod (62) also showed experimentally that there was a constant relationship between the growth of a culture and substrate utilization

$$\frac{dX}{dt} = -Y\frac{ds}{dt}$$

where Y is termed the "yield factor." Over any finite period of time during the exponential growth phase

$$Y = \frac{\text{weight of bacteria formed}}{\text{weight of substrate consumed}}$$

Therefore, if the values of μ_m, K_s, and Y are known, a completely quantitative description of the events occurring during the growth cycle of a batch culture can be obtained. For detailed accounts on growth and growth limitations, as well as the yield factor or molar growth yield determinations, several excellent textbooks or review articles can be consulted (19, 70, 77, 82).

Continuous Culture

The bacterial growth kinetics of a batch culture now having been considered, another important type will be treated. Although there are a number of different types of continuous cultures in operation (55), the basic principles of all these systems are very similar. All continuous cultures start as batch cultures. The medium, which is contained in a growth vessel, is inoculated with an organism that then proceeds to grow and divide. If during the exponential growth phase fresh medium is added to the culture at a rate sufficient to maintain the culture population density at a fixed and submaximal value, then growth cannot cease as it does in batch cultures but will continue indefinitely. However, the medium input rate and the culture volume would have to increase exponentially with biomass, unless provision is made for the continuous removal of culture at a rate equal to that of the medium flow. The material enters the system of constant volume at the same rate as it leaves the system, possibly being transformed. The exchange of energy is constant (ΔF = const.) and the biophase shows minimum entropy production.

Let us first consider a type of continuous culture apparatus, called the "chemostat" (71), that allows continuous growth. In a chemostat, the organism is grown at specific growth rates smaller than μ_m. For this purpose a culture of fixed volume (V) is contained in a suitably constructed growth vessel (65) to which medium is pumped at a constant rate (F). This medium is compounded so that all substances for the organism's growth, except one, are present in the culture at concentrations that are in excess of the growth requirements. The one, so-called growth-limiting, nutrient is present in the medium at a concentration sufficient to obtain the required growth. It has now become customary to separate the growth-limiting substrate from the rest of the medium and feed them separately into the growth vessel, thus obtaining a better controllable growth (96, 97).

This type of single-step open system is characterized by the inflow of fresh medium into the culture vessel and the overflow, at the same rate, of the medium modified by the metabolic activity of the organisms together with part of the grown culture. The ratio of the inflowing amount of nutrient medium per hour to the volume of the culture is called the dilution rate, D

$$D = \frac{F}{V}$$

With this experimental arrangement, the growth rate of organisms will depend not merely on the medium flow rate, but on the dilution rate, i.e., the number of culture volumes of medium passing through the growth vessel per unit time, the dimensions being reciprocal time (hr^{-1}). If the dilution rate is kept constant, the concentration of the so-called growth-limiting substrate gives a value equal to the specific growth rate μ. Therefore

$$\mu = D$$

and we obtain a steady-state culture. This steady state may be maintained for an infinitely long period, provided no change of the dilution rate or substrate concentration occurs.

The effective substrate concentration can be expressed as

$$S_x = S_0 - S_1$$

where S_0 is the concentration of the main substrate in the inflowing medium and S_1 is the concentration of the main substrate in the cultivating vessel. At $S_x = 0$, the substrate is not utilized, and the biophase does not grow. If the biomass is to grow, S_x must be positive. With the S_x concentration increasing toward S_0, the specific growth rate μ will change asymptotically to its limit value μ_m, the maximum specific growth rate. The specific growth rate thus becomes a function of the concentration of the substrate.

In order to fully appreciate the kinetics of bacterial growth in a chemostat culture, it is essential to understand the exact relationship between dilution rate, growth rate, and substrate concentration (see also 18).

If the assumptions are made (*a*) that the culture vessel is perfectly agitated, (*b*) that the culture is homogenous, (*c*) that the properties of the culture are practically constant at high concentration of the substrate limiting growth (i.e., at $S \gg K_s$), and (*d*) that all other substances are present in excess, the specific growth rate μ can be presumed to approach μ_m. From Eq. (2.14) is obtained

$$X = X_0 e^{\mu t} \tag{2.18}$$

By its derivation with time, the relation determining the instantaneous increment of microorganisms during cultivation is obtained

$$\frac{dX}{dt} = \mu X \tag{2.19}$$

If μX is now compensated by the outflow of biomass with medium DX, the net change in concentration of biomass with time will be determined by the relation

$$\text{increase} = \text{growth} - \text{outflow}$$

$$dX/dt = \mu X - DX$$

or

$$dX/dt = X(\mu - D) \tag{2.20}$$

It follows that if $\mu > D$, dX/dt will be positive and the concentration of organisms in the culture will increase with time. If, however, $\mu < D$, then dX/dt will have a negative value, and the cell concentration will diminish with time: the culture is washed out from the growth vessel. Only when

$\mu = D$ will $dX/dt = 0$ and the concentration of organisms in the culture remains constant with time. This is the stage when one terms a culture as being in a "steady state":

$$D = \mu = \mu_m \left(\frac{s}{K_s + s}\right) = \frac{\ln 2}{g} \tag{2.21}$$

This steady state is not difficult to achieve because the growth rate of the organism in the culture, being limited by the rate of supply of a growth-limiting substance, must be proportional to the dilution rate. Provided the dilution rate is maintained constant, the system is inherently self-balancing.

This relationship between the growth rate of the organism and the dilution rate makes it possible to adjust the growth rate of the organism, within certain limits, to any value desired. When the dilution rate is increased above the specific growth rate, $D > \mu$, then dX/dt becomes negative, the biomass decreases, and the substrate is not utilized and increases. With $\mu < \mu_m$, the increased substrate concentration positively influences the specific growth rate up to its limit, where the specific growth rate equals the maximum growth rate ($\mu = \mu_m$). However, the specific growth rate cannot be made to exceed μ_m and therefore steady-state conditions cannot be obtained at dilution rates above a critical value (D_c), which is nearly equal to μ_m. If the dilution rate is set to a value greater than D_c, the culture will be progressively washed out from the fermenter.

It has been mentioned above that the bacterial growth is also dependent on the substrate concentration. Therefore, it is necessary to consider not only the effect of the dilution rate on the specific growth rate, but the effect of the dilution rate on the concentration of growth-limiting substrate (s) and biomass (X) in the culture. The substrate enters the growth vessel at concentration S_0, is consumed by the organisms, and emerges in the overflow at a concentration S. Therefore, the net change in substrate concentration is

$$\begin{aligned} \text{Increase of substrate} &= \text{input} - \text{outflow} - \text{consumption} \\ dS/dt &= DS_0 - DS - (\text{growth/yield}) \\ &= DS_0 - DS - (\mu X/Y) \end{aligned}$$

Rearranging the equation and substituting μ by Eq. (2.16)

$$\frac{dS}{dt} = D(S_0 - S) - \frac{\mu_m X}{Y}\left(\frac{s}{K_s + s}\right) \tag{2.22}$$

Similarly μ can be substituted for in Eq. (2.20)

$$\frac{dX}{dt} = \mu_m X \left(\frac{s}{K_s + s}\right) - DX$$

or

$$\frac{dX}{dt} = X\mu_m \left(\frac{s}{K_s + s}\right) - D \qquad (2.23)$$

The Eqs. (2.21), (2.22), and (2.23) quantitatively define the behavior of the culture in the chemostat. They also show that the continuous culture, irrespective of the initial state of the culture, should ultimately establish a steady state. These equations express the self-regulating capacity of the continuous system.

Because Eqs. (2.22) and (2.23) equal zero in a steady state, the values of X and S can be calculated for the conditions of the steady state. For the calculations of X, Eq. (2.22) is used

$$D(S_0 - S) = \frac{X}{Y} \mu_m \left(\frac{s}{K_s + s}\right)$$

Since in the steady state $D = \mu_m(s/K_s + s)$,

$$X = Y(S_0 - S) \qquad (2.24)$$

The concentration of the substrate S is then determined from Eq. (2.23):

$$D = \mu_m \left(\frac{s}{K_s + s}\right)$$

or

$$S = K_s \left(\frac{D}{\mu_m - D}\right) \qquad (2.25)$$

Substituting S from Eq. (2.25) in Eq. (2.24), it follows that

$$X = Y\left[S_0 - K_s\left(\frac{D}{\mu_m - D}\right)\right]$$

As the values K_s, μ_m, and Y are determined experimentally in a batch culture and thus become constants in a continuous culture, it is possible to plot the self-regulating capacity of the continuous process. By varying

S_0 or D experimentally, a great number of steady states can be obtained, restricted only by the concentration of S remaining at a level where it is the limiting factor in the growth of the bacterial culture.

Modifications of the basic theory with regard to variations occurring in yield values or for cultures of low viability are described in the excellent review by Tempest (90).

These considerations have so far all been made with respect to the chemostat. This chemostat, as was mentioned above, controls the growth rate and thus the biomass with the help of dilution rate and a growth-limiting substance. An experimental microbiologist would find, however, that each increase in dilution rate would lead to a lower stable population density. Finally, at a certain dilution rate, the population density could no longer reach a stable value but would decrease steadily (67). When the growth rate is no longer dependent on population density, a constant density can only be maintained by a system that monitors some property that depends on population density rather than growth rate and that adjusts the dilution rate accordingly. Such a device is called a "turbidostat" (8, 66). A turbidostat is a continuous culture apparatus in which a photoelectric monitor detects deviations from some desired culture turbidity and passes a signal calling for a compensatory increase or decrease in dilution rate. All other components function as in the chemostat. Details about the turbidostat can be found in the review by Munson (67). The turbidostat is mainly used in conjunction with the chemostat, or alone should the substrate not be a growth-limiting factor for experimental purposes.

References

1. Abiko, Y. (1967). Investigations on pantothenic acid and its related compounds. IX. Biochemical studies. (4). Separation and substrate specificity of pantothenate kinase and phosphopantothenoylcysteine synthetase. *J. Biochem. (Tokyo)* **61,** 290.
2. Abiko, Y. (1967). Investigations on pantothenic acid and its related compounds. X. Biochemical studies. (5). Purification and substrate specificity of phosphopantothenoylcysteine decarboxylase from rat liver. *J. Biochem. (Tokyo)* **61,** 300.
3. Abiko, Y., Suzuki, T., and Shimizu, M. (1967). Investigation on pantothenic acid and its related compounds. VI. Biochemical studies. (2). Determination of Coenzyme A by the phosphotransacetylase system of *Escherichia coli. J. Biochem. (Tokyo)* **61,** 10.
4. Abiko, Y., Suzuki, T., and Shimizu, M. (1967). Investigation on pantothenic acid and its related compounds. XI. Biochemical studies. (6). A final stage in the biosynthesis of CoA. *J. Biochem. (Tokyo)* **61,** 309.
5. Bartsch, R. G., and Kamen, M. D. (1960). Isolation and properties of two soluble heme proteins in extracts of the photoanaerobe *Chromatium. J. Biol. Chem.* **235,** 825.
6. Bartsch, R. G. (1968). Bacterial cytochromes. *Annu. Rev. Microbiol.* **22,** 181.

7. Benson, A. M., and Yasunobu, K. T. (1969). Nonheme iron proteins. X. The amino acid sequence of ferredoxins from *Leucaena glauca*. *J. Biol. Chem.* **244,** 955.
8. Bryson, V. (1952). The turbidostatic selectro—a device for automatic selection of bacterial variants. *Science* **116,** 48.
9. Buchanan, B. B., and Rabinowitz, J. C. (1964). Some properties of *Methanobacterium omelianskii*-ferredoxin. *J. Bacteriol.* **88,** 806.
10. Buchanan, B. B., Matsubara, H., and Evans, M. C. W. (1969). Ferredoxin from the photosynthetic bacterium, *Chlorobium thiosulfatophilum*. A link to ferredoxins from nonphotosynthetic bacteria. *Biochim. Biophys. Acta* **189,** 46.
11. Buchanan, B. B., and Arnon, D. I. (1970). Ferredoxins: Chemistry and function in photosynthesis, nitrogen fixation and fermentative metabolism. *Advan. Enzymol.* **33,** 114.
12. Bullock, M. W., Brockman, J. A., Jr., Patterson, E. L., Pierce, J. V., and Stockstadt, E. L. R. (1952). Synthesis of compounds in the thioctic acid series. *J. Amer. Chem. Soc.* **74,** 3455.
13. Castor, L. N., and Chance, B. (1959). Photochemical determinations of the oxidases of bacteria. *J. Biol. Chem.* **234,** 1587.
14. Chaykin, S. (1967). Nicotinamide coenzymes. *Annu. Rev. Biochem.* **36,** 149.
15. Cole, J. A., and Wimpenny, J. W. T. (1966). The interrelationships of low redox potential cytochrome c-552 and hydrogenase in facultative anaerobes. *Biochim. Biophys. Acta* **128,** 419.
16. Colowick, S., Lazarow, A., Racker, E., Schwarz, D. R., Stadtman, E. R., and Waelsch, H. (1954). "Glutathione." Academic Press, New York.
17. Conrad, H. E., Lieb, K., and Gunsalus, I. C. (1965). Mixed function oxidation. III. An electron transport complex in camphor ketolactonization. *J. Biol. Chem.* **240,** 4029.
18. Contois, D. E. (1959). Kinetics of bacterial growth: Relationship between population density and specific growth rate of continuous cultures. *J. Gen. Microbiol.* **21,** 40.
19. Dean, A. C. R., and Hinshelwood, C. (1966). "Growth, Function and Regulation in Bacterial Cells." Oxford Univ. Press (Clarendon), London and New York.
20. De Klerk, H., Bartsch, R. G., and Kamen, M. D. (1965). Atypical soluble heme proteins from a strain of *Rhodopseudomonas palustris sp. Biochim. Biophys. Acta* **97,** 275.
21. Devanathan, T., Akagi, J. M., Hersh, R. T., and Himes, R. H. (1969). Ferredoxin from two thermophilic clostridia. *J. Biol. Chem.* **244,** 2846.
22. Dixon, M., and Webb, E. C. (1964). "Enzymes," 2nd ed. Longmans, Green, New York.
23. Dubourdieu, M., Le Gall, J., and Leterrier, F. (1968). *C. R. Acad. Sci., Ser. D* **267,** 1653 [as cited by Neims and Hellerman (69)].
24. Fasella, P. (1967). Pyridoxal phosphate. *Annu. Rev. Biochem.* **36,** 185.
25. Fujita, A., and Kodama, T. (1934). *Biochem. Z.* **273,** 186 [as cited by Kamen and Horio (42)].
26. Fujita, T. (1966). Studies on soluble cytochromes in Enterobacteriaceae. II. Cytochrome b-562 and c-550. *J. Biochem.* (*Tokyo*) **60,** 329.
27. Fujita, T., and Sato, R. (1966). Studies on soluble cytochromes in Enterobacteriaceae. IV. Possible involvement of cytochrome c-552 in anaerobic nitrite metabolism. *J. Biochem.* (*Tokyo*) **60,** 691.
28. Ginsburg, A., and Stadtman, E. R. (1970). Multienzyme systems. *Annu. Rev. Biochem.* **39,** 429.

29. Gray, C. T., Wimpenny, J. W. T., Hughes, D. E., and Ranlett, M. (1963). A soluble c-type cytochrome from anaerobically grown *Escherichia coli* and various Enterobacteriaceae. *Biochim. Biophys. Acta* **67,** 157.
30. Gray, C. T., Wimpenny, J. W. T., Hughes, D. E., and Mosman, M. R. (1966). Regulation of metabolism in facultative bacteria. I. Structural and functional changes in *Escherichia coli* associated with shifts between the aerobic and anaerobic states. *Biochim. Biophys. Acta* **117,** 22.
31. Guirard, B. M., Snell, E. E., and Williams, R. J. (1946). The nutritional role of acetate for lactic acid bacteria. I. The response to substances related to acetate. *Arch. Biochem. Biophys.* **9,** 361.
32. Guirard, B. M., Snell, E. E., and Williams, R. J. (1946). The nutritional role of acetate for lactic acid bacteria. II. Fractionation of extracts of natural materials. *Arch. Biochem. Biophys.* **9,** 381.
33. Gunsalus, I. C., Chapman, P. J., and Kuo, J. F. (1965). Control of catabolic specificity and metabolism. *Biochem. Biophys. Res. Commun.* **18,** 924.
34. Gutfreund, H. (1965). "An Introduction to the Study of Enzymes." Blackwell, Oxford.
35. Hatefi, Y. (1959). Studies on the electron transport system. XXIII. Coenzyme Q oxidase. *Biochim. Biophys. Acta* **34,** 183.
36. Henderson, R. W., and Nankiville, D. D. (1966). Electrophoretic and other studies on heme pigments from *Rhodopseudomonas palustris:* cytochrome 552 and cytochromoid c. *Biochem. J.* **98,** 587.
37. Horio, T., Kamen, M. D., and De Klerk, H. (1961). Relative oxidation–reduction potentials of heme groups in two soluble double-heme proteins. *J. Biol. Chem.* **236,** 2783.
38. Horio, T., and Kamen, M. D. (1962). Observations on the respiratory system of *Rhodospirillum rubrum. Biochemistry* **1,** 1141.
39. Horio, T., and Kamen, M. D. (1970). Bacterial cytochromes. II. Functional aspects. *Annu. Rev. Microbiol.* **24,** 399.
40. International Union of Biochemistry Commission. (1972). "Enzyme Nomenclature: Recommendations of the International Union of Biochemistry on the Nomenclature and Classification of Enzymes, Together with their Units and the Symbols of Enzyme Kinetics." Elsevier, Amsterdam.
41. Iwasaki, H. (1966). Lactate oxidation system in *Acetobacter suboxydans* with special reference to carbon monoxide-binding pigment. *Plant Cell Physiol.* **7,** 199.
42. Kamen, M. D., and Horio, T. (1970). Bacterial cytochromes. I. Structural aspects. *Annu. Rev. Biochem.* **39,** 673.
43. Karlson, P. (1970). "Kurzes Lehrbuch der Biochemie," 7th ed. Thieme, Stuttgart.
44. Katagiri, M., Ganguli, B. N., and Gunsalus, I. C. (1968). A soluble cytochrome P-450 functional in methylene hydroxylation. *J. Biol. Chem.* **243,** 3543.
45. Keresztes-Nagy, S., and Margoliash, E. (1966). Preparation and characterization of alfalfa ferredoxin. *J. Biol. Chem.* **241,** 5955.
46. Knappe, J. (1970). Mechanism of biotin action. *Annu. Rev. Biochem.* **39,** 757.
47. Knight, E., Jr., D'Eustachio, A. J., and Hardy, R. W. F. (1966). Flavodoxin: A flavoprotein with ferredoxin activity from *Clostridium pasteurianum. Biochim. Biophys. Acta* **113,** 626.
48. Knight, E., Jr., and Hardy, R. W. F. (1966). Isolation and characteristics of flavodoxin from nitrogen-fixing *Clostridium pasteurianum. J. Biol. Chem.* **241,** 2752.

49. Knight, E., Jr., and Hardy, R. W. F. (1967). Flavodoxin. Chemical and biological properties. *J. Biol. Chem.* **242,** 1370.
50. Le Galle, J., Mazza, G., and Dragoni, N. (1965). Le cytochrome c_3 de *Desulfovibrio gigas*. *Biochim. Biophys. Acta* **99,** 385.
51. Lehninger, A. L. (1965). "Bioenergetics." Benjamin, New York.
52. Lovenberg, W., Buchanan, B. B., and Rabinowitz, J. C. (1963). Studies on the chemical nature of clostridial ferredoxin. *J. Biol. Chem.* **238,** 3899.
53. Lovenberg, W., and Sobel, B. E. (1965). Characterization of a new electron transfer protein from *Clostridium pasteurianum*. *Fed. Proc., Fed. Amer. Soc. Exp. Biol.* **24,** 233.
54. Lovenberg, W., and Sobel, B. E. (1965). Rubredoxin: A new electron transfer protein from *Clostridium pasteurianum*. *Proc. Nat. Acad. Sci. U.S.* **54,** 193.
55. Málek, I., and Fencl, Z. (1966). "Theoretical and Methodological Basis of Continuous Culture of Microorganisms." Academic Press, New York.
56. Malkin, R., and Rabinowitz, J. C. (1967). Nonheme iron electron transfer proteins. *Annu. Rev. Biochem.* **36,** 113.
57. Matsubara, H. (1968). Purification and some properties of *Scenedesmus* ferredoxin. *J. Biol. Chem.* **243,** 370.
58. Mayhew, S. G., and Peel, J. L. (1966). Rubredoxin from *Peptostreptococcus elsdenii*. *Biochem. J.* **100,** 80P.
59. Mayhew, S. G., and Massey, V. (1969). Purification and characterization of flavodoxin from *Peptostreptococcus elsdenii*. *J. Biol. Chem.* **244,** 794.
60. Mayhew, S. G., Foust, G. P., and Massey, V. (1969). Oxidation–reduction properties of flavodoxin from *Peptostreptococcus elsdenii*. *J. Biol. Chem.* **244,** 803.
61. Mayhew, S. G. (1971). Properties of two clostridial flavodoxins. *Biochim. Biophys. Acta* **235,** 276.
62. Monod, J. (1942). "Recherches sur la croissance bactérienne." Masson, Paris.
63. Mortenson, L. E., Valentine, R. C., and Carnaham, J. E. (1962). An electron transport factor from *Clostridium pasteurianum*. *Biochem. Biophys. Res. Commun.* **7,** 448.
64. Morton, R. L. (1953). As cited by Dixon and Webb (22).
65. Moss, F. J., and Bush, F. (1967). Working design for a 5-liter controlled continuous culture apparatus. *Biotechnol. Bioeng.* **9,** 585.
66. Munson, R. J., and Jeffery, A. (1964). Reversion rate in continuous cultures of an *Escherichia coli* auxotroph exposed to gamma rays. *J. Gen. Microbiol.* **35,** 191.
67. Munson, R. J. (1970). Turbidostats. In "Methods in Microbiology" (J. R. Norris and D. W. Ribbons, eds.), Vol. 2, p. 349. Academic Press, New York.
68. Negelein, E., and Gerischer, W. (1934). *Biochem. Z.* **268** [as cited by Kamen and Horio (42)].
69. Neims, A. H., and Hellerman, L. (1970). Flavoenzyme catalysis. *Annu. Rev. Biochem.* **39,** 867.
70. Nineteenth Symposium for General Microbiology. (1969). "Microbial Growth." Cambridge Univ. Press, London and New York.
71. Novick, A., and Szilard, L. (1950). Description of the chemostat. *Science* **112,** 715.
72. Painter, P. R., and Marr, A. G. (1968). Mathematics of microbial populations. *Annu. Rev. Microbiol.* **22,** 519.
73. Postgate, J. R. (1956). Cytochrome c_3 and desulfoviridin; pigments of the anaerobe *Desulfovibrio desulfuricans*. *J. Gen. Microbiol.* **14,** 545.
74. Powell, E. O., Evans, C. G. T., Strange, R. E., and Tempest, D. W. (1967). "Microbial Physiology and Continuous Culture," 3rd Int. Symp. HM Stationery Office, London.

75. Rao, K. K., and Matsubara, H. (1970). The amino acid sequence of Taro ferredoxin. *Biochem. Biophys. Res. Commun.* **38,** 500.
76. Reed, L. J., de Busk, B. G., Gunsalus, I. C., and Hornberger, C. S., Jr. (1957). Crystalline α-lipoic acid: A catalytic agent associated with pyruvate dehydrogenase. *Science* **114,** 93.
77. Rose, A. H. (1969). "Chemical Microbiology," 3rd ed. Butterworth, London.
78. Sasaki, R. M., and Matsubara, H. (1967). Molecular weight and amino acid composition of *Chromatium* ferredoxin. *Biochem. Biophys. Res. Commun.* **28,** 467.
79. Schmidt, U., Grafen, P., Altland, K., and Goedde, H. W. (1969). Biochemistry and chemistry of lipoic acid. *Advan. Enzymol.* **32,** 423.
80. Sih, C. J. (1969). Enzymatic mechanism of steroid hydroxylation. *Science* **163,** 1297.
81. Sokatch, J. R. (1969). "Bacterial Physiology and Metabolism." Academic Press, New York.
82. Stouthamer, A. H. (1969). Determination and significance of molar growth yields. In "Methods in Microbiology" (J. R. Norris and D. W. Ribbons, eds.), Vol. 1, p. 629. Academic Press, New York.
83. Sugeno, K., and Matsubara, H. (1968). The amino acid sequence of *Scenedesmus* ferredoxin. *Biochem. Biophys. Res. Commun.* **32,** 951.
84. Suzuki, H., and Mori, T. (1962). Studies on denitrification. V. Purification of denitrifying enzyme by means of electrophoresis. *J. Biochem.* (*Tokyo*) **52,** 190.
85. Suzuki, H., and Iwasaki, H. (1962). Studies on denitrification, VI. Preparations and properties of crystalline blue protein and cryptocytochrome c, and role of copper in denitrifying enzyme from a denitrifying bacterium. *J. Biochem.* (*Tokyo*) **52,** 193.
86. Tagawa, K., and Arnon, D. I. (1962). Ferredoxins as electron carriers in photosynthesis and in the biological production and consumption of hydrogen gas. *Nature* (*London*) **195,** 537.
87. Tagawa, K., and Arnon, D. I. (1968). Oxidation–reduction potentials and stoichiometry of electron transfer in ferredoxins. *Biochim. Biophys. Acta* **153,** 602.
88. Tamiya, H., and Yamaguchi, S. (1933). *Acta Phytochim.* **7,** 233 [as cited by Kamen and Horio (42)].
89. Tanaka, M., Hanin, M., Yasunobu, K. T., Mayhew, S., and Mayhew, V. (1971). Amino acid sequence of the *Peptostreptococcus elsdenii* flavodoxin. *Biochem. Biophys. Res. Commun.* **44,** 886.
90. Tempest, D. W. (1970). The continuous cultivation of microorganisms. I. Theory of the chemostat. In "Methods in Microbiology" (J. R. Norris and D. W. Ribbons, eds.), Vol. 2, p. 259. Academic Press, New York.
91. Trudgill, P. W., Du Bus, R., and Gunsalus, I. C. (1966). Mixed function oxidation. VI. Purification of a tightly coupled electron transport complex in camphor lactonization. *J. Biol. Chem.* **241,** 4288.
92. Tsunoda, J. N., and Yasunobu, K. T. (1968). Nonheme iron proteins. IX. The amino sequence of ferredoxin from *Micrococcus aerogenes*. *J. Biol. Chem.* **243,** 6262.
93. Van Gelder, B. F. (1966). On cytochrome c oxidase. I. The extinction coefficients of cytochrome a and cytochrome a_3. *Biochim. Biophys. Acta* **118,** 36.
94. Van Gelder, B. F., and Muijsers, A. O. (1966). On cytochrome c oxidase. II. The ratio of cytochrome a to cytochrome a_3. *Biochim. Biophys. Acta* **118,** 47.
95. Van Lin, B., and Bothe, H. (1972). Flavodoxin from *Azotobacter vinelandii*. *Arch. Mikrobiol.* **82,** 155.
96. Van Uden, N. (1967). Transport-limited growth in the chemostat and its competitive inhibition: A theoretical treatment. *Arch. Mikrobiol.* **58,** 145.

97. Van Uden, N. (1969). Kinetics of nutrient-limited growth. *Annu. Rev. Microbiol.* **23,** 473.
98. Vernon, L. P., and Kamen, M. D. (1954). Hematin compounds in photosynthetic bacteria. *J. Biol. Chem.* **211,** 643.
99. Warburg, O., and Negelein, E. (1933). As cited by Kamen and Horio (42).
100. Warburg, O., and Negelein, E. (1933). *Biochem. Z.* **266** [as cited by Kamen and Horio (42)].
101. White, A., Handler, P., and Smith, E. L. (1964). "Principles of Biochemistry," 3rd ed. McGraw-Hill, New York.
102. Wimpenny, J. W. T., Ranlett, P., and Gray, C. T. (1963). Repression and derepression of cytochrome c biosynthesis in *Escherichia coli. Biochim. Biophys. Acta* **73,** 170.
103. Yoch, D. C., Benemann, J. R., Valentine, R. C., and Arnon, D. I. (1969). The electron transport system in nitrogen fixation by *Azotobacter.* II. Isolation and function of a new type of ferredoxin. *Proc. Nat. Acad. Sci. U.S.* **64,** 1404.

Supplementary Readings

Bleecken, S. (1973). Dynamik von Zellpopulationen und ihre mathematische Beschreibung. *Ztschr. Allgem. Mikrobiol.* **13,** 3.

Colleran, E. M., and Jones, O. T. G. (1973). Studies on the biosynthesis of cytochrome c. *Biochem. J.* **134,** 89.

Johnson, P. W., and Canale-Parola, E. (1973). Properties of rubredoxin and ferredoxin isolated from spirochetes. *Arch. Mikrobiol.* **89,** 341.

Lemberg, R., and Barrett, J. (1973). "Cytochromes." Academic Press, New York.

Orme-Johnson, W. H. (1973). Iron-sulfur proteins: structure and function. *Ann. Rev. Biochem.* **42,** 159.

Yoch, D. C., and Valentine, R. C. (1972). Ferredoxins and flavodoxins in bacteria. *Ann. Rev. Microbiol.* **26,** 139.

Questions

1. What is the function of an enzyme?
2. What is an enzyme?
3. How does an enzyme act on its substrate?
4. What does the Michaelis-Menten constant (K_m) represent?
5. Explain the formula

$$v = V\left(\frac{s}{K_s + s}\right)$$

6. What is the importance of the Lineweaver-Burk plot?
7. What is understood by one international enzyme unit and the specific activity of an enzyme?
8. What is the difference between a coenzyme and a prosthetic group?
9. Explain the action of a hydrogen-transferring coenzyme in enzyme catalysis.
10. What is the difference between NAD^+, FAD^+, lipoic acid, and ubiquinone?
11. Name the four major groups of cytochromes and the way they are differentiated.

12. What is the difference between cytochromes, ferredoxins, rubredoxins, and flavodoxins?

13. Discuss the role of tetrahydrofolic acid in comparison to that of thiamine pyrophosphate as coenzymes.

14. "Acetyl-CoA and ATP are regarded as the most important energy-rich compounds." Discuss this statement together with their mode of action as coenzymes.

15. Give the formulas for determining specific growth rates, generation time, multiplication rate, and number of generations of a batch culture.

16. Explain the difference between specific growth rate and growth rate of a batch culture.

17. What is the mathematical expression for determining the specific growth rate graphically?

18. Explain and discuss the relationship between the kinetics of an enzyme system and that of bacterial growth.

19. *Escherichia coli* is grown for 9 hours in 100 ml of a mineral salts medium containing 0.25 gm glucose and an unknown quantity of galactose. If glucose is used preferentially and to completion, determine by graphical methods and calculation the:
   (a) specific growth rates
   (b) generation times
   (c) multiplication rates
   (d) growth rate at 0.5, 4.5, and 7.5 hours
   (e) lag period required for adaptation to galactose
   (f) Y_{glucose}, given the average weight of a cell is 10^{-12} gm, using the results in the following tabulation:

Age (hr)	Viable counts	Age (hr)	Viable count
0	1.778×10^3	4.5	2.820×10^6
1	1.870×10^3	5.0	3.176×10^6
1.5	3.162×10^3	6.0	1.780×10^7
2.0	1.778×10^4	7.0	1.010×10^8
2.5	3.162×10^4	7.5	1.995×10^8
3.0	9.988×10^4	8.0	1.585×10^8
3.5	3.090×10^5	9.0	7.943×10^6
4.0	3.981×10^5		

20. During the batch culture cultivation of a microorganism, stationary phase was reached 8 hours after inoculation. The lag phase was 1.5 hours and the specific growth rate 0.2 hr^{-1}. Determine during the log phase of growth the following:
   (a) generation time
   (b) the number of generations
   (c) the multiplication rate

21. What is the difference between a chemostat and a turbidostat?

22. What is the importance of the formula

$$X = Y\left[S_o - K_s\left(\frac{D}{\mu_m - D}\right)\right]$$

23. What is meant by the self-regulating capacity of a continuous culture?

3

Photosynthesis and Photometabolism

Terminology in Bacterial Metabolism

All the anabolic and catabolic reactions that occur during the lifetime of a microorganism are included under the term "metabolism." Anabolism is the biosynthetic buildup of cell material from simple inorganic or organic compounds, whereas catabolism supplies all the energy, and in many cases the building blocks or precursors, for these essential biosynthetic reactions. Many of the terms used in microbial chemistry are unfortunately used as generalizations and have caused great confusion in terminology. In order to rectify this, a number of the major terms that are essential in dealing with this subject matter are defined below.

Metabolism	Represents the overall chemical reactions that occur in microorganisms
Anabolism	Represents the biosynthetic reactions that lead to the buildup of such cell material as polymers, DNA, RNA, and lipids
Catabolism	Represents all chemical reactions that are involved in the breakdown of inorganic and organic material for the purpose of supplying energy and precursors for the biosynthesis of cell material
Autotroph	A microorganism that is able to use CO_2 as sole carbon source for growth
Heterotroph	A microorganism that requires carbon sources more reduced than CO_2; the majority of microorganisms fall within this category

Photosynthetic microorganisms	A microorganism that derives its energy from light quanta; it may be autotrophic or heterotrophic
Anaerobic or anoxybiontic respiration	The chemical energy-yielding reactions in which inorganic compounds other than oxygen act as the terminal electron acceptor
Aerobic or oxybiontic respiration	The chemical energy-yielding reactions that require molecular oxygen as the terminal electron acceptor
Fermentation	The chemical energy-yielding reactions that require organic compounds as electron acceptors
Photolithotroph	A microorganism that derives its energy from light and uses inorganic compounds as electron donors
Photoorganotroph	A microorganism that derives its energy from light and uses organic compounds as electron donors
Chemolithotroph	A microorganism that derives its energy from biochemical reactions and uses inorganic compounds as electron donors
Chemoorganotroph	A microorganism that derives its energy from biochemical reactions and uses organic compounds as electron donors

In regard to the relationship between bacteria and their oxygen requirement the definitions as set out by McBee *et al.* (182) as recommendations should be taken into serious consideration. Here, two distinct groupings are suggested: (1) a description of the environment or atmosphere in which the bacteria can live, for which the terms "aerobic" and "anaerobic" are adequate; and (2) a description of the metabolic use of gaseous oxygen by living bacteria, for which the terms "oxybiontic" (oxybiotic) and "anoxybiontic" (anoxybiotic) should be introduced.

There are several reasons for this separation. Some organisms grow in the presence of oxygen but do not use it. Some cannot grow in the presence of molecular oxygen but are not killed by it. To some oxygen is lethal as a gas. Some organisms generally recognized as anaerobes, e.g., *Clostridium perfringens,* can not only tolerate oxygen at partial pressures below that of the normal atmosphere but can actually metabolize it.

Liquid media are most unsuitable for defining oxygen relationships because of the difficulties associated with maintaining uniform distribution of oxygen throughout the fluid and exposed to air in the stationary state. Liquid media can often be internally deoxygenated.

Attempts to grow organisms on the surface of solid media, where cells are freely exposed to air in the initial stages of growth, is therefore a more suitable and reliable index of oxygen sensitivity.

The present recommendations are as follows: The terms "aerobic," "anaerobic," and "facultative anaerobic" should be applied only to the description of practical cultural conditions used for the cultivation of bacteria or to the growth of bacteria under these conditions. Aerobic bacteria will grow on the surface of a solid medium exposed to air. Anaerobic bacteria will not grow on the surface of a solid medium freely exposed to air. Facultatively anaerobic bacteria are those aerobic bacteria which have the ability to grow anaerobically. The word "facultative" should not be used by itself. In order to cover these microaerophilic bacteria, the term "aerotolerant" was proposed as a term additional to anaerobic. All bacteria that are described as "microaerophilic" would therefore be termed "anaerobic–aerotolerant." "Aerobic incubation" refers to incubation in the atmosphere of the laboratory and the medium used should not contain reducing substances added for the sole and specific purpose of reducing the oxidation–reduction potential.

The requirements for gaseous oxygen in the growth of an organism should be considered apart from the conditions of culture because they are not necessarily related. The term "oxybiontic" could be applied to those bacteria capable of using atmospheric oxygen in their growth, whereas "anoxybiontic" would apply to those bacteria not capable of using atmospheric oxygen in their growth. Many bacteria have not been studied adequately to permit their classification on the basis of oxygen utilization. However, the utilization of oxygen should be included in the description of each new species and in studies that reexamine a taxonomic group of bacteria. Exceptions that are not adequately covered under these terms need further description. Those clostridia, for example, that will grow to a limited extent under aerobic conditions should probably be defined as "aerotolerant–anaerobic–anoxybiontic." This would give a better description than "aerobic" or "microaerophilic."

For the so-called microaerophilic organisms, which actually require increased carbon dioxide tensions rather than reduced oxygen tension for growth, the term "microaerophilic" should not be used. Consideration should be given to the word "capneic" to describe these microorganisms.

The terms "oxybiontic" and "anoxybiontic" may be used without confusion to describe cultural habits as well as metabolism. For example, *Pseudomonas aeruginosa* is aerobic with an oxybiontic metabolism. *Escherichia coli* is facultatively anaerobic and may be either oxybiontic or anoxybiontic, depending on conditions of growth. *Streptococcus lactis* is facultatively anaerobic with an anoxybiontic metabolism. *Clostridium histolyticum* is anaerobic–aerotolerant with an anoxybiontic metabolism, and *Clostridium tetani* is anaerobic with an anoxybiontic metabolism.

These major terms indicate that the nutritional classification in microbial chemistry depends on two main factors: (a) the source of energy and (b) the nature of the electron donor and acceptor.

A number of these terms undoubtedly come from mammalian biochemistry, but others had to be introduced in order to cope with the significant differences occurring between mammalian tissues and bacteria.

Photosynthesis

Historical Development

Rabinowitch (226, 227), Bassham (15), Arnon (4), and Lascelles (171a) have ably condensed the early history of the study of photosynthesis. It commences in 1772 when Priestley discovered that green plants did not respire in the same way as animal cells but seemed to use the reverse method (226); he thus discovered the capacity of plants to produce free oxygen. It was Ingenhousz, however, in 1779, who suggested a connection between sunlight and oxygen development and therefore discovered the necessity of light and of the green pigment chlorophyll in the leaves to photosynthesis. The importance of carbon dioxide in the air emerged a few years later, when Ingenhousz and Senebier indicated that green plants exposed to light absorbed carbon dioxide and liberated oxygen. Saussure corrected this suggestion, indicating that water must enter into the photosynthetic production of organic matter.

$$CO_2 + H_2O \xrightarrow{\text{light}} \text{organic matter} + O_2$$

This suggestion that air is absorbed, oxygen thrown out, and the carbon kept for its cell synthesis became a fixed principle of biology for the next 100 years.

Later investigators therefore divided all living organisms into two groups (4): (*a*) green plants, which were regarded as the only carbon dioxide assimilators; and (*b*) all other forms of life, which must consume the organic products of the photosynthetic group. This led to the assumption that CO_2 assimilation was in fact photosynthesis. The role of light in photosynthesis became obvious as soon as von Mayer (ca. 1845) discovered the principle of conservation of energy, particularly the conversion of light energy into chemical energy. A few years later (ca. 1880), Winogradsky started to shatter the firm belief that CO_2 assimilation was, in fact, photosynthesis when he discovered that certain microorganisms were also able to produce organic material by carbon dioxide assimilation without the influence of light. This finding was supported by Lebedev, who suggested further that all cells in one way or another possess the ability to assimilate carbon dioxide. Engelman also showed that purple bacteria perform a type of photosynthesis but do not evolve oxygen. These bacteria require light to metabolize sulfur compounds.

From measuring the rate of photosynthesis in plants under different conditions, Blackman, in 1905, concluded that photosynthesis could not be a single photochemical reaction. He proposed that a "dark reaction"—a reaction that is not affected by light—must occur and therefore divided photosynthesis into two steps: a photochemical step and a dark step. The former produces unstable intermediates, which are stabilized by conversion into the final products, oxygen and cellular material, by the dark reaction.

The first work was done on the light or photochemical step in photosynthesis. It was known that in plant respiration carbon chains were broken down and hydrogen atoms were transferred to oxygen, producing water. In photosynthesis, therefore, the same type of reactions must be involved, only in reverse. Hydrogen must therefore be transferred from water to carbon dioxide for the building of carbon chains. Because this transfer of hydrogen in respiration liberates energy, it must be the one that results in the ultimate storage of energy in photosynthesis. Because the energy came from light, the light reaction was in all probability a hydrogen transfer from water to carbon against the gradient of chemical potential. The impact of light quanta adsorbed by chlorophyll would provide the necessary energy. This is approximately 112 kcal, the combustion heat of carbohydrate. In 1913, Bohr and Einstein first showed that light is adsorbed by atoms or molecules in the form of quanta of definite energy content, which is proportional to the wavelength of the light. The first attempt to measure the quanta required for photosynthesis was made in 1923 by Warburg, who stated that four quanta of light energy were required per molecule oxygen produced. However, this value is considered too low; the true value is still unsettled. The absorbing molecule was assumed to be chlorophyll, a theory strongly supported by the fact that this pigment absorbs red light. However, other pigments are also able to absorb light energy (e.g., carotenoids).

The process at this period appeared to proceed in two separate sequences: (1) the oxidation of water, which releases free oxygen, while hydrogen becomes attached to some acceptor; and (2) the hydrogenation of carbon dioxide to produce carbohydrates.

A great step forward was made independently in the 1930's by Hill and van Niel and later by Calvin and his associates. In 1937, Hill (226) found that dried powdered leaves were still able to oxidize water and liberate oxygen but could not reduce carbon dioxide to carbohydrates. This discovery separated the first sequence from the second. Van Niel and his associates obtained the converse, namely, organisms capable of reducing carbon dioxide in light, anaerobically, were dependent on hydrogen sulfide. From the H_2S, sulfur was deposited and a strict stoichiometric relationship obtained between the amount of CO_2 reduced and H_2S oxidized. Van Niel was struck by the similarity of this process to plant photosynthesis. In the latter O_2 is liberated, and in the former S is liberated. At the same time, Roelofson (235) and Gaffron (99) showed that H_2 could replace H_2S, and Foster (89) demonstrated that hydrogen could be removed from alcohols by purple bacteria. The reader is referred to Foster's excellent review (89) on developments in photosynthesis research to 1951. According to van Niel's ideas, water was split into [H^+] and [OH^-] radicals; [H^+] supplies the reductant for the conversion of carbon dioxide to carbohydrates and oxygen is derived from the [OH^-]. He suggested that bacteria lack the responsible enzyme involved in the photolytic process, and [OH^-] together with [H^+] from an outside source reform water. In the photosynthetic bacteria the reductant is supplied by a compound other than water. Van Niel therefore proposed the following equation for photosynthesis

$$2H_2A + CO_2 \xrightarrow{\text{light}} (CH_2O) + 2A^- + H_2O$$

If A is taken to be an oxygen atom, the process is plant photosynthesis; if it represents a sulfur atom, the process is the photosynthesis performed by sulfur bacteria. With this great discovery, van Niel removed photosynthesis by green plants from its unique position and placed it alongside other types of photosynthetic processes. He envisaged all photosynthetic processes as requiring some source of hydrogen, therefore, released by light energy with the concomitant production of relatively oxidized substances (e.g., O_2 and S).

It appeared that the origin of oxygen (and sulfur) in photosynthesis had been resolved and it became necessary to trace the path of the reductant. It was mentioned above that carbon dioxide reduction was considered the reverse of respiration. The reductant H^+ would then be responsible for the backward drive of the respiration. This type of synthesis does not require any light and therefore represents the dark step of photosynthesis. The energy was assumed to come from the adsorption of light quanta. The reduced pyridine nucleotides, however, were a problem because it was already known that these compounds were powerful biological reductants that could force their hydrogen atoms on other molecules and so participate in oxidation–reduction reactions of the living cell. Studies on the dark reaction were undertaken by Calvin and his research group, who confirmed that the carbon dioxide assimilation was the reverse of respiration and traced the pathway of carbon in photosynthesis along the "Calvin cycle" or "autotrophic CO_2 fixation mechanism" (15), which required the participation of ATP and reduced pyridine nucleotide. They also proved that ATP and reduced pyridine nucleotide formation requires light. This automatically suggested that the function of light lay in the production of ATP and reduced pyridine nucleotides.

Meanwhile, Arnon and his associates were directing their attention to the light reaction. The results of their investigation may be integrated with Calvin's findings under the name "phosphorylation."

BASIC PRINCIPLES OF PHOTOSYNTHETIC PHOTOPHOSPHORYLATION. The basic requirements for photosynthesis are twofold: (1) there must be a production of energy (ATP) with the help of light quanta, which is called "photophosphorylation"; and (2) provision must be made for the formation of a reductant that is able to reduce high-energy compounds into cellular material. Extensive research over the past decade, mainly by Arnon and co-workers (3, 7–9), has made it quite clear that two different types of photophosphorylation exist. These involve different pathways of electron transfer: (a) cyclic photophosphorylation and (b) noncyclic photophosphorylation.

Cyclic photophosphorylation. This type is predominant in plants and is used in a minor fashion in bacteria. All photosynthetic organisms, whether bacteria, algae, or plants, have one common denominator: ATP generation (Fig. 3.1). With the proposed model of the Calvin cycle (CO_2 fixation in the dark) the essence of photosynthesis—the conversion of light energy into physiologically useful chemical work or energy—must lie in the photochemical reaction that generates ATP. This involves cyclic phosphorylation. The basic difference between this and respiration is that in the former electrons are not removed, whereas they are in respiration, in which they are finally accepted by an electron acceptor. A light-absorbing molecule (chlorophyll or bacteriochlorophyll) becomes "excited" and acquires a

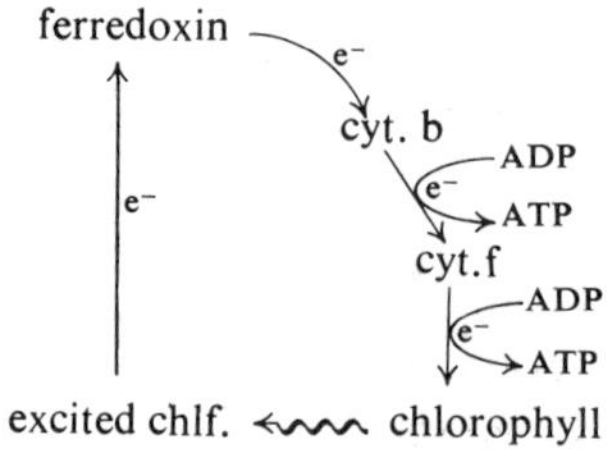
ferredoxin
e⁻
cyt. b
ADP
e⁻
ATP
e⁻
cyt.f
ADP
e⁻
ATP
excited chlf.
chlorophyll

Fig. 3.1. Scheme for cyclic photophosphorylation (8) [reprinted with permission of Macmillan (Journals) Pty, Ltd.]; chlf, chlorophyll.

tendency to expel an electron as well as a capacity to accept one as replacement for the expelled one. The expelled electron is raised to a high energy potential, as described in Chapter 2 and is transferred to the special redox system, ferredoxin. Ferredoxin is an iron-containing protein (150) with the unusual negative redox potential of -0.432 V, which is $\sim$100 mV less than that of the $NAD^+/NADH + H^+$ system. This redox potential is similar to that of the hydrogen electrode at pH 7.0 (H_2/H^+ -0.42 V). From ferredoxin an electron pair travels through a number of redox systems before it finally returns to the chlorophyll, where it is brought back to its ground state. The system is now again ready to accept a light quantum and to repeat the process.

It can be shown experimentally that 2 moles of ATP are generated in the course of this reaction. The catalytic action of ferredoxin in cyclic photophosphorylation, however, can only occur under strictly anaerobic conditions inside the cell.

$$n\text{ADP} + n\text{P}_i \xrightarrow[h\nu]{\text{Fd}} n\text{ATP}$$

The formation of ATP can therefore depend either on the availability of inorganic phosphate or pyrophosphate (10, 11) or on the availability of ADP (139). Whereas two possible mechanisms for the formation of inorganic pyrophosphate were described, an ADP-dependent, high-energy intermediate in photosynthetic ATP formation has been postulated in the photophosphorylation with chromatophores from *Rhodospirillum rubrum* (139). Figure 3.1 gives two sites for ATP generation in this electron transport chain, but the possibility of additional phosphorylation sites is not excluded (120a). Ferredoxin-catalyzed cyclic photophosphorylation will occur only under conditions such that noncyclic photophosphorylation is excluded (8). This cycle can therefore generate the necessary energy for use in the Calvin cycle, but the system lacks the necessary reducing power. For the operation of the Calvin cycle $NADH + H^+$ is required. The cyclic phosphorylation, however, cannot generate $NADH + H^+$ and for this dependence lies on noncyclic phosphorylation.

Noncyclic photophosphorylation. Van Niel's original concept was that CO_2 was reduced ultimately by hydrogen originating from the photolysis of water. However, CO_2 has been shown to be incorporated by fixation in the Calvin cycle. Therefore, any reduction effected by photolysis of water must be indirectly through another substance, since shown to be NAD^+, involved in the Calvin cycle. Van Niel's concept was that photolysis of water was common to all photosynthetic organisms, including plants, and

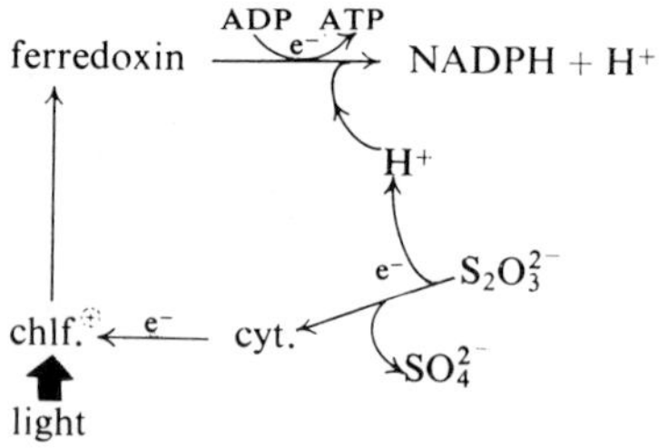
ADP ATP
e⁻
ferredoxin
NADPH + H+
H+
e⁻
S2O3 2−
chlf.
e⁻
cyt.
SO4 2−
light

Fig. 3.2. Scheme for the photosynthetic $NADP^+$ reduction in sulfur bacteria; chlf, chlorophyll.

that the [OH^-] released was enzymatically converted to molecular oxygen in plants and reduced to water by an external electron donor in bacteria. However, this conflicts with a unified concept for all photosynthetic mechanisms.

Losada *et al.* (181) showed that photolysis of water occurs in plants but not in bacteria. It was assumed that the hydrogen evolved reduced the NAD^+. Arnon (5) therefore distinguishes between a noncyclic photophosphorylation of the plant type and one of the bacterial type.

A proposal that in bacteria a light-independent reduction of NAD^+ was effected by an external electron donor was supported by Ogata and co-workers (3). According to these researchers, electrons resulting from quantum-activated chlorophyll in plant photosynthesis are required for (a) the production of ATP and (b) the reduction of NAD^+. The proton for the latter arises from the photolysis of water. In bacteria, electrons arising from chlorophyll are likewise involved in the production of ATP and the reduction of NAD^+; the proton for the latter, however, is donated by the external reducing agent (Fig. 3.2).

The illuminated chromatophore reacts not directly with NAD^+ or $NADP^+$ but first with ferredoxin, as explained in the section on cyclic phosphorylation (8). The reduction of $NADP^+$ is the second of two dark reactions following the photochemical reduction of ferredoxin. The reoxidation of reduced ferredoxin occurs with the help of a flavoprotein enzyme, called "ferredoxin–NADP reductase" (reduced-NADP:ferredoxin oxidoreductase, EC 1.6.99.4), and the reduced reductase is finally reoxidized by $NADP^+$. This flavoprotein enzyme, the flavin of which is probably FAD, is very similar to the corresponding green plant enzyme. Apart from the ferredoxin–NADP reductase activity, cell free extracts of *Rhodopseudomonas palustris* also exhibited a reduced NADP–cytochrome c-552 reductase (EC 1.6.2.4 or 1.6.2.5), reduced NADP–DCIP (dichlorophenolindophenol) reductase (EC 1.6.99.2), and a pyridine nucleotide transhydrogenase (EC 1.6.1.1) activity (319). The presence of all these enzymes could be very important for biosynthesis regulation. A vigorous pyridine nucleotide transhydrogenase transfers the hydrogen from the reduced NADP formed to NAD^+ producing reduced NAD, depleting the reduced NADP pool necessary for the biosynthesis of cellular material. This reoxidation of $NADP^+$ together with the inhibition of ferredoxin-NADP reductase by $NADP^+$ (319) explains the suggestion that cytochrome c-552 could take the place of ferredoxin under certain conditions in the Athiorhodaceae to make certain of the reduced NAD supply during photosynthesis. A pyridine nucleotide transhydrogenase has been observed in *Rhodopseudomonas spheroides* (208), where the trapping of reduced NAD completely inhibited the reduction of $NADP^+$, although $NADP^+$ was preferred in the reduction. The oxidation–reduction reaction employing ferredoxin–NADP reductase transfers one electron. This enzyme can also form complexes with bacterial ferredoxin, rubredoxin, and flavodoxin (90), which could be important in the interactions between the chromatophores and these complex flavoproteins.

No cyclic electron flow is possible under strongly reducing conditions, because the components of the cyclic chain will be kept in the reduced state. Electrons will flow unidirectionally via a portion of the cyclic chain to ferredoxin and to $NADP^+$. As long as this flow is maintained through a cytochrome, a phosphorylation at a cyclic site will be induced (Fig. 3.3).

As long as electrons flow from an external hydrogen donor to ferredoxin (via chlorophyll), and from there to NADP, cyclic photophosphorylation cannot proceed. Photosynthetic bacteria, in other words, are able to regulate ATP and reducing power production. If there is a great need for reducing power, the noncyclic photophosphorylation will operate with the help of an exogenous hydrogen donor. When the bacterium does not require reducing power, cyclic phosphorylation supplies the ATP required for cell synthesis.

This noncyclic photophosphorylation of the bacterial type (4) explains

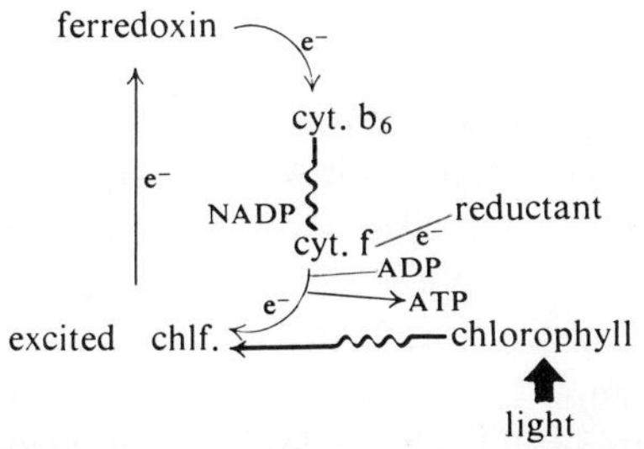
ferredoxin
e⁻
cyt. b₆
e⁻
NADP
reductant
cyt. f
e⁻
ADP
e⁻
ATP
excited chlf.
chlorophyll
light

Fig. 3.3. Noncyclic photophosphorylation of the bacterial type (8) (reprinted with permission of Macmillan (Journals) Pty, Ltd.); chlf, chlorophyll.

(*1*) the ATP formation at a cyclic site and (*2*) the dependence for the formation of reducing power on a noncyclic electron flow, as is indicated by the requirement for a terminal acceptor (NADP).

Studies in chloroplasts with ascorbate–dichloroindophenol as electron donor for $NADP^+$ reduction revealed no indication of a sharp transition from a nonphosphorylating to a phosphorylating system. Whenever a phosphorylation is observed in this system, cyclic photophosphorylation systems operate simultaneously with two different sites of ATP formation. One site is located on the noncyclic pathway, prior to the point of entry of electrons from the outside donor, and the other on the pathway that is exclusive for the cyclic system and not shared by the noncyclic ones (123). These studies were confirmed by the stoichiometric calculation of ATP formation, which gave a $P/2\ e^-$ ratio of 2 on chloroplasts (87). Although this theory explains the simultaneous operation of both photophosphorylation systems, it requires two different systems (system I and system II) of light-absorbing pigments; one that excites the electrons from the photolysis of water and a second that absorbs and excites the electrons from the reductant (46). It is possible that system I and system II are similar, if not identical, to the short-wavelength-sensitized and long-wavelength-sensitized phosphorylation processes found in chloroplasts (254). A similar system has been indicated in *Chromatium* D (112) [now *Chromatium vinosum* (221a)].

Systematics of Photosynthetic Bacteria

A closer look into the primary and secondary electron acceptors, as well as the actual reaction center of bacteriochlorophyll, revealed that the schemes presented on cyclic and noncyclic photophosphorylation are certainly oversimplified. However, they should demonstrate what is meant by cyclic and noncyclic photophosphorylation and why one could regard both as separate pathways, although neither of them could function alone in photosynthesis. Measurements of the $P/2\ e^-$ ratio of noncyclic photophosphorylation revealed that only one of the pathways operates at a time. This ratio of 1 obtained with bacteria (61) certainly differed from the higher ratios obtained from chloroplasts (87). The number of electron transport particles necessary also indicates that variations among the various families of photosynthetic bacteria are most likely.

To discuss the more detailed information available, therefore, it is necessary to first learn something about the systematics of photosynthetic bacteria. Three major families (269, 310) of photosynthetic bacteria are distinguishable by the nature of the pigment and the nature of the utilizable substrate.

1. The green sulfur bacteria, or Chlorobacteriaceae, include the genera *Chlorobium, Chloropseudomonas, Pelodictyon* (221), *Clathrochloris, Chlorobacterium, Chlorochromatium,* and *Cylindrogloea.* Of these genera, *Chlorobium* and *Chloropseudomonas* are the only ones being investigated more or less extensively. All these bacteria contain chlorobium chlorophyll, which has an absorption band at 750 nm. Bacteriochlorophyll a (absorption maximum at 590 nm) is present only in small amounts. The function of a second chlorobium chlorophyll, with an absorption band at 660 nm, is not fully understood, although its biosynthesis has been described (233). The genus *Chlorobium* can use four different types of inorganic hydrogen donors (174).

a. Sulfide. *Chlorobium thiosulfatophilum* will always oxidize sulfide to sulfate.

$$CO_2 + 2H_2S \xrightarrow{h\nu} (CH_2O) + 2S + H_2O$$

$$3CO_2 + 2S + 5H_2O \xrightarrow{h\nu} 3(CH_2O) + 2H_2SO_4$$

It was found that the rate of CO_2 assimilation diminished when all the sulfide had been converted to sulfur and the production of sulfate had began. *Chlorobium limicola* converts sulfide to sulfur only, which then accumulates outside the cell.

b. Thiosulfate. *Chlorobium thiosulfatophilum* can oxidize thiosulfate to sulfate.

$$2CO_2 + Na_2S_2O_3 + 3H_2O \xrightarrow{h\nu} 2(CH_2O) + Na_2SO_4 + H_2SO_4$$

c. Hydrogen. Both *Chlorobium thiosulfatophilum* and *Chlorobium limicola* can use hydrogen as their hydrogen donor.

$$2H_2 + CO_2 \xrightarrow{h\nu} (CH_2O) + H_2O$$

d. Organic compounds. *Chlorobium thiosulfatophilum* is also able to utilize organic compounds under certain conditions as electron donors.

Of these electron donors, thiosulfate metabolism is the best studied and it is assumed that the other inorganic electron donors may serve in the same way (7) (see p. 312). The DNA base ratio of Chlorobacteriaceae is 48.5–58.1% guanosine plus cytosine (183).

2. The photosynthetic purple and red sulfur bacteria, or Thiorhodaceae, behave similarly to the green sulfur bacteria in their use of inorganic com-

pounds as hydrogen donors (272). The best known genera are *Chromatium* and *Thiospirillum,* which have DNA base ratios of 61–63% guanosine plus cytosine (183).

The separation of the green from the red and purple sulfur bacteria rests on the pigmentation of the respective organisms. The Thiorhodaceae contain mainly bacteriochlorophyll a or bacteriochlorophyll b. Spectroscopically they show bands of absorption at 375 and 590 nm in the visible and 800, 850, and 890 nm in the infrared region for their chlorophylls. They also contain a large amount of carotenoid pigments that exhibit absorption peaks between 400 and 600 nm, often obscuring the 590 nm band for bacteriochlorophyll (269).

3. The nonsulfur bacteria, or Athiorhodaceae, with the genera *Rhodopseudomonas, Rhodospirillum, Rhodomicrobium* and *Vannielia,* also contain bacteriochlorophylls a and b and are primarily dedicated to the metabolism of organic compounds. They are capable of growth anaerobically when exposed to light and are also able to grow in the dark when exposed to air. DNA base ratios vary among the different genera and species: *Rhodomicrobium vannielia,* 62–64%; *Rhodopseudomonas palustris, R. acidophila, R. capsulata,* and *Rhodospirillum,* 63–67%; *Rhodopseudomonas viridis* 66–68%; *Rhodopseudomonas spheroides,* 68–70%; and *Rhodopseudomonas gelatinosa* 71–72% (183). The average contents of 60.5–65% for *Rhodospirillum* and 64.4–70.5% for *Rhodopseudomonas* (266) indicate a distinct difference between the two genera.

The Thiorhodaceae are separated from the Athiorhodaceae because the former are predominantly autotrophic and the latter heterotrophic (304). Metabolically, the Athiorhodaceae do not oxidize H_2S and therefore do not deposit sulfur internally or externally. The differentiation between these two families, however, is not sharp. Some purple sulfur bacteria can also grow in media devoid of H_2S but supplied instead with organic compounds, thus resembling the purple nonsulfur bacteria, and vice versa. Moreover, *Rhodomicrobium,* which is classified with the Athiorhodaceae, does not appear to require any vitamins for growth and can readily grow on purely mineral media containing H_2S, whereas *Chromatium* and *Thiospirillum,* of the Thiorhodaceae, seem to depend on an external supply of vitamin B_{12}. This distinction between the families is not above criticism, however, although morphological differences can still support a differentiation between the two groups of purple bacteria.

Purple bacteria in general grow readily when illuminated under anaerobic conditions in synthetic media with malate and an ammonium salt as sole carbon and nitrogen sources. With a suitable organic compound available, CO_2 or inorganic donors are also necessary (212). Purple bacteria seem to grow much more rapidly with organic carbon sources, such as C_4-dicarboxylic acids, than with CO_2. On media containing a more oxidized substrate, excellent growth is obtained in the absence or with evolution of CO_2 (107). These organic hydrogen donors do not undergo a simple one-step oxidation in order to furnish the cell with CO_2 for photochemical reduction (303) but generally supply carbon intermediates, other than CO_2, that are directly used by the cell for synthetic purposes (107). This, of course, led to the conclusion that when organic compounds supply intermediates for direct assimilation, photosynthetic CO_2 fixation is not obligatory and may, in fact, be bypassed or suppressed (106, 149, 261, 263–265). The anaerobic utilization of organic compounds is therefore essentially of a heterotrophic character.

Photosynthetic Phosphorylation

THE GREEN SULFUR BACTERIA OR CHLOROBACTERIACEAE. It was mentioned above that bacteria of this family (*Chlorobium* and *Chloropseudomonas*) possess predominantly a chlorophyll, known as chlorobium chlorophyll, that is characteristic for this family. Measurements of light-induced absorbancy changes revealed a photochemical reaction center P-840 (284, 285), which is the site of a primary photochemical reaction. The above-described cyclic and noncyclic photophosphorylation, often referred to as "cyclic" and "open" electron transport chains (283), can be found here in their purest form.

Chlorobium thiosulfatophilum possesses a ferredoxin that is very similar to the ferredoxins from nonphotosynthetic anaerobic bacteria (35) and that very closely resembles the *Chromatium* ferredoxin. Another member of the electron transport chain is the quinone group. Three quinones were isolated from *Chlorobium thiosulfatophilum* and *Chloropseudomonas ethylicum* (224): menaquinone-7, chlorobium quinone, and the polar menaquinone. Because the chlorobium quinone is present only in sulfide-adapted cells, the suggestion of its participation in the cyclic photophosphorylation pathway is valid. Polar menaquinone breaks down to chlorobium quinone *in vitro* and is therefore thought to be a precursor to this quinone. Apart from ferredoxin and quinones, three c-type cytochromes (c-551, c-553, and c-555), have been isolated and characterized from *Chlorobium thiosulfatophilum* (187). As no protoheme was detected in the cells, no cytochrome b could be found. The redox potentials of all these cytochromes are close to each other. Chlorobium cytochrome c-555 is very similar to the algal c-type cytochrome (the f-type cytochromes) but differs from the nonsulfur purple bacterial c-type cytochrome (c_2 type) (320). Cytochrome c-553 seems to act as a sulfide–cytochrome c reductase (166a,b) and cytochrome c-551 as a thiosulfate–cytochrome c reductase (166b). It therefore appears that the green sulfur bacteria possess a unique chlorophyll, ferredoxin, menaquinones, and three c-type cytochromes.

Light-induced UV absorption changes revealed that under anaerobic conditions the UV absorption between 260 and 290 nm increased on illumination and that of wavelengths shorter than 260 nm decreased. On illumination 60% of the total menaquinone was oxidized (289). This photooxidation was abolished completely by the addition of air and could be restored after making the suspension anaerobic. The role of ferredoxin, however, appears to be confined solely to the reduction of NAD^+ to obtain the reducing power necessary for CO_2 fixation. This reduction could be carried out by chlorophyll-containing particles, ferredoxin, and a soluble protein fraction of *Chlorobium thiosulfatophilum*. The soluble protein fraction was replaceable by purified ferredoxin–NADP reductase from spinach chloroplasts (34, 90). This soluble chromatophore fraction contains a hydrogenase that can reduce NAD^+ in the presence of ferredoxin (83). The second hydrogenase, which can be solubilized by Triton X-100 treatment of the particulate chromatophore fraction, could not perform this reaction. The rate by which NAD^+ was reduced was comparable to the overall rate of CO_2 assimilation. Ferredoxin itself was photoreducible by the chlorophyll-containing particles with sodium sulfide or 2-mercaptoethanol as the electron donor. Considering that cytochrome c-551 (= cytochrome c-419) exhibited a fast dark reaction following photooxidation, whereas cytochrome c-555 (= cytochrome c-422) exhibited a slow dark reduction, pointing to a closer link to the substrate (94), the following electron transport system could be envisaged (Fig. 3.4).

The problem of the primary electron acceptor is still not solved. It is not certain whether ferredoxin is photoreduced or whether this reduction

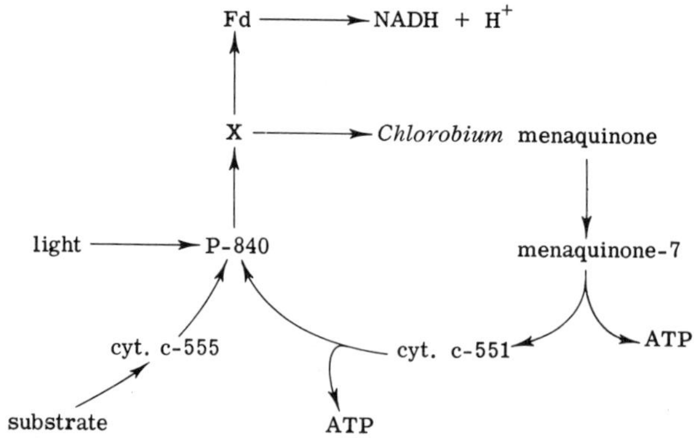
Fd
NADH + H+
X
Chlorobium menaquinone
light
P-840
menaquinone-7
cyt. c-555
cyt. c-551
ATP
substrate
ATP

Fig. 3.4. Hypothetical electron transport system in the green sulfur bacteria, or Chlorobacteriaceae.

occurs via an intermediate (X), which can be either a common intermediate for the open as well as the cyclic phosphorylation system or two compounds, one for each system. This close interaction between both systems and the possible existence of a photoreduction of ferredoxin may be one of the reasons why green photosynthetic bacteria are less resistant to the action of oxygen. Although they may retain viability during constant contact of the cells with air for several hours (124), no change to an oxidative respiratory system seems to be possible.

The isolation of a rubredoxin with properties like those of the clostridial type raises the question as to its role in Chlorobacteriaceae. Rubredoxin was only found in *Chlorobium thiosulfatophilum* and *Chloropseudomonas ethylicum*, but not in the Thiorhodaceae or Athiorhodaceae (188). It is possibly an important link to the plant ferredoxin and may serve as a phylogenetic guide, as it appears to bridge the bacterial and plant ferredoxins (313). In contrast, rhodanese (thiosulfate sulfurtransferase, EC 2.8.1.1) has been observed in all three taxonomic families of photosynthetic bacteria (326). The observation that cytochrome c-551 may be a cytochrome c_3-class enzyme (189), differing only in one missing heme group from the *Desulfovibrio vulgaris* cytochrome c_3, brings forward the problem and importance of redox potential in photochemical processes or photophosphorylation. This cytochrome has so far only been found in *Chloropseudomonas ethylicum* and not in *Chlorobium* or any other photosynthetic organism. It could replace cytochrome c-555 in the scheme in Fig. 3.4 and cytochrome c-555 could take the place of cytochrome c-551.

PURPLE AND RED SULFUR BACTERIA OR THIORHODACEAE. Investigations into the light-induced electron transport system have been focused around the light-induced absorbancy changes associated with bacteriochlorophyll (48) and around variations in oxidation states of cytochromes (206). The illumination of the purple photosynthetic bacteria caused an absorbancy change at 890 nm, which was accompanied by a second shift in the spectrum at around 800 nm (72). Of these two absorbancy changes, the first one, around 890 nm, was equivalent to the one at 870 nm (890 nm in *Chromatium*) occurring in other photosynthetic strains. This P-870 nm oxidation was found to be the primary photochemical reaction center in *Chromatium* chromatophores (214). The primary photochemical reaction of bacterial photosynthesis, in other words, is the oxidation of a special bacteriochlorophyll trimer (243), P-870, which has already been isolated and characterized from the bulk bacteriochlorophyll (229). The present evidence allows the interpretation that P-800 and P-870 are separate molecules, with P-890 (in *Chromatium*) being the "active center" bacteriochlorophyll, which initiates light-induced electron transport, and P-800 being the "light-harvesting" bacteriochlorophyll in close association with P-890 or P-870. The discovery of a new bacteriochlorophyll component, "P" (91), awaits confirmation.

In addition to these two bacteriochlorophyll components, a high-potential cytochrome c-553 was found (59). This cytochrome has properties like those of several isolated algal cytochromes of the f type. It may be related to cytochrome c-555 and appears to be identical to the cytochrome c-555 described from *Chlorobium.* A correlation of P-890 and cytochrome c-555 has been reported by Clayton (47), who also showed that the c-555 and P-890 oxidation could be correlated with the reduction of coenzyme Q-7 in chromatophores with a 1:1:1 stoichiometry. Therefore, coenzyme Q-7 should be included in the P-890 reaction center. In contrast is the stoichiometry of P-870 and the cytochrome c-552 (= c-422), as each P-870 reaction center is responsible for the oxidation of two c-552 hemes (216, 265). Therefore, cytochrome c-552 would not be expected to participate in the cyclic electron flow. Cytochrome cc′ is an additional cytochrome found to complex with P-870, which would result in the following flow system:

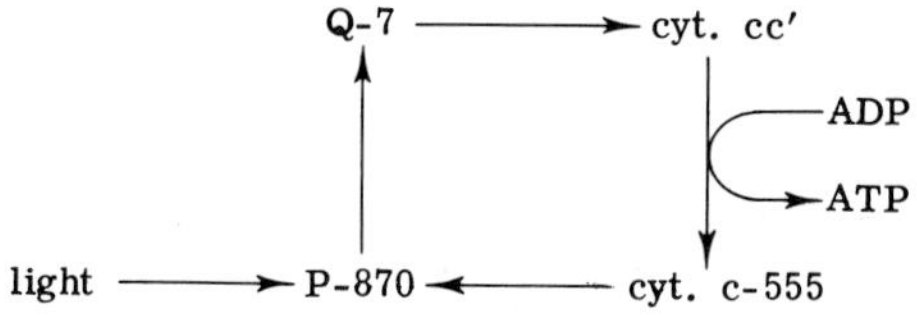
Q-7
cyt. cc′
ADP
ATP
light
P-870
cyt. c-555

The cytochrome c-552 of *Chromatium* has been purified and characterized, and chromatography showed that it must be of a FAD type (129).

The above scheme, by which quinones were confirmed as the first electron acceptors from the photochemical center P-870 (151), being the primary photochemical reaction in *Chromatium,* produces oxidized P-870 and reduced ubiquinone. If the cyclic electron flow of *Chromatium* is compared with the one of *Chlorobium,* it can be seen that both genera differ not only in their chlorophyll constituents but also in whether they have menaquinone or ubiquinone as their first electron acceptor. The possession of ubiquinone (288) by *Chromatium* forms a link to the Athiorhodaceae, as will be seen below (p. 103).

The cyclic electron flow having been elaborated on, the question now arises as to whether the Thiorhodaceae have one photochemical center serving the cyclic and noncyclic electron flow, as has been seen in the green sulfur bacteria, or whether they possess two such centers for the oxidation of both the high- and the low-potential cytochromes. The reports are still conflicting.

It was originally thought that the presence of three c-type cytochromes must reflect distinct electron pathways, as all the cytochromes reacted with P-890 (191, 206, 207). Cusanovich and associates (57, 58) confirmed the idea of the presence of three pathways. To the above-mentioned cyclic flow system they attached a noncyclic electron transport system, which would function under more reduced conditions or at high light intensities, and including cytochrome c-552 and a P-905 reaction center. The third pathway would include both systems operating in a transition region of redox potentials. The optimum level of photophosphorylation is obtained when the apparent redox potential of the system is between 50 and 100 mV (58).

This suggestion of three pathways depending on the redox potential is of great interest when it is considered that cell-free extracts of *Chromatium* are able to carry out an aerobic metabolism (113). Air in itself is not lethal to *Chromatium*, because aerated suspensions survived as well if not better in the presence of thiosulfate as did control suspensions in nitrogen. Although the rate of oxygen uptake by malate-grown *Chromatium* cells was only 0.2–0.3 times that of *Rhodopseudomonas spheroides*, the low rate of oxygen uptake does not account for the lack of aerobic growth. Measurements of ATP contents after the admission of air indicated a limited capacity for oxidative phosphorylation. A value of approximately 1 for the P/O ratio has been obtained. The addition of light inhibits the rate of oxygen consumption by suspensions equilibrated with excess substrate but may stimulate the rate in previously starved suspensions. It may be that once the substrate concentration in the cell exceeds enzyme saturation, the oxidant produced in the light reaction then competes effectively with oxygen for the available supply of electrons (137). The actual failure of aerobic growth or *Chromatium* cells is probably caused by a combination of factors, such as low Q_{O_2} and low phosphorylation efficiency.

The more oxidizing conditions require low light intensities and the "P-890" pathway, whereas the "P-905" pathway functions under more reducing conditions and/or high light intensities. The latter pathway can only operate under anaerobic conditions. This "open" or noncyclic phosphorylation pathway would be gradually replaced by an oxidative respiratory pathway once aerobic conditions prevailed. The P-890 pathway, however, can function far longer under semiaerobic conditions and thus makes air not lethal. However, once the redox potential conditions exceed 400 mV (e.g., fully aerobic conditions), the respiration is not efficient enough to tolerate growth of the whole cell. Cell-free extracts are able to take up oxygen, however, as appropriate experiments indicate the occurrence of a noncyclic phosphorylation with oxygen as a terminal electron acceptor (58).

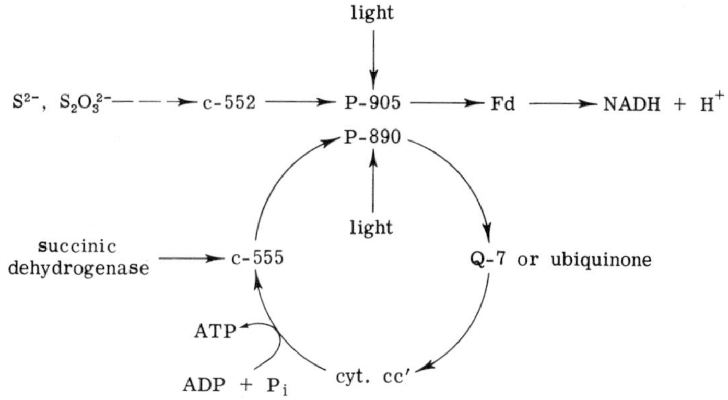

light
S^{2-}, $S_2O_3^{2-}$
c-552
P-905
Fd
NADH + H$^+$
P-890
light
succinic dehydrogenase
c-555
Q-7 or ubiquinone
ATP
ADP + P$_i$
cyt. cc'

Fig. 3.5. Hypothetical two-system scheme of the electron transport chain in Thiorhodaceae.

Another feature of the two systems is that the organism is able to operate at different redox levels in the light (94). When cells are grown photoautotrophically, reduced ferredoxin (Fd) or NADH + H^+ required for CO_2 reduction could be supplied by the "P-905" pathway. The relationship of cellular ATP and chlorophyll levels to sulfur metabolism also gives support for the existence of two photosystems (248). When cells are grown photoheterotrophically with succinate, less reducing power is needed and the cells could almost entirely operate with the "P-890" pathway. Light intensity has the effect of shifting the redox potential of the system, and phosphorylation activity shifts the redox potential into the transition region, where both systems are operative. This is certainly a versatile system for the production of ATP and reducing power. The combined system would therefore be as outlined in Fig. 3.5.

This two-system scheme has its critics, however. Although the observations on the effect of oxidation–reduction potentials were confirmed, the existence of two distinct photochemical centers has been questioned (70, 217, 255). In all cases, the P-905 center is the disputable one, and Seibert and Devault (255) proposed three different possible models to support their claims for the existence of only one single photochemical center.

It is also not as yet known whether ferredoxin and ubiquinone are the first electron acceptors for the noncyclic and the cyclic photophosphorylation system, respectively (152, 215). A low-potential light reaction has been observed, in chromatophores of *Chromatium* D (256) [now *C. vinosum* (221a)], that is thermodynamically able to donate electrons to NAD^+. Redox potentials of the primary electron donor (P-870) and this primary electron acceptor (X) have been established in *Chromatium* (54, 57, 70,

256, 257), *Rhodospirillum rubrum* (54, 231), and *Rhodopseudomonas spheroides* (54, 231). The so reduced primary electron acceptor (X^-) is claimed to transfer an electron to a secondary acceptor (Y) (215), and the kinetics of this transfer have been established (40). It is not certain, however, whether this system represents a new reaction center or a different form of P-883. The observation of a component responsible for the oxidation at 836 nm in chromatophores of *Chromatium vinosum* strain D with a redox potential of 320 mV (247) indicates that the electron transport system of the Thiorhodaceae is still far from being solved.

THE PURPLE NONSULFUR BACTERIA OR ATHIORHODACEAE. In contrast to the other two photosynthetic families, Athiorhodaceae are able to grow either anaerobically in the light (photosynthesis) or aerobically in the dark (oxidative respiration). The organisms therefore must be able to switch from photophosphorylation to oxidative phosphorylation. It is therefore expected that both electron transport systems are present in the various organisms.

The evidence in the literature that P-890 in *Rhodospirillum rubrum* or P-870 in *Rhodopseudomonas spheroides* are the primary photochemical centers is overwhelming (230, 310). The light is absorbed by another bacteriochlorophyll molecule, P-800 (48, 102, 311), and is transferred to the reaction center for the initial photochemical event. Both molecules form a very close association with each other, and the ratio appears to be 2:1 in favor of P-800 (50, 51, 232, 270).

All species of the Athiorhodaceae contain ubiquinone, but no menaquinone (184). Ubiquinone-10 appears to be the predominant member of the quinones in the Athiorhodaceae (220), as it is found almost in abundancy (204). Apart from ubiquinone-10 or ubiquinone-8 (*Rhodopseudomonas gelatinosa*) (39), rhodoquinone has also been found in most of the Athiorhodaceae. It is assumed that rhodoquinone is able to function in a way similar to ubiquinone-10 (318). As in the Thiorhodaceae, the ubiquinones are very readily reduced by the photochemical center P-890 (20, 141, 310). However, because oxidized P-890 in turn also oxidizes a cytochrome, it is postulated that an unknown electron acceptor X must be reduced when P-890 is oxidized, possibly being followed by a second electron acceptor Y (40, 215). The primary electron acceptor X is possibly an iron–sulfur protein (71, 176). It is not known whether or not a connection exists between X and the two different ferredoxins (type I and type II) isolated from *Rhodospirillum rubrum* (258, 259). Ferredoxin type II was found regularly in cells grown heterotrophically in the dark and light, whereas type I only accompanied type II when cells were grown in the light.

In addition to the ubiquinones, of course, there are ferredoxin and a number of cytochromes present in the Athiorhodaceae (94, 310). The main cytochromes found are of the c type (c-550.5, c-553, and c-558) and a b-type cytochrome. Cytochrome c-553 was mainly oxidized at low light intensities, whereas the other two cytochromes were oxidized almost equally well at high light intensities.

As do all the other photosynthetic families, the Athiorhodaceae possess a cyclic electron transport system (305) that includes ubiquinone and the c-type and b-type cytochromes. In studying this cyclic photophosphorylation a photoreduction of NAD^+ that occurred only in the presence of succinate was observed (306). It was therefore suggested that the very active succinic dehydrogenase must interact with the closed system in freeing the photoreducing system to react with NAD^+. A similar stimulation was obtained on adding fumarate. It is thought that succinate and fumarate interact in the transfer of electrons from the photoreducing site to the photooxidizing site of the chromatophore, which would cause an oxidation–reduction imbalance of some components (305). This concept of a light-dependent noncyclic electron transfer could also be explained in a different way (26). The interference of succinate and the resulting evolution of hydrogen needed for NAD^+ reduction could occur through the promotion of dark oxidation–reduction reactions by energy-rich intermediates, which are produced by the photochemical apparatus. This "accessory hydrogen donor" can therefore only be provided if light energy is available. ADP and inorganic phosphate (Pi) inhibited this system (138, 305). From these results two distinct general metabolic patterns, an autoheterotrophic and an autotrophic, have been postulated (110).

Investigations with *Rhodopseudomonas capsulata* (156), however, revealed that ADP did not inhibit the reduction of NAD^+ but in some cases stimulated this process. In *Rhodopseudomonas capsulata* NAD photoreduction and phosphorylation occurred simultaneously. As these results could not be explained by competition with the photochemical reaction, it was suggested that the photoreduction of NAD^+ with succinate or molecular hydrogen as electron donors must occur via a noncyclic photophosphorylation. The existence of such a separate noncyclic photophosphorylation was supported by the occurrence of ferredoxin, NADPH reductase, and a NADH–DCIP (dichlorophenolindophenol) reductase, together with a pyridine nucleotide transhydrogenase (84, 85, 209, 319). A very striking property of the NADPH reductase is that its activity is strongly inhibited by $NADP^+$. The presence of this enzyme could be important for biosynthesis regulation in photosynthetic bacteria. It suggests, however, that NAD^+ is not the exclusive electron acceptor in a noncyclic photophosphorylation. The evidence for a light-dependent reduction of NAD^+ by energy-dependent, reversed electron flow was also confirmed for *Rhodospirillum rubrum* (145).

A possible decisive step forward was taken when Knobloch and associates (162, 163) could prove for the first time that NAD reduction can take place as a secondary energy-dependent process in the dark and does not require an integrated chromatophore structure. *Rhodopseudomonas palustris* is able to couple the energy-dependent NAD reduction in the dark to the oxidation of thiosulfate. Electrons from this thiosulfate oxidation must therefore enter the electron transport chain at the level of cytochrome c and the energy-linked reverse electron flow to NAD^+ is therefore similar to the aerobic respiration of *Thiobacillus* (see p. 339). This would mean that under light and anaerobic conditions, *Rhodopseudomonas palustris* utilizes a cyclic electron flow for the ATP generation and a photoreduction of NAD^+, whereas under aerobic and dark conditions, an aerobic respiration with reversed electron flow (see p. 319) operates—a perfect intermediate between photosynthesis and aerobic respiration. Possible support for this comes from the observation (154) that the photooxidation of anaerobic c-type cytochrome under anaerobic conditions is nonexistant in aerobic samples. Instead, cytochrome c_2 appears to be a sort of branching point, as this cytochrome is able to interact with both cyclic and respiratory electron flow pathways (145). Cytochrome c_2 was also found in form of a NADPH–cytochrome reductase (239), which does not reduce cytochrome c-553 but reduces 2,6-dichlorophenolindophenol and $K_3Fe(CN)_6$ in light-grown *Rhodopseudomonas spheroides*. The enzyme is not associated with $NADP^+$ reductase or $NADP^+$-transhydrogenase (EC 1.6.1.1), but NADH functions as a competitive inhibitor. Apart from the NADPH–cytochrome c reductase there appears also a NADH oxidase (322) in both dark-grown and light-grown cells. Both types of cells exhibit terminal oxidases (241), although *Rhodospirillum rubrum* possessed only cytochrome o and *Rhodopseudomonas spheroides* cytochrome o and cytochrome a. The additional terminal oxidase in *Rhodopseudomonas spheroides* may be indicative of the previously suggested link of *Rhodopseudomonas* to aerobically respiring thiobacilli.

In addition, there are reports of two distinctly different NADH dehydrogenases (23, 24, 142) that are both activated by flavins, one having a heme protein as prosthetic group and the other a quinone. A possible function for these two dehydrogenases was suggested as shown in Scheme 3.1. The NADH oxidation is tightly coupled to phosphorylation and is significantly stimulated by the addition of Pi and Mg^{2+}, or a complex system consisting of Pi, Mg^{2+}, ATP, and the hexokinase–glucose system in aerobically dark-grown cultures of *Rhodospirillum rubrum* (321, 323).

Aerobically dark-grown and anaerobically light-grown cells of *Rhodo-*

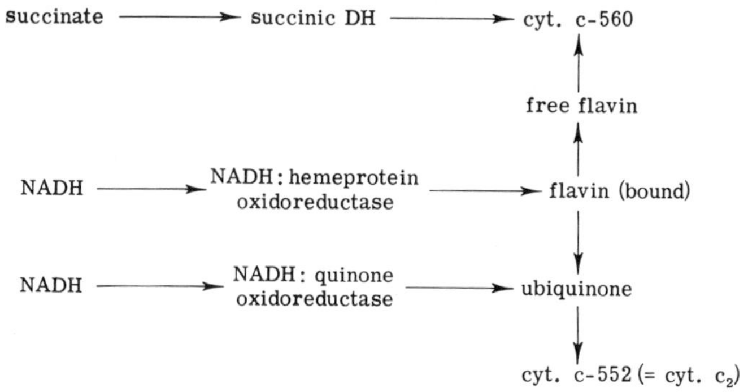
succinate
succinic DH
cyt. c-560
free flavin
NADH
NADH: hemeprotein oxidoreductase
flavin (bound)
NADH
NADH: quinone oxidoreductase
ubiquinone
cyt. c-552 (= cyt. c_2)

Scheme 3.1.

spirillum rubrum did not show any difference in the specific activities of NADH dehydrogenase and succinate–cytochrome c reductase (25).

With the establishment of the two light-dependent systems, the actual effect of oxygen was investigated. In the absence of an external donor, such as succinate, cyclic phosphorylation was strongly inhibited by oxygen (111, 157, 178). When the oxygen concentrations are varied from a high to a low partial pressure, the level of NADH increases greatly as soon as the pO_2 drops to 0.5 mm Hg (250). If this change from aerobic to anaerobic conditions is made rapidly, an overreduction of NAD^+ occurs, whereas in reverse from anaerobic to aerobic conditions an overoxidation of reduced NAD is the result. Cytochromes b and c_2 follow the same kinetics. These changes correlate with the breakdown of oxidative phosphorylation.

If all the available information is taken into consideration, an electron transport scheme as outlined in Fig. 3.6 could be evolved; this does not mean, however, that this scheme is actually operating in that way. This scheme outlines the cyclic photophosphorylation together with a photoreduction of NAD^+ via the reversed electron flow system. It also includes the possibility of a noncyclic phosphorylation from thiosulfate via ferredoxin with the help of the pyridine nucleotide transhydrogenase. The role of ferredoxin is still disputed (309), however, and could be mainly in reductive CO_2 assimilation (see p. 130). This would agree with the finding that hydrogenase activities depend strongly on the culture conditions (158) and are lowest under photoheterotrophic conditions. Under aerobic conditions, the electron flows via cytochrome c_2 and cytochrome b to oxygen, whereas NAD^+ reduction occurs via the same reversible electron flow system. The importance of the two NADH dehydrogenases could be for low-potential redox systems in the case of heterotrophic growth. This scheme should only be seen as a summary and not as a functional hypothesis.

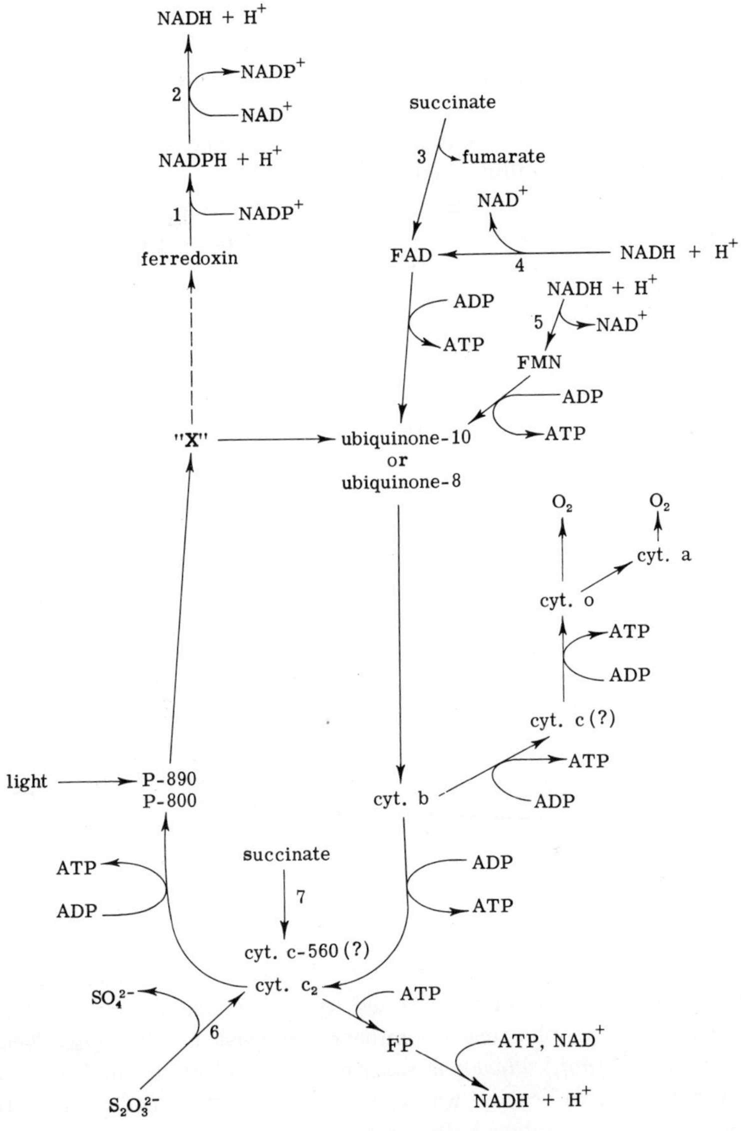
NADH + H+
NADP+
2
NAD+
NADPH + H+
1
NADP+
ferredoxin
"X"
succinate
3
fumarate
NAD+
FAD
4
NADH + H+
ADP
ATP
NADH + H+
5
NAD+
FMN
ADP
ATP
ubiquinone-10
or
ubiquinone-8
O2
O2
cyt. a
cyt. o
ATP
ADP
cyt. c (?)
ATP
ADP
light
P-890
P-800
cyt. b
ATP
ADP
succinate
7
ADP
ATP
cyt. c-560 (?)
cyt. c2
ATP
SO4 2-
6
FP
ATP, NAD+
S2O3 2-
NADH + H+

Fig. 3.6. A generalized electron transport scheme for photosynthetic bacteria under aerobic and anaerobic conditions. 1, ferredoxin-NADP reductase; 2, NADPH transhydrogenase; 3, succinic dehydrogenase; 4, NADH: heme protein oxidoreductase or NADH: cyt. c reductase; 5, NADH: ubiquinone oxidoreductase; 6, thiosulfate: cytochrome c reductase; 7, succinate: cytochrome c reductase.

Actual measurements of ATP contents in intact cells of *Rhodospirillum rubrum* and *Rhodopseudomonas spheroides* (228) revealed that the rate of decrease in ATP in the intact bacteria, in the dark and in the absence of oxygen, is low. On illumination or oxygenation, the formation of ATP increases rapidly from a similar initial rate. However, the total content of ATP is higher during illumination. In order to obtain the same level, aerobic cells must be illuminated. This illumination, however, inhibits the respiration of the bacteria. These observations agree best with the existence of two independent pathways, as both could indirectly compete for the phosphate acceptor (308). The inhibition of respiration by illumination would then be the result of an increased ATP formation that causes the depletion of the phosphate. A very similar observation was obtained with *Rhodospirillum rubrum* (202).

Another reason for the inhibition could be the high ATP/ADP ratio in aerated illuminated cultures (252). This ratio could function as a regulatory mechanism on the enzymes of the respiratory chain. This hypothesis certainly conforms with observations that the ATP/ADP ratio is strongly influenced by culture conditions. Sojka and Gest (273) suggest the inclusion of AMP in this ratio and propose that ATP/ADP/AMP forms a signal system that is able to influence the synthesis of certain macromolecules, including the bacteriochlorophyll. Attention has just recently been drawn to the possible inhibitory effect of cyclic AMP on NAD^+ photoreduction and energy-linked transhydrogenation (44). This inhibition could be an internal effect that occurs only in the actively photosynthesizing cell.

Cyclic photophosphorylation is not the only energy-conserving process in photosynthetic nonsulfur bacteria, however (14). *Rhodospirillum rubrum* is able to form pyrophosphate via the photosynthetic electron transport (11, 13). The pyrophosphatase necessary to make the conserved energy available to the cell was the first found example of a bacterial inorganic pyrophosphatase subject to allosteric inhibition (159). At low Zn^{2+} concentrations, the enzyme is inhibited by NADH + H^+, NADPH + H^+, and Mg-ATP. The effect could be reversed by high concentrations of zinc. A second type of pyrophosphatase appears to exist in *Rhodopseudomonas capsulata* and *Rhodopseudomonas spheroides*; it exhibits normal Michaelis-Menten reaction kinetics and was stabilized by Co^{2+} instead of zinc (160). The existence of pyrophosphate formation is important insofar that the activated phosphate intermediate does not necessarily require ADP but can be transferred to a phosphate molecule. It was also observed (153) that pyrophosphate drives the synthesis of ATP from ADP and inorganic phosphate with the help of the pyrophosphatase action. The formation of pyrophosphate also had a stimulatory effect on CO_2 fixation in a light-independent reaction (222), wherein pyrophosphate could replace ATP to 60% during autotrophic and up to 90% during photoheterotrophic growth on acetate. Even AMP, a potent inhibitor of the ATP-dependent CO_2 fixation, stimulates in the presence of pyrophosphates.

A third route of ATP formation was observed in light-grown cells of *Rhodospirillum rubrum*. Light causes an uptake of protons (73, 310). Applying the chemoosmotic hypothesis for photophosphorylation to the observed proton movement, Edward and Borell (73) proposed that the light-dependent proton extrusion creates an electrochemical potential that drives phosphorylation. The electric field ("plus" inside the chromatophore) was found (144, 249) to be the motive force for ion transfer through the chromatophore membrane against a concentration gradient. With NAD^+ photoreduction, noncyclic photophosphorylation, and the energy-dependent anion transport, then, four sites of energy coupling exist in the chromatophore redox chain:

$$NADPH \rightarrow NAD^+; \quad NADH \rightarrow \text{cytochrome b}; \quad \text{cytochrome b} \rightarrow \text{cytochrome c}$$

and fourth the region after cytochrome c. Each of these coupling sites can provide energy for the generation of a membrane potential. First experiments have shown that adenylate nucleotides in cellular pools may operate as energy charge controls (327).

The existence of a separate noncyclic photophosphorylation is still disputed (12, 42, 60, 114, 128, 147, 213, 223, 242), particularly as $P/2\ e^-$ ratios of 1 (61, 240) indicate the presence of only one type of electron transport. Evidence suggests, however, that *Rhodospirillum rubrum* would have the cyclic photophosphorylation and NAD^+ reduction, whereas *Rhodopseudomonas* appears to have a second noncyclic phosphorylating pathway.

Primary Photosynthetic Reactions

One of the outstanding features of the process of photosynthesis is the ability of the organism to utilize quanta of energy to accomplish an ordered chemical transformation with high efficiency.

THE PHOTOCHEMICAL PROCESS OF THE PRIMARY REACTION. Electron microscopic studies of *Chlorobium* revealed (98) no vesicular or lamellar structure. The pigments seem to be readily dissociated and appear to be the equivalent of subunits of chromatophore fragments of a more highly organized system. It is therefore not possible to regard them as chromatophores, although the particles contain carotenoids, chlorophyll, cytochromes, quinones, and phospholipids. These particles are referred to as "thylakoid structures" (66). These observations are somewhat in contrast to investigations on

Chlorobium thiosulfatophilum by Cohen-Bazire and co-workers (52, 53, 56) and on *Chloropseudomonas ethylicum* (135). In both these species the photosynthetic pigment is located in large peripheral vesicles. Electron micrographs of *Chloropseudomonas ethylicum* revealed vesicles 1300–1500 Å long and 300–500 Å wide along the cell periphery. All these vesicles seem to be interconnected at the apex (202a). There is strong evidence that the chlorophyll in the green bacteria is contained as part of a structure that may be differentiated both structurally and functionally from the bacterial cytoplasmic membrane (195a). It is possible that ballistic treatment of the cells causes disruption of the vesicles at their apparent interconnections (135), which may be the cause of the contrasting results obtained.

Studies on the effect of light intensity on the formation of the photochemical apparatus in *Chloropseudomonas ethylicum* (136) supported the presence of vesicles and revealed that the formation of the photosynthetic vesicles is an inverse function of the light intensity at which cells were grown. However, not only the formation of these vesicles but also the specific chlorophyll content of isolated vesicles varied with light intensity, which has also been shown to be the case in *Rhodomicrobium vanielii* (295). These effects mean that in fact the regulation of chlorophyll content in *Chloropseudomonas ethylicum* in response to light intensity changes is achieved both by a change in the specific chlorophyll content of the vesicles and by a change in the number of vesicles formed.

In green plants, chlorophylls a and b are contained in plastids. The

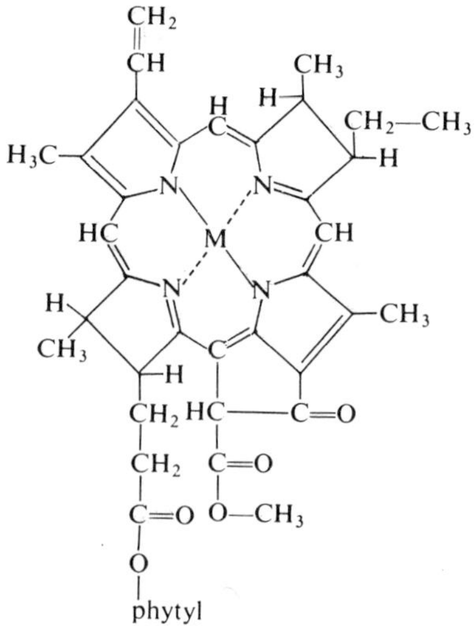
CH2
CH
CH3
H
C
H
CH2—CH3
H3C
H
N
N
HC
M
CH
N
N
H
CH3
CH3
C
H
CH2 HC
C=O
CH2
C=O
C=O
O—CH3
O
phytyl

Fig. 3.7. Chlorophyll a.

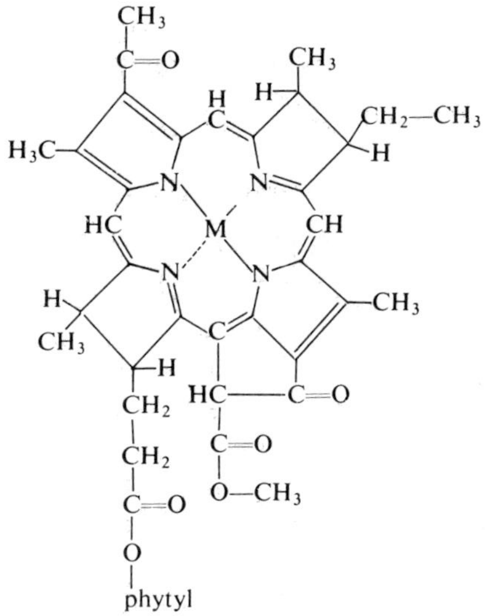

CH3
C=O
CH3
H
H
C
CH2—CH3
H3C
H
N
N
HC
M
CH
N
N
H
CH3
CH3
C
H
CH2
HC
C=O
C=O
CH2
O—CH3
C=O
O
phytyl

Fig. 3.8. Bacteriochlorophyll.

absorption bands for chlorophyll a are 683 and 695 nm and for chlorophyll b, 650 and 480 nm. *In vitro* absorption measurements of the chlorophyll of the green sulfur bacteria revealed a maximum at 750 nm, which was different from the bacteriochlorophyll found in the other two groups of photosynthetic bacteria and which showed maxima at 590 nm and in the infrared region between 800 and 900 nm (49, 304). The pigment of the green sulfur bacteria was therefore named "*Chlorobium* chlorophyll." *Chlorobium thiosulfatophilum* also accumulates large amounts of polymetaphosphates (143) in discrete intracellular granules, from which inorganic phosphate is released by an ADP- and light-dependent reaction (see section on photophosphorylation). The structural formulas for chlorophyll a and bacteriochlorophyll are shown in Figs. 3.7 and 3.8.

The purple sulfur bacteria possess definitely circular vesicular chromatophores, as has been shown in *Chromatium*, but the structure of these in a single species might be dependent on growth conditions and physiological age of the cell (28, 93).

In lysates of *Rhodospirillum rubrum* prepared by osmotic shock of protoplasts, it was found that all pigments sediment at a lower centrifugal force than the chromatophores after mechanical disruption (302). This result could be explained either by the size of the chromatophore or by the location of the photosynthetic apparatus in the membrane. Marr (185) and Stanier (277) postulated that the photosynthetic apparatus is part of the intracytoplasmic membrane because the differential release of chlorophyll during sonic and ballistic disruption of the cells is not in agreement with the location of the chlorophyll in independent vesicles.

Direct electron microscopy of sonically disrupted cells of *Rhodospirillum rubrum*, as well as stereoelectron microscopy of osmotically shocked cells, revealed a network of tubules attached to the envelope (132). The micrographs suggest that the internal membranes containing the photosynthetic pigments originate from the peripheral membrane in the form of spherical vesicles that later develop into bulged and sometimes flattened tubes. Holt and Marr (132) supported the idea that chromatophores are really fragments of these tubular internal membranes, which represent the intracytoplasmic membrane. The isolation and purification of this membrane (133) revealed further that the fine structure of the "chromatophores" resembles the fine structure of mitochondria and chloroplasts, or quantosomes. The striking difference, however, is the small size of the units (approximately 50 Å in diameter) compared with those in chloroplasts (185 × 155 × 100 Å). It has been speculated by Holt and Marr (133) that these small units are the loci of the photosynthetic pigments.

A further support to the location of the photosynthetic pigments in intracytoplasmic membranes comes again from Holt and Marr (134), who studied the effect of light on the formation of the intracytoplasmic membrane in *Rhodospirillum rubrum*. They demonstrated that highly purified membranes (chromatophores) from cells grown at low to moderate light intensity had a constant content of chlorophyll. It could be concluded, therefore, that the regulation of the chlorophyll content in response to changes in light depends on the formation of greater or lesser amounts of membrane.

In *Rhodospirillum rubrum* it appears that protein and chlorophyll synthesis (169, 267) are coupled with an increased invagination of the intracytoplasmic membrane into the cytoplasm of the cell. This is not the case in *Chloropseudomonas ethylicum*, however, which seems to couple its chlorophyll synthesis with an increase or decrease in the number and size of the vesicles formed (136). However, the change in the specific chlorophyll content in *C. ethylicum* with changes in light intensity does not occur in purple bacteria and seems also not in agreement with the idea of a constant relationship between the amount of internal membrane and the amount of pigment. Investigations are difficult, as the photochemical pigment appears to change its spectrophotometer absorbance bands (2, 115, 116, 120) and its chemical constitution (21, 43, 65, 67, 170, 268) on isolation under different environmental conditions.

The above-mentioned effect of light intensities on the synthesis of bacteriochlorophyll and the morphogenesis of thylakoids indicates that both are light-independent processes (290). However, they are induced by lowering the oxygen partial pressure in the culture to 5 mm Hg. The thylakoid formation starts with an invagination of the cytoplasmic membrane. The arising protuberances grow plainlike and parallel to the cytoplasmic membrane, producing the flat vesicles of the thylakoids. These thylakoids are able to branch and to fuse with one another and with the cytoplasmic membrane (167). The formation of the photosynthetic pigment, therefore, is not just an insertion of the pigment into preformed membraneous structures, but a production of increased amounts of membrane with constant pigment concentration (22). Fractionation of cell components during the adaptation from aerobic dark to anaerobic light cultivation showed that a preferential synthesis of protein occurs at the membrane site (121, 260). No shift was observed (122) in nucleotide composition during this adaptation period, indicating that no differential synthesis of RNA takes place (316b), although a difference in the stability of an RNA component was observed under aerobic and anaerobic conditions (316a).

The synthesis of bacteriochlorophyll during this adaptation period is carried on from reserve material and can be speeded up by the addition of exogenous carbon sources, such as fructose (253) or malate (251). In other words, the Athiorhodaceae are able to carry out a fermentation. The requirement for an N source correlates with the new and extensive protein synthesis mentioned above.

Oxidative metabolism is not changed with the start of the bacteriochlorophyll synthesis and the formation of thylakoids. The limitation of the growth rate and aerobic respiration does not start before the pO_2 drops below 3–4 mm Hg. At this partial pressure the production of bacteriochlorophyll comes to a maximum. Thus, the production rate of bacteriochlorophyll is greatly enhanced when the oxygen partial pressure is suddenly lowered (41). As the growth rate is not affected above 4 mm Hg and the bacteriochlorophyll synthesis is relatively high below 20 mm Hg, the regulation of the pigment synthesis must be independent of the growth rate (68). Comparing the effect of oxygen on photophosphorylation and bacteriochlorophyll, both effects certainly follow entirely different kinetics. The synthesis of bacteriochlorophyll first starts at below 50 mm Hg and is followed by the photosynthetic apparatus at below 5 mm Hg. This inhibitory effect of molecular oxygen on bacteriochlorophyll synthesis was reinvestigated by Marrs and Gest (185a) using a respiratory mutant of *Rhodopseudomonas capsulata.* Their results led to the proposal that molecular oxygen directly inactivates a "factor," that is required for bacteriochlorophyll synthesis. This regulatory factor, F^{bchl}, may exist in an active and an inactive form and could also be one or more of the enzymes directly involved in bacteriochlorophyll synthesis. Apart from an enzyme, this factor could also be a regulatory compound that interacts with such an enzyme. Such a hypothesis agrees also with suggestions made by Oelze and Drews in 1972 (202a).

The higher content of bacteriochlorophyll under low illumination can be explained by the fact that after induction the rate of synthesis of thylakoids is higher than that of the cytoplasmic membrane (198). Once a certain ratio has been obtained, the rate of further synthesis is constant; i.e., the bacteriochlorophyll content per thylakoid stays constant and is independent of the bacteriochlorophyll content per unit cell. It is very difficult indeed, if not impossible, to separate thylakoids from the cytoplasmic membrane, as it was found that both form a continuous membrane system (118, 119, 196, 197). It is also of interest that *Rhodopseudomonas capsulata* responds to energy restrictions or energy flux by preferentially synthesizing excess bacteriochlorophyll–membrane (274, 278).

Cells precultured at low light intensities (400 lux) on [2-^{14}C]acetate showed an equal distribution between thylakoids and cytoplasmic membranes. As soon as the cells were incubated under moderate light intensities (4000 lux), a preferential labeling occurred in the cytoplasmic membrane (199). The specific bacteriochlorophyll content of thylakoids, however, decreased with the higher light intensity. With increasing light intensity, the thylakoid proteins approach a pattern typical for cytoplasmic membranes, which could be observed using radioactively labeled amino acids. If the light intensity is lowered, the modified membrane system becomes again typical for thylakoids. It therefore appears that thylakoids and the cytoplasmic membrane represent reversible modifications of a dynamic membrane system.

This idea was further supported by investigations into the alterations of protein patterns in purified thylakoids (200). Specific membrane proteins were found in light- and dark-grown cells under high oxygen tension. These proteins were also characteristic for anaerobically light- and aerobically dark-grown cells. Under these conditions the bacteriochlorophyll content of thylakoids decreased, whereas the NADH oxidase activity increased. If these cells were incubated anaerobically at low light intensities, protein was newly synthesized that was typical of thylakoids from anaerobic illuminated cells. These proteins had not been found in the presence of oxygen. Also, the amount of bacteriochlorophyll increased and the NADH oxidase activity decreased. There appears to be a correlation between variations in catalytic functions and protein patterns of thylakoids. In membrane differentiation, oxygen appears to be more important than light. Oxygen appears not to influence the photochemical system (201), which suggests that oxygen dilutes the bacteriochlorophyll content and the respiratory activity thus becomes dominant.

The high amount of ubiquinone present in photosynthetic bacteria

(140) could also put this compound, or any other energy-rich intermediate (294), into a regulatory role. A further possibility would be that the presence of oxygen could stimulate heme production, which in turn has been found to influence α-aminolevulinate synthase, the first enzyme in chlorophyll or heme formation (172, 194). *Rhodopseudomonas spheroides* has only one form (173) of this enzyme, which provides α-aminolevulinate for bacteriochlorophyll and heme synthesis. A further possible step of interaction could be at the conversion of coproporphyrinogen to protoporphyrin (286), as this reaction occurs only under anaerobic and semianaerobic light conditions.

The presently available information shows that photosynthetic bacteria, in common with plants, have the capacity to regulate chlorophyll synthesis according to the environmental conditions (171, 202). The key factors in the Chlorobacteriaceae and Thiorhodaceae are light intensity and in the case of the Athiorhodaceae light intensity and, particularly, oxygen partial pressure.

THE PHYSICAL PROCESS OF THE PRIMARY REACTION. Light in general is a form of electromagnetic radiation and consists of a wave of particles known as "photons." It can be divided into three regions; namely, the ultraviolet, visible, and infrared. When light impinges on the surface of certain substances, electrons are ejected from the surface atoms. Einstein first proposed that photons are actually units of energy of light quanta and established the relation

$$E = hv - h(c/\lambda)$$

in which E is the energy of a light quantum, v is the frequency in vibrations per second, h is the Planck constant, c is the velocity of light, and λ is the wavelength (279).

The absorption of light and its energy varies greatly from substance to substance, particularly according to the number of surrounding electrons. It is, in fact, the electron of an atom that absorb the photons. When these photons strike an atom or molecule that can absorb light, an electron in one of the orbitals may absorb it and thus gain its energy (38). This energy may be sufficient to move the electron farther away from the nucleus to an outer orbital with a higher energy level. The atom is then said to be in the "excited state." This process can only happen when the electron and the photon have the exact energy to equal the energy difference between the initial orbital and the outer orbital to which the electron has been moved (175). Because the energy of a photon can only be used on an all-or-none basis, the term "quantum" was introduced. The excited atoms are very unstable.

Such a process seems certain to happen in the pigments of photosynthetic bacteria or plants. Chlorophyll is not the only light-absorbing molecule; carotenoids are also able to absorb light. Purple bacteria and red algae, which are known for their high carotenoid content, owe much of their color to such pigments. It seems certain that these carotenoids, although they possess a different absorption spectrum from chlorophyll (284), are able to supplement chlorophyll. However, the light energy absorbed by carotenoids has first to pass through chlorophyll before it can be used to do photochemical work (175). For further detailed reading it is suggested that the excellent books by Thomas (292) and by Fogg (86) be obtained.

Photosynthetic bacteria can grow anaerobically under two markedly different conditions: (1) CO_2 can serve as the sole carbon source, provided an inorganic electron donor is present; (2) simple organic compounds, such as malate, can serve as the sole carbon source instead of CO_2, provided an "accessory" electron donor is present. There are certain species, of course, such as *Chromatium*, that can grow under both conditions. Fixation of CO_2 liberated during metabolic conversions of added organic substrates also occurs, but the extent of this process varies considerably depending on the nutritional conditions (81, 82, 109).

Photometabolism

In order to avoid repetition, all metabolic events are shown under this section, whether the organism grows photosynthetically in the light and anaerobically or aerobically in the dark.

Hydrogen and Inorganic Sulfur Compounds

There are two distinct photoreactions that can occur in photosynthetic bacteria, both involving molecular hydrogen (131): (1) the photoreduction of carbon dioxide, and (2) the photoreduction of hydrogen with subsequent carbon dioxide reduction. Both reactions can be carried out by *Chlorobium*, *Chromatium*, and *Rhodomicrobium*. The significance of both reactions is not fully understood as yet.

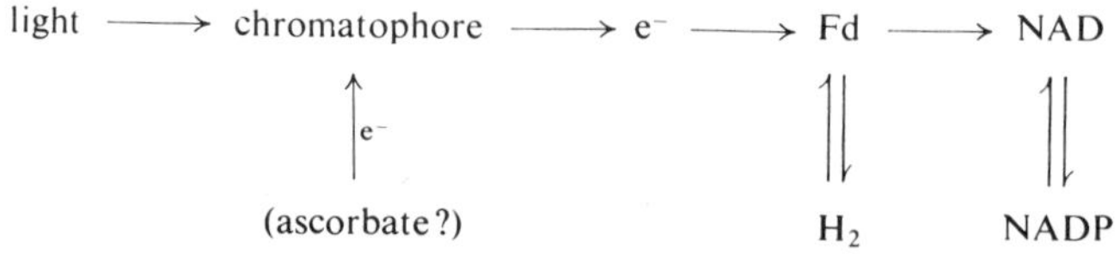
light
chromatophore
e⁻
Fd
NAD
e⁻
(ascorbate?)
H₂
NADP

Fig. 3.9. Hypothetical electron flow pattern in photosynthetic bacteria leading to pyridine nucleotide reduction (131).

If molecular hydrogen serves as electron donor, it can be used as a useful reductant in the reduction of NAD^+ (312). Extracts of *Chromatium* and *Chlorobium* showed a ferredoxin-dependent reduction of NAD^+. The enzyme responsible for this step is called "ferredoxin–NAD reductase" and couples ferredoxin (Fd) with NAD^+ (see Figure 3.9). This ferredoxin–NAD reaction is a so-called "dark" reaction (29, 82). The rate of NAD^+ reduction is three to four times that of $NADP^+$, which indicates that NAD^+ is the primary, but not the sole, electron acceptor. It may therefore be that the enzyme ferredoxin–NAD reductase could well play a key role in photochemical NAD reduction.

The photoproduction of hydrogen, however, requires such inorganic electron donors as H_2S or thiosulfate. *Chromatium* cytochromes are reduced by thiosulfate and can therefore serve as a point of entry for electrons from this source. With the help of light and a hydrogenase, hydrogen was evolved in addition to the reduction of thiosulfate to sulfate. This evolution of hydrogen occurred only in the light and is dependent on the thiosulfate concentration. With 1 mole of thiosulfate, two electrons and two protons combine with the help of the hydrogenase and hydrogen is evolved. This photoproduction of hydrogen can be viewed as a reduction of H^+ by a hydrogenase with the aid of electrons expelled from the excited chlorophyll molecules. The cytochrome system would thus become a gateway for the entry of electrons and for their transport to chlorophyll. The electron flow mechanism for the photoproduction of hydrogen from thiosulfate is shown by the electron transfer given in Fig. 3.10. In other words, *Chromatium* raises electrons supplied by thiosulfate to chlorophyll via cytochromes to a reducing potential at least equal to that of molecular hydrogen.

Whereas the CO_2 reduction with molecular hydrogen as electron donor is very much independent of light, the photoproduction is a function of the light intensity plus substrate concentration. At low light intensities the rates of fixation in the presence of molecular hydrogen and thiosulfate are

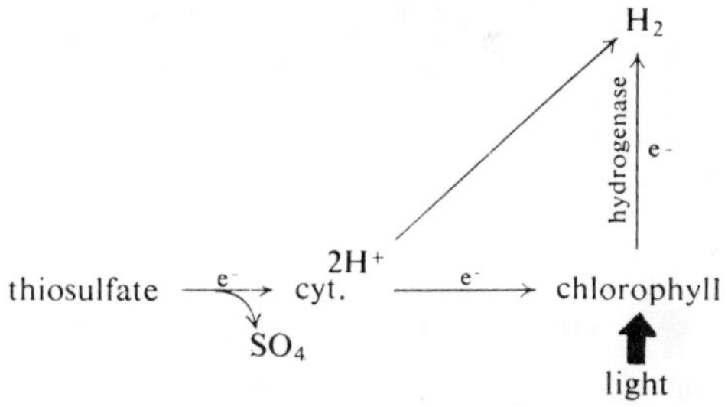

H2
hydrogenase
e-
2H+
thiosulfate
e-
cyt.
e-
chlorophyll
SO4
light

Fig. 3.10. The electron flow mechanism for the photoproduction of hydrogen from thiosulfate; cyt., cytochrome.

equal. The light saturation of CO_2 fixation with hydrogen is reached at about 100–150 lux (179), whereas the saturation with thiosulfate as electron donor is not reached before 700 lux. The higher the light intensity, the lower the CO_2 fixation with molecular hydrogen.

The hydrogen evolution is inhibited by nitrogen gas and ammonium ions as well as by organic electron donors (104, 105). However, CO_2 fixation was not inhibited either by N_2 or NH_4 as long as thiosulfate was the electron donor, which indicates that only the flow of electrons to H^+, but not the flow of electrons for CO_2 fixation via NAD(P), is inhibited. The reaction sequence (see also Chapter 6) would therefore be

$$S_2O_3^{2-} + 2\,OH^- \rightarrow S + SO_4^{2-} + 2e^- + H_2O$$

$$S + 8\,OH^- \rightarrow SO_4^{2-} + 6e^- + 4\,H_2O$$

$$10\,H_2O \longrightarrow 10\,H^+ + 10\,OH^-$$

$$\overline{S_2O_3^{2-} + 5\,H_2O \rightarrow 2\,SO_4^{2-} + 8e^- + 10\,H^+}$$

This evolution of hydrogen is quite common in photosynthetic bacteria (104, 105; see Fig. 3.11).

The detailed properties of the hydrogenase, however, are still not well understood (212), particularly if the possibilities of photoreduction of CO_2 with molecular hydrogen are considered (92, 166, 186, 219, 238, 315).

The presence of adenosine-5′-phosphosulfate reductase (APS reductase) in *Thiocapsa floridana* strain 6311; *Chromatium* strains 1611, 2811, and 6412 (291); and *Thiocapsa roseopericina* (298), as well as adenosine-5′-diphosphate sulfurylase in the *Thiocapsa floridana* strain, are not explicable at the present time. Both enzymes are known to be part of the oxidative metabolism of sulfide in the thiobacilli (see p. 348). The reaction of APS formation in phototrophic bacteria, however, is inhibited by oxygen,

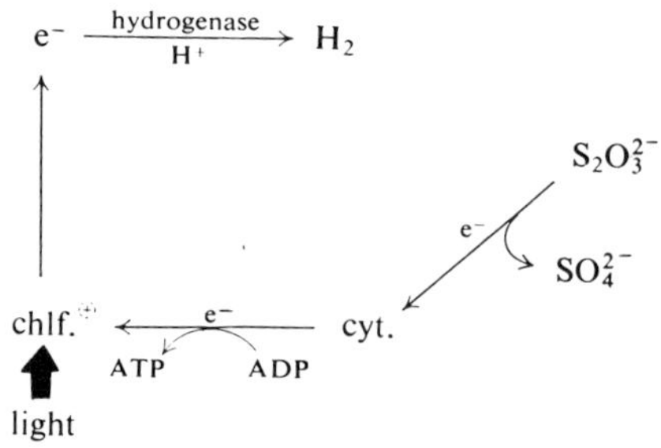
e⁻
hydrogenase
H⁺
H₂
S₂O₃²⁻
e⁻
SO₄²⁻
chlf.
e⁻
cyt.
ATP
ADP
light

Fig. 3.11. The mechanism postulated by Stanier (276) to account for photohydrogen evolution from thiosulfate by purple bacteria (reprinted with permission of the American Society of Microbiologists); chlf., chlorophyll; cyt., cytochrome.

molecular hydrogen, or the addition of elemental sulfur (297). The presence of APS reductase in the Chlorobiaceae and Thiorhodaceae and its absence in the Athiorhodaceae could develop into a valuable taxonomic characteristic.

Rhodopseudomonas palustris is also able to oxidize thiosulfate to tetrathionate (236). Cell yield measurements revealed that thiosulfate addition to photoheterotrophically grown cells with pyruvate as carbon source was stimulatory, whereas no effect was observed on cultures grown aerobically in the dark (237). In both cases thiosulfate was oxidized.

Acetate Photometabolism

Rhodospirillum rubrum is able to assimilate acetate directly in the absence of CO_2 (275). Acetate is converted to acetyl-CoA by an acetate-activating enzyme (76, 77) with acetyl adenylate as the anhydride intermediate (18)

$$\text{acetyl} + \text{ATP} \rightleftharpoons \text{acetyl adenylate} + \text{pyrophosphate}$$

$$\text{acetyl adenylate} + \text{CoA} \rightleftharpoons \text{acetyl-CoA} + \text{AMP}$$

The main assimilatory product formed is poly-β-hydroxybutyric acid. Because this synthesis is reductive, some acetate must be oxidized to provide the necessary reducing power. The conversion of acetate to the polymer will therefore compete with CO_2 fixation for the limiting reducing power available. In the absence of CO_2, poly-β-hydroxybutyric acid can be formed as a reserve product (Fig. 3.12). The supply of reducing power is an anaerobic process also and probably proceeds via the tricarboxylic acid cycle, as will be discussed below (p. 127).

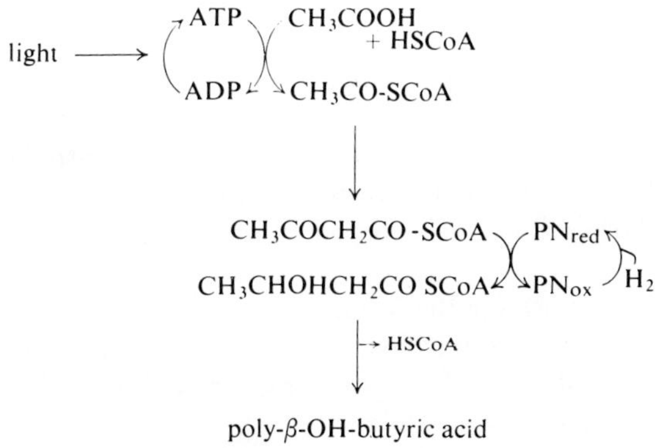
light
ATP
ADP
CH₃COOH
+ HSCoA
CH₃CO-SCoA
CH₃COCH₂CO-SCoA
CH₃CHOHCH₂CO SCoA
PNred
PNox
H₂
HSCoA
poly-β-OH-butyric acid

Fig. 3.12. The mechanism postulated by Stanier (276) to account for the acetate–hydrogen reaction in purple bacteria discovered by Gaffron in 1935 (reprinted with permission of the American Society of Microbiologists).

The formation of poly-β-hydroxybutyric acid depends on the amount of CO_2 present, because CO_2 will swing the synthesis toward polysaccharide formation. The formation, economics, and usefulness of polymers as reserve products is quite remarkable, because the bacterial cell cannot accumulate free fatty acids to any extent without causing serious damage to itself. In forming the polymer, these acids are neutralized and made osmotically inert and the cell is able to build up a reserve of reducing power that can be used for subsequent CO_2 fixation.

In the presence of bicarbonate in addition to acetate, *Rhodospirillum rubrum* (234) performs a metabolism similar to that of *Chlorobium* (130). Studies with *R. rubrum* indicate that glutamate is synthesized from acetate and bicarbonate probably via a light-dependent anaerobic TCA cycle or reductive CO_2 fixation. *Rhodospirillum* can therefore assimilate acetate in the absence of bicarbonate to form poly-β-hydroxybutyrate and in the presence of bicarbonate to synthesize glutamate. *Chlorobium thiosulfatophilum,* although a strict autotroph, assimilates acetate provided that bicarbonate and a source of reducing power (i.e., hydrogen gas) are present. There is no formation of the polymer poly-β-hydroxybutyrate. The pathways leading to glutamate are not known as yet, although a hypothesis has been put forward (see Fig. 3.15).

With the finding that even autotrophs may be able to assimilate organic compounds, it can be seen that the limitations of autotrophs may reside not in their capacity to assimilate or metabolize organic compounds, but in their capacity to oxidize them and so derive reducing power for biosynthetic reactions leading to cell growth. *Chloropseudomonas ethylicum* could play an intermediary role because this green, obligately photosynthetic bacterium has a nutritional requirement for a two-carbon compound (37) and most of the TCA cycle enzymes are present. Polymer synthesis from acetate requires only 1 mole of ATP per mole acetate. If, therefore, all the electrons needed for the reductive steps are transferred from hydrogen (reducing power) at the pyridine nucleotide level, no ATP can be formed as a result of hydrogen oxidation. The only function that could be attributed to light is the generation of ATP by cyclic photophosphorylation (275, 276). The two additional ways of ATP formation (see the section on photophosphorylation), however, could provide the cell with additional ATP.

The two key enzymes for the formation and degradation of the polymer have been identified and purified as 3-hydroxybutyrate dehydrogenase (19) (D-hydroxybutyrate:NAD oxidoreductase, EC 1.1.1.30), which catalyzes the reaction

$$\text{acetoacetate} + \text{NADH} + \text{H}^+ \rightarrow \text{3-hydroxybutyrate} + \text{NAD}^+$$

and 3-hydroxy acid dehydrogenase (271), which can catalyze the reversible reaction of L(+)-3-hydroxybutyrate into acetoacetate.

Under anaerobic conditions in the light, *Rhodopseudomonas spheroides* is able to convert acetate via malate to pyruvate. This route enables the continuous generation of malate from acetate through the action of malate synthase [L-malate glyoxylate-lyase (CoA-acetylating), EC 4.1.3.2]:

$$\text{Acetate} \rightarrow \text{acetyl-CoA} \rightarrow \text{malate} \rightarrow \text{pyruvate}$$

This subsequent oxidation of malate to pyruvate would then become a site of acetate oxidation, which does not pass through the stage of citrate (301). This mechanism allows the organism to avoid the use of the enzyme citrate synthase (EC 4.1.3.7), which was found to be a regulatory enzyme in *Rhodopseudomonas capsulata*. NADH + H^+ strongly inhibits this enzyme (74), whereas AMP relieves the inhibition. The activity of this enzyme varies very much from carbon source to carbon source and is particularly high in acetate-grown cells under aerobic, dark conditions.

Chromatium and *Rhodospirillum rubrum* have a further pathway of acetate transformation, with the end products being citramalate and possibly glutamate (16, 180):

$$\begin{array}{ccccc} \text{acetate} & \longrightarrow & \text{acetyl-CoA} & \longrightarrow & \text{citramalate} \\ & & & & \downarrow \\ & & \text{glutamate} & \longleftarrow & \text{itaconate} \end{array}$$

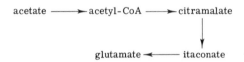

Citramalate and itaconate were found to strongly inhibit the metabolism of acetate and glutamate, but not the metabolism of succinate and fumarate (317). The addition of most of the C_4-dicarboxylic acids could release the inhibition. In *Rhodospirillum rubrum*, glutamate is metabolized via the TCA cycle (117).

Pyruvate Photometabolism

Pyruvate can be metabolized in various ways. *Rhodospirillum rubrum* not only metabolizes pyruvate but uses it for the synthesis of cell components. If enough light energy is being supplied and the substrate is in excess, storage products, such as poly-β-hydroxybutyrate and polysaccharides, are formed (27). The storage product poly-β-hydroxybutyrate prefers anaerobic light conditions with molecular hydrogen, whereas the polysaccharide formation prefers anaerobic conditions with molecular nitrogen as the gas phase. It is assumed that the enzyme pyruvate kinase

(ATP:pyruvate phosphotransferase, EC 2.7.1.40) may play a regulatory role (160a).

The isolation of a CoA-dependent pyruvate carboxylase (EC 6.4.1.1) from *Rhodopseudomonas spheroides* (218) indicates that pyruvate can be decarboxylated straight to acetyl-CoA and acetate. This information suggests that the metabolism of pyruvate may encounter a dismutation reaction (203), with the carboxylation reaction dominating under anaerobic, light conditions. The dismutation produces not only poly-β-hydroxybutyrate but also CO_2 and H^+ via the TCA cycle. During anaerobic dark growth, *Rhodospirillum rubrum* synthesizes poly-β-hydroxybutyric acid in fermenting pyruvate with a pyruvate ferredoxin-oxidoreductase (EC 1.2.7.1; 302a) and/or a pyruvate formate-lyase reaction (302b, 193a). These enzymes are responsible for the end products acetate, H_2 and CO_2 or poly-β-hydroxybutyric acid. The formation of H_2 therefore is light independent during anaerobic dark and light-dependent during anaerobic light conditions. In addition to these end products, Schön and Biedermann (253a–c) also found the production of propionate. In this case, pyruvate was metabolized to propionate, formate and H_2 (253c), whereby H_2 formation was inversely correlated with the production of propionate.

The observations that *Rhodospirillum rubrum* is able to grow equally well in anaerobic light or dark conditions certainly provides a unique model system for the study of differentiating between photosynthetic and fermentative metabolism (253c).

Formate Photometabolism

Rhodopseudomonas palustris utilizes formate as organic carbon source (280). Over 96% of the formate carbon is photoassimilated via CO_2 and the autotrophic CO_2 assimilation (281). The photoassimilation of formate occurs in light, anaerobic culture conditions. The organism contained an inducible formic hydrogenlyase system (225), which consists of two enzymes, a soluble formic dehydrogenase and a particulate hydrogenase. The electron carrier mediating between the two enzymes is NAD^+ (324). As the $HCOOH/CO_2$ redox potential is −400 mV, this reaction does not require light energy:

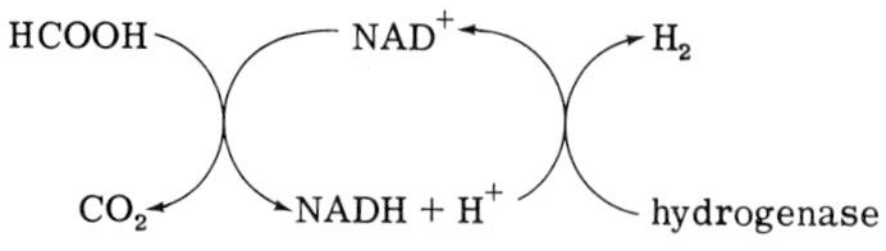
HCOOH
NAD⁺
H₂
CO₂
NADH + H⁺
hydrogenase

No ferredoxin was found to be involved in this reaction. The metabolism of formate is therefore very similar to the one carried out by *Pseudomonas oxalaticus* (see p. 414).

Succinate Photometabolism

The photometabolism of succinate is slightly more complicated than that of butyrate and acetate. Succinate as hydrogen donor transfers electrons at a potential that is lower than that of $NAD^+/NADH + H^+$. These electrons are therefore not able to reduce NAD^+ directly. The bacterial chromatophore, however, is capable of performing a light-dependent reduction of NAD^+ in order to overcome this problem. This reduction is coupled with the oxidation of either succinate or reduced flavin mononucleotides (95). Since the photochemical reactions by *Rhodospirillum rubrum* chromatophores are concluded in terms of only one electron transfer system (307), part of the electron transport scheme must be operative in the respiratory reaction which takes place in the dark with this particular organism. In principle, the result of the metabolism is the synthesis of polysaccharides. The substrate is oxidized via fumarate (Fig. 3.13) to pyruvate and a continuation of the reverse of the Embden-Meyerhof pathway. A small amount of substrate is also oxidized to acetate and forms the polymer poly-β-hydroxybutyric acid.

A stoichiometric relationship between the amount of succinate and the NAD^+ reduced was found (79). Moreover, fumarate was found to be the only product of succinate oxidation and succinate the first detectable product of propionate oxidation, followed by fumarate and malate (161). Therefore, a pathway was proposed (80) leading from succinate to hexose (Fig. 3.13). The enzymes involved in this pathway are

EC 1.3.99.1	succinate dehydrogenase [succinate: (acceptor) oxidoreductase]
EC 4.2.1.2	fumarate hydratase (L-malate hydro-lyase)
EC 1.1.1.37	malate dehydrogenase (L-malate:NAD oxidoreductase)
EC 4.1.1.32	phosphoenolpyruvate carboxykinase [GTP:oxalacetate carboxy-lyase (transphosphorylating)]
EC 4.2.1.11	enolase (2-phospho-D-glycerate hydro-lyase)
EC 5.4.2.1	phosphoglycerate phosphomutase (D-phosphoglycerate 2,3-phosphomutase)
EC 2.7.2.3	phosphoglycerate kinase (ATP:3-phospho-D-glycerate 1-phosphotransferase)
EC 1.2.1.12	triosephosphate dehydrogenase [D-glyceraldehyde 3-phosphate:NAD oxidoreductase (phosphorylating)]
EC 5.3.1.1	triosephosphate isomerase (D-glyceraldehyde-3-phosphate ketol-isomerase)
EC 4.1.2.13	fructose-bisphosphate aldolase (D-fructose-1,6-bisphosphate D-glyceraldehyde-3-phosphate-lyase)

A comparison of this pathway with the acetate metabolism and carbohydrate cycle in *Chromatium* (see Fig. 3.15) shows the similarities between the two systems. Earlier observations on the propionate fermentation (164) in *Rhodospirillum rubrum* suggest that pyruvate is an intermediate,

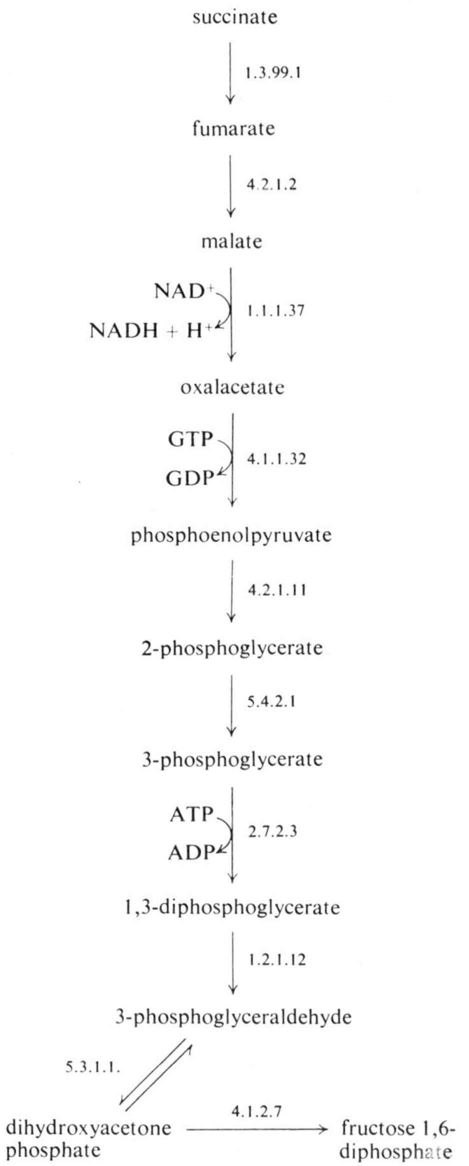
succinate
1.3.99.1
fumarate
4.2.1.2
malate
NAD+
NADH + H+
1.1.1.37
oxalacetate
GTP
GDP
4.1.1.32
phosphoenolpyruvate
4.2.1.11
2-phosphoglycerate
5.4.2.1
3-phosphoglycerate
ATP
ADP
2.7.2.3
1,3-diphosphoglycerate
1.2.1.12
3-phosphoglyceraldehyde
5.3.1.1.
dihydroxyacetone phosphate
4.1.2.7
fructose 1,6-diphosphate

Fig. 3.13. Photometabolism of succinate by *Rhodospirillum rubrum*.

possibly instead of oxalacetate. The difference in the photometabolism of succinate and propionate is an additional carboxylation step that converts propionate to succinate (161), wherein propionyl-CoA is the substrate for the carboxylation and methylmalonyl-CoA the product of the reaction. As this propionate carboxylation resembles the ones in *Propionibacterium shermanii*, the detailed mechanism will be outlined in Chapter 9 under propionic acid metabolism.

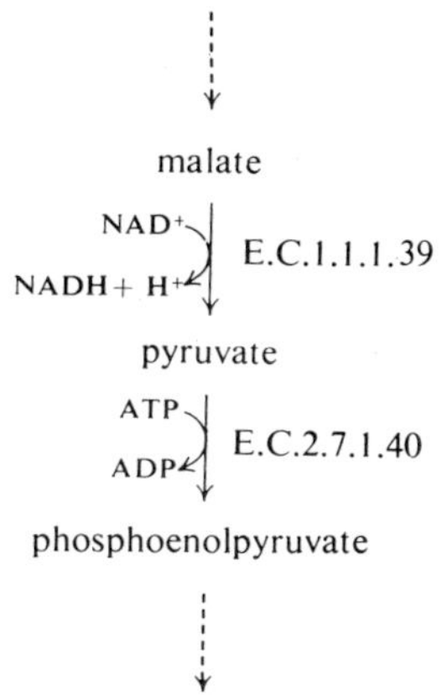
malate
NAD+
E.C.1.1.1.39
NADH+ H+
pyruvate
ATP
E.C.2.7.1.40
ADP
phosphoenolpyruvate

Fig. 3.14. A variation of succinate photometabolism in *Rhodospirillum rubrum.*

If pyruvate is an intermediate, then part of the proposed pathway would have to be changed as outlined in Fig. 3.14. Malate would be converted to pyruvate instead of to oxalacetate by a reaction catalyzed by malate dehydrogenase [L-malate:NAD oxidoreductase (decarboxylating), EC 1.1.1.39] and phosphoenolpyruvate would be formed with the help of pyruvate kinase (ATP:pyruvate phosphotransferase, EC 2.7.1.40). This has been shown to be the case in enzyme fractions of *Chlorobium thiosulfatophilum*, *Chromatium*, and *Rhodospirillum rubrum* (32).

It can therefore be assumed that substrates that are convertible to acetyl units without the formation of pyruvate, e.g., acetate and butyrate, yield mostly poly-β-hydroxybutyrate, whereas substrates that can be converted to pyruvate with an accompanying generation of reducing power, e.g., succinate, malate, and propionate, yield mostly polysaccharides. Synthesis of the latter follows a pathway similar to that of CO_2 fixation, which also leads to polysaccharide formation, but mainly via the Calvin cycle.

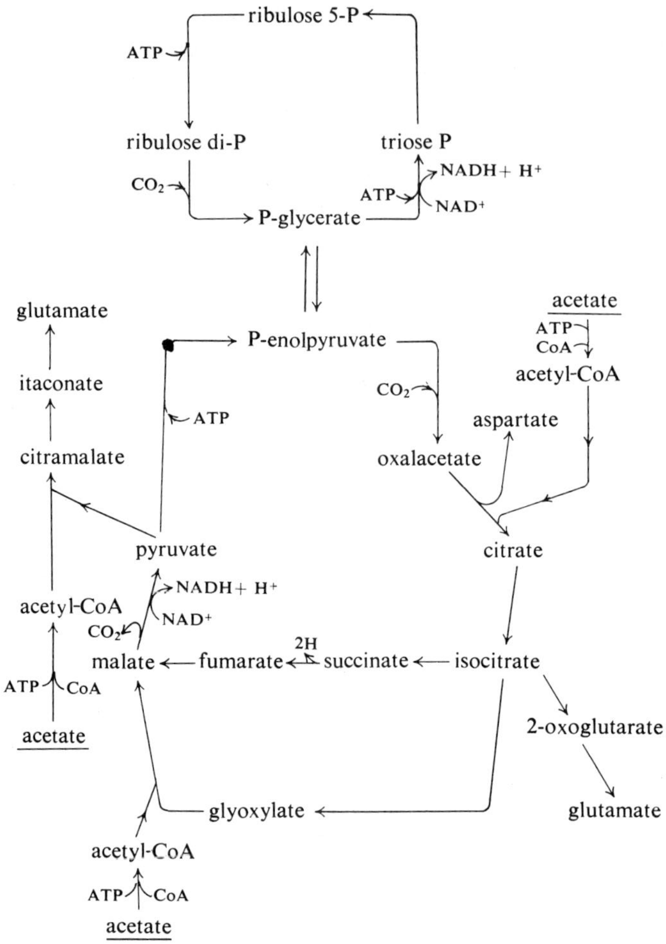
ribulose 5-P
ATP
ribulose di-P
triose P
NADH+ H+
CO2
ATP
NAD+
P-glycerate
glutamate
P-enolpyruvate
acetate
ATP
CoA
acetyl-CoA
itaconate
CO2
ATP
aspartate
citramalate
oxalacetate
pyruvate
citrate
NADH+ H+
acetyl-CoA
NAD+
CO2
2H
malate
fumarate
succinate
isocitrate
ATP
CoA
acetate
2-oxoglutarate
glyoxylate
glutamate
acetyl-CoA
ATP
CoA
acetate

Fig. 3.15. Acetate metabolism and carbohydrate cycle in *Chromatium* (180) (reprinted with permission of Macmillan (Journals) Pty, Ltd.).

Butyrate Photometabolism

Butyrate is assimilated to poly-β-hydroxybutyric acid without CO_2 evolution (210). The CO_2 that is fixed photosynthetically with molecular hydrogen is converted to polysaccharides as indicated above:

$$2nC_4H_8O_2 \rightarrow 2(C_4H_6O_2)_n + 4n[H]$$

$$4n[H] + nCO_2 \rightarrow (CH_2O)_n + nH_2O$$

$$2nC_4H_8O_2 + nCO_2 \rightarrow 2(C_4H_6O_2)_n + (CH_2O)_n$$

Anaerobic Tricarboxylic Acid Cycle and Reductive CO_2 Fixation

The photometabolism of the organic acids mentioned above revealed that the role of ATP is twofold. It is required in forming either an "activated" substrate to bring about CO_2 fixation into the Calvin cycle or an "activated" carbon source, such as acetyl-CoA from acetate (146) and CoA. The activated compounds are then ready to participate in synthetic reactions that are catalyzed by a specific enzyme system, all of which function in the dark (180). It can therefore be stated that CO_2 assimilation is fundamentally only a special case of the use of light energy. The photosynthetic transformation of energy is more closely linked with phosphorus than with carbon assimilation.

It was mentioned above that the photometabolism of C_4-dicarboxylic acids, such as malate and succinate, form significant quantities of CO_2 and H_2. Because of this and the discovery of almost every enzyme of the TCA cycle in the cell-free extracts, it is now believed that the photosynthetic bacteria use for their assimilation part of the TCA cycle, the glyoxylate cycle (Fig. 3.15).

In *Rhodospirillum palustris,* one of the pathways from α-oxoglutarate via succinate, fumarate, malate, and pyruvate to cell material was confirmed, but the metabolism proceeds anaerobically in light, and only aerobically in the dark (190):

$$\text{2-Oxoglutarate} + 1.5\ O_2 \rightarrow 2.5\ (CH_2O) + 2.5\ CO_2 + 0.5\ H_2O\ \text{(light)}$$

$$\text{2-Oxoglutarate} + 2.0\ O_2 \rightarrow 2.0\ (CH_2O) + 3\ CO_2 + H_2O\ \text{(dark)}$$

$$\text{Succinate} + O_2 \rightarrow 2.5\ (CH_2O) + 1.5\ CO_2 + 0.5\ H_2O\ \text{(light)}$$

$$\text{Succinate} + 1.5\ O_2 \rightarrow 2.0\ (CH_2O) + 2\ CO_2 + H_2O\ \text{(dark)}$$

$$\text{Fumarate} + 0.5\ O_2 + 0.5\ H_2O \rightarrow 2.5\ (CH_2O) + 1.5\ CO_2\ \text{(light)}$$

$$\text{Fumarate} + O_2 + 0.5\ H_2O \rightarrow 2.0\ (CH_2O) + 2\ CO_2\ \text{(dark)}$$

$$\text{Malate} + 0.5\ O_2 \rightarrow 2.5\ (CH_2O) + 1.5\ CO_2 + 0.5\ H_2O$$

$$\text{Malate} + O_2 \rightarrow 2.0\ (CH_2O) + 2\ CO_2 + H_2O$$

$$\text{Lactate} + 0.5\ O_2 \rightarrow 2.5\ (CH_2O) + 0.5\ CO_2 + 0.5\ H_2O\ \text{(light)}$$

$$\text{Lactate} + O_2 \rightarrow 2.0\ (CH_2O) + CO_2 + H_2O\ \text{(dark)}$$

$$\text{Pyruvate} + 0.5\ O_2 \rightarrow 2.5\ (CH_2O) + 0.5\ CO_2\ \text{(light)}$$

$$\text{Pyruvate} + 0.5\ O_2 \rightarrow 2.0\ (CH_2O) + CO_2$$

The molar amounts of oxygen uptake for 1 mole of each substrate consumed are shown in the accompanying tabulation.

	Light	Dark
2-Oxoglutarate	1.13	1.51
Succinate	0.73	1.45
Malate	0.45	0.92

The higher rate of oxygen uptake in the dark reaction results in a higher rate of CO_2 formation [see also Stoppani *et al.* (282) for *R. capsulatus*].

The oxidation of citric, isocitric, and α-oxoglutaric acids did not occur in living cells but did in dried cells or cell-free extracts (75). It is assumed that permeability barriers, rather than lack of enzymes, are the reason for this situation. The presence of all the required enzymes, however, indicates that this TCA cycle may be the pathway for the terminal oxidation of the organic acid in a number of purple bacteria such as *Chromatium* and *Rhodospirillum rubrum.* The metabolism of organic acids under anaerobic conditions via the TCA cycle appears to be best in the dark. Light–anaerobic conditions, however, seem to suppress somewhat the activity of the TCA cycle having it only partly functioning (155). This suppression could be caused indirectly by a stimulation of the citrate cleavage reaction, which increases the levels of ATP and NADH + H^+ in the cell (299).

The observations on the photometabolism of acetate and the various dicarboxylic acids with *Rhodospirillum rubrum* (164) and *Chromatium* (180, 296) could be summarized by the comprehensive scheme shown in Figure 3.16.

Coupled oxidation–reduction reactions between carboxylic acids of oxidatively different levels are favored as was indicated when the photometabolism of succinate in *Rhodospirillum rubrum,* for example, increased threefold when CO_2 was present, suggesting that CO_2 played the role of a hydrogen acceptor (78). The CO_2 evolution was dependent on the oxidation level of the acids to be metabolized. In the case of *Chromatium* sp. strain D, 2-oxoglutarate dehydrogenase [2-oxoglutarate:lipoate oxidoreductase (acceptor-acylating), EC 1.2.4.2] and malate dehydrogenase (EC 1.1.1.37) are absent and substituted by malate dehydrogenase (decarboxylating) (EC 1.1.1.40) and pyruvate carboxylase (EC 6.4.1.1.) (96, 97), which would give the modified glyoxylate cycle. The Athiorhodaceae, however, possess a complete TCA cycle, which functions oxidatively in the dark (45, 55, 75, 78, 282) as well as anaerobically in the light (108).

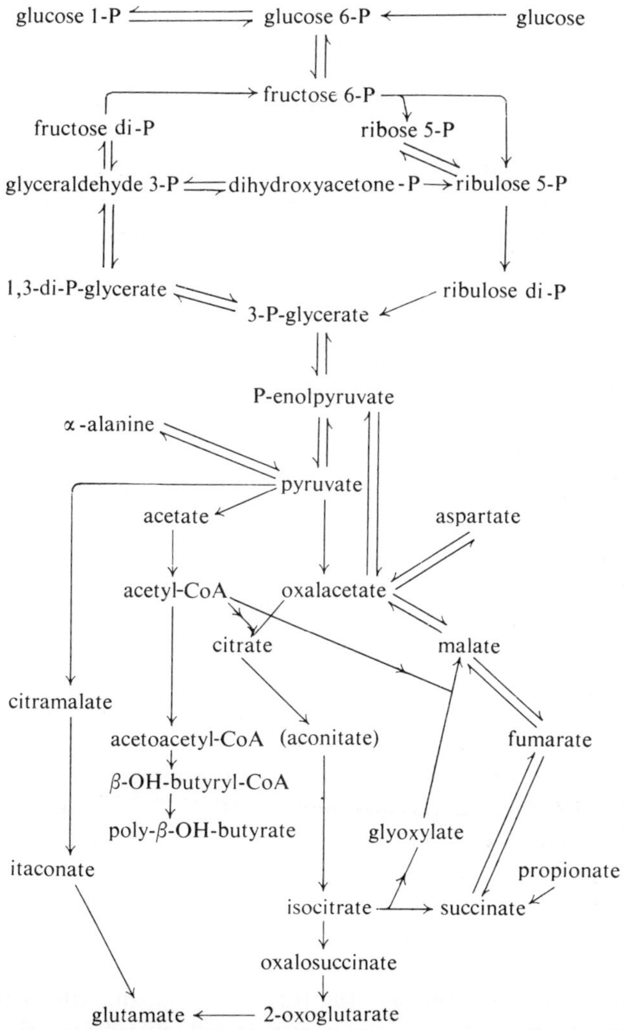
glucose 1-P
glucose 6-P
glucose
fructose 6-P
fructose di-P
ribose 5-P
glyceraldehyde 3-P
dihydroxyacetone-P
ribulose 5-P
1,3-di-P-glycerate
ribulose di-P
3-P-glycerate
P-enolpyruvate
α-alanine
pyruvate
acetate
aspartate
acetyl-CoA
oxalacetate
citrate
malate
citramalate
acetoacetyl-CoA
(aconitate)
fumarate
β-OH-butyryl-CoA
poly-β-OH-butyrate
glyoxylate
itaconate
propionate
isocitrate
succinate
oxalosuccinate
glutamate
2-oxoglutarate

Fig. 3.16. Photometabolism of acetate and dicarboxylic acids in *Rhodospirillum rubrum* and *Chromatium.*

All the previously mentioned dicarboxylic acids are members of the TCA cycle. It could therefore be expected that their photometabolism functions without hydrogen production if a suitable electron acceptor is provided or generated by the metabolism of the added substrate. Even CO_2 itself may act to some extent as an electron acceptor. Photoheterotrophically growing purple bacteria possess high levels of ribulose-1,5-diphosphate carboxylase [3-phospho-D-glycerate carboxy-lyase (dimerizing), EC 4.1.1.39], the key enzyme for CO_2 fixation in the Calvin cycle for autotrophs (168). It can therefore be assumed that under certain conditions purple bacteria are able to simultaneously fix CO_2 via the autotrophic scheme (carboxylase reaction) and to photoassimilate fragments derived from added organic compounds (296).

It has been suggested that CO_2 has the potential of influencing the direction of electron transfer (262). The presence of a certain steady-state concentration of CO_2 appears to be a major requirement for H_2 evolution also. A decreased yield of H_2 can certainly be correlated with the overall oxidation level of the substrate; i.e., the higher the oxidation level, the smaller the effect. The mechanism that is involved in the CO_2 stimulation of H_2 evolution is still obscure, however.

In the presence of light as an energy source and substrates more reduced than CO_2, it may be assumed that the major function of the various metabolic cycles of photosynthetic bacteria is the synthesis of new cellular material rather than the production of energy (17). In *Chromatium* there exists a mechanism whereby an adequate supply of oxalacetate can be produced independently of the TCA cycle. The existence of enzymes that would bring about phosphoroclastic cleavage of pyruvate to acetyl phosphate is now well established in anaerobes. This cleavage requires coenzyme A, inorganic phosphate, and ferredoxin and results in the formation of a two-carbon unit, H_2, and CO_2, without formate as an intermediate. It was also shown that N_2 assimilation takes place at the expense of hydrogen evolution and is stimulated by pyruvate (Fig. 3.17). Hydrogen evolution was strong in the dark (165), whereas in the light N_2 assimilation was favored. An enzyme was recently found (30) that can use ferredoxin directly as reductant in carbon assimilation. This enzyme was called "pyruvate synthase" [pyruvate:ferredoxin oxidoreductase (CoA-acetylating,) EC 1.2.7.1] and seems to catalyze a reductive synthesis of pyruvate from CO_2 and acetyl-CoA. This mechanism can also function for the assimilation of acetate as well as of CO_2.

Evidence is accumulating for the role of ferredoxin, which is approximately equal to H_2 in its reducing power, in reductive CO_2 assimilation by photosynthetic bacteria. The pyruvate synthase can utilize this reducing power in synthesizing pyruvate from acetyl-CoA and CO_2

$$\text{Acetyl-CoA} + CO_2 + Fd_{red} \rightarrow \text{pyruvate} + \text{CoA} + Fd_{ox}$$

In the 2-oxoglutarate synthase (EC 1.2.7.3) reaction, α-ketoglutarate is formed from succinyl-CoA and CO_2 (31):

$$\text{Succinyl-CoA} + CO_2 + Fd_{red} \rightarrow \alpha\text{-ketoglutarate} + CO_2 + Fd_{ox}$$

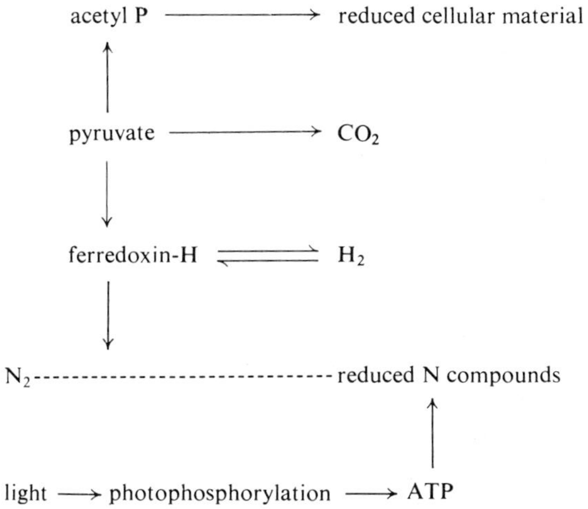
acetyl P
reduced cellular material
pyruvate
CO2
ferredoxin-H
H2
N2
reduced N compounds
light
photophosphorylation
ATP

Fig. 3.17. The relationship between N_2 assimilation and hydrogen evolution (17) (reprinted with permission of the National Academy of Sciences).

Rhodospirillum rubrum appears to possess both reactions (33), whereas *Chromatium* and *Chlorobium thiosulfatophilum* replace the α-ketoglutarate synthase reaction with a third reaction, involving the synthesis of phenylacetyl-CoA and bicarbonate (103). This reaction requires not only the reducing power of ferredoxin, but thiamine pyrophosphate as well.

Propionic acid is also able to function as an endogenous CO_2 acceptor in *Rhodospirillum rubrum*, forming methylmalonic acid and succinic acid. This reaction occurred only in cell-free extracts from acetate-grown cells (205).

A very similar carboxylation reaction is the one involving propionyl-CoA, which again is ferredoxin dependent. The enzyme catalyzing this reaction is 2-oxobutyrate synthase (36) (EC 1.2.7.2):

$$\text{Propionyl-CoA} + CO_2 + Fd_{red} \rightarrow \alpha\text{-ketobutyrate} + \text{CoA} + Fd_{ox}$$

This reaction is very important as α-ketobutyrate is a precursor in the biosynthesis of isoleucine and γ-aminobutyrate. It is of interest that this enzyme was also isolated from *Clostridium pasteurianum* and *Desulfovibrio desulfuricans*.

All these reductive CO_2 assimilation reactions are certain evidence for the existence of a reversed TCA cycle. This reductive carboxylic acid cycle constitutes a mechanism for utilizing ATP and the reducing power of ferredoxin. The acetyl-CoA becomes available for the synthesis of lipids and carbohydrates.

The existence of such a reductive carboxylic acid cycle next to the reductive pentose cycle (or autotrophic mechanism of CO_2 fixation) leaves open speculation as to whether photosynthetic bacteria are able to switch from one cycle to the other depending on the substrate. Lithotrophic photosynthetic bacteria could utilize the reductive pentose cycle, whereas organotrophic photosynthetic bacteria could prefer the reductive carboxylic acid cycle.

Illuminated cells of *Rhodospirillum rubrum* photometabolize 1 mole of malate to 3 moles of CO_2 and 3.8 moles of H_2. When H_2 is not produced, malate photodissimilation suggests that this C_4-dicarboxylic acid is converted to CO_2 and a C_3 fragment, which is then assimilated to carbohydrates (211). Hydrogen formation may represent an alternative electron transfer mechanism that comes into play when the metabolic energy balance is such that the ADP level has decreased to a critical level or, in other words, when the rate of ATP formation has exceeded the rate of utilization. The photoevolution of hydrogen via a terminal hydrogenase system can therefore be interpreted as the manifestation of an alternative pathway of electron transfer, which serves as a regulatory function in energy metabolism.

Acetone and Alcohol Photometabolism

A number of the Athiorhodaceae, or nonsulfur purple bacteria, are able to utilize various alcohols for the photosynthetic reduction of CO_2 (88). Some species seem to specialize on primary, others on secondary alcohols. In many cases induced enzymes are required. The general reactions are

$$CH_3CH_2OH + 3H_2O \rightarrow 2CO_2 + 12H^+$$

$$3CO_2 + 12H^+ \rightarrow 3(CH_2O) + 3H_2O$$

$$CH_3CH_2CH_2OH + 5H_2O \rightarrow 3CO_2 + 18H^+$$

The photometabolism of acetone to products other than cell material is possible in *Rhodopseudomonas gelatinosa* (261). Because there were no intermediates found to accumulate, the following mechanism is postulated (reprinted with the permission of the American Society of Microbiologists):

isopropanol
↓
(I) acetone → acetoacetate → acetate → cell material
↓
(II) acetol → methylglyoxal → pyruvate → cell material
↓
(III) dihydroxyacetone → glyceraldehyde → pyruvate
↓
cell material

This scheme indicates that acetone condenses with CO_2 to form acetoacetate which is governed by induced enzymes. Photophosphorylation, however, seems to be strongly inhibited by these lower aliphatic alcohols as shown with chromatophores of *Rhodospirillum rubrum* (293).

Glycine Photometabolism

Glycine is metabolized to acetyl-CoA by *Rhodopseudomonas spheroides*. The detailed metabolic pathway, however, is still disputed. It has been demonstrated that glycine is exclusively converted to glyoxylate mainly by a transamination reaction, wherein pyruvate and oxalacetate act as amino acceptors (300). Glyoxylate itself is converted to pyruvate via the glyoxylate cycle. The more recent claims are (101, 287) that a glycine decarboxylase, together with methylene tetrahydrofolate, metabolizes glycine via serine and pyruvate to acetyl-CoA. The latter system would be identical with the one carried out by *Diplococcus glycinophilus, Micrococcus anaerobiosis,* and *M. variabilis* under strict anaerobic conditions (see p. 653). With this system functioning, the methylene carbon atom of glycine would be converted to acetyl-CoA, whereas the carboxyl carbon atom appears in the newly formed protein. ATP has a marked stimulatory effect on this reaction.

Methane Photometabolism

A recent report (314) brought the first evidence that a strain of *Rhodopseudomonas gelatinosa* is able to utilize methane. This bacterium was found to grow with methane as the sole electron donor and can incorporate methane carbon into cellular material as well as oxidize methane to carbon dioxide. Details of the pathway are not available as yet.

Aromatic Compound Metabolism

Recent reports (69, 126, 127) indicate that *Rhodopseudomonas palustris* is able to metabolize aromatic compounds aerobically and anaerobically. Benzoate, for example, supports growth only under photosynthetic conditions (126), but not aerobically in the dark. Other compounds, such as *p*-hydroxybenzoate, however, can be metabolized under both conditions. Aerobically in the dark, *Rhodopseudomonas palustris* uses a similar pathway to the pseudomonads (see p. 510) for their *p*-hydroxybenzoate utilization. *p*-Hydroxybenzoate is metabolized to protocatechuate, which itself is cleaved by a protocatechuate 4,5-oxygenase to form γ-carboxy-α-hydroxymuconic semialdehyde. The presence of oxygen was obligatory for this pathway.

By isotope dilution experiments under anaerobic photosynthetic conditions, a novel reduction mechanism was discovered that finally leads to the formation of pimelic acid (69). The key difference between the aerobic and anaerobic pathways is the reduction of the aromatic nucleus to an aliphatic cyclic acid first, before ring fission occurs under anaerobic conditions. Benzoate is first reduced to cyclohex-1-ene-1-carboxylic acid. The

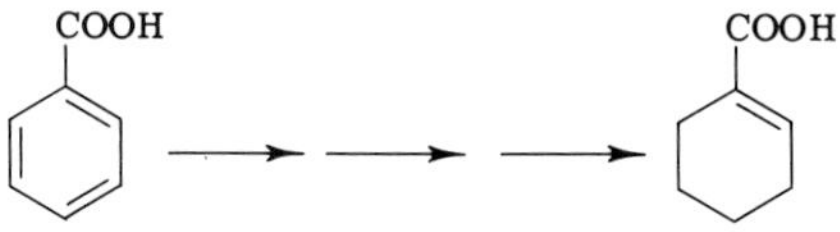
COOH
COOH

latter compound is now hydrated to the intermediate 2-hydroxycyclo-

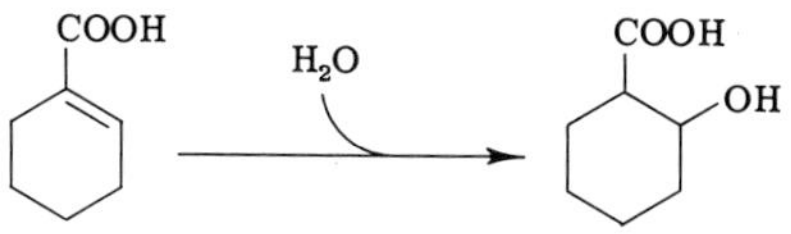
COOH
H₂O
COOH
OH

hexanecarboxylic acid, which is followed by a dehydrogenation to 2-oxo-

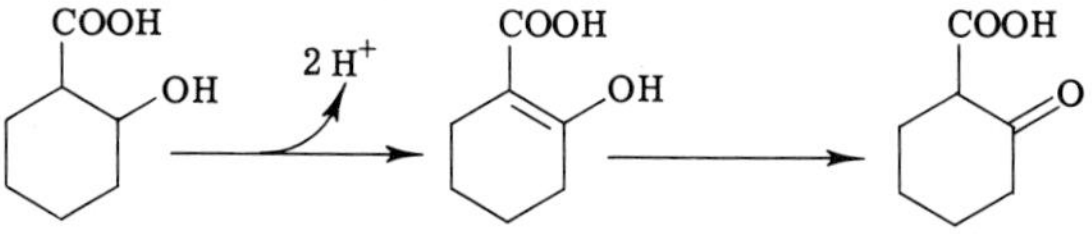
COOH
OH
2 H⁺
COOH
OH
COOH
O

cyclohexanecarboxylic acid. A further hydration of this aliphatic cyclic

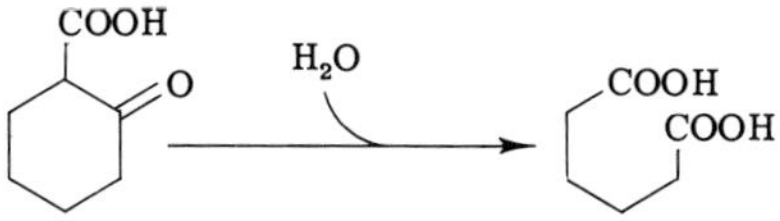
COOH
O
H₂O
COOH
COOH

acid results in the ring fission and formation of pimelic acid. It has been suggested that this reduction of aromatic compounds may be catalyzed by reductases coupled possibly to ferredoxin.

Photochemical Nitrogen Fixation

The finding that molecular nitrogen acts as a repressor to hydrogen evolution (148) led to the discovery that both green and purple bacteria are capable of fixing nitrogen under anaerobic conditions in the light (177). This nitrogen fixation by photosynthetic organisms (148, 195) could therefore be the result of a noncyclic electron flow in which electrons pass from the substrate via cytochromes to the chlorophyll (7, 62), as was described for hydrogen evolution. Evidence for this system was obtained with illuminated cells and thiosulfate or succinate as substrate or electron donor (6). The similarity to hydrogen evolution goes even further, as nitrogen assimilation was greatly increased by the addition of ATP (125). Reduced ferredoxin was also necessary (63, 192) as the immediate reductant but could be replaced by dithionate or H_2 in the presence of catalytic amounts of methyl- or benzylviologen (316). The nitrogenase system of *Chromatium* therefore appears to have the typical properties of other nitrogenases (*Azotobacter*, etc.) from nonphotosynthetic bacteria, as it reduces nitrogen to ammonia and acetylene to ethylene. The energy requirements for the nitrogen fixation are now also known (325) to be qualitatively the same as for CO_2 assimilation, allowing the statement that the nitrogenase activity is coupled to photochemically generated ATP and reductant.

Quantitative measurements of nitrogen fixation in *Rhodospirillum rubrum* revealed (193, 244) that about 6 moles of L-malate, fumarate, or succinate or 10 moles of pyruvate are consumed per mole of molecular nitrogen. There also exists a stoichiometric relationship between the amount of NH_4Cl added and the amount of gas evolved. About 8 H_2 and 4 CO_2 were taken up per mole of NH_3 (246). The fixation was quickly suppressed by small amounts of ammonia (244, 245).

This nitrogen-fixing capability is another way of utilizing the electrons expelled from chlorophyll by light (6). This interrelationship between the photoproduction of hydrogen and photosynthetic nitrogen fixation does not seem to be confined to the photosynthetic bacteria. For example, blue-green algae, which are known for their photosynthetic nitrogen-fixing ability, can be adapted to hydrogen formation (1, 100).

The Electron Acceptor in Purple Bacteria

Throughout this chapter the main emphasis has been on the metabolism of the various photosynthetic bacteria with regard to their ATP formation and hydrogen donors, but little has been said about the hydrogen or electron acceptors. It was stated above that ATP plays a dual function and that there exists a relation between CO_2 concentration and H_2 evolution as well as a competition between H_2 evolution and nitrogen fixation. Assuming that the electrons (or hydrogens) required for the net generation of reduced NAD are derived from an accessory inorganic or organic compound, which would therefore serve as hydrogen donor, the mechanism

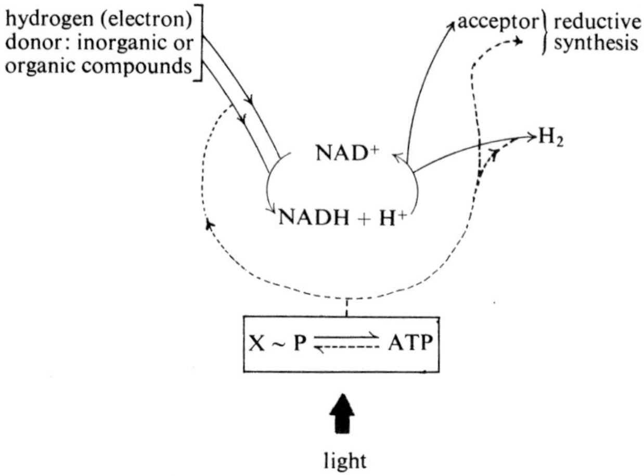
hydrogen (electron)
donor: inorganic or
organic compounds
acceptor
reductive
synthesis
H_2
NAD^+
$NADH + H^+$
X ~ P
ATP
light

Fig. 3.18. Scheme for hydrogen (electron) flow from donors to acceptors and photoproduction of bacterial photosynthesis (109) (reprinted with permission of The Antioch Press).

shown in Fig. 3.18 would be possible (109). Depending on the redox potential of the donor, or on the steady-state concentration of the reduced and oxidized form of donor and NAD, the formation of reduced NAD may be promoted by energy-rich intermediates created by the action of light on the photochemical apparatus. Reduced NAD would then be available for a great variety of reductive syntheses, such as the conversion of CO_2 to cell material, the transformation of C_2 and C_3 compounds to reserve materials, or even reductive aminations to provide the protein synthesis with amino acid precursors. In the case of an excess production of ATP and reduced NAD in relation to the biosynthetic machinery, reduced NAD can be reoxidized through the liberation of molecular hydrogen by an energy-dependent process. The evolution of H_2 can therefore be interpreted as a sort of regulatory device that maintains ATP and reduced NAD at levels consistent with the overall rate of biosynthetic activity (212).

The only common event left in all types of photosynthesis is the primary photochemical reaction that leads to ATP formation, including the transfer of electrons through a closed carrier system coupled to chlorophyll. The use that a photosynthetic cell makes of the ATP generated is determined in part by its inherited enzymatic constitution and in part by environmental conditions. In green plants, green bacteria, and some of the purple sulfur bacteria, major portions of ATP go into CO_2 reduction to drive cell synthesis. In purple nonsulfur bacteria and some of the purple sulfur bacteria, it is used to drive cell synthesis from externally provided organic substrates.

The use of CO_2 as a sole source in photosynthesis, in other words, is not a fundamental feature of this process but is restricted to those organisms that are called "autotrophs."

Evolution of Photosynthesis

Arnon and co-workers (7) proposed a very good theory on the evolution and development of photosynthesis. The first step is considered to be an emergence of a porphyrin that gave rise to chlorophyll. Primitive photosynthesis was probably limited to anaerobic cyclic photophosphorylation because there was no need for a photochemically formed reductant as long as H_2 was present. *Chromatium* still has this type of photosynthesis. As H_2 vanished from the atmosphere, the photosynthetic cell probably became dependent on the photosynthetic apparatus for photochemically generating a strong reductant from such electron donors as succinate or thiosulfate. In organisms that contain or can adaptively form hydrogenase or nitrogenase, this place in photosynthesis can be served today by photoproduction of molecular hydrogen or a photofixation of nitrogen. Water became an electron donor only with the emergence of plant photosynthesis. The difference between bacterial and plant photosynthesis seems to center on the electron donors that are consumed in the reduction of NAD.

References

1. Allen, M. B. (1956). Photosynthetic nitrogen fixation by blue-green algae. *Sci. Mon.* **83,** 100 [as cited by Stanier (276)].
2. Amesz, J., and Vredenberg, W. J. (1966). Absorbance changes of photosynthetic pigments in various purple bacteria. *Curr. Photosyn., Proc. West-Eur. Conf., 2nd, 1965* p. 75. *Chem. Abstr.* **65,** 19018, (1966).
3. Arnon, D. I. (1959). Conversion of light into chemical energy in photosynthesis. *Nature (London)* **184,** 10.
4. Arnon, D. I. (1965). The role of light in photosynthesis. *In* "Readings from Scientific American: The Living Cell," p. 97. Freeman, San Francisco, California.
5. Arnon, D. I. (1966). The photosynthetic energy conversion process in isolated chloroplasts. *Experientia* **22,** 273.
6. Arnon, D. I., Losada, M., Nozaki, M., and Tagawa, K. (1960). Photofixation of nitrogen and photoproduction of hydrogen by thiosulfate during bacterial photosynthesis. *Biochem. J.* **77,** 23P.
7. Arnon, D. I., Losada, M., Nosaki, M., and Tagawa, K. (1961). Photoproduction of hydrogen, photofixation of nitrogen, and a unified concept of photosynthesis. *Nature (London)* **190,** 601.
8. Arnon, D. I., Tsujimoto, H. Y., and McSwain, B. D. (1965). Photosynthetic phosphorylation and electron transport. *Nature (London)* **207,** 1367.
9. Arnon, D. I. (1971). The light reactions of photosynthesis. *Proc. Nat. Acad. Sci. U.S.* **68,** 2883.

10. Baltscheffsky, H., and von Stedingk, L. V. (1966). Bacterial photophosphorylation in the absence of added nucleotide. A second intermediate stage of energy transfer in light-induced formation of ATP. *Biochem. Biophys. Res. Commun.* **22,** 722.
11. Baltscheffsky, H., von Stedingk, L. V., Heldt, H. W., and Klingenberg, M. (1966). Inorganic pyrophosphate: Formation in bacterial photophosphorylation. *Science* **153,** 1120.
12. Baltscheffsky, M., and Baltscheffsky, H. (1963). Photophosphorylation in *Rhodospirillum rubrum.* About the electron transport chain and the phosphorylation reactions. *In* "Bacterial Photosynthesis" (H. Gest, A. S. Pietro, and L. P. Vernon, eds.), p. 195. Antioch Press, Yellow Springs, Ohio.
13. Baltscheffsky, M., Baltscheffsky, H., and von Stedingk, L. V. (1966). Light-induced energy conversion and the inorganic pyrophosphatase reaction in chromatophores from *Rhodospirillum rubrum. Brookhaven Symp. Biol.* **19,** 246.
14. Baltscheffsky, M. (1967). Inorganic pyrophosphate and ATP as energy donors in chromatophores from *Rhodospirillum rubrum. Nature* (*London*) **216,** 241.
15. Bassham, J. A. (1962). The path of carbon on photosynthesis. *Sci. Amer.* **206,** 88.
16. Benedict, C. R. (1962). Early products of [^{14}C]acetate incorporation in resting cells of *Rhodospirillum rubrum. Biochim. Biophys. Acta* **56,** 620.
17. Bennett, H. R., Rigopoulos, N., and Fuller, R. C. (1964). The pyruvate phosphoroclastic reaction and light-dependent N_2 fixation in bacterial photosynthesis. *Proc. Nat. Acad. Sci. U.S.* **52,** 762.
18. Berg, P. (1955). Participation of adenyl acetate in the acetate-activating system. *J. Amer. Chem. Soc.* **77,** 3163.
19. Bergmeyer, H. W., Gawehn, K., Klotzsch, H., Krebs, H. A., and Williamson, D. H. (1967). Purification and properties of crystalline 3-hydroxybutyrate dehydrogenase from *Rhodopseudomonas spheroides. Biochem. J.* **102,** 423.
20. Beugeling, T. (1968). Photochemical activities of $K_3Fe(CN)_6$-treated chromatophores from *Rhodospirillum rubrum. Biochim. Biophys. Acta* **153,** 143.
21. Biedermann, M., Drews, G., Marx, R., and Schröder, J. (1967). Der Einfluss des Sauerstoffpartialdruckes und der Antibiotica Actinomycin and Puromycin auf das Wachstum, die Synthese von Bakteriochlorophyll und die Thylakoidmorphogenese in Dunkelkulturen von *Rhodospirillum rubrum. Arch. Mikrobiol.* **56,** 133.
22. Biedermann, M., and Drews, G. (1968). Trennung der Thylakoidbausteine einiger Athiorhodaceae durch Gelelektrophorese. *Arch. Mikrobiol.* **61,** 48.
23. Boll, M. (1968). Oxydation von reduziertem Nicotinamid-Adenin-Dinucleotid in *Rhodospirillum rubrum.* I. Charakterisierung einer löslichen NADH-Dehydrogenase. *Arch. Mikrobiol.* **62,** 94.
24. Boll, M. (1968). Oxydation von reduziertem Nicotinamid-Adenin-Dinucleotid in *Rhodospirillum rubrum.* II. Über eine reversible, temperaturabhängige Aktivierung der ApoNADH-Dehydrogenase. *Arch. Mikrobiol.* **62,** 349.
25. Boll, M. (1968). Enzyme der Elektronentransportpartikel aus *Rhodospirillum rubrum:* Eigenschaften von NADH- und Succinat-Cytochrome c-Reduktase. *Arch. Mikrobiol.* **64,** 85.
26. Bose, S. K., and Gest, H. (1962). Electron transport systems in purple bacteria. Hydrogenase and light-stimulated electron transfer reactions in photosynthetic bacteria. *Nature* (*London*) **195,** 1168.
27. Bosshard-Hear, E., and Bachofen, R. (1969). Synthese von Speicherstoffen aus Pyruvat durch *Rhodospirillum rubrum. Arch. Mikrobiol.* **65,** 61.
28. Bril, C. (1963). Studies on bacterial chromatophores. II. Energy transfer and photooxidative bleaching of bacteriochlorophyll in relation to structure in normal and carotinoid-depleted *Chromatium. Biochim. Biophys. Acta* **66,** 50.

29. Buchanan, B. B., Bachofen, R., and Arnon, D. I. (1964). Role of ferredoxin in the reductive assimilation of CO_2 and acetate by extracts of the photosynthetic bacterium, *Chromatium. Proc. Nat. Acad. Sci. U.S.* **52,** 829.

30. Buchanan, B. B., and Arnon, D. I. (1965). Ferredoxin-dependent synthesis of labeled pyruvate from labeled acetyl-CoA and CO_2. *Biochem. Biophys. Res. Commun.* **20,** 163.

31. Buchanan, B. B., and Evans, M. C. W. (1965). The synthesis of α-ketoglutarate from succinate and carbon dioxide by a subcellular preparation of a photosynthetic bacterium. *Proc. Nat. Acad. Sci. U.S.* **54,** 1212.

32. Buchanan, B. B., and Evans, M. C. W. (1966). The synthesis of phosphoenol pyruvate from pyruvate and ATP by extracts of photosynthetic bacteria. *Biochem. Biophys. Res. Commun.* **22,** 484.

33. Buchanan, B. B., Evans, M. C. W., and Arnon, D. I. (1967). Ferredoxin-dependent carbon assimilation in *Rhodospirillum rubrum. Arch. Mikrobiol.* **59,** 32.

34. Buchanan, B. B., and Evans, M. C. W. (1969). Photoreduction of ferredoxin and its use in $NAD(P)^+$ reduction by a subcellular preparation from the photosynthetic bacterium *Chlorobium thiosulfatophilum. Biochim. Biophys. Acta* **180,** 123.

35. Buchanan, B. B., Mitsubara, H., and Evans, M. C. W. (1969). Ferredoxin from the photosynthetic bacterium, *Chlorobium thiosulfatophilum.* A link to ferredoxins from nonphotosynthetic bacteria. *Biochim. Biophys. Acta* **189,** 46.

36. Buchanan, B. B. (1969). Role of ferredoxin in the synthesis of α-ketobutyrate from propionyl coenzyme A and carbon dioxide by enzymes from photosynthetic and nonphotosynthetic bacteria. *J. Biol. Chem.* **244,** 4218.

37. Callely, A. G., and Fuller, R. C. (1967). Carboxylic acid cycle enzymes in *Chloropseudomonas ethylicum. Biochem. J.* **103,** 74P.

38. Calvin, M., and Androes, G. M. (1962). Primary quantum conversion in photosynthesis. *Science* **138,** 867.

39. Carr, N. G., and Exell, G. (1965). Ubiquinone concentrations in Athiorhodaceae grown under various environmental conditions. *Biochem. J.* **96,** 688.

40. Case, G. D., and Parson, W. W. (1971). Thermodynamics of the primary and secondary photochemical reactions in *Chromatium. Biochim. Biophys. Acta* **253,** 187.

41. Cellarius, R. A., and Peters, G. A. (1969). Photosynthetic membrane development in *Rhodopseudomonas spheroides:* Incorporation of bacteriochlorophyll and development of energy transfer and photochemical activity. *Biochim. Biophys. Acta* **189,** 234.

42. Chance, B., Horio, T., Kamen, M. D., and Taniguchi, S. (1965). Kinetic studies on the oxidase systems of photosynthetic bacteria. *Biochim. Biophys. Acta* **112,** 1.

43. Chance, B., Nishimura, M., Avron, M., and Baltscheffsky, M. (1966). Light-induced intravesicular changes pH in *Rhodospirillum rubrum* chromatophores. *Arch. Biochem. Biophys.* **117,** 158.

44. Chaudhary, A. H., and Frenkel, A. W. (1970). Effects of adenosine 3′,5′-cyclic monophosphoric acid on certain light-induced reactions and on ATPase activity of isolated chromatophores from *Rhodospirillum rubrum. Biochem. Biophys. Res. Commun.* **39,** 238.

45. Cheruyad'ev, I. I., and Doman, N. G. (1970). The pathway of carbon in the photosynthesis of *Rhodopseudomonas palustris. Biokhimiya* **35,** 968.

46. Clayton, R. K. (1962). Symposium on autotrophy. III. Recent developments in photosynthesis. *Bacteriol. Rev.* **26,** 151.

47. Clayton, R. K. (1962). Evidence for the photochemical reaction of coenzyme Q in chromatophores of photosynthetic bacteria. *Biochem. Biophys. Res. Commun.* **9,** 49–53.
48. Clayton, R. K. (1962). Primary reactions in bacterial photosynthesis. I. The nature of light-induced absorbancy changes in chromatophores; evidence for a special bacteriochlorophyll component. *Photochem. Photobiol.* **1,** 201.
49. Clayton, R. K. (1963). Toward the isolation of a photochemical reaction center in *Rhodopseudomonas spheroides. Biochim. Biophys. Acta* **75,** 312.
50. Clayton, R. K., and Sistrom, W. R. (1966). An absorption band near 800 mμ associated with P_{870} in photosynthetic bacteria. *Photochem. Photobiol.* **5,** 661.
51. Clayton, R. K. (1966). The bacterial photosynthetic reaction center. *Brookhaven Symp. Biol.* **19,** 62.
52. Cohen-Bazire, G., Pfennig, N., and Kunisawa, R. (1964). The fine structure of green bacteria. *J. Cell Biol.* **22,** 207.
53. Cohen-Bazire, G., and Sistrom, W. R. (1966). The procaryotic photosynthetic apparatus. *In* "The Chlorophylls" (L. P. Vernon and G. R. Seeley, eds.), p. 328. Academic Press, New York.
54. Cramer, W. A. (1969). Low potential titration of the fluorescence yield changes in photosynthetic bacteria. *Biochim. Biophys. Acta* **189,** 54.
55. Crook, P. G., and Lindstrom, E. S. (1956). A comparison of the oxidative metabolism of light- and dark-grown *Rhodospirillum rubrum. Can. J. Microbiol.* **2,** 427.
56. Cruden, D. L., Cohen-Bazire, G., and Stanier, R. Y. (1970). *Chlorobium* vesicles: The photosynthetic organelles of green bacteria. *Nature (London)* **228,** 1345.
57. Cusanovich, M. A., Bartsch, R. G., and Kamen, M. D. (1968). Light-induced electron transport in *Chromatium* strain D. II. Light-induced absorbance changes in *Chromatium* chromatophores. *Biochim. Biophys. Acta* **153,** 397.
58. Cusanovich, M. A., and Kamen, M. D. (1968). Light-induced electron transfer in *Chromatium* strain D. III. Photophosphorylation by *Chromatium* chromatophores. *Biochim. Biophys. Acta* **153,** 418.
59. Cusanovich, M. A., and Bartsch, R. G. (1969). A light potential cytochrome c from *Chromatium* chromatophores. *Biochim. Biophys. Acta* **189,** 245.
60. De Klerk, H., and Kamen, M. D. (1966). A high-potential nonheme iron protein from the facultative photoheterotroph *Rhodopseudomonas gelatinosa. Biochim. Biophys. Acta* **112,** 175.
61. Del Campo, F. F., Ramirez, J. M., and Arnon, D. I. (1968). Stoichiometry of photosynthetic phosphorylation. *J. Biol. Chem.* **243,** 2805.
62. Demina, N. S., Ivanov, I. D., and Medvedev, G. A. (1969). Interrelation between the processes of nitrogen fixation, molecular hydrogen evolution and phosphorylation in photosynthesizing bacteria. *Mikrobiologiya* **38,** 428.
63. D'Eustachio, A. J., and Hardy, R. W. F. (1964). Reductants and electron transport in nitrogen fixation. *Biochem. Biophys. Res. Commun.* **15,** 319.
64. Dixon, M., and Webb, E. C. (1964). "Enzymes," 2nd ed. Longmans, Green, New York.
65. Drews, G. (1965). Untersuchungen zur Regulation der Bakteriochlorophyll-Synthese bei *Rhodospirillum rubrum. Arch. Mikrobiol.* **51,** 186.
66. Drews, G., and Giesebrecht, P. (1965). Die Thylakoidstrukturen von *Rhodopseudomonas* sp. *Arch. Mikrobiol.* **52,** 242.
67. Drews, G., and Oelze, J. (1966). Regulation of bactriochlorophyll synthesis in *Rhodospirillum rubrum. Zentralbl. Bakteriol., Parasitenk., Infektionskr. Hyg., Abt.* **2, 120,** 1; *Chem. Abstr.* **65,** 4298 (1966).

68. Drews, G., Lampe, H. H., and Ladwig, R. (1969). Die Entwicklung des Photosyntheseapparates in Dunkelkulturen von *Rhodopseudomonas capsulata*. *Arch. Mikrobiol.* **65,** 12.
69. Dutton, P. L., and Evans, W. C. (1969). The metabolism of aromatic compounds by *Rhodopseudomonas palustris*. A new, reductive method of aromatic ring fission. *Biochem. J.* **113,** 525.
70. Dutton, P. L. (1971). Oxidation–reduction potential dependence of the interaction of cytochromes, bacteriochlorophyll, and carotenoids at 77°K in chromatophores of *Chromatium* D and *Rhodopseudomonas gelatinosa*. *Biochim. Biophys. Acta* **226,** 63.
71. Dutton, P. L., Leigh, J. S., and Seibert, M. (1972). Primary processes in photosynthesis: *In situ* ESR studies on the light-induced oxidized and triplet state of reaction center bacteriochlorophyll. *Biochem. Biophys. Res. Commun.* **46,** 406.
72. Duysens, L. N. M., Juniskamp, W. J., Vos, J. J., and van der Hart, J. M. (1956). Reversible changes in bacteriochlorophyll in purple bacteria upon illumination. *Biochim. Biophys. Acta* **19,** 188.
73. Edwards, G. E., and Borell, C. R. (1969). Characteristics of a light-dependent proton transport in cells of *Rhodospirillum rubrum*. *Biochim. Biophys. Acta* **172,** 126.
74. Eidels, L., and Preiss, J. (1970). Citrate synthase. A regulatory enzyme from *Rhodopseudomonas capsulata*. *J. Biol. Chem.* **245,** 2937.
75. Eisenberg, M. A. (1953). The tricarboxylic acid cycle in *Rhodospirillum rubrum*. *J. Biol. Chem.* **203,** 815.
76. Eisenberg, M. A. (1955). The acetate-activating enzyme of *Rhodospirillum rubrum*. *Biochim. Biophys. Acta* **16,** 58.
77. Eisenberg, M. A. (1957). The acetate-activating enzyme of *Rhodospirillum rubrum*. *Biochim. Biophys. Acta* **23,** 327.
78. Elsden, S. R., and Ormerod, J. G. (1956). The effect of monofluoroacetate on the metabolism of *Rhodospirillum rubrum*. *Biochem. J.* **63,** 691.
79. Evans, M. C. W. (1965). The photooxidation of succinate by chromatophores of *Rhodospirillum rubrum*. *Biochem. J.* **95,** 661.
80. Evans, M. C. W. (1965). The photoassimilation of succinate to hexose by *Rhodospirillum rubrum*. *Biochem. J.* **95,** 669.
81. Evans, M. C. W., and Buchanan, B. B. (1965). Photoreduction of ferredoxin and its use in carbon dioxide by a subcellular system from a photosynthetic bacterium. *Proc. Nat. Acad. Sci. U.S.* **53,** 1420.
82. Evans, M. C. W., Buchanan, B. B., and Arnon, D. I. (1966). A new ferredoxin-dependent carbon reduction cycle in a photosynthetic bacterium. *Fed. Proc., Fed. Amer. Soc. Exp. Biol.* **25,** 226.
83. Feigenblum, E., and Krasna, A. I. (1970). Solubilization and properties of the hydrogenase of *Chromatium*. *Biochim. Biophys. Acta* **189,** 157.
84. Fisher, R. R., and Guillory, R. J. (1971). Resolution of enzymes catalyzing energy-linked transhydrogenation. II. Interaction of transhydrogenase factor with the *Rhodospirillum rubrum* chromatophore membrane. *J. Biol. Chem.* **246,** 4679.
85. Fisher, R. R., and Guillory, R. J. (1971). Resolution of enzymes catalyzing energy-linked transhydrogenation. III. Preparation and properties of *Rhodospirillum rubrum* transhydrogenase factor. *J. Biol. Chem.* **246,** 4687.
86. Fogg, G. E. (1968). "Photosynthesis." English Univ. Press, London.
87. Forti, G. (1968). The stoichiometry of NAD-dependent photosynthetic phosphorylation. *Biochem. Biophys. Res. Commun.* **32,** 1020.

88. Foster, J. W. (1944). Oxidation of alcohols by nonsulfur photosynthetic bacteria. *J. Bacteriol.* **47,** 355.
89. Foster, J. W. (1951). Autotrophic assimilation of carbon dioxide. *In* "Bacterial Physiology" (C. H. Werkman and P. W. Wilson, eds.), p. 361. Academic Press, New York.
90. Foust, G. P., Mayhew, S. G., and Massey, V. (1969). Complex formation between ferredoxin triphosphopyridine nucleotide reductase and electron transfer protein. *J. Biol. Chem.* **244,** 964.
91. Fowler, C. F., and Sybesma, C. (1970). Light- and chemical induced oxidation–reduction reactions in chromatophore fractions of *Rhodospirillum rubrum. Biochim. Biophys. Acta* **197,** 276.
92. French, C. S. (1937). The quantum yield of hydrogen and carbon dioxide assimilation in purple bacteria. *J. Gen. Physiol.* **20,** 711 [as cited by Ormerod and Gest (212)].
93. Frenkel, A. W., and Hickman, D. D. (1959). *J. Biophys. Biochem. Cytol.* **6,** 284 [as cited by Fuller *et al.* (98)].
94. Frenkel, A. W. (1970). Multiplicity of electron transport reactions in bacterial photosynthesis. *Biol. Rev. Cambridge Phil. Soc.* **45,** 569.
95. Frenkel, H. (1958). *Brookhaven Symp. Biol.* **11,** 276 [as cited by Stanier (276)].
96. Fuller, R. C., and Kornberg, H. L. (1961). A possible route for malate oxidation by *Chromatium. Biochem. J.* **79,** 8P.
97. Fuller, R. C., Smillie, R. M., Sisler, E. C., and Kornberg, H. L. (1961). Carbon metabolism in *Chromatium. J. Biol. Chem.* **236,** 2140.
98. Fuller, R. C., Conti, S. F., and Mellin, D. B. (1963). The structure of the photosynthetic apparatus in the green and purple sulfur bacteria. *In* "Bacterial Photosynthesis" (H. Gest, A. S. Pietro, and L. P. Vernon, eds.), p. 71. Antioch Press, Yellow Springs, Ohio.
99. Gaffron, H. (1935). Über den Stoffwechsel der Purpurbakterien. II. *Biochem. Z.* **275,** 301 [as cited by Foster (89)].
100. Gaffron, H., and Rubin, J. (1942). *J. Gen. Physiol.* **26,** 129 [as cited by Stanier (276)].
101. Gajdos, A., Gajdos-Torok, M., Gorsheim, A., Neuberger, A., and Tait, G. H. (1968). The effect of adenosine triphosphate on porphyrin excretion and on glycine metabolism in *Rhodopseudomonas spheroides. Biochem. J.* **106,** 185.
102. Garcia, A., Vernon, L. P., Ke, B., and Mollnehauer, H. (1968). Some structural and photochemical properties of *Rhodopseudomonas palustris* subchromatophore particles obtained by treatment with Triton X-100. *Biochemistry* **7,** 319.
103. Gehring, U., and Arnon, D. I. (1971). Ferredoxin-dependent phenyl pyruvate synthesis by cell-free preparations of photosynthetic bacteria. *J. Biol. Chem.* **246,** 4518.
104. Gest, H., and Kamen, M. D. (1949). Photoproduction of molecular hydrogen by *Rhodospirillum rubrum. Science* **109,** 558.
105. Gest, H., and Kamen, M. D. (1949). Studies on the metabolism of photosynthetic bacteria. IV. Photochemical production of H_2 by growing cultures of photosynthetic bacteria. *J. Bacteriol.* **58,** 239.
106. Gest, H., Kamen, M. D., and Bregoff, H. M. (1950). Studies on the metabolism of photosynthetic bacteria. V. Photoproduction of hydrogen and nitrogen fixation by *Rhodospirillum rubrum. J. Biol. Chem.* **182,** 153.
107. Gest, H. (1951). Metabolic pattern in photosynthetic bacteria. *Bacteriol. Rev.* **15,** 183.

108. Gest, H., Ormerod, J. G., and Ormerod, K. S. (1962). Photometabolism of *Rhodospirillum rubrum:* Light-dependent dissimilation of organic compounds to carbon dioxide and molecular hydrogen by an anaerobic citric acid cycle. *Arch. Biochem. Biophys.* **97,** 21.
109. Gest, H. (1963). Metabolic aspects of bacterial photosynthesis. *In* "Bacterial Photosynthesis" (H. Gest, A. S. Pietro, and L. P. Vernon, eds.), p. 129. Antioch Press, Yellow Springs, Ohio.
110. Gest, H. (1966). Comparative biochemistry of photosynthetic processes. *Nature (London)* **209,** 879.
111. Gibbs, M. (1970). The inhibition of photosynthesis by oxygen. *Amer. Sci.* **58,** 634.
112. Gibson, J., and Morita, S. (1967). Changes in adenine nucleotides of intact *Chromatium* D by illumination. *J. Bacteriol.* **93,** 1544.
113. Gibson, J. (1967). Aerobic metabolism of *Chromatium* sp. strain D. *Arch. Mikrobiol.* **59,** 104.
114. Gibson, K. D. (1955). Nature of the insoluble pigmented structures (chromatophores) in extracts and lyzates of *Rhodopseudomonas spheroides. Biochemistry* **4,** 2027.
115. Gibson, K. D. (1965). Isolation and characterization of chromatophores from *Rhodopseudomonas spheroides. Biochem. J.* **4,** 2042.
116. Gibson, K. D. (1965). Structure of chromatophores of *Rhodopseudomonas spheroides:* Removal of a nonpigmented outer layer of lipid. *Biochemistry* **4,** 2052.
117. Gibson, M. S., and Wang, C. H. (1968). Utilization of fructose and glutamate by *Rhodospirillum rubrum. Can. J. Microbiol.* **14,** 493.
118. Gorsheim, A. (1968). The relation between the pigment content of isolated chromatophores and that of the whole cell in *Rhodopseudomonas spheroides. Proc. Roy. Soc., Ser. B* **170,** 247.
119. Gorsheim, A., Neuberger, A., and Tait, G. H. (1968). Metabolic turnover of the lipids of *Rhodopseudomonas spheroides. Proc. Roy. Soc., Ser. B* **170,** 311.
120. Gould, E. S., Kuntz, I. D., Jr., and Calvin, M. (1965). Absorption changes in bacterial chromatophores. II. A new chlorophyll-like pigment from the oxidation of chromatophores from *Rhodospirillum rubrum. Photochem. Photobiol.* **4,** 482; *Chem. Abstr.* **63,** 13617 (1965).
120a. Gould, J. M., and Winget, G. D. (1972). The mechanism of photophosphorylation. I. Inhibition of the light-induced proton translocation by inorganic phosphate. *Biochem. Biophys. Res. Commun.* **47,** 309.
121. Gray, E. D. (1967). Studies on the adaptive formation of photosynthetic structures in *Rhodopseudomonas spheroides.* I. Synthesis of macromolecules. *Biochim. Biophys. Acta* **138,** 550.
122. Gray, E. D. (1967). Studies on the adaptive formation of photosynthetic structures in *Rhodopseudomonas spheroides.* II. Nucleotide composition during adaptation. *Biochim. Biophys. Acta* **138,** 564.
123. Gromet-Elhanan, Z. (1967). The relationship of cyclic and noncyclic electron flow patterns with reduced indophenols to photophosphorylation. *Biochim. Biophys. Acta* **131,** 526.
124. Guswr, M. V., Shenderova, L. V., and Kondrat'eva, E. N. (1970). Influence of the oxygen concentration on the growth and survival of photosynthetic bacteria. *Mikrobiologiya* **39,** 562.
125. Hardy, R. W. F., and D'Eustachio, A. J. (1964). The dual role of pyruvate and the energy requirement in nitrogen fixation. *Biochem. Biophys. Res. Commun.* **15,** 314.

126. Hawk, A., and Leadbetter, E. R. (1965). Aromatic acid utilization by Athiorhodaceae. *Bacteriol. Proc.* p. 22.
127. Hegeman, G. D. (1967). The metabolism of *p*-hydroxybenzoate by *Rhodopseudomonas palustris* and its regulation. *Arch. Mikrobiol.* **59,** 143.
128. Henderson, R. W., and Nankiville, D. D., (1966). Electrophoretic and other studies on heme pigments from *Rhodopseudomonas palustris:* Cytochrome 552 and cytochromoid c. *Biochem. J.* **98,** 587.
129. Hendriks, R., and Cronin, J. R. (1971). The flavin of *Chromatium* cytochrome c-552. *Biochem. Biophys. Res. Commun.* **44,** 313.
130. Hoare, D. S., and Gibson, J. (1964). Photoassimilation of acetate and the biosynthesis of amino acids by *Chlorobium thiosulfatophilum. Biochem. J.* **91,** 546.
131. Hoare, D. S., and Hoare, S. L. (1969). Hydrogen metabolism by *Rhodomicrobium vannielii. J. Bacteriol.* **100,** 1124.
132. Holt, S. C., and Marr, A. G. (1965). Location of chlorophyll in *Rhodospirillum rubrum. J. Bacteriol.* **89,** 1402.
133. Holt, S. C., and Marr, A. G. (1965). Isolation and purification of the intracytoplasmic membrane of *Rhodospirillum rubrum. J. Bacteriol.* **89,** 1413.
134. Holt, S. C., and Marr, A. G. (1965). Effect of light intensity on the formation of intracytoplasmic membrane in *Rhodospirillum rubrum. J. Bacteriol.* **89,** 1421.
135. Holt, S. C., Conti, S. F., and Fuller, R. C. (1966). Photosynthetic apparatus in the green bacterium *Chloropseudomonas ethylicum. J. Bacteriol.* **91,** 311.
136. Holt, S. C., Conti, S. F., and Fuller, R. C. (1966). Effect of light intensity on the formation of the photochemical apparatus in the green bacterium *Chloropseudomonas ethylicum. J. Bacteriol.* **91,** 349.
137. Horio, T., and Kamen, M. D. (1962). Observations on the respiratory system of *Rhodospirillum rubrum. Biochemistry* **1,** 1141.
138. Horio, T., Yamashita, J., and Nishikawa, K. (1963). Photosynthetic adenosine triphosphate formation and photoreduction of diphosphopyridine nucleotide with chromatophores of *Rhodospirillum rubrum. Biochim. Biophys. Acta* **66,** 37.
139. Horio, T., Nishikawa, K., and Yamashita, J. (1966). Synthesis and possible character of a high-energy intermediate in bacterial photophosphorylation. *Biochem. J.* **98,** 321.
140. Horio, T., Nishikawa, K., Horiuti, Y., and Kakuno, T. (1968). Mode of coupling of the phosphorylating system of the electron transport system in *Rhodospirillum rubrum* chromatophores. *In* "Comparative Biochemistry and Biophysics of Photosynthesis" (K. Shibata *et al.*, eds.), p. 408. Univ. of Tokyo Press, Tokyo.
141. Horio, T., Nishikawa, K., Okayama, S., Horiuti, Y., Yamamoto, N., and Kakutani, Y. (1968). The requirement of ubiquinone-10 for an ATP-forming system and an ATPase system of chromatophores from *Rhodospirillum rubrum. Biochim. Biophys. Acta* **153,** 913.
142. Horio, T., Bartsch, R. G., Kakuno, T., and Kamen, M. D. (1969). Two reduced nicotinamide adenine dinucleotide dehydrogenases from the photosynthetic bacterium *Rhodospirillum rubrum. J. Biol. Chem.* **244,** 5899.
143. Hughes, D. E., Conti, S. F., and Fuller, R. C. (1963). Inorganic polyphosphate metabolism in *Chlorobium thiosulfatophilum. J. Bacteriol.* **85,** 577.
144. Isaer, P. I., Liberman, E. A., Samuilov, V. D., Skulachev, V. P., and Tsofina, L. M. (1970). Conversion of biomembrane-produced energy into electric form. III. Chromatophores of *Rhodospirillum rubrum. Biochim. Biophys. Acta* **216,** 22.
145. Jackson, J. B., and Crofts, A. R. (1968). Energy-linked reduction of nicotinamide adenine dinucleotides in cells of *Rhodospirillum rubrum. Biochem. Biophys. Res. Commun.* **32,** 908.

146. James, A. T., Harris, R. V., Hitchcock, C., Wood, B. J. B., and Nichols, B. W. (1965). The biosynthesis and degradation of unsaturated fatty acids in higher plants and photosynthetic bacteria. *Fette, Seifen, Anstrichm.* **67,** 393; *Chem. Abstr.* **63,** 15154 (1965).
147. Jones, C. W., and Vernon, L. P. (1969). Nicotinamide-adenine dinucleotide photoreduction in *Rhodospirillum rubrum* chromatophores. *Biochim. Biophys. Acta* **180,** 149.
148. Kamen, M. D., and Gest, H. (1949). Evidence for a nitrogenase system in the photosynthetic bacterium *Rhodospirillum rubrum. Science* **109,** 560.
149. Kamen, M. D. (1950). Hydrogenase activity and photoassimilation. *Fed. Proc., Fed. Amer. Soc. Exp. Biol.* **9,** 543.
150. Karlson, P. (1968). "Introduction to Modern Biochemistry," 3rd ed. Academic Press, New York.
151. Ke, B., Vernon, L. P., Garcia, A., and Ngo, E. (1968). Coupled photooxidation of bacteriochlorophyll P_{890} and photoreduction of ubiquinone in a photochemically active subchromatophore particle derived from *Chromatium. Biochemistry* **7,** 311.
152. Ke, B. (1969). Nature of the primary electron acceptor in bacterial photosynthesis. *Biochim. Biophys. Acta* **172,** 583.
153. Keister, D. L., and Minton, N. J. (1971). ATP synthesis driven by inorganic pyrophosphate in *Rhodospirillum rubrum* chromatophores. *Biochem. Biophys. Res. Commun.* **42,** 932.
154. Kihara T., and Dutton, P. L. (1970). Light-induced reactions of photosynthetic bacteria. I. Reactions in whole cells and in cell-free extracts at liquid nitrogen temperature. *Biochim. Biophys. Acta* **205,** 196.
155. Kikuchi, G., Abe, S., and Muto, A. (1961). Carboxylic acids metabolism and its relation to porphyrin biosynthesis in *Rhodopseudomonas spheroides* under light-anaerobic conditions. *J. Biochem.* (*Tokyo*) **49,** 570.
156. Klemme, H.-J., and Schlegel, H. G. (1967). Photoreduktion von Pyridinnucleotid durch Chromatophoren aus *Rhodopseudomonas capsulata* mit molekularem Wasserstoff. *Arch. Mikrobiol.* **59,** 185.
157. Klemme, H.-J., and Schlegel, H. G. (1968). Cyclic photophosphorylation by chromatophores of the facultative phototroph, *Rhodopseudomonas capsulata. Arch. Mikrobiol.* **63,** 154.
158. Klemme, H.-J. (1968). Untersuchungen zur Photoautotrophie mit molekularem Wasserstoff bei neuisolierten schwefelfreien Purpurbakterien. *Arch. Mikrobiol.* **64,** 29.
159. Klemme, H.-J., and Schlegel, H. G. (1971). Regulatory properties of an inorganic pyrophosphatase from the photosynthetic bacterium *Rhodospirillum rubrum. Proc. Nat. Acad. Sci. U.S.* **68,** 721.
160. Klemme, H.-J., Klemme, B., and Gest, H. (1971). Catalytic properties and regulatory diversity of inorganic pyrophosphatases from photosynthetic bacteria. *J. Bacteriol.* **108,** 1122.
160a. Klemme, J. H. (1973). Allosteric control of pyruvate kinase from *Rhodospirillum rubrum* by inorganic phosphate and sugar phosphate esters. *Arch. Mikrobiol.* **90,** 305.
161. Knight, M. (1962). The photometabolism of propionate by *Rhodospirillum rubrum. Biochem. J.* **84,** 170.
162. Knobloch, K., Eley, J. H., and Aleem, M. I. H. (1971). Thiosulfate-linked ATP-dependent NAD^+ reduction in *Rhodopseudomonas palustris. Arch. Mikrobiol.* **80,** 97.

163. Knobloch, K., Eley, J. H., and Aleem, M. I. H. (1971). Generation of reducing power in bacterial photosynthesis. *Rhodopseudomonas palustris. Biochem. Biophys. Res. Commun.* **43,** 834.
164. Kohlmiller, E. F., and Gest, H. (1951). A comparative study of the light and dark fermentations of organic acids by *Rhodospirillum rubrum. J. Bacteriol.* **61,** 269.
165. Kondratieva, E. N., and Gogotov, I. N. (1969). Production of hydrogen by green photosynthetic bacteria. *Nature (London)* **221,** 83.
166. Korkes, S. (1955). Enzymatic reduction of pyridine nucleotides by molecular hydrogen. *J. Biol. Chem.* **216,** 737.
166a. Kusai, K., and Yamanaka, T. (1973). Cytochrome c (553, *Chlorobium thiosulfatophilum*) is a sulphide-cytochrome c reductase. *FEBS Lettrs.* **34,** 235.
166b. Kusai, K., and Yamanaka, T. (1973). The oxidation mechanisms of thiosulfate and sulphide in *Chlorobium thiosulfatophilum.* Roles of cytochrome c-551 and cytochrome c-553. *Biochim. Biophys. Acta* **325,** 304.
167. Lampe, H. H., Oelze, J., and Drews, G. (1972). Die Fraktionierung des Membransystems von *Rhodopseudomonas capsulata* und seine Morphogenese. *Arch. Mikrobiol.* **83,** 78.
168. Lascelles, J. (1960). The formation of ribulose 1,5-diphosphate carboxylase by growing cultures of Athiorhodaceae. *J. Gen. Microbiol.* **23,** 499.
169. Lascelles, J. (1962). The chromatophores of photosynthetic bacteria. *J. Gen. Microbiol.* **29,** 47.
170. Lascelles, J. (1966). The accumulation of bacteriochlorophyll precursors by mutant and wild-type strains of *Rhodopseudomonas spheroides. Biochem. J.* **100,** 175.
171. Lascelles, J. (1968). The bacterial photosynthetic apparatus. *Advan. Microbial Physiol.* **2,** 1–42.
171a. Lascelles, J. (1973). Microbial Photosynthesis. Benchmark Papers in Microbiology (W. W. Umbreit, ed.) Dowden, Hutchison & Ross, Inc.
172. Lascelles, J., and Hatch, T. P. (1969). Bacteriochlorophyll and heme synthesis in *Rhodopseudomonas spheroides:* Possible role of heme in regulation of the branched biosynthetic pathway. *J. Bacteriol.* **98,** 712.
173. Lascelles, J., and Altshuler, T. (1969). Mutant strains of *Rhodopseudomonas spheroides* lacking α-aminolevulinate synthase: Growth, heme, and bacteriochlorophyll synthesis. *J. Bacteriol.* **98,** 721.
174. Lees, H. (1955). The photosynthetic bacteria. *In* "Biochemistry of Autotrophic Bacteria," (H. Lees, ed.), p. 61. Butterworth, London.
175. Lehninger, A. L. (1965). "Bioenergetics." Benjamin, New York.
176. Leigh, J. S., Jr., and Dutton, P. L. (1972). The primary electron acceptor in photosynthesis. *Biochem. Biophys. Res. Commun.* **46,** 414.
177. Lindstrom, E. S., Burris, R. H., and Wilson, P. W. (1949). Nitrogen fixation by photosynthetic bacteria. *J. Bacteriol.* **58,** 313.
178. Lippert, K.-D., and Klemme, H.-J. (1968). Untersuchungen zum Mechanismus der Photoreduktion von Pyridinnukleotid durch Chromatophoren aus *Rhodospirillum rubrum. Arch. Mikrobiol.* **62,** 307.
179. Lippert, K.-D., and Pfennig, N. (1969). Die Verwertung von molekularem Wasserstoff durch *Chlorobium thiosulfatophilum.* Wachstum und CO_2-Fixierung. *Arch. Mikrobiol.* **65,** 29.
180. Losada, M., Trebst, A. V., Ogata, S., and Arnon, D. I. (1960). Equivalence of light and ATP in bacterial photosynthesis. *Nature (London)* **186,** 753.
181. Losada, M., Whatley, F. R., and Arnon, D. I. (1961). Separation of two light

reactions in noncyclic photophosphorylation of green plants. *Nature* (*London*) **190,** 606.

182. McBee, R. H., Lamanna, C., and Weeks, O. B. (1955). Definitions of bacterial oxygen relationship. *Bacteriol. Rev.* **19,** 45.
183. Mandel, M., Leadbetter, E. R., Pfennig, N., and Truper, H. G. (1971). Deoxyribonucleic acid base compositions of phototrophic bacteria. *Int. J. Syst. Bacteriol.* **21,** 222.
184. Marock, J., deKlerk, H., and Kamen, M. D. (1968). Quinones of Athiorhodaceae. *Biochim. Biophys. Acta* **162,** 621.
185. Marr, A. G. (1960). Localization of enzymes in bacteria. *In* "The Bacteria" (I. C. Gunsalus and R. Y. Stanier, eds.), Vol. 1, p. 443. Academic Press, New York.

185a. Marrs, B., and Gest, H. (1973). Regulation of bacteriochlorophyll synthesis by oxygen in respiratory mutants of *Rhodopseudomonas capsulata. J. Bacteriol.* **114,** 1052.

186. Mechalas, B. J., and Rittenberg, S. C. (1960). Energy coupling in *Desulfovibrio desulfuricans. J. Bacteriol.* **80,** 501.
187. Meyer, T. E., Bartsch, R. G., Cusanovich, M. A., and Mathewson, J. H. (1968). The cytochromes of *Chlorobium thiosulfatophilum. Biochim. Biophys. Acta* **153,** 854.
188. Meyer, T. E., Sharp, J. J., and Bartsch, R. G. (1971). Isolation and properties of rubredoxin from the photosynthetic green sulfur bacteria. *Biochim. Biophys. Acta* **234,** 266.
189. Meyer, T. E., Bartsch, R. G., and Kamen, M. D. (1971). Cytochrome c_3. A class of electron transfer heme proteins found in both photosynthetic and sulfur-reducing bacteria. *Biochim. Biophys. Acta* **245,** 453.
190. Morita, S. (1961). Metabolism of organic acids in *Rhodopseudomonas palustris* in light and dark. *J. Biochem.* (*Tokyo*) **50,** 190.
191. Morita, S. (1968). Evidence for three photochemical systems in *Chromatium* D. *Biochim. Biophys. Acta* **153,** 241.
192. Mortenson, L. E. (1964). Ferredoxin requirement for nitrogen fixation by extracts of *Clostridium pasteurianum. Biochim. Biophys. Acta* **81,** 473.
193. Munson, T. O., and Burris, R. H. (1969). Nitrogen fixation by *Rhodospirillum rubrum* grown in nitrogen-limited continuous culture. *J. Bacteriol.* **97,** 1093.

193a. Nakayama, H., Midwinter, G. G., and Krampitz, L. O. (1971). Properties of the pyruvate formate–lyase reaction. *Archs. Biochem. Biophys.* **143,** 526.

194. Naudi, D. L., and Shemin, D. (1968). α-Aminolevulinic dehydratase of *Rhodopseudomonas spheroides.* III. Mechanism of porphobilinogen synthesis. *J. Biol. Chem.* **243,** 1236.
195. Newton, J. W., and Wilson, P. W. (1953). Nitrogen fixation and photoproduction of molecular hydrogen by Thiorhodaceae. *Antonie van Leeuwenhoek; J. Microbiol. Serol.* **19,** 71.

195a. Niederman, R. A. (1974). Membranes of *Rhodopseudomonas spheroides:* Interactions of chromatophores with the cell envelope. *J. Bacteriol.* **117,** 19.

196. Oelze, J., Biedermann, M., and Drews, G. (1969). Die Morphogenese des Photosyntheseapparatus von *Rhodospirillum rubrum.* I. Die Isolierung und Charakterisierung von zwei Membransystemen. *Biochim. Biophys. Acta* **173,** 436.
197. Oelze, J., and Drews, G. (1969). Die Morphogenese des Photosyntheseapparatus von *Rhodospirillum rubrum.* II. Die Kinetik der Thykaloidsynthese nach Markierung der Membranen mit (2-^{14}C)Azetat. *Biochim. Biophys. Acta* **173,** 448.

198. Oelze, J., Biedermann, M., Freund-Molbert, E., and Drews, G. (1969). Bakteriochlorophyllgehalt und Proteinmuster der Thykaloide von *Rhodospirillum rubrum* während der Morphogenese des Photosynthese-Apparates. *Arch. Mikrobiol.* **66,** 154.
199. Oelze, J., and Drews, G. (1970). Der Einfluss der Lichtintensität und der Sauerstoffspannung auf die Differenzierung der Membranen von *Rhodospirillum rubrum. Biochim. Biophys. Acta* **203,** 189.
200. Oelze, J., and Drews, G. (1970). Variations of NADH oxidase activity and bacteriochlorophyll contents during membrane differentiation in *Rhodospirillum rubrum. Biochim. Biophys. Acta* **219,** 131.
201. Oelze, J., and Weaver, P. (1971). The adjustment of photosynthetically grown cells of *Rhodospirillum rubrum* to aerobic light conditions. *Arch. Mikrobiol.* **79,** 108.
202. Oelze, J., and Kamen, M. D. (1971). Adenosine triphosphate cellular levels in *Rhodospirillum rubrum* during transition from aerobic to anaerobic metabolism. *Biochim. Biophys. Acta* **234,** 137.
202a. Oelze, J., and Drews, G. (1972). Membranes of photosynthetic bacteria. *Biochim. Biophys. Acta* **265,** 209.
203. Ohashi, A., Ishihara, N., and Kikuchi, G. (1967). Pyruvate metabolism in *Rhodopseudomonas spheroides* under light-anaerobic conditions. *J. Biochem.* (*Tokyo*) **62,** 497.
204. Okayama, S., Yamamoto, N., Nishikawa, K., and Horio, T. (1968). Roles of ubiquinone-10 and rhodoquinone in photosynthetic formation of adenosine triphosphate by chromatophores from *Rhodospirillum rubrum. J. Biol. Chem.* **243,** 2995.
205. Oken, I., and Merrick, J. M. (1968). Identification of propionate as an endogenous CO_2 acceptor in *Rhodospirillum rubrum* and properties of purified propionyl-coenzyme A carboxylase. *J. Bacteriol.* **95,** 1774.
206. Olson, J. M., and Chance, B. (1960). Oxidation–reduction reactions in the photosynthetic bacterium *Chromatium.* I. Absorption spectrum changes in whole cells. *Arch. Biochem. Biophys.* **88,** 26.
207. Olson, J. M., and Chance, B. (1960). Oxidation–reduction reactions in the photosynthetic bacterium *Chromatium.* II. Dependence of light reactions on intensity of irradiation and quantum efficiency of cytochrome oxidation. *Arch. Biochem. Biophys.* **88,** 40.
208. Orlando, J. A., Sabo, D., and Curnyn, C. (1966). Photoreduction of pyridine nucleotide by subcellular preparations from *Rhodopseudomonas spheroides. Plant Physiol.* **41,** 937.
209. Orlando, J. A. (1968). Light-dependent reduction of nicotinamide adenine dinucleotide phosphate by chromatophores of *Rhodopseudomonas spheroides. Arch. Biochem. Biophys.* **124,** 413.
210. Ormerod, J. G. (1956). The use of radioactive CO_2 in the measurement of CO_2 fixation in *Rhodospirillum rubrum. Biochem. J.* **64,** 373.
211. Ormerod, J. G., Ormerod, K. S., and Gest, H. (1961). Light-dependent utilization of organic compounds and photoproduction of molecular hydrogen by photosynthetic bacteria; relationships with nitrogen metabolism. *Arch. Biochem. Biophys.* **94,** 449.
212. Ormerod, J. G., and Gest, H. (1962). Symposium on metabolism of inorganic compounds. IV. Hydrogen photosynthesis and alternative pathways metabolic in photosynthetic bacteria. *Bacteriol. Rev.* **26,** 51.

213. Park, C., and Berger, L. R. (1967). Fatty acids of extractable and bound lipids of *Rhodomicrobium vannielii*. *J. Bacteriol.* **93,** 230.
214. Parson, W. W. (1968). The role of P_{870} in bacterial photosynthesis. *Biochim. Biophys. Acta* **153,** 248.
215. Parson, W. W. (1969). The reaction between primary and secondary electron acceptors in bacterial photosynthesis. *Biochim. Biophys. Acta* **189,** 384.
216. Parson, W. W. (1969). Cytochrome photooxidations in *Chromatium* chromatophores. Each P_{870} oxidizes two cytochrome c-422 hemes. *Biochim. Biophys. Acta* **189,** 397.
217. Parson, W. W., and Case, G. D. (1970). In *Chromatium*, a single photochemical reaction center oxidizes both cytochrome c-552 and cytochrome c-555. *Biochim. Biophys. Acta* **205,** 232.
218. Payne, J., and Morris, J. G. (1969). Pyruvate carboxylase in *Rhodopseudomonas spheroides*. *J. Gen. Microbiol.* **59,** 97.
219. Peck, H. D., Jr. (1960). Evidence for oxidative phosphorylation during the reduction of sulfate with hydrogen by *Desulfovibrio desulfuricans*. *J. Biol. Chem.* **235,** 2734.
220. Peters, G. A., and Cellarius, R. A. (1972). The ubiquinone homolog of the green mutant of *Rhodopseudomonas spheroides*. *Biochim. Biophys. Acta* **256,** 544.
221. Pfennig, N., and Cohen-Bazire, G. (1967). Some properties of the green bacterium *Pelodictyon clathratiforma*. *Arch. Mikrobiol.* **59,** 226.
221a. Pfennig, N., and Trüper, H. G. (1971). Type and neotype strains of the species of phototrophic bacteria maintained in pure culture. *Int. J. Syst. Bacteriol.* **21,** 21.
222. Pfluger, V. N., and Bachofen, R. (1971). Effect of inorganic pyrophosphate on carbon dioxide fixation in photosynthetic bacteria. *Arch. Mikrobiol.* **77,** 36.
223. Porva, R. J., and Lascelles, J. (1965). Hemoproteins and heme synthesis in facultative photosynthetic and denitrifying bacteria. *Biochem. J.* **94,** 120.
224. Powls, R., and Redfearn, E. R. (1969). Quinones of the Chlorobacteriaceae. Properties and possible function. *Biochim. Biophys. Acta* **172,** 429.
225. Quadri, S. M. H., and Hoare, D. S. (1968). Formic hydrogenlyase and the photoassimilation of formate by a strain of *Rhodopseudomonas palustris*. *J. Bacteriol.* **95,** 2344.
226. Rabinowitch, E. I. (1948). Photosynthesis. *Sci. Amer.* **179,** 24.
227. Rabinowitch, E. I. (1971). An unfolding discovery. *Proc. Nat. Acad. Sci. U.S.* **68,** 2875.
228. Ramirez, J., and Smith, L. (1968). Synthesis of adenosine triphosphate in intact cells of *Rhodospirillum rubrum* and *Rhodopseudomonas spheroides* on oxygenation or illumination. *Biochim. Biophys. Acta* **153,** 466.
229. Reed, D. W., and Clayton, R. K. (1968). Isolation of a reaction center fraction from *Rhodopseudomonas spheroides*. *Biochem. Biophys. Res. Commun.* **30,** 471.
230. Reed, D. W. (1969). Isolation and composition of a photosynthetic reaction center complex from *Rhodopseudomonas spheroides*. *J. Biol. Chem.* **244,** 4936.
231. Reed, D. W., Zankel, K. L., and Clayton, R. K. (1969). The effect of redox potential on P-870 fluorescence in reaction centers from *Rhodopseudomonas spheroides*. *Proc. Nat. Acad. Sci. U.S.* **63,** 42.
232. Reiss-Husson, F., and Jolchine, G. (1972). Purification and properties of a photosynthetic reaction center isolated from various chromatophore fractions of *Rhodopseudomonas spheroides* Y. *Biochim. Biophys. Acta* **256,** 440.
233. Richards, W. R., and Rapoport, H. (1967). The biosynthesis of *Chlorobium* chlorophyll-660. *Biochemistry* **6,** 3830.

234. Rinne, R. W., Buckman, R. W., and Benedict, R. C. (1965). Acetate and bicarbonate metabolism in photosynthetic bacteria. *Plant Physiol.* **40,** 1066.

235. Roelofson, P. A. (1934). On the metabolism of the purple sulfur bacteria. *Proc., Kon. Ned. Akad. Wetensch.* **37,** 660 [as cited by Foster (89)].

236. Rolls, J. P., and Lindstrom, E. S. (1967). Induction of a thiosulfate-oxidizing enzyme in *Rhodopseudomonas palustris. J. Bacteriol.* **94,** 784.

237. Rolls, J. P., and Lindstrom, E. S. (1967). Effect of thiosulfate on the photosynthetic growth of *Rhodopseudomonas palustris. J. Bacteriol.* **94,** 860.

238. Rose, I. A., and Ochoa, S. (1956). Phosphorylation by particulate preparations of *Azotobacter vinelandii. J. Biol. Chem.* **220,** 307.

239. Sabo, D. J., and Orlando, J. A. (1968). Isolation, purification, and some properties of reduced nicotinamide adenine dinucleotide phosphate-cytochrome c_2 reductase from *Rhodopseudomonas spheroides. J. Biol. Chem.* **243,** 3742.

240. Saha, S., and Good, N. E. (1970). Products of the photophosphorylation reaction. *J. Biol. Chem.* **245,** 5017.

241. Sasaki, T., Motokawa, Y., and Kikuchi, G. (1970). Occurrence of both a-type and o-type cytochromes as the functional terminal oxidases in *Rhodopseudomonas spheroides. Biochim. Biophys. Acta* **197,** 284.

242. Sato, H., Takahashi, K., and Kikuchi, G. (1966). Inhibition studies of photophosphorylation by *Rhodospirillum rubrum* chromatophores with particular concerns to antimycin-resistant photophosphorylation in the presence of artificial electron carriers. *Biochim. Biophys. Acta* **112,** 8.

243. Sauer, K., Dratz, E. A., and Coyne, L. (1968). Circular dichroism spectra and the molecular arrangement of bacteriochlorophyll in the reaction centers of photosynthetic bacteria. *Proc. Nat. Acad. Sci. U.S.* **61,** 17.

244. Schick, H. J. (1971). Substrate- and light-dependent fixation of molecular nitrogen in *Rhodospirillum rubrum. Arch. Mikrobiol.* **75,** 89.

245. Schick, H. J. (1971). Interrelationship of nitrogen fixation, hydrogen evolution, and photoreduction in *Rhodospirillum rubrum. Arch. Mikrobiol.* **75,** 102.

246. Schick, H. J. (1971). Regulation of photoreduction in *Rhodospirillum rubrum* by ammonia. *Arch. Mikrobiol.* **75,** 110.

247. Schmidt, G. L., and Kamen, M. D. (1971). Redox properties of the "P-836" pigment complex of *Chromatium. Biochim. Biophys. Acta* **234,** 70.

248. Schmidt, G. L., and Kamen, M. D. (1971). Control of chlorophyll synthesis in *Chromatium vinosum. Arch. Mikrobiol.* **76,** 57.

249. Scholes, P., Mitchell, P., and Moyle, J. (1969). The polarity of proton translocation in some photosynthetic microorganisms. *Eur. J. Biochem.* **8,** 450.

250. Schön, G., and Drews, G. (1968). Der Redoxzustand des NAD(P) und der Cytochrome b und c_2 in Abhängigkeit vom pO_2 bei einigen Athiorhodaceae. *Arch. Mikrobiol.* **61,** 317.

251. Schön, G. (1968). Fruktoseverwertung und Bacteriochlorophyllsynthese in anaeroben Dunkel- und Lichtkulturen von *Rhodospirillum rubrum. Arch. Mikrobiol.* **63,** 362.

252. Schön, G. (1969). Der Einfluss der Kulturbedingungen auf den ATP-, ADP-, und AMP-spiegel bei *Rhodospirillum rubrum. Arch. Mikrobiol.* **66,** 348.

253. Schön, G., and Ladwig, R. (1970). Bakteriochlorophyllsynthese und Thykaloidmorphogenese in anaerober Dunkelkultur von *Rhodospirillum rubrum. Arch. Mikrobiol.* **74,** 356.

253a. Schön, G. (1972). Substratverwertung und Bacteriochlorophyll-Synthese in anaerober Dunkelkultur von *Rhodospirillum rubrum.* I. Abhängigkeit der Bacteriochlorophyll-Synthese von der Substratkonzentration und dem Elektronenacceptor. *Zentralbl. Bakteriol., I. Abtlg. A* **220,** 380.

253b. Schön, G., and Biedermann, M. (1972). Bildung flüchtiger Säure bei der Vergärung von Pyruvat und Fructose in anaerober Dunkelkultur von *Rhodospirillum rubrum. Arch. Mikrobiol.* **85,** 77.

253c. Schön, G., and Biedermann, M. (1973). Growth and adaptive hydrogen production of *Rhodospirillum rubrum* (F 1) in anaerobic dark cultures. *Biochim. Biophys. Acta* **304,** 65.

254. Schwartz, M. (1967). Wavelength-dependent quantum yields of chloroplast phosphorylation catalyzed by phenazine methosulfate. *Biochim. Biophys. Acta* **131,** 548.

255. Seibert, M., and Devault, D. (1970). Relations between the laser-induced oxidations of the high- and low-potential cytochromes of *Chromatium* D. *Biochim. Biophys. Acta* **205,** 220.

256. Seibert, M., Dutton, P. L., and Devault, D. (1971). A low-potential photosystem in *Chromatium* D. *Biochim. Biophys. Acta* **226,** 189.

257. Seibert, M., and Devault, D. (1971). Photosynthetic reaction center transients, P_{435} and P_{424} in *Chromatium* D. *Biochim. Biophys. Acta* **253,** 396.

258. Shanmugan, K. T., Buchanan, B. B., and Arnon, D. I. (1972). Ferredoxins in light- and dark-grown photosynthetic cells with special reference to *Rhodospirillum rubrum. Biochim. Biophys. Acta* **256,** 477.

259. Shanmugan, K. T., and Arnon, D. I. (1972). Effect of ferredoxin on bacterial photophosphorylation. *Biochim. Biophys. Acta* **256,** 487.

260. Shaw, M. A., and Richards, W. R. (1971). Evidence for the formation of membranous chromatophore precursor fractions in *Rhodopseudomonas spheroides. Biochem. Biophys. Res. Commun.* **45,** 863.

261. Siegel, J. M. (1950). The metabolism of acetone by the photosynthetic bacterium *Rhodopseudomonas gelatinosa. J. Bacteriol.* **60,** 595.

262. Siegel, J. M., and Kamen, M. D. (1951). Studies on the metabolism of photosynthetic bacteria. VII. Comparative studies on the photoproduction of hydrogen by *Rhodopseudomonas gelatinosa* and *Rhodospirillum rubrum. J. Bacteriol.* **61,** 215.

263. Siegel, J. M. (1954). The photosynthetic metabolism of acetone by *Rhodopseudomonas gelatinosa. J. Biol. Chem.* **208,** 205.

264. Siegel, J. M., and Smith, A. M. (1955). The dark aerobic metabolism of acetone by the photosynthetic bacterium *Rhodopseudomonas gelatinosa. J. Biol. Chem.* **214,** 475.

265. Siegel, J. M. (1957). The dark anaerobic metabolism of acetone and acetate by the photosynthetic bacterium *Rhodopseudomonas gelatinosa. J. Biol. Chem.* **228,** 41.

266. Silver, M., Friedman, S., Guay, R., Couture, J., and Tanguay, R. (1971). Base composition of deoxyribonucleic acid isolated from Athiorhodaceae. *J. Bacteriol.* **107,** 368.

267. Sistrom, W. R. (1962). Observations on the relationships between formation of photopigments and the synthesis of protein in *Rhodopseudomonas spheroides. J. Gen. Microbiol.* **28,** 599.

268. Sistrom, W. R. (1965). Effect of oxygen on growth and the synthesis of bacteriochlorophyll in *Rhodospirillum molischianum. J. Bacteriol.* **89,** 403.

269. Skerman, V. B. D. (1967). "A Guide to the Identification of the Genera of Bacteria," 2nd ed., pp. 96–105. Williams & Wilkins, Baltimore, Maryland.

270. Slooten, L. (1972). Reaction center preparations of *Rhodopseudomonas spheroides:* Energy transfer and structure. *Biochim. Biophys. Acta* **256,** 452.

271. Smiley, J. D., and Ashwell, G. (1961). Purification and properties of β-L-hydroxy acid dehydrogenase. II. Isolation of β-keto-L-gluconic acid, an intermediate in L-xylulose biosynthesis. *J. Biol. Chem.* **236,** 357.

272. Smith, A. J., and Lascelles, J. (1966). Thiosulfate metabolism and rhodanese in *Chromatium* sp. strain D. *J. Gen. Microbiol.* **42,** 357.

273. Sojka, G. A., and Gest, H. (1968). Integration of energy conversation and biosynthesis in the photosynthetic bacterium *Rhodopseudomonas capsulata. Proc. Nat. Acad. Sci. U.S.* **61,** 1468.

274. Sojka, G. A., Baccarini, A., and Gest, H. (1969). Energy flux and membrane synthesis in photosynthetic bacteria. *Science* **166,** 113.

275. Stanier, R. Y., Doudoroff, M., Kunisawa, R., and Contopoulou, R. (1959). The role of organic substrates in bacterial photosynthesis. *Proc. Nat. Acad. Sci. U.S.* **45,** 1246.

276. Stanier, R. Y. (1961). Photosynthetic mechanisms in bacteria and plants. Development of a unitary concept. *Bacteriol. Rev.* **25,** 1.

277. Stanier, R. Y. (1963). The organization of photosynthetic apparatus in purple bacteria. *In* "The General Physiology of Cell Specialization" (D. Mazie and A. Tyler, eds.), p. 242. McGraw-Hill, New York.

278. Steiner, S., Sojka, G. A., Conti, S. F., Gest, H., and Lester, R. L. (1970). Modification of membrane composition in growing photosynthetic bacteria. *Biochim. Biophys. Acta* **203,** 571.

279. Stern, H., and Nanney, D. L. (1965). "The Biology of Cells." Wiley, New York.

280. Stokes, J. E., Quadri, S. M. H., and Hoare, D. S. (1968). Formic hydrogenlyase and the photoautotrophic growth of *Rhodopseudomonas palustris. Bacteriol. Proc.* p. 115.

281. Stokes, J. E., and Hoare, D. S. (1969). Reductive pentose cycle and formate assimilation in *Rhodopseudomonas palustris. J. Bacteriol.* **100,** 890.

282. Stoppani, A. O. M., Fuller, R. C., and Calvin, M. (1955). Carbon dioxide fixation by *Rhodopseudomonas capsulata. J. Bacteriol.* **61,** 491.

283. Sybesma, C. (1967). Light-induced cytochrome reactions in the green photosynthetic bacterium *Chloropseudomonas ethylicum. Photochem. Photobiol.* **6,** 261.

284. Sybesma, C., and Beugeling, T. (1967). Light-induced absorbance changes in the green photosynthetic bacterium *Chloropseudomonas ethylicum. Biochim. Biophys. Acta* **131,** 357.

285. Sybesma, C., and Vredenberg, W. J. (1964). Kinetics of light-induced cytochrome oxidation and P_{840} bleaching in green photosynthetic bacteria under various conditions. *Biochim. Biophys. Acta* **88,** 205.

286. Tait, G. H. (1969). Coproporhyrinogenase activities in extracts from *Rhodopseudomonas spheroides. Biochem. Biophys. Res. Commun.* **37,** 116.

287. Tait, G. H. (1970). Glycine decarboxylase in *Rhodopseudomonas spheroides* and in rat liver mitochondria. *Biochem. J.* **118,** 819.

288. Takamiya, K., and Takamiya, A. (1970). Nature of the photochemical reaction in chromatophores of *Chromatium* D. I. Effects of isooctane extraction on the photochemical reactions of P_{890} and ubiquinone in chromatophores of *Chromatium* D. *Biochim. Biophys. Acta* **205,** 72.

289. Takamiya, K. (1971). The light-induced oxidation–reduction reactions of menaquinone in intact cells of a green photosynthetic bacterium, *Chloropseudomonas ethylicum. Biochim. Biophys. Acta* **234,** 390.

290. Tauschel, H. D., and Drews, G. (1967). Thykaloidmorphogenese bei *Rhodopseudomonas palustris. Arch. Mikrobiol.* **59,** 381.

291. Thiele, H. H. (1968). Sulfur metabolism in Thiorhodaceae. V. Enzymes of sulfur metabolism in *Thiocapsa floridana* and *Chromatium* species. *Antonie van Leeuwenhoek; J. Microbiol. Serol.* **34,** 350.
292. Thomas, J. B. (1965). "Primary Photoprocesses in Biology." North-Holland Publ., Amsterdam.
293. Thore, A., and Baltscheffsky, H. (1965). Inhibitory effect of lower aliphatic alcohols on electron transport phosphorylation systems. I. Straight chain primary alcohols. *Acta Chem. Scand.* **19,** 1591; *Chem. Abstr.* **64,** 5316 (1966).
294. Thore, A., Keister, D. L., and San Pietro, A. (1969). Studies of the respiratory system of aerobically (dark) and anaerobically (light) grown *Rhodospirillum rubrum. Arch. Mikrobiol.* **67,** 378.
295. Trentini, W. C., and Starr, M. P. (1967). Growth and ultrastructure of *Rhodomicrobium vannielii* as a function of light intensity. *J. Bacteriol.* **93,** 1699.
296. Trüper, H. G. (1964). CO_2-fixierung und Intermediärstoffwechsel bei *Chromatium okenii* Perty. *Arch. Mikrobiol.* **49,** 23.
297. Trüper, H. G., and Peck, H. D., Jr. (1970). Formation of adenyl sulfate in phototrophic bacteria. *Arch. Mikrobiol.* **73,** 125.
298. Trüper, H. G., and Rogers, L. A. (1971). Purification and properties of adenylylsulfate reductase from the phototrophic sulfur bacterium *Thiocapsa roseopericina. J. Bacteriol.* **108,** 1112.
299. Tsuboi, S., and Kikuchi, G. (1966). Regulation by illumination of the citric acid cycle activity in *Rhodopseudomonas spheroides. J. Biochem.* (*Tokyo*) **59,** 456.
300. Tsuiki, S., and Kikuchi, G. (1962). Catabolism of glycine by *Rhodopseudomonas spheroides. Biochim. Biophys. Acta* **64,** 514.
301. Tsuiki, S., Muto, A., and Kikuchi, G. (1963). A possible route of acetate oxidation in *Rhodopseudomonas spheroides. Biochim. Biophys. Acta* **69,** 181.
302. Tuttle, A. L., and Gest, H. (1959). Subcellular particulate systems and the photochemical apparatus of *Rhodospirillum rubrum. Proc. Nat. Acad. Sci. U.S.* **45,** 1261.
302a. Uffen, R. L., Sybesma, C., and Wolfe, R. S. (1971). Mutants of *Rhodospirillum rubrum* obtained after long-term anaerobic, dark growth. *J. Bacteriol.* **108,** 1348.
302b. Uffen, R. L. (1973). Growth properties of *Rhodospirillum rubrum* mutants and fermentation of pyruvate in anaerobic, dark conditions. *J. Bacteriol.* **116,** 874.
303. Van Niel, C. B. (1941). The bacterial photosynthesis and their importance for the general problem of photosynthesis. *Advan. Enzymol.* **1,** 263.
304. Van Niel, C. B. (1963). A brief survey of the photosynthetic bacteria. *In* "Bacterial Photosynthesis' (H. Gest, A. S. Pietro, and L. P. Vernon, eds.), p. 459. Antioch Press, Yellow Springs, Ohio.
305. Vernon, L. P., and Ash, O. K. (1959). The photoreduction of pyridine nucleotides by illuminated chromatophores of *Rhodospirillum rubrum* in the presence of succinate. *J. Biol. Chem.* **234,** 1878.
306. Vernon, L. P. (1959). Photooxidations catalyzed by chromatophores of *Rhodospirillum rubrum* under anaerobic conditions. *J. Biol. Chem.* **234,** 1883.
307. Vernon, L. P. (1963). Photooxidation and photoreduction reactions catalyzed by chromatophores of purple photosynthetic bacteria. *In* "Bacterial Photosynthesis" (H. Gest, A. S. Pietro, and L. P. Vernon, eds.), p. 235. Antioch Press, Yellow Springs, Ohio.
308. Vernon, L. P. (1964). Bacterial photosynthesis. *Annu. Rev. Plant Physiol.* **15,** 73.
309. Vernon, L. P., and Ke, B. (1966). Photochemistry of chlorophyll *in vivo*. *In* "The Chlorophylls" (L. P. Vernon and G. R. Seely, eds.), p. 569. Academic Press, New York.

310. Vernon, L. P. (1968). Photochemical and electron transport reactions of bacterial photosynthesis. *Bacteriol. Rev.* **32,** 243.
311. Vredenberg, W. J., and Duysens, L. N. M. (1963). Transfer of energy from bacteriochlorophyll to a reaction center during bacterial photosynthesis. *Nature* (*London*) **197,** 355.
312. Weaver, P., Tinker, K., and Valentine, R. C. (1965). Ferredoxin-linked DPN reduction by the photosynthetic bacteria *Chromatium* and *Chlorobium. Biochem. Biophys. Res. Commun.* **12,** 195.
313. Weinstein, B. (1969). An archetype correlation between bacterial rubredoxin and both bacterial and plant ferredoxin. *Biochem. Biophys. Res. Commun.* **35,** 109.
314. Wertlieb, D., and Vishniac, W. (1967). Methane utilization by a strain of *Rhodopseudomonas gelatinosa. J. Bacteriol.* **93,** 1722.
315. Whittenberger, C. C., and Repaske, R. (1961). Studies on hydrogen oxidation in cell-free extracts of *Hydrogenomonas eutropha. Biochim. Biophys. Acta* **47,** 542.
316. Winter, H. C., and Arnon, D. I. (1970). The nitrogen fixation system of photosynthetic bacteria. I. Preparation and properties of a cell-free extract from *Chromatium. Biochim. Biophys. Acta* **197,** 170.
316a. Witkin, S. S., and Gibson, K. D. (1972). Changes in ribonucleic acid turnover during aerobic and anaerobic growth in *Rhodopseudomonas spheroides. J. Bacteriol.* **110,** 677.
316b. Witkin, S. S., and Gibson, K. D. (1972). Ribonucleic acid from aerobically and anaerobically grown *Rhodopseudomonas spheroides:* Comparison by hybridization to chromosomal and satellite deoxyribonucleic acid. *J. Bacteriol.* **110,** 684.
317. Yamada, T., and Kikuchi, G. (1968). Inhibition of the metabolism of carboxylic acids by citramalate and other related compounds in *Rhodopseudomonas spheroides. J. Biochem* (*Tokyo*) **63,** 462.
318. Yamamoto, N., Hatakeyama, H., Nishikawa, K., and Horio, T. (1970). Function of ubiquinone-10 both in the electron transport system and in the energy conservation system of chromatophores from *Rhodospirillum rubrum. J. Biochem.* (*Tokyo*) **67,** 587.
319. Yamanaka, T., and Kamen, M. D. (1967). An NADP reductase and a nonheme iron protein isolated from a facultative photoheterotroph, *Rhodopseudomonas palustris. Biochim. Biophys. Acta* **131,** 317.
320. Yamanaka, T., and Okuniki, K. (1968). Comparison of *Chlorobium thiosulfatophilum* cytochrome c-555 with c-type cytochromes derived from algae and nonsulfur purple bacteria. *J. Biochem.* (*Tokyo*) **63,** 341.
321. Yamashita, J., Yoshimura, S., Matuo, Y., and Horio, T. (1967). Relationship between photosynthetic and oxidative phosphorylation in chromatophores from light-grown cells of *Rhodospirillum rubrum. Biochim. Biophys. Acta* **143,** 154.
322. Yamashita, J., and Kamen, M. D. (1969). Observations on distribution of NADH oxidase in particles from dark-grown and light-grown *Rhodospirillum rubrum. Biochem. Biophys. Res. Commun.* **34,** 418.
323. Yamashita, J., Kamen, M. D., and Horio, T. (1969). Effect of oligomycin on NADH oxidation and its coupled phosphorylation with the particulate fraction from dark aerobically grown *Rhodospirillum rubrum. Arch. Mikrobiol.* **66,** 304.
324. Yoch, D. C., and Lindstrom, E. S. (1969). Nicotinamide adenine dinucleotide-dependent formate dehydrogenase from *Rhodospirillum palustris. Arch. Mikrobiol.* **67,** 182.
325. Yoch, D. C., and Arnon, D. I. (1970). The nitrogen fixation system of photosynthetic bacteria. II. *Chromatium* nitrogenase activity linked to photochemically generated assimilatory power. *Biochim. Biophys. Acta* **197,** 180.

326. Yoch, D. C., and Lindstrom, E. S. (1971). Survey of the photosynthetic bacteria for rhodanese (thiosulfate:cyanide sulfur transferase) activity. *J Bacteriol.* **106,** 700.
327. Zilinsky, J. W., Sojka, G. A., and Gest, H. (1971). Energy charge regulation in photosynthetic bacteria. *Biochem. Biophys. Res. Commun.* **42,** 955.

Supplementary Readings

Gest, H. (1972). Energy conversion and generation of reducing power in bacterial photosynthesis. *Adv. Microbiol. Physiol.* **7,** 243.

Questions

1. Briefly define the terms
   (a) autotroph and heterotroph
   (b) oxybiontic and anoxybiontic respiration
   (c) fermentation
   (d) photolithotroph and photoorganotroph
   (e) chemolithotroph and chemoorganotroph
2. What is the difference between "aerobic" and "oxybiontic"?
3. Explain van Niel's equation for photosynthesis.
4. Discuss the basic principles of photosynthetic phosphorylation.
5. Explain the major differences between cyclic and noncyclic photophosphorylation and the need for both during photosynthesis.
6. What are the basic differences between plant and bacterial photosynthesis?
7. Name the three major families of photosynthetic bacteria together with at least two major genera in each. What characteristics are important for classification in the families and genera?
8. Draw and discuss the electron transport system of the Chlorobacteriaceae.
9. Draw and discuss the electron transport system of the Thiorhodaceae.
10. Draw and discuss the electron transport system of the Athiorhodaceae
11. Discuss briefly the major differences in the electron transport systems of the three major families of photosynthetic bacteria.
12. What role does oxygen play in photosynthetic phosphorylation of bacteria?
13. What is meant by the terms "light reaction" and "dark reaction" in photosynthesis?
14. Athiorhodaceae are able to grow either anaerobically in the light (photosynthesis) or aerobically in the dark (oxidative respiration). As the organisms must be able to switch from photophosphorylation to oxidative phosphorylation, discuss (a) the effect of light on oxidative phosphorylation, and (b) the effect of oxygen on photophosphorylation.
15. What other energy-conserving processes are known to occur in Athiorhodaceae apart from cyclic photophosphorylation?
16. What is meant by the term "primary photosynthetic reaction"?
17. Discuss the basic differences in the photochemical apparatus of the three families of photosynthetic bacteria.
18. Discuss the effect of light intensity on the photochemical apparatus and bacteriochlorophyll synthesis.

19. Discuss the effect of oxygen on bacteriochlorophyll synthesis.

20. *Chromatium* either produces hydrogen or uses hydrogen as the electron donor during photometabolism. Discuss both metabolic events with the help of the electron flow mechanisms.

21. Discuss the formation of poly-β-hydroxybutyric acid during acetate photometabolism in *Rhodospirillum rubrum*.

22. Describe the photometabolism of succinate by *Rhodospirillum rubrum*.

23. Discuss the importance of the anaerobic TCA cycle and reductive CO_2 fixation for photoorganotrophs.

24. Outline the relationship between N_2 assimilation and hydrogen evolution in photosynthetic bacteria.

25. Write an essay on whether or not the present system of classification of the photosynthetic bacteria into three major families is justified on metabolic grounds. Take into consideration not only the various aspects of photometabolism, but also photophosphorylation.

4

Anaerobic Respiration

It has been shown in the preceding chapter that photosynthetic bacteria obtain their energy from light, transfer it via cytochromes, and finally store it as ATP. Their reducing power comes from the substrate, however; i.e., the exogenous hydrogen donor. The majority of microorganisms, however, obtain their energy from chemical reactions and are called, accordingly, "chemosynthetic bacteria." We shall now deal with the various kinds of chemosynthesis.

On comparative biochemical grounds, the energy system of the chemosynthetic bacteria is very much like that of the photosynthetic bacteria and consists of a cytochrome system coupled to phosphorylation, which generates ATP. Light, as the source of the highly energized electrons, is replaced by a great number of chemical compounds, the utilization of which is extensively used for classification purposes. In chemolithotrophs, the whole system is driven by the oxidation of inorganic compounds, whereas in chemoorganotrophs this system is driven by the oxidation of organic compounds. As in photosynthesis, three general processes must be considered in the study of biological oxidations (51): (*1*) the dehydrogenation of a substrate, followed by the transfer of the hydrogen or electron to an ultimate acceptor; (*2*) conservation of the energy released in step (*1*); and (*3*) the subsequent metabolism of the dehydrogenated (oxidized) substrate.

In biological oxidations, the energy present in an organic substrate is released by successive dehydrogenations of the carbon chain. The reducing equivalents are thereby removed, normally two at a time, and transferred to a final acceptor, which may be oxygen in the case of aerobic respiration, inorganic compounds except oxygen in the case of anaerobic respiration, and organic compounds in the case of fermentation. This transfer proceeds via a potentially graded series of reversible oxidation–reduction systems known as electron transport systems, the basic scheme of which has been outlined in Chapter 1.

The first group of chemosynthetic bacteria includes all those bacteria that perform energy-yielding reactions in which inorganic compounds other than oxygen act as terminal acceptor. This process is generally referred to as "anaerobic" or "anoxybiontic respiration." As the name indicates, it is an anaerobic process, which means that all bacteria carrying out this type of respiration are either strictly or facultatively anaerobes.

Three main groups of bacteria are known to use an inorganic compound as terminal electron acceptor:

a. *Desulfovibrio* or *Desulfotomaculum*, using sulfate (SO_4^{2-}) as terminal electron acceptor.

b. The heterogenous group of denitrifying bacteria, using nitrate (NO_3^-) as terminal electron acceptor.

c. *Methanobacterium*, using carbon dioxide (CO_2) as terminal electron acceptor.

The first two groups of bacteria are very closely related in their electron transport systems but differ in that the sulfate-reducing bacteria do not possess NAD (arrows indicate the direction of electron flow):

$$2\,[\mathrm{H}] \rightarrow \text{cytochrome(s)} \rightarrow \begin{cases} \text{sulfate reductase} \rightarrow SO_4^{2-} \\ \text{thiosulfate reductase} \rightarrow S_2O_3^{2-} \end{cases}$$

$$\text{NAD} \rightarrow \text{FD} \rightarrow \text{cytochrome(s)} \rightarrow \begin{cases} \text{nitrate reductase} \rightarrow NO_3^- \\ \text{nitrite reductase} \rightarrow NO_2^- \\ \text{nitric oxide reductase} \rightarrow NO^- \end{cases}$$

This difference in the electron transport systems also reflects that between the strictly anaerobic sulfate reducers and the facultative anaerobic group of denitrifiers.

Although the nature of the reductases is not as yet precisely known (51), they represent the cytochrome oxidases of anaerobic respiration.

Sulfur Compounds as Electron Acceptors

The number of microorganisms utilizing sulfate as the terminal electron acceptor is very small. They are divided into the following groups (37):

1. Nonsporing sulfate-reducing bacteria
   a. *Desulfovibrio desulfuricans*

b. *Desulfovibrio gigas* (129)
c. *Desulfovibrio africanus* (38, 111a)
2. Spore-forming, sulfate-reducing bacteria (193)
a. *Desulfotomaculum nigrificans* [syn. *Clostridium nigrificans, Desulfovibrio thermodesulfuricans* (160), *Sporovibrio thermodesulfuricans*]
b. *Desulfotomaculum orientis* (syn. *Desulfovibrio orientis*)
c. *Desulfotomaculum ruminis* (Coleman's organism).

Most studies of the mechanism of sulfate reduction, however, have been carried out on the classic sulfate reducer *Desulfovibrio desulfuricans.*

Desulfovibrio desulfuricans is an obligately anaerobic organism that reduces sulfate in the presence of hydrogen as electron donor, carbon dioxide, and yeast extract (35), as well as thiosulfate, bisulfite or sulfate, and an organic hydrogen (electron) donor. The ability to use hydrogen as electron donor and the requirement for carbon dioxide has given the impression for a long time that these microorganisms are potentially capable of autotrophic growth. However, autotrophic cultivation resulted in extremely scanty growth. The addition of yeast extract greatly enhanced growth (35, 192), making it obvious that *Desulfovibrio desulfuricans* has a requirement for organic carbon (148). This requirement was found to be acetate (226), which *D. desulfuricans* required in addition to carbon dioxide for its biosynthesis of cell material. Therefore, the sulfate reducers are classified as facultative autotrophs (180) or chemolithotrophic heterotrophs (201). Whatever medium is used, the presence of sulfate is obligatory for growth, whether hydrogen or any organic substrate functions as hydrogen (electron) donor, except in the case of pyruvate (190).

Sulfate Reduction with Hydrogen as Electron Donor

Whole cells of *Desulfovibrio desulfuricans* reduce sulfate very rapidly in the presence of hydrogen

$$4\,H_2 + SO_4^{2-} \rightarrow S^{2-} + 4\,H_2O$$

Stoichiometric and thermodynamic studies with whole cells and cell-free extracts revealed that *Desulfovibrio desulfuricans* must reduce sulfate in a number of steps, using such organic sulfur compounds as adenosine 5′-phosphosulfate (APS) (Fig. 4.1) intermediates. The involvement of such a high-energy organic sulfur compound means that sulfate must require some sort of activation energy. This requirement was found to be satisfied by ATP (20). Although the activation energy required is very minute and at the catalytic level, it revealed a complex initial step in the sulfate

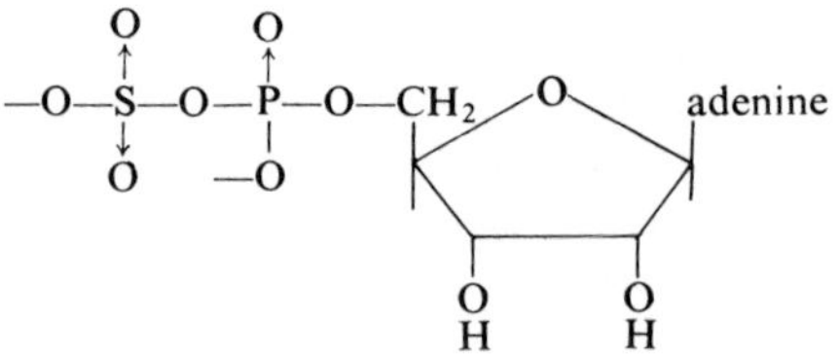
adenine

Fig. 4.1. Adenosine 5′-phosphosulfate (APS).

reduction of *Desulfovibrio desulfuricans.* Segal (217) indicated that pyrophosphate (PPi) is eliminated from ATP during this activation reaction and is rapidly converted to inorganic phosphate (Pi). This liberation of inorganic phosphate is very rapid (254). Thus, the initial step in sulfate

$$ATP + SO_4^{2-} \rightleftharpoons APS + PP_i \rightleftharpoons 2P_i$$

reduction (Fig. 4.2) first creates a high-energy compound in APS and, second, causes a cleavage of PPi from ATP. This cleavage ultimately causes the hydrolysis of two high-energy phosphate bonds of ATP (133), which evidently confers in a much larger thermodynamic pull (178) on such coupled reactions than is the case for reactions in which only the terminal phosphate group of ATP is lost. The split of pyrophosphate into two inorganic phosphate molecules in the presence of pyrophosphatase is therefore equivalent to another ATP → ADP step as far as energy is concerned. The enzyme catalyzing this reaction (i) from sulfate to APS (see Fig. 4.2) is sulfate adenylyltransferase (EC 2.7.7.4) (223c). The

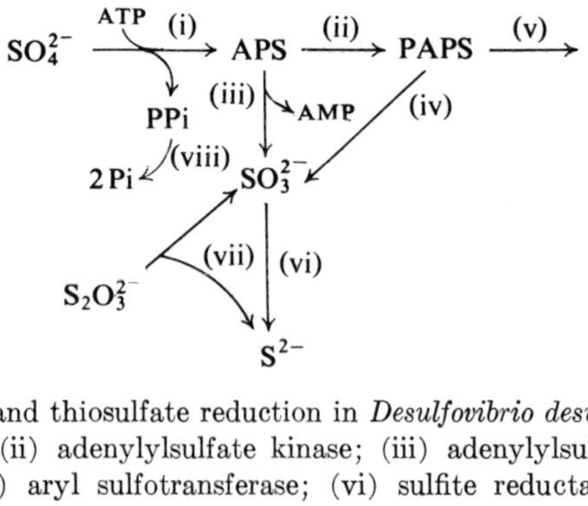
SO₄²⁻
ATP
(i)
APS
(ii)
PAPS
(v)
PPi
(iii)
AMP
(iv)
2 Pi
(viii)
SO₃²⁻
(vii)
(vi)
S₂O₃²⁻
S²⁻

Fig. 4.2. Sulfate and thiosulfate reduction in *Desulfovibrio desulfuricans.* (i) Sulfate adenylyltransferase; (ii) adenylylsulfate kinase; (iii) adenylylsulfate reductase; (iv) PAPS reductase; (v) aryl sulfotransferase; (vi) sulfite reductase; (vii) thiosulfate reductase; (viii) inorganic pyrophosphatase.

second step (ii) in sulfate reduction involves the oxidation of molecular hydrogen (174). A very active hydrogenase, which requires Fe^{2+} in most of the strains studied so far, led to the discovery of cytochrome c_3, which was in fact the first cytochrome discovered in a nonphotosynthetic, anaerobic microorganism. This cytochrome c_3 is a soluble, autoxidizable hemeprotein (191) that has reduction bands at 553, 525, and 419 nm. It also has a very low redox potential, −250 mV, and a high isoelectric point (IEP) at >pH 10; it contains 0.9% Fe. Cytochrome c_3 can act as a carrier in reactions that have oxygen as the terminal acceptor and it is therefore assumed that it provides a mechanism by which the organism could remove traces of oxygen from its environment. The E_m of cytochrome c_3 of −250 mV is related to the free energy of its oxidation at pH 7.0

$$Fe_2^{4+} + 2H^+ = Fe_2^{6+} + H_2 \qquad \Delta F = -9500 \text{ cal}$$

Cytochrome c_3 may also be responsible for the low oxidation–reduction potential required for the growth of *Desulfovibrio desulfuricans*. This organism therefore conducts its oxidation–reduction at the strongly reducing potential of about −200 mV. Cytochrome c_3, of course, mainly functions as an electron carrier in this second step

$$\text{APS} + \text{cytochrome } c_3 \text{ (red.)} \rightarrow \text{AMP} + SO_3^{2-} + \text{cytochrome } c_3 \text{ (oxid.)}$$

which is catalyzed by the enzyme adenylylsulfate reductase (EC 1.8.99.2). In the absence of an electron acceptor, such as cytochrome c_3, adenylylsulfate reductase catalysis involves three steps (149). The first step involves the reversible association of sulfite with enzyme-bound FAD to form a flavin–sulfite intermediate. The second step is a transfer of the sulfur moiety from FAD to a mononucleotide acceptor to yield APS, and the third step involves an internal reduction of an as yet unknown chromophore, yielding oxidized FAD.

The final reduction of sulfate to sulfide (step vi) therefore occurs in the presence of cytochrome c_3, hydrogenase, and sulfite reductase (EC 1.8.99.1). The hydrogenase (83, 132) is responsible for the reduction of the oxidized cytochrome c_3, whereas the sulfite reductase reduces sulfite to sulfide

$$H_2 + \text{cyt. } c_{3\,(\text{oxid})} \longrightarrow \text{cyt. } c_{3\,(\text{red})} + 2\,H^+$$

$$SO_3^{2-} + 6\,H^+ \longrightarrow S^{2-} + 3\,H_2O$$

If all the sequential steps of sulfate reduction with hydrogen as electron donor are combined, the following reactions can be given:

$$SO_4^{2-} + ATP \underset{}{\overset{\text{sulfate adenylyl transferase}}{\rightleftharpoons}} APS + PP_i$$

$$APS + \text{cyt. } c_{3\,(red)} + 2\,H^+ \xrightarrow[\text{reductase}]{\text{adenylylsulfate}} AMP + SO_3^{2-} + H_2O + \text{cyt. } c_{3\,(oxid)}$$

$$H_2 + \text{cyt. } c_{3\,(oxid)} \xrightarrow{\text{hydrogenase}} \text{cyt. } c_{3\,(red)} + 2\,H^+$$

$$SO_3^{2-} + 6\,H^+ \xrightarrow{\text{sulfite reductase}} S^{2-} + 3\,H_2O$$

This summary demonstrates not only that the metabolism of sulfate combines phosphorylation and electron transport via cytochrome c_3 (175) but also why it has been considered a generation of high-energy phosphate coupled with electron transport occurs in the oxidation of hydrogen with sulfate (174). A menaquinone-6 found in *Desulfovibrio vulgaris* and *D. gigas* (144, 249) may be involved as an additional oxidant–reductant in the electron transport pathway. The reduction of sulfate requires a source of ATP ($2P/SO_4^{2-}$) for the activation of sulfate (176) that neither of the substrates, sulfate or hydrogen, can provide by means of substrate phosphorylation. The only alternative source for the required ATP is an oxidative phosphorylation coupled with the oxidation of hydrogen. The sulfate-reducing bacterium *Desulfovibrio gigas* has been demonstrated to mediate exactly such a phosphorylation of ADP coupled to hydrogen oxidation, with sulfite or fumarate as electron acceptor (19, 179) whereby menaquinone-6 is involved in the electron transport between hydrogenase and fumarate reductase (85a). The recent demonstration of a dinitrophenol-stimulated ATPase in *Desulfovibrio gigas* (80) finally confirmed the existence of this oxidative phosphorylation in these strict nonphotosynthetic, anaerobic, sulfate reducers, as the association of ATPase activity with oxidative phosphorylation and photophosphorylation in bacterial extracts has been well documented (10, 28, 93, 95). The cytochrome c_3 involved in these reactions appears to vary slightly from species to species (56). It is different from cytochrome c-553 (32) and is probably not the only cytochrome in these sulfate reducing bacteria (131), although the functions of the additional c-type cytochromes found are unknown. It therefore appears that *Desulfovibrio* is able to carry out an oxidative phosphorylation equivalent to the one in aerobic bacteria, but with only a single cytochrome. This organism is therefore unique.

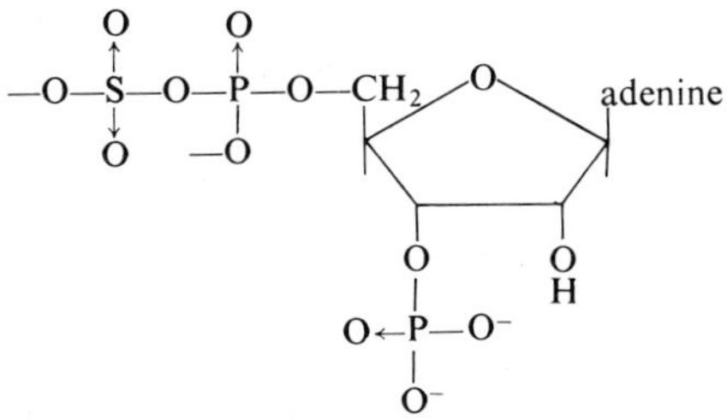

Fig. 4.3. 3'-Phosphoadenosine 5'-phosphosulfate (PAPS).

If a parallel is drawn to yeasts, which work on the same substrates, the difference becomes quite clear, for the pathway proceeds via 3'-phosphoadenosine 5'-phosphosulfate (PAPS) (Fig. 4.3). This yeast-type of sulfate reduction requires, in addition to the enzyme and ATP, reduced NADP or reduced lipoic acid as electron donor (89). The enzyme catalyzing the reaction is a kinase, adenylylsulfate kinase, which transfers a second phosphate group to APS (175). The final step to sulfite is a reduction step, with reduced NADP as electron donor and the enzyme PAPS-reductase. This

$$(NADPH + H^+) + PAPS \rightarrow NADP^+ + PAP + HSO_3^-$$

reduction of PAPS is thought to involve a thiolytic split by dihydrolipoic acid (88) via the following proposed mechanism (89):

$$PAP-O-SO_3^- + lip\genfrac{}{}{0pt}{}{SH}{SH} \rightleftharpoons PAP-OH + lip\genfrac{}{}{0pt}{}{S-SO_3^-}{SH}$$

$$lip\genfrac{}{}{0pt}{}{S-SO_3^-}{SH} \rightleftharpoons lip\genfrac{}{}{0pt}{}{SH}{SH} + HSO_3^-$$

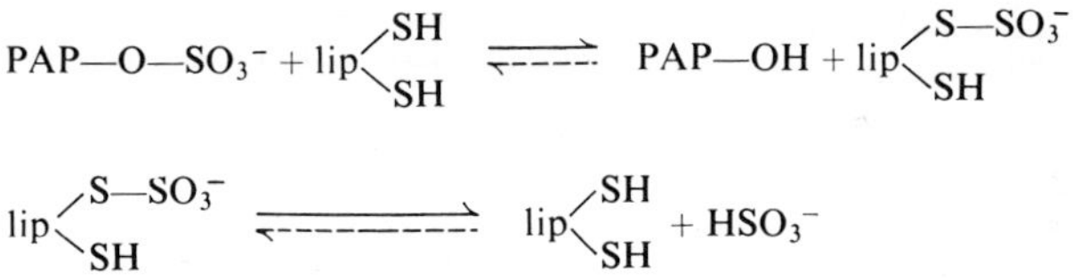

The postulated intermediate "lipothiosulfate," however, has not yet been isolated.

This reduction of PAPS is paralleled in the case of *Desulfovibrio desulfuricans* to a certain extent, for APS is reduced (99) with adenylylsulfate reductase (27, 178) as the catalyzing enzyme but there is no lipoic acid involvement. Recent chromatographic observations led to the discovery of yet another organic sulfur compound, which was tentatively named "adenosine 5'-phosphodithionate" (APSS) (26). It was detected when crude extracts of *Thiobacillus denitrificans* were incubated in the presence of AMP, sulfite, and ferricyanide. Autoradiography also showed the production of APSS from APS and sulfite in the presence and absence of ferricyanide (45).

The existence of these two different pathways of sulfate reduction (88, 255) appears to separate all those organisms that use sulfate reduction as dissimilatory pathway (192) from all those that use it as an assimilatory pathway (128b). The dissimilatory pathway is taken by those organisms which reduce sulfate in great excess of nutritional requirements and which therefore produce large amounts of sulfide by using sulfate as terminal electron acceptor. The amount of sulfide produced is proportional to the amount of H_2 or organic material dissimilated (231). Yeasts, *Escherichia coli* (40), and *Salmonella typhimurium* (55, 135), however, have the ability to grow on sulfate as their sole source of sulfur (216) and use the sulfite or sulfide produced for assimilation. The H_2S produced by these organisms is far in excess of the sulfate reduced and must therefore come from the metabolism of sulfur-containing amino acids (177). The formation of PAPS separates these two pathways clearly and distinctly.

Thiosulfate Reduction

Thiosulfate can be reduced in various ways, as is outlined in Fig. 4.4. *Desulfovibrio vulgaris, Desulfovibrio desulfuricans,* and *Desulfotomaculum* reduce thiosulfate via a dismutation reaction. Thiosulfate reductase, in conjunction with cytochrome c_3 (96–98), reduces thiosulfate to H_2S and sulfite (157); the outer sulfur atom of thiosulfate is reduced to H_2S and the inner sulfur atom accumulates as sulfite (84). The sulfite formed in this way can then be reduced to H_2S, as described above, with the help of sulfite reductase (EC 1.8.99.1) or can be recycled to thiosulfate (63) with the aid of a thiosulfate-forming system (235). The existence of this enzyme system allows thiosulfate to be an intermediate in sulfite reduction. This thiosulfate-forming system consists of two components, cytochrome c_3, ferredoxin, or flavodoxin (57), and hydrogenase. The dismutation reaction and the thiosulfate-forming system, not only make thiosulfate appear as

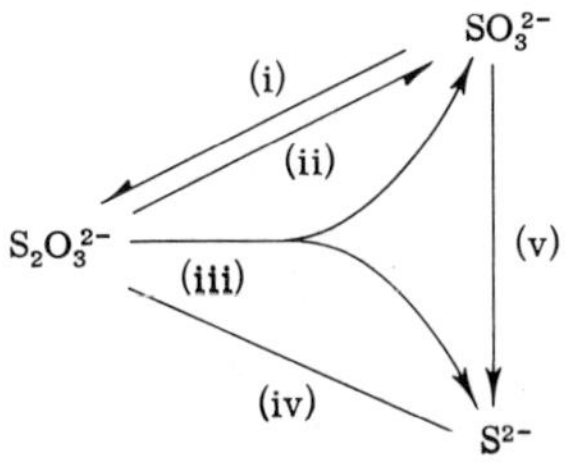

SO_3^{2-}
(i)
(ii)
$S_2O_3^{2-}$
(iii)
(v)
(iv)
S^{2-}

Fig. 4.4. Thiosulfate reduction and formation by *Desulfovibrio, Desulfotomaculum,* and *Thiobacillus denitrificans.* (i) Thiosulfate-forming system (bisulfite reductase?); (ii) thiosulfate sulfurtransferase; (iii) thiosulfate reductase; (iv) thiosulfate reduction system; (v) sulfite reductase.

an intermediate in the sulfite reduction, but also make it possible for the cell to control the intracellular level of sulfite. Sulfite itself inhibits thiosulfate reductase at relatively low levels and also inhibits sulfite reductase at higher concentrations. The characterization of the two components of the thiosulfate-forming system must be awaited. Using labeled sulfite as substrate, thiosulfate with equal distributions of radioactivity in both sulfur atoms was formed in *Desulfovibrio vulgaris* (64). The ionic species acting as the substrate for the thiosulfate formation was concluded to be bisulfite (235). Whether or not this enzyme is identical with the bisulfite reductase (128) that catalyzes the reductive formation of trithionate from sulfite is not established as yet. The bisulfite reductase forms trithionate as the sole product from bisulfite and contains the green pigment desulfoviridin in *Desulfovibrio gigas* (119a, 128, 239). In *Desulfovibrio vulgaris* the same enzyme was found to accumulate thiosulfate (119a). *Desulfotomaculum nigrificans* contains, instead of desulfoviridin, a carbon monoxide-binding pigment P582 that is claimed to reduce sulfite directly to sulfide (239a), although there appears to be a likelihood that this pigment may also be identical with the bisulfite reductase of this organism as trithionate and thiosulfate were found as intermediates (5a). A third pigment, desulforubidin, has been isolated from a *Desulfovibrio desulfuricans* strain, which lacked desulfoviridin, and has also been suggested to represent the bisulfite reductase (128a), as it reduces sulfite mainly to trithionate. The present methods available (223a,b,c) do not allow as yet to differentiate the direct conversion from sulfite to sulfide from those using trithionate and thiosulfate as intermediates.

A direct reduction of thiosulfate by molecular hydrogen was also observed with cell-free extracts of *Desulfovibrio gigas* (79). This system, similarly to the above-mentioned thiosulfate reductase, requires ferredoxin (270) or flavodoxin as electron carrier. It is not known whether or not the enzyme is a thiosulfate reductase similar or identical to the one described above. As molecular hydrogen is involved, it is almost certain that hydrogenase and cytochrome c_3 are part of the thiosulfate reduction system.

Anaerobically grown *Thiobacillus denitrificans* exhibits another route of thiosulfate reduction. An enzyme known as "rhodanese" or thiosulfate

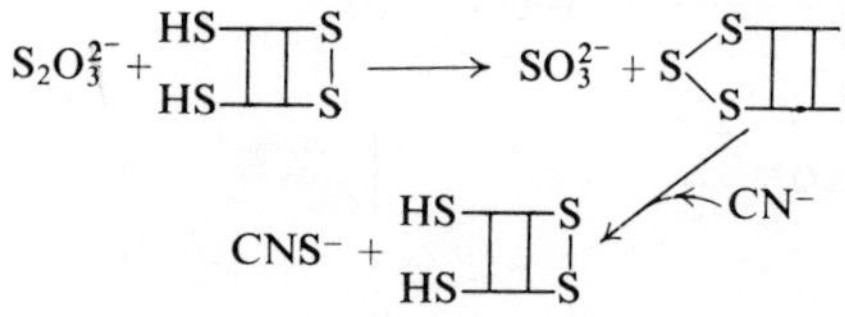

$S_2O_3^{2-}$ +
HS
HS
S
S
SO_3^{2-} + S
S
S
SH
SH
CN⁻
CNS⁻ +
HS
HS
S
S

Fig. 4.5. The reduction of thiosulfate by anaerobically grown *Thiobacillus denitrificans* (reprinted with permission of *The Biochemical Journal*, London).

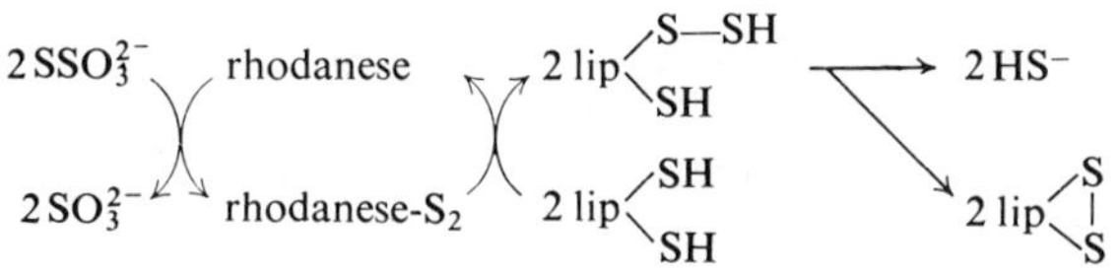
2SSO$_3^{2-}$
2SO$_3^{2-}$
rhodanese
rhodanese-S$_2$
2 lip S—SH SH
2 lip SH SH
2HS$^-$
2 lip S S

Fig. 4.6. The general mechanism of thiosulfate reduction by anaerobically grown *Thiobacillus denitrificans* (reprinted with permission of the American Society of Biological Chemists).

sulfurtransferase (thiosulfate:cyanide sulfurtransferase, EC 2.8.1.1) has been isolated (210) and found to reduce thiosulfate to sulfite (Fig. 4.5) in a stereospecific reaction (243), that is coupled with lipoate or lipoamide (242). This mechanism very much resembles that involving the corresponding mammalian enzyme. The bacterial rhodanese, however, has been regarded as a distinct enzyme in regard to its unique combination of properties (25) and may therefore occupy a transitional position among sulfurtransferases. Thiosulfate sulfurtransferase reacts with dihydrolipoate and sulfite is produced. The overall reaction mechanism can be summarized as is shown in Fig. 4.6. In this reaction sequence the outer sulfur atom of thiosulfate is transferred to dihydrolipoate (244) via an enzyme–sulfur intermediate. This intermediate transfers its sulfur to dihydrolipoate with the formation of lipoate persulfide, which in turn is highly active and breaks down immediately to HS^- and lipoate-S_2. Apart from being a double displacement mechanism (78, 122, 134, 250), the reaction is very similar to the one known for the oxidative decarboxylation of pyruvate and 2-oxoglutarate (see Chapter 7), with acetyl-CoA being substituted for rhodanese-S_2.

The discovery of thiosulfate sulfurtransferase in *Desulfotomaculum nigrificans* (34) as a separate protein from thiosulfate reductase may

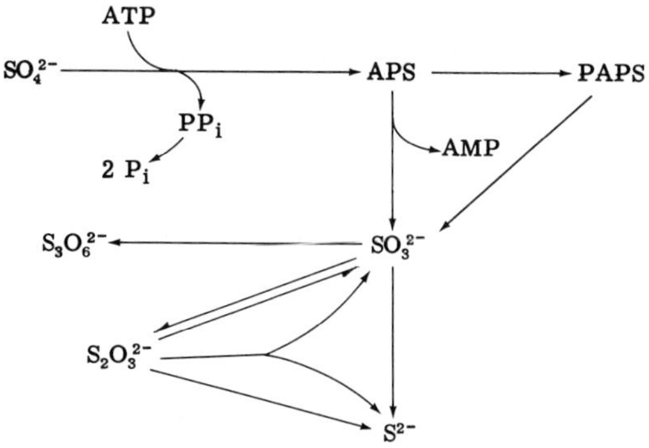
ATP
SO$_4^{2-}$
APS
PAPS
PP$_i$
2 P$_i$
AMP
S$_3$O$_6^{2-}$
SO$_3^{2-}$
S$_2$O$_3^{2-}$
S^{2-}

Fig. 4.7. Sulfate, thiosulfate, and sulfite reduction in the sulfate-reducing bacteria *Desulfovibrio*, *Desulfotomaculum*, and *Thiobacillus denitrificans* and yeast.

indicate that the sulfurtransferase-catalyzed reaction exists in addition to the thiosulfate reductase mechanism.

As molecular hydrogen is the electron donor in all these systems of sulfate and thiosulfate reduction, the electron transport system is the same as the one given for the sulfate reduction.

In summarizing the present knowledge of sulfate and thiosulfate reduction, the general scheme given in Fig. 4.1 could be expanded as shown in Fig. 4.7, indicating the most central position of sulfite.

Sulfate Reduction with Organic Compounds as Electron Donor

Hydrogen can be replaced by numerous organic compounds (193, 226), provided that inorganic sulfate and yeast extract or its equivalent are present in the medium.

GLUCOSE METABOLISM. Although the majority of the sulfate-reducing bacteria are unable to utilize glucose, *Desulfotomaculum nigrificans* (37), which is a thermophilic member of the sulfate-reducing bacteria (36), can be adapted to utilize glucose as an energy source (5). *Desulfotomaculum nigrificans* metabolizes glucose primarily through the Embden-Meyerhof pathway (see Chapter 5). It is one of the few species that utilize the Entner-Doudoroff pathway in addition to the Embden-Meyerhof pathway. As will be seen in Chapter 5 this pathway combination is quite unusual, although the Entner-Doudoroff pathway itself has been found operative in a number of anaerobic bacteria (49, 69, 225). *Desulfotomaculum nigrificans* metabolizes glucose to acetate, CO_2, and ethanol. The oxidation of pyruvate coupled to sulfite reduction (2) will be given below.

LACTATE AND PYRUVATE METABOLISM. The metabolism of glucose, lactate, and pyruvate produces acetate, carbon dioxide, and hydrogen sulfide (174) as end products. There is also one report by Postgate (126) claiming methane as a minor product of pyruvate metabolism. Tests in D_2O indicated that the methane came from the methyl carbon of pyruvate in some *Desulfovibrio* species and in *Desulfotomaculum ruminis.*

The generalized pathway in Fig. 4.8 should indicate roughly the interconnection between the pyruvate metabolism and sulfate reduction. The dissimilation of the organic compound produces the electrons required for the reduction of sulfate and also the energy required for growth and sulfate activation. The most important metabolic step is the utilization of pyruvate.

Pyruvate is decarboxylated to acetyl phosphate, carbon dioxide, and hydrogen in many anaerobes (151). The pyruvate–carbon dioxide exchange reaction in cell-free extracts of *Desulfovibrio desulfuricans* showed a definite

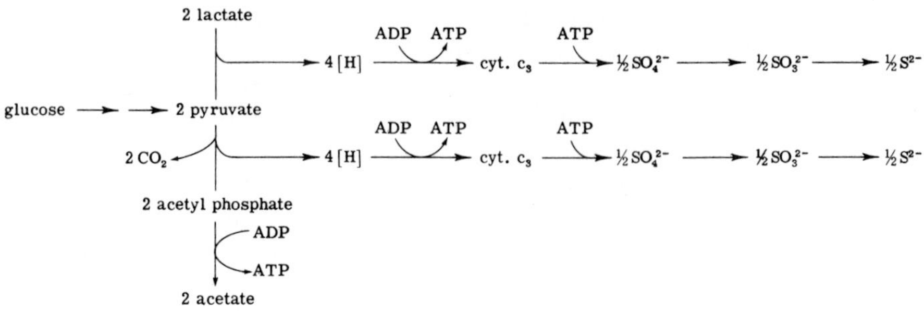
2 lactate
ADP ATP
ATP
4 [H]
cyt. c3
½ SO4 2-
½ SO3 2-
½ S2-
glucose
2 pyruvate
ADP ATP
ATP
2 CO2
4 [H]
cyt. c3
½ SO4 2-
½ SO3 2-
½ S2-
2 acetyl phosphate
ADP
ATP
2 acetate

Fig. 4.8. Lactate and pyruvate metabolism of the sulfate-reducing bacteria (reprinted with permission of the American Society of Microbiologists).

requirement for inorganic phosphate and coenzyme A (234). As this requirement, together with the products formed, is very similar to those in the clostridial type of phosphoroclastic reaction (see Chapter 8), it was assumed (151) that thiamine diphosphate is also required. The initial ATP expenditure in sulfate reduction is finally regenerated by the conversion of acetyl phosphate to acetate and ATP. The enzyme acetokinase involved in this reaction has been purified (31) and appears to be no different from the one in *Escherichia coli*, as it is not dependent on coenzyme A. This enzyme is specific for acetate and is inactive against formate, propionate, butyrate, and succinate. The detailed reaction sequence would therefore be

$$2CH_3CHOHCOOH \rightarrow 2CH_3COCOOH + 4H^+$$

$$2CH_3COCOOH + 2HOPO_3^{2-} \rightarrow 2CH_3COOPO_3 + 2CO_2 + 4H^+$$

$$2CH_3COOPO_3 + AMP + 2H^+ \rightarrow 2CH_3COOH + ATP$$

$$SO_4^{2-} + ATP + 8H^+ \rightarrow S^{2-} + 2H_2S + AMP + 2HOPO_3^{2-} + 2H^+$$

$$2CH_3CHOHCOOH + SO_4^{2-} \rightarrow 2CH_3COOH + 2CO_2 + S^{2-} + 2H_2S$$

Ferredoxin has been claimed necessary for the phosphoroclastic decarboxylation (4) as well as stimulation for sulfite reduction (2). The metabolism of lactate via pyruvate could therefore be as illustrated in Figure 4.9, which indicates that the hydrogen evolved would be used for the reduction of sulfate, as described in the previous section of this chapter.

This metabolic scheme also illustrates why ATP has a stimulatory effect on the pyruvate breakdown (268). This stimulation might be attributed to the formation of ADP, which by simple mass reaction accelerates the removal of acetyl phosphate and pulls the whole reaction sequence.

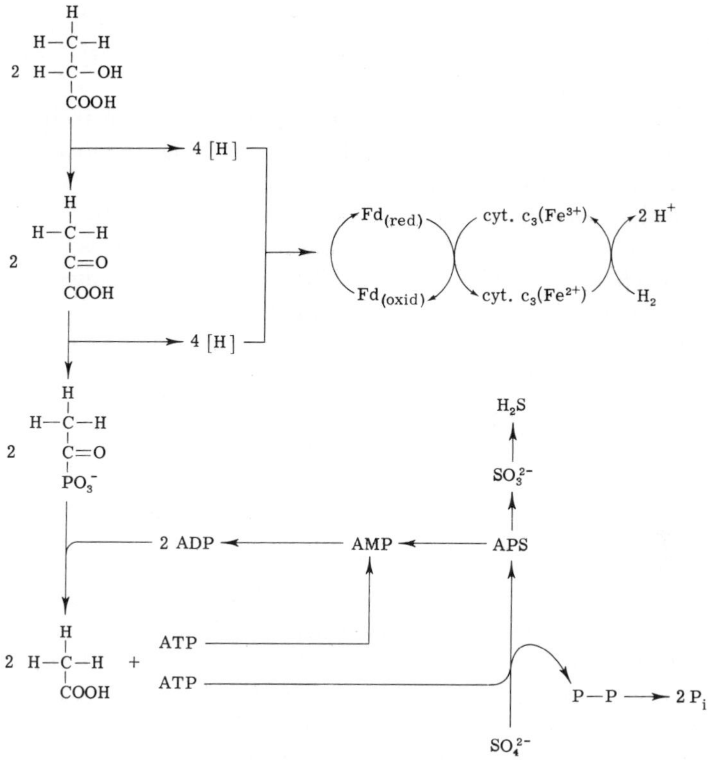
4 [H]
Fd(red)
cyt. c3(Fe3+)
2 H+
Fd(oxid)
cyt. c3(Fe2+)
H2
4 [H]
H2S
SO3 2-
2 ADP
AMP
APS
ATP
ATP
P—P
2 Pi
SO4 2-

Fig. 4.9. Metabolism of lactate and the reduction of sulfate in sulfate-reducing bacteria.

The overall ATP balance (245) is very similar to the one in sulfate reduction with hydrogen as electron donor. The four electron pairs originating from the oxidation of two molecules of acetyl phosphate are utilized for the conversion of one molecule of sulfate to sulfide. Because there would be no net production of ATP from substrate phosphorylation, the organism obtains its energy for growth from phosphate esterification (218) or oxidative phosphorylation coupled with the electron transport from lactate and pyruvate to sulfate.

If pyruvate replaces lactate as the sole energy source, *Desulfovibrio desulfuricans* can grow without sulfate but evolves hydrogen. The finding of ferredoxin (154) in *Desulfovibrio desulfuricans* (237) as well as cytochrome c_3 would explain the process of molecular hydrogen evolution (4). Electrons released from pyruvate are accepted by ferredoxin and transferred to cytochrome c_3. The latter donates the electrons finally to a hydrogenase (173). However, doubts have arisen about whether the same system of electron transport exists in this system as was outlined above. With lactate as carbon source, sulfate was absolutely necessary for growth and this carbon source mainly served as a hydrogen donor to the sulfate reduction system. Cytochrome c_3 and ferredoxin are known to be closely linked to the sulfite reductase system (130). In addition, *Desulfotomaculum* also evolves hydrogen without cytochrome c_3.

The question arose as to whether *Desulfovibrio* can substitute its sulfate reduction by utilizing the classical phosphoroclastic reaction as it occurs in *Escherichia coli* (76) (see Chapter 8). This mechanism produces formate in addition to acetyl phosphate, but formate has never been isolated as an intermediate in the pyruvate metabolism of *Desulfovibrio desulfuricans* (234). It is possible that the formate exchange proceeds via cytochrome c_3-mediated breakdown to H_2 and CO_2, the latter subsequently exchanging with pyruvate. The discovery of a membrane-bound formate dehydrogenase in *Desulfovibrio vulgaris* (265) strengthened the evidence for the existence of a classical phosphoroclastic reaction together with a formate–hydrogenlyase system. Although most of the known formate dehydrogenases are linked to NAD^+ (196), $NADP^+$ (138), ubiquinone, and ferredoxin (29), the formate dehydrogenase of *D. vulgaris* is linked to a c-type cytochrome, tentatively called "cytochrome c-553." The name "formate:ferricytochrome c-553 oxidoreductase" is being suggested. The observation, however, that the formation of hydrogen and acetyl phosphate is proportional to the ferredoxin concentration (2) leaves open speculation as to whether *D. desulfuricans* has a formate dehydrogenase linked to ferredoxin.

If *Desulfovibrio desulfuricans* utilizes pyruvate without the reduction of sulfate, this organism must then use additional oxidative phosphorylation steps to obtain its energy for growth. So far, the conversion of acetyl phosphate to acetate is the only known energy-producing step. This additional phosphorylation was found to be at the fumarase level. *Desulfovibrio gigas* possesses a fumarase and a malate enzyme (85). These enzymes catalyze the hydration of fumarate into malate and the oxidative decarboxylation of malate to pyruvate. The malate enzyme was specific for NADP and lacked oxalacetate decarboxylase activity. The existence of a malate dismutation of malate to succinate, fumarate, and acetate, as well as the presence of succinate dehydrogenase (150), indicates that a number of *Desulfovibrio* species are able to utilize part of the anaerobic TCA cycle. However, malate as the only carbon source again requires the presence of sulfate as terminal electron acceptor.

The present knowledge of their metabolism indicates that the sulfate-reducing bacteria are more versatile than originally thought. Sulfate is reduced by hydrogen, which can be replaced by a number of organic compounds. The latter produce enough ATP to stimulate sulfate reduction and to otherwise serve as hydrogen donors for sulfate reduction. The organisms, however, can also grow on pyruvate without sulfate, substituting the oxidative phosphorylation of the APS pathway, possibly with a reductive or anaerobic TCA cycle. The ability of sulfate reducers to oxidize pyruvate via the clostridial type of phosphoroclastic reaction in the case of sulfate reduction appears to be unique; without a sulfate requirement, the coli-type, or classical phosphoroclastic, reaction appears to be functional.

Apart from growth investigations (7, 92, 100, 143) studies on the deoxyribonucleic acid (DNA) base composition of *Desulfovibrio* (211) and *Desulfotomaculum* (212) showed the diversity of these sulfate-reducing bacteria. The thirty strains of *Desulfovibrio desulfuricans* fall into three groups, judged on their DNA base composition:

Group 1 contained 60–62% guanine + cytosine

Group 2 contained 54–56% guanine + cytosine

Group 3 contained 46–47% guanine + cytosine

The sulfate-reducing strains of *Desulfotomaculum* showed a base composition of 42–46% guanine + cytosine.

There is also some evidence (15) that *Desulfovibrio desulfuricans* is able to metabolize choline in the absence of sulfate. However, it is not as yet clear whether this organism is *Desulfovibrio desulfuricans* or *Vibrio cholinicus* (86).

Nitrate as Electron Acceptor

A small number of bacteria are able to reduce nitrate by using molecular hydrogen as the electron donor

$$5H_2 + 2NO_3^- \rightarrow N_2 + 4H_2O + 2OH^-$$

The end product of nitrate reduction is molecular nitrogen. Certain species of hydrogen bacteria are able to use this reaction as a source for their anaerobic growth in a strictly mineral medium. The great majority of nitrate-reducing bacteria, however, are chemoorganotrophs.

There are two types of microbial enzymes that reduce nitrate to nitrite: (*1*) the assimilatory enzymes, which contain flavin and molybdenum and usually utilize reduced NAD or reduced NADP as a hydrogen donor; and (*2*) the dissimilatory enzyme, which has an additional iron requirement. During assimilation, nitrate is reduced and incorporated into cellular material as cellular nitrogenous material, whereas in the dissimilatory process or nitrate respiration, nitrate serves as an alternative hydrogen acceptor to oxygen. Only the dissimilatory section or anaerobic respiration will be dealt with here.

Anaerobic respiration that results in the conversion of nitrate to molecular nitrogen, nitric oxide, or a mixture of these two gases is known as "denitrification" (203). Altogether, there are three microbiological reaction sequences in the dissimilatory nitrate reduction (6): (*1*) a complete reduction to ammonia, frequently with the transitory appearance of nitrite; (*2*) an incomplete reduction and an accumulation of nitrite in the medium; and (*3*) a reduction of nitrite followed by the evolution of gaseous compounds, or denitrification. Regardless of their other physiological characteristics, microorganisms that use nitrate as a nitrogen source carry out reaction (1), whereas cultures incapable of the complete reduction must be supplied with ammonia or other reduced nitrogen compounds for growth to proceed.

The suitability of nitrite for N_2 formation and its utilization in ammonia formation suggests that it is an intermediate in both cases. On the basis of current knowledge the following biochemical pathway is likely to operate:

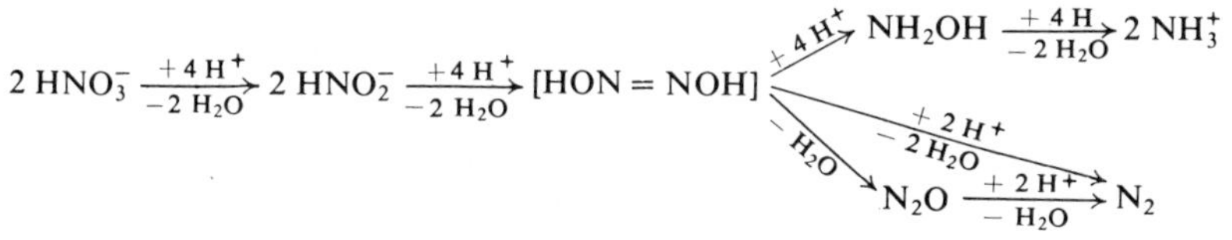
2 HNO3−
+4 H+
−2 H2O
2 HNO2−
+4 H+
−2 H2O
[HON = NOH]
+4 H+
NH2OH
+4 H
−2 H2O
2 NH3+
+2 H+
−2 H2O
− H2O
N2O
+2 H+
− H2O
N2

This pathway assumes that the reduction of nitrate to ammonia proceeds by a sequence of two-electron changes from the 5^+ oxidation state of nitrate to the 3^- of ammonia, a shift of eight electrons. The enzymes responsible for accepting these electrons are the nitrate, nitrite, hyponitrite, and hydroxylamine reductases. The pathway indicates that of all these reductases, nitrate reductase is the most common and important enzyme in nitrate reduction.

The electron transport system in denitrifying bacteria is very obscure. Over the past years, many suggestions have been put forward for different microorganisms (159); these will be dealt with according to the taxonomical groups below.

The first step in denitrification is the addition of two electrons to the nitrate with the formation of nitrite. All denitrifying bacteria can use nitrite in place of nitrate for denitrification. Molecular nitrogen is the usual major end product of denitrification, but under certain conditions, large amounts of N_2O can be formed from nitrite, which represents the addition of two further electrons per nitrogen atom

$$2NO_2^- + 4H^+ \rightarrow N_2O + H_2O + 2OH^-$$

Nitrous oxide, however, can also be reduced by denitrifying bacteria to molecular nitrogen

$$N_2O + 2H^+ \rightarrow N_2 + H_2O$$

It is therefore possible that denitrifying bacteria can form either N_2O or N_2 directly from nitrite via nitrous oxide. Because of this, it has been suggested that an intermediate of the elementary composition of $N_2O_2H_2$ may be involved (117)

$$N_2O_2H_2 \rightarrow N_2O + H_2O \quad \text{or} \quad N_2O_2H_2 + 2H^+ \rightarrow N_2 + H_2O$$

This intermediate can be formed relatively easily from nitrite via a free radical

$$NO_2^- + 2H^+ \rightarrow NOH + H_2O$$

$$2NOH \rightarrow N_2O_2H_2$$

Chemolithotrophic Reduction of Nitrate

One prominent representative for chemolithotrophic nitrate reduction is *Thiobacillus denitrificans*. Elemental sulfur or thiosulfate, both of which are converted to sulfate (see Chapter 6), are the energy sources for this organism. It is therefore a sulfur-oxidizing chemoautotroph (1):

$$5S + 6NO_3^- + 2H_2O \rightarrow 5SO_4^{2-} + 3N_2 + 4H^+$$

Chemoorganotrophic Reduction of Nitrate

Micrococcus denitrificans, a facultative autotroph, grows in air or anaerobically, with either organic compounds or molecular hydrogen as the source of energy, and with molecular oxygen or nitrate as electron acceptor (16). *Micrococcus denitrificans* can utilize a number of organic compounds to supply the reductant H_2 for nitrate reduction. Nitrate, in other words, functions as a terminal electron acceptor, replacing oxygen under aerobic conditions. The problem in such a replacement of the terminal acceptor is always to find the phosphorylation site, which gives the organism the required energy in the form of ATP for growth.

Before the problem of the electron transport system is dealt with, one

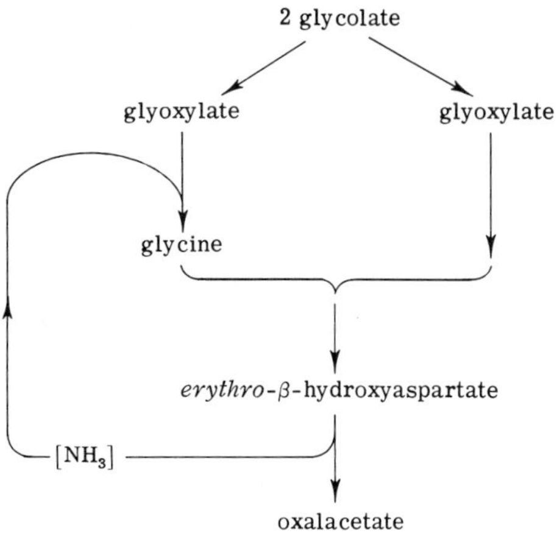
2 glycolate
glyoxylate
glyoxylate
glycine
erythro-β-hydroxyaspartate
[NH3]
oxalacetate

Fig. 4.10. Metabolism of glycolate by *Micrococcus denitrificans.*

degrading pathway of an organic substrate unique to *Micrococcus denitrificans* should be considered. In general, it can be said that the utilization of the carbon source by *Micrococcus* is very similar to that of the chemo-organotrophic sulfate reducer *Desulfotomaculum* (3). In addition to lactate (169) and pyruvate, *M. denitrificans* also utilizes glycolate, succinate, etc., as sole carbon source. With glycolate as substrate, this denitrifying bacterium is able to synthesize a C_4-dicarboxylic acid by the condensation of two C_2 compounds without a prior activation to CoA derivatives (121), yielding a precursor of the tricarboxylic acid cycle. This new pathway goes via *erythro*-β-hydroxyaspartate, which plays a key role in glycolate utilization in *Micrococcus denitrificans* (Fig. 4.10).

Glycolate is first converted to glyoxylate by the NAD-dependent glyoxylate reductase (glycolate:NAD oxidoreductase, EC 1.1.1.26). The

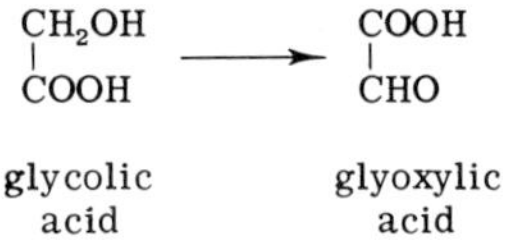
CH2OH
COOH
COOH
CHO
glycolic acid
glyoxylic acid

amination of glyoxylate in the following step to form glycine is not yet fully elucidated, but is thought that a specific aminotransferase might be in-

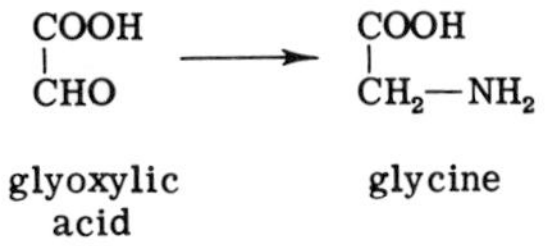
COOH
CHO
COOH
CH2—NH2
glyoxylic acid
glycine

volved (71). The succeeding cleavage reaction of glyoxylate and glycine,

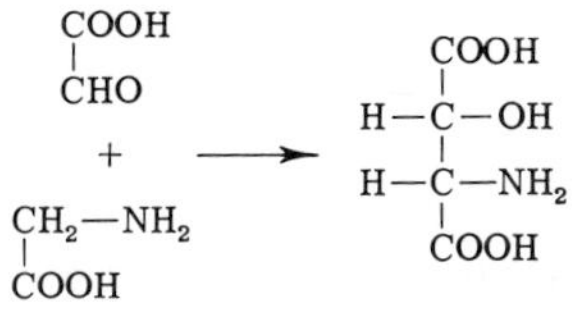

erythro-β-hydroxy-
aspartic acid

producing *erythro*-β-hydroxyaspartate, is very similar to the known aldolase cleavages (see Chapter 5) and the name "*erythro*-β-hydroxyaspartate aldolase" has been given to the enzyme catalyzing this reaction. The final

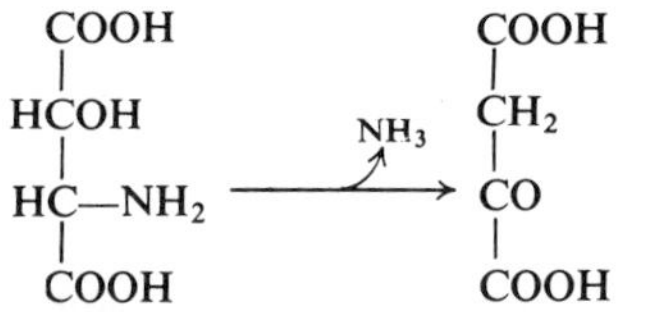

erythro-β-OH-aspartic acid   oxalacetic acid

step toward the tricarboxylic acid (TCA) cycle precursor, oxalacetate, is a dehydration reaction. The enzyme catalyzing this reaction has been named "*erythro*-β-hydroxyaspartate dehydratase" [*erythro*-3-hydroxyaspartate hydrolyase (deaminating)] (70); it has also been purified. The liberated ammonia is thought to reenter the metabolism in the conversion of glyoxylate to glycine.

The length of the electron transport and the sites of oxidative phosphorylation are still under dispute. It has been found that *M. denitrificans*, a facultative anaerobe, possesses cytochrome c, cytochrome $a + a_3$, and cytochrome o (214). However, under anaerobic conditions no phosphorylation takes place in the cytochrome c to oxygen region (11, 12, 93, 94), although it has been observed that phosphorylation occurs if the organism is grown on succinate as the electron donor and nitrate as the electron acceptor (155, 251). Phosphorylation occurred with NADH- or NAD-linked substrates by nitrate or nitrite. After having carefully broken cells with treatment of lysozymes, John and Whatley (111) were able to demonstrate a $P:NO_3$ ratio of 1.0 with $NADH + H^+$ and of 0.4 with succinate as the electron donor. They also obtained good evidence that the nitrate reductase interacted with the respiratory chain in the region of the b-type cytochrome and showed that the c-type cytochrome was not involved in the nitrate reduction. It therefore appears that the electron transport system would be

$$[H] \rightarrow NAD^+ \rightarrow FP \rightarrow cyt.b \rightarrow NO^{3-}$$

This preference of nitrate reductase for cytochrome b oxidation was confirmed by Lam and Nicholas (124). Bacteria grown anaerobically in media containing nitrate have a much higher content than those grown aerobically. The electron transport system of cells grown aerobically with molecular oxygen and grown anaerobically with nitrate as electron acceptors appear to have some electron acceptors in common

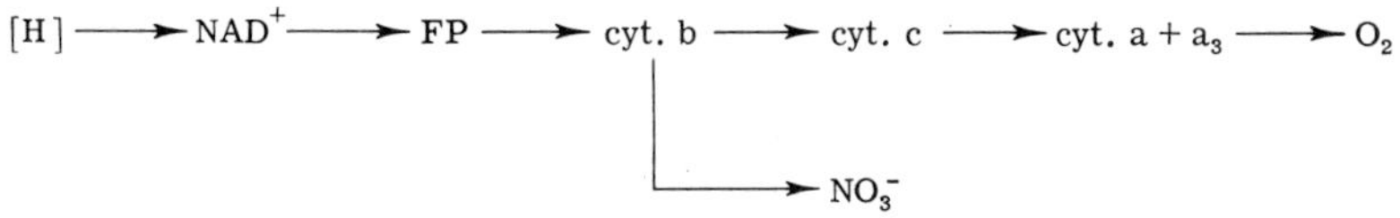
[H] → NAD+ → FP → cyt. b → cyt. c → cyt. a + a3 → O2
NO3-

Whether or not nitrate reductase operates at the cytochrome b level is still in dispute, however. The increase of cytochrome c under anaerobic conditions (214), as well as the finding that autotrophically grown *Micrococcus denitrificans* is capable of catalyzing phosphate esterification coupled to the oxidation of H_2, NADH + H^+, succinate, and ascorbate (119), indicates an entry of electrons at the level of cytochrome b and cytochrome c. Although all reports agree that the terminal phosphorylation site is lost as soon as *M. denitrificans* is grown anaerobically with nitrate, it is not known whether this site is cytochrome a + a_3 alone or cytochrome c plus the cytochrome oxidase. Evidence is accumulating in favor of cytochrome b, however, which would be in agreement with the observation that 3 moles of ATP are formed per atom of oxygen, but only 2 moles of ATP are formed per mole of nitrate (227).

The most common product of nitrate reduction is nitrite. Nitrite, however, is toxic to a number of bacteria and consequently a reduction of nitrate with the accumulation of nitrite is not a reaction that supports extensive growth (24). This almost proves that nitrite reductase is an inducible, not a constitutive, enzyme. Using a partially purified enzyme preparation of a halotolerant *Micrococcus* (*strain 203*), Asano (8, 9) showed that the electron transport system functioning in the reduction of nitrite was similar to that for oxygen. The *Micrococcus* spp. all converted nitrite to nitrogen with 1 mole of molecular nitrogen being formed for every 2 moles of nitrite reduced (Fig. 4.11). The preparation possessed a reduced NAD–nitrite reductase activity in the presence of 0.6 *M* NaCl and was stimulated by FAD and menadione. A cytochrome b_4 seemed to function as electron carrier. To date, five chromoproteins have been isolated: cyto-

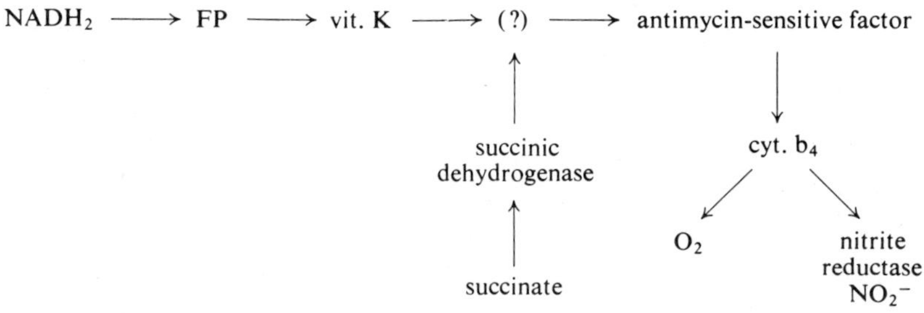
NADH2
FP
vit. K
(?)
antimycin-sensitive factor
succinic dehydrogenase
succinate
cyt. b4
O2
nitrite reductase
NO2−

Fig. 4.11. Proposed electron transport pathway in the reduction of nitrite by *Micrococcus* sp. (159) (reprinted with permission of the American Society of Microbiology).

chrome b_4(I), cytochrome b_4(II), cytochrome c-625,553 (HR), cytochrome c-551, and a "brown protein" (90, 91). Whereas cytochrome c-560, cytochrome b_4(I), and cytochrome b_4(II) can function in both anaerobic and aerobic cells, it is thought that cytochrome c-551 may function in a role similar to that of mammalian cytochrome c and that brown protein may function as a direct electron donor to the nitrate reductase (90, 91) (Fig. 4.12). Evidence suggests (161) that the brown protein is, in fact, a cytochrome pigment containing both heme c and an (a_2)-like heme and might function as a nitrate reductase.

No definite conclusion about phosphorylation coupled to electron trans-

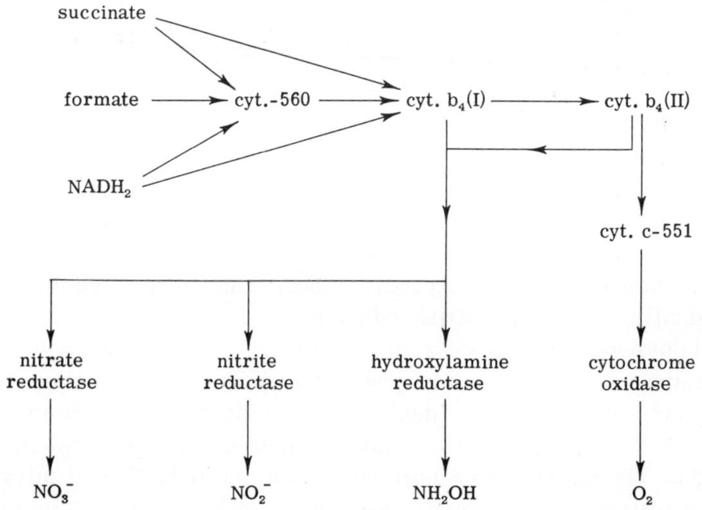
succinate
formate
cyt.-560
cyt. b4(I)
cyt. b4(II)
NADH2
cyt. c-551
nitrate reductase
nitrite reductase
hydroxylamine reductase
cytochrome oxidase
NO3−
NO2−
NH2OH
O2

Fig. 4.12. Electron transport sequence of anaerobic and aerobic respiration by *Pseudomonas aeruginosa* and *Micrococcus denitrificans* (91) (reprinted with permission of the Japanese Biochemical Society).

TABLE 4.1
Properties of Cytochrome b_4(I) and Cytochrome b_4(II)[a]

	Cytochrome b_4(I)	Cytochrome b_4(II)
$s_{20;20}$ (S)	2.20	2.17
$D_{20;10}$ (cm^2/second)	10.2×10^{-7}	—
Minimum molecular weight	9,300	14,800
Molecular weight	18,000	18,000
Electrophoretic mobility (cm^2/second/V)	-9.6×10^{-5}	-12.7×10^{-5}
$E_{0'9}$ (pH 7) (V)	+0.113	+0.180
Iron content (%)	0.61	0.38

[a] From (91). Reprinted with permission of the Japanese Biochemical Society.

port in these organisms can yet be reached. If it is assumed that the starting potential for nitrate reduction is near the NAD^+/NADH + H^+ system and that the potential for the NO^{3-}/NO^{2-} system is +0.5 V, then the potential change is reduced to approximately 0.9 V. The differences between the two cytochromes b_4 are given in Table 4.1.

Recent observations, however, appear to favor cytochrome c instead of cytochrome b_4. It was found that nitrite reductase consists of a two-heme cytochrome (162), containing a c-type and an a_2(d)-like heme. This two-heme cytochrome appears only in cells grown anaerobically with nitrate as electron acceptor, as it is induced not only by nitrate but also by nitrite. As even highly purified nitrite reductase still contains cytochrome c oxidase activity, it was suggested that cytochrome c and a_2 may be involved in the reduction process (125). Apart from the two-heme cytochrome nitrite reductase, a second nitrite reductase, with *p*-phenylenediamine as cofactor, has been found in *M. denitrificans* (186). No further details could be found, however, apart from its being observed only in denitrifying bacteria. These characteristics clearly indicate that nitrate reductase acts independently (206) from nitrite reductase.

The halotolerant *Micrococcus* sp. (strain 203) has in addition to the nitrate and nitrite reductases a very active hydroxylamine reductase, which is even active under strongly denitrifying conditions (120). Similarly to the nitrite reductase, the hydroxylamine reductase also reacts with molecular oxygen. The electron transport system for the reduction of nitrate to molecular nitrogen or ammonia is been shown in Fig. 4.13. The branching point of nitrate reductase is cytochrome b, whereas nitrite and hydroxylamine reductase require in addition cytochrome c. The other difference

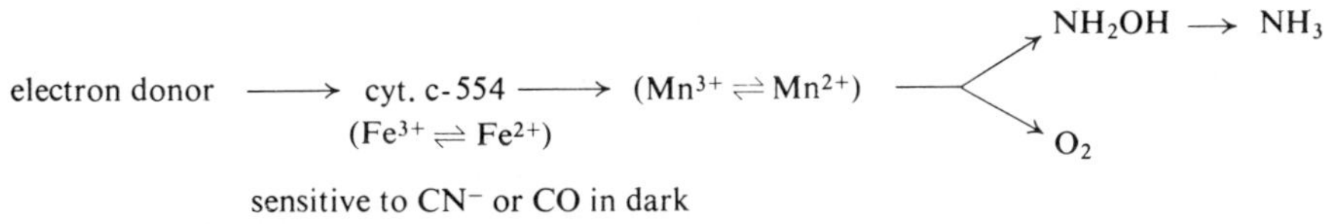
NH₂OH → NH₃
electron donor
cyt. c-554
(Fe³⁺ ⇌ Fe²⁺)
(Mn³⁺ ⇌ Mn²⁺)
O₂
sensitive to CN⁻ or CO in dark

Fig. 4.13. Electron transport of hydroxylamine reductase in *Micrococcus* sp. (strain 203) (159) (reprinted with permission of the American Society of Microbiology).

is undoubtedly the two-heme cytochrome nitrite reductase, which indicates the independence of each of the reductase from each other. A similar arrangement functions in *Pseudomonas aeruginosa* (246) and *Bacterium anitratum* (112).

PSEUDOMONAS DENITRIFICANS. Some genera from the *Pseudomonas* family are also able to reduce nitrate to molecular nitrogen under anaerobic conditions. *Pseudomonas denitrificans* is the main representative. The nitrate-reducing system, however, appears to be slightly different from that in *Micrococcus denitrificans,* as no gaseous products could be observed with *Pseudomonas denitrificans* (198). Nevertheless, a formation of ATP was coupled with nitrate respiration (167). The best electron donors appear to be reduced NAD and formate. With the latter compound as substrate, formate dehydrogenase increased ninefold (198). All observations of the electron transport chain suggest that the nitrate reductase derives its electrons from the cytochrome chain prior to cytochrome c. Therefore, it appears that the branching point for the nitrate reductase is at the same level as in *Micrococcus denitrificans.* The competition between nitrate and oxygen therefore seems to be caused by the withdrawal of electrons at the cytochrome b level for the reduction of nitrate.

The nitrite-reducing system in *Pseudomonas denitrificans* (102) seems to be completely different from that in *Micrococcus denitrificans,* because there was no phosphorylation observed with nitrite as terminal acceptor with *Pseudomonas denitrificans* (197). Iwasaki (101) purified two fractions

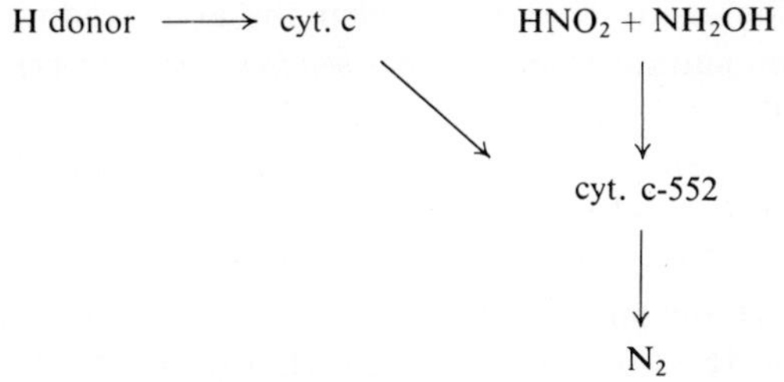
H donor → cyt. c
HNO₂ + NH₂OH
cyt. c-552
N₂

Fig. 4.14. Nitrite reductase system in *Pseudomonas denitrificans* (159) (reprinted with permission of the American Society of Microbiology).

from *Pseudomonas denitrificans*. One of these fractions had the typical absorption spectrum of a cytochrome c type. It was considered to be an electron carrier to the denitrifying or "cytochrome c" system. The latter also mediates the conversion of nitrite and hydroxylamine to molecular nitrogen (Fig. 4.14). Cytochrome c-553 was crystallized and found able to replace cytochrome c-552 in the denitrifying process (103). As flavin nucleotides stimulate the reaction (152), a flavoprotein must be involved between reduced NAD and cytochrome c. Whether or not the enzyme also reduces hydroxylamine is still under dispute (197). The alternative electron transport would be

$$\text{NADH} + \text{H}^+ \rightarrow \text{flavoprotein} \rightarrow \text{metal protein} \rightarrow \text{NO}_2^-$$

This latter suggestion is gaining support, as intact cells of *Pseudomonas denitrificans* were found to reduce NO to N_2O (153). The nitric oxide reductase appears to require no flavoprotein. The reduction of N_2O to N_2 (145) occurred only in the presence of lactate as the hydrogen donor.

The close relationship between cytochrome oxidase and nitrite oxidase led to the suggestion (267) that both are identical in *Pseudomonas*. However, there is a possibility that there are two types of cytochrome oxidases (13):

1. Ferrocytochrome c_2:oxygen oxidoreductase, EC 1.9.3.2 (syn. *Pseudomonas* cytochrome oxidase; cytochrome cd; cytochrome oxidase/nitrite reductase; *Pseudomonas* cytochrome c-551; nitrite, oxygen, oxidoreductase), which does not act on mammalian cytochrome c.
2. Ferrocytochrome c:oxygen oxidoreductase, EC 1.9.3.1 (syn. cytochrome a_3; cytochrome oxidase; *Pseudomonas* cytochrome c-551/O_2 oxidoreductase), which can reduce mammalian cytochrome c, but does not reduce nitrite.

A similar finding was also observed in *Micrococcus denitrificans* (241).

PSEUDOMONAS STUTZERI. The *Pseudomonas stutzeri* group (168) is another group of pseudomonads that have been found to reduce nitrite. Using crude ammonium sulfate fractions of *Pseudomonas stutzeri* extract, Najjar and Chung (156) showed that reduced NADP and reduced NAD served as electron donors and FAD and FMN stimulated nitrite reduction. As the nitrite reductase and the nitric oxide reductase exhibited almost identical properties and both led to the formation of molecular nitrogen, the electron transport scheme given in Fig. 4.15 could be constructed (159). The separation of three complex fractions that utilize $NADH_2$ as the source of electrons for the reduction of nitrite to nitric oxide, of nitric oxide to nitrous oxide, and of nitrous oxide to molecular nitrogen in *Pseudomonas*

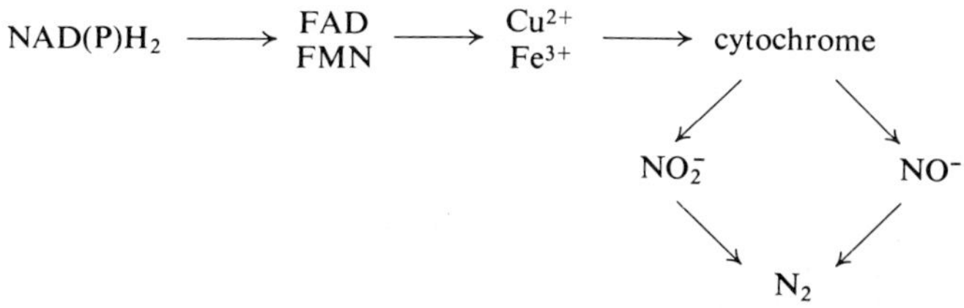
NAD(P)H2
FAD
FMN
Cu2+
Fe3+
cytochrome
NO2-
NO-
N2

Fig. 4.15. Electron transport systems of nitrite reduction by *Pseudomonas stutzeri* (reprinted with permission of the American Society of Microbiology).

perfectomarium (171) may lead to further knowledge in this area of electron transport systems.

The main differences between the electron transport systems for nitrate reductase of *Achromobacter* (Fig. 4.16) and *Pseudomonas aeruginosa* (Fig. 4.17) appear in the electron carrier immediately before the enzyme and in the requirement for molybdenum (164, 165, 208).

In the case of nitrite reductase, *Achromobacter fisheri* possessed a heme-containing protein with only a c-type spectrum (194), whereas *Pseudomonas aeruginosa* (62) has a cytochrome c and a cytochrome oxidase (266, 267). In contrast, *Neurospora* (158, 163), soybean leaves (158), *Escherichia coli* (113), and *Azotobacter agile* (228) are believed to possess metalloflavoproteins. *Achromobacter liquefaciens* reduces only nitrite, but not nitrate (223).

ENTEROBACTERIACEAE. The members of the family Enterobacteriaceae are facultative anaerobes and are therefore able to perform not only aerobic respiration and fermentation, but also anaerobic respiration if supplied with the corresponding substrate. Depending on the inorganic substrate supplied, they are able to form nitrate reductase (47, 181, 182, 185, 232), chlorate reductase, thiosulfate reductase, and tetrathionate reductase

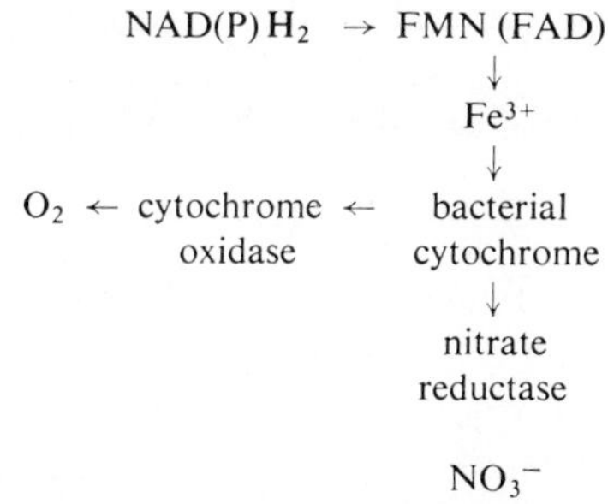
NAD(P)H2
FMN (FAD)
Fe3+
O2
cytochrome
oxidase
bacterial
cytochrome
nitrate
reductase
NO3-

Fig. 4.16. Electron transport system of nitrate reductase by *Achromobacter fisheri* (reprinted with permission of the authors and the Elsevier Publishing Co.).

$$NADH_2 \rightarrow FAD \rightarrow cyt.c \rightarrow Mo \rightarrow NO_3^-$$
$$\downarrow$$
$$\text{cytochrome oxidase} \rightarrow O_2$$

Fig. 4.17. Electron transport system of nitrate reductase by *Pseudomonas aeruginosa* (reprinted with permission of the authors and the Elsevier Publishing Co.).

(47). For nitrate reduction, formate, lactate, pyruvate, and reduced NAD are the most effective donors (43). The formation of these end products can only occur under fermentation or anaerobic conditions (see Chapter 8). Increasing concentrations of nitrate in the medium not only increased the reduced NAD(P):nitrate oxidoreductase (EC 1.6.6.2) activity but also the synthesis of cytochrome c-552 (42). However, this cytochrome appears not to function with nitrate reductase. The localization of the enzyme in the particulate fraction (220, 243a), as well as evidence for the existence of a redox-sensitive repressor that mediates nitrate reductase regulation, led to the thought that the anaerobic formic hydrogen-lyase system may be interconnected with the nitrate reductase system (166, 204, 233). Further evidence is the recent observation that molybdate and selenite significantly increased not only the formate hydrogen-lyase and nitrate reductase systems but also the aerobic formate oxidation (137). The presence of cytochrome b_1 in nitrate-induced cells then led to the electron transport chain shown in Scheme 4.1. It is not as yet clear whether the two cytochrome b-555 components are identical to the previously described cytochrome b_4 components of *Pseudomonas aeruginosa*. However, the transport sequence certainly is very similar. Ruiz-Herrera and co-workers (204, 205) suggest that the formate dehydrogenase–cytochrome b-555 (I and II) complex belongs to the formate oxidase system, which in connection with the additional complexing with nitrate reductase reduces nitrate to nitrite. The above-mentioned increase in cytochrome c may be caused by its requirement as an electron carrier in the formic hydrogen-lyase system. Molar growth yields almost doubled when nitrate was added to an anaerobic glucose–mineral medium, which indicates (82) that 3

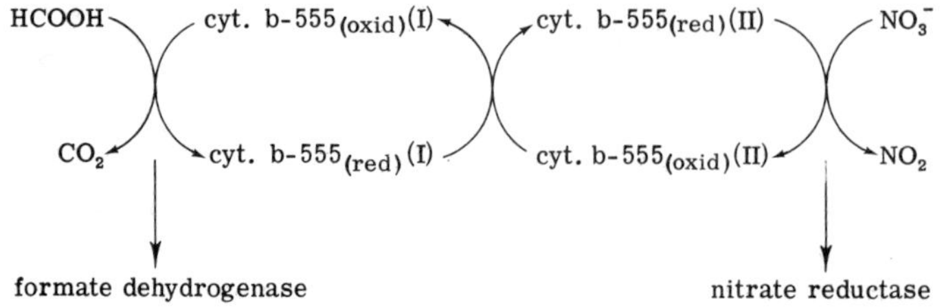
HCOOH
cyt. b-555(oxid)(I)
cyt. b-555(red)(II)
NO3-
CO2
cyt. b-555(red)(I)
cyt. b-555(oxid)(II)
NO2
formate dehydrogenase
nitrate reductase

Scheme 4.1. Nitrate reduction by *Pseudomonas aeruginosa* with formate as electron donor.

moles ATP were produced per mole nitrate reduced. The indications are that all the ATP produced comes from the nitrate to nitrite reduction step.

As described for *Micrococcus denitrificans,* the nitrite reductase in the Enterobacteriaceae is linked to the cytochrome c. The increase in cytochrome c-552 during the addition of nitrate was found to be necessary to induce the nitrite reductase. However, the nitrite reductase activity was not consistently proportional to the amount of cytochrome c-552 present in the cells (44), whereas cytochrome c reductase was proportional to nitrite reductase in most cases. It is therefore suggested that a second nitrite reductase may exist that would be cytochrome c-552 linked (66, 67). Some evidence has recently been given for the existence of a reduced NAD-linked (114) and a cytochrome c-552-linked nitrite reductase (248):

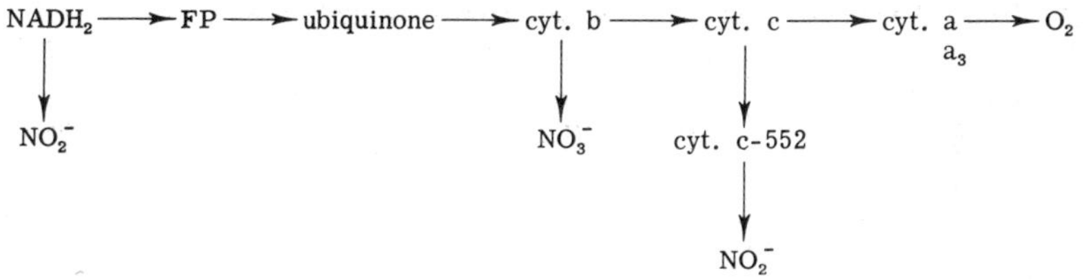
NADH2
FP
ubiquinone
cyt. b
cyt. c
cyt. a
a3
O2
NO2-
NO3-
cyt. c-552
NO2-

The involvement of the low-redox potential cytochrome c-552 in nitrite reduction tends to agree with the suggestion by Cole and Wimpenny (42, 43) that the redox potential of the medium rather than the concentration of a particular electron acceptor regulates the formation of many of the enzymes. This concept also takes into account the second hypothesis, made by Pichinoty (185), that a single metabolic redox couple (e.g., $NAD^+/NADH + H^+$), or a range of such metabolites, is functional as regulator molecules (45a, 48).

In *Bacillus licheniformis* (215) and *Bacillus stearothermophilus* (53, 54), nitrate reductase is also an inducible enzyme. The most striking feature of the regulation of nitrate reductase formation in *B. licheniformis* is the induction by anaerobiosis, even in the absence of nitrate. The intracellular redox potential is of prime importance. In *Bacillus stearothermophilus,* however, nitrate reductase is only formed in the presence of nitrate and low oxygen tension. Whether or not the increase in cytochrome c reflects an association of nitrate reductase with this cytochrome is not known as yet. Nitrate itself preferentially oxidizes cytochrome b of the membrane (115). In *Bacillus cereus* nitrate and nitrite reductase are simultaneously induced by nitrate, nitrite, and some other monovalent anions (81).

A very similar nitrate reductase induction pattern was found in *Spirillum itersonii.* As in *Bacillus licheniformis,* nitrate reductase was observed in cells grown with low aeration and without nitrate (68). The addition of nitrate increased the activity about twofold. The increase of cytochrome c was related to the nitrite reductase and an electron transport scheme similar to the one described for the Enterobacteriaceae may therefore exist.

Even *Corynebacterium nephridii* was reported to be able to reduce nitrate, nitrite, and nitric oxide to nitrous oxide (200). Two moles of nitrite were required for each mole of nitrous oxide produced. Molecular nitrogen was only produced by resting cells, with hydroxylamine as substrate and in the presence of nitrite.

Denitrification plays an important role in soil metabolism (6, 195). A high rate of denitrification occurs only when oxygen tension is very low. About 1% oxygen will suppress denitrification to about 12%. It is therefore suggested that the amount of oxygen present serves to control the denitrification rate. The first gaseous product is nitric oxide (approximately 5% of nitrate in soil), with nitrous oxide being next. Finally, molecular nitrogen appears and accounts for 83–95% of the nitrate present in soil. Quastel (195) states that only in the presence of appreciable concentrations of organic matter and low oxygen levels (7%) will an appreciable reduction of nitrate to nitrite occur.

The respiratory nitrate reductase is bound to subcellular particles that are themselves intimately involved in aerobic respiration. Nitrate respiration is competitively inhibited by molecular oxygen. This competition between nitrate and oxygen reduction in cells of *Pseudomonas denitrificans*, for example, led to the conclusion that nitrate reduction can only occur when the oxygen concentration is below the critical level at which the oxygen-utilizing enzymes are saturated (90). It appears that the competition between oxygen and nitrate for the donor electron favors the oxygen so much that nitrate reduction only occurs when the oxygen supply is completely inadequate to meet the demand (221, 222). However, there exists a direct relationship between nitrate reduction and oxygen solution rate. No reduction was detectable at an oxygen concentration above 0.2 ppm. Similar results were obtained with *Achromobacter* sp. (139). This competition seems to be restricted to the nitrate reductase. Nitrite reduction was observed with *Achromobacter liquefaciens*, which reduces nitrite but not nitrate (213) in solutions in which the oxygen concentration was in equilibrium with air. With *Pseudomonas stutzeri*, however, Sachs and Barker (207) recorded a high oxygen sensitivity of nitrite reduction.

Anaerobic respiration of nitrate and nitrite appears to be widespread among facultative anaerobic bacteria. Nitrate replaces oxygen as terminal electron acceptor and cytochrome a disappears as soon as oxygen tension drops to zero. Cytochrome b is the branch point for nitrate and cytochrome c for nitrite reductase. The regulation of reductase formation is therefore an oxygen effect. This is because oxygen itself is a corepressor (185), and because a low molecular weight metabolite, the presence of which depends on the redox state of the cell, is the corepressor, or because oxygen inhibits the removal of those enzymes from protein-synthesizing sites. The present hypothesis is that a combination of two factors are involved. The redox potential of the cell certainly plays an important role (256), probably together with a low molecular weight metabolite the presence of which depends on the redox state of the cell. This hypothesis would explain the formation of nitrate reductase under anaerobic conditions, even in the absence of nitrate. Studies on chlorate-resistant mutants have led to the suggestion (14) that nitrate reductase formation is controlled by a mechanism that also regulates the biosynthesis of membrane elements. This apparent oxygen toxicity in facultative anaerobic bacteria may represent a disturbance of metabolism, because it can be reversed nutritionally (72).

There is no further information available as to the effect of glucose on the formation of cytochromes (253).

The difficulties in obtaining a regulatory function may be seen in that there exist four different nitrate reductases (50):

1. Reduced-NAD:nitrate oxidoreductase, EC 1.6.6.1, which catalyzes the reaction:

$$NADH + H^+ + NO_3^- \rightleftharpoons NAD^+ + NO_2^- + H_2O$$

and requires metalloflavoprotein for its function.

2. Reduced-NAD(P):nitrate oxidoreductase, EC 1.6.6.2, which catalyzes the reaction:

$$NAD(P)H + H^+ + NO_3^- \rightleftharpoons NAD(P)^+ + NO_2^- + H_2O$$

and requires only a simple flavoprotein.

3. Reduced-NADP:nitrate oxidoreductase, EC 1.6.6.3, which catalyzes the reaction:

$$NADPH + H^+ + NO_3^- \rightleftharpoons NADP^+ + NO_2^- + H_2O$$

and requires a flavoprotein containing molybdenum.

4. Ferrocytochrome:nitrate oxidoreductase, EC 1.9.6.1, which catalyzes the reaction:

$$\text{Ferrocytochrome} + NO_3^- \rightleftharpoons \text{ferricytochrome} + NO_2^-$$

The existence of these four nitrate reductases gives some indication of the variation that can be expected in the electron transport systems of bacteria that use nitrate as the final hydrogen (electron) acceptor. Whether or not nitrate reductase is an adaptive or constitutive enzyme is still under dispute. Although it has been claimed that it is adaptive in *Escherichia coli* (141, 142), *Aerobacter aerogenes*, and *Bacillus stearothermophilus* (52), Nicholas and co-workers (165) claim it to be a constitutive enzyme for *Achromobacter fisheri* [*Photobacterium sepia* (sic)].

A major reason that different nitrate reductases exist may be that oxygen may not always influence nitrate reductase but instead influences the formation of cytochromes, which form the electron carrier system to nitrate reductase. Extracts of *Micrococcus denitrificans* showed a higher cytochrome content of b and c types if the organism was grown anaerobically on nitrate than those from aerobically grown cells (189). In contrast, some strains of *Staphylococcus epidermidis* (105) produced nitrite from nitrate when grown in air but failed to do so during growth under anaerobic conditions; although nitrate reductase was present, it remained unfunctional under anaerobic conditions. These observations clearly indicate that some electron carrier of the respiratory chain must be affected by the anaerobic conditions (41, 213, 224). Although it is not known what role oxygen plays in the development of the respiratory systems (107) in these facultative anaerobes, a parallel could be drawn to the yeast *Saccharomyces cerevisiae.* It has been shown quite clearly here that oxygen induces the formation of respiratory enzymes, cytochrome pigments, and even mitochondrial structures (188, 247). *Staphylococcus epidermidis* shows a decreased cytochrome content as a result of anaerobiosis; this was also found in *Pasteurella pestis* (59, 60), *Proteus vulgaris* (123), and *Staphylococcus aureus* (73, 126). This deficiency of nitrate reduction could be compensated for by the addition of hemin (104). Because nitrate reductase does not contain a heme component (39), the effect of hemin in nitrate reduction must result from its conversion into a cytochrome that mediates in electron transfer to nitrate. It can therefore be said that all those bacteria which reduce nitrate under aerobic conditions (assimilatory nitrate reduction) do possess the nitrate reductase and require oxygen not as terminal electron acceptor but for the biosynthesis of a heme-containing cytochrome that mediates in the electron carrier system to nitrate reduction. Whereas in other facultative anaerobes, such as *Saccharomyces cerevisiae, Pasteurella pestis,* and *Bacillus cereus,* oxygen is required for the synthesis of the complete respiratory system (61), the major role oxygen seems to play in *Staphylococcus epidermidis* and *Staphylococcus aureus* (236) is as a requirement for heme biosynthesis (106).

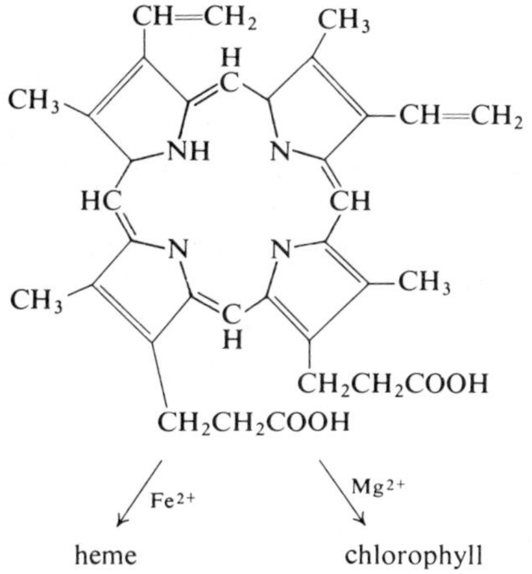
CH=CH2
CH3
H
C
CH3
CH=CH2
NH
N
HC
CH
N
N
CH3
CH3
C
H
CH2CH2COOH
CH2CH2COOH
Fe2+
Mg2+
heme
chlorophyll

Fig. 4.18. Protoporphyrin IX.

One possible explanation for this requirement is suggested by the finding in beef liver mitochondria that one of the oxidative steps in heme biosynthesis exhibits a requirement for molecular oxygen and no other electron acceptor will substitute (209). In the heme biosynthesis, protoporphyrin IX (Fig. 4.18) is the key porphyrin, from which two branches of the biosynthetic chain are postulated, an "iron branch" and a "magnesium branch." The former arises by the insertion of Fe^{2+} into the ring of protoporphyrin IX to form heme. This is the prosthetic group of hemoglobin, catalase, peroxidase, and some cytochromes. The "magnesium branch," however, leads toward chlorophyll. The details of heme biosynthesis will not be given here but can be followed up in a series of publications (18, 22, 23, 73–75, 116, 127, 136, 146, 147, 170, 209, 219, 252). The enzyme required for the molecular oxygen incorporation is not as yet known, although it could be an oxygenase (110). This, however, does not explain the fact that chlorophyll synthesis, as well as heme synthesis, occurs under anaerobic conditions.

It has been concluded that hemin combines with the protein component of cytochrome b to form cytochrome b_1 (39, 46), which participates in the electron transfer to nitrate, mediated by nitrate reductase.

Among the heterotrophic bacteria, *Escherichia coli* can synthesize cytochromes when grown either aerobically or anaerobically (77), which suggests that not all bacteria exhibit an oxygen requirement for heme biosynthesis (77). In these bacteria it seems likely that an alternative electron acceptor, which has not yet been identified, can substitute for oxygen in the conversion of coproporphyrinogen to protoporphyrin IX. However, certain other facultative anaerobes, in the genera *Staphylococcus* and *Bacillus*, exhibit a markedly diminished heme content and accumulate large quantities of coproporphyrin under anaerobic conditions (65, 87, 108, 109). So far only one report exists (58) which claims that soluble extracts from an aerobically grown species of *Pseudomonas* could convert coproporphyrinogen to protoporphyrin IX under both aerobic and anaerobic conditions. Every attempt to support this finding with *Pseudomonas denitrificans* and a number of other denitrifying bacteria failed (110), leaving the question open as to how the conversion could occur in the absence of molecular oxygen.

SUMMARY. According to our present knowledge, facultative anaerobes can be divided into two groups. The first group includes all those which use nitrate as terminal electron acceptor under aerobic and anaerobic conditions without heme requirements. Their cytochrome systems consist of metalloflavoproteins; e.g., *Escherichia coli* and *Pseudomonas*. The second group includes those facultative anaerobes that use only nitrate as their final electron acceptor but that require either oxygen or heme for their growth. The cytochrome system of this group contains heme as prosthetic group for at least one of their cytochromes; e.g., *Pasteurella pestis*, *Staphylococcus aureus*, *Staphylococcus epidermidis*, and *Hemophilus parainfluenzae*. This division into two groups could fit very well into the hypothesis that the redox potential is the prime effector for the formation of nitrate reductase (group 1?) together with a low molecular weight metabolite (group 2?). It could also explain the fact that only one nitrate reductase may be necessary for the assimilatory and respiratory function under the appropriate conditions (240).

Aerobacter aerogenes utilizes nitrate as a sole source of nitrogen for the synthesis of cell material (181), and also as hydrogen acceptor under anaerobic conditions (183), and is thought to have one nitrate reductase (185). This organism would therefore belong to the group that has metalloflavoprotein-containing cytochromes. This one nitrate reductase may be unique because it can use molecular hydrogen as an electron source for nitrate reduction to nitrite as well as to ammonia. The indication that *Micrococcus denitrificans* possesses two nitrate reductases (184) suggests that its cytochromes require heme when it is grown under anaerobic conditions. *Vibrio sputorum* (140) also seems to possess a nitrate reductase because it reduces nitrate beyond nitrite under anaerobic conditions. For detailed information the excellent review by Taniguchi (238) should be consulted.

Carbon Dioxide as Electron Acceptor

Carbon dioxide is used by a small group of bacteria as their final electron acceptor, the reduction product being methane. These are the "methane bacteria." These bacteria are entirely different from the aerobic methane-oxidizing bacteria, as will be seen in Chapter 7. The former produce methane from various organic and inorganic compounds, whereas the latter oxidize methane to carbon dioxide and water.

The methane bacteria have not been studied as extensively as most other groups of bacteria. The reason for this is most likely the extreme difficulty of isolating and maintaining pure cultures (172, 257). All of the species are strict anaerobes and therefore develop only in the absence of oxygen and in the presence of a suitable reducing agent. The methane bacteria are much more sensitive to oxygen or certain other oxidizing agents, such as nitrate, than are most other anaerobic bacteria. Their energy metabolism is specialized for a process that produces methane as the only major product. They also specialize with respect to the type of substrates utilized for energy and carbon sources. They have not been reported to utilize carbohydrates or amino acids. The substrates utilized by the methane bacteria fall into three categories:

1. The lower fatty acids, containing from one to six carbon atoms
2. The normal and isoalcohols, containing from one to five carbon atoms
3. The three inorganic gases hydrogen, carbon monoxide, and carbon dioxide

Each species therefore is restricted to the use of a few compounds.

The methane bacteria are represented by *Methanobacterium*, *Methanobacillus*, *Methanococcus*, and *Methanosarcina* (17, 264, 269).

Barker (17) reviewed the literature on methane production. Early theories have proposed that organic compounds metabolized by methane bacteria are oxidized completely to carbon dioxide and that this oxidation is coupled with a reduction of some or all of the carbon dioxide to methane:

$$\begin{array}{lrcl} \text{oxidation} & CH_3COOH + 2\,H_2O & \rightarrow & 2\,CO_2 + 8\,[H] \\ \text{reduction} & 8\,[H] + CO_2 & \rightarrow & CH_4 + 2\,H_2O \\ \hline \text{net} & CH_3COOH & \rightarrow & CH_4 + CO_2 \end{array}$$

Coupled reactions were cited for the following species.

1. *Methanobacterium omelianski*, which is now being regarded as a

symbiotic association of two species of bacteria (33). One of these is a gram-negative, motile, anaerobic rod that ferments ethanol with the production of hydrogen and acetate (199), whereas the second species is a gram-variable, nonmotile, anaerobic rod that utilizes H_2 but not ethanol for growth and methane formation:

(1)
$$2\,CH_3CH_2OH + 2\,H_2O \rightarrow 2\,CH_3COOH + 8\,[H]$$
$$8\,[H] + CO_2 \rightarrow CH_4 + 2\,H_2O$$
$$\overline{2\,CH_3CH_2OH + CO_2 \rightarrow 2\,CH_3COOH + CH_4}$$

(2)
$$4\,H_2 \rightarrow 8\,[H]$$
$$8\,[H] + CO_2 \rightarrow CH_4 + 2\,H_2O$$
$$\overline{4\,H_2 + CO_2 \rightarrow CH_4 + 2\,H_2O}$$

(3)
$$4\,CH_3OH + 4\,H_2O \rightarrow 4\,CO_2 + 24\,[H]$$
$$24\,[H] + 3\,CO_2 \rightarrow 3\,CH_4 + 6\,H_2O$$
$$\overline{4\,CH_3OH \rightarrow 3\,CH_4 + 2\,H_2O + CO_2}$$

The first reaction was quite distinct from the third reaction. The oxidation of ethanol was strictly CO_2 dependent. When the CO_2 was completely consumed, the oxidation of ethanol stopped (118). During the third reaction CO_2 was formed first and was then reduced to methane. It was therefore necessary not to supply the medium with CO_2 continuously.

2. *Methanobacterium formicicum*

$$CO + H_2O \rightarrow CO_2 + H_2$$
$$CO_2 + 4\,H_2 \rightarrow CH_4 + 2\,H_2O$$
$$\overline{CO + 3\,H_2 \rightarrow CH_4 + H_2O}$$

3. *Methanobacillus suboxydans*

$$2\,CH_3CH_2CH_2COOH + 4\,H_2O \rightarrow 4\,CH_3COOH + 8\,[H]$$
$$8\,[H] + CO_2 \rightarrow CH_4 + 2\,H_2O$$
$$\overline{2\,CH_3CH_2CH_2COOH + 2\,H_2O + CO_2 \rightarrow 4\,CH_3COOH + CH_4}$$

4. *Methanobacterium propionicum*

$$4\,CH_3CH_2COOH + 8\,H_2O \rightarrow 4\,CH_3COOH + 4\,CO_2 + 24\,[H]$$
$$3\,CO_2 + 24\,[H] \rightarrow 3\,CH_4 + 6\,H_2O$$
$$\overline{4\,CH_3CH_2COOH + 2\,H_2O \rightarrow 4\,CH_3COOH + CO_2 + 3\,CH_4}$$

In cases 1(a), 2, and 3, acetate was an additional end product. The substrates were not converted completely into methane as in the other coupled reactions. These are called the "fermentative type" because the substrate is not completely oxidized.

Tracer experiments on the metabolism of acetate and methanol by methane bacteria suggested that the carbon dioxide reduction theory was not tenable in at least two cases. Stadtman and Barker (229, 230) showed that the methane formed is entirely from the methyl carbon and the carbon dioxide is exclusively from the carboxyl carbon of acetate, and Pine and Barker (187) found similar results for methanol. These results suggested then that the methane formation from methanol and acetate was independent of carbon dioxide. To provide for the various modes of methane production, Barker (17) postulated possible pathways wherein XH represents an unknown carrier (see Fig. 4.19). The reduction steps require energy, which probably comes from ATP. The detailed mechanism of CO_2 reduction is therefore still obscure (258), but a hydrogen atmosphere was found essential for optimal methane formation. Cell-free extracts of *Methanobacterium omelianski* produced methane when H_2, CO_2, CoA, and

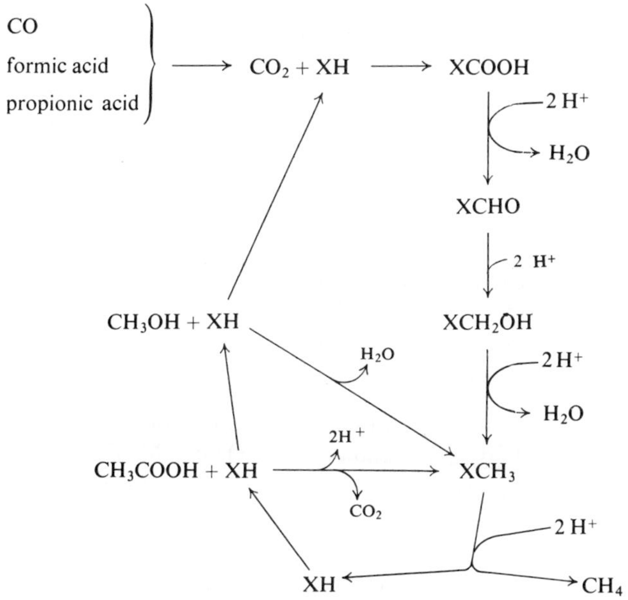
CO
formic acid
propionic acid
CO2 + XH
XCOOH
2 H+
H2O
XCHO
2 H+
XCH2OH
2 H+
H2O
CH3OH + XH
H2O
2H+
CH3COOH + XH
XCH3
CO2
2 H+
XH
CH4

Fig. 4.19. The formation of methane by the methane bacteria (17) (reprinted with permission of John Wiley and Sons).

ATP were in the reaction mixture. This suggests that the oxidation of hydrogen, coupled with the reduction of carbon dioxide to methane involves CoA and ATP (258). Pyruvate, serine, and *o*-phosphoserine could substitute for ATP and CO_2. The search for a methyl transfer enzyme in methane formation has led to comparisons with a cobamide enzyme that is involved in a similar step during methionine biosynthesis of *Escherichia coli* (30). It was found that methyl cobalamine was an excellent substrate for the methane-forming reaction (258, 259). This formation of methane from methyl cobalamine by extracts of *Methanobacterium omelianski* was ATP dependent; in extracts of *Methanosarcina barkeri* (21) it was also CoA dependent. With the purification of the methane-forming system of *M. omelianski,* it was possible to support the concept that a cobamide moiety of the methane-forming enzyme exists (263). The site of participation of ATP is still obscure, although it is known that 1 μmole of ATP is required for each micromole of methane evolved. The strong adenosine triphosphatase activity seems to indicate that the products ADP and AMP somehow regulate methane formation (262). This discovery was interesting insofar as the presence or absence of ATP regulates C_1 transfer in the methane formation from N^5,N^{10}-methylenetetrahydrofolate in extracts of *Methanobacterium omelianski* (260, 261).

Our present state of knowledge concerning the methane formation from carbon dioxide and such substrates as pyruvate or serine under hydrogen atmosphere could be summarized as follows:

a. That with some substrates (e.g., ethanol, butyrate, and hydrogen) methane is formed by CO_2 reduction if CO_2 is added to the atmosphere

b. That methane is formed by CO_2 reduction if CO_2 is formed during the oxidation of the substrate (e.g., CO and propionate)

c. That methane is formed via a methyl transfer using either methyl cobalamine or N^5,N^{10}-methylenetetrahydrofolate as cofactors, but without carbon dioxide (e.g., acetate and methanol)

In the last case there is a requirement for 1 μmole of ATP for each micromole of methane formed. As the depletion of ATP by a hexokinase trap did not stop methane synthesis (202), it is assumed that there may be a way of producing energy in form of an energy-rich compound different from ATP, i.e., $I \sim X$. The ATP produced makes the functioning of the reductive pentose phosphate cycle or reductive dicarboxylic acid cycle for CO_2 assimilation impossible. It is also possible that further ATP may be required to generate a reducing potential. This role of ATP and the presence of an active adenosine triphosphatase may lead to the assumption that ADP and AMP as products of this enzyme could regulate methane formation.

References

1. Adams, C. A., Warnes, G. M., and Nicholas, D. J. D. (1971). A sulfite-dependent nitrate reductase from *Thiobacillus denitrificans*. *Biochim. Biophys. Acta* **235,** 398.
2. Akagi, J. M. (1965). The participation of a ferredoxin of *Clostridium nigrificans* in sulfite reduction. *Biochem. Biophys. Res. Commun.* **21,** 72.
3. Akagi, J. M. (1965). Phosphoroclastic reaction of *Clostridium nigrificans*. *J. Bacteriol.* **88,** 813.
4. Akagi, J. M. (1967). Electron carriers for the phosphoroclastic reaction of *Desulfovibrio desulfuricans*. *J. Biol. Chem.* **242,** 2478.
5. Akagi, J. M., and Jackson, G. (1967). Degradation of glucose by proliferating cells of *Desulfotomaculum nigrificans*. *Appl. Microbiol.* **15,** 1427.

5a. Akagi, J. M., and Adams, V. (1973). Isolation of a bisulfite reductase from *Desulfotomaculum nigrificans* and its identification as the carbon monoxide-binding pigment P582. *J. Bacteriol.* **116,** 392.

6. Alexander, M. (1964). "Introduction to Soil Microbiology." Wiley, New York.
7. Alico, R. K., and Liegey, R. W. (1966). Growth of *Desulfovibrio desulfuricans* under heterotrophic and anaerobic conditions. *J. Bacteriol.* **91,** 1112.
8. Asano, A. (1959). Studies on enzymic nitrite reduction. I. Properties of the enzyme system involved in the process of nitrite reduction. *J. Biochem.* (*Tokyo*) **46,** 781.
9. Asano, A. (1959). Studies on enzymic nitrite reduction. II. Separation of nitrite reductase to particulate and soluble components. *J. Biochem.* (*Tokyo*) **46,** 1235.
10. Asano, A., and Brodie, A. F. (1963). Oxidative phosphorylation in fractionated bacterial systems. XI. Separation of soluble factors necessary for oxidative phosphorylation. *Biochem. Biophys. Res. Commun.* **13,** 416.
11. Asano, I., and Imai, K., and Sato, R. (1967). Oxidative phosphorylation in *Micrococcus denitrificans*. II. The properties of pyridine nucleotide transhydrogenase. *Biochim. Biophys. Acta* **143,** 477.
12. Asano, I., Imai, K., and Sato, R. (1967). Oxidative phosphorylation in *Micrococcus denitrificans*. III. ATP supported reduction of NAD^+ by succinate. *J. Biochem.* (*Tokyo*) **62,** 210.
13. Azoulay, E. (1964). Influence des conditions de culture sur la respiration de *Pseudomonas aeruginosa*. *Biochim. Biophys. Acta* **92,** 458.
14. Azoulay, E., Puig, J., and Martins Rosado de Sousa, M. L. (1969). Régulation de la nitrate-reductase chez *Escherichia coli* K 12. *Ann. Inst. Pasteur, Paris* **117,** 474.
15. Baker, F. D., Papiska, H. R., and Campbell, L. L. (1962). Choline fermentation by *Desulfovibrio desulfuricans*. *J. Bacteriol.* **84,** 973.
16. Bannerje, A. K. (1966). Physiologische Untersuchungen an *Micrococcus denitrificans* Beijerinck und auxotrophen Mutanten. Isolierung auxotropher Mutanten und Spaltung des Cystathionins. *Arch. Mikrobiol.* **53,** 107.
17. Barker, H. A. (1956). "Bacterial Fermentations," CIBA Lectures in Microbial Biochemistry. Wiley, New York.
18. Barrett, J. (1956). The prosthetic group of cytochrome a_2. *Biochem. J.* **64,** 626.
19. Barton, L. L., LeGall, J., and Peck, H. D., Jr. (1970). Phosphorylation coupled to oxidation of hydrogen with fumarate in extracts of the sulfate-reducing bacterium *Desulfovibrio gigas*. *Biochem. Biophys. Res. Commun.* **41,** 1036.
20. Bernstein, S., and McGilvery, R. W. (1952) Substrate activation in the synthesis of phenyl sulfate. *J. Biol. Chem.* **199,** 745.

21. Blaylock, B. A., and Stadtman, T. C. (1966). Methane biosynthesis by *Methanosarcina barkeri*. Properties of the soluble enzyme system. *Arch. Biochem. Biophys.* **116,** 138.
22. Bogorad, L. (1958). The enzymatic synthesis of porphyrins from porphobilinogen I, uroporphyrin I. *J. Biol. Chem.* **233,** 501.
23. Bogorad, L. (1958). The enzymatic synthesis of porphyrins from porphobilinogen II, uroporphyrin II. *J. Biol. Chem.* **233,** 510.
24. Bovell, C. (1967). The effect of sodium nitrite on the growth of *Micrococcus denitrificans*. *Arch. Mikrobiol.* **59,** 13.
25. Bowen, T. J., Butler, P. J., and Happold, F. C. (1965). Some properties of the rhodanese system of *Thiobacillus denitrificans*. *Biochem. J.* **97,** 651.
26. Bowen, T. J., and Cook, W. T. (1966). Detection of the polythionates on paper chromatograms. *J. Chromatogr.* **22,** 488 [as cited by Cook (45)].
27. Bowen, T. J., Happold, F. C., and Taylor, B. F. (1966). Studies on adenosine 5′-phosphosulfate reduction from *Thiobacillus denitrificans*. *Biochim. Biophys. Acta* **118,** 566.
28. Bragg, P. D., and Hou, C. (1968). Oxidative phosphorylation in *Escherichia coli*. *Can. J. Biochem.* **46,** 631.
29. Brill, W. J., Wolin, E. A., and Wolfe, R. S. (1964). Anaerobic formate oxidation: A ferredoxin-dependent reaction. *Science* **144,** 297.
30. Brot, N., and Weissbach, H. (1966). The role of cobamides in methionine synthesis. Enzymatic formation of holoenzyme. *J. Biol. Chem.* **241,** 2024.
31. Brown, M. S., and Akagi, J. M. (1966). Purification of acetokinase from *Desulfovibrio desulfuricans*. *J. Bacteriol.* **92,** 1273.
32. Bruschi, M., LeGall, J., and Dus, K. (1970). c-Type cytochromes of *Desulfovibrio vulgaris*. Amino acid composition and end groups of cytochrome c-553. *Biochem. Biophys. Res. Commun.* **38,** 607.
33. Bryant, M. P., Wolin, E. A., Wolin, M. J., and Wolfe, R. S. (1967). *Methanobacillus omelianski*, a symbiotic association of two species of bacteria. *Arch. Mikrobiol.* **59,** 20.
34. Burton, C. P., and Akagi, J. M. (1971). Observations on the rhodanese activity of *Desulfotomaculum nigrificans*. *J. Bacteriol.* **107,** 375.
35. Butlin, K. R., Adams, M. E., and Thomas, M. (1949). The isolation and cultivation of sulfate-reducing bacteria. *J. Gen. Microbiol.* **3,** 46.
36. Campbell, L. L., Jr., Frank, H. A., and Hall, E. R. (1957). Studies on thermophilic sulfate-reducing bacteria. I. Identification of *Sporovibrio desulfuricans* as *Clostridium nigrificans*. *J. Bacteriol.* **73,** 516.
37. Campbell, L. L., and Postgate, J. R. (1965). Classification of the spore-forming, sulfate-reducing bacteria. *Bacteriol. Rev.* **29,** 359.
38. Campbell, L. L., Kasprzycki, M. A., and Postgate, J. R. (1966). *Desulfovibrio africanus* sp.n., a new dissimilatory sulfate-reducing bacterium. *J. Bacteriol.* **92,** 1122.
39. Chang, J. P., and Lascelles, J. (1963). Nitrate reductase on cell-free extracts of a hemin-requiring strain of *Staphylococcus aureus*. *Biochem. J.* **89,** 503.
40. Chaste, J., and Pierfitte, M. (1965). Sulfur metabolism of *Escherichia coli*. *Bull. Soc. Pharm. Nancy* **64,** 13; *Chem. Abstr.* **63,** 13723 (1965).
41. Chin, C. H. (1950). Effect of aeration on the cytochrome systems of the resting cells of brewer's yeast. *Nature* (*London*) **165,** 926.
42. Cole, J. A., and Wimpenny, J. W. T. (1966). The interrelationships of low-redox

potential cytochrome c-552 and hydrogenase in facultative anaerobes. *Biochim. Biophys. Acta* **128,** 419.

43. Cole, J. A., and Wimpenny, J. W. T. (1968). Metabolic pathways for nitrate reduction in *Escherichia coli. Biochim. Biophys. Acta* **162,** 39.
44. Cole, J. A. (1968). Cytochrome c-552 and nitrite reduction in *Escherichia coli. Biochim. Biophys. Acta* **162,** 356.
45. Cook, W. K. T. (1967). Aspects of dithionate metabolism in *Thiobacillus denitrificans. Biochem. J.* **102,** 5P.

45a. Cornish-Bowden, A. J., Ward, F. B., and Cole, J. A. (1973). Product activation and 'redox control' of nitrite reductase activity in *Escherichia coli. J. Gen. Microbiol.* **75,** xi.

46. Deeb, S. S., and Hager, L. P. (1964). Crystalline cytochrome b_1 from *E. coli. J. Biol. Chem.* **239,** 1024.
47. DeGroot, G. N., and Stouthamer, A. H. (1969). Regulation of reductase formation in *Proteus mirabilis.* I. Formation of reductases and enzymes of the formic hydrogenlyase complex in the wild type and in chlorate-resistant mutants. *Arch. Mikrobiol.* **66,** 220.
48. DeGroot, G. N. (1970). Regulation of reductase formation in *Proteus mirabilis.* Ph.D. Thesis, Free University of Amsterdam.
49. DeMoss, R. D., Bard, R. C., and Gunsalus, I. C. (1951). The mechanism of the heterolactic fermentation: A new route of ethanol formation. *J. Bacteriol.* **62,** 499.
50. Dixon, M., and Webb, E. C. (1964). "Enzymes," 2nd ed. Longmans, Green, New York.
51. Dolin, M. I. (1961). Survey of microbial electron transport mechanism. *In* "The Bacteria" (I. C. Gunsalus and R. Y. Stanier, eds.), Vol. 2, p. 341. Academic Press, New York.
52. Downey, R. J. (1966). Nitrate reductase and respiratory adaption in *Bacillus stearothermophilus. J. Bacteriol.* **91,** 634.
53. Downey, R. J., and Nuner, J. N. (1968). Regulation of nitrate reductase in *Bacillus stearothermophilus. J. Cell Biol.* **39,** 37a.
54. Downey, R. J., and Kiszkiss, D. F. (1969). Oxygen- and nitrate-induced modification of the electron transfer system of *Bacillus stearothermophilus. Microbios* **1,** 145.
55. Dreyfuss, J., and Monty, K. J. (1963). Coincident repression of the reduction of 3′-phosphoadenosine 5′-phosphosulfate, sulfite, and thiosulfate in the cysteine pathway of *Salmonella typhimurium. J. Biol. Chem.* **238,** 3781.
56. Drucker, H., and Campbell, L. L. (1969). Electrophoretic and immunological differences between the cytochrome c_3 of *Desulfovibrio desulfuricans* and that of *Desulfovibrio vulgaris. J. Bacteriol.* **100,** 358.
57. Dubourdieu, M., and LeGall, J. (1970). Chemical study of two flavodoxins extracted from sulfate-reducing bacteria. *Biochem. Biophys. Res. Commun.* **38,** 965.
58. Ehfeshamuddin, A. F. M. (1968). Anaerobic formation of protoporphyrin IX from coproporphyrinogen III by bacterial preparations. *Biochem. J.* **107,** 446.
59. Englesberg, E., Gibor, A., and Levy, J. B. (1954). Adaptive control of terminal respiration in *Pasteurella pestis. J. Bacteriol.* **68,** 146.
60. Englesberg, E., Levy, J. B., and Gibor, A. (1954). Some enzymatic changes accompanying the shift from anaerobiosis to aerobiosis in *Pasteurella pestis. J. Bacteriol.* **68,** 178.
61. Ephrussi, B., and Slonimski, P. P. (1950). La synthèse adaptive des cytochrome chez la levure de boulangerie. *Biochim. Biophys. Acta* **6,** 256.

62. Fewson, C. A., and Nicholas, D. J. D. (1961). Nitrate reductase from *Pseudomonas aeruginosa*. *Biochim. Biophys. Acta* **49,** 335.
63. Findley, J. E., and Akagi, J. M. (1969). Evidence for thiosulfate formation during sulfite reduction by *Desulfovibrio vulgaris*. *Biochem. Biophys. Res. Commun.* **36,** 266.
64. Findley, J. E., and Akagi, J. M. (1970). Role of thiosulfate in bisulfite reduction as catalyzed by *Desulfovibrio vulgaris*. *J. Bacteriol.* **103,** 741.
65. Frerman, F. F., and White, D. C. (1967). Membrane lipid changes during formation of a functional electron transport system in *Staphylococcus aureus*. *J. Bacteriol.* **94,** 1868.
66. Fujita, T., and Sato, R. (1966). Studies on soluble cytochromes in Enterobacteriaceae. IV. Possible involvement of cytochrome c-552 in anaerobic nitrite metabolism. *J. Biochem.* (*Tokyo*) **60,** 691.
67. Fujita, T., and Sato, R. (1967). Studies on soluble cytochromes in Enterobacteriaceae. V. Nitrite-dependent gas evolution in cells containing cytochrome c-552. *J. Biochem.* (*Tokyo*) **62,** 230.
68. Gauthier, D. K., Clark-Walker, G. D., Garrard, W. T., Jr., and Lascelles, J. (1970). Nitrate reductase and soluble cytochrome c in *Spirillum itersonii*. *J. Bacteriol.* **102,** 797.
69. Gibbs, M., and DeMoss, R. D. (1954). Anaerobic dissimilation of ^{14}C-labeled glucose and fructose by *Pseudomonas lindneri*. *J. Biol. Chem.* **207,** 689.
70. Gibbs, R. G., and Morris, J. G. (1965). Purification and properties of erythro-β-hydroxyaspartate dehydratase from *Micrococcus denitrificans*. *Biochem. J.* **97,** 547.
71. Gibbs, R. G., and Morris, J. G. (1966). Formation of glycine from glyoxylate in *Micrococcus denitrificans*. *Biochem. J.* **99,** 27P.
72. Gottlieb, S. F. (1966). Bacterial nutritional approach to mechanisms of oxygen toxicity. *J. Bacteriol.* **92,** 1021.
73. Granick, S. (1958). Porphyrin biosynthesis in erythocytes. I. Formation of α-aminolevulinic acid in erythrocytes. *J. Biol. Chem.* **232,** 1101.
74. Granick, S., and Mauzerall, D. (1958). Porphyrin biosynthesis in erythrocytes. II. Enzymes converting α-aminolevulinic acid to coproporphyrinogen. *J. Biol. Chem.* **232,** 1119.
75. Granick, S., and Mauzerall, D. (1961). The metabolism of heme and chlorophyll. *Metab. Pathways* **3,** 525.
76. Gray, C. T., and Gest, H. (1965). Biological function of molecular hydrogen. *Science* **148,** 186.
77. Gray, C. T., Wimpenny, J. W. T., Hughes, D. E., and Mossman, M. R. (1966). Regulation of metabolism in facultative bacteria. I. Structural and functional changes in *Escherichia coli* associated with shifts between the aerobic and anaerobic state. *Biochim. Biophys. Acta* **117,** 22.
78. Green, J. R., and Westley, J. (1961). Mechanism of rhodanese action: Polarographic studies. *J. Biol. Chem.* **236,** 3047.
79. Guarraia, L. J., Laishley, E. J., Forget, N., and Peck, H. D., Jr. (1968). The role of ferredoxin and flavodoxin in the sulfur metabolism of *Desulfovibrio gigas*. *Bacteriol. Proc.* p. 133.
80. Guarraia, L. J., and Peck, H. D., Jr. (1971). Dinitrophenol-stimulated adenosine triphosphatase activity in extracts of *Desulfovibrio gigas*. *J. Bacteriol.* **106,** 890.
81. Hackenthal, E. (1966). Die parallele Induktion von Nitrat-reduktase bei *Bacillus cereus* durch verschiedene Anionen. *Biochem. Pharmacol.* **15,** 1119.

82. Hadjipetron, L. P., and Stouthamer, A. H. (1965). Energy production during nitrate respiration by *Aerobacter aerogenes*. *J. Gen. Microbiol.* **38,** 29.
83. Haschke, R. H., and Campbell, L. L. (1971). Purification and properties of a hydrogenase from *Desulfovibrio vulgaris*. *J. Bacteriol.* **105,** 249.
84. Haschke, R. H., and Campbell, L. L. (1971). Thiosulfate reductase of *Desulfovibrio vulgaris*. *J. Bacteriol.* **106,** 603.
85. Hatchikian, E. C., LeGall, J., LeGuen, M., and Forget, N. (1970). Metabolism of dicarboxylic acids and pyruvate in sulfate-reducing bacteria. I. Enzymic oxidation of fumarate to acetate. *Ann. Inst. Pasteur, Paris* **118,** 125.
85a. Hatchikian, E. C. (1974). On the role of menaquinone-6 in the electron transport of hydrogen: fumarate reductase system in the strict anaerobe *Desulfovibrio gigas*: *J. Gen. Microbiol.* **81,** 261.
86. Hayward, H. R., and Stadtman, T. C. (1959). Anaerobic degradation of choline. I. Fermentation of choline by an anaerobic, cytochrome-producing bacterium, *Vibrio cholinicus* n.sp. *J. Bacteriol.* **78,** 557.
87. Heady, R. E., Jacobs, N. J., and Deibel, R. H. (1964). Effect of hemin supplementation on porphyrin accumulation and catalase synthesis during anaerobic growth of *Staphylococcus*. *Nature (London)* **203,** 1285.
88. Hilz, H., Kittler, M., and Knape, G. (1959). Die Reduktion von Sulfat in der Hefe. *Biochem. Z.* **332,** 151.
89. Hilz, H., and Kittler, M. (1960). Reduction of active sulfate (PAPS) by dihydrolipoic acid. *Biochem. Biophys. Res. Commun.* **3,** 140.
90. Hori, K. (1961). Electron transport components participating in nitrate and oxygen respiration from a halo-tolerant *Micrococcus*. I. Purification and properties of cytochromes b_4(I) and b_4(II). *J. Biochem. (Tokyo)* **50,** 440.
91. Hori, K. (1961). Properties of cytochrome c_{551} and brown protein. *J. Biochem. (Tokyo)* **50,** 481.
92. Hutchinson, M., Johnstone, K. I., and White, D. (1967). Taxonomy of anaerobic thiobacilli. *J. Gen. Microbiol.* **47,** 17.
93. Imai, K., Asano, A., and Sato, R. (1967). Oxidative phosphorylation in *Micrococcus denitrificans*. I. Preparation and properties of phosphorylating membrane fragments. *Biochim. Biophys. Acta* **143,** 462.
94. Imai, K., Asano, A., and Sato, R. (1970). Oxidative phosphorylation in *Micrococcus denitrificans*. IV. Further characterization of electron-transfer pathway and phosphorylation activity in NADH oxidation. *J. Biochem. (Tokyo)* **63,** 207.
95. Ishikawa, S. (1970). Properties of an oxidative phosphorylation system reconstituted from coupling factors in *Micrococcus lysodeikticus*. *J. Biochem. (Tokyo)* **67,** 297.
96. Ishimoto, M., and Koyama, J. (1955). On the role of a cytochrome in thiosulfate reduction by a sulfate-reducing bacterium. *Bull. Chem. Soc. Jap.* **28,** 231.
97. Ishimoto, M., Koyama, J., and Nagal, Y. (1955). Biochemical studies on sulfate-reducing bacteria. IV. Reduction of thiosulfate by cell-free extracts. *J. Biochem. (Tokyo)* **42,** 41.
98. Ishimoto, M., and Koyama, J. (1957). Biochemical studies on sulfate-reducing bacteria. VI. Separation of hydrogenase and thiosulfate reductase and partial purification of cytochrome and green pigment. *J. Biochem. (Tokyo)* **44,** 233.
99. Ishimoto, M. (1959). Sulfate reduction in cell-free extracts of *Desulfovibrio*. *J. Biochem. (Tokyo)* **46,** 105.
100. Iverson, W. P. (1966). Growth of *Desulfovibrio* on the surface of agar media. *Appl. Microbiol.* **14,** 529.

101. Iwasaki, H. (1960). Studies on denitrification. IV. Participation of cytochromes in the denitrification. *J. Biochem. (Tokyo)* **47,** 174.
102. Iwasaki, H., Shidara, S., Suzuki, H., and Mori, T. (1963). Studies on denitrification. VII. Further purification and properties of denitrifying enzyme. *J. Biochem. (Tokyo)* **53,** 299.
103. Iwasaki, H., and Shidava, S. (1969). Crystallization of cytochrome c-553 in aerobically grown *Pseudomonas denitrificans. J. Biochem. (Tokyo)* **66,** 775.
104. Jacobs, N. J., and Deibel, R. H. (1963). Effect of anaerobic growth on nitrate reduction by staphylococci. *Bacteriol. Proc.* p. 124.
105. Jacobs, N. J., Hohantges, J., and Deibel, R. H. (1963). Effect of anaerobic growth on nitrate reduction by *Staphylococcus epidermidis. J. Bacteriol.* **85,** 782.
106. Jacobs, N. J., and Conti, S. (1965). Effect of hemin on the formation of the cytochrome system of anaerobically grown *Staphylococcus epidermis. J. Bacteriol.* **89,** 675.
107. Jacobs, N. J., Maclosky, E. R., and Conti, S. F. (1967). Effects of oxygen and heme on the development of a microbial respiratory system. *J. Bacteriol.* **93,** 278.
108. Jacobs, N. J., Maclosky, E. R., and Jacobs, J. M. (1967). Role of oxygen and heme in heme synthesis and the development of hemoprotein activity in anaerobically grown *Staphylococcus. Biochim. Biophys. Acta* **148,** 645.
109. Jacobs, N. J., Jacobs, J. M., and Sheng, G. S. (1969). Effect of oxygen on heme and porphyrin accumulation from delta aminolevulinic acid by suspensions of anaerobically grown *Staphylococcus epidermidis. J. Bacteriol.* **99,** 37.
110. Jacobs, J. N., Jacobs, J. M., and Brent, P. (1970). Formation of protoporphyrin from coproporphyrinogen in extracts of various bacteria. *J. Bacteriol.* **102,** 398.
111. John, P., and Whatley, F. R. (1970). Oxidative phosphorylation coupled to oxygen uptake and nitrate reduction in *Micrococcus denitrificans. Biochim. Biophys. Acta* **216,** 342.
111a. Jones, H. E. (1971). A re-examination of *Desulfovibrio africanus. Arch. Mikrobiol.* **80,** 78.
112. Jyssum, K., and Joner, P. E. (1966). Hydroxylamine as a possible intermediate in nitrate reduction by *Bacterium anitratum* (B5W). *Acta Pathol. Microbiol. Scand.* **67,** 139.
113. Kemp, J. D., Atkinson, D. E., Ehret, A., and Lazzarini, R. A. (1963). Evidence for the identity of the nicotinamide adenine dinucleotide phosphate-specific sulfite and nitrate reduction of *Escherichia coli. J. Biol. Chem.* **238,** 3466.
114. Kemp, J. D., and Atkinson, D. E. (1966). Nitrite reductase of *Escherichia coli* specific for $NADH_2$. *J. Bacteriol.* **92,** 628.
115. Kiczkiss, D. F., and Downey, R. J. (1972). Localization and solubilization of the respiratory nitrate reductase of *Bacillus stearothermophilus. J. Bacteriol.* **109,** 803.
116. Kikuchi, G., Kumar, A., Talmage, P., and Shemin, D. (1958). The enzymatic synthesis of α-aminolevulinic acid. *J. Biol. Chem.* **233,** 1214.
117. Kluyver, A. J., and Verhoeven, W. (1954). Studies on two dissimilatory nitrate reductions. II. The mechanism of denitrification. *Antonie van Leeuwenhoek; J. Microbiol. Serol.* **20,** 241.
118. Knight, M., Wolfe, R. S., and Elsden, S. R. (1966). The synthesis of amino acids by *Methanobacterium omelianski. Biochem. J.* **99,** 76.
119. Knobloch, K., Ishaque, M., and Aleem, M. I. H. (1971). Oxidative phosphorylation in *Micrococcus denitrificans* under autotrophic growth conditions. *Arch. Mikrobiol.* **76,** 114.

119a. Kobayashi, K., Takashashi, E., and Ishimoto, M. (1972). Biochemical studies on sulfate-reducing bacteria xi. Purification and some properties of sulfite reductase, desulfoviridin. *J. Biochem. (Tokyo)* **72,** 879.

120. Kono, M., and Taniguchi, S. (1960). Hydroxylamine reductase of a halotolerant *Micrococcus. Biochim. Biophys. Acta* **43,** 419.

121. Kornberg, H. L., and Morris, J. G. (1965). The utilization of glycolate by *Micrococcus denitrificans:* β-Hydroxyaspartate pathway. *Biochem. J.* **95,** 577.

122. Koshland, D. E., Jr. (1954). *In* "The Mechanism of Enzyme Action" (W. D. McElroy and B. Glass, eds.), p. 608. Johns Hopkins Press, Baltimore, Maryland [as cited by Green and Westley (78)].

123. Krasna, A. I., and Rittenberg, D. (1954). Reduction of nitrate with molecular hydrogen by *Proteus vulgaris. J. Bacteriol.* **68,** 53.

124. Lam, Y., and Nicholas, D. J. D. (1969). Aerobic and anaerobic respiration in *Micrococcus denitrificans. Biochim. Biophys. Acta* **172,** 450.

125. Lam, Y., and Nicholas, D. J. D. (1969). A nitrite reductase with cytochrome oxidase activity from *Micrococcus denitrificans. Biochim. Biophys. Acta* **180,** 459.

126. Lascelles, J. (1956). An assay of iron protoporphyrin based on the reduction of nitrate by a variant strain of *Staphylococcus aureus:* Synthesis of iron protoporphyrin by suspensions of *Rhodopseudomonas spheroides. J. Gen. Microbiol.* **15,** 404.

127. Lascelles, J. (1960). The synthesis of enzymes concerned in bacteriochlorophyll formation in growing cultures of *Rhodopseudomonas spheroides. J. Gen. Microbiol.* **23,** 487.

128. Lee, J. P., and Peck, H. D., Jr. (1971). Purification of the enzyme reducing bisulfite to trithionate from *Desulfovibrio gigas* and its identification as desulfoviridin. *Biochem. Biophys. Res. Commun.* **45,** 583.

128a. Lee, J.-P., Yi, C.-S., LeGall, J., and Peck, Jr., H. D. (1973). Isolation of a new pigment, desulforubidin, from *Desulfovibrio desulfuricans* (Norway strain) and its role in sulfite reduction. *J. Bacteriol.* **115,** 453.

128b. Lee, J.-P., LeGall, J., and Peck, Jr., H. D. (1973). Isolation of assimilatory- and dissimilatory-type sulfite reductases from *Desulfovibrio vulgaris. J. Bacteriol.* **115,** 529.

129. LeGall, J. (1963). A new species of *Desulfovibrio. J. Bacteriol.* **86,** 1120.

130. LeGall, J., and Dragoni, N. (1966). Dependence of sulfite reduction on a crystallized ferredoxin from *Desulfovibrio gigas. Biochem. Biophys. Res. Commun.* **23,** 145.

131. LeGall, J., Bruschi-Heriand, M., and Dervartanian, D. V. (1971). Electron paramagnetic resonance and light absorption studies in c-type cytochromes of the anaerobic sulfate reducer *Desulfovibrio. Biochim. Biophys. Acta* **234,** 499.

132. LeGall, J., Dervartanian, D. V., Spilker, E., Lee, J., and Peck, H. D., Jr. (1971). Evidence for the involvement of nonheme iron in the active site of hydrogenase from *Desulfovibrio vulgaris. Biochim. Biophys. Acta* **234,** 525.

133. Lehninger, A. C. (1965). "Bioenergetics." Benjamin, New York.

134. Leinniger, V. R., and Westley, J. (1968). The mechanisms of the rhodanese-catalyzed thiosulfate–cyanide reaction. *J. Biol. Chem.* **243,** 1892.

135. Leinweber, F. J., and Monty, K. J. (1963). The metabolism of thiosulfate in *Salmonella typhimurium. J. Biol. Chem.* **238,** 3775.

136. Lemberg, R. (1961). Cytochromes of group a and their prosthetic groups. *Advan. Enzymol.* **23,** 265.

137. Lester, R. L., and DeMoss, J. A. (1971). Effects of molybdate and selenite on formate and nitrate metabolism in *Escherichia coli. J. Bacteriol.* **105,** 1006.

138. Li, L. F., Ljungdahl, L., and Wood, H. G. (1966). Properties of nicotinamide adenine dinucleotide phosphate-dependent formate dehydrogenase from *Clostridium thermoaceticum*. *J. Bacteriol.* **92,** 405.
139. Lindeberg, G., Lode, A., and Somme, R. (1963). Effect of oxygen on formation and activity of nitrate reductase in a halophilic *Achromobacter* species. *Acta Chem. Scand.* **17,** 232.
140. Loesche, W. J., Gibbons, R. J., and Socransky, S. S. (1965). Biochemical characteristics of *Vibrio sputorum* and relationship to *Vibrio bubulus* and *Vibrio fetus*. *J. Bacteriol.* **89,** 1109.
141. McNall, E. G., and Atkinson, D. E. (1956). Nitrate reduction. I. Growth of *Escherichia coli* with nitrate as sole source of nitrogen. *J. Bacteriol.* **72,** 226.
142. McNall, E. G., and Atkinson, D. E. (1957). Nitrate reduction. II. Utilization of possible intermediates as nitrogen sources and as electron acceptors. *J. Bacteriol.* **74,** 60.
143. Macpherson, R., and Miller, J. D. A. (1963). Nutritional studies on *Desulfovibrio desulfuricans* using chemically defined media. *J. Gen. Microbiol.* **31,** 365.
144. Maroc, J., Azerad, R., Kamen, M. D., and LeGall, J. (1970). Menaquinone (MK-6) in the sulfate-reducing obligate anaerobe, *Desulfovibrio*. *Biochim. Biophys. Acta* **197,** 87.
145. Matsubara, T., and Mori, T. (1968). Studies on denitrification. IX. Nitrous oxide, its production and reduction to nitrogen. *J. Biochem.* (*Tokyo*) **64,** 863.
146. Mauzerall, D., and Granick, S. (1958). Porphyrin synthesis in erythrocytes. III. Uroporphyrin and its decarboxylases. *J. Biol. Chem.* **232,** 1141.
147. Mauzerall, D. (1960). The thermodynamic stability of the porphyrinogens. *J. Amer. Chem. Soc.* **82,** 2601.
148. Mechalas, B. J., and Rittenberg, S. C. (1960). Energy coupling in *Desulfovibrio desulfuricans*. *J. Bacteriol.* **80,** 501.
149. Michaelis, G. B., Davidson, J. T., and Peck, H. D., Jr. (1970). A flavin–sulfite adduct as an intermediate in the reaction catalyzed by adenylylsulfate reductase from *Desulfovibrio vulgaris*. *Biochem. Biophys. Res. Commun.* **39,** 321.
150. Miller, J. D. A., Neumann, P. M., Elford, L., and Wakerley, D. S. (1970). Malate dismutation by *Desulfovibrio*. *Arch. Mikrobiol.* **71,** 214.
151. Millet, I. (1954). *C. R. Acad. Sci.* **238,** 408 [as cited by Peck (174)].
152. Miyata, M., and Mori, T. (1969). Studies on denitrification. X. The "denitrifying enzyme" as a nitrite reductase and the electron donating system for denitrification. *J. Biochem.* (*Tokyo*) **66,** 463.
153. Miyata, M., Matsubara, T., and Mori, T. (1969). Studies on denitrification. XI. Some properties of nitric oxide reductase. *J. Biochem.* (*Tokyo*) **66,** 759.
154. Mortensen, L. E., Valentine, R. C., and Carnahan, J. E. (1962). An electron transport factor from *Clostridium pasteurianum*. *Biochem. Biophys. Res. Commun.* **7,** 448.
155. Naik, M. S., and Nicholas, D. J. D. (1966). Phosphorylation associated with nitrate and nitrite reduction in *Micrococcus denitrificans* and *Pseudomonas denitrificans*. *Biochim. Biophys. Acta* **113,** 490.
156. Najjar, V. A., and Chung, C. W. (1956). Enzymatic steps in denitrification. *In* "Inorganic Nitrogen Metabolism" (W. D. McElroy and B. Glass, eds.), p. 260. Johns Hopkins Press, Baltimore, Maryland [as cited by Nason (159)].
157. Nakatsukasa, W., and Akagi, J. M. (1969). Thiosulfate reductase isolated from *Desulfotomaculum nigrificans*. *J. Bacteriol.* **98,** 429.

158. Nason, A., Abraham, R. C., and Averbach, B. C. (1954). The enzymic reduction of nitrite to ammonia by reduced pyridine nucleotides. *Biochim. Biophys. Acta* **15,** 159 [as cited by Prakash *et al.* (194)].
159. Nason, A. (1962). Symposium on metabolism of inorganic compounds. II. Enzymatic pathways of nitrate, nitrite, and hydroxylamine metabolism. *Bacteriol. Rev.* **26,** 16.
160. Newton, J. W., and Kamen, M. D. (1961). Cytochrome systems in anaerobic electron transport. *In* "The Bacteria" (I. C. Gunsalus and R. Y. Stanier, eds.), Vol. 2, p. 397. Academic Press, New York.
161. Newton, N. (1967). A soluble cytochrome containing c-type and a_2-type heme groups from *Micrococcus denitrificans. Biochem. J.* **105,** 21C.
162. Newton, N. (1969). The two-heme nitrite reductase of *Micrococcus denitrificans. Biochim. Biophys. Acta* **185,** 316.
163. Nicholas, D. J. D., Medina, A., and Jones, O. T. G. (1960). A nitrite reductase from *Neurospora crassa. Biochim. Biophys. Acta* **37,** 468 [as cited by Prakash *et al.* (194)].
164. Nicholas, D. J. D., Redmond, W. J., and Wright, M. A. (1963). Mo and Fe requirements for nitrate reductase in *Photobacterium sepia. Nature (London)* **200,** 1125.
165. Nicholas, D. J. D., Redmond, W. J., and Wright, M. A. (1964). Effects of cultural conditions in *Photobacterium sepia. J. Gen. Microbiol.* **35,** 401.
166. O'Hara, J., Gray, C. T., Puig, J., and Pichinoty, F. (1967). Defects in formate hydrogenlyase in nitrate-negative mutants of *Escherichia coli. Biochem. Biophys. Res. Commun.* **28,** 951.
167. Ohmishi, T. (1963). Oxidative phosphorylation coupled with nitrate respiration with cell-free extracts of *Pseudomonas denitrificans. J. Biochem. (Tokyo)* **53,** 71.
168. Palleroni, N. J., Doudoroff, M., Stanier, R. Y., Solanes, R. E., and Mandel, M. (1970). Taxonomy of the aerobic pseudomonads: the properties of the *Pseudomonas stutzeri* group. *J. Gen. Microbiol.* **60,** 215.
169. Pascal, M. C., Pichinoty, F., and Bruno, V. (1965). Sur les lactate-déhydrogènases d'une bactérie dénitrifiante. *Biochim. Biophys. Acta* **99,** 543.
170. Paul, K. G. (1951). The porphyrin component of cytochrome c and its linkage to the protein. *Acta Chem. Scand.* **5,** 389.
171. Payne, W. J., Riley, P. S., and Cox, C. D., Jr. (1971). Separate nitrite, nitric oxide, and nitrous oxide reducing fractions from *Pseudomonas perfectomarium. J. Bacteriol.* **106,** 356.
172. Paynter, M. J. B., and Hungate, R. E. (1968). Characterization of *Methanobacterium mobilis,* sp.n., isolated from the bovine rumen. *J. Bacteriol.* **95,** 1943.
173. Peck, H. D., Jr., and Gest, H. (1956). A new procedure for assay of bacterial hydrogenases. *J. Bacteriol.* **71,** 70.
174. Peck, H. D., Jr. (1960). Evidence for oxidative phosphorylation during the reduction of sulfate with hydrogen by *Desulfovibrio desulfuricans. J. Biol. Chem.* **235,** 2734.
175. Peck, H. D., Jr. (1961). Enzymatic basis for assimilatory and dissimilatory sulfate reduction. *J. Bacteriol.* **82,** 933.
176. Peck, H. D., Jr. (1962). The role of adenosine 5'-phosphosulfate to sulfite by *Desulfovibrio desulfuricans. J. Biol. Chem.* **237,** 198.
177. Peck, H. D., Jr. (1962). Symposium on metabolism on inorganic compounds. V. Comparative metabolism of inorganic sulfur compounds in microorganisms. *Bacteriol. Rev.* **26,** 67.

178. Peck, H. D., Jr., Deacon, T. E., and Davidson, J. T. (1965). Studies on adenosine-5'-phosphosulfate reductase from *Desulfovibrio desulfuricans* and *Thiobacillus thioparus* I. The assay and purification. *Biochim. Biophys. Acta* **96,** 429.
179. Peck, H. D., Jr. (1966). Phosphorylation coupled with electron transfer in extracts of the sulfate-reducing bacterium *Desulfovibrio gigas. Biochem. Biophys. Res. Commun.* **22,** 112.
180. Peck, H. D., Jr. (1968). Energy-coupling mechanisms in chemolithotrophic bacteria. *Annu. Rev. Microbiol.* **22,** 489.
181. Pichinoty, F. (1960). Reduction assimilative du nitrate par les cultures aerobies d'*Aerobacter aerogenes.* Influence de la nutrition azotee sur la croissance. *Folia Microbiol.* (Prague) **5,** 165.
182. Pichinoty, F., and d'Ornano, L. (1961). Sur le mécanisme de l'inhibition par l'oxygène de la dénitrification bacterienne. *Biochim. Biophys. Acta* **52,** 386.
183. Pichinoty, F., and d'Ornano, L. (1961). Sur le mécanisme de l'inhibition par l'oxygène de la dénitrification bactérienne. *Biochim. Biophys. Acta* **52,** 386.
184. Pichinoty, F. (1964). A propos des nitrate-réductases d'une bactérie dénitrificante. *Biochim. Biophys. Acta* **89,** 378.
185. Pichinoty, F. (1965). L'effect oxygène de la biosynthèse des enzymes d'oxydoreduction bactériens. *In* "Mécanisme de regulation des activités cellulaires chez les microorganismes," pp. 507–520. CNRS, Paris [as cited in Stouthamer (232)].
186. Pichinoty, F., Bagliardi-Rouvier, J., and de Rimassa, R. (1969). La denitrification bactérienne. I. Utilisation des amines aromatique domeuses d'electrons dans la reduction du nitrite. *Arch. Mikrobiol.* **69,** 314.
187. Pine, M. S., and Barker, H. A. (1954). *Bacteriol. Proc.* p. 98.
188. Polakis, G. S., Bartley, W., and Meek, G. A. (1964). Changes in the structure and enzyme activity of *Saccharomyces cerevisiae* in response to changes in the environment *Biochem. J.* **90,** 369.
189. Porva, R. J., and Lascelles, J. (1965). Hemoproteins and hemesynthesis in facultative photosynthetic and denitrifying bacteria. *Biochem. J.* **94,** 120.
190. Postgate, J. R. (1952). *J. Res.* **5,** 189 [as cited by Peck (174)].
191. Postgate, J. R. (1956). Cytochrome c_3 and desulfoviridin; pigments of the anaerobe *Desulfovibrio desulfuricans. J. Gen. Microbiol.* **14,** 545.
192. Postgate, J. R. (1959). Sulfate reduction by bacteria. *Annu. Rev. Microbiol.* **13,** 505.
193. Postgate, J. R. (1965). Recent advances in the study of the sulfate-reducing bacteria. *Bacteriol. Rev.* **29,** 425.
194. Prakash, O., Rao, R. R., and Sadana, J. C. (1966). Purification and characterization of nitrite reductase from *Achromobacter fisheri. Biochim. Biophys. Acta* **118,** 426.
195. Quastel, J. H. (1965). Soil metabolism. *Annu. Rev. Plant Physiol.* **16,** 217.
196. Quayle, J. R. (1966). Formate dehydrogenase. *In* "Methods in Enzymology" (W. A. Wood, ed.), Vol. 9, p. 360. Academic Press, New York.
197. Radcliffe, B. C., and Nicholas, D. J. D. (1968). Some properties of a nitrite reductase from *Pseudomonas denitrificans. Biochim. Biophys. Acta* **153,** 545.
198. Radcliffe, B. C., and Nicholas, D. J. D. (1970). Some properties of a nitrite reductase from *Pseudomonas denitrificans. Biochim. Biophys. Acta* **205,** 273.
199. Reddy, C. A., Bryant, M. P., and Wolin, M. J. (1972). Characteristics of S organism isolated from *Methanobacterium omelianskii. J. Bacteriol.* **109,** 539.
200. Renner, E. D., and Becker, G. E. (1970). Production of nitric oxide and nitrous oxide during denitrification by *Corynebacterium nephridii. J. Bacteriol.* **101,** 821.

201. Rittenberg, S. C. (1969). The roles of exogenous organic matter in the physiology of chemolithotrophic bacteria *Advan. Microbial Physiol.* **3,** 159.
202. Roberton, A. M., and Wolfe, R. S. (1969). ATP requirement for methanogenesis in cell extracts of *Methanobacterium strains M.o.H. Biochim. Biophys. Acta* **192,** 420.
203. Rose, A. H. (1965). "Chemical Microbiology," p. 108. Butterworth, London.
204. Ruiz-Herrera, J., Showe, M. K., and DeMoss, J. A. (1969). Nitrate reductase complex of *Escherichia coli* K-12: Isolation and characterization of mutants unable to reduce nitrate. *J. Bacteriol.* **97,** 1291.
205. Ruiz-Herrera, J., and DeMoss, J. A. (1969). Nitrate reductase complex of *Escherichia coli* K-12: Participation of specific formate dehydrogenase and cytochrome b_1 components in nitrate reduction. *J. Bacteriol.* **99,** 720.
206. Sabater, F. (1966). Phosphorylation coupled with nitrite and nitrate reductase in *Micrococcus denitrificans. Rev. Espan. Fisiol.* **22,** 1; *Chem. Abstr.* **65,** 10827 (1966).
207. Sachs, L. E., and Barker, H. A. (1949). The influence of oxygen on nitrate and nitrite reduction. *J. Bacteriol.* **58,** 11.
208. Sadana, J. C., and McElroy, W. D. (1957). Nitrate reductase from *Achromobacter fisheri.* Purification and properties: Function of flavines and cytochromes. *Arch. Biochem. Biophys.* **67,** 16.
209. Sano, S., and Granick, S. (1961). Mitochondrial coproporphyrinogen oxidase and protoporphyrin formation. *J. Biol. Chem.* **236,** 1173.
210. Sargeant, K., Buck, P. W., Ford, J. W. S., and Yeo, R. G. (1967). Anaerobic production of *Thiobacillus denitrificans* for the enzyme rhodanese. *Appl. Microbiol.* **14,** 998.
211. Saunders, G. F., Campbell, L. L., and Postgate, J. R. (1964). Base composition of DNA of sulfate-reducing bacteria deduced from buoyant density measurements in cesium chloride. *J. Bacteriol.* **87,** 1073.
212. Saunders, G. F., and Campbell, L. L. (1966). Deoxyribonucleic acid base composition of *Desulfotomaculum nigrificans J. Bacteriol.* **92,** 515.
213. Schaeffer, P. (1952). Recherches sur la métabolisme bactérien des cytochromes et des porphyrins. I. Disparition partielle des cytochromes par culture anaérobie chez certaines bactéries aérobies facultative. *Biochim. Biophys. Acta* **9,** 261.
214. Scholes, P. B., and Smith, L. (1968). The isolation and properties of cytoplasmic membrane of *Micrococcus denitrificans. Biochim. Biophys. Acta* **153,** 350.
215. Schulp, J. A., and Stouthamer, A. H. (1970). The influence of oxygen, glucose, and nitrate upon the formation of nitrate reductase and the respiratory system in *Bacillus licheniformis. J. Gen. Microbiol.* **64,** 195.
216. Schultz, A. S., and McManus, D. K. (1949). Amino acids and inorganic sulfur as sulfur source for the growth of yeasts. *Arch. Biochem. Biophys.* **25,** 401.
217. Segal, H. L. (1956). Sulfate-dependent exchange of pyrophosphate with nucleotide phosphate. *Biochim. Biophys. Acta* **21,** 194.
218. Senez, J. C. (1962). Some considerations of the energetics of bacterial growth. *Bacteriol. Rev.* **26,** 95.
218a. Sekiguchi, T., and Noso, Y. (1973). Pyruvate-supported acetylene and sulfate reduction by cell free extracts of *Desulfovibrio desulfuricans. Biochem. Biophys. Res. Commun.* **51,** 331.
219. Shennin, D. (1970). On the synthesis of heme. *Naturwissenschaften* **57,** 185.
220. Showe, M. K., and DeMoss, J. A. (1968). Localization and regulation of synthesis of nitrate reduction in *Escherichia coli. J. Bacteriol.* **95,** 1305.

221. Skerman, V. B. D., and MacRae, I. C. (1957). The influence of oxygen on the reduction of nitrate by adapted cells of *Pseudomonas denitrificans*. *Can. J. Microbiol.* **3,** 215.

222. Skerman, V. B. D., and MacRae, I. C. (1957). The influence of oxygen availability in the degree of nitrate reduction by *Pseudomonas denitrificans*. *Can. J. Microbiol.* **3,** 506.

223. Skerman, V. B. D., Carey, B. J., and MacRae, I. C. (1958). The influence of oxygen on the reduction of nitrite by washed suspensions of adapted cells of *Achromobacter liquefaciens*. *Can. J. Microbiol.* **4,** 243.

223a. Skyring, G. W., and Trudinger, P. A. (1972). A method for the electrophoretic characterization of sulfite reductases in crude preparations from sulfate-reducing bacteria using polyacrylamide gels. *Can. J. Biochem.* **50,** 1145.

223b. Skyring, G. W., and Trudinger, P. A. (1973). A comparison of the electrophoretic properties of the ATP-sulfurylases, APS-reductases, and sulfite reductases from cultures of dissimilatory sulfate-reducing bacteria. *Can. J. Microbiol.* **19,** 375.

223c. Skyring, G. W., and Trudinger, P. A. (1972). Electrophoretic characterization of ATP-sulfate adenylyltransferase (ATP-sulfurylase) using acrylamide gels. *Analyt. Biochem.* **48,** 259.

224. Slonimski, P. P. (1953). "La formation des enzymes respiratoires chez la levure." Masson, Paris [as cited by Jacobs *et al.* (107)].

225. Sokatch, J. T., and Gunsalus, I. C. (1957). Aldonic acid metabolism. I. Pathway of carbon in an inducible gluconate fermentation by *Streptococcus faecalis*. *J. Bacteriol.* **73,** 452.

226. Sorokin, Y. I. (1966). Role of carbon dioxide and acetate in biosynthesis by sulfate-reducing bacteria. *Nature (London)* **210,** 551.

227. Spangler, W. J., and Gilmour, C. H. (1966). Biochemistry of nitrate respiration in *Pseudomonas stutzeri*. I. Aerobic and nitrate respiration routes of carbohydrate catabolism. *J. Bacteriol.* **91,** 245.

228. Spencer, D., Takahashi, H., and Nason, A. (1957). Relationship of nitrite and hydroxylamine reductases to nitrate assimilation and nitrogen fixation in *Azotobacter agile*. *J. Bacteriol.* **73,** 553.

229. Stadtman, T. C., and Barker, H. A. (1949). Studies on the methane fermentation. VII. Tracer experiments on the mechanism of methane formation. *Arch. Biochem.* **21,** 256.

230. Stadtman, T. C., and Barker, H. A. (1951). Studies on the methane fermentation. IX. The origin of methane in the acetate and methanol fermentation by *Methanosarcina*. *J. Bacteriol.* **61,** 81.

231. Starkey, L. R. (1960). Sulfate-reducing bacteria—physiology and practical significance. "Lectures on Theoretical and Applied Aspects of Modern Microbiology." University of Maryland, College Park, [as cited by Peck (175)].

232. Stouthamer, A. H. (1967). Nitrate reduction in *Aerobacter aerogenes*. I. Isolation and properties of mutant strains blocked in nitrate assimilation and resistant against chlorate. *Arch. Mikrobiol.* **56,** 68.

233. Stouthamer, A. H. (1967). Nitrate reduction in *Aerobacter aerogenes*. II. Characterization of mutants blocked in the reduction of nitrate and chlorate. *Arch. Mikrobiol.* **56,** 76.

234. Suh, B., and Akagi, J. M. (1966). Pyruvate–carbon dioxide exchange reaction of *Desulfovibrio desulfuricans*. *J. Bacteriol.* **91,** 2281.

235. Suh, B., and Akagi, J. M. (1969). Formation of thiosulfate from sulfite by *Desulfovibrio vulgaris*. *J. Bacteriol.* **99,** 210.
236. Taber, H. W., and Morrison, M. (1964). Electron transport in staphylococci. Properties of a particle preparation from exponential phase *Staphylococcus aureus*. *Arch. Biochem. Biophys.* **105,** 367.
237. Tagawa, K., and Arnon, D. I. (1962). Ferredoxins as electron carriers in photosynthesis and in the biological production and consumption of hydrogen gas. *Nature (London)* **195,** 537.
238. Taniguchi, S. (1961). Comparative biochemistry of nitrate metabolism. *Z. Allg. Mikrobiol.* **1,** 341.
239. Trudinger, P. A. (1969). Assimilatory and dissimilatory metabolism of inorganic sulfur compounds by microorganisms. *Advan. Microbial Physiol.* **3,** 111.
239a. Trudinger, P. A., and Chambers, L. A. (1973). Reactions of P582 from *Desulfotomaculum nigrificans* with substrates, reducing agents and carbon monoxide. *Biochim. Biophys. Acta* **293,** 26.
240. Van't Riet, J., Stouthamer, A. H., and Planta, R. J. (1968). Regulation of nitrate assimilation and nitrate respiration in *Aerobacter aerogenes*. *J. Bacteriol.* **96,** 1455.
241. Vernon, L. P., and White, F. G. (1957). Terminal oxidases of *Micrococcus denitrificans*. *Biochim. Biophys. Acta* **25,** 321 [as cited by Dolin (51)].
242. Villarejo, M., and Westley, J. (1963). Rhodanese-catalyzed reduction of thiosulfate by reduced lipoic acid. *J. Biol. Chem.* **238,** PC1185.
243. Villarejo, M., and Westley, J. (1963). Mechanism of rhodanese catalyses of thiosulfate–lipoate oxidation–reduction. *J. Biol. Chem.* **238,** 4016.
243a. Villarreal-Moguel, E. I., Ibarra, V., Ruiz-Herrera, J., and Gitler, C. (1973). Resolution of the nitrate reductase complex from the membrane of *Escherichia coli*. *J. Bacteriol.* **113,** 1264.
244. Volini, M., and Westley, J. (1966). The mechanism of the rhodanese-catalyzed thiosulfate–lipoate reaction. Kinetic analysis. *J. Biol. Chem.* **241,** 5168.
245. Vosjan, J. H. (1970). ATP generation by electron transport in *Desulfovibrio desulfuricans*. *Antonie van Leeuwenhoek; J. Microbiol. Serol.* **36,** 584.
246. Walker, G. C., and Nicholas, D. J. D. (1961). Hydroxylamine reductase from *Pseudomonas aeruginosa*. *Biochim. Biophys. Acta* **49,** 361.
247. Wallace, P. G., and Linnane, A. W. (1964). Oxygen-induced synthesis of yeast mitochondria. *Nature (London)* **201,** 1191.
248. Ward, F. B., and Cole, J. A. (1971). Nitrite reductases of *Escherichia coli*. *J. Gen. Microbiol.* **68,** xiii.
249. Weber, M. M., Matschiner, J. T., and Peck, H. D., Jr. (1970). Menaquinone-6 in the strict anaerobes *Desulfovibrio vulgaris* and *Desulfovibrio gigas*. *Biochem. Biophys. Res. Commun.* **38,** 197.
250. Westley, J., and Nakamoto, T. (1962). Mechanism of rhodanese action: Isotopic tracer studies. *J. Biol. Chem.* **237,** 547.
251. Whateley, F. R. (1962). Phosphorylation accompanying nitrate respiration in cell-free extracts of *Micrococcus denitrificans*. *Plant Physiol.* **37,** Suppl., viii.
252. White, D. C., and Granick, S. (1963). Hemin biosynthesis in *Hemophilus*. *J. Bacteriol.* **85,** 842.
253. White, D. C. (1967). Effect of glucose on the formation of the membrane-bound electron transport system in *Hemophilus parainfluenzae*. *J. Bacteriol.* **93,** 567.
254. Wilson, L. G., and Bandurski, R. S. (1958). Enzymatic reactions involving sulfate, sulfite, selenate, and molybdate. *J. Biol. Chem.* **233,** 975.

255. Wilson, L. G., and Bandurski, R. S. (1958). Enzymatic reduction of sulfate. *J. Amer. Chem. Soc.* **80,** 5576.
256. Wimpenny, J. W. T., and Cole, J. A. (1967). Regulation of metabolism in facultative bacteria: The effect of nitrate. *Biochem. J.* **103,** 20P.
257. Wolfe, R. S. (1971). Microbial formation of methane. *Advan. Microbial Physiol.* **6,** 107.
258. Wolin, M. J., Wolin, E. A., and Wolfe, R. S. (1963). ATP-dependent formation of methane from methylcobalamin by extracts of *Methanobacillus omelianski. Biochem. Biophys. Res. Commun.* **12,** 464.
259. Wolin, M. J., Wolin, E. A., and Wolfe, R. S. (1964). The cobalamin product of the conversion of methylcobalamin to methane by extracts of *Methanobacillus omelianski. Biochem. Biophys. Res. Commun.* **15,** 420.
260. Wood, J. M., and Wolfe, R. S. (1965). The formation of methane from N^5-methyltetrahydrofolate monoglutamate by cell-free extracts of *Methanobacillus omelianskii. Biochem. Biophys. Res. Commun.* **19,** 306.
261. Wood, J. M., Allan, A. M., Brill, W. J., and Wolfe, R. S. (1965). Formation of methane from serine by cell-free extracts of *Methanobacillus omelianskii. J. Biol. Chem.* **240,** 4564.
262. Wood, J. M., and Wolfe, R. S. (1966). Components required for the formation of CH_4 from methylcobalamin by extracts of *Methanobacillus omelianski. J. Bacteriol* **92,** 696.
263. Wood, J. M., and Wolfe, R. S. (1966). Alkylation of an enzyme in the methane-forming system of *Methanobacillus omelianskii. Biochem. Biophys. Res. Commun.* **22,** 119.
264. Wood, W. A. (1961). Fermentation of carbohydrates and related compounds. *In* "The Bacteria" (I. C. Gunsalus and R. Y. Stanier, eds.), Vol. 2, p. 59. Academic Press, New York.
265. Yagi, T. (1969). Formate:cytochrome oxidoreductase of *Desulfovibrio vulgaris. J. Biochem.* (*Tokyo*) **66,** 473.
266. Yamanaka, T., Ota, A., and Okumuki, K. (1961). A nitrite reducing system reconstructed with purified cytochrome components of *Pseudomonas aeruginosa. Biochim. Biophys. Acta* **53,** 294 [as cited by Prakash *et al.* (194)].
267. Yamanaka, T. (1964). Identity of *Pseudomonas* cytochrome oxidase with *Pseudomonas* nitrite reductase. *Nature* (*London*) **204,** 253.
268. Yates, M. G. (1967). Stimulation of the phosphoroclastic system of *Desulfovibrio* by nucleotide triphosphate. *Biochem. J.* **103,** 32C.
269. Zhilina, T. N. (1971). The fine structure of *Methanosarcina. Mikrobiologiya* **40,** 674.
270. Zubieta, J. A., Mason, R., and Postgate, J. R. (1973). A four-iron ferredoxin from *Desulfovibrio desulfuricans. Biochem. J.* **133,** 851.

Supplementary Readings

LeGall, J., and Postgate, J. R. (1973). The physiology of sulphate-reducing bacteria. *Adv. Microbial Physiol.* **10,** 82.

Barton, L. L., LeGall, J., and Peck, Jr., H. D. (1972). Oxidative phosphorylation in the obligate anaerobe, *Desulfovibrio gigas. In* "Horizons of Bioenergetics" (A. San Pietro and H. Gest, eds.) p. 33–51. Academic Press Inc., New York.

Questions

1. What is anaerobic respiration?
2. Name the three main groups of bacteria that use inorganic compounds as terminal electron acceptors.
3. Draw the generalized scheme of sulfate and thiosulfate reduction by *Desulfovibrio desulfuricans*, using molecular hydrogen as electron donor.
4. Explain the thermodynamics of sulfate reduction in the presence of hydrogen.
5. What is the difference between sulfate reduction in *Desulfovibrio* and that in yeasts?
6. Explain why *Desulfovibrio* was originally thought to be an autotroph and why thiosulfate was thought to be an intermediate in sulfate reduction.
7. Explain the differences between thiosulfate reduction in *Desulfovibrio desulfuricans* and that in *Thiobacillus denitrificans*.
8. Draw the generalized scheme of sulfate reduction by *Desulfovibrio desulfuricans*, using lactate as electron donor.
9. What are the differences between lactate and pyruvate as energy sources for the growth of *Desulfovibrio*?
10. Discuss the roles of cytochrome c_3, ferredoxin, and APS in sulfate reduction by *Desulfovibrio desulfuricans*.
11. What are the reasons for the sulfate requirement in the metabolism of organic compounds by *Desulfovibrio desulfuricans*?
12. Explain why oxidative phosphorylation is possible under anaerobic conditions in the case of *Desulfovibrio desulfuricans*.
13. What is meant by denitrification in contrast to nitrification?
14. Draw the presently conceived biochemical pathway of denitrification.
15. Discuss the thermodynamics of nitrate reduction.
16. Describe the metabolism of glycolate by *Micrococcus denitrificans*.
17. Explain the differences in the electron transport sequences of nitrate reduction by *Escherichia coli* and *Pseudomonas aeruginosa*.
18. Discuss the regulation of nitrate reduction in such facultative anaerobic bacteria as *Escherichia coli*.
19. Why do microorganisms employ at least four different nitrate reductases?
20. Explain the importance of heme in nitrate reduction.
21. Name the four major genera of methane bacteria.
22. Discuss whether or not it is justified to call the methane bacteria "carbon dioxide reducers."
23. Discuss methane formation.
24. What cofactors would be required in the reduction of CO_2 to methane via a methyl transfer?
25. Discuss the thermodynamics of methane formation.

5

Carbohydrate Metabolism

Pyruvic acid has been established as the key intermediate substance in the metabolism of carbohydrates by bacteria. Almost all six-, five-, and four-carbon compounds are converted initially to pyruvate, from which substance further catabolic or synthetic reactions proceed. Because of its key position, it was thought best to deal first with the origin of pyruvic acid.

Glucose Metabolism

The main carbohydrate compound that serves as carbon source for bacteria is glucose. Glucose is converted to pyruvic acid mainly via four different pathways, which have been named after those researchers who discovered and established them or according to their main components:

1. Embden-Meyerhof-Parnas (EMP) pathway
2. Warburg-Dickens or hexose monophosphate (HMP) pathway
3. Entner-Doudoroff (ED) pathway
4. Phosphoketolase (PK) pathway

The first two pathways also function in mammalian tissue and in yeasts. However, there are significant differences between the terminologies used by biochemists and by microbiologists. The EMP pathway is very often referred to as "glycolysis," or the "glycolytic" or "anaerobic" pathway. As used by mammalian biochemists, the term "glycolysis" refers to the pathway by which glycogen or glucose is converted anaerobically via pyruvic acid to lactic acid. Only in the homofermentative lactobacilli is the complete pathway, including the conversion of pyruvic acid to lactic acid, found. Many other microorganisms use this pathway to pyruvic acid, but the subsequent metabolism of pyruvic acid varies from one group of bacteria to another. The EMP pathway is used not only by anaerobic bacteria, but also by facultative anaerobic bacteria and, as is shown in the iron bacteria (see Chapter 6), even by aerobic bacteria. It is therefore not an exclusive anaerobic pathway. Some anaerobic bacteria, such as *Zymomonas mobilis*, metabolize glucose via the ED pathway, making the EMP pathway again not an exclusive pathway for the anaerobic utilization of glucose. The term "glycolysis" would therefore best be restricted to its mammalian biochemical usage.

Another term that needs clarification is "fermentation." Anaerobiosis is often considered as synonymous with fermentation and all the different anaerobic pathways as "fermentative" pathways (3). The term "ferment" has an ancient history and with progress in biochemical knowledge it has assumed various meanings. As defined in this book (see Chapter 8), "fermentative" is restricted to those pathways in which the terminal electron acceptor is an organic compound. Where the terminal acceptor is oxygen or an inorganic compound, the pathway is regarded as oxidative and the process one of respiration. The latter is either aerobic (using oxygen) or anaerobic (using inorganic compounds other than oxygen).

Disregarding for the moment whether the various pathways are functional under aerobic or anaerobic conditions, each will be traced to the level of the formation of pyruvic acid, which is an intermediate common to them all (89). As the problem of glucose transport will be dealt with separately (see p. 241), this aspect is also disregarded for the time being.

Embden-Meyerhof-Parnas Pathway

The EMP pathway, outlined in Fig. 5.1, is widely distributed among bacteria (Enterobacteriaceae, Lactobacillaceae, saccharolytic clostridia, etc.) and shows the overall reaction

$$\text{glucose} + 2\text{ATP} + 2\text{NAD}^+ \rightleftharpoons 2\ \text{pyruvate} + 4\text{ATP} + 2(\text{NADH} + \text{H}^+)$$

The first step (i) in the glucose breakdown is a phosphorylation step that requires 1 mole of ATP and the enzyme hexokinase (ATP:D-hexose 6-phosphotransferase, EC 2.7.1.1)

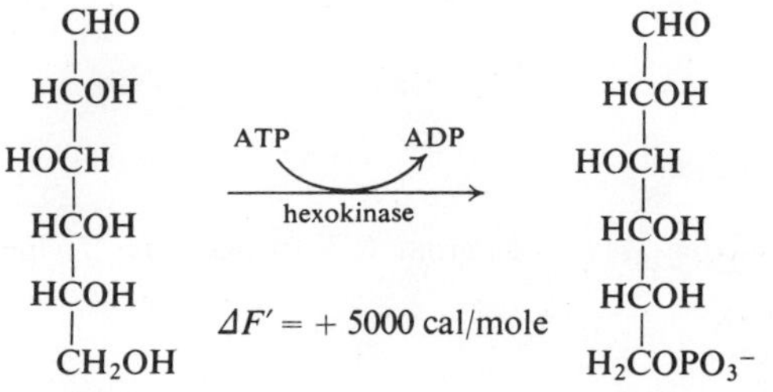
CHO
HCOH
HOCH
HCOH
HCOH
CH2OH
ATP
ADP
hexokinase
ΔF′ = + 5000 cal/mole
CHO
HCOH
HOCH
HCOH
HCOH
H2COPO3−

glucose glucose 6-phosphate

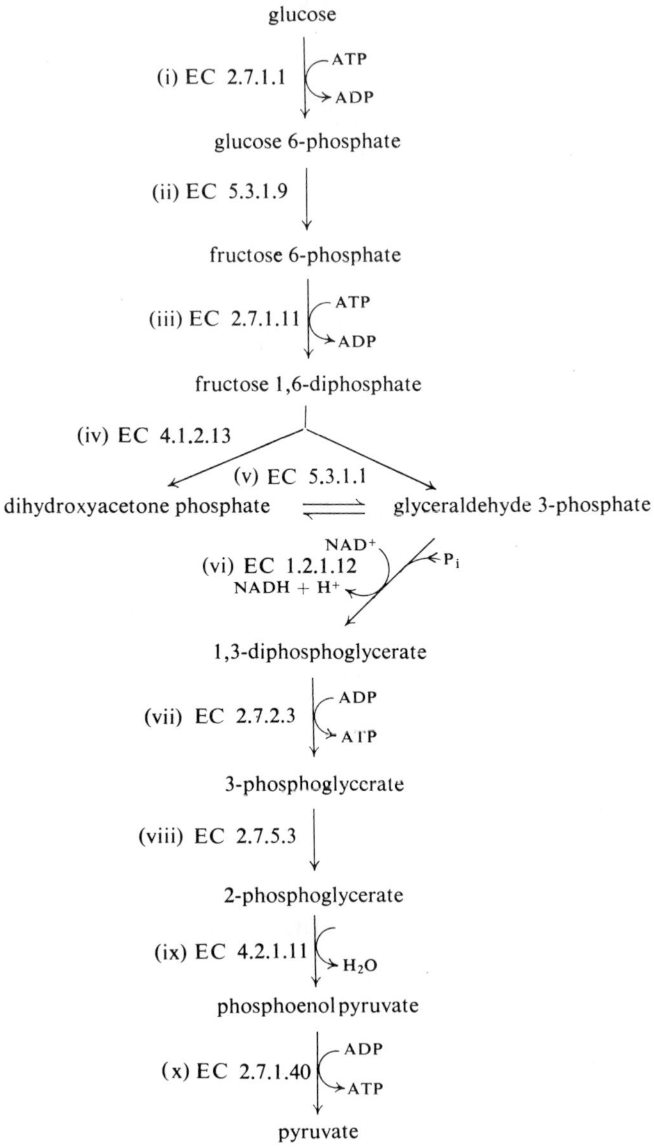
glucose
(i) EC 2.7.1.1
ATP
ADP
glucose 6-phosphate
(ii) EC 5.3.1.9
fructose 6-phosphate
(iii) EC 2.7.1.11
ATP
ADP
fructose 1,6-diphosphate
(iv) EC 4.1.2.13
(v) EC 5.3.1.1
dihydroxyacetone phosphate
glyceraldehyde 3-phosphate
NAD+
(vi) EC 1.2.1.12
NADH + H+
Pi
1,3-diphosphoglycerate
(vii) EC 2.7.2.3
ADP
ATP
3-phosphoglycerate
(viii) EC 2.7.5.3
2-phosphoglycerate
(ix) EC 4.2.1.11
H2O
phosphoenol pyruvate
(x) EC 2.7.1.40
ADP
ATP
pyruvate

Fig. 5.1. The Embden-Meyerhof-Parnas (EMP) pathway of glucose utilization.

The terms "hexokinase" and "pentokinase" include all enzymes that catalyze the transfer of the terminal phosphoryl residue of ATP to any one of the hydroxyl groups of a C-5 or C-6 monosaccharide. In actual fact, however, only the terminal —OH group of the free sugar seems to be involved in the reaction (72). Enzyme-catalyzed phosphorylations of hydroxyl groups occupying other positions on the sugar molecule have been observed only in the ribose moiety of certain coenzymes and cofactors.

Hexokinase, the prototype of this group of enzymes, was first described in yeasts. The distribution of these enzymes among microorganisms is shown in Table 5.1. The hexokinases of bacterial origin appear to have a far greater substrate specificity than those of mammalian or yeast origin. The yeast hexokinase is capable of phosphorylating a number of different sugars and appears to be specific for the 3,4,5,6 region of the molecule but not for the 1,2 part, provided, of course, that no larger groups are attached to them (94). Brain and *Aspergillus* hexokinases are broadly similar but differences occur between the two; e.g., brain enzyme is much less exacting with regard to substitution on C-5, and *Aspergillus* enzyme with regard to the C-4 atom. Because of this, galactose can serve almost as well as

TABLE 5.1
Hexokinases and Pentokinases of Microbial Origin[a]

Enzyme	Substrate	Product	Source
Hexokinase	Glucose, mannose, fructose, 2-deoxyglucose, etc.	Hexose 6-phosphate	Yeast, *Neurospora crassa*, *Pseudomonas putrefaciens*, *Borrelia recurrentis*
Glucokinase	Glucose	Glucose 6-phosphate	*P. saccharophila*, *Staph. aureus*
Mannokinase	Mannose	Mannose 6-phosphate	*P. saccharophila*
Galactokinase	Galactose	Galactose 6-phosphate	*S. fragilis*
Fructokinase	Fructose	Fructose 6-phosphate	*P. saccharophila*
Ribokinase	Ribose, 2-deoxyribose	Ribose 5-phosphate	Bacteria, yeast
Ribulokinase	L-Ribulose, arabitol	Ribulose 5-phosphate	Bacteria
Xylulokinase	D-Xylulose	Xylulose 5-phosphate	Bacteria
N-Acetylglucosamine kinase	*N*-Acetyl glucosamine	*N*-Acetyl glucosamine 6-phosphate	General bacterial distribution

[a] From Crane (72).

glucose as a substrate. The phosphorylation of hexoses and pentoses in bacteria, however, is governed by more specific enzymes; glucose by glucokinase (EC 2.7.1.2), fructose by ketohexokinase (EC 2.7.1.3), etc. In many cases, kinases are quite specific for a particular sugar structure; e.g., the ribulokinase (EC 2.7.1.16) of *Aerobacter aerogenes* phosphorylates only D- and L-ribulose.

All kinases require the participation of ATP as the phosphate donor and also a metal, normally Mg^{2+}. The actual number of discovered hexokinases continues to grow, as can well be expected from the almost universal occurrence of glucose throughout the living forms.

Hexokinases, with the exception of glucokinase, are inhibited by glucose 6-phosphate (385, 386). A mechanism for ADP control of hexokinases in yeast was proposed (60) that invoked an ATP compartment. Adenosine diphosphate enters the compartment and is also phosphorylated there, which means that ATP must be compartmented either with or without hexokinase.

The rate of glucose utilization in a cell that has rapid glucose transport and that has little glucose 6-phosphate hydrolysis has been shown to depend on the hexokinase step, which has been located in soluble and mitochondrial fractions of mammalian cells (248) and is strongly inhibited by glucose 6-phosphate (71). This inhibition is competitive (132) with ATP in the presence of excess Mg^{2+}. However, Van Thiedemann and Born (384) observed that inorganic phosphate offsets the inhibitory effect of glucose 6-phosphate in Ehrlich ascites cell homogenates and concluded that the specific inhibition by glucose 6-phosphate is overcome by inorganic phosphate. They also suggested that this might have an important relation to the Pasteur effect. Rose *et al.* (331) traced the effect of inorganic phosphate in stimulating glucose utilization in red blood cells to this property of inorganic phosphate.

It is possible, however, that glucose 6-phosphate, as a regulatory substance of hexokinase, is a consequence rather than a cause of metabolic

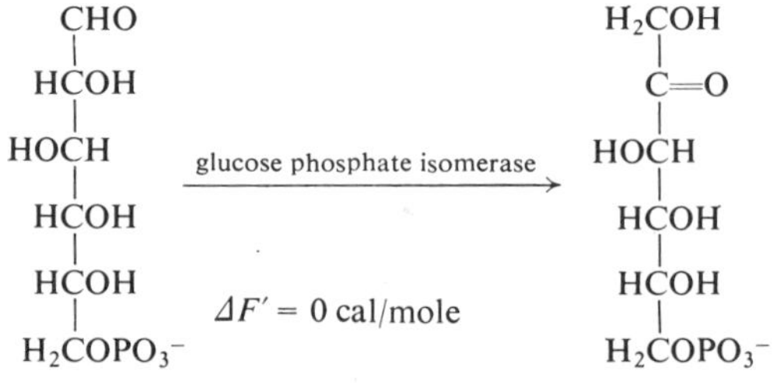
glucose phosphate isomerase
ΔF′ = 0 cal/mole

glucose 6-phosphate fructose 6-phosphate

control. Therefore, hexokinase could effect a limitation on the size of the glucose 6-phosphate pool only.

The second step (ii) in the EMP pathway is the isomerization of glucose 6-phosphate to fructose 6-phosphate (350a) by the enzyme glucosephosphate isomerase (D-glucose-6-phosphate ketol-isomerase, EC 5.3.1.9).

The second phosphorylation step (iii) requires a second mole of ATP and is catalyzed by the most important enzyme in the pathway, phosphofructokinase (ATP:D-fructose-6-phosphate 1-phosphotransferase, EC 2.7.1.11)

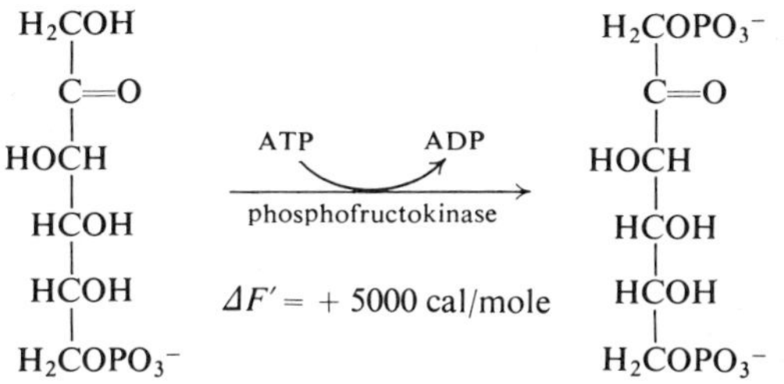
H₂COH
C=O
HOCH
HCOH
HCOH
H₂COPO₃⁻
ATP
ADP
phosphofructokinase
ΔF′ = + 5000 cal/mole
H₂COPO₃⁻
C=O
HOCH
HCOH
HCOH
H₂COPO₃⁻

fructose 6-phosphate fructose 1,6-diphosphate

This enzyme is a most complex enzyme, in terms of both kinetics and physical structure, for it is inhibited by ATP (226), by citrate (303, 304, 313), and to a lesser degree by Mg^{2+}. In contrast, phosphofructokinase is activated or reactivated by NH_4^+ (2), K^+, inorganic phosphate, ADP, 3′-AMP, 5′-AMP, and fructose 6-phosphate (303, 304). There is strong evidence that each of these activators acts at a different site on the enzyme molecule. In considering possible structures that might give the kinetic properties of phosphofructokinase, it is important to keep in mind the fact that citrate and ATP are synergistic in action, each lowering the inhibition constant of the other.

Phosphofructokinase is a unique enzyme of the EMP pathway in facultative anaerobes in that it is not involved in any of the other types of carbohydrate degradations. This enzyme therefore plays an important function in glucose utilization (217). Evidence indicates that phosphofructokinase may be the rate-limiting factor of the EMP pathway. Its suppression by oxidation has led to the proposals that the enzyme is the mediator of the Pasteur effect (see p. 609).

There is also other evidence indicating that phosphofructokinase is endowed with many properties that make it well adapted for regulation in the cell. Three such properties are recognized: (*a*) kinetic properties (allostery) of the enzyme (17); (*b*) reversible dissociation of the active enzyme to an inactive enzyme (178, 259); and (*c*) aggregation of the enzyme in a monomer–polymer system in equilibrium. These three properties depend on adenylic nucleotides and on the hexose phosphates that induce a primary change in the molecular configuration of the enzyme (258). After such structural change, the enzyme may become more susceptible to the formation of inactive subunits (407). A further phenomenon observed in yeast (386) is (*d*) the transformation of the enzyme from the active form (α) to an active form (β). Only the active form (α) is susceptible to ATP inhibition. A similar transformation has been observed in *Escherichia coli* K 12, in which the enzyme from strictly aerobically grown cells exhibited no ATP sensitivity and the enzyme from anaerobically grown cells was ATP sensitive (377).

Phosphofructokinase, however, does not exhibit allosteric kinetics in every microorganism. The enzyme from *Arthrobacter crystallopoietes* (117), as well as that from lactobacilli (97), has been found to follow simple Michaelis-Menten kinetics with respect to ATP and fructose 6-phosphate.

Different temperatures also appear to effect the kinetics of the enzyme. Thermostable phosphofructokinase from an extreme thermophilic bacteria appears to change its substrate effect (408). Whereas the enzyme followed simple Michaelis-Menten kinetics with respect to fructose 6-phosphate and ATP, it is phosphoenol pyruvate that turns the normal saturation curve for fructose 6-phosphate into a sigmoidal one independent of the ATP concentration. In the cold, in contrast, the enzyme undergoes inactivation caused by dissociation (216). This makes any suggestion for a mechanism difficult, although attempts have been made (116).

In summary, phosphofructokinase is influenced by no less than nine metabolites, eight of which (viz., ATP, ADP, AMP, F6-P, FDP, Pi, citrate, PEP) are all affected, or capable of being affected, by phosphofructokinase activity. Many of these in turn determine through equilibria the level of other metabolites. Consequently, cell composition may be determined to a considerable extent by the kinetic parameter of the particular brand of phosphofructokinase present in the cell (see also Pasteur effect, p. 609 and regulation on p. 256).

The second key enzyme of the EMP pathway catalyzes reaction (iv), which is a cleavage of fructose 1,6-diphosphate to 2 moles of triose phosphates. The enzyme is fructose-diphosphate aldolase (fructose-1,6-diphosphate:D-glyceraldehyde-3-phosphate-lyase, EC 4.1.2.13) or in the old literature also "zymohexase." This enzyme is even more widely distributed than phosphofructokinase, as it is also a key enzyme in gluconeogenesis. Meyerhof and Lohmann discovered this enzyme in 1934 (275) and proposed the above reaction; they thus confirmed, experimentally, a view expressed

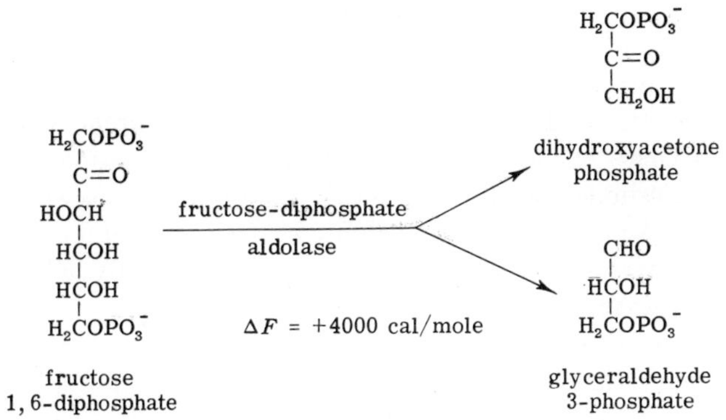
H$_2$COPO$_3^-$
C=O
HOCH
HCOH
HCOH
H$_2$COPO$_3^-$
fructose 1,6-diphosphate
fructose-diphosphate
aldolase
ΔF = +4000 cal/mole
H$_2$COPO$_3^-$
C=O
CH$_2$OH
dihydroxyacetone phosphate
CHO
HCOH
H$_2$COPO$_3^-$
glyceraldehyde 3-phosphate

by Embden in 1913 (98, 133). The term "zymohexase" has been used to connote the combined activity of two independent catalysts, fructose-diphosphate aldolase and triosephosphate isomerase (277), although the name has been used to refer to aldolase free of isomerase (174).

Warburg and Christian (390), who first reported the purification and crystallization of the aldolase from rat muscle, brought the first evidence that there exist differences between the mammalian and the yeast aldolases. Investigations into the nature of bacterial aldolases indicated that these enzymes were inhibited by chelating agents (e.g., EDTA) and that this inhibition could be overcome with metal ions (23, 26, 139, 249, 368, 396).

TABLE 5.2
Some Differences between Class I and Class II Aldolases

Source	Chelation inhibition	Metal ion reversal of inhibition	Optimal pH range
Class I			
Muscle	No	—	6.7–8.5
Liver	No	—	
bovine			7.2–9.2
rabbit			7.5–8.0
Green pea	No	—	5.9–8.8
Class II			
Anacystis nidulans	Yes	Fe^{2+}	8.0–8.5
Clostridium perfringens	Yes	Fe^{2+}; Co^{2+}	7.5–7.9
Brucella suis	Yes	Fe^{2+}; Mn^{2+}	7.0–7.5
Aspergillus niger	Yes	Fe^{2+}; Co^{2+} Zn^{2+}; Mn^{2+}	7.6–7.9
Bacillus stearothermophilus	No	—	7.2–7.5

The accumulation of these and other differences led Richards and Rutter (324) to consider aldolase the prototype of metal ion-independent (Class I) enzymes, whereas yeast aldolase was considered the prototype of metal ion-dependent (Class II) aldolases (Table 5.2). Differences between the Class I and Class II aldolase were considered to be owing either to the existence of dissimilarities in the overall protein molecule, with the active site in each case remaining equivalent, or to the existence of different active sites. It was also found that yeast aldolase had only half the molecular weight of the muscle aldolase.

A concise idea of the active site of aldolase has been possible as a result of the discovery that a stable, inert dihydroxyacetone-phosphate aldolase compound could be formed by treating rabbit muscle or *Candida utilis* aldolases with borohydride in the presence of dihydroxyacetone phosphate

$$\begin{array}{c} CH_2OH \\ | \\ C=O \\ | \\ CH_2-O-\textcircled{P} \end{array} + H_2N-(CH_2)_4-\begin{array}{c} R_1 \\ | \\ O=C \\ | \\ CH \\ | \\ NH \\ | \\ R_2 \end{array} \rightleftharpoons \left[\begin{array}{cc} CH_2OH & C=O \\ | & | \\ C=N-(CH_2)_4-CH \\ | & | \\ CH_2-O-\textcircled{P} & NH \end{array} \right]$$

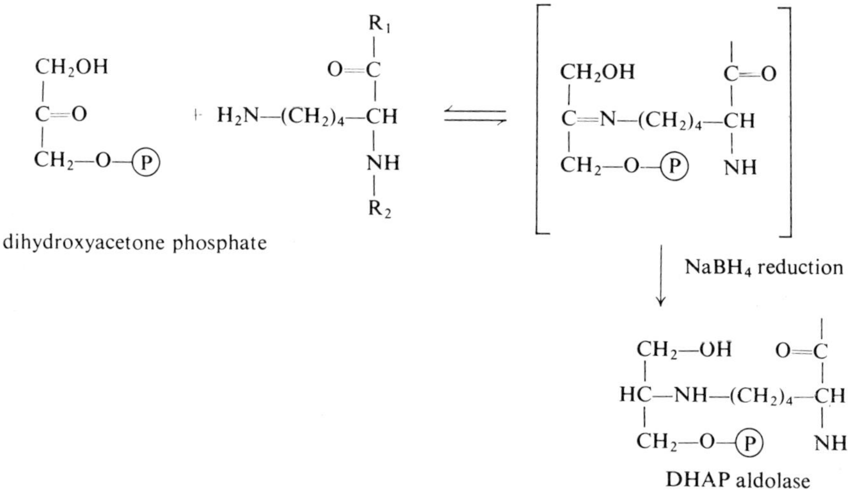

dihydroxyacetone phosphate

$\downarrow$ NaBH$_4$ reduction

$$\begin{array}{cc} CH_2-OH & O=C \\ | & | \\ HC-NH-(CH_2)_4-CH \\ | & | \\ CH_2-O-\textcircled{P} & NH \end{array}$$

DHAP aldolase

(152, 153). It was shown that the linkage of the substrate carbonyl group to the enzyme involved the formation of a Schiff base intermediate with the ϵ-amino group of lysine residue of the enzyme protein. Grazi *et al.* (152, 153) concluded that the lysine was located at the active site of the protein and that the Schiff base formation occurs during the enzyme-catalyzed aldol condensation and the transfer reactions.

Both yeast and muscle aldolase have been shown to proceed almost exclusively by an ordered reaction sequence

enzyme + FDP $\rightleftharpoons$ GAP + enzyme − DHAP (lacking a proton on α-carbon)
enzyme + DHAP (332)

Details of this reaction sequence have indicated the following steps

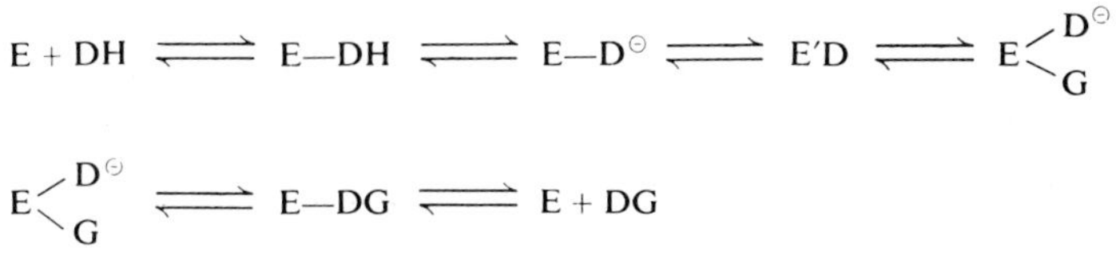

where DH indicates dihydroxyacetone phosphate; E, enzyme; E–DH, Schiff base intermediate; and E–D⁻, intermediate after loss of proton on α carbon; E′D, product following isomerization step; G, glyceraldehyde 3-phosphate; and DG, condensation product, i.e., FDP.

Rose *et al.* (332) considered that the rate-limiting step involves the —C—H bond formation, i.e., the proton-exchange reaction.

The enzyme is generally intracellular and is found in the soluble fraction of the cytoplasm (329), although there have been a number of reports indicating the association of the enzyme with certain intracellular structures (47, 355). In microorganisms, the enzyme has been observed in the soluble fraction (23, 104, 149). At high concentrations of fructose 1,6-diphosphate, a sixfold augmentation of aldolase activity was observed (223), which may result from a specific interaction of ionized FDP and the enzyme molecule rather than from a nonspecific anion stimulation.

The greatest difference between the mammalian and bacterial aldolase, however, is the existence of large numbers of isozymes of the former (13, 320). Attempts are being made to use this enzyme as a model of biochemical evolution (245a, 336). This attempt may be particularly useful also for the evolution of the bacteria, as the Class II aldolases turn out to be much more heterogenous in regard to stability, inhibition, and reactivation compared to the Class I aldolases. As no isozymes have been reported for Class II aldolases as yet, to the author's knowledge, this heterogeneity may indicate a similar diversity. Fructose-diphosphate aldolase was found to be different from spores and vegetative cells of *Bacillus cereus* (337), and aldolase from *Escherichia coli* reacts differently from the one isolated from *Lactobacillus casei* var. *rhamnosus* (96). Whereas the enzyme from *Escherichia coli* was extremely unstable in a buffer containing mercaptoethanol, the *Lactobacillus* enzyme depended on this sulfhydryl compound for its activity. The fructose-diphosphate aldolases from *Thermus aquaticus* and *Bacillus stearothermophilus* are extremely heat stable (130). Despite all these variations, all bacterial aldolases belong to the Class II aldolases, although there are indications of the existence of multiple forms in *Lactobacillus casei* (195a, 195b) and *Escherichia coli* K-12 (98a) and of a Class I aldolase in *Escherichia coli* (Crookes' strain) (375a). However, it would be more helpful if the Class II aldolases could be subgrouped to make comparisons easier.

The cleavage of fructose 1,6-diphosphate to 2 moles of triose phosphate would lead to accumulation of one of the triose phosphates. However, a triosephosphate isomerase (D-glyceraldehyde-3-phosphate ketol-isomerase, EC 5.3.1.1) (53, 276) ensures that neither of the two products do accumulate.

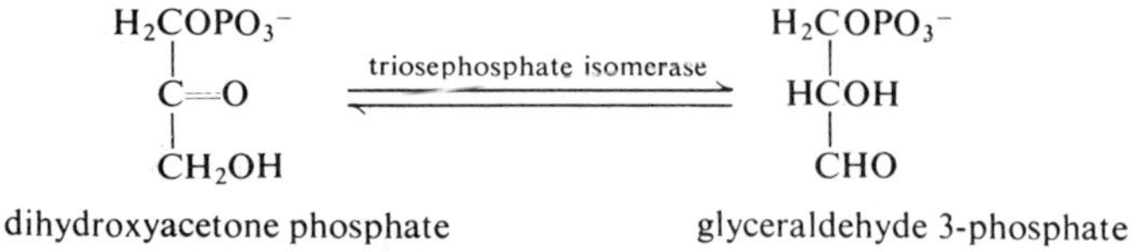
triosephosphate isomerase
dihydroxyacetone phosphate
glyceraldehyde 3-phosphate

The equilibrium lies toward the right (glyceraldehyde 3-phosphate) as long as the EMP pathway functions in the proper way.

The following step (vi) is a combined oxidation and phosphorylation step, which is catalyzed by the enzyme glyceraldehyde-phosphate de-

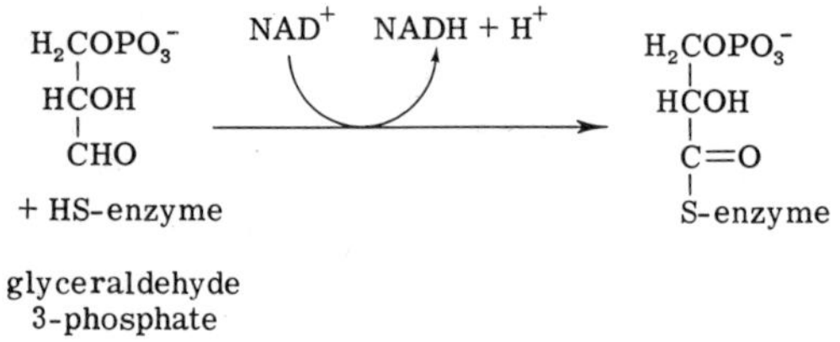
NAD+
NADH + H+
+ HS-enzyme
S-enzyme
glyceraldehyde 3-phosphate

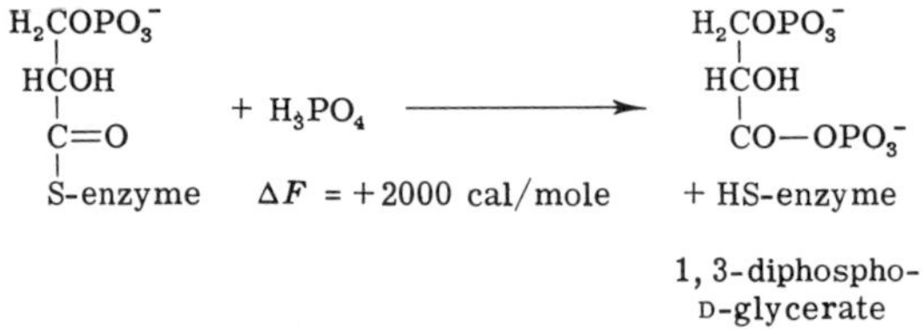
+ H3PO4
ΔF = +2000 cal/mole
+ HS-enzyme
1, 3-diphospho-D-glycerate

hydrogenase [D-glyceraldehyde-3-phosphate: NAD oxidoreductase (phosphorylating), EC 1.2.1.12]. This enzyme is unique insofar as it also transfers acyl groups (137). It is composed of four chemically similar subunits, of which each is able to bind 1 mole of NAD^+, and it also contains four SH groups, one of which is highly reactive and participates in the enzymatic reaction (393). The reactivity of the thiol group is brought about by the binding of NAD^+, as this would lower the pK of the particular SH group (379). The importance of the coenzyme NAD^+ therefore lies in the activation of the thiol group and not in the normal redox step. Adenine nucleotides, such as ATP (127, 128), and such substrates as L-serine (365) inhibit the enzyme activity strongly. Whether or not this inhibition could result in the participation of the enzyme in the Pasteur effect or in general metabolic control will be discussed below. The complexity of the reaction, however, together with the ATP inhibition, certainly suggests some participation in the control mechanism. The enzyme was also found to be thermostable in *Bacillus stearothermophilus* (364), although the mechanism of this thermophily is still unknown.

The high-energy compound 1,3-diphospho-D-glycerate now releases (step vii) one phosphate group to form 1 mole of ATP under the catalytic

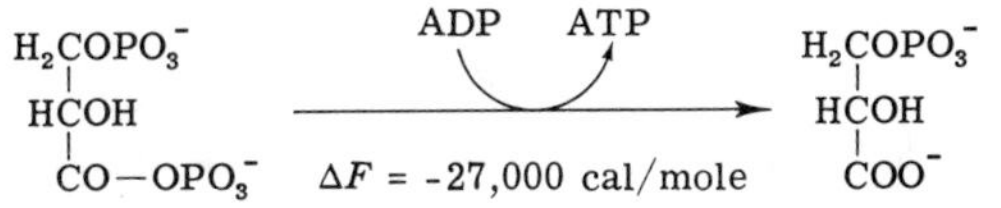
ADP
ATP
ΔF = -27,000 cal/mole

1, 3-diphospho-D-glycerate — 3-phosphoglycerate

action of phosphoglycerate kinase (ATP:3-phospho-D-glycerate 1-phosphotransferase, EC 2.7.2.3).

The conversion of 3-phosphoglycerate to 2-phosphoglycerate (reaction viii) results from the interaction of 2,3-diphosphoglycerate and the enzyme

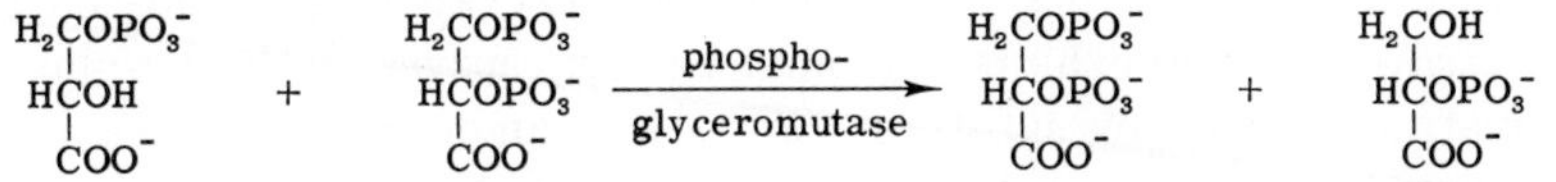
phospho-
glyceromutase

3-phosphoglyceric acid — 2, 3-diphospho-D-glyceric acid — 2, 3-diphospho-D-glyceric acid — 2-phosphoglyceric acid

phosphoglyceromutase (2,3-diphospho-D-glycerate:2-phospho-D-glycerate phosphotransferase EC 2.7.5.3). The phosphate group is transferred to the 2 position in the molecule. Because 2,3-diphospho-D-glycerate is regenerated during this reaction, it is used in a cycling process by this particular reaction.

The formation of phosphoenolpyruvate (reaction ix) is obtained by the action of enolase (2-phospho-D-glycerate hydro-lyase, EC 4.2.1.11) (370).

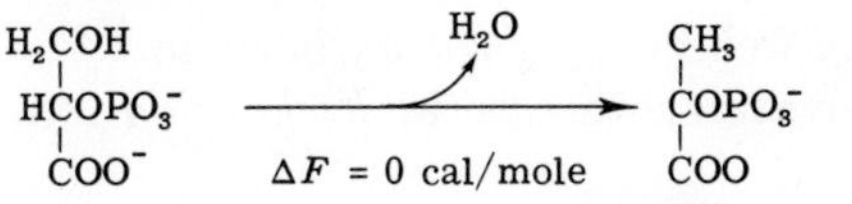
H2O
ΔF = 0 cal/mole

2-phosphoglyceric acid — phosphoenolpyruvic acid

This reaction is connected with an intramolecular electron shift, which is very often referred to as an "intramolecular oxidation–reduction reaction." The electrons are concentrated around C-2 of phosphoenolpyruvate. This concentration and the double-bond formation result in a closer contact between the negative loaded phosphate and carboxyl groups, thus increasing the inner energy tension of the molecule. This rearrangement is the reason phosphoenolpyruvate is a more energy-rich compound than 2-phosphoglycerate. Phosphoenolpyruvate is one of the key intermediates in the anaplerotic sequence system, as will be seen in Chapter 7.

The phosphate group of phosphoenolpyruvate can now, in the final reaction (reaction x), be transferred to ADP with the enzyme pyruvate

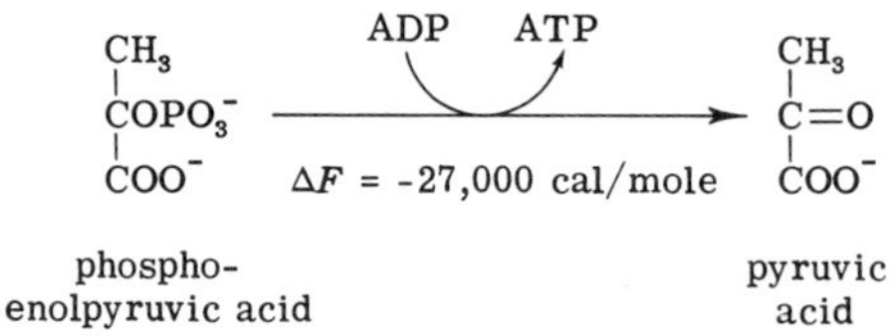
CH3
COPO3-
COO-
ADP
ATP
ΔF = -27,000 cal/mole
CH3
C=O
COO-
phospho-
enolpyruvic acid
pyruvic
acid

kinase (ATP:pyruvate phosphotransferase, EC 2.7.1.40). This reaction completes the pathway, with the formation of pyruvic acid and an additional mole of ATP. Because of the central position of phosphoenolpyruvate as far as anaplerotic and catabolic metabolism is concerned, the enzyme pyruvate kinase underlies certain controls. The kinetics of this enzyme from *Escherichia coli* exhibit sigmoidal velocity curves, with phosphoenolpyruvate as the variable substrate (253), that can be converted to a normal hyperbola on addition of AMP. The specific substrate required is Mg^{2+}–ADP^- (271), with ATP being a strong inhibitor (41) and fructose 1,6-diphosphate a good activator (253). Pyruvate kinase, in other words, is the second allosteric enzyme in the EMP pathway. It is not likely to play any role in gluconeogenesis, as the reverse reaction from pyruvate to phosphoenolpyruvate is almost impossible because of its unfavorable equilibrium under physiological conditions (9). The organism overcomes this barrier by inducing a number of enzymes, such as phosphoenolpyruvate synthetase (30–32), pyruvatephosphate dikinase (9, 323), and a system that uses oxalacetate as intermediate. Although these enzymes are used mainly in gluconeogenesis (glucose synthesis) of acetate- or dicarboxylic acid-grown cells (28), they are mentioned here to show that the pyruvate kinase reaction may well be a regulatory enzyme.

Phosphoenolpyruvate synthetase in *Escherichia coli* (31) catalyzes the reaction

$$\text{ATP} + \text{pyruvate} \rightleftharpoons \text{AMP} + \text{PEP} + \text{Pi}$$

The enzyme has been purified. [^{32}P]-Labeled ATP experiments indicated that two types of transfer reaction are equally probable (30):

1. A mechanism involving a pyrophosphoryl transfer

$$\text{ATP} + \text{pyruvate} \rightarrow \text{pyrophosphorylenolpyruvate} + \text{AMP}$$

$$\text{Pyrophosphorylenolpyruvate} + H_2O \rightarrow \text{PEP} + \text{Pi}$$

2. A mechanism involving an adenosine pyrophosphoryl transfer

$$\text{ATP} + \text{pyruvate} \rightarrow \text{ADP-enolpyruvate} + \text{Pi}$$

$$\text{ADP-enolpyruvate} + H_2O \rightarrow \text{AMP} + \text{PEP}$$

Although evidence seems to favor the first reaction, the mechanism has not been fully elucidated.

Pyruvate, orthophosphate dikinase (ATP:pyruvate, orthophosphate phosphotransferase, EC 2.7.9.1) catalyzes the reaction

$$\text{Pyruvate} + \text{ATP} + \text{Pi} \rightleftharpoons \text{PEP} + \text{AMP} + \text{PPi}$$

and has only been purified as yet from *Bacteroides symbiosus* (323).

The third way PEP can be synthesized from pyruvate is via oxalacetate, whereby pyruvate is first carboxylated to form oxalacetate, which in turn is decarboxylated to phosphoenolpyruvate (PEP). The two enzymes involved in this system are pyruvate carboxylase and PEP-carboxykinase (9). Both reactions require the presence of ATP (254) and/or free fatty acids and CoA (187) and are therefore not favored.

The EMP pathway thus produces two molecules of ATP and 2 moles of $NADH + H^+$, which are used for biosynthesis and as hydrogen donor, respectively, in the further breakdown of pyruvate. The 2 moles of $NADH + H^+$ are formed during the production of 2 moles of pyruvate from 1 mole of glucose at the 1,3-diphosphoglyceric acid level. This metabolic breakdown of glucose to pyruvate can be used by most of the anaerobic microorganisms, such as yeast and fermentative bacteria, because no oxygen is necessary. We know, however, that it is not restricted to these particular groups (23, 309, 340).

Warburg-Dickens or Hexose Monophosphate (HMP) Pathway

The hexose monophosphate pathway of glucose metabolism carries a number of names, e.g., "shunt" and "pentose cycle." It is made up of a rather complicated series of reactions, which can be carried out by many microorganisms that metabolize glucose via the EMP or ED pathway (54, 64, 142, 266, 309). *Escherichia coli,* for example, uses the EMP pathway mainly during anaerobic glucose utilization but also uses the HMP pathway to 20–30% (see regulation on p. 255) of the glucose metabolized and *Agrobacterium tumefaciens* uses the combination ED and HMP in a ratio of 55:44 (15a).

The process of glucose conversion to glucose 6-phosphate (reaction i) is identical to reaction (i) of the EMP pathway. From here, however, the HMP pathway departs from the EMP pathway (Fig. 5.2). Glucose 6-phosphate is not isomerized to fructose 6-phosphate but is oxidized instead

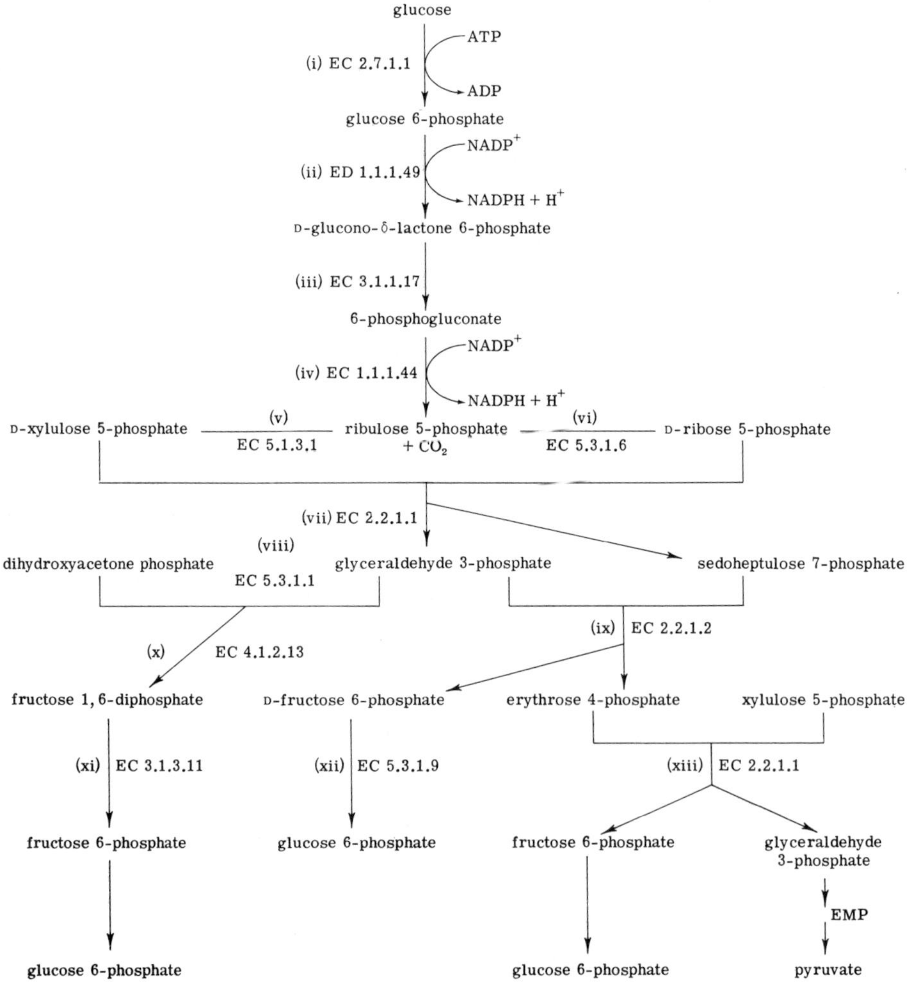
glucose
ATP
(i) EC 2.7.1.1
ADP
glucose 6-phosphate
NADP+
(ii) ED 1.1.1.49
NADPH + H+
D-glucono-δ-lactone 6-phosphate
(iii) EC 3.1.1.17
6-phosphogluconate
NADP+
(iv) EC 1.1.1.44
NADPH + H+
D-xylulose 5-phosphate
(v)
EC 5.1.3.1
ribulose 5-phosphate
+ CO2
(vi)
EC 5.3.1.6
D-ribose 5-phosphate
(vii) EC 2.2.1.1
(viii)
dihydroxyacetone phosphate
EC 5.3.1.1
glyceraldehyde 3-phosphate
sedoheptulose 7-phosphate
(ix) EC 2.2.1.2
(x)
EC 4.1.2.13
fructose 1, 6-diphosphate
D-fructose 6-phosphate
erythrose 4-phosphate
xylulose 5-phosphate
(xi) EC 3.1.3.11
(xii) EC 5.3.1.9
(xiii) EC 2.2.1.1
fructose 6-phosphate
glucose 6-phosphate
fructose 6-phosphate
glyceraldehyde
3-phosphate
EMP
glucose 6-phosphate
glucose 6-phosphate
pyruvate

Fig. 5.2. Hexose monophosphate (HMP) pathway of glucose utilization.

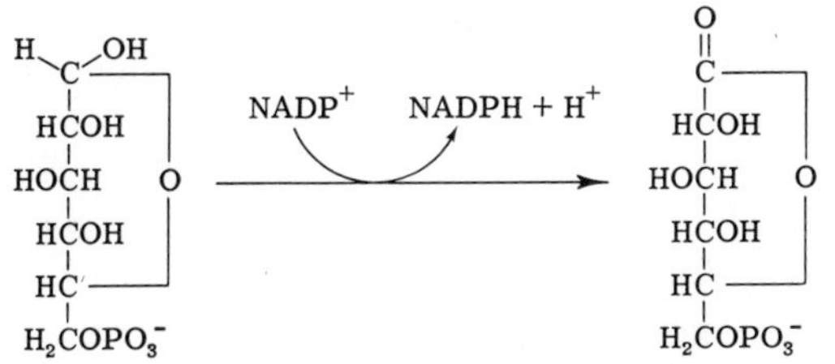
NADP+
NADPH + H+

glucose 6-phosphate   D-glucono-δ-lactone 6-phosphate

with the NADP-linked glucose-6-phosphate dehydrogenase (D-glucose-6-phosphate:NADP oxidoreductase, EC 1.1.1.49). This reaction (ii) produces D-glucono-δ-lactone 6-phosphate. The enzyme glucose-6-phosphate dehydrogenase is becoming known as an important regulatory enzyme as it is inhibited not only by NADH + H^+, in the case of *Escherichia coli* (see p. 255), but also by ATP, as in the case of the pseudomonads (377). This key role was exposed in studies of the metabolism of *Hydrogenomonas* and other mixotrophic bacteria. Here, glucose-6-phosphate dehydrogenase regulates the use of the autotrophic carbon dioxide fixation or heterotrophic metabolism. Although this enzyme is mainly $NADP^+$ linked, it can also occur as the NAD^+-linked enzyme (44).

Glucose-6-phosphate dehydrogenase is localized exclusively in the soluble fraction of the cell and has been detected in a wide variety of cells and tissues in the plant and animal kingdoms (yeast; *Escherichia coli; Bacillus subtilis; Bacillus megaterium; Pseudomonas fluorescens; Zymomonas mobilis; Leuconostoc mesenteroides;* red, blue-green, and green algae; and higher plants). The enzyme of yeast, *Escherichia coli,* and mammalian tissues origin is specific for $NADP^+$. However, it has been reported (28a, 69, 405) that the enzyme lacks coenzyme specificity and is also active with NAD^+. This lack of specificity could also be owing to a transhydrogenase (69), which catalyzes the reaction

$$NADPH + H^+ + NAD^+ \rightarrow NADP^+ + NADH + H^+$$

as is the case in *Pseudomonas fluorescens* (405) or two different dehydrogenases with distinctly different functional roles as in *Acetobacter xylinum* (28a). Although the physiological function of transhydrogenases are not fully elucidated as yet (316), they appear to exist wherever $NADP^+$-linked dehydrogenases are functioning. The enzyme in *Aspergillus flavus-oryzae* is specific for both $NADP^+$ and NAD^+. In *Bacterium anitratum* (165) and *Bacillus megaterium* (44), glucose dehydrogenase can metabolize glucose without the intervention of pyridine or flavin nucleotides. It is therefore assumed that a novel prosthetic group is involved.

The product of the glucose-6-phosphate dehydrogenase reaction is almost immediately hydrolyzed (reaction iii) to 6-phosphogluconate by

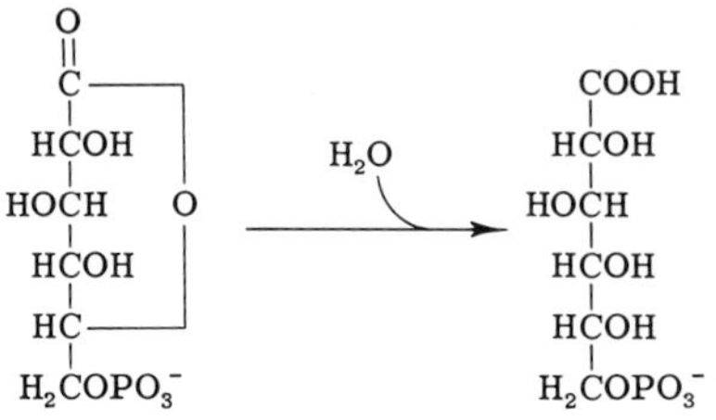
H_2O

D-glucono-δ-lactone 6-phosphate — 6-phospho-gluconic acid

gluconolactonase (D-glucono-δ-lactone hydrolase, EC 3.1.1.17). Very little is known about this enzyme because of its quick catalytic reaction.

A second $NADP^+$-linked oxidation occurs in reaction (iv), which leads to the formation of D-ribulose 5-phosphate. This reaction, catalyzed by

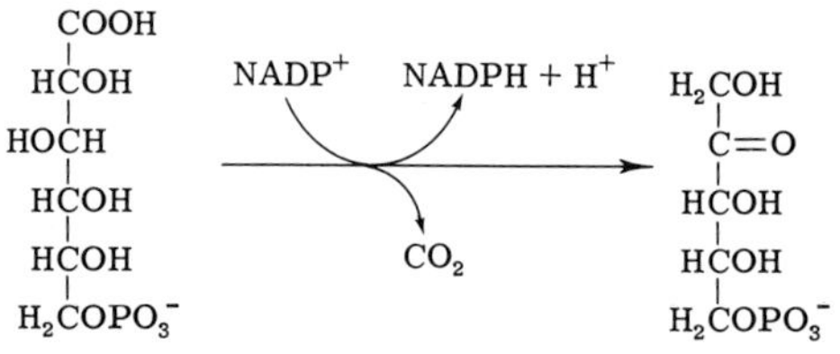
$NADP^+$
$NADPH + H^+$
CO_2

6-phospho-gluconic acid — ribulose 5-phosphoric acid

phosphogluconate dehydrogenase [6-phospho-D-gluconate:NADP oxido-reductase (decarboxylating), EC 1.1.1.44], was first demonstrated in yeast by Warburg and Christian in 1933 (389). They obtained evidence that the product was a C_5 compound, most likely a pentose phosphate, which we now know to be ribulose 5-phosphate.

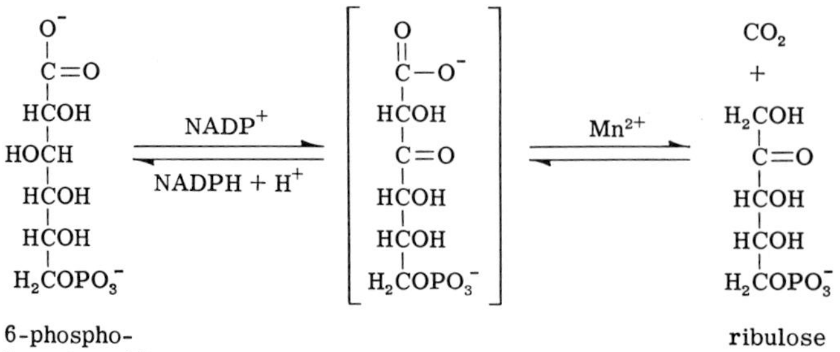
$NADP^+$
$NADPH + H^+$
Mn^{2+}
CO_2 +

6-phospho-gluconic acid — ribulose 5-phosphoric acid

Recent investigations suggest that 3-keto-6-phosphogluconate (239) may be an intermediate in the reaction, but this compound has not yet been isolated or even detected. Although the equilibrium of 6-phosphogluconate oxidation favors oxidative decarboxylation, the reaction is reversible, as demonstrated by the enzymatic fixation of $^{14}CO_2$ into the C-1 atom of 6-phosphogluconate and by the reductive carboxylation of ribulose 5-phosphate in the presence of NADPH + H^+ and carbon dioxide.

Although the majority of workers reported 6-phosphogluconate dehydrogenase to be exclusively in the soluble fraction of the cell, there is evidence that the enzyme is associated with a particulate fraction in *Pseudomonas fluorescens* (403). 6-Phosphogluconate dehydrogenase is similar in its distribution in the animal and plant kingdoms to glucose-6-phosphate dehydrogenase. The mammalian and *Escherichia coli* (367) enzymes are specific for $NADP^+$, but the enzyme obtained from *Leuconostoc mesenteroides* reduces NAD^+ at 25 times the rate of $NADP^+$. As seen with glucose-6-phosphate dehydrogenase, *Aspergillus flavus-oryzae* contains two 6-phosphogluconate dehydrogenases that are specific for either NAD^+ or $NADP^+$. The occurrence of $NADP^+$- and NAD^+-linked 6-phosphogluconate dehydrogenase in *Rhizobium* (264) led to the suggestion of an enzymatic differentiation of rhizobia.

6-Phosphogluconate dehydrogenase, like glucose-6-phosphate dehydrogenase, is an important regulatory enzyme, for it is inhibited by fructose 1,6-diphosphate, ATP, and possibly NADPH + H^+, depending on the

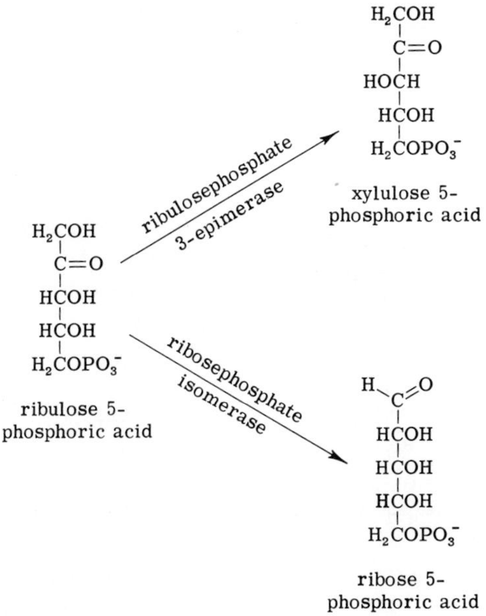
ribulosephosphate 3-epimerase
xylulose 5-phosphoric acid
ribulose 5-phosphoric acid
ribosephosphate isomerase
ribose 5-phosphoric acid

organism (393a). It could therefore play an important role in the Pasteur effect and in glucose regulation (see p. 255).

D-Ribulose 5-phosphate is attacked partly by two different enzymes. The action of ribulosephosphate 3-epimerase (D-ribulose-5-phosphate 3-epimerase, EC 5.1.3.1) converts ribulose 5-phosphate into xylulose 5-phosphate (reaction v), whereas the enzyme ribose-5-phosphate isomerase (D-ribose-5-phosphate ketol-isomerase, EC 5.3.1.6) is responsible for the formation of ribose 5-phosphate (reaction vi). This type of conversion makes it possible for the organism to replenish ribose 5-phosphate, as this intermediate forms the basic precursor for the purine, pyrimidine, and aromatic amino acid biosynthesis.

Both intermediates, xylulose 5-phosphate and ribose 5-phosphate, are required for the cleavage reaction (vii) of the enzyme transketolase (se-

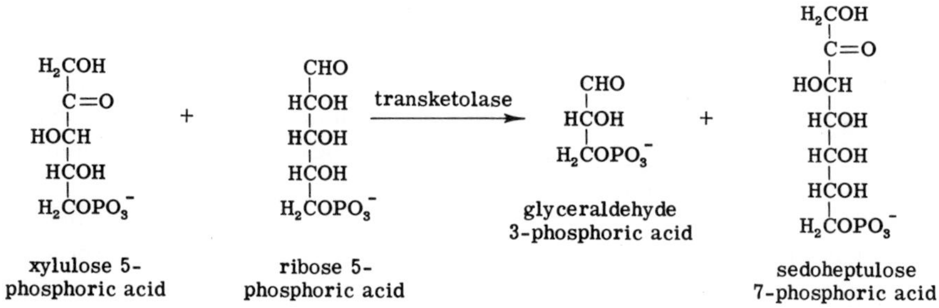
transketolase
xylulose 5-phosphoric acid
ribose 5-phosphoric acid
glyceraldehyde 3-phosphoric acid
sedoheptulose 7-phosphoric acid

doheptulose-7-phosphate: D-glyceraldehyde-3-phosphate glycolaldehydetransferase, EC 2.2.1.1), which yields glyceraldehyde 3-phosphate and sedoheptulose 7-phosphate.

The second cleavage reaction of this pathway (ix) cleaves both intermediates from the transketolase reaction and produces D-fructose 6-phosphate and D-erythrose 4-phosphate.

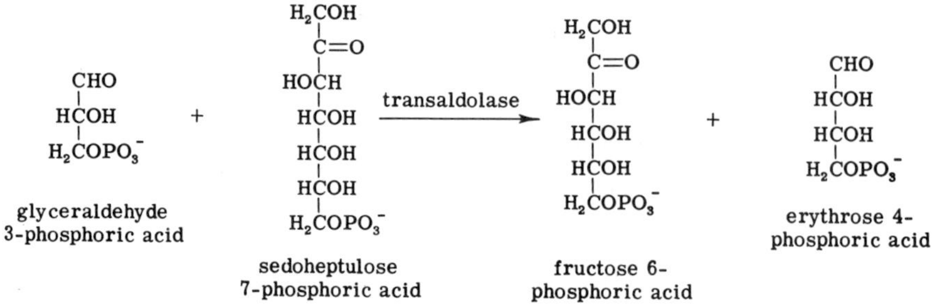
transaldolase
glyceraldehyde 3-phosphoric acid
sedoheptulose 7-phosphoric acid
fructose 6-phosphoric acid
erythrose 4-phosphoric acid

This reaction is catalyzed by transaldolase (sedoheptulose-7-phosphate: D-glyceraldehyde-3-phosphate dihydroxyacetonetransferase, EC 2.2.1.2).

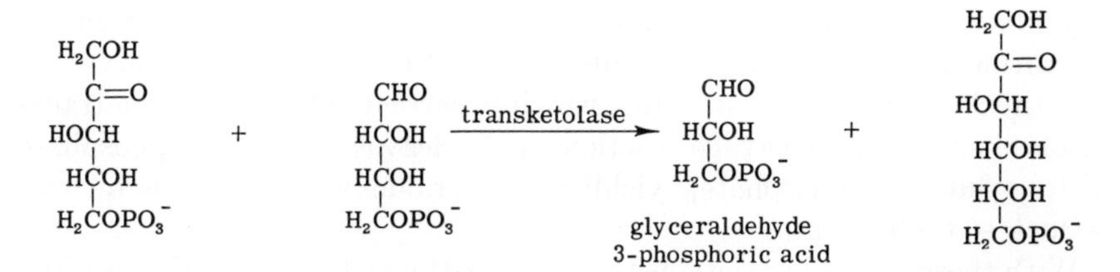
transketolase

xylulose 5-phosphoric acid + erythrose 4-phosphoric acid → glyceraldehyde 3-phosphoric acid + fructose 6-phosphoric acid

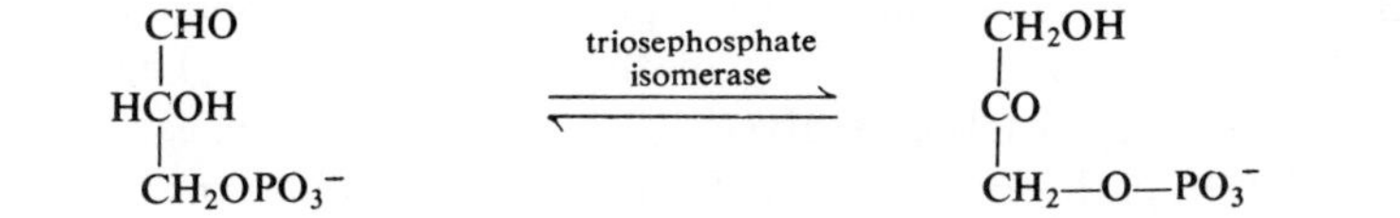
triosephosphate isomerase

glyceraldehyde 3-phosphate ⇌ dihydroxyacetone phosphate

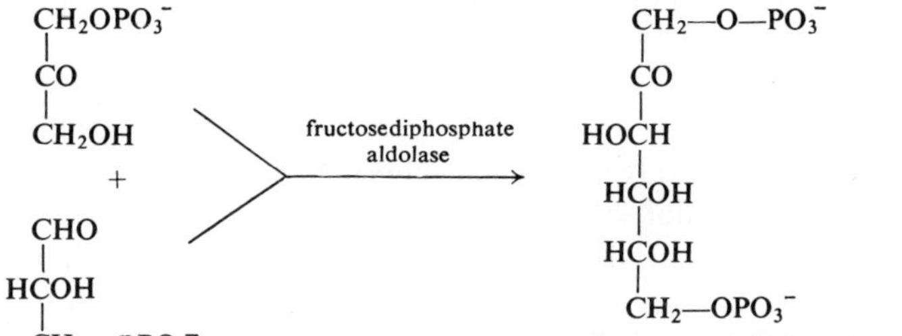
fructosediphosphate aldolase

fructose 1,6-diphosphate

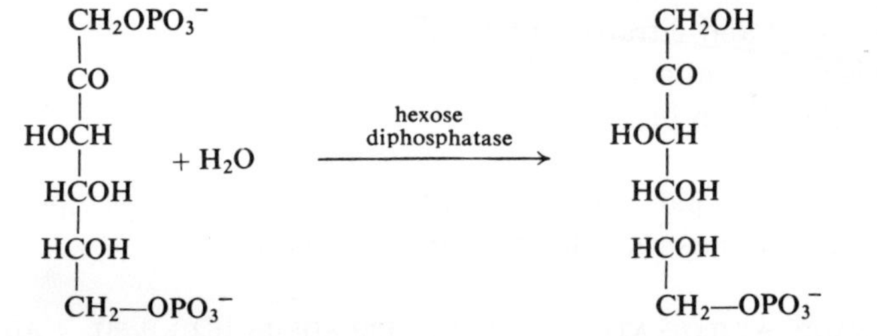
hexose diphosphatase

fructose 1,6-diphosphate → fructose 6-phosphate

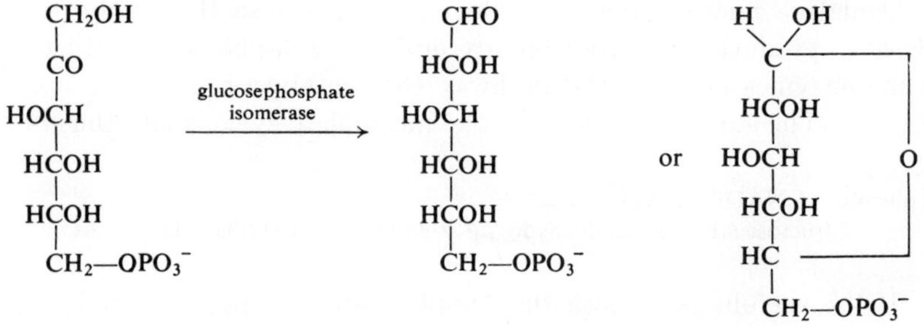
glucosephosphate isomerase
or

fructose 6-phosphate → glucose 6-phosphate

Erythrose 4-phosphate, like ribose 5-phosphate, is an important precursor for purine and pyrimidine and in aromatic amino acid biosynthesis.

A third cleavage reaction (xiii), which is carried out by the same transketolase as the first cleavage reaction (vii), cleaves erythrose 4-phosphate with xylulose 5-phosphate, yielding glyceraldehyde 3-phosphate and D-fructose 6-phosphate.

With these cleavage reactions the HMP pathway links up with the EMP pathway at the fructose 6-phosphate and glyceraldehyde 3-phosphate level. The cycle can be complete, because fructose 6-phosphate can be converted to D-glucose 6-phosphate with a glucosephosphate isomerase, an enzyme occurring also in reaction (ii) of the EMP pathway. The other product of the cleavage reaction, D-glyceraldehyde 3-phosphate, can follow the reverse EMP pathway and also form D-glucose 6-phosphate. The formation of dihydroxyacetone phosphate by triosephosphate isomerase is identical to reaction (v). The cleavage of both triose phosphates to fructose 1,6-diphosphate by fructose-diphosphate aldolase is also identical with reaction (iv) of the EMP pathway. In order to form fructose 6-phosphate, the organism must produce a separate enzyme, as reaction (iii) of the EMP pathway catalyzed by phosphofructokinase is an irreversible reaction. Fructose 1,6-diphosphate can be converted to fructose 6-phosphate by the enzyme hexosediphosphatase (D-fructose-1,6-diphosphate 1-phosphohydrolase, EC 3.1.3.11) (121, 122), and this is also an irreversible reaction. This enzyme has been reported to exist in two forms (247) in rabbit liver. A recent purification from *Acinetobacter* (288) indicated an allostery to ATP and citrate. The possible importance of this enzyme for regulation is not as yet certain. The "end product" glucose 6-phosphate is obtained by the action of glucosephosphate isomerase (EC 5.3.1.9), as noted above.

This complete HMP pathway represents what is called the "pentose cycle" or "shunt" and has the following sum of reactions:

$$\text{glucose} + 12\,\text{NADP}^+ + 7\,\text{H}_2\text{O} + \text{ATP} \rightleftharpoons 6\,\text{CO}_2 + 12(\text{NADPH} + \text{H}^+) + \text{H}_3\text{PO}_4 + \text{ADP}$$

Oxidative microorganisms, however, can also use the partly complete cycle to produce pyruvate from glyceraldehyde 3-phosphate, utilizing the same enzymes of the EMP pathway that catalyze reactions (vi–x) (see p. 219). The sum of reactions for the incomplete cycle would thus be:

$$3\,\text{glucose} + 6\,\text{NADP}^+ + \text{ATP} \rightleftharpoons 2\,\text{fructose 6-P} + \text{glyceraldehyde 3-P} + 3\,\text{CO}_2 + 6(\text{NADPH} + \text{H}^+) + \text{ADP} + \text{H}_3\text{PO}_4$$

If the usefulness of both the complete and the partly complete HMP pathway are compared, it could be envisaged that oxidative microorganisms most certainly would utilize the partly complete pathway to obtain energy via pyruvate and the TCA cycle. However, those microorganisms which metabolize glucose mainly through the EMP pathway under anaerobic conditions would utilize the complete HMP pathway only for the purpose of producing the precursors for their purine, pyrimidine, and aromatic amino acid biosynthesis. Such microorganisms as the lactobacilli, which utilize solely the EMP pathway for their glucose degradation, require complex media to compensate for the loss of biosynthetic ability.

The incomplete HMP cycle requires 1 mole of ATP and produces 2 moles of ATP in the formation of pyruvate. Reduced pyridine nucleotides are also formed and could function as hydrogen donors for the reduction

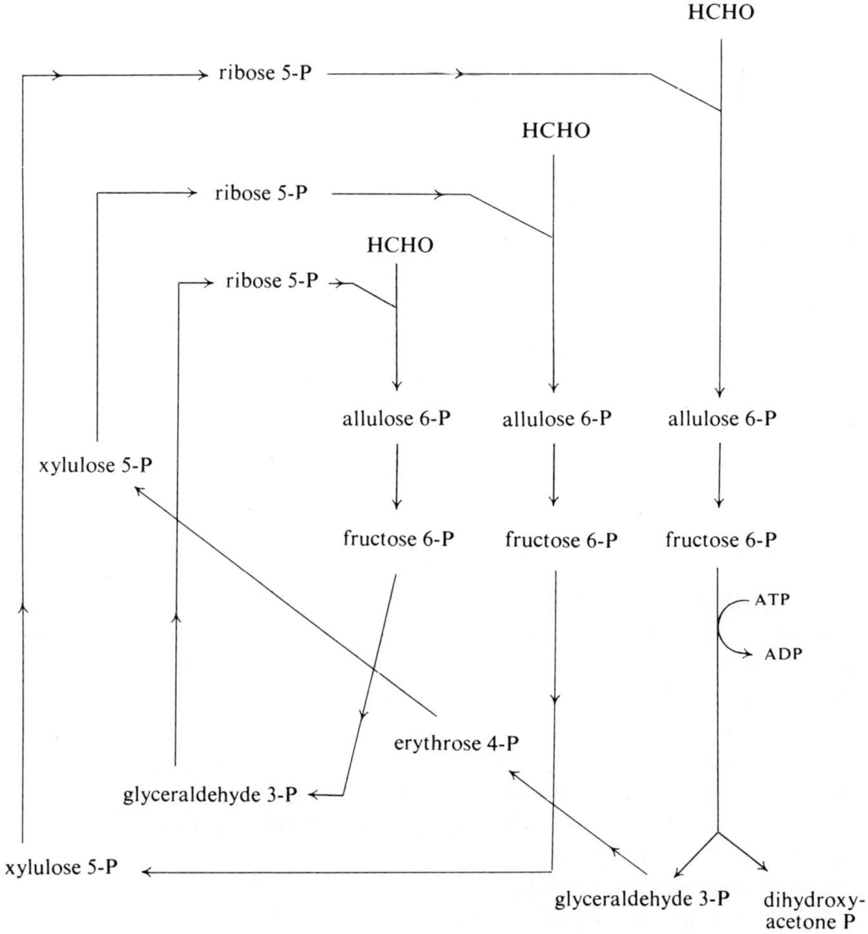
HCHO
ribose 5-P
HCHO
ribose 5-P
HCHO
ribose 5-P
allulose 6-P
allulose 6-P
allulose 6-P
xylulose 5-P
fructose 6-P
fructose 6-P
fructose 6-P
ATP
ADP
erythrose 4-P
glyceraldehyde 3-P
xylulose 5-P
glyceraldehyde 3-P
dihydroxy-
acetone P

Scheme 5.1. Allulose 6-phosphate pathway of *Pseudomonas methanica.*

of an organic compound. The pathway therefore does not necessarily have to be dependent on oxygen in order to function. As the pathway also has no oxygen-sensitive enzyme, it can function both aerobically and anaerobically. The functioning of the HMP pathway is also of great importance to photosynthetic and chemosynthetic autotrophs because all their cellular carbon is derived by condensing carbon dioxide with ribulose 1,5-diphosphate, which is derived from ribulose 5-phosphate.

A slight variation of the HMP pathway has been found to occur in *Pseudomonas methanica*. As will be mentioned when aerobic respiration is considered (see Chapter 6), *Pseudomonas methanica* oxidizes methane stepwise via methanol, formaldehyde, and formate to carbon dioxide. It was thought that methane and methanol were rapidly assimilated by growing cell suspensions into sugar phosphates, whereas carbon dioxide was incorporated mainly into C_4-dicarboxylic acids (189). Tracer element work, however, revealed that it is the formaldehyde that is incorporated into phosphorylated compounds, whereas formate is assimilated into serine and malate (207).

It appears that high fixation occurs between the oxidation levels of methane and formaldehyde and low fixation between formate and carbon dioxide. This means that formaldehyde must be the oxidation level at which most of the carbon is assimilated into cell constituents. With the observation that all the pentose phosphate cycle intermediates occur in *Pseudomonas methanica* and with the identification of a new hexose phosphate, allulose 6-phosphate (206), it has been possible to suggest a mechanism for formaldehyde incorporation. The first reaction involves a hydroxymethylation of ribose 5-phosphate and formaldehyde to allulose 6-phosphate (Scheme 5.1). With this mechanism functioning, *Pseudomonas methanica* is able to produce most of the required intermediates for its cell material biosynthesis.

Entner-Doudoroff Pathway

This pathway is one of the most recently described pathways, discovered by Entner and Doudoroff (114) during metabolic studies of *Pseudomonas saccharophila*. It has since been found in a number of other bacteria (26a, 64, 84, 145, 218, 219, 373, 391) (see Fig. 5.3).

Reactions (i), (ii), and (iii) are identical to the ones of the HMP pathway, although it is not as yet certain whether or not the enzymes themselves differ in their kinetic characteristics. The marked differences of this pathway from the other two is the dehydration of 6-phosphogluconate (reaction iv) (218) to form 2-keto-3-deoxy-6-phosphogluconate (KDPG) by a phosphogluconate dehydratase (6-phosphogluconate hydro-lyase, EC 4.2.1.12). It has been proposed that 6-phosphogluconate is dehydrated

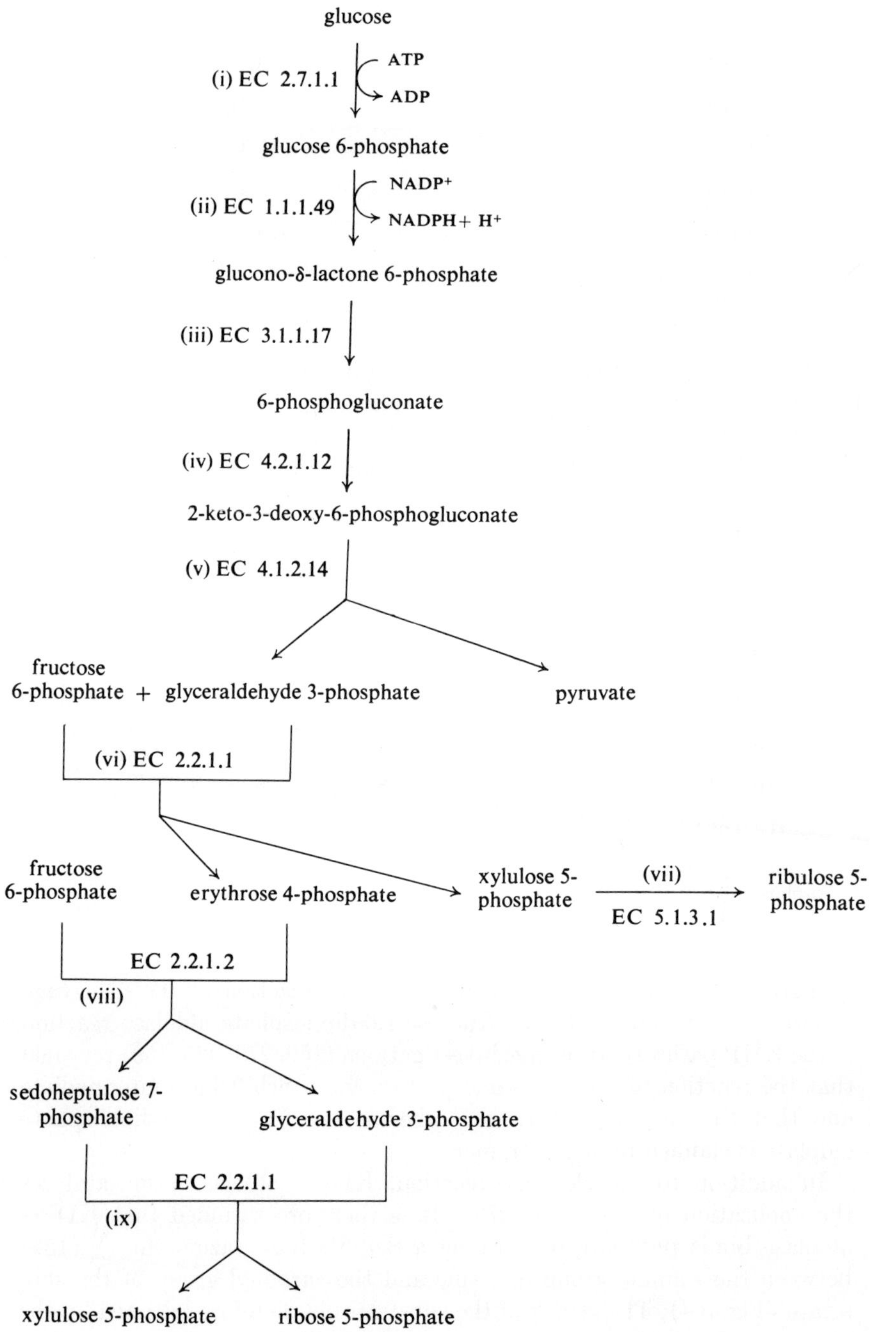
glucose
(i) EC 2.7.1.1
ATP
ADP
glucose 6-phosphate
(ii) EC 1.1.1.49
NADP+
NADPH+ H+
glucono-δ-lactone 6-phosphate
(iii) EC 3.1.1.17
6-phosphogluconate
(iv) EC 4.2.1.12
2-keto-3-deoxy-6-phosphogluconate
(v) EC 4.1.2.14
fructose 6-phosphate + glyceraldehyde 3-phosphate
pyruvate
(vi) EC 2.2.1.1
fructose 6-phosphate
erythrose 4-phosphate
xylulose 5-phosphate
(vii)
EC 5.1.3.1
ribulose 5-phosphate
EC 2.2.1.2
(viii)
sedoheptulose 7-phosphate
glyceraldehyde 3-phosphate
EC 2.2.1.1
(ix)
xylulose 5-phosphate
ribose 5-phosphate

Fig. 5.3. Entner-Doudoroff pathway of glucose utilization.

COOH
HCOH
HOCH
HCOH
HCOH
$H_2COPO_3^-$
COOH
C=O
CH_2
HCOH
HCOH
$H_2COPO_3^-$

6-phosphogluconic acid

2-keto-3-deoxy-6-phosphogluconic acid

first to form enol-2-keto-3-deoxy-6-phosphogluconate, which spontaneously ketonizes (272) to 2-keto-3-deoxy-6-phosphogluconate.

A subsequent cleavage reaction by phospho-2-keto-3-deoxy-gluconate aldolase (6-phospho-2-keto-3-deoxy-D-gluconate D-glyceraldehyde-3-phosphate-lyase, EC 4.1.2.14) cleaves the 2-keto-3-deoxy-6-phosphogluconate

COOH
C=O
HCH
HCOH
HCOH
$H_2COPO_3^-$
COOH
C=O
CH_3
pyruvic acid
+
CHO
HCOH
$H_2COPO_3^-$

2-keto-3-deoxy-6-phosphogluconic acid

glyceraldehyde 3-phosphoric acid

into glyceraldehyde 3-phosphate and pyruvate (reaction v). This cleavage reaction is very similar to the fructose-1,6-diphosphate aldolase reaction in the EMP pathway. Intensive investigations (159, 273, 325, 359) revealed that the reaction follows the same pattern via a Schiff base intermediate and that the enzyme is a trimer, whereas the fructose-1,6-diphosphate aldolase is claimed to be a tetramer.

In addition to the cleavage reaction, KDPG aldolase also catalyzes the enolization of pyruvate (273). It is therefore assumed that KDPG aldolase binds pyruvate by forming a Schiff's base (azomethine) (154) between the ε-amino group of lysine and the carbonyl group of the substrate (Fig. 5.4). The attack of the substrates is therefore initiated by the electron pair of an amino nitrogen, forming a KDPG–lysine azomethine,

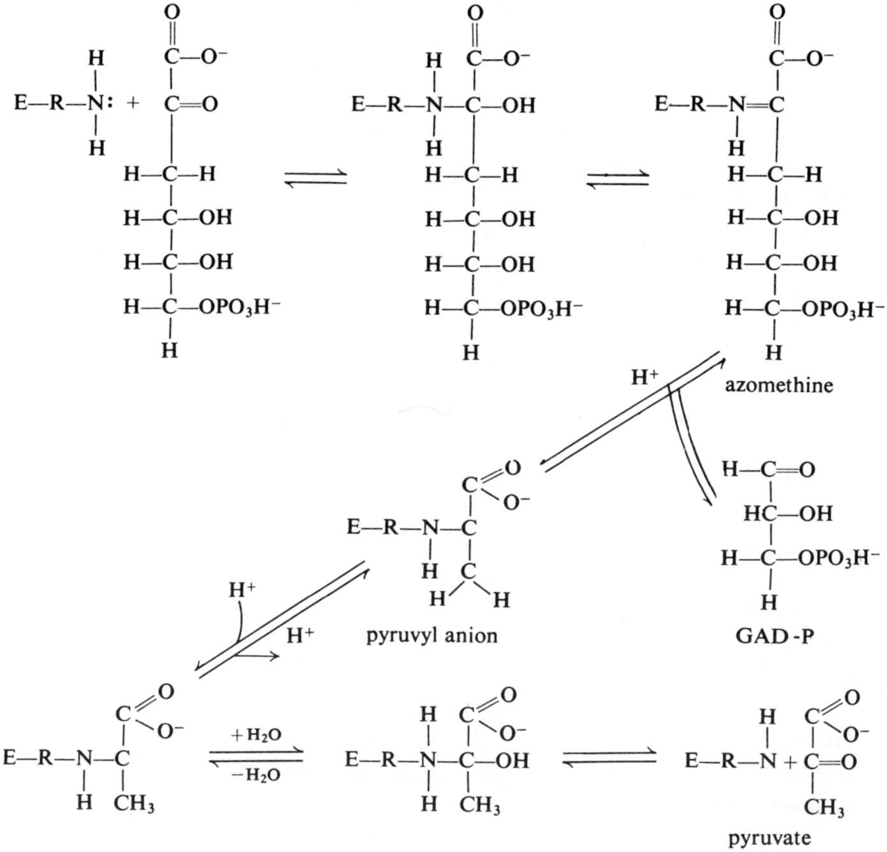
azomethine
pyruvyl anion
GAD-P
pyruvate

Fig. 5.4. Proposed mechanism of α-keto-3-deoxy-6-phosphogluconic acid aldolase [Meloche and Wood (273); reprinted with the permission of *Journal of Biological Chemistry*].

which is rearranged with the elimination of glyceraldehyde 3-phosphate and yields a pyruvyl anion. The pyruvyl anion can then be attacked either by a proton or by a tautomeric form of glyceraldehyde 3-phosphate, yielding pyruvate in the former and KDPG in the latter case. The overall system is completely reversible. *Acetobacter melanogenum* and *Azotobacter vinelandii* also show KDPG aldolase activity as well as that of phosphogluconate dehydratase. KDPG aldolase was not detected in *Leuconostoc mesenteroides* 8081, *Corynebacterium creatinovorans* 7562, *Microbacterium lacticum* 8081, *Lactobacillus arabinosus* 8014, or brewer's yeast.

The triose phosphate formed during the cleavage of glyceraldehyde 3-phosphate is now free for use in the EMP pathway to pyruvate. Where this pathway is followed it yields 2 moles of ATP per mole substrate (glucose), and 1 mole each of reduced $NADP^+$ and NAD^+.

The more important feature of this pathway, however, is that it enables organisms to produce their pentose precursors for pyridine and pyrimidine as well as for aromatic amino acid biosynthesis by a reversed hexose monophosphate pathway utilizing the same enzymes as discussed in the HMP pathway. Glyceraldehyde 3-phosphate may condense (reaction vi) with fructose 6-phosphate, yielding erythrose 4-phosphate and xylulose 5-phosphate. The reaction is catalyzed by transketolase (EC 2.2.1.1). Ribulose-phosphate 3-epimerase (EC 5.1.3.1) converts xylulose 5-phosphate into ribulose 5-phosphate (reaction vii), whereas a transaldolase (EC 2.2.1.2) forms glyceraldehyde 3-phosphate and sedoheptulose 7-phosphate from erythrose 4-phosphate and fructose 6-phosphate (reaction viii). Ribose 5-phosphate and xylulose 5-phosphate are the products from a second transketolase (EC 2.2.1.1) reaction (reaction ix). Depending on the environmental conditions, microorganisms are able to use the reversed pentose cycle to meet their pyridine and pyrimidine requirement or, alternatively, to use this cycle only partially to form pyruvate from glyceraldehyde 3-phosphate. Pyruvate can then be metabolized in ways to be considered below.

Each of the three different pathways could serve certain aspects of metabolism. The EMP pathway provides the greatest amount of ATP, 2 moles, but does not produce the most important precursors for purine and pyrimidine biosynthesis, ribose 5-phosphate and erythrose 4-phosphate. Therefore, it could be assumed that microorganisms using this pathway must have specific growth requirements for pentoses, purines, or pyrimidines in order to build, for example, their nucleic acids. In general, organisms that grow only on complex media, e.g., media containing meat extract and yeast extract, use this particular pathway (257, 326).

The HMP pathway, in contrast, produces all the precursors necessary for purine and pyrimidine biosynthesis but produces only half the amount of ATP. This pathway does not produce pyruvate directly, so the organism must have at least part of the EMP pathway enzyme complex, from glyceraldehyde 3-phosphate onward, to form pyruvate. It is therefore not surprising that both pathways may be present in those organisms that do possess the HMP pathway (see Table 5.3) and that are therefore independent in regard to their purine and pyrimidine requirements. The ratio of usage of the EMP and HMP pathways can vary greatly, depending on environmental conditions.

The Entner-Doudoroff pathway is at the other side of the HMP, where the reverse HMP pathway may be present but the direct formation of pyruvate makes it possible for it to be independent of the other two pathways. The production of 1 mole of ATP and all necessary pentoses makes

TABLE 5.3
Estimation of Catabolic Pathways of Glucose Pathway Estimation[a]

Microorganisms	EMP	HMP	ED
Saccharomyces cerevisiae	88	12	—
Candida utilis	66–81	19–34	—
Streptomyces griseus	97	3	
Penicillium chrysogenum	77	23	
Escherichia coli	72	28	
Sarcina lutea	70	30	
Bacillus subtilis	74	26	
Pseudomonas aeruginosa	—	29	71
Acetomonas oxydans	—	100	—
Zymomonas mobilis	—	—	100

[a] From (62). (Reprinted with permission of John Wiley and Sons.)

it very similar to the HMP pathway, particularly as a number of enzymes are the same. There are some groups of bacteria, particularly gram-negative bacteria with a deoxynucleic acid (DNA) base composition of between 52 and 70% guanine plus cytosine (G + C) (211), that are able to use this pathway alone. Table 5.3 indicates the percentage distribution of these pathways in some microorganisms. The reasons for the individual choices are not yet quite clear.

It was thought that oxygen may play an important role in the selection of pathway usage, since the majority of anaerobic bacteria were found to contain the EMP pathway, e.g., clostridia (12, 14, 157, 234, 263, 307, 360, 402), enteric bacteria (46), spirochetes (135, 136, 175), and sarcinae (56). Those bacteria which are facultative aerobes were found to contain a combination of EMP and HMP pathway, e.g., *Escherichia coli,* and strict aerobes were found to contain almost exclusively the ED pathway, e.g., pseudomonads and *Rhizobium* (203). However, the findings that the strict anaerobes *Zymomonas mobilis* and *Zymomonas anaerobia* (250) can use only the ED pathway, that *Clostridium aceticum* can use a modified ED pathway (12) with gluconate as carbon source, and that homofermentative lactobacilli (154a, 278) can use the EMP pathway even under aerobic conditions, as *Ferrobacillus ferrooxidans* does, cast doubt on the early assumptions.

A few examples of the participation of mixed pathways and the problems they involve for metabolic studies and evaluation in general will now be given.

The pathway of D-glucose metabolism in *Salmonella typhimurium* (120) can follow either a combination of the EMP and HMP pathways or the ED pathway. Glucose strongly inhibited catalase synthesis (222) and catalase activity therefore changed throughout the growth cycle. A similar suppression has been observed in *Mycoplasma pneumoniae* (246). Investigations on a wild and a mutant strain of *S. typhimurium* showed that the wild type utilized 20% of the glucose via the HMP pathways and metabolized the remainder via the EMP pathway. Although the mutant was deficient in the enzyme phosphoglucose isomerase, it was not possible to adapt the strain to the HMP to a greater extent than the wild type, despite the greater need of the mutant for this particular pathway. On gluconate as substrate, however, both the wild type and the mutant utilized gluconate via the ED pathway and not via the EMP or HMP pathway. Gluconokinase and gluconate-6-phosphate dehydratase were found to be inducible.

A similar arrangement of glucose utilization was found in *Leptospira* (25), *Myxococcus xanthus* (392), and *Haemophilus parainfluenzae* (394), where the EMP and HMP are functional. Evidence has been given for the need of a functioning electron transport system. The growth of *Haemophilus parainfluenzae* depends on the presence of electron acceptors, such as oxygen, nitrate, pyruvate, fumarate, NAD^+, and $NADP^+$, in the media. Such an obligatory requirement of oxygen for glucose catabolism via the EMP pathway has also been demonstrated for *Pasteurella tularensis, Agrobacterium tumefaciens* (177), and *Streptomyces coelicolor* (66). Whether or not oxygen is required for the biosynthesis of cofactors, an example of which has been given in Chapter 4 for heme biosynthesis, must be left for further studies. A good example of the dependence of end products on cultural conditions has been demonstrated with the facultative anaerobe *Actinomyces naeslundii* (49). This organism is pathogenic. It ferments sugar but requires carbon dioxide for maximal growth. Without the addition of carbon dioxide, growth is limited and the fermentation product is lactic acid. Addition of carbon dioxide causes a change in end product formation, and formate, acetate, and succinate are found in addition to lactate in a ratio that depends on the carbon dioxide concentration (48, 311). The major pathway used is the EMP pathway, although the presence of most of the enzymes of the HMP pathway indicates a small participation of this pathway as well.

Carbohydrate metabolism can also occur when an organism is unable to grow in the presence of nonphosphorylated sugars as sole source of carbon and energy but it requires permeable intermediates of the TCA cycle (27). The parent strain of *Mima polymorpha*, for example, lacks some kinases and therefore requires additional energy. The enzymes found in this organism (265), glucose dehydrogenase, 6-phosphogluconate dehydrogenase, and phospho-2-keto-3-deoxy-6-gluconate aldolase, indicate the functioning of the HMP and ED pathway.

Radiorespirometric studies (319) revealed that no EMP pathway exists in *Leucothrix mucor* but that glucose is mainly metabolized via the ED pathway with a 20–25% participation of the HMP pathway.

Another combination of pathways appears to exist in *Microbacterium lacticum* (382), where the HMP and EMP seem to function under anaerobic conditions and the phosphoketolase pathway (see p. 244) under aerobic conditions.

Most changes in pathways appear to be attributed to oxygen availability. The most interesting materials for this type of study are certainly the facultative anaerobes (227). The available evidence suggests that these bacteria preferentially use an oxygen-linked electron transport mechanism and therefore develop control mechanisms triggered by oxygen for repressing the synthesis of systems unnecessary for aerobic growth. In other words, facultative aerobes may retain anaerobic mechanisms for the synthesis of certain metabolites but use aerobic energy-yielding mechanisms preferentially to anaerobic ones. Very little is known at present about whether the available oxygen or the nutritional changes influence the carbohydrate utilization. Oxygen availability certainly changes the end products formed (see Chapter 7), but it is uncertain whether it influences the pathways.

Studies with *Bacillus subtilis* revealed that oxygen has no effect on the ratio of participation of EMP and HMP during glucose catabolism (148) when the radiorespirometric method (388) was employed in the study. In resting cells of *Bacillus cereus* (39), only 1–2% of the glucose was utilized by the HMP and the remainder was used by the EMP pathway. During sporulation, however, this ratio changed (99, 147, 158), although it eventually returned to the previous level. The EMP pathway appears to occupy the key position of glucose catabolism at all stages of development of *Bacillus cereus* (149). Similar studies have been carried out with *Clostridium botulinum* (360).

The only genus of bacteria that has been classified according to the pathway of glucose breakdown is *Arthrobacter* (410). *Arthrobacter ureofaciens* and *Arthrobacter globiformis* (284) use primarily the EMP and to some extent the HMP pathway, whereas *A. simplex*, *A. pascens*, and *A. atrocyaneus* catabolize glucose mainly via the ED and HMP pathways. The presence and absence of the ED pathway has been used as the differential criterion. A similar distinction could be made between *Mycobacterium phlei* and *M. tuberculosis* strain H_{37}RA (232). Whereas the latter uses only the EMP pathway, *M. phlei* and *M. tuberculosis* strain H_{37}Rv (321, 322) also use the HMP pathway, as they possess glucose-6-phosphate dehydrogenase and 6-phosphogluconate dehydrogenase. Much more work must be done in this field to find out whether the use of one or another

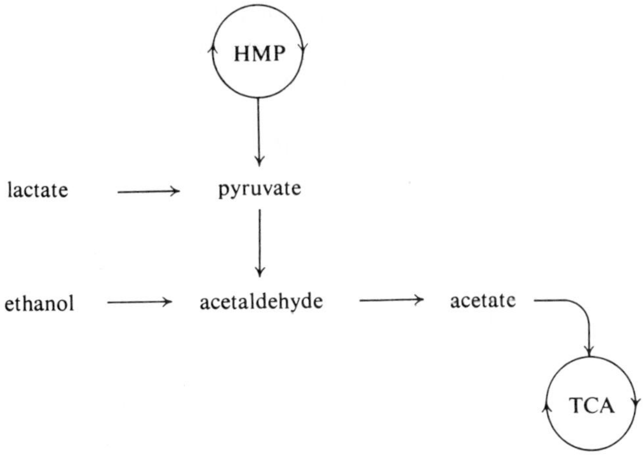
HMP
lactate
pyruvate
ethanol
acetaldehyde
acetate
TCA

Fig. 5.5. Carbohydrate metabolism of the *Acetobacter peroxydans* group.

pathway is dependent on nutritional requirements (24, 300, 301) or on oxygen availability.

A further family of bacteria is classified according to their carbohydrate metabolism. Although the acetic acid bacteria use solely the HMP pathway for carbohydrate utilization, they are known for their oxidative versatility.

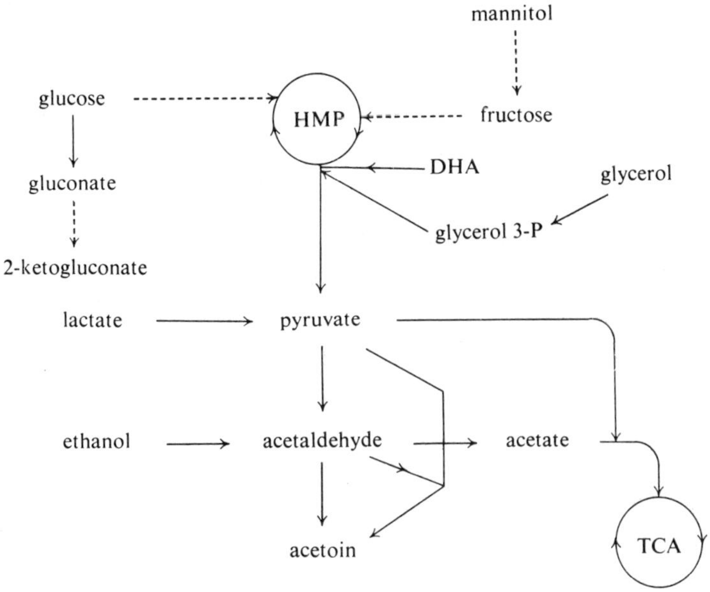
mannitol
glucose
HMP
fructose
DHA
glycerol
gluconate
glycerol 3-P
2-ketogluconate
lactate
pyruvate
ethanol
acetaldehyde
acetate
acetoin
TCA

Fig. 5.6. Carbohydrate metabolism of the *Acetobacter oxydans* group (375) (reprinted with permission of the *Antonie van Leeuwenhoek Journal for Microbiology and Serology*).

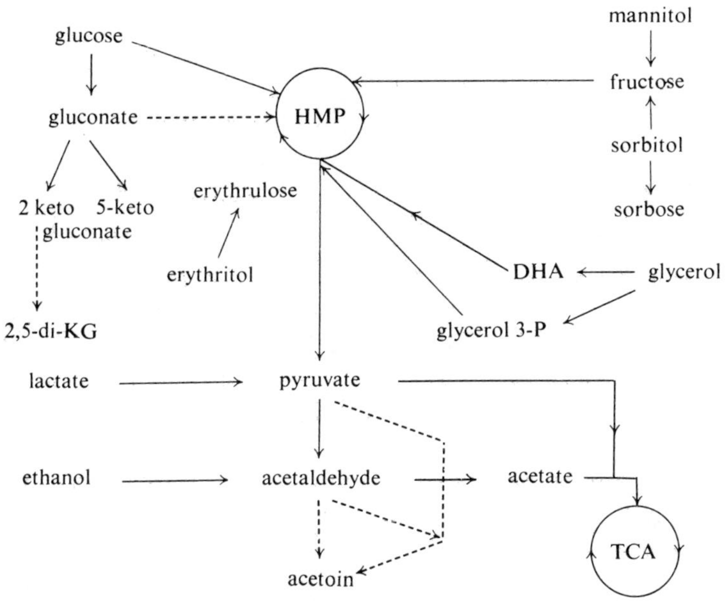
mannitol
glucose
fructose
gluconate
HMP
sorbitol
erythrulose
2 keto 5-keto
gluconate
sorbose
erythritol
DHA
glycerol
2,5-di-KG
glycerol 3-P
lactate
pyruvate
ethanol
acetaldehyde
acetate
TCA
acetoin

Fig. 5.7. Carbohydrate metabolism of the *Acetobacter mesoxydans* group (375) (reprinted with permission of the *Antonie van Leeuwenhoek Journal for Microbiology and Serology*).

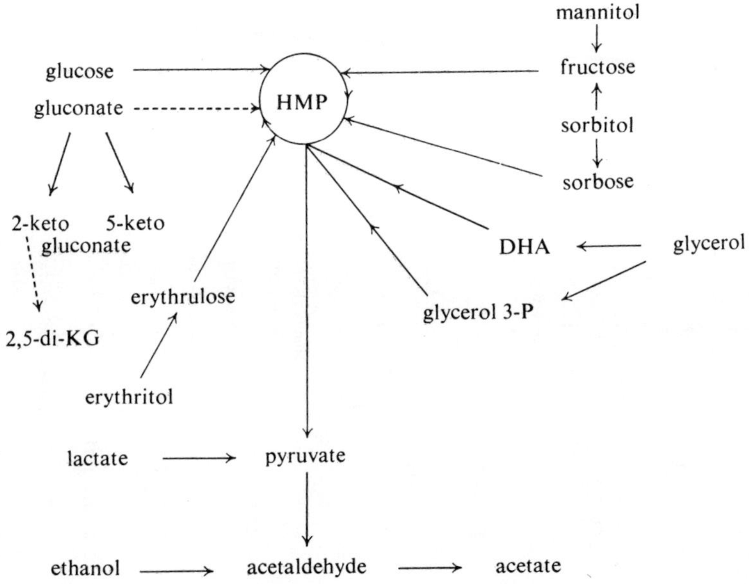
mannitol
glucose
fructose
gluconate
HMP
sorbitol
sorbose
2-keto 5-keto
gluconate
DHA
glycerol
erythrulose
glycerol 3-P
2,5-di-KG
erythritol
lactate
pyruvate
ethanol
acetaldehyde
acetate

Fig. 5.8. Carbohydrate metabolism of *Acetomonas* (*Gluconobacter*) (123) (reprinted with permission of the *Antonie van Leeuwenhoek Journal for Microbiology and Serology*).

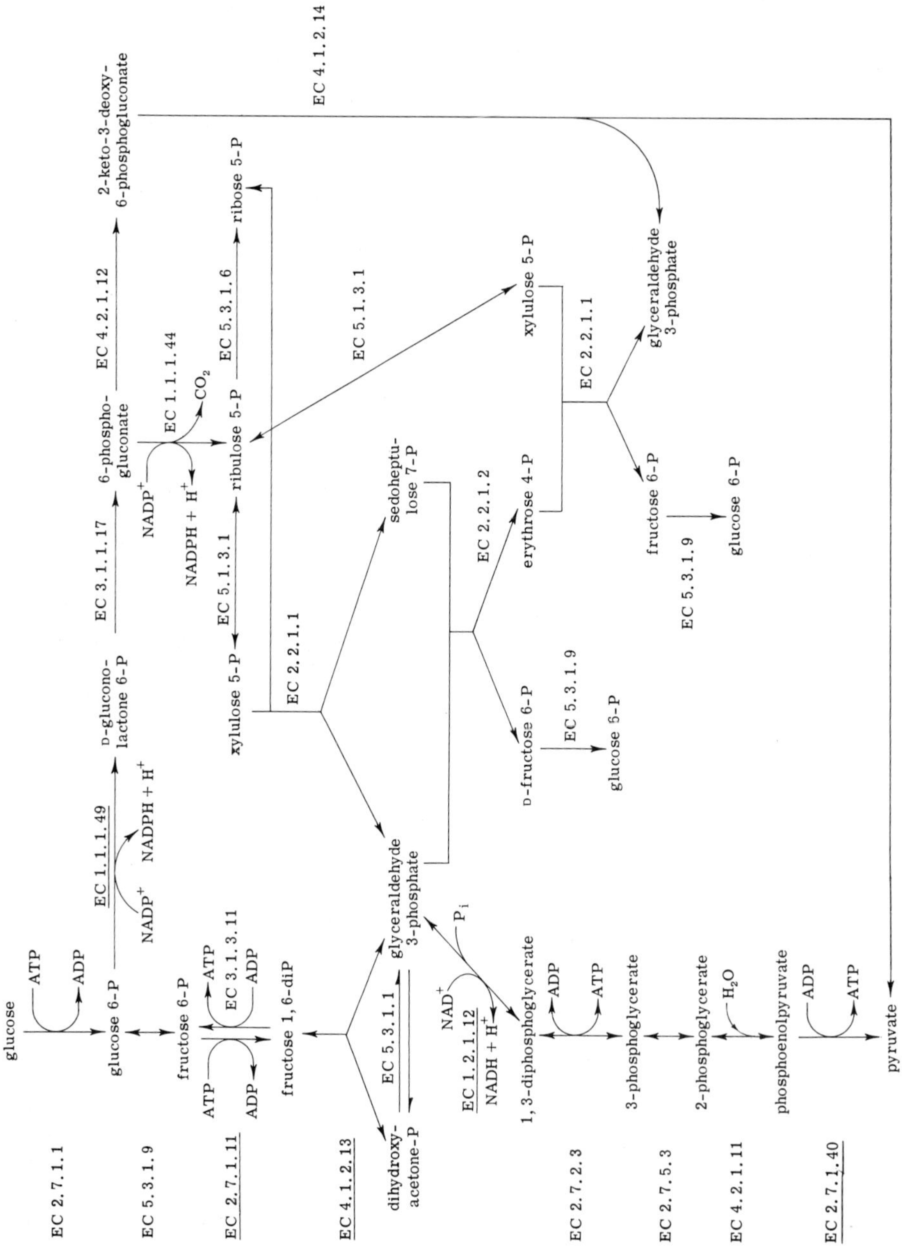

EC 2.7.1.1
glucose
ATP
ADP
glucose 6-P
EC 5.3.1.9
fructose 6-P
EC 2.7.1.11
EC 3.1.3.11
fructose 1, 6-diP
EC 4.1.2.13
dihydroxy-acetone-P
EC 5.3.1.1
glyceraldehyde 3-phosphate
P_i
NAD^+
EC 1.2.1.12
$NADH + H^+$
1, 3-diphosphoglycerate
EC 2.7.2.3
3-phosphoglycerate
EC 2.7.5.3
2-phosphoglycerate
EC 4.2.1.11
H_2O
phosphoenolpyruvate
EC 2.7.1.40
pyruvate
EC 1.1.1.49
$NADP^+$
$NADPH + H^+$
D-glucono-lactone 6-P
EC 3.1.1.17
6-phospho-gluconate
EC 1.1.1.44
CO_2
ribulose 5-P
EC 4.2.1.12
2-keto-3-deoxy-6-phosphogluconate
EC 4.1.2.14
xylulose 5-P
EC 5.1.3.1
EC 5.3.1.6
ribose 5-P
EC 2.2.1.1
sedoheptu-lose 7-P
EC 2.2.1.2
erythrose 4-P
D-fructose 6-P
glucose 6-P

The separation into two main genera (85, 237), the polarly flagellated *Acetomonas* Leifson 1954 (syn. *Gluconobacter* Asai 1935) and the peritrichous *Acetobacter* Beijerinck 1898, is based on the fact that *Acetobacter* oxidizes ethanol to CO_2 and water via the TCA cycle and *Acetomonas* to acetate only (357, 358). Moreover a subdivision into four species is possible by comparing the versatility of carbohydrate utilization (129). Stouthamer summarized the carbohydrate metabolism of *Acetobacter oxydans, A. mesoxydans,* and *Acetomonas suboxydans* and de Ley (85) that of the *Acetobacter peroxydans* group, as illustrated in Figs. 5.5–5.8.

This separation into two genera (85) is further supported by a comparison of the infrared spectra of some 22 strains of the acetic acid bacteria (352). It is emphasized by the distinction between the various strains as lactophilic and glycophilic (70, 397). These terms signify a preference for the metabolism of lactate (*Acetobacter;* see p. 421) or glycols (*Acetomonas;* see p. 284) (318).

The complexity of glucose metabolism via the EMP, HMP, and ED pathways is well reflected in the interlinkage of all three pathways (Fig. 5.9). Our present knowledge of the usage of the different pathways no longer allows one or the other pathway to be regarded as for exclusive use under aerobic or anaerobic conditions, although one may predominate under certain environmental conditions. In order to utilize glucose most efficiently under the set environment, the organisms must possess an elaborate control system, which will be dealt with separately (see Chapter 7). However, as glucose cannot be utilized without entering the cell itself, unless it is in a derivative form, some consideration should be given at this time to the glucose transport system.

Glucose Transport Mechanism

Glucose added to a medium must cross the cell membrane to be metabolized. The cell membrane itself is known to be a very selective barrier for inorganic and organic compounds. Because of the importance of the transport of these compounds via the cell membrane into the cell, the major systems will be considered here with special emphasis on glucose transport. It is intended to cover the permeation problem as comprehensively as possible but a further detailed reading of the excellent reviews written on this subject over the past two decades is suggested (68, 193, 193a, 279, 281, 372).

Fig. 5.9. Interlinkage of the three major pathways of glucose metabolism, EMP, HMP, and ED. The underlined enzyme numbers indicate a possible regulation at that particular enzyme level. With permission of Dowden, Hutchinson, and Ross, Inc.

Four major processes are known at the present moment that could be responsible for the transport of molecules across the cell membrane (330, 334): (*1*) simple diffusion, (*2*) facilitated diffusion, (*3*) active transport, and (*4*) group translocation.

In the first case, the solutes can pass across the cell membrane by simple diffusion. Neither the membrane itself nor any of its components act as catalysts. The energy necessary for this diffusion is derived from the thermal agitation of the molecules. The process depends mainly on the existence of an electrochemical gradient across the membranes. The number of solutes that can utilize this process is very small, e.g., water, glycerol, and urea.

Facilitated diffusion has a number of similarities to simple diffusion, for it also is a passive transport mechanism and the energy required comes from the thermal agitation of the molecules. This process also requires an electrochemical gradient across the membrane. The difference to the simple diffusion mechanism, however, is that a component in the cell membrane (C) acts as catalyst, which results in a more rapid transport across it. This particular compound (C) is specific for stereo- and optical isomers of the solute. The rate of transport resembles Michaelis-Menten kinetics.

Active transport is a second type of carrier mechanism. It is identical to facilitated diffusion, with the additional requirement of energy. This process can move molecules across the membrane against a concentration gradient. This active transport mechanism has been studied in great detail and is identical with a permease system. Very extensive studies have been carried out, particularly with permeases connected with galactosidase. This enzyme has been isolated and characterized (52, 73, 262, 327) and was found to be bound to ribosomes in *Escherichia coli* (412). Because ATP (1) or related energy-rich compounds (299) are able to act as catabolite repressors of β-galactosidase, which in turn reduces the transport of galactosides, it is thought that β-galactosidase must be closely connected with the permease system (108, 294, 346) in the cell membrane. The occurrence of ATPase activities in the particulate fractions led to the suggestion that the permease may be a membrane-bound galactoside ATPase. In the absence of this permease, ATP could suppress β-galactosidase and thus regulate galactoside utilization. The active transport works with facilitated diffusion in a great number of cases, as it was found that metabolite transport is closely linked with cation transport (170). For further details see the excellent book written by Stein (374).

The most favored mechanism for the transport of sugars across the bacterial membranes is now believed to be the group translocation process. The basic principle of this process is that the solute undergoes a chemical reaction at the cell membrane that results in the formation of a derivative, e.g., glucose → glucose 6-phosphate. Such a system was first postulated by Mitchell in 1959 (280), but it did not explain the movement within the membrane. The first intensively studied sugar transport mechanism was that in *Staphylococcus aureus*. It was found that a phosphoenolpyruvate (PEP)-dependent phosphotransferase system is necessary (171, 172, 208, 228, 229) to bring about the phosphorylation of glucose in the cell membrane and the subsequent release of glucose 6-phosphate into the cell. This system was found to consist of four protein components (355, 361), which were shown (173) to be a heat stable, histidine-containing, low molecular weight protein (HPr) or an inducible soluble protein called "K_m-factor" (387a) and a soluble protein enzyme (E I), which is constitutively produced, and a membrane-bound component (E II), which is carbohydrate specific. The general mechanism proposed was

$$\text{PEP} + \text{HPr} \overset{\text{E I}}{\rightleftharpoons} \text{HPr-P} + \text{pyruvate}$$

$$\text{HPr-P} + \text{galactoside} \overset{\text{E II, F III}}{\rightleftharpoons} \text{galactoside 6-P} + \text{HPr}$$

The fourth component was an additional protein component, called factor III. It is completely specific for the phosphorylation of the sugar used for growth of the cells for which it was isolated. This factor could thus be involved in the second reaction, together with E II.

In the first step, enzyme I catalyzes the transfer of phosphate from PEP to the protein HPr. This phosphor transfer results in the formation of pyruvate and phospho-HPr. During the second step, enzyme II, together with factor III, catalyzes the transfer of the phosphate from phospho-HPr to the corresponding sugar, regenerating HPr. The protein is therefore a phosphate carrier.

This PEP–phosphotransferase system has not only been found in gram-positive but also in gram-negative bacteria, such as *Escherichia coli* (5). The difference between both bacterial groups is that gram-negative bacteria appear not to possess factor III and their system therefore includes the three components, HPr, enzyme I, and enzyme II (220, 221, 333). Here again it was found that HPr (5) and enzyme I are soluble and that enzyme II was located in the membrane fraction. Investigations with mutant strains of *Salmonella typhimurium* lacking enzyme I revealed that these organisms lost their ability to transport any carbohydrate (338). This suggests that enzyme I is responsible for the mobilization of the sugar transport in general.

Evidence is also accumulating that this system (156) exists for glucose transport and glucosidases (240) in saccharolytic clostridia.

The question arises, of course, as to why certain bacteria, e.g., *Escherichia coli,* have both permease– and PEP–phosphotransferase systems. The answer is partly given by the survey conducted of the distribution of the PEP–phosphotransferase system (328). This survey indicated that the PEP–phosphotransferase system was mainly found in representatives of genera that are characteristically facultative anaerobes. It apparently plays no role in the sugar transport of strictly aerobic bacteria. As the facultative anaerobes utilize more sugar anaerobically to obtain the energy required for biosynthesis than they do under aerobic conditions (see Pasteur effect, p. 609), the rapid glucose transport is facilitated by the PEP–phosphotransferase system.

This system also appears to be regulated by its product, glucose 6-phosphate (192). It is only the glucose 6-phosphate formed by this system, and not the endogenously formed one (93, 251a), that induces or inhibits the transport system. It is therefore possible that under anaerobic conditions the utilization of metabolic energy for the permease system is curtailed (302), as less ATP but more PEP is produced. Under aerobic conditions, however, more energy but less PEP is available, which possibly favors the permease system. The close relationship between the PEP:sugar phosphotransferase system and permease system was demonstrated with *Salmonella typhimurium* strains defective in enzyme I of the PEP–phosphotransferase system (339). The accumulating sugar substrates repressed not only the synthesis of the enzyme systems required for melibiose, maltose, and glycerol metabolism but also the corresponding permease systems.

Phosphoketolase Pathway

In addition to the well-known EMP, HMP, and ED pathways, there are two other pathways that are possessed by only a small group of bacteria (88, 90, 91, 95, 233), the heterofermentative lactobacilli and the Bifidobacteria. These pathways can be regarded as a branch or a variation of the HMP pathway, because a part of both these pathways is identical with the HMP pathway. Because of this, they are referred to as the "pentose phosphoketolase" and the "hexose phosphoketolase" pathways.

PENTOSE PHOSPHOKETOLASE PATHWAY. Glucose, as well as the pentoses,

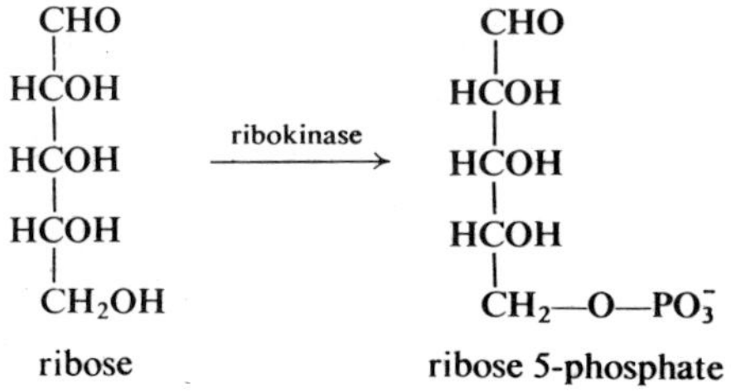
CHO
HCOH
HCOH
HCOH
CH₂OH
ribose
ribokinase
CHO
HCOH
HCOH
HCOH
CH₂—O—PO₃⁻
ribose 5-phosphate

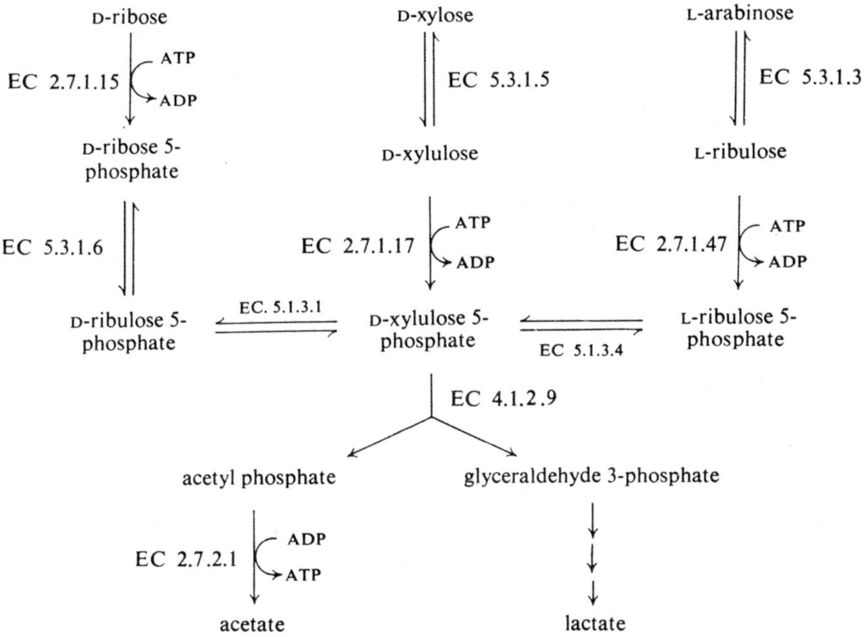
D-ribose
EC 2.7.1.15
ATP
ADP
D-ribose 5-phosphate
EC 5.3.1.6
D-ribulose 5-phosphate
EC. 5.1.3.1
D-xylose
EC 5.3.1.5
D-xylulose
EC 2.7.1.17
ATP
ADP
D-xylulose 5-phosphate
L-arabinose
EC 5.3.1.3
L-ribulose
EC 2.7.1.47
ATP
ADP
L-ribulose 5-phosphate
EC 5.1.3.4
EC 4.1.2.9
acetyl phosphate
glyceraldehyde 3-phosphate
EC 2.7.2.1
ADP
ATP
acetate
lactate

Scheme 5.2. Pentose phosphoketolase pathway of lactic acid bacteria.

can serve as a carbon source, in which case D-ribose 5-phosphate, or D-xylulose 5-phosphate are formed via the HMP pathway (Scheme 5.2). Bacteria using this pathway lack the enzyme transketolase (EC 2.2.1.1). If supplied with the pentose D-ribose, a ribokinase (ATP:D-ribose 5-phosphotransferase, EC 2.7.1.15) transfers one phosphate group from ATP to ribose, forming D-ribose 5-phosphate.

The product of this reaction is then isomerized to D-ribulose 5-phosphate by ribosephosphate isomerase (D-ribose-5-phosphate ketolisomerase, EC 5.3.1.6). Ribulosephosphate 3-epimerase (D-ribulose-5-phosphate 3-epimerase, EC 5.1.3.1) converts the D-ribulose 5-phosphate to D-xylulose 5-phosphate. This newly formed phosphate ester plays a key role in this

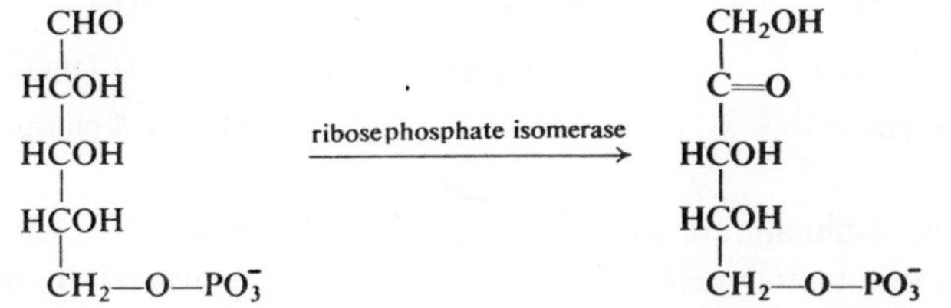
CHO
HCOH
HCOH
HCOH
CH2—O—PO3
ribosephosphate isomerase
CH2OH
C=O
HCOH
HCOH
CH2—O—PO3

ribose 5-phosphate ribulose 5-phosphate

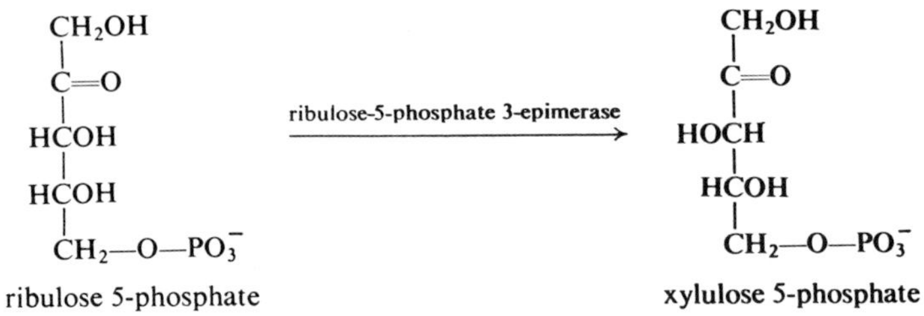
CH₂OH
C=O
HCOH
HCOH
CH₂—O—PO₃⁻
ribulose 5-phosphate
ribulose-5-phosphate 3-epimerase
CH₂OH
C=O
HOCH
HCOH
CH₂—O—PO₃⁻
xylulose 5-phosphate

pathway. All other pentoses, such as D-xylose or L-arabinose, have first to be converted to D-xylulose 5-phosphate, because the key enzyme of this pathway, phosphoketolase [D-xylulose-5-phosphate D-glyceraldehyde-3-phosphate-lyase (phosphate-acetylating), EC 4.1.2.9] reacts only on this compound. D-Xylose, for example, undergoes first an isomerization cata-

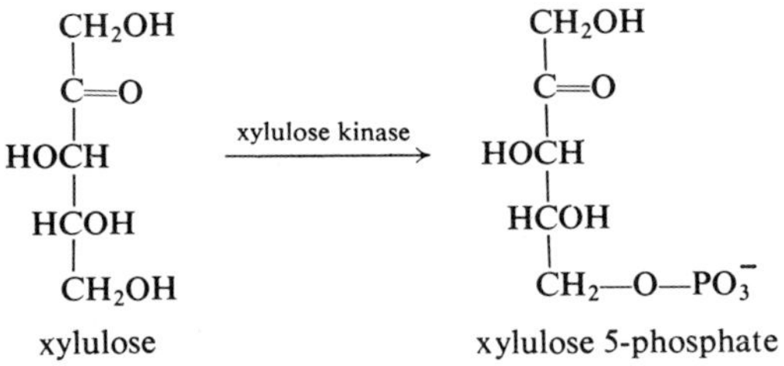
CH₂OH
C=O
HOCH
HCOH
CH₂OH
xylulose
xylulose kinase
CH₂OH
C=O
HOCH
HCOH
CH₂—O—PO₃⁻
xylulose 5-phosphate

lyzed by D-xylose isomerase (D-xylose ketol-isomerase, EC 5.3.1.5) to D-xylulose. After this conversion, D-xylulokinase (ATP:D-xylulose 5-phosphotransferase, EC 2.7.1.17) phosphorylates D-xylulose to D-xylulose 5-phosphate.

Another well-known pentose attacked by heterofermentative lactobacilli is L-arabinose, which also undergoes first an isomerization (D-arabinose keto-isomerase, EC 5.3.1.3) to D-ribulose and, second, a phosphorylation

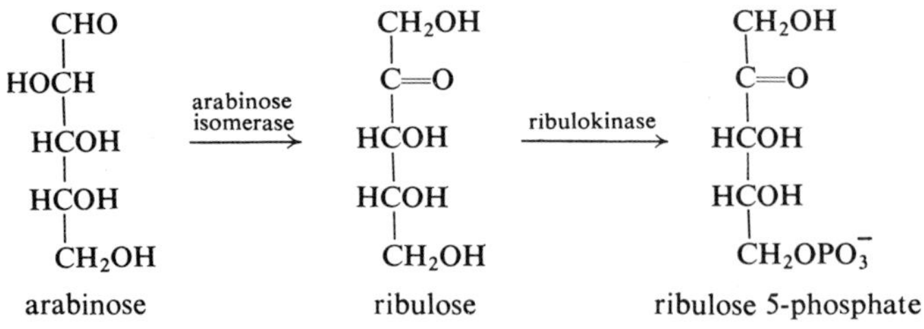
CHO
HOCH
HCOH
HCOH
CH₂OH
arabinose
arabinose isomerase
CH₂OH
C=O
HCOH
HCOH
CH₂OH
ribulose
ribulokinase
CH₂OH
C=O
HCOH
HCOH
CH₂OPO₃⁻
ribulose 5-phosphate

to D-ribulose 5-phosphate with the help of ribulokinase (50) (ATP:D-ribulose 5-phosphotransferase, EC 2.7.1.47). The conversion of ribulose

5-phosphate to xylulose 5-phosphate is catalyzed by ribulosephosphate 4-epimerase (51) (L-ribulose-5-phosphate 4-epimerase, EC 5.1.3.4)

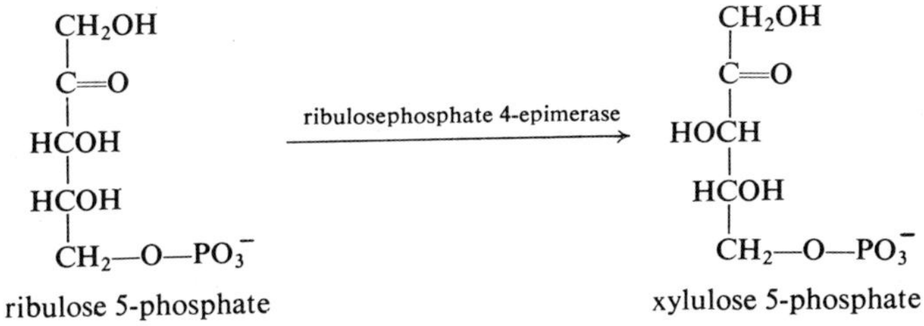
ribulosephosphate 4-epimerase
ribulose 5-phosphate
xylulose 5-phosphate

Having formed D-xylulose 5-phosphate, phosphoketolase splits the pentose molecule into acetyl phosphate and glyceraldehyde 3-phosphate. This reaction requires thiamine pyrophosphate as well as inorganic phosphate.

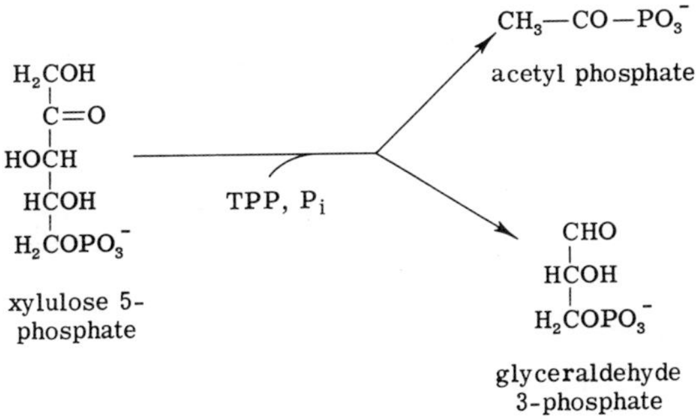
acetyl phosphate
TPP, Pi
xylulose 5-phosphate
glyceraldehyde 3-phosphate

The reaction of this enzyme is the simplest of a series known as "phosphoroclastic" cleavages (168, 169, 350). When *Lactobacillus plantarum* is grown in the presence of xylose, cells contained an enzyme that cleaves xylulose 5-phosphate phosphorolytically to glyceraldehyde 3-phosphate and acetyl phosphate. The action of this enzyme phosphoketolase is dependent on TPP and Mg^{2+}, is stimulated by SH compounds, and is irreversible (387). As acetyl phosphate is the product, the reaction resembles the phosphoroclastic split with pyruvate; as xylulose 5-phosphate is the substrate and glyceraldehyde 3-phosphate is a product, the reaction also resembles a transketolase reaction. The reaction described for *Lactobacillus plantarum* was found to be identical to that in *Leuconostoc mesenteroides* (167). In these microorganisms the usual pathways, i.e., EMP, HMP, and ED, are absent (167). Phosphoketolase is produced by *Lactobacillus plantarum* when it is grown in a xylose-containing medium (induced enzyme) or

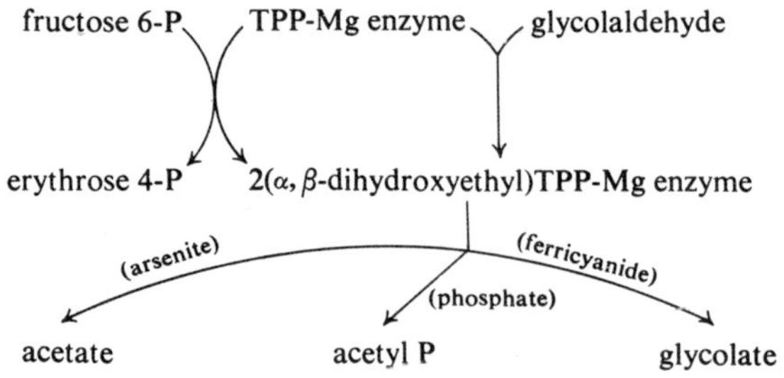

fructose 6-P
TPP-Mg enzyme
glycolaldehyde
erythrose 4-P
2(α, β-dihydroxyethyl)TPP-Mg enzyme
(arsenite)
(ferricyanide)
(phosphate)
acetate
acetyl P
glycolate

Fig. 5.10. Phosphoketolase reaction.

the enzyme is constitutive, as in *Leuconostoc mesenteroides* and *Acetobacter xylinum* (350). In essence, the reaction is none other than a substrate level phosphorylation, one of the carbon of the substrate becoming "more" oxidized, while the other carbon of the substrate becomes "more" reduced. The conservation of energy is at the carbon that becomes more oxidized.

Another reaction catalyzed by phosphoketolase is given in Fig. 5.10. The first step of this reaction is the formation of 2-(α,β-dihydroxyethyl)TPP–Mg^{2+} enzyme intermediate (179). This reaction is followed by an active interaction of the glycolaldehyde enzyme with arsenate, which yields acetate; or with phosphate, which yields acetyl phosphate; or with ferricyanide, which yields glycolate. Other details of this mechanism can be obtained from articles by Breslow (42) and Breslow and McNelis (43).

It was mentioned above that *Acetobacter xylinum* has a functional phosphoketolase, which is, however, less apparent. It is thought that a cyclic process is involved, which can be called the "phosphoketolase shunt" and is summarized as shown in Fig. 5.11. The formulation of this cycle represents a short circuit pathway for the production of acetate in *Acetobacter xylinum*. The function may be important, for the organism lacks phosphofructokinase. The energy yield of the short circuit in terms of ATP is 3 moles/mole of fructose 6-phosphate or 2 moles of ATP/mole of glucose. This yield is identical with that of the EMP system and the cycle is considerably simpler. The energy yield is low, however, compared to that obtained by oxidation of acetate in the tricarboxylic acid cycle. The phosphoketolase shunt therefore probably functions as a short circuit to acetate rather than as an important contributor to the energy budget of this microorganism.

Whereas glyceraldehyde 3-phosphate continues to be metabolized via the EMP pathway to pyruvate and lactate, acetyl phosphate is converted to acetate with acetokinase

$$\underset{\text{acetyl phosphate}}{CH_3COPO_3^-} \xrightarrow{\text{acetokinase}} \underset{\text{acetate}}{CH_3COOH} + H_3PO_4$$

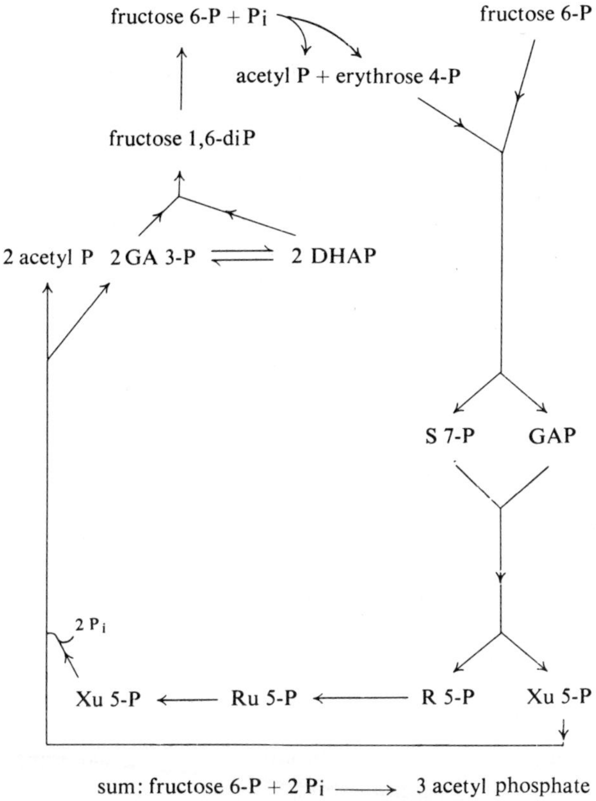
fructose 6-P + Pi
fructose 6-P
acetyl P + erythrose 4-P
fructose 1,6-diP
2 acetyl P
2 GA 3-P
2 DHAP
S 7-P
GAP
2 Pi
Xu 5-P
Ru 5-P
R 5-P
Xu 5-P
sum: fructose 6-P + 2 Pi ⟶ 3 acetyl phosphate

Fig. 5.11. Phosphoketolase shunt in *Acetobacter xylinum.*

(ATP:acetate phosphotransferase, EC 2.7.2.1) and the formation of 1 mole of ATP. From the point of view of ATP formation, the phosphoketolase pathway behaves similarly to the HMP pathway. The net production is 1 mole ATP. The main difference, of course, is that the phosphoketolase pathway substitutes a phosphoketolase reaction for the transketolase–transaldolase reaction of the HMP pathway. The other difference is that the cycling mechanism of the HMP pathway, according to the carbon balance sheet, seems not to occur.

HEXOSE PHOSPHOKETOLASE PATHWAY. The occurrence of a specific catabolic route in the genus *Bifidobacterium* (*Lactobacillus bifidus*) has been reported (92). This group of bacteria was found to lack fructosediphosphate aldolase as well as glucose-6-phosphate dehydrogenase. They ferment glucose via a pathway different from those found in homo- and heterofermentative lactic acid bacteria, which means that none of the EMP, HMP, ED, or pentose phosphoketolase pathways could operate. The key

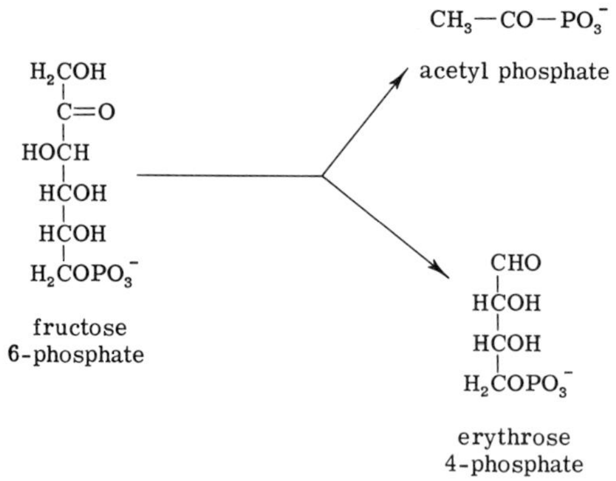
CH3—CO—PO3⁻
acetyl phosphate
fructose 6-phosphate
erythrose 4-phosphate

reaction in the fermentation of glucose appears to be a phosphoketolase cleavage of fructose 6-phosphate into acetyl phosphate and erythrose 4-phosphate. Pentose phosphates are thus formed in a reverse HMP pathway by the action of transaldolase (EC 2.2.1.2) and transketolase (EC 2.2.1.1). The product of this sequence, xylulose 5-phosphate, can then be split into acetyl phosphate and glyceraldehyde 3-phosphate as described above. Hexose phosphoketolase was not found in *Lactobacillus*, *Ramibacterium*, *Arthrobacter*, *Propionibacterium*, *Corynebacterium*, or *Actinomyces* (345). All hexose phosphoketolase-positive strains of *Bifidobacterium* form more acetic than lactic acid.

All five pathways of glucose metabolism have a great number of intermediates and enzymes in common. There are some enzymes, however, that are characteristic for the individual pathways, as they occur only in that particular pathway and not in any of the other pathways. These so-called "key enzymes" are, for example, phosphofructokinase (EMP pathway), 6-phosphogluconate dehydrogenase (HMP pathway), phospho-2-keto-3-deoxy-6-gluconate aldolase (ED pathway), xylulose 5-phosphoketolase (PK pathway), and fructose 6-phosphoketolase (hexose phosphoketolase pathway). These enzymes play an important role in the differentiation between the pathways.

Glucose Metabolism of Acetic Acid Bacteria

In the early stages of investigation, it was thought that the EMP pathway was present in acetic acid bacteria. This was challenged by deLey (85), and present indications are that the EMP pathway is completely absent (318). The main metabolic pathway of glucose metabolism is the HMP pathway (see p. 221).

Acetobacter and *Acetomonas* are well known for their oxidative versatility, substrates ranging from carbohydrates to a great number of straight-chain polyols (317). The oxidation of polyols and reduction of ketones are carried out by specific dehydrogenases that follow the Bertrand-Hudson rule. Bertrand (34) found with *Acetobacter xylinum* that a generalization could be made concerning the structure of polyols susceptible to oxidation. According to his established rule, a secondary group of a polyol is oxidized to ketone if it has the cis configuration with respect to the primary alcohol group:

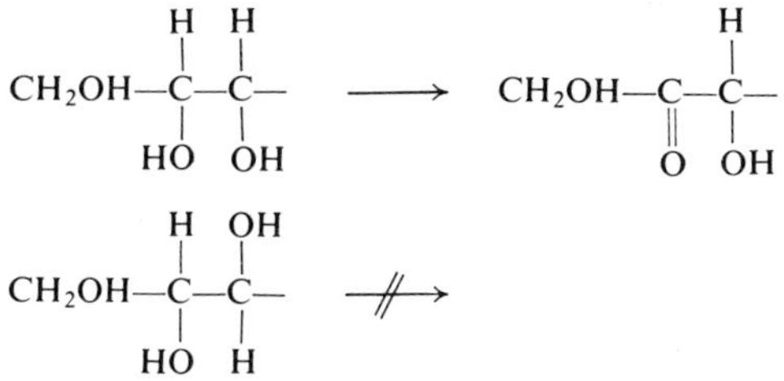

With glucose, the terminal aldehyde group is oxidized to a carboxyl group. This was explained by considering glucose as a cyclic polyol, to which the Bertrand rule may not apply.

The most versatile microorganism, as far as carbohydrate metabolism is concerned, is undoubtedly *Acetomonas suboxydans*. It oxidizes a great number of polyols in one- or two-step oxidations. It has been suggested that *A. suboxydans* is a "metabolic cripple" (63), for in many instances it is unable to oxidize the compounds beyond the first or second step. Cell-free extracts of *Acetomonas suboxydans* were shown to possess two kinds of glucose dehydrogenases (138), one 2,6-dichlorophenolindophenol (DPI) linked and a second $NADP^+$ linked. It is very likely that the DPI-linked dehydrogenase is identical with the glucose oxidase (β-D-glucose:oxygen oxidoreductase, EC 1.1.3.4) which is a flavoprotein. The NADP-linked one is identical to the glucose dehydrogenase [β-D-glucose:NAD(P) oxidoreductase, EC 1.1.1.47] that catalyzes the reaction:

$$\text{glucose} + O_2 \rightarrow \text{D-glucono-}\delta\text{-lactone} + H_2O_2$$

$$\text{glucose} + NADP^+ \rightarrow \text{D-glucono-}\delta\text{-lactone} + NADPH + H^+$$

Phosphate does not participate in this glucose oxidation, although the reaction is strictly $NADP^+$ dependent in *Acetomonas suboxydans* (295). A 5-keto-D-gluconate 5-reductase (D-gluconate:$NAD(P)^+$ 5-oxidoreductase, EC 1.1.1.69) metabolizes the gluconate further to 5-ketogluconate (see p. 264).

Investigations into the dissimilation of glucose with the radiorespirometric method (214) revealed that, in the presence of oxygen, the HMP pathway accounts for most of the glucose metabolized (to CO_2 and water); the remainder is directly oxidized to 2-ketogluconate. This exclusive use of the HMP pathway by all acetic acid bacteria is a clear-cut difference between them and the related genera of the family Pseudomonadaceae.

The connection between the glucose → 5-ketogluconate reaction and the HMP pathway was established by de Ley and Stouthamer (83) when they discovered the following three enzymes:

1. A soluble NADP-specific dehydrogenase yielding 2-ketogluconate, which could be gluconate dehydrogenase [D-gluconate:NAD(P)$^+$ oxidoreductase, EC 1.1. group];
2. A soluble NADP-specific dehydrogenase yielding 5-ketogluconate, which is possibly 5-keto-D-gluconate 5-reductase [D-gluconate:NAD(P)$^+$ 5-oxidoreductase, EC 1.1.1.69];
3. A particulate, possibly cytochrome-linked, gluconate dehydrogenase (D-gluconate:(acceptor)2-oxidoreductase, EC 1.1.99.3) yielding 2-ketogluconate.

As there was no 5-ketogluconate detectable, both 2- and 5-ketogluconate are metabolized by a dehydrogenase to gluconate. This was followed by a gluconokinase (ATP:D-gluconate 6-phosphotransferase, EC 2.7.1.12) and 6-phosphogluconate dehydrogenase [6-phospho-D-gluconate:NADP oxidoreductase (decarboxylating), EC 1.1.1.44], which catalyzed the same reactions as described in the HMP pathway.

Acetobacter melanogenum, however, does not possess a gluconokinase and therefore oxidizes 2-ketogluconate further to 2,5-diketogluconate by means of a ketogluconate dehydrogenase [2-keto-D-gluconate:(acceptor) oxidoreductase, EC 1.1.99.4] (see p. 265).

Acetobacter xylinum combines the activities of *A. melanogenum* and *Acetomonas suboxydans* in being able to (155, 348):

(1) phosphorylate glucose → glucose 6-phosphate

(2) oxidize glucose → gluconate → 2-keto- and 5-ketogluconate → α-oxoglutarate + CO_2

Only a minor fraction of glucose undergoes phosphorylation as such. The bulk of glucose carbon can be introduced into the hexose phosphate pool directly via gluconate, which is phosphorylated to 6-phosphogluconate. Pyruvate and acetate may also originate from 5-oxogluconate in *Acetomonas suboxydans* (see Fig. 5.12).

Inhibition studies with iodoacetate in *A. xylinum* have revealed that a further pathway must exist that does not involve glyceraldehyde-3-phos-

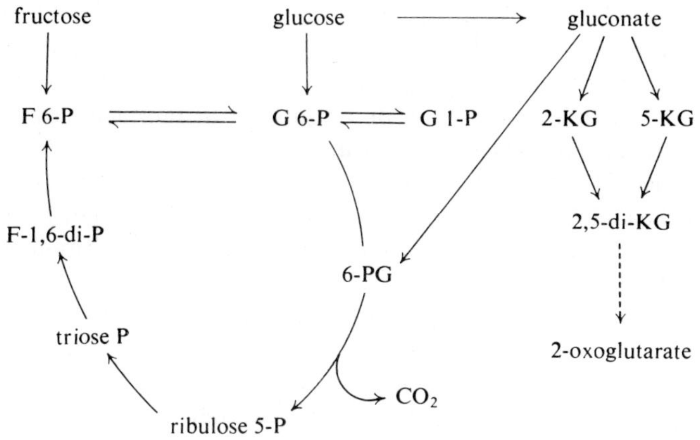
fructose
glucose
gluconate
F 6-P
G 6-P
G 1-P
2-KG
5-KG
F-1,6-di-P
2,5-di-KG
6-PG
triose P
2-oxoglutarate
CO₂
ribulose 5-P

Fig. 5.12. Carbohydrate metabolism of *Acetobacter xylinum.*

phate dehydrogenase (349). *Acetobacter xylinum* is still able to synthesize cellulose after the addition of the inhibitor. It was also observed that inorganic phosphate was converted to acyl phosphate, which could be linked to a hexose phosphorylation system

$$\begin{aligned}
\text{fructose 6-P} + P_i &\rightarrow \text{acetyl P} + \text{erythrose 4-P}\\
\text{erythrose 4-P} + \text{fructose 6-P} &\rightarrow \text{sedoheptulose 7-P} + \text{glyceraldehyde 3-P}\\
\text{sedoheptulose 7-P} + \text{glyceraldehyde 3-P} &\rightarrow \text{ribose 5-P} + \text{xylulose 5-P}\\
\text{ribose 5-P} &\rightarrow \text{xylulose 5-P}\\
2\ \text{xylulose 5-P} + 2\ P_i &\rightarrow 2\ \text{acetyl P} + \text{glyceraldehyde 3-P}
\end{aligned}$$

$$\begin{aligned}
\text{net: fructose 6-P} + 2\ P_i &\rightarrow 2\ \text{acetyl P}\\
3\ \text{acetyl P} + 3\ \text{ADP} &\rightarrow 3\ \text{ATP} + 3\ \text{acetate}
\end{aligned}$$

The enzyme catalyzing the initial step was called "fructose-6-phosphate phosphoketolase" (350). The presence of this enzyme may indicate a "short circuit" pathway (356) for the case when no phosphofructokinase or no glucose-6-phosphate dehydrogenase is present. The detailed description of this pathway was given above. It is interesting that apart from *Acetobacter xylinum* only species of the genus *Bifidobacterium* possess this pathway to our present knowledge.

Regulation of Carbohydrate Pathways

The diverse activities of microorganisms with regard to their energy and biosynthetic activities has brought about the very artificial separation of metabolism into catabolic and anabolic pathways. It should be realized, however, that most of the pathways described can have more than one function, for a great number of the enzymatic reactions are reversible reactions. The EMP pathway, for example, can function not only as a catabolic route of carbohydrate utilization, but also as a biosynthetic pathway, called "gluconeogenesis," for those organisms utilizing smaller molecular compounds, such as dicarboxylic acids. Whenever pathways carry out such dual functions, they are referred to as "amphibolic pathways" (80).

At this time only the regulation of the carbohydrate pathways will be considered. Before these control mechanisms are dealt with, however, a clarification of the terminology may be necessary.

"Catabolite repression" is the control exerted by metabolites on the rate of synthesis of certain enzymes.

"Transient repression" is a transient control exerted by metabolites to which the organism is exposed. This effect may last only one generation and is very often referred to as "adaptation."

"Catabolite inhibition" is the control exerted by glucose or other metabolites on the activity of certain enzymes. This type of control is analogous to the feedback inhibition of biosynthetic pathways.

"Allosteric enzyme control" mechanisms are carried out by enzymes the protein of which is assumed to possess two stereospecifically different receptor sites (283), an active and an allosteric site. The active site binds the substrate and the allosteric site the allosteric effector. Whereas the active site is responsible for the biological activity of the enzyme protein, the enzyme–allosteric effector complex brings about a reversible alteration of the protein structure (allosteric transition) that modifies the properties of the active site. These modifications, in turn, bring about changes in the kinetic parameters of the enzyme, which are reflected in the sigmoidal shape of the enzyme–substrate velocity curve. As NADH + H^+ and ATP can function as allosteric effectors during catabolism, the activity of these enzymes (e.g., phosphofructokinase) could serve as indicators of the energy state of a cell.

It was demonstrated above that glucose is mainly metabolized via three major pathways, the Embden-Meyerhof-Parnas pathway, the hexose monophosphate pathway, and the Entner-Doudoroff pathway. Whereas the Enterobacteriaceae, as facultative anaerobes, mainly use the first two, the Pseudomonadaceae favor the Entner-Doudoroff pathway. How do these organisms regulate the distribution of the glucose carbon between these pathways?

The first control mechanism most certainly will be a transient repression in the glucose transport mechanism across the cell membrane, if the organism encounters a new carbon compound (251, 381). This mechanism requires the phosphoenolpyruvate phosphotransferase and β-galactosidase systems. The availability of these systems is responsible for the first step in the glucose metabolism. Enterobacteriaceae cultivated on glucose medium have this system preformed and no repression occurs. It is suggested that the actual passage of the compound through the cell membrane is responsible for the repression (251a). Permanent and transient catabolite repressions of the β-galactosidase synthesis in *Escherichia coli* are, for example, abolished by the addition of 5 m*M* 3′,5′-cyclic AMP (287, 305).

The major intermediate for the distribution of the glucose carbon between the EMP and HMP pathway is undoubtedly glucose 6-phosphate (315, 341, 371). It has been well documented (198, 199) that 20–30% of glucose is utilized by the Enterobacteriaceae through the HMP pathway under anaerobic conditions. Evidence is accumulating (58, 105, 200, 282) that these microorganisms are not using this pathway mainly for its pentose supply but rather for the generation of reduced NADP for reductive biosynthesis. If this is the case, one would immediately look for key enzymes in the respective pathways and their catalytic properties. The first key enzyme in the hexose monophosphate pathway is glucose-6-phosphate dehydrogenase. The enzyme from all sources is specific for the substrate glucose 6-phosphate, although a few cases of low activity have been reported for closely related substrates (109, 245, 274, 366). Its importance in carbohydrate metabolism became known with the observation that NADH + H^+ inhibits glucose-6-phosphate dehydrogenase of *Escherichia coli* in an allosteric manner (342) that cannot be reversed by NAD^+, AMP, or spermidine. ATP, acetyl-CoA, and CoA itself, however, did not inhibit the enzyme. With this observation, NADH + H^+ becomes almost a universal inhibitor of most of the NADP-dependent enzymes in the Enterobacteriaceae. The enzyme from *E. coli* is not affected by ATP. This is in contrast to the effect of ATP on the same enzyme from *Hydrogenomonas* and *Pseudomonas* (347) as well as *Acetobacter* (28a, 393b), three genera that do not use the EMP pathway under any known conditions. Such an inhibition of NADH + H^+ would curtail the capacity of the pentose phosphate pathway. This inhibition by NADH + H^+, however, has been disputed (58a, 393a) and an inhibition by NADPH + H^+ found instead (393a), that caused competitive inhibition with respect to $NADP^+$ and mixed type inhibition with regard to glucose 6-phosphate. This type of inhibition is similar to the type reported to occur in rat liver (108a) and therefore awaits further investigation. The same enzyme has also been reported to control the HMP pathway in yeast (20, 182, 298).

The second important enzyme in the HMP pathway is the NADP-dependent 6-phosphogluconate dehydrogenase, which has been reported inhibited by FDP in *Streptococcus faecalis* (45), rat liver (252), sheep liver (103), and in *Escherichia coli* (393a). This inhibition also suggests that FDP could control the HMP pathway.

As far as the EMP pathway is concerned, it has already been mentioned that a number of enzyme could cause certain control effects. The phosphofructokinase of the Enterobacteriaceae is well known for its sensitivity to ATP under anaerobic conditions. After NADH + H^+ or NADPH + H^+ and FDP, ATP is the third compound that could influence the distribution of the glucose carbon in either of the two pathways under consideration. It should be mentioned, however, that not all bacteria utilizing the EMP pathway possess an ATP-sensitive phosphofructokinase; at least the lactic acid bacteria appear to be an exception (97). Fructose-diphosphate aldolase is an enzyme that has been neglected in the past but that exhibits a very high instability toward β-mercaptoethanol in *E. coli* (118) in comparison to the same enzyme from lactic acid bacteria (256). Glucose stimulated the activity in yeast (180) and the same effect was found recently with the *E. coli* enzyme (98a).

Another important step in the EMP pathway is the formation of reduced NAD by the catalytic action of glyceraldehyde-3-phosphate dehydrogenase. Investigations into the interaction of 3-phosphoglyceraldehyde dehydrogenase with NAD^+ and adenine nucleotides revealed that ATP inhibited almost every step in the multifunctional enzyme catalysis (128), which makes this enzyme a metabolic control point similar to phosphofructokinase.

A summary of the enzymatic activities and their inhibitors, is given in the regulatory system shown in Scheme 5.3 (broken line indicates inhibition).

Under strict aerobic conditions and low glucose concentration (0.1%), fructose-diphosphate aldolase activity in yeasts (180) and *E. coli* (98a) is very low, which would indicate use of the HMP pathway. However, the accumulation of FDP because of a low phosphofructokinase activity would inhibit 6-phosphogluconate dehydrogenase, making this enzyme the pacemaker of the HMP pathway. As the glucose concentration increases, FDP aldolase activity increases, opening the EMP pathway for the production of NADH + H^+, which in turn inhibits glucose-6-phosphate dehydrogenase. As the FDP aldolase of yeasts increases about 60-fold (from 0.6 to 20% glucose) against a twofold increase of glucose-6-phosphate dehydrogenase (180), the inhibition of the latter enzyme can certainly be severe. At the same time, the FDP inhibition is taken off the 6-phosphogluconate dehydrogenase, making glucose-6-phosphate dehydrogenase the pacemaker of the HMP pathway. Under these highly aerobic conditions, phosphofructokinase of *E. coli* exhibits and maintains its relatively low activity

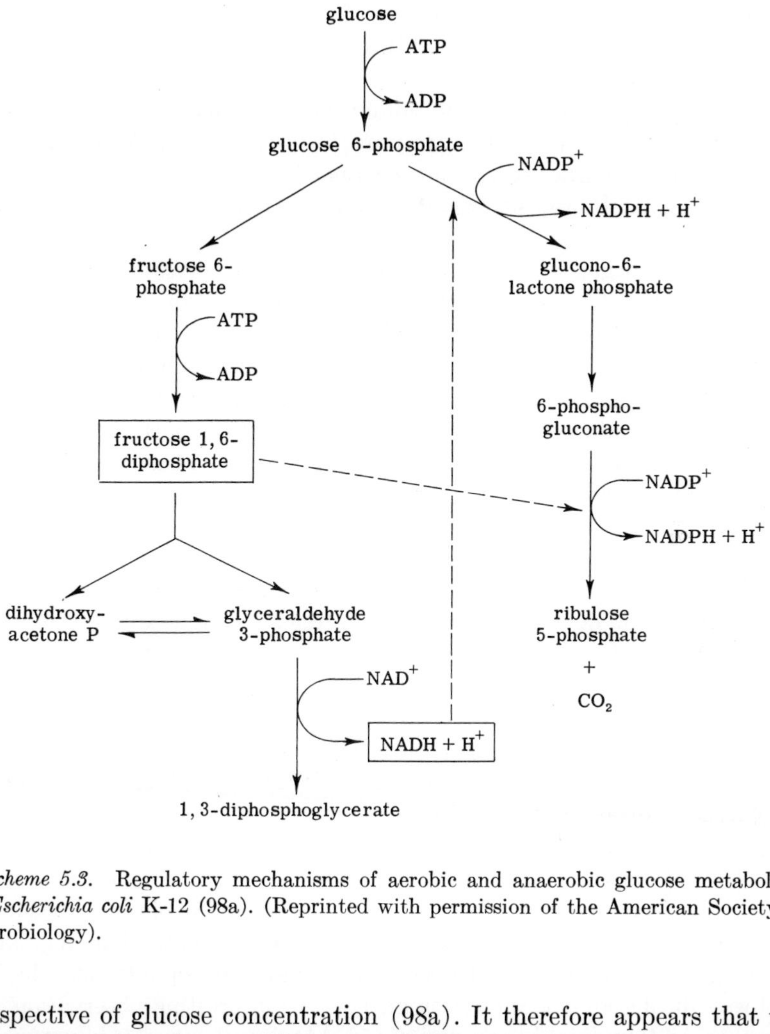
glucose
ATP
ADP
glucose 6-phosphate
NADP$^+$
NADPH + H$^+$
fructose 6-phosphate
glucono-6-lactone phosphate
ATP
ADP
6-phosphogluconate
fructose 1, 6-diphosphate
NADP$^+$
NADPH + H$^+$
dihydroxyacetone P
glyceraldehyde 3-phosphate
ribulose 5-phosphate
+
CO_2
NAD$^+$
NADH + H$^+$
1, 3-diphosphoglycerate

Scheme 5.3. Regulatory mechanisms of aerobic and anaerobic glucose metabolism in *Escherichia coli* K-12 (98a). (Reprinted with permission of the American Society of Microbiology).

irrespective of glucose concentration (98a). It therefore appears that the glucose concentration could be responsible for the selection of either pathway.

The negative control of 6-phosphogluconate dehydrogenase by FDP under anaerobic conditions in *Streptococcus faecalis* is more difficult to explain, for these organisms produce lactate mainly via the EMP pathway (146, 312), with a special requirement of FDP for lactate dehydrogenase (398, 401). Brown and Wittenberger (45) therefore suggest that an intracellular FDP pool must exist to activate lactate dehydrogenase. Such an FDP pool could also exert control on the HMP pathway activity. The latter is certainly necessary as the accumulation of biosynthetic intermediates would occur at the expense of glucose carbon that is actually required for the generation of ATP through the EMP pathway. However, the reduced NAD formed via the EMP pathway could also cause a competitive effect between the inhibition of glucose-6-phosphate dehydrogenase and the reduction of pyruvate to lactic acid. An energy control at the phosphofructokinase level cannot be considered, as this enzyme appears to be ATP insensitive and is comparable to the one in lactobacilli (97).

In the gram-negative microorganisms, such as *Escherichia coli*, phosphofructokinase is ATP sensitive under anaerobic conditions and could play a major role. It seems possible that at high ATP concentrations, the enzyme will certainly be repressed and thus would relieve the inhibition caused by $NADH + H^+$ on glucose-6-phosphate dehydrogenase and of FDP on 6-phosphogluconate dehydrogenase, opening the HMP pathway for glucose utilization. As soon as the ATP pool diminishes, phosphofructokinase would regain its activity and thus again impose its control on the HMP pathway.

The role of glyceraldehyde-3-phosphate dehydrogenase has not been elucidated as yet. The ATP inhibition of this enzyme (75) indicates, however, that it could be responsible for regulating the complete and incomplete HMP pathway. In the case of high ATP concentrations, the pentose shunt could operate, for the inhibition of phosphofructokinase does not effect the reverse reaction catalyzed by fructose-1,6-diphosphatase.

The Pseudomonadaceae metabolize hexoses via the ED pathway (101, 113, 114). The major reactions are catalyzed by glucose-6-phosphate dehydrogenase, 6-phosphogluconate dehydratase, and phospho-2-keto-3-deoxy-6-gluconate aldolase (see p. 230). It is of interest that glucose-6-phosphate dehydrogenase is not inhibited by $NADH + H^+$ but is by ATP (238) and that the other two enzymes are inducible by exposure of the organism to glucose, fructose, mannose, or gluconate. This regulation by ATP certainly prevents drainage of glucose 6-phosphate into the ED pathway during an oversupply on energy. An interesting observation was also that ATP inhibited the activity of NAD-linked enzymes in preference to the NADP-linked enzymes in *Pseudomonas multivorans* (383), which would keep a balance between NAD^+ and $NADP^+$ reduction in the cell. The induction of enzymes even extends to glyceraldehyde-3-phosphate dehydrogenase (335), but the uncertainty of the presence of FDP aldolase in pseudomonads (238, 378) leaves open the question of whether glyceraldehyde-3-phosphate dehydrogenase plays a similar role for gluconeogenesis as was described above. Pseudomonads appear not to possess the key enzymes of the EMP and HMP pathway and therefore are confined to the ED pathway. This selection must therefore be a genetic one.

Fructose Metabolism

The utilization of fructose is known to occur particularly in *Alcaligenes* (100) and *Aerobacter* (160), but was also found in *Escherichia coli* (125), *Zymomonas mobilis* (250), *Clostridium aceticum* (243), and *Clostridium thermocellum* (308). The pathway for this utilization depends on whether the free hexose is supplied extracellularly and must therefore be transported across the cell membrane or whether it is supplied as a product of sucrose utilization inside the cell (204).

The exogenously supplied fructose is mediated across the cell membrane by a phosphoenolpyruvate:D-fructose 1-phosphotransferase system (125, 161) (ATP:D-fructose 1-phosphotransferase, EC 2.7.1.3) which is induced by D-fructose and enters the cell as D-fructose 1-phosphate. An inducible

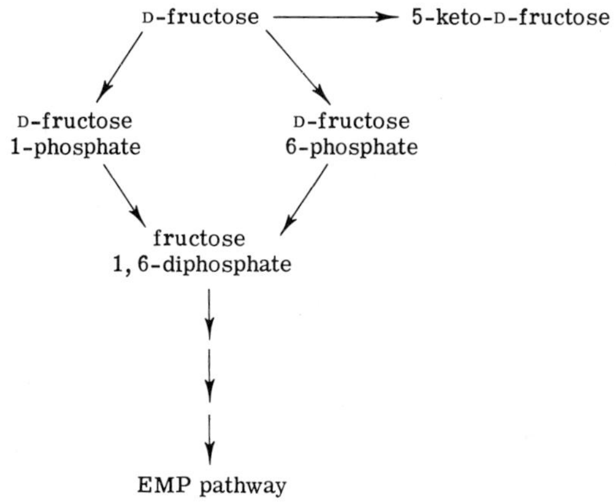
D-fructose
5-keto-D-fructose
D-fructose 1-phosphate
D-fructose 6-phosphate
fructose 1, 6-diphosphate
EMP pathway

enzyme, which has been named D-fructose-1-phosphate kinase (ATP:D-fructose-1-phosphate 6-phosphotransferase) (160) phosphorylates fructose 1-phosphate to fructose 1,6-diphosphate. This enzyme, in contrast to phosphofructokinase (see p. 213), is a nonallosteric enzyme (343) and is activated by potassium and ammonium ions (344). This pathway is also found in *Escherichia coli* (125), *Zymomonas* (250), and *Clostridium* (243).

The second pathway is mediated by a specific sucrose-induced D-fructokinase (ATP:D-fructose 6-phosphotransferase, EC 2.7.1.4) and phosphorylates D-fructose to fructose 6-phosphate. Both pathways continue via the EMP pathway.

A third pathway exists in *Acetobacter cerinus* (18, 111), a strictly aerobic organism. The oxidation of D-fructose to 5-keto-D-fructose conforms to

Bertrand's rules (see p. 251). This reaction is catalyzed by a NADPH-dependent dicarbonylhexose reductase (D-fructose:$NADP^+$ 5-oxidoreductase, EC 1.1.1.124) (112), which is extremely substrate specific (19) and strongly favors the reduction of 5-keto-D-fructose at equilibrium.

Lactose Metabolism

Lactose is very readily metabolized (Scheme 5.4) to galactose and glucose by Enterobacteriaceae and Group N streptococci. It is used as one of the main classification criteria (140, 141, 188), together with genetics, because of the importance of the *lac* operon (107). Glucose is further metabolized via the EMP pathway (see p. 209).

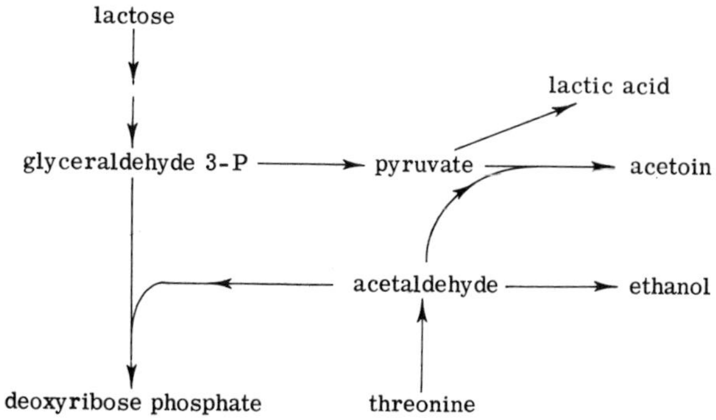
lactose
lactic acid
glyceraldehyde 3-P
pyruvate
acetoin
acetaldehyde
ethanol
deoxyribose phosphate
threonine

Scheme 5.4. Lactose metabolism of Group N streptococci.

The utilization of lactose by Group N streptococci, however, has been found to differ from that by the Enterobacteriaceae. These bacteria, namely *Streptococcus diacetilactis*, *S. lactis*, and *S. cremoris*, grow on lactose and casein hydrolyzate. They utilize threonine with the formation of glycine and acetaldehyde. The enzyme catalyzing the utilization of threonine is threonine aldolase (L-threonine acetaldehyde-lyase, EC 4.1.2.5) and pyridoxal phosphate is required as cofactor. Acetaldehyde itself can be metabolized further via three different paths, depending on the availability

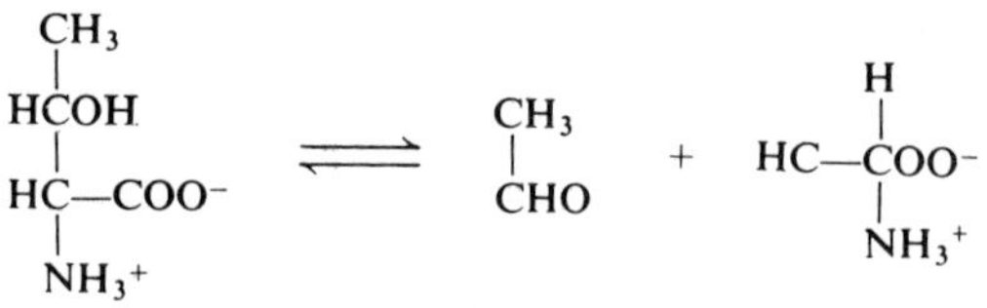

L-threonine glycine

of three enzymes as well as the utilization of lactose via the EMP pathway:

1. Acetaldehyde and glyceraldehyde 3-phosphate are cleaved under the catalytic action of deoxyriboaldolase (2-deoxy-D-ribose-5-phosphate acetaldehyde-lyase, EC 4.1.2.4) to form deoxyribose phosphate:

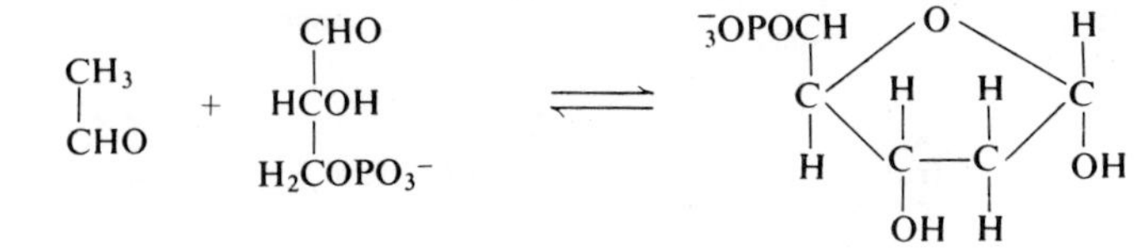

acetaldehyde D-glyceraldehyde 3-phosphoric acid 2-deoxyribose 5-phosphoric acid

2. The Group N streptococci also possess an enzyme that cleaves pyruvate with acetaldehyde with the formation of acetoin:

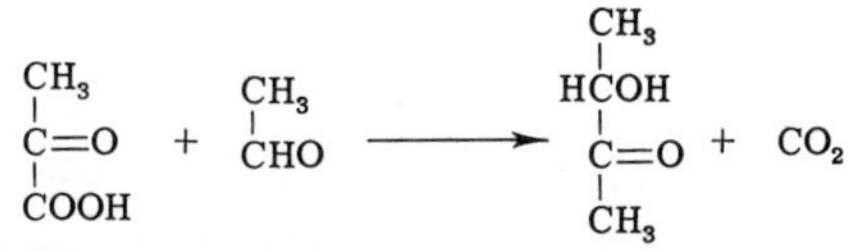

pyruvic acid acetaldehyde acetoin

This enzyme has been called "acetoin synthetase" by Lees and Jago (personal communication).

3. The third possible acetaldehyde metabolism is directed by an active alcohol dehydrogenase (alcohol:NAD oxidoreductase, EC 1.1.1.1) that converts acetaldehyde to ethanol:

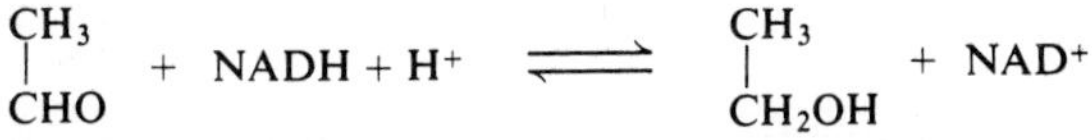

ethanol

Of course, there may be a number of Group N streptococci that do not follow this metabolism, as was shown with the glycine-requiring *Streptococcus cremoris* Z 8. This strain lacked the threonine aldolase.

Mannose Metabolism

The catabolism of mannose can follow two different mechanisms, a cyclic and a noncyclic mechanism for the D-isomer, whereas only the noncyclic system of utilization appears to exist when L-mannose is the substrate

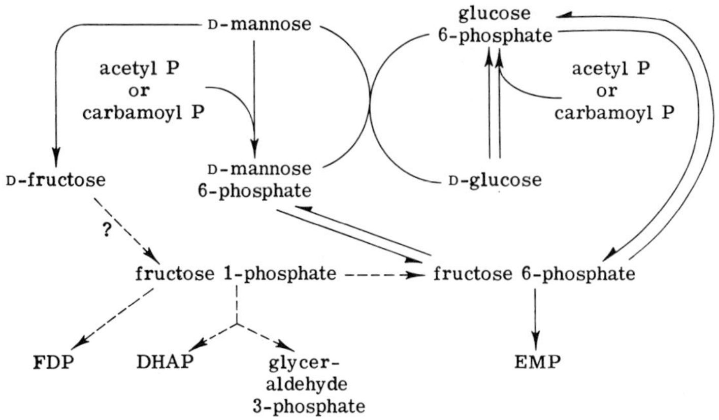
D-mannose
glucose
6-phosphate
acetyl P
or
carbamoyl P
acetyl P
or
carbamoyl P
D-fructose
D-mannose
6-phosphate
D-glucose
?
fructose 1-phosphate
fructose 6-phosphate
FDP
DHAP
glycer-
aldehyde
3-phosphate
EMP

Fig. 5.13. Mannose metabolism (reprinted with permission of the American Society of Biological Chemists).

(Fig. 5.13). *Aerobacter aerogenes* (196) phosphorylates D-mannose with glucose 6-phosphate and phosphotransferase to mannose 6-phosphate. This phosphate transfer from glucose 6-phosphate to D-mannose also yields glucose. The same enzyme is also able to transfer the phosphate group from acetyl phosphate or carbamyl phosphate. There is no D-mannokinase present. Glucose 6-phosphate can now be regenerated by the isomerization of D-mannose 6-phosphate to fructose 6-phosphate, which is catalyzed by mannosephosphate isomerase (D-mannose-6-phosphate ketol-isomerase, EC 5.3.1.8) (151). Fructose 6-phosphate is now converted to glucose 6-phosphate by a second isomerase (D-glucose-6-phosphate ketol-isomerase, EC 5.3.1.9). *Aerobacter aerogenes* PRL-R3 can also regenerate glucose 6-phosphate by phosphorylating the formed glucose with a stereospecific glucokinase. Fructose 6-phosphate is further metabolized via the EMP pathway.

The apparent epimerization of D-mannose to D-glucose may occur via a cyclic process involving D-mannose-6-phosphate isomerase, D-glucose-6-phosphate isomerase, and acylphosphate:hexose phosphotransferase (196).

The purification of a D-mannose isomerase from *Mycobacterium smegmatis* (176) suggests the conversion of D-mannose to D-fructose. Whether or not D-fructose can be metabolized further via fructose 1-phosphate to fructose 6-phosphate or fructose diphosphate, as previously described with fructose metabolism, or cleaved by an aldolase as in the L-mannose utilization, is not known.

The utilization of L-mannose involves a noncyclic mechanism (269, 270). L-Mannose is converted to L-fructose by a cobalt-activated isomerase.

The product of this reaction is phosphorylated at C-1 with ATP by a kinase reaction and L-fructose 1-phosphate is formed. An aldolase-type cleavage produces dihydroxyacetone phosphate and L-glyceraldehyde and thus connects this pathway with the EMP pathway.

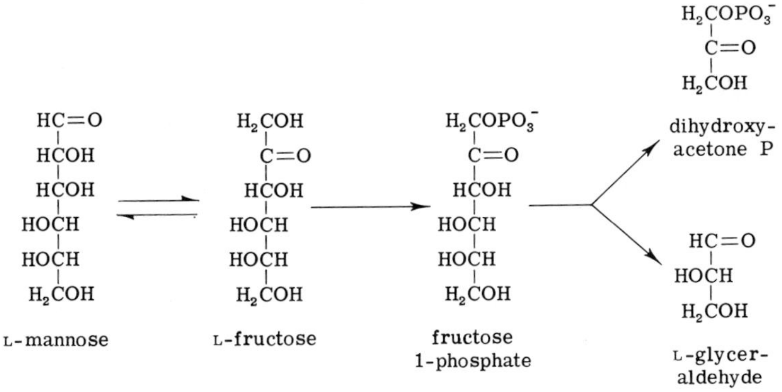
L-mannose
L-fructose
fructose 1-phosphate
dihydroxy-acetone P
L-glycer-aldehyde

Allose Metabolism

The catabolism of glucose and allose follow independent paths until they merge and join the EMP pathway at the fructose 6-phosphate level (144):

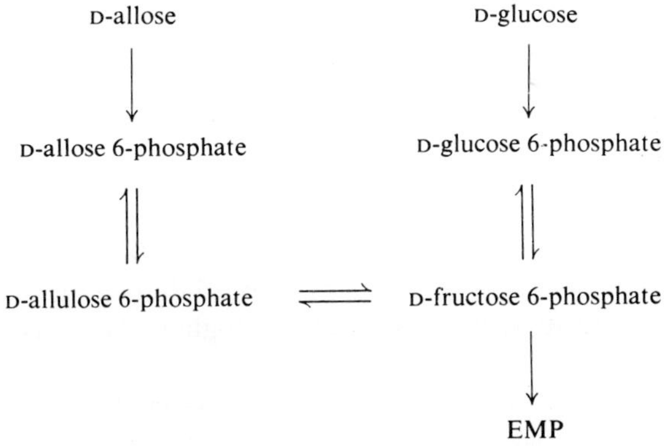
D-allose
D-glucose
D-allose 6-phosphate
D-glucose 6-phosphate
D-allulose 6-phosphate
D-fructose 6-phosphate
EMP

Because *Aerobacter aerogenes* metabolizes allose (4) under both aerobic and anaerobic conditions, most of the study of allose metabolism has been conducted with this microorganism. D-Allose is phosphorylated in the first step by a specific kinase, allose-6-kinase, to D-allose 6-phosphate. The kinase has been purified and characterized (143). The second step is performed by an inducible isomerase, D-allose-6-phosphate ketol-isomerase (268), and D-allulose 6-phosphate is formed. This inducible enzyme also isomerizes ribulose 5-phosphate nonspecifically to ribose 5-phosphate but does not act on other phosphorylated aldoses. The interconversion of D-allulose 6-phosphate to D-fructose 6-phosphate is thought to be catalyzed by a 3-epimerase. *Aerobacter aerogenes* is also able to metabolize glucose, for it possesses the glucosephosphate isomerase (EC 5.3.1.9) (267) as well as a stereospecific D-glucokinase (EC 2.7.1.2) (197).

Gluconate Metabolism

A number of microorganisms are not able to phosphorylate glucose, as they lack either the hexokinase or the PEP–phosphotransferase system. Cell-free extracts of *Acetomonas suboxydans* were shown to possess two kinds of glucose dehydrogenases (138), one DPI (2,6-dichloroindophenol) linked and a second NADP linked. It is very likely that the DPI-linked dehydrogenase is identical with the glucose oxidase (β-D-glucose:oxygen oxidoreductase, EC 1.1.3.4), which is a flavoprotein (see p. 251). The presence of a gluconokinase (ATP:D-gluconate 6-phosphotransferase, EC 2.7.1.12), together with the 6-phosphogluconate dehydrogenase [6-phospho-D-gluconate:NADP oxidoreductase (decarboxylating), EC 1.1.1.44], links the gluconate metabolism to the metabolism of the HMP pathway in the acetic acid bacteria (214).

Acetic acid bacteria have a second possible path for gluconate metabolism. A 5-ketogluconate reductase (D-gluconate:NADP oxidoreductase, EC 1.1.1.group) metabolizes gluconate to 5-ketogluconate in *Acetomonas suboxydans* (Scheme 5.5). A pH optimum of 7.5 favors the forward reaction, whereas one of 9.5 favors the reverse reaction (296).

Acetobacter melanogenum, however, does not possess a gluconokinase (EC 2.7.1.12) and is therefore not able to form 6-phosphogluconate. It therefore oxidizes 2-ketogluconate further, to 2,5-diketogluconate, by means of a ketogluconate dehydrogenase [2-keto-D-gluconate:(acceptor) oxido-

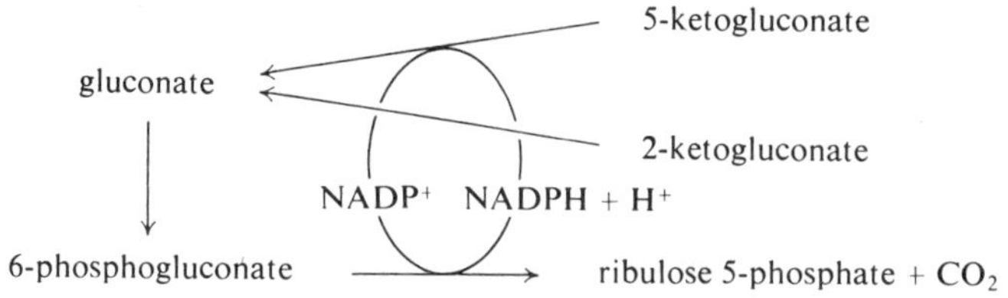
5-ketogluconate
gluconate
2-ketogluconate
NADP+
NADPH + H+
6-phosphogluconate
ribulose 5-phosphate + CO2

Scheme 5.5. The metabolism of gluconate in acetic acid bacteria.

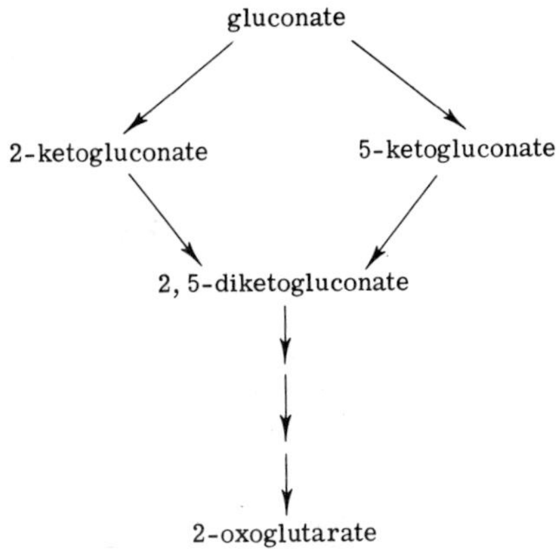
gluconate
2-ketogluconate
5-ketogluconate
2, 5-diketogluconate
2-oxoglutarate

Scheme 5.6. The metabolism of gluconate in *Acetobacter melanogenum.*

reductase, EC 1.1.99.4] (Scheme 5.6). The end product of this reaction is very unstable (201), especially above pH 4.5, and gives rise to brown decomposition products. It therefore seems that *Acetobacter melanogenum* possesses two completely separate systems for the oxidation of glucose and hexose phosphates (202). Probably because of this, *A. melanogenum* is able to oxidize 2-ketogluconate further to 2,5-diketogluconate and through a series of unknown intermediates to 2-oxoglutarate (76). This organism has no phosphohexokinase activity. The generalized scheme of gluconate metabolism in acetic acid bacteria could be as presented in Fig. 5.14.

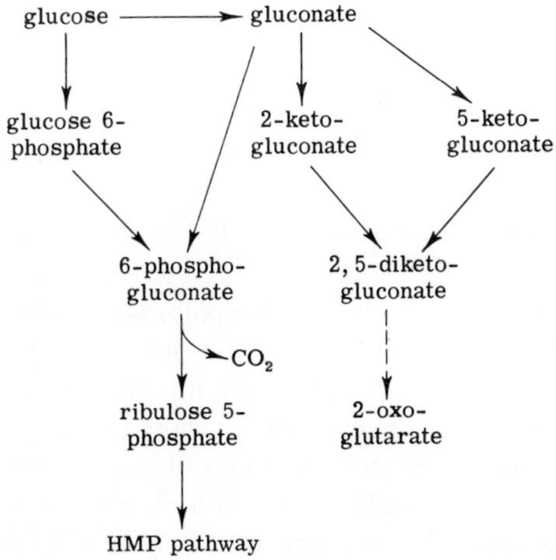
glucose
gluconate
glucose 6-phosphate
2-keto-gluconate
5-keto-gluconate
6-phospho-gluconate
2, 5-diketo-gluconate
CO2
ribulose 5-phosphate
2-oxo-glutarate
HMP pathway

Fig. 5.14. Glucose and gluconate metabolism in acetic acid bacteria.

In the lactic acid bacteria, the fermentation of hexonic acids resembles the fermentation of glucose:

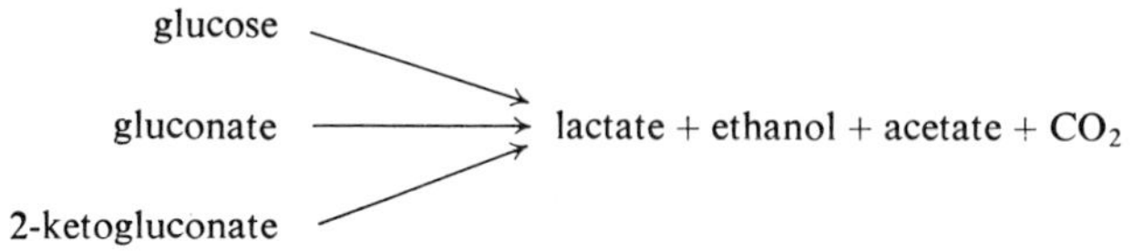
glucose
gluconate
2-ketogluconate
lactate + ethanol + acetate + CO_2

Cells of *Streptococcus faecalis* grown on gluconate, however, ferment glucose (369), gluconate, and 2-ketogluconate, but glucose-grown cells were unable to attack gluconate. It was therefore assumed that a multiple pathway for gluconate fermentation must exist (131) and it was therefore postulated that after the phosphorylation of gluconate to 6-phosphogluconate had taken place, two pathways function equally for the further dissimilation of 6-phosphogluconate:

1. Oxidation of 6-phosphogluconate via the HMP and EMP pathways

$$3\ \text{6-PG} \rightarrow 3\ \text{pentose phosphate} + 3\ CO_2 + 6\ H^+\ (\text{HMP})$$

$$3\ \text{Pentose phosphate} \rightarrow 2\ \text{F6-P} + \text{glyceraldehyde 3-P}\ (\text{HMP})$$

$$2\ \text{F6-P} + \text{glyceraldehyde 3-P} \rightarrow 5\ \text{lactate}\ (\text{EMP})$$

2. Oxidation of 6-phosphogluconate via the ED and EMP pathways:

$$3\ \text{6-PG} \rightarrow 3\ \text{pyruvate} + 3\ \text{glyceraldehyde 3-P}\ (\text{ED})$$

$$3\ \text{Pyruvate} + 6\ H^+ \rightarrow 3\ \text{lactate}\ (\text{EMP})$$

$$3\ \text{Glyceraldehyde 3-P} \rightarrow 3\ \text{lactate}\ (\text{EMP})$$

The overall stoichiometry of these pathways would therefore be

$$\text{6-Phosphogluconate} \rightarrow 1.83\ \text{lactate} + 0.5\ CO_2$$

This has been worked out only for *Streptococcus faecalis*, which is considered a homofermentative organism. The details of the 2-ketogluconate metabolism have not been elucidated. By analogy to *Enterobacter cloacae* (40) and *Pseudomonas fluorescens* (119), it would be expected that 2-ketogluconate is first phosphorylated to 2-keto-6-phosphogluconate, followed by a reduction to 6-phosphogluconate. Whether the 6-phosphogluconate is metabolized as above or heterofermentatively is still obscure.

2-Keto-6-phosphogluconate is also an intermediate in the gluconate metabolism of *Aerobacter cloaceae* (81).

The Pseudomonadaceae use the ED pathway for their gluconate utilization. *Pseudomonas fluorescens* (291) phosphorylates not only gluconate (67), but also 2-ketogluconate via 2-keto-6-phosphogluconate to 6-phosphogluconate (55, 82, 106, 114, 126, 215, 255, 290, 293, 374, 404). The latter compound is then metabolized via the ED pathway.

Investigations with mutants lacking both phosphoglucose isomerase and glucose-6-phosphate dehydrogenase showed that Enterobacteriaceae, e.g., *Escherichia coli* (124, 217a) and *Pasteurella pseudotuberculosis* (46), appear not to possess a glucose dehydrogenase to convert glucose to gluconate. However, they metabolize gluconate via 6-phosphogluconate and the HMP or ED (409) pathway. They also do not possess the corresponding reductases for the formation of 2-keto- or 5-ketogluconate.

To summarize the utilization of gluconate by the different bacterial families, it seems that gluconate metabolism does not involve the EMP pathway, but only the ED or HMP pathway (110). In most cases, the diversion from glucose to gluconate metabolism is caused by the lack of either glucokinase, which phosphorylates glucose, or phosphoglucoisomerase, which converts glucose 6-phosphate to fructose 6-phosphate. *Escherichia coli* (123) and *Salmonella typhimurium* (120) are able to (*a*) utilize gluconate as substrate and, in the case of *E. coli*, preferentially via the ED pathway (409) and (*b*) utilize glucose via the HMP pathway. Acetic acid bacteria and lactic acid bacteria utilize gluconate solely via the HMP pathway, whereas pseudomonads prefer the ED pathway.

Mannitol Metabolism

Bacterial species can be divided into two groups with respect to their mode of mannitol catabolism (376): (a) those which initiate it by a phosphorylation, and (b) those which initiate it by a dehydrogenation. In both cases, the mannitol metabolism joins the EMP pathway at the fructose 6-phosphate level. The division into these two groups can be made on the basis of the organism's ability to form an NAD-dependent mannitol-1-phosphate dehydrogenase (D-mannitol-1-phosphate:NAD oxidoreductase, EC 1.1.1.17). With this criterion *Aerobacter aerogenes* (244, 399), *Bacillus subtilis* (183), *Diplococcus pneumoniae* (261), *Escherichia coli* (399, 400), *Salmonella typhimurium* (29), *Lactobacillus plantarum* (59), and *Staphylococcus aureus* (289) would follow Scheme 5.7, I, whereas *Acetobacter suboxydans* (15, 16, 310), *Azotobacter agilis* (260), *Cellvibrio*

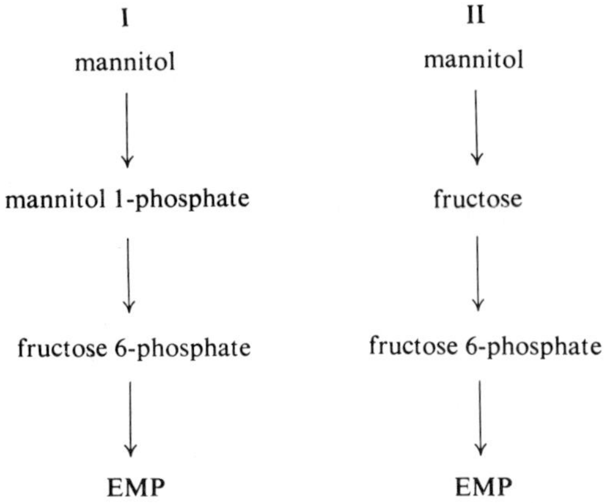
I
mannitol
mannitol 1-phosphate
fructose 6-phosphate
EMP
II
mannitol
fructose
fructose 6-phosphate
EMP

Scheme 5.7. Metabolism of mannitol.

polyoltrophicus (351), and *Pseudomonas fluorescens* (353) possess a mannitol dehydrogenase and follow Scheme 5.7, II.

The first step in Scheme 5.7, I, is the phosphorylation of mannitol by a mannitol kinase (ATP:mannitol 1-phosphotransferase, EC 2.7.1.57) that is phosphoenolpyruvate dependent (376) and converts mannitol to mannitol 1-phosphate, which can accumulate (22). The latter compound is now converted to the EMP pathway intermediate fructose 6-phosphate

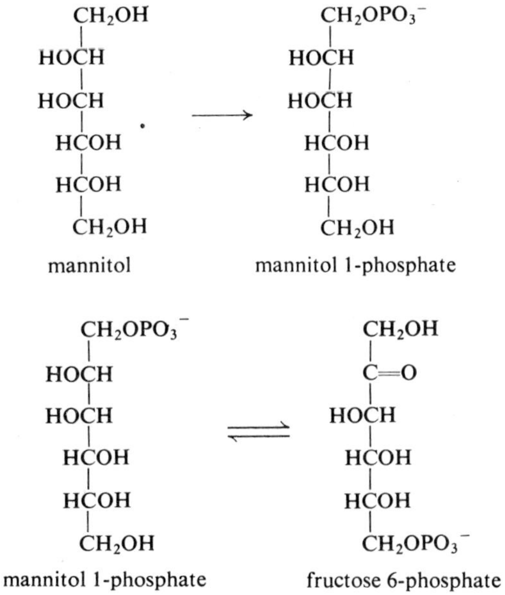
mannitol
mannitol 1-phosphate
mannitol 1-phosphate
fructose 6-phosphate

with the enzyme D-mannitol-1-phosphate dehydrogenase (EC 1.1.1.17), which is NAD linked, as catalyst.

Those microorganisms that metabolize D-mannitol to D-fructose at a nonphosphorylated level (15, 354) possess an NAD-linked mannitol dehydrogenase and carry out an oxidative dissimilation of mannitol (353).

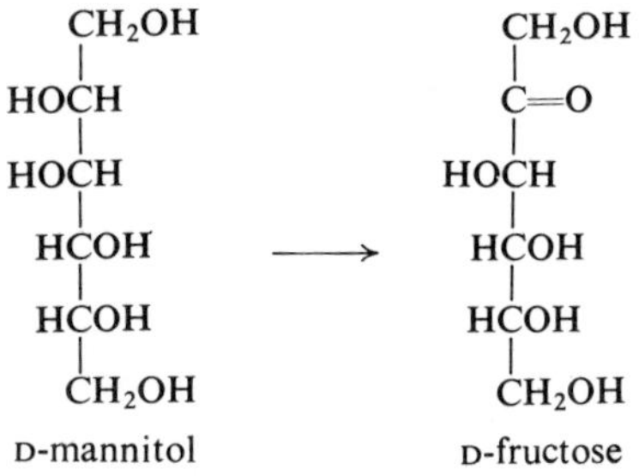
D-mannitol
D-fructose

This enzyme has been reported to be NAD linked. Whether this enzyme is identical with the cytochrome-linked mannitol dehydrogenase (D-mannitol:ferricytochrome oxidoreductase, EC 1.1.2.2) or with the D-mannitol: NAD^+ 2-oxidoreductase (EC 1.1.1.67) is not known. The similarity with the former is striking, however, because it oxidizes D-sorbitol, as has been reported for the NAD-linked enzyme (353). Fructose can either be phosphorylated by fructokinase (EC 2.7.1.4) and join the EMP pathway at the fructose 1,6-diphosphate level, as described under fructose metabolism, or be converted to glucose and gluconic acid (353, 380), which is characteristic for the Pseudomonadaceae.

Sorbitol Metabolism

The metabolism of sorbitol is almost identical to that for mannitol because only the arrangement on C-2 differentiates between mannitol and sorbitol. The only difference occurs in the phosphorylation of sorbitol with a sorbitol kinase to sorbitol 6-phosphate (244). Subsequently, a NAD-linked sorbitol-6-phosphate dehydrogenase (EC 1.1.1.140) converts sorbitol 6-phosphate to fructose 6-phosphate. Sorbitol phosphorylation is mediated by a PEP:sorbitol 6-phosphotransferase system in *Aerobacter aerogenes* PRL-R3 (205). Both enzymes, mannitol-1-phosphate dehydrogenase and sorbitol-6-phosphate dehydrogenase, have been purified and found to be completely different enzymes (244). In *Clostridium pasteurianum* (102), sorbitol-6-phosphate dehydrogenase exhibited specificity for NAD^+, $NADH + H^+$, sorbitol 6-phosphate, and fructose 6-phosphate. The equilibrium of the reaction lies heavily on the side of sorbitol 6-phosphate formation. Sorbitol itself, however, can induce the formation of mannitol-1-phosphate dehydrogenase in *Bacillus subtilis* (183) although sorbitol 6-phosphate is inactive as substrate.

Acetic acid bacteria not only oxidize D-sorbitol to D-fructose in the presence of NAD+, but also produce L-sorbose in the presence of NADP+

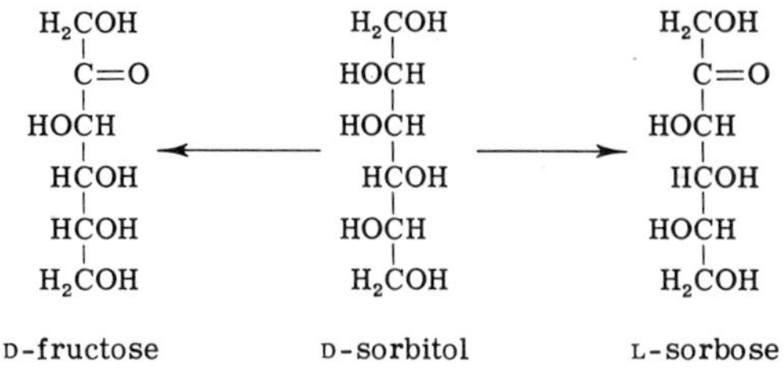
D-fructose
D-sorbitol
L-sorbose

(16). As the NADP-linked sorbitol dehydrogenase is very unstable (395), the fructose formation appears more likely to occur. This would agree with the known fructose metabolism, wherein fructose is phosphorylated to fructose 6-phosphate and metabolized via the HMP pathways.

A number of acetic acid bacteria, however, appear able to oxidize L-sorbose very rapidly with the oxidation products being 5-ketofructose or kojic acid. The latter compound could also be formed from the former by *Acetobacter suboxydans* var. *nonaceticum* 1 FO 3254. This indicated that 5-ketofructose is an intermediate in the formation of kojic acid, not only from D-fructose but also from L-sorbose (see Scheme 5.8).

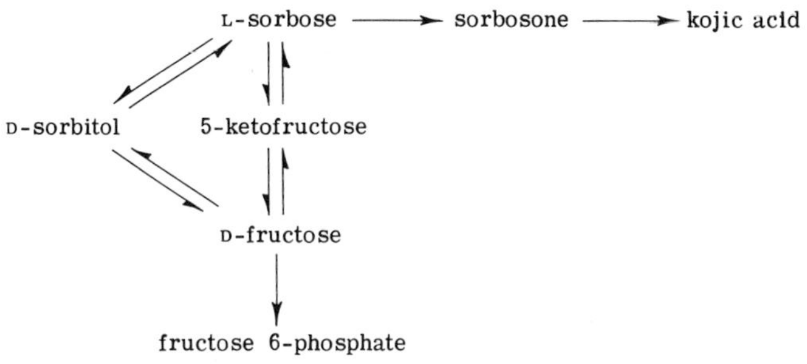
L-sorbose
sorbosone
kojic acid
D-sorbitol
5-ketofructose
D-fructose
fructose 6-phosphate

Scheme 5.8. Sorbitol metabolism.

Inositol Metabolism

The degradation of myoinositol has been elucidated in *Aerobacter aerogenes* (10, 11, 33, 150) (Scheme 5.9). Myoinositol is oxidized by an NAD-

myoinositol
↓
2-ketomyoinositol
↓
D-2, 3-diketo-4-deoxy-*epi*-inositol
|
[1, 3-diketo-4-deoxyinositol]
↓
2-deoxy-5-keto-D-gluconic acid
↓
2-deoxy-5-keto-D-6-phosphogluconic acid
↙ ↘
dihydroxyacetone phosphate | malonic semialdehyde → acetyl-CoA + CO_2

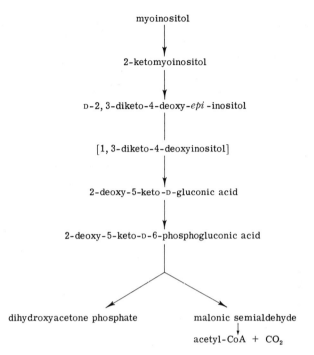

Scheme 5.9. General scheme of inositol metabolism in *Aerobacter aerogenes.*

linked dehydrogenase (inositol:NAD oxidoreductase) to 2-ketomyoinositol. The enzyme is strongly inhibited by its product. The oxidation occurs on C-2, where two hydrogen atoms are transferred to NAD^+. The second step

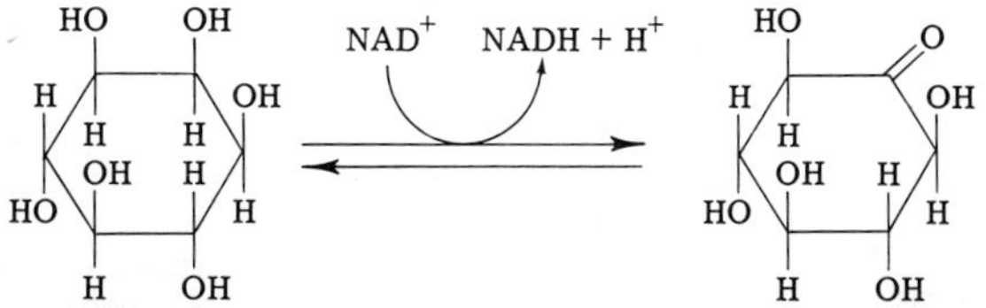
NAD+
NADH + H+

myoinositol 2-ketomyoinositol

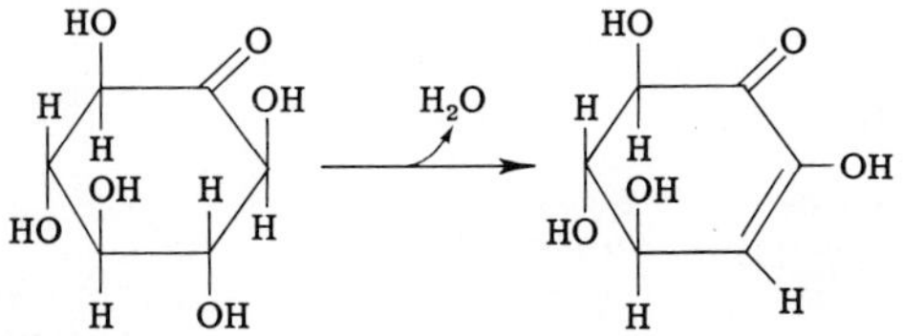
H2O

2-ketomyoinositol D-2, 3-diketo-4-deoxy-*epi*-inositol

is a dehydration (keto inositol dehydrase) to D-2,3-diketo-4-deoxy-*epi*-inositol. It is assumed that this dehydration is followed by an epimerization to 1,3-diketo-4-deoxyinositol, a compound that has not been isolated as

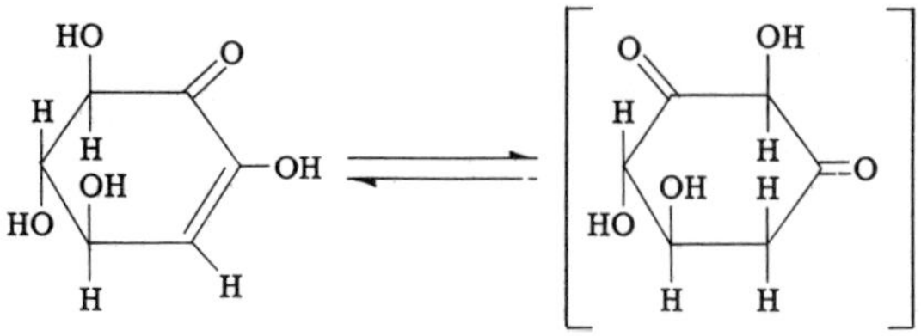

D-2, 3-diketo-4-deoxy-*epi*-inositol

1, 3-diketo-4-deoxyinositol

yet and that is still hypothetical A hydrolytic cleavage of the cyclitol ring produces 2-deoxy-5-ketogluconic acid. A phosphorylation followed by

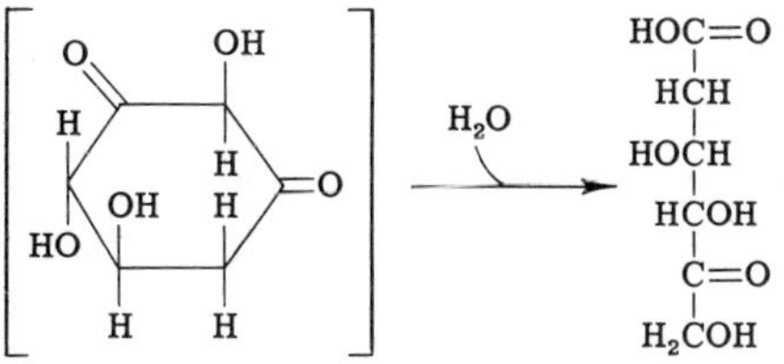

1, 3-diketo-4-deoxyinositol

2-deoxy-5-keto-gluconic acid

an aldolase cleavage yields dihydroxyacetone phosphate and malonic semialdehyde.

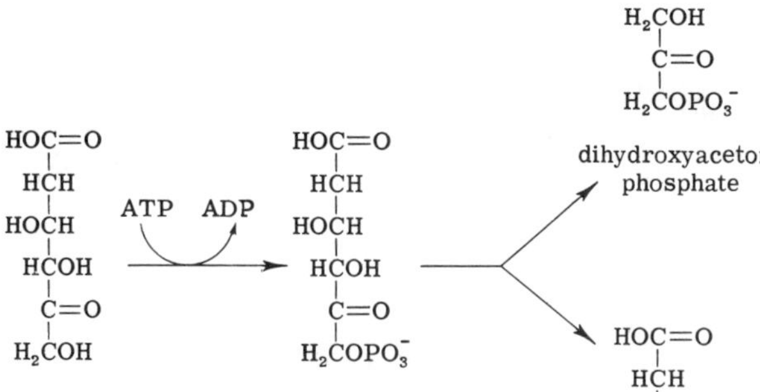

2-deoxy-5-keto-gluconic acid

2-deoxy-5-keto-6-phosphoglu-conic acid

dihydroxyacetone phosphate

malonic semialdehyde

An oxidative decarboxylation with NAD^+ and coenzyme A converts malonic semialdehyde into acetyl-CoA and carbon dioxide.

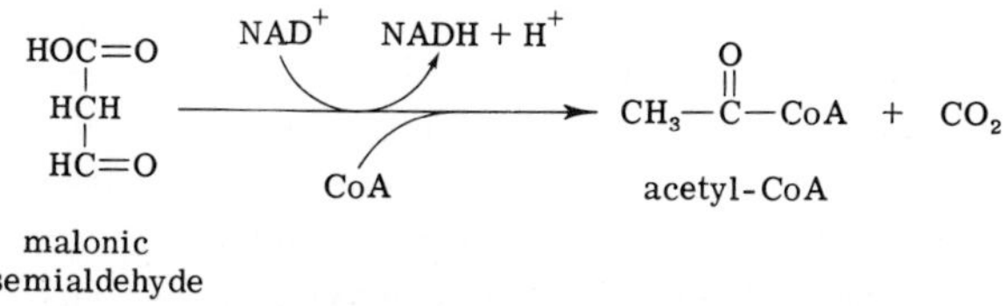
HOC=O
HCH
HC=O
malonic
semialdehyde
NAD+
NADH + H+
CoA
CH3—C—CoA + CO2
acetyl-CoA

The acetyl-CoA formed can thus enter the TCA cycle under aerobic conditions.

Hexuronic Acid Metabolism

Hexuronic acids, such as glucuronic acid, galacturonic acid, mannuronic acid, and polygalacturonic acid, are metabolized by a few microorganisms

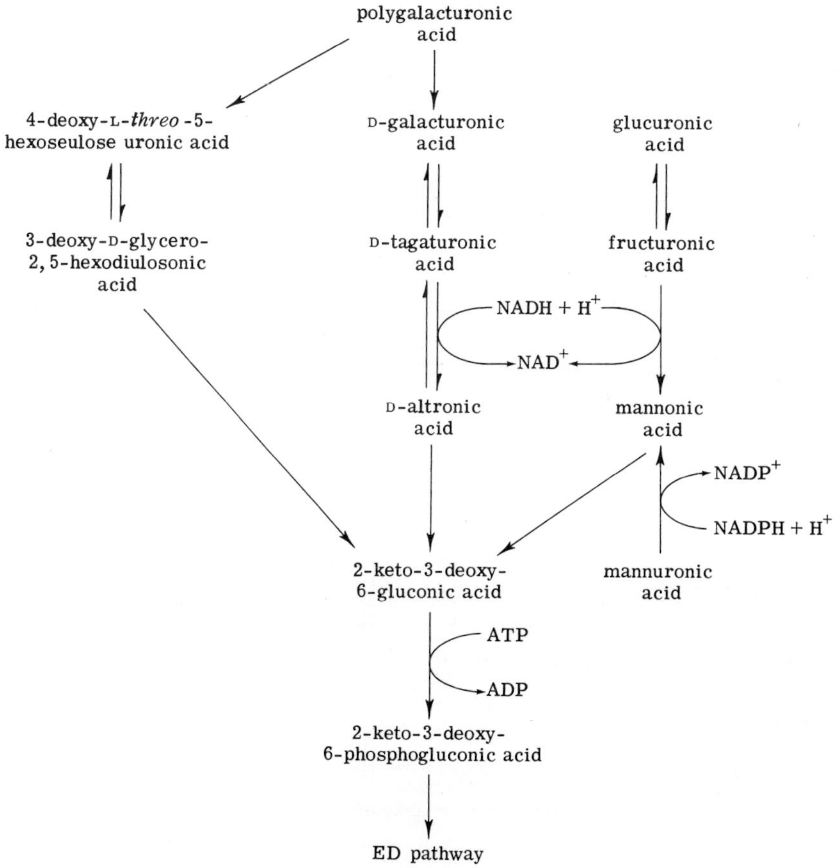
polygalacturonic
acid
4-deoxy-L-threo-5-
hexoseulose uronic acid
D-galacturonic
acid
glucuronic
acid
3-deoxy-D-glycero-
2, 5-hexodiulosonic
acid
D-tagaturonic
acid
fructuronic
acid
NADH + H+
NAD+
D-altronic
acid
mannonic
acid
NADP+
NADPH + H+
2-keto-3-deoxy-
6-gluconic acid
mannuronic
acid
ATP
ADP
2-keto-3-deoxy-
6-phosphogluconic acid
ED pathway

Fig. 5.15. Hexuronic acid metabolism. (Reprinted with permission of the authors and the American Society of Microbiology.)

of the genera *Pseudomonas* (86), *Aeromonas* (115), and *Agrobacterium* (88). As all hexuronic acids are metabolized via the ED pathway, the general scheme as outlined in Fig. 5.15 emerges. Glucuronate and galacturonate are first converted to fructuronate or tagaturonate by the respective isomerases. These products, as well as mannuronate, are now metabolized by the function of hexuronic acid dehydrogenase (D-aldohexuronic acid:NAD oxidoreductase) (61), which is NAD linked in the case of fructuronic acid and tagaturonic acid and NADP linked with mannuronic acid. After the dehydration of these hexuronic acids to 2-keto-3-deoxygluconic acid, a phosphorylation by ATP takes place, with the formation of the key intermediate of the ED pathway. As these organisms are strictly aerobes, pyruvate can be metabolized via the tricarboxylic acid cycle.

Pentose and Pentitol Metabolism

Pentose and pentitol metabolism has been extensively studied in the Enterobacteriaceae and Lactobacillaceae, the main difference being that xylulose 5-phosphate is either further degraded via phosphoketolase (EC

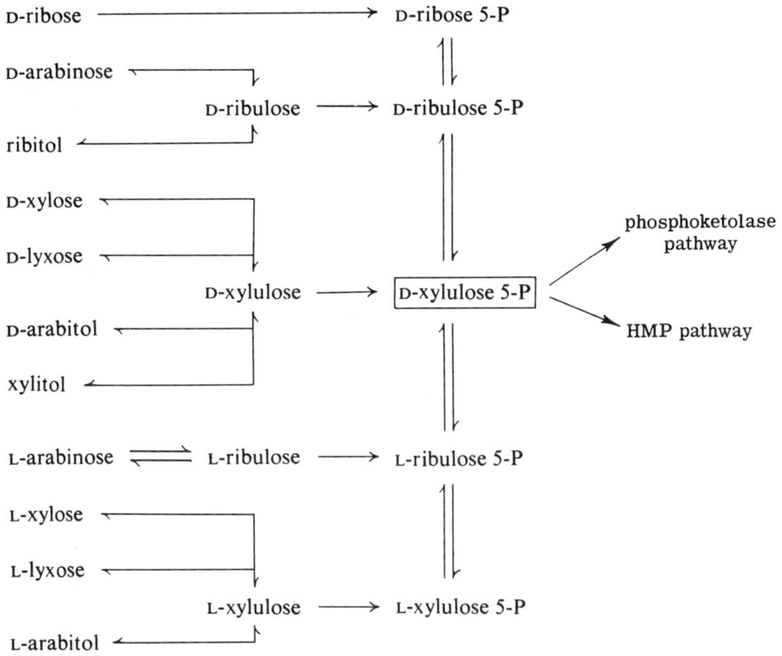
D-ribose
D-ribose 5-P
D-arabinose
D-ribulose
D-ribulose 5-P
ribitol
D-xylose
D-lyxose
D-xylulose
D-xylulose 5-P
phosphoketolase pathway
HMP pathway
D-arabitol
xylitol
L-arabinose
L-ribulose
L-ribulose 5-P
L-xylose
L-lyxose
L-xylulose
L-xylulose 5-P
L-arabitol

Scheme 5.10. General scheme of pentose and pentitol metabolism.

4.1.2.9) (168) or the HMP pathway, as in *Escherichia coli* and *Aerobacter aerogenes*. The overall scheme of the catabolism could be summarized in Scheme 5.10 (285, 286).

Ribose Metabolism

The reaction of D-ribose to ribose 5-phosphate and further to D-xylulose 5-phosphate has been described in detail in the section on the phosphoketolase pathway (see p. 244) and functions in the same way in the Enterobacteriaceae. Kinetic analysis, however, indicated (77) that two distinct ribose-specific permease systems exist, one being constitutive and the second being inducible only at high external ribose concentrations (approximately 1 mM). Another interesting observation was that only the inducible permease underlies catabolic repression and that it cannot be induced in the presence of glucose.

In addition to the two different permease systems, *Escherichia coli* possesses two distinct ribose 5-phosphate isomerases (78), one being heat stable and the other heat labile. Apparently only the heat-labile enzyme functions in the conversion of D-ribose 5-phosphate to D-ribulose 5-phosphate. It is assumed that the second, heat-stable isomerase regulates the intracellular concentrations of ribose 5-phosphate, as the latter compound inhibits the ribokinase activity. As glucose 6-phosphate also inhibits the heat-labile isomerase, this enzyme seems to play an important regulatory function. It was therefore proposed that two independently regulated pathways must exist in *Escherichia coli:* (a) an anabolic pathway, consisting of the constitutive permease and inducible kinase; and (b) a catabolic pathway, consisting of both ribose permeases, the ribokinase, and the heat-labile isomerase. If ribose is not required as a carbon source, glucose 6-phosphate concentration increases and inhibits the heat-labile ribose-5-phosphate isomerase. In this case, ribose 5-phosphate is converted mainly anabolically to 5-phosphoribosyl-1-pyrophosphate. With D-ribose as the main carbon source, glucose 6-phosphate concentration would be low and the heat-labile isomerase could convert D-ribose 5-phosphate catabolically to D-xylulose 5-phosphate. These mechanisms have been confirmed by *in vitro* investigations (79). A biosynthetic utilization of D-ribose has also been found in *Veillonella* (194, 195).

D-*Arabinose Metabolism*

The metabolism of D-arabinose is a variable one, for not less than three pathways are known depending on the bacterial species in question.

In *Aerobacter aerogenes,* D-arabinose is attacked by an arabinose isomerase (D-arabinose ketol-isomerase, EC 5.3.1.3) (297) that converts D-arabinose to D-ribulose, which in turn is phosphorylated by a specific D-ribulokinase (ATP:D-ribulose 5-phosphotransferase, EC 2.7.1.47) to D-ribulose 5-phosphate. This specific D-ribulokinase has also been found in *Escherichia coli* (207), although it appears that the latter organism phosphorylates C-1, as will be seen below. This specific ribulokinase appears to be different also from the other ribulokinases (EC 2.7.1.16). The conversion of D-ribulose 5-phosphate to D-xylulose 5-phosphate is the same as that described for D-ribose metabolism. The other two pentoses that undergo isomerizations are xylose and lyxose (6).

In *Escherichia coli*, D-arabinose is isomerized to D-ribulose as in *Aerobacter aerogenes*, although a requirement of adenosine-3′,5′-cyclic monophosphate (cyclic AMP) for the synthesis of L-arabinose isomerase appears to exist (289a). The D-ribulokinase of *Escherichia coli* (230) does not phosphorylate C-5, however, but C-1 instead, thus forming D-ribulose 1-phosphate. An aldolase (231) cleaves D-ribulose 1-phosphate to dihydroxyacetone phosphate and glycolaldehyde, which in turn is metabolized to glycolate.

The D-arabinose metabolism in pseudomonads is completely different, although one of the end products is also glycolic acid. D-Arabinose is first converted to its lactone by the enzyme arabinose dehydrogenase (D-arabinose:NAD^+ 1-oxidoreductase, EC 1.1.1.116) (65). A lactonase (D-arabinono-γ-lactone hydrolase, EC 3.1.1.30) metabolizes the lactone to the corresponding acid, arabonic acid. The similarity to the ED pathway is

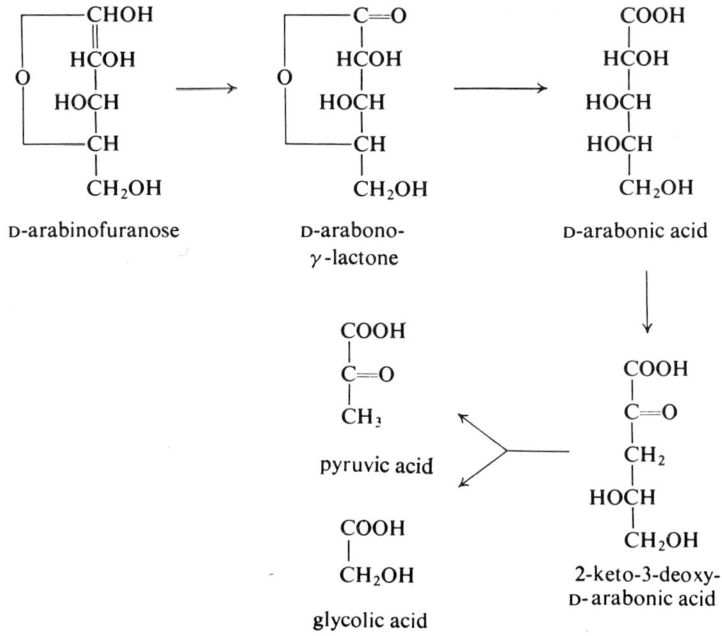
CHOH
HCOH
HOCH
CH
CH₂OH
O
D-arabinofuranose
C=O
HCOH
HOCH
CH
CH₂OH
O
D-arabono-γ-lactone
COOH
HCOH
HOCH
HOCH
CH₂OH
D-arabonic acid
COOH
C=O
CH₃
pyruvic acid
COOH
CH₂OH
glycolic acid
COOH
C=O
CH₂
HOCH
CH₂OH
2-keto-3-deoxy-D-arabonic acid

continued as an arabinonate dehydratase (D-arabinonate hydro-lyase, EC 4.2.1.5) forms 2-keto-3-deoxyarabinonate. An NAD^+-linked dehydrogenase-type reaction forms a 2,4-ketoarabinonate (COOH—CO—CH_2CO—CH_2OH) followed by a hydrolysis, which leads to the formation of pyruvate and glycolic acid. When this is compared with the ED pathway and the *Escherichia coli* metabolism of D-arabinose, this last reaction looks more like an aldolase (EC 4.1.2.18) cleavage reaction, as it was observed in the L-arabinose metabolism of *Pseudomonas* sp. strain MSU-1.

L-*Arabinose Metabolism*

Like D-arabinose, L-arabinose can be metabolized in three different ways.

Escherichia coli, together with *Aerobacter aerogenes*, *Lactobacillus plantarum*, *Bacillus subtilis*, and *Salmonella typhimurium*, first isomerize L-arabinose to L-ribulose (362) by an isomerase action (L-arabinose ketol-isomerase, EC 5.3.1.4) (35, 306). This is then followed by a phosphorylation step. The

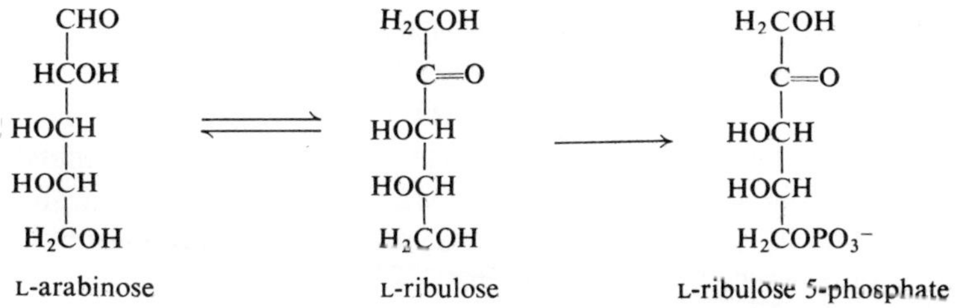
CHO
HCOH
HOCH
HOCH
H₂COH
L-arabinose
H₂COH
C=O
HOCH
HOCH
H₂COH
L-ribulose
H₂COH
C=O
HOCH
HOCH
H₂COPO₃⁻
L-ribulose 5-phosphate

acting enzyme in this reaction is L-ribulokinase (235, 363) (ATP:L-ribulose 5-phosphotransferase, EC 2.7.1.16) and the product formed is L-ribulose 5-phosphate. The latter is now acted upon by a ribulosephosphate 4-epimerase (236) (L-ribulose-5-phosphate 4-epimerase, EC 5.1.3.4), which

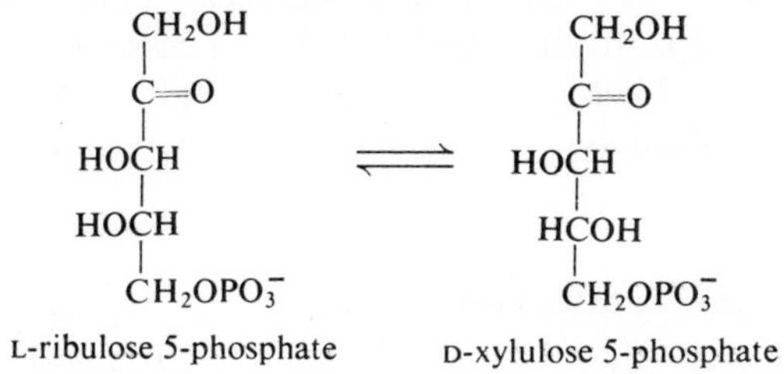
CH₂OH
C=O
HOCH
HOCH
CH₂OPO₃⁻
L-ribulose 5-phosphate
CH₂OH
C=O
HOCH
HCOH
CH₂OPO₃⁻
D-xylulose 5-phosphate

converts L-ribulose 5-phosphate to the key intermediate D-xylulose 5-phosphate.

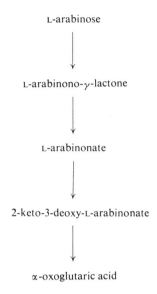

L-arabinose

↓

L-arabinono-γ-lactone

↓

L-arabinonate

↓

2-keto-3-deoxy-L-arabinonate

↓

α-oxoglutaric acid

Pseudomonas saccharophila and *Pseudomonas fragi* convert L-arabinose in like manner to the D-arabinose pathway. L-Arabinose is first attacked by a dehydrogenase (L-arabinose:NAD oxidoreductase, EC 1.1.1.46), forming the corresponding lactone. This reaction is NAD^+ linked and produces 1 mole of reduced NAD. Arabinonolactonase (L-arabinono-γ-lactone hydrolase, EC 3.1.1.15) hydrolyzes the lactone to L-arabinonate, which undergoes a reaction together with L-arabinonate dehydratase (L-arabinonate hydro-lyase, EC 4.2.1.25) producing 2-keto-3-deoxy-L-arabinonate. Whereas in the ED pathway a split of the molecule occurs, this is not the case here. 2-Keto-3-deoxy-L-arabinonate with the addition of 2 H^+, forms 2-oxoglutarate. *Pseudomonas* species are therefore able to produce this dicarboxylic acid straight from L-arabinose without the formation of pyruvate.

The third pathway of L-arabinose metabolism differs only in the last step. *Pseudomonas* sp. strain MSU-1 (74) possesses a 2-keto-3-deoxy-L-arabonate aldolase (2-keto-3-deoxy-L-arabonate glycolaldehyde-lyase, EC 4.1.2.18), which carried out a cleavage reaction forming pyruvate and glycolaldehyde.

Xylose and Lyxose Metabolism

The conversion of D-xylose to D-xylulose is catalyzed by xylose isomerase (D-xylose ketol-isomerase, EC 5.3.1.5), whereas D-lyxose is isomerized by

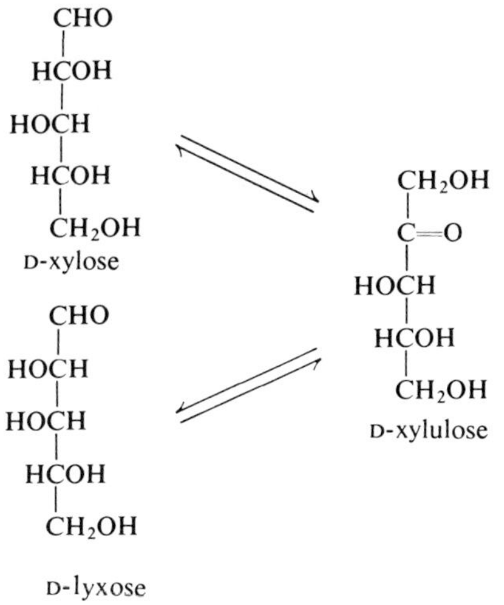

CHO
HCOH
HOCH
HCOH
CH₂OH
D-xylose
CHO
HOCH
HOCH
HCOH
CH₂OH
D-lyxose
CH₂OH
C=O
HOCH
HCOH
CH₂OH
D-xylulose

D-lyxose isomerase (D-lyxose ketol-isomerase, EC 5.3.1.15), which has been purified (8). The common product of both isomerization reactions, D-xylulose, is subsequently phosphorylated by xylulokinase (ATP:D-xylulose

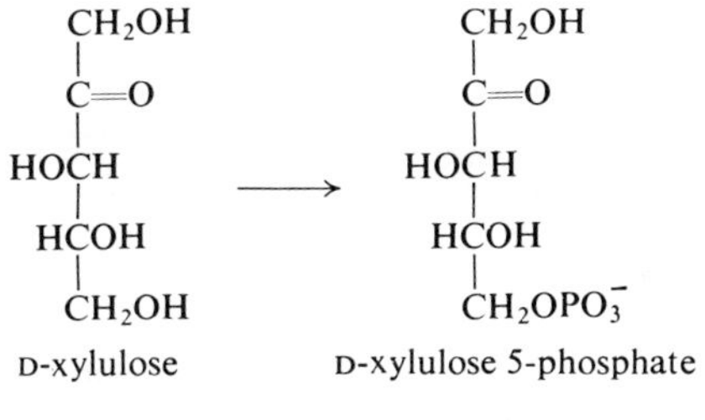

CH₂OH
C=O
HOCH
HCOH
CH₂OH
D-xylulose
CH₂OH
C=O
HOCH
HCOH
CH₂OPO₃⁻
D-xylulose 5-phosphate

5-phosphotransferase, EC 2.7.1.17) (36). The pathways of the L-isomers of these two pentoses have also been evaluated (6), but none of the isomerases is reported in the Enzyme Commission's report. It appears that the main difference between the isomerases of the D and L isomers lies in the cofactor requirements. Whereas the D-isomerases require Mg^{2+}, the L-isomerases are reported to require Co^{2+} for their activity.

The isomerization of L-xylose and L-lyxose to L-xylulose is followed by a phosphorylation reaction, which requires ATP and Mg^{2+}, to L-xylulose 5-phosphate. The enzyme involved in this reaction is named "L-xylulokinase" (ATP:L-xylulose-5-phosphotransferase, EC 2.7.1.53) (7). The conversion of L-xylulose 5-phosphate to D-xylulose 5-phosphate includes two enzymatic steps, involving a 3- and a 4-epimerase. The first step in-

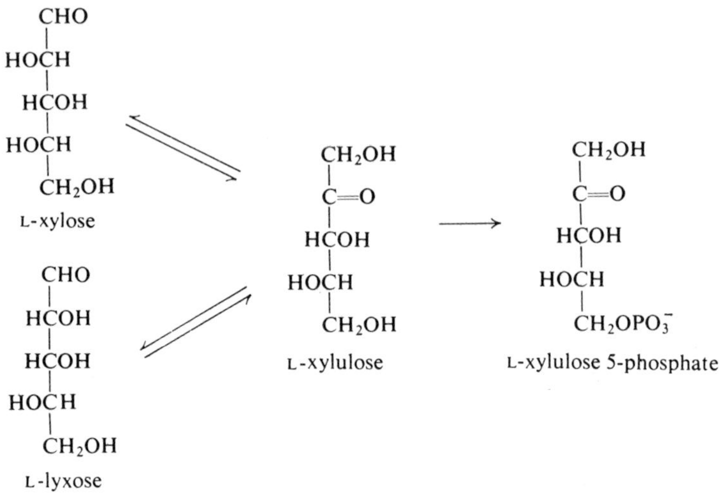

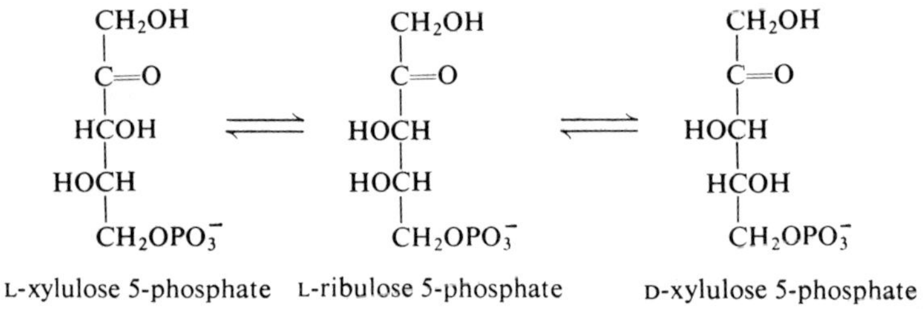

volves a ribulose phosphate 3-epimerase, which converts L-xylulose 5-phosphate to L-ribulose 5-phosphate. Whether or not this enzyme is identical to EC 5.1.3.1 could not be clarified. The second stage with the formation of D-xylulose 5-phosphate is the action of ribulosephosphate 4-epimerase (EC 5.1.3.4).

Pentitol Metabolism

If the metabolism of pentoses involves mainly isomerases, the metabolism of pentitols requires NAD^+-linked dehydrogenases. Ribitol is acted upon by an NAD^+-specific ribitol dehydrogenase (ribitol:NAD^+ 2-oxido-reductase, EC 1.1.1.56) (185, 406), which converts ribitol to D-ribulose. The enzyme appears to be specific for the substrate (131), whereas D-arabitol dehydrogenase, which converts D-arabitol to D-xylulose, also acted upon D-mannitol to form D-fructose (242).

Ribitol dehydrogenase and D-ribulokinase are coordinately induced in *Aerobacter aerogenes* (37), whereas this is not the case with D-arabitol de-

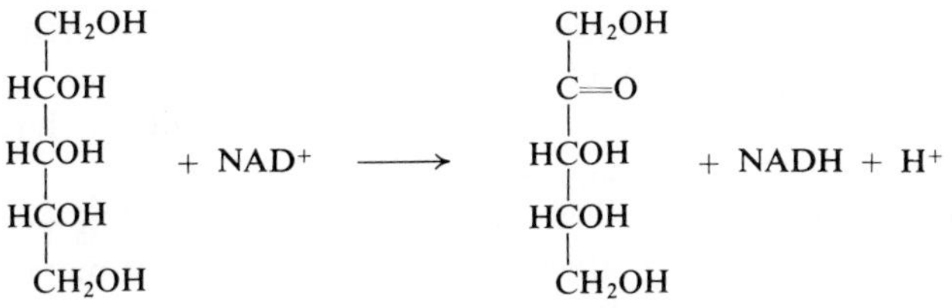

$$\begin{array}{c}\text{CH}_2\text{OH}\\|\\\text{HCOH}\\|\\\text{HCOH}\\|\\\text{HCOH}\\|\\\text{CH}_2\text{OH}\\\text{ribitol}\end{array} + \text{NAD}^+ \longrightarrow \begin{array}{c}\text{CH}_2\text{OH}\\|\\\text{C}=\text{O}\\|\\\text{HCOH}\\|\\\text{HCOH}\\|\\\text{CH}_2\text{OH}\\\text{D-ribulose}\end{array} + \text{NADH} + \text{H}^+$$

hydrogenase and D-xylulokinase. It also appears that D-ribulose and not ribitol acts as the inducer in ribitol metabolism (38).

In xylitol metabolism, only D-xylulose, and never L-xylulose, could be identified as a reaction product of an NAD^+-dependent dehydrogenase reaction (119).

The NAD^+-dependent dehydrogenases of L- and D-arabitol oxidation (119, 406) have also been identified. The reaction products have been identified as L- and D-xylulose, respectively. The key intermediate, D-xylulose 5-phosphate, is able to undergo transketolase and transaldolase reactions as described in the discussion of the HMP pathway to produce fructose 6-phosphate. From here, most of the bacteria, in particular *Aerobacter*, follow the Embden-Meyerhof-Parnas or the hexose monophosphate pathway in further metabolism.

D-Apiose Metabolism

Aerobacter aerogenes PRL-R3 is able to metabolize D-apiose to D-apiitol with a newly discovered enzyme, D-apiose reductase (D-apiitol:NAD^+

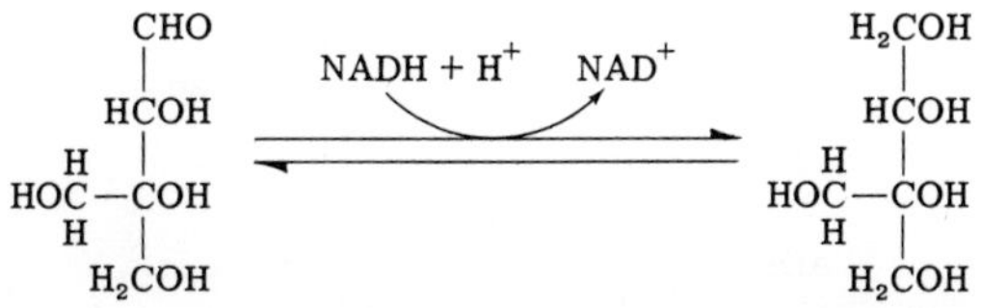

$$\begin{array}{c}\text{CHO}\\|\\\text{HCOH}\\\text{H}\\\text{HOC}-\text{COH}\\\text{H}\\\text{H}_2\text{COH}\\\text{D-apiose}\end{array} \underset{\text{NAD}^+}{\overset{\text{NADH} + \text{H}^+}{\rightleftharpoons}} \begin{array}{c}\text{H}_2\text{COH}\\|\\\text{HCOH}\\\text{H}\\\text{HOC}-\text{COH}\\\text{H}\\\text{H}_2\text{COH}\\\text{D-apiitol}\end{array}$$

1-oxidoreductase, EC 1.1.1.114) (292). D-Apiose and D-apiitol did not serve as substrates for ribitol dehydrogenase or D-arabitol dehydrogenase.

Glycerol Metabolism

The metabolism of glycerol has been observed in *Clostridium butyricum* (21), Enterobacteriaceae, lactobacilli, and acetic acid bacteria.

Members of the Enterobacteriaceae are able to metabolize glycerol either to trimethylene glycol or to glyceraldehyde 3-phosphate and subsequently via the EMP pathway. As these metabolic events are fermentative, they will be dealt with in Chapter 8.

Acetomonas suboxydans, which does not possess a TCA cycle, is also able to grow on and utilize glycerol (162). This glycerol metabolism proceeds along two pathways (163):

1. Glycerol is oxidized to dihydroxyacetone, which proceeds at pH 6.0 and is independent of ATP. This reaction is catalyzed by a glycerol dehydrogenase (glycerol:NAD oxidoreductase, EC 1.1.1.6). A kinase reaction would then phosphorylate dihydroxyacetone, for this reaction has been found to be ATP dependent and also requires Mg^{2+}

Glycerol → dihydroxyacetone → dihydroxyacetone phosphate

2. Glycerol is first phosphorylated by glycerol kinase (ATP:glycerol phosphotransferase, EC 2.7.1.30) to L-glycerol 3-phosphate. Glycerolphosphate dehydrogenase [L-glycerol-3-phosphate: (acceptor) oxidoreductase, EC 1.1.99.5] finally converts glycerol 3-phosphate to dihydroxyacetone phosphate.

The overall scheme for glycerol metabolism in acetic acid bacteria is therefore

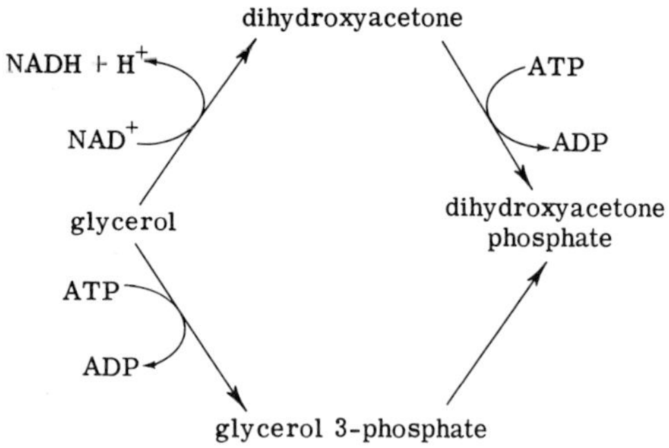
dihydroxyacetone
NADH + H+
NAD+
ATP
ADP
glycerol
dihydroxyacetone phosphate
ATP
ADP
glycerol 3-phosphate

Dihydroxyacetone phosphate is then isomerized by the triosephosphate isomerase (D-glyceraldehyde-3-phosphate ketol-isomerase, EC 5.3.1.1.) to glyceraldehyde 3-phosphate and the fructose-diphosphate aldolase (EC 4.1.2.13) equilibrium forms fructose 1,6-diphosphate (164). The actions of hexosediphosphatase (EC 3.1.3.11) and glucose-6-phosphate isomerase (EC 5.3.1.9) produce glucose 6-phosphate. The oxidation of glucose 6-phosphate to 6-phosphogluconate and further to ribose 5-phosphate is the only way for the glycerol metabolism to function. The amount of phosphate present in the medium appears to be important (356).

Homofermentative lactobacilli metabolize glycerol via mannose phosphate to pyruvic acid (57).

The facultative anaerobic bacterium *Escherichia coli* also possesses the enzyme glycerol kinase (ATP:glycerol phosphotransferase, EC 2.7.1.30)

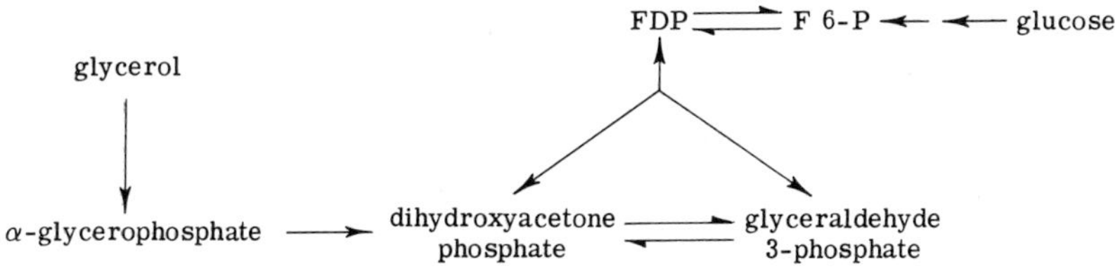
FDP
F 6-P
glucose
glycerol
α-glycerophosphate
dihydroxyacetone phosphate
glyceraldehyde 3-phosphate

(166) as well as a glycerolphosphate dehydrogenase [L-glycerol-3-phosphate:(acceptor) oxidoreductase, EC 1.1.99.5]. As glycerol enters the cell by simple diffusion, it can be converted to glyceraldehyde 3-phosphate via α-glycerophosphate and dihydroxyacetone phosphate. The enzyme glycerol kinase appears to be the rate-limiting enzyme, as it is feedback controlled by FDP (413).

A further pathway of glycerol metabolism was suggested with the detection of a second soluble α-glycerophosphate dehydrogenase, which is FAD dependent and seems to function only under anaerobic conditions (213). This enzyme is very similar to the one found in *Aerobacter aerogenes* (241), which converts glycerol to dihydroxyacetone. An inducible dihydroxyacetonekinase phosphorylates dihydroxyacetone to its phosphate derivative.

Polyol Metabolism of Acetic Acid Bacteria

Investigations into the polyol oxidation of *Acetomonas suboxydans* (210) showed that this organism is able to oxidize a great number of acyclic polyols and related substances according to the rules of Bertrand (see p. 251). The secondary OH group involved in the oxidation must have a D configuration with respect to the primary alcohol group adjacent to the site of oxidation. *Acetomonas suboxydans* therefore contains six different polyol dehydrogenases and each of these has a unique structural specificity: (a) $NADP^+$-xylitol dehydrogenase [xylitol:NADP oxidoreductase (L-xylulose-forming) EC 1.1.1.10]; (b) NAD^+-D-mannitol dehydrogenase (D-mannitol:NAD^+ 2-oxidoreductase, EC 1.1.1.67); (c) NAD^+-D-*erythro*-dehydrogenase; (d) NAD^+-D-*xylo*-(D-sorbitol) dehydrogenase (411) (D-sorbitol-6-phosphate:NAD^+ 2-oxidoreductase, EC 1.1.1.140); (e) $NADP^+$-D-xylo-dehydrogenase [xylitol:NAD^+ 2-oxidoreductase (D-xylulose forming), EC 1.1.1.9]; and (f) $NADP^+$-D-*lyxo*-(D-mannitol) dehydrogenase (D-mannitol:$NADP^+$ 2-oxidoreductase, EC 1.1.1.138).

Glycol Oxidation

Glycerol is not only metabolized in the HMP cycle but also can function as an inducer for glycol dehydrogenases in cells originally grown on glucose or lactate medium (225). The formation of these enzymes seems to be especially dependent on glycerol, although a nitrogen source as well as an energy source, such as glucose, is required (224). The glycol oxidation of the *Acetobacter* and *Acetomonas* group appears to involve three basic enzymes (209):

1. A soluble NAD^+-linked primary alcohol dehydrogenase, which does not react with $NADP^+$ (87) but seems to require the

$$>CH—CH_2OH \text{ group}$$

because methanol is not oxidized and an OH group or a second CH_2OH group at C-2 decreases or inhibits enzymatic action. The further the second CH_2OH group is removed in position from the one to be attacked, the greater is its susceptibility to enzymatic action. The end products are most likely to be the corresponding aldehydes before strong NAD^+- and $NADP^+$-linked dehydrogenases were found to be present (212) in *Acetomonas suboxydans.*

2. A soluble NAD^+-linked secondary alcohol dehydrogenase that attacks glycols with a secondary OH group. This enzyme seems to be less specific. The presence of an adjacent OH group improves the enzymatic activity and the end products appear to be the corresponding ketones. The presence of a third OH group, a C=O group, or a COOH group in the molecule decreases the enzymatic activity.

3. A particulate oxidative system by which the oxidation of glycols proceeds with the uptake of oxygen. It is very likely that an electron transport system via cytochromes occurs. This oxidative system not only oxidizes all primary and secondary alcohols but many other compounds as well, e.g., hexoses, pentoses, polyols, and aldehydes. The oxidation of diols depends on the distance between the two terminal CH_2OH groups (209). In short-chain diols, i.e., where these groups are situated close together, as in 1,2-ethanediol, only one of them can be oxidized. The resulting carboxyl group apparently prevents further enzymatic action. With larger molecules the negative influence of the carboxyl group decreases, and the oxidation proceeds with both types of molecules at once to the dicarboxylic acids. It appears that cytochrome 553 is tightly bonded to the enzyme aggregate. There is no evidence for the participation of NAD, NADP, flavins, ubiquinones, or free heavy metals in this system.

Further details of glycol oxidations in acetic acid bacteria are presented in a review article by DeLey and Kersters (87).

Oxidation of the Primary Alcohol Group

Acetobacter xylinum, A. aceti, A. acetosus, A. kuetzingianus, A. pasteurianus, A. acetigenus, A. ascendens, Acetomonas suboxydans (oxydans), and *Acetomonas industrium* are able to oxidize ethylene glycol to glycolic acid, with glycolic acetaldehyde as a possible intermediate:

$$\begin{array}{c} CH_2OH \\ | \\ CH_2OH \end{array} \longrightarrow \begin{array}{c} COOH \\ | \\ CH_2OH \end{array}$$

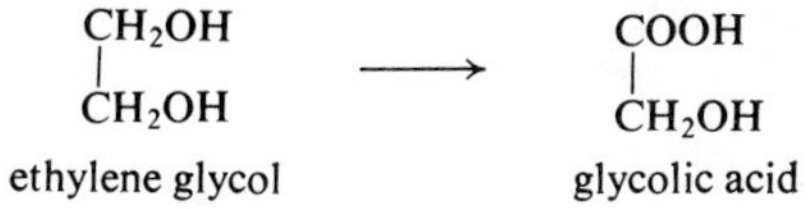

ethylene glycol → glycolic acid

The mechanism by which this C-2 substrate is converted into cell material, however, is still unknown. Ethylene glycol monomethyl ether (Methyl Cellosolve) is quickly oxidized by resting cells of *Acetomonas suboxydans*

$$\begin{array}{c} CH_2OH \\ | \\ CH_2-O-CH_3 \end{array} \longrightarrow \left[\begin{array}{c} COOH \\ | \\ CH_2-O-CH_3 \end{array} \right]$$

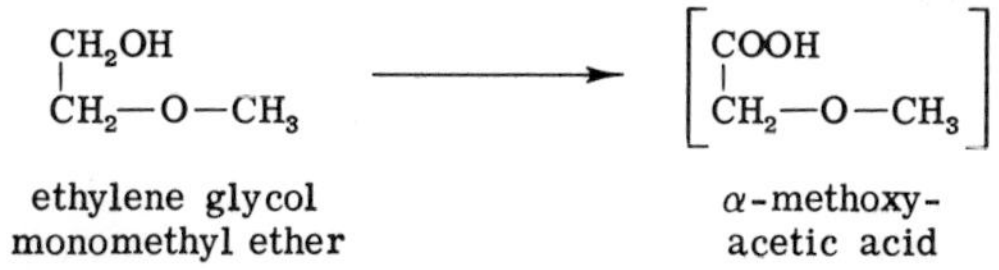

ethylene glycol monomethyl ether → α-methoxy-acetic acid

with the uptake of 1 mole of O_2 per mole substrate and with the possible formation of α-methoxyacetic acid. The monoethyl ether of ethylene glycol (Cellosolve) is oxidized very slowly, and therefore this oxidation never comes to completion with submerged cultures of *Acetomonas suboxydans*

$$\begin{array}{c} CH_2OH \\ | \\ CH_2-O-CH_2CH_3 \end{array} \longrightarrow \begin{array}{c} COOH \\ | \\ CH_2-O-CH_2CH_3 \end{array}$$

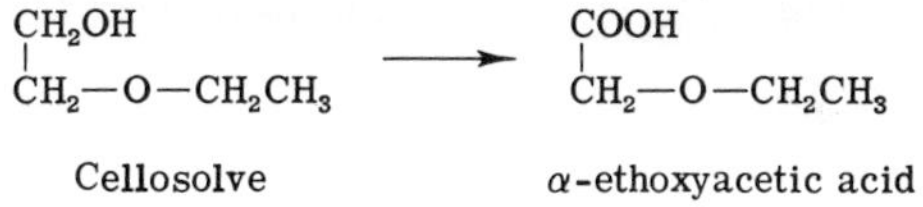

Cellosolve → α-ethoxyacetic acid

(184, 314). The α-ethoxyacetic acid is formed. An incomplete oxidation occurs also with diethylene glycol in aerated, submerged cultures of *Acetomonas suboxydans* with the formation of α-hydroxyethoxyacetic

$$\begin{array}{c} CH_2OH \\ | \\ CH_2 \\ | \\ O \\ | \\ CH_2 \\ | \\ H_2COH \end{array} \longrightarrow \begin{array}{c} COOH \\ | \\ CH_2 \\ | \\ O \\ | \\ CH_2 \\ | \\ H_2COH \end{array} \longrightarrow \begin{array}{c} COOH \\ | \\ CH_2 \\ | \\ O \\ | \\ CH_2 \\ | \\ COOH \end{array}$$

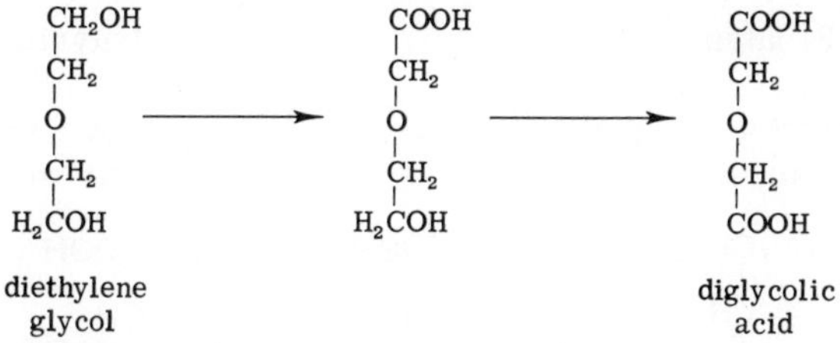

diethylene glycol → → diglycolic acid

acid as an intermediate and the corresponding dicarboxylic acid (diglycolic acid) as end product.

The monomethyl ether of diethylene glycol can be oxidized by *Acetomonas suboxydans* as well (184), but the yield of 2-(2-methoxy)ethoxyacetic acid

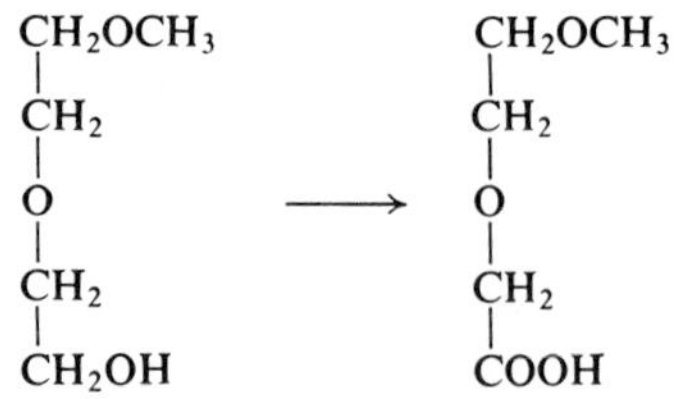

monomethyl ether of diethylene glycol — 2-(2-methoxy)-ethoxy acid

is only 36%. Triethylene glycol is oxidized very rapidly by resting cells of *Acetomonas suboxydans*, with an uptake of 1 mole of O_2 per mole substrate, probably to a carboxyl group on one end of the molecule. This is followed by a slower oxidation to another carboxyl group at the other end (209). *Acetomonas melanogenum* grows rapidly on trimethylene glycol (1,3-propanediol) and oxidizes this compound to malonic acid with hydra-

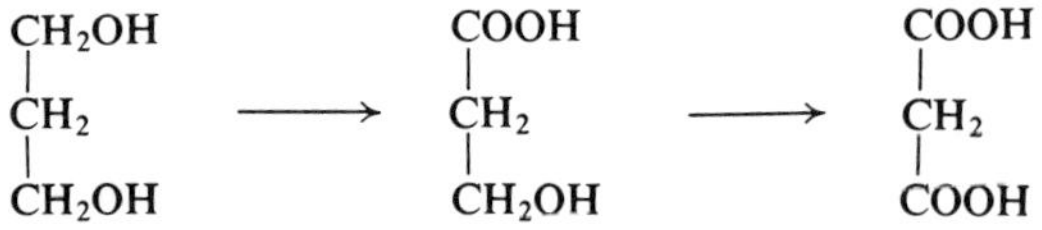

1,3-propanol — malonic acid

crylic acid (β-hydroxypropionic acid) as an intermediate. *Acetomonas suboxydans* is only able to grow on and utilize this compound very slowly.

DL-1,3-Butanediol seems to be a good substrate for all acetic acid bacteria because all strains studied to date are able to oxidize it to DL-β-hydroxybutyric acid.

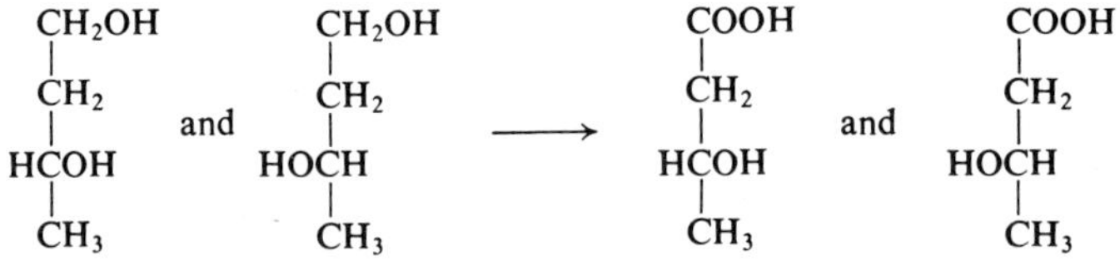

1,3-butanediol — β-hydroxybutyric acid

Acetomonas strains are able to oxidize 1,4-butanediol (tetramethylene glycol) to succinic acid with γ-hydroxybutyric acid as intermediate.

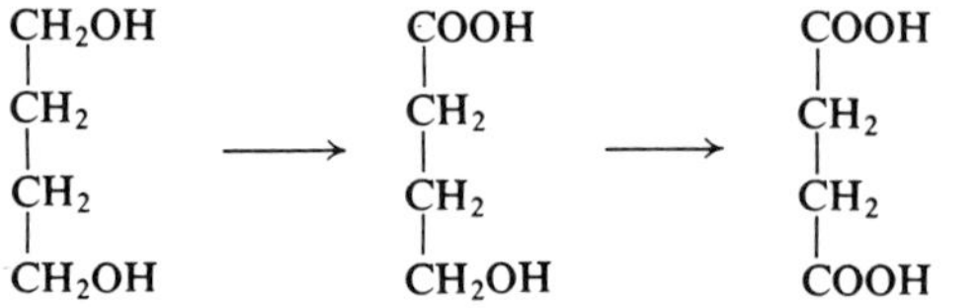

1,4-butanediol — γ-hydroxybutyric acid — succinic acid

Strains of the *Acetobacter mesoxydans*, *A. oxydans*, and *A. peroxydans* group can oxidize succinic acid even further through the TCA cycle.

Glutaric acid is produced in two steps from 1,5-pentanediol by *Acetomonas*

$$\begin{array}{c} CH_2OH \\ | \\ (CH_2)_3 \\ | \\ CH_2OH \end{array} \longrightarrow \begin{array}{c} COOH \\ | \\ (CH_2)_3 \\ | \\ CH_2OH \end{array} \longrightarrow \begin{array}{c} COOH \\ | \\ (CH_2)_3 \\ | \\ COOH \end{array}$$

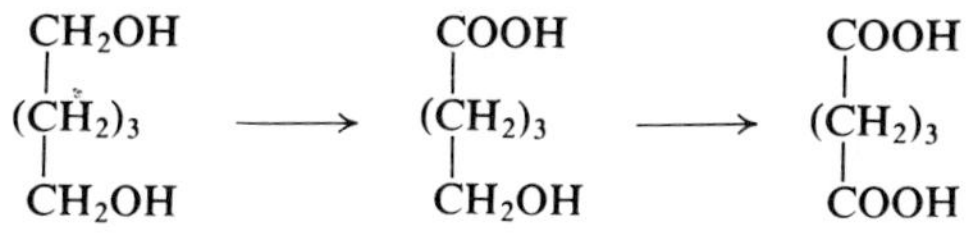

1,5-pentanediol $\qquad\qquad\qquad$ glutaric acid

suboxydans, forming β-hydroxyvaleric acid as intermediate. The same oxidation applies for 1,6-hexanediol, which oxidizes to adipic acid via a

$$\begin{array}{c} CH_2OH \\ | \\ (CH_2)_4 \\ | \\ CH_2OH \end{array} \longrightarrow \begin{array}{c} COOH \\ | \\ (CH_2)_4 \\ | \\ CH_2OH \end{array} \longrightarrow \begin{array}{c} COOH \\ | \\ (CH_2)_4 \\ | \\ COOH \end{array}$$

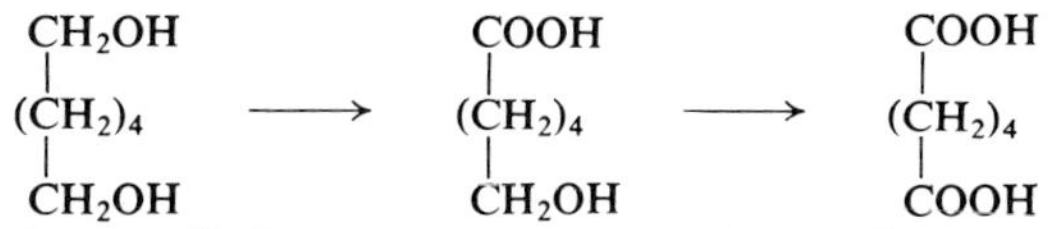

1,6-hexanediol $\qquad\qquad\qquad$ adipic acid

$$\begin{array}{c} CH_2OH \\ | \\ (CH_2)_5 \\ | \\ CH_2OH \end{array} \longrightarrow \begin{array}{c} COOH \\ | \\ (CH_2)_5 \\ | \\ CH_2OH \end{array} \longrightarrow \begin{array}{c} COOH \\ | \\ (CH_2)_5 \\ | \\ COOH \end{array}$$

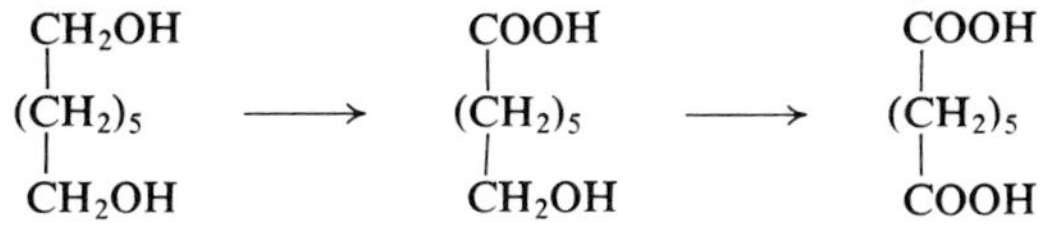

1,7-heptanediol $\qquad\qquad\qquad$ pimelic acid

hydroxycaproic acid, and for 1,7-heptanediol, which forms pimelic acid with 7-hydroxyheptylic acid as intermediate.

Oxidation of Secondary Alcohol Groups

Experiments have shown that the mode of attack on secondary alcohol groups depends on the distance between both groups in the molecule. When groups are adjacent, as in 1,2-propanediol, only secondary alcohol groups are oxidized:

$$\begin{array}{c} CH_3 \\ | \\ HCOH \\ | \\ CH_2OH \end{array} \rightleftharpoons \begin{array}{c} CH_3 \\ | \\ C=O \\ | \\ CH_2OH \end{array} \longleftarrow \begin{array}{c} CH_3 \\ | \\ HOCH \\ | \\ CH_2OH \end{array}$$

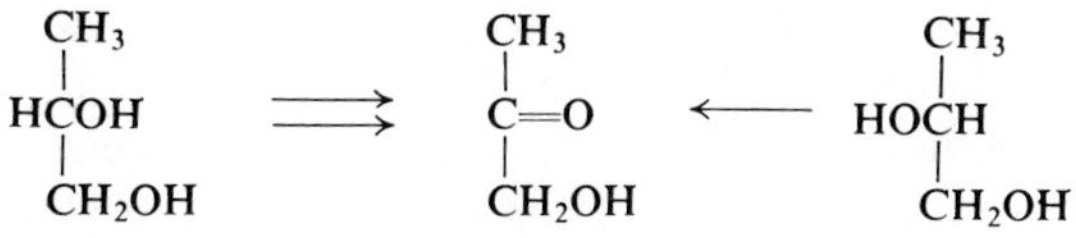

$\quad$ D(−) $\qquad\qquad$ acetol $\qquad\qquad$ L(+)

When both groups are separated by a CH_2 group, as in 1,3-butanediol, the opposite happens. The primary alcohol group is oxidized but the secondary CHOH group is not attacked at all.

Acetobacter rancens (181) is able to attack and oxidize acetol further to acetoin and acetate:

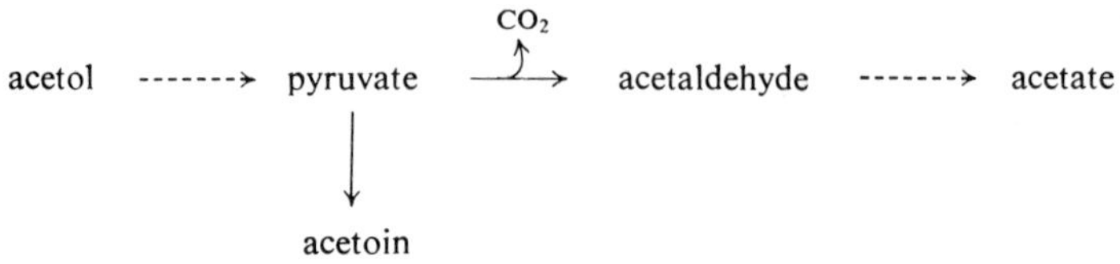
acetol
pyruvate
CO_2
acetaldehyde
acetate
acetoin

2,3-Butanediol Metabolism

As will be discussed under fermentation, 2,3-butanediol is a major end product of carbohydrate metabolism in *Aerobacter* and *Bacillus* and can be broken down aerobically by pseudomonads. Almost all other oxidation reactions involving 2,3-butanediol can also take place anaerobically (190).

Butanediol dehydrogenase (2,3-butanediol:NAD oxidoreductase, EC 1.1.1.4) oxidizes 2,3-butanediol to acetoin. This is followed by a second oxidation step, catalyzed by acetoin dehydrogenase (acetoin:NAD oxidoreductase, EC 1.1.1.5).

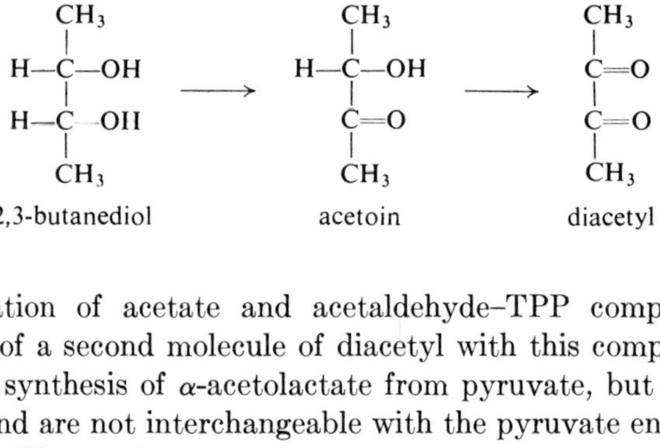

2,3-butanediol acetoin diacetyl

The formation of acetate and acetaldehyde–TPP complex and the combination of a second molecule of diacetyl with this complex are analogous to the synthesis of α-acetolactate from pyruvate, but the enzymes are specific and are not interchangeable with the pyruvate enzyme system of *Aerobacter* (Fig. 5.16).

Diacetyl, together with acetaldehyde–TPP, forms diacetyl methylcarbinol and a diacetylmethylcarbinol reductase that is probably very similar to butanediol dehydrogenase (EC 1.1.1.4) because it is NAD^+-linked (191), reduces diacetyl methylcarbinol to acetylbutanediol. Hydrolysis now occurs and, in a step similar to that of acetaldehyde–TPP formation, a further molecule of acetate is formed, together with butanediol. In this cyclic process, two molecules of acetate are formed from two mole-

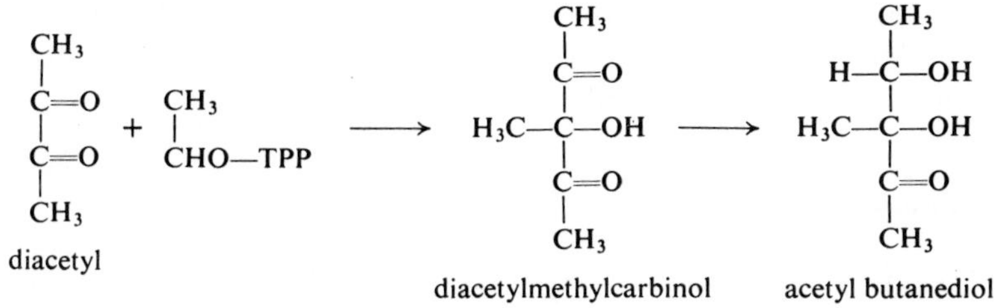
CH₃
C=O
C=O
CH₃
diacetyl
+
CH₃
CHO—TPP
H₃C—C—OH
diacetylmethylcarbinol
H—C—OH
acetyl butanediol

cules of 2,3-butanediol. It seems probable that the same cycle is operative in *Pseudomonas fluorescens* (190) and *Pseudomonas* sp. (186). The acetate produced is used for synthesizing cell material in processes involving the TCA and glyoxylate cycles.

The mechanism of oxidation of 2,3-butanediol and acetoin in acetic acid bacteria is not as yet clear. According to Bertrand's rule, only the meso form should be oxidized. However, experiments revealed that both

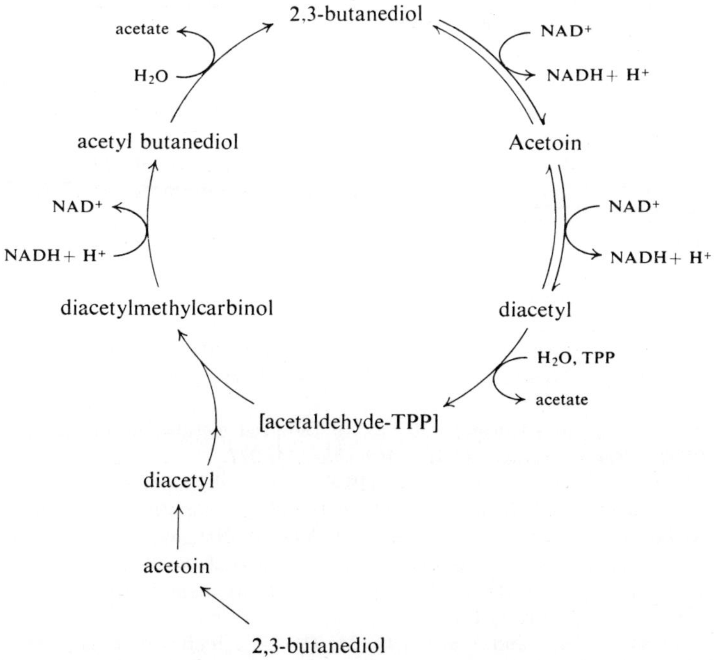
2,3-butanediol
acetate
H₂O
NAD⁺
NADH+ H⁺
acetyl butanediol
Acetoin
NAD⁺
NADH+ H⁺
NAD⁺
NADH+ H⁺
diacetylmethylcarbinol
diacetyl
H₂O, TPP
acetate
[acetaldehyde-TPP]
diacetyl
acetoin
2,3-butanediol

Fig. 5.16. Metabolism of 2,3-butanediol by *Aerobacter*.

the meso and D(−) forms are readily oxidizable, whereas the L(+) form is very slowly oxidized. *Acetobacter xylinum*, *A. aceti*, and *Acetomonas suboxydans* produce L(+)-acetoin from *meso*-2,3-butanediol.

References

1. Abond, M., and Burger, M. (1971). Adenosine triphosphate and catabolite repression of β-galactosidase in *Escherichia coli*. *Biochem. Biophys. Res. Commun.* **45,** 190.
2. Abrahams, S. L., and Younathan, E. S. (1970). Modulation of the kinetic properties of phosphofructokinase by ammonium ions. *J. Biol. Chem.* **246,** 2464.
3. Aiba, S., Humphrey, A. E., and Millis, N. F. (1973). "Biochemical Engineering," 2nd ed. p. 61ff. Academic Press, Inc., New York.
4. Altermatt, H. A., Simpson, F. J., and Neish, A. C. (1955). The fermentation of D-allose and D-glucose by *Aerobacter aerogenes*. *Can. J. Microbiol.* **1,** 473.
5. Anderson, B., Weigel, N., Kundig, W., and Roseman, S. (1971). Sugar transport. III. Purification and properties of a phosphocarrier protein (HPr) of the phosphoenolpyruvate-dependent phosphotransferase system of *Escherichia coli*. *J. Biol. Chem.* **246,** 7023.
6. Anderson, R. L., and Wood, W. A. (1962). Pathway of L-lyxose and L-lyxose degradation in *Aerobacter aerogenes*. *J. Biol. Chem.* **237,** 296.
7. Anderson, R. L., and Wood, W. A. (1962). Purification and properties of L-xylulokinase. *J. Biol. Chem.* **237,** 1029.
8. Anderson, R. L., and Allison, D. P. (1965). Purification and characterization of D-lyxose isomerase. *J. Biol. Chem.* **240,** 2367.
9. Anderson, R. L., and Wood, W. A. (1969). Carbohydrate metabolism in microorganisms. *Annu. Rev. Microbiol.* **23,** 539.
10. Anderson, W. A., and Magasanik, B. (1971). The pathway of myoinositol degradation in *Aerobacter aerogenes*. Identification of the intermediate 2-deoxy-5-keto-D-gluconic acid. *J. Biol. Chem.* **246,** 5653.
11. Anderson, W. A., and Magasanik, B. (1971). The pathway of myoinositol degradation in *Aerobacter aerogenes*. Conversion of 2-deoxy-5-keto-D-gluconic acid to glycolytic intermediates. *J. Biol. Chem.* **246,** 5662.
12. Andreesen, J. R., and Gottschalk, G. (1969). The occurrence of a modified Entner-Doudoroff pathway in *Clostridium aceticum*. *Arch. Mikrobiol.* **69,** 160.
13. Anstall, H. B., Lapp, C., and Trujillo, J. M. (1966). Isozymes of aldolase. *Science* **154,** 657.
14. Anthony, C., and Guest, J. R. (1968). Deferred metabolism of glucose by *Clostridium tetanomorphum*. *J. Gen. Microbiol.* **54,** 277.
15. Arcus, A. C., and Edson, N. L. (1956). Polyol dehydrogenases. 2. The polyol dehydrogenases of *Acetobacter suboxydans* and *Candida utilis*. *Biochem. J.* **64,** 385.
15a. Arthur, L. O., Bulla, Jr., L. A., Julian, G. St., and Nekamura, L. K. (1973). Carbohydrate metabolism in *Agrobacterium tumefaciens*. *J. Bacteriol.* **116,** 304.
16. Asai, T. (1968). "Acetic Acid Bacteria. Classification and Biochemical Activities." Univ. of Tokyo Press, Tokyo.
17. Atkinson, D. E., and Walton, G. M. (1965). Kinetics of regulatory enzymes. *Escherichia coli* phosphofructokinase. *J. Biol. Chem.* **240,** 757.
18. Avigad, G., and Englard, S. (1965). 5-Keto-D-fructose. I. Chemical characterization and analytical determination of the dicarboxyl hexose produced by *Gluconobacter cerinus*. *J. Biol. Chem.* **240,** 2290.

19. Avigad, G., Englard, S., and Pifko, S. (1966). 5-Keto-D-fructose. IV. A specific nicotinamide adenine dinucleotide phosphate-linked reductase from *Gluconobacter cerinus*. *J. Biol. Chem.* **241,** 373.

20. Avigad, G. (1966). Inhibition of glucose-6-phosphate dehydrogenase by adenosine 5′-triphosphate. *Proc. Nat. Acad. Sci. U.S.* **56,** 1543.

21. Azova, L. G. (1967). Utilization of glycerol by some strains of butyric acid bacteria. *Dokl. Akad. Nauk SSSR* **172,** 1434.

22. Baddiley, J., Buchanan, J. G., Carss, B., Mathies, A. P., and Sanderson, A. R. (1956). The isolation of cytidine diphosphate glycerol, cytidine diphosphate ribitol, and mannitol 1-phosphate from *Lactobacillus arabinosus*. *Biochem. J.* **64,** 599.

23. Bard, R. C., and Gunsalus, I. C. (1950). Glucose metabolism of *Clostridium perfringens*. Existence of metalloaldolase. *J. Bacteriol.* **59,** 387.

24. Baschnagel-De Pamphilis, J., and Hanson, R. S. (1969). Relationship between glucose utilization and growth rate in *Bacillus subtilis*. *J. Bacteriol.* **98,** 222.

25. Baseman, J. B., and Cox, C. D. (1969). Intermediate energy metabolism of *Leptospira*. *J. Bacteriol.* **97,** 992.

26. Bastarracea, F., Anderson, D. G., and Goldman, D. S. (1961). Enzyme systems in mycobacteria. XI. Evidence for a functional glycolytic system. *J. Bacteriol.* **82,** 94.

26a. Baumann, L., and Baumann, P. (1973). Enzymes of glucose catabolism in cell free extracts of non-fermentative marine Eubacteria. *Can. J. Microbiol.* **19,** 302.

27. Bell, E. J., and Marus, A. (1966). Carbohydrate catabolism of *Mima polymorpha*. I. Supplemental energy from glucose added to a growth medium. *J. Bacteriol.* **91,** 2223.

28. Benziman, M. (1969). Role of phosphoenolpyruvate carboxylation in *Acetobacter xylinum*. *J. Bacteriol.* **98,** 1005.

28a. Benziman, M., and Mazova, A. (1973). NAD- and NADP-specific glucose-6-phosphate dehydrogenase of *Acetobacter xylinum* and their role in the regulation of the pentose cycle. *J. Biol. Chem.* **248,** 1603.

29. Berkowitz, D. (1971). D-Mannitol utilization in *Salmonella typhimurium*. *J. Bacteriol.* **105,** 232.

30. Berman, K. M., Itada, N., and Cohn, M. (1967). On the mechanism of ATP cleavage in phosphoenolpyruvate synthase reaction of *Escherichia coli*. *Biochim. Biophys. Acta* **141,** 214.

31. Berman, K. M., and Cohn, M. (1970). Phosphoenolpyruvate synthetase of *Escherichia coli*. Purification, some properties, and the role of divalent metal ions. *J. Biol. Chem.* **245,** 5309.

32. Berman, K. M., and Cohn, M. (1970). Phosphoenolpyruvate synthetase of *Escherichia coli*. Partial reactions studied with adenosine triphosphate analogs and the inorganic phosphate–$H_2{}^{18}O$ exchange reaction. *J. Biol. Chem.* **245,** 5319.

33. Berman, T., and Magasanik, B. (1966). The pathway of myoinositol degradation in *Aerobacter aerogenes*. Dehydrogenation and dehydration. *J. Biol. Chem.* **241,** 801.

34. Bertrand, G. (1904). *Ann. Chim. Phys.* **8,** 181 [as cited by Rainbow (317)].

35. Bhattacharya, A. K., and Chakravorty, M. (1971). Induction and repression of L-arabinose isomerase in *Salmonella typhimurium*. *J. Bacteriol.* **106,** 107.

36. Bhuyan, B. K., and Simpson, F. J. (1962). Some properties of the D-xylulokinase of *Aerobacter aerogenes*. *Can. J. Microbiol.* **8,** 737.

37. Bisson, T. M., and Mortlock, R. P. (1968). Regulation of pentitol metabolism by *Aerobacter aerogenes*. I. Coordinate control of ribitol dehydrogenase and D-ribulokinase activities. *J. Bacteriol.* **95,** 925.

38. Bisson, T. M., and Mortlock, R. P. (1968). Regulation of pentitol metabolism by *Aerobacter aerogenes*. II. Induction of ribitol pathway. *J. Bacteriol.* **95,** 932.
39. Blumenthal, H. J. (1961). Discussion of Chapter VI (Biochemical changes occurring in *Bacillus cereus* during sporulation by K. G. Gollakota and H. O. Halvorson). *In* "Spore II" (H. O. Halvorson, ed.), p. 120–126. Burgess, Minneapolis, Minnesota.
40. Boehringer, C. F. (1964). Aldolase. Colorimetric determinations. *In* "General Instructions," TC-O Art. No. 15961. Boehringer, Mannheim, Germany.
41. Boyer, P. D. (1969). The inhibition of pyruvate kinase by ATP:Mg^{2+} buffer system for use in enzyme studies. *Biochem. Biophys. Res. Commun.* **34,** 702.
42. Breslow, R. (1959). Discussion part in "Aldol and Ketol condensation" by B. L. Horecker. *J. Cell. Comp. Physiol.* **54,** 100.
43. Breslow, R., and McNelis, E. (1960). On the mechanism of thiamine action. VI 2-Acetylthrazolium salts as "active acetate." *J. Amer. Chem. Soc.* **82,** 2394.
44. Broberg, P. L., Welsch, M., and Smith, L. (1969). Glucose dehydrogenase of *Bacillus megaterium* KM. Coupling of the cytoplasmic enzyme with membrane-bound cytochromes. *Biochim. Biophys. Acta* **172,** 205.
45. Brown, A. T., and Wittenberger, C. L. (1971). Mechanism for regulating the distribution of glucose carbon between the Embden-Meyerhof and hexose monophosphate pathways in *Streptococcus faecalis*. *J. Bacteriol.* **106,** 456.
46. Brubaker, P. R. (1968). Metabolism of carbohydrates by *Pasteurella pseudotuberculosis*. *J. Bacteriol.* **95,** 1698.
47. Brumgraber, E., and Abood, G. (1960). Mitochondrial glycolysis of rat brain to its relationship to the remainder of cellular glycolysis. *J. Biol. Chem.* **235,** 1847.
48. Buchanan, B. B., and Pine, L. (1963). Factors influencing the fermentation and growth of an atypical strain of *Actinomyces naeslundii*. *Sabouraudia* **3,** 26.
49. Buchanan, B. B., and Pine, L. (1967). Path of glucose breakdown and cell yields of a facultative anaerobe, *Actinomyces naeslundii*. *J. Gen. Microbiol.* **46,** 225.
50. Burma, D. P., and Horecker, B. L. (1958). Pentose fermentation by *Lactobacillus plantarum*. III. Ribulokinase. *J. Biol. Chem.* **231,** 1039.
51. Burma, D. P., and Horecker, B. L. (1958). Pentose fermentation by *Lactobacillus plantarum*. IV. L-Ribulose-5-phosphate-4-epimerase. *J. Biol. Chem.* **231,** 1053.
52. Burstein, C., and Kepes, A. (1971). The β-galactosidase from *Escherichia coli* K12. *Biochim. Biophys. Acta* **230,** 52.
53. Burton, P. M., and Waley, S. G. (1966). The active center of triose phosphate isomerase. *Biochem. J.* **100,** 702.
54. Campbell, J. J. R., and Norris, F. C. (1950). The intermediary metabolism of *Pseudomonas aeruginosa*. IV. The absence of an Embden-Meyerhof system as evidenced by phosphorus distribution. *Can. J. Res. Sect. C* **28,** 203.
55. Campbell, J. J. R., Ramakrishnan, T., Linnes, A. G., and Eagles, B. (1956). Evaluation of the energy gained by *Pseudomonas aeruginosa* during the oxidation of glucose to 2-ketogluconate. *Can. J. Microbiol.* **2,** 304.
56. Canale-Parola, E. (1970). Biology of the sugar-fermenting sarcinae. *Bacteriol. Rev.* **34,** 82.
57. Cantoni, C., and Molnax, M. R. (1967). Investigations on the glycerol metabolism of lactobacilli. *J. Appl. Bacteriol.* **30,** 197.
58. Caprioli, R., and Rittenberg, D. (1969). Pentose synthesis in *Escherichia coli*. *Biochemistry* **8,** 3375.
58a. Cavalieri, R. L., and Sable, H. Z. (1973). Enzymes of pentose biosynthesis. II. Evidence that the proposed control of glucose-6-phosphate dehydrogenase by reduced DPN is an instrumental artifact. *J. Biol. Chem.* **248,** 2815.

59. Chakravorty, M. (1964). Metabolism of mannitol and induction of mannitol-1-phosphate dehydrogenase in *Lactobacillus plantarum*. *J. Bacteriol.* **87,** 1246.
60. Chance, B., and Hess, B. (1959). Spectroscopic evidence of metabolic control. *Science* **129,** 700.
61. Chang, Y. F., and Feingold, D. S. (1969). Hexuronic acid dehydrogenase of *Agrobacterium tumefaciens*. *J. Bacteriol.* **99,** 667.
62. Cheldelin, V. H. (1961). Evaluation of metabolic pathways. *In* "Metabolic Pathways in Microorganisms," p. 61. E. R. Squibb Lectures on Chemistry of Microbial Products. Wiley, New York.
63. Cheldelin, V. H. (1961). The acetic acid bacteria. *In* "Metabolic Pathways in Microorganisms," pp. 1–29. E. R. Squibb Lectures on Chemistry of Microbial Products. Wiley, New York.
64. Claridge, C. A., and Werkman, C. H. (1954). Evidence for alternate pathways for the oxidation of glucose by *Pseudomonas aeruginosa*. *J. Bacteriol.* **68,** 77.
65. Cline, A. L., and Hu, A. S. L. (1965). The isolation of three sugar dehydrogenases from a pseudomonad. *J. Biol. Chem.* **240,** 4488.
66. Cochrane, V. W. (1955). The metabolism of species of streptomyces. VIII. Reactions of the Embden-Meyerhof-Parnas sequence in *Streptomyces coelicolor*. *J. Bacteriol.* **69,** 256.
67. Coffee, C. J., and Ui, A. S. L. (1970). The purification and partial characterization of gluconokinase from a pseudomonad. *J. Biol. Chem.* **245,** 3703.
68. Cohen, G. N., and Monod, J. (1957). Bacterial permeases. *Bacteriol. Rev.* **21,** 169.
69. Colowick, S. P., Kaplan, N. O., Neufeld, E. F., and Ciotti, M. M. (1952). Pyridine nucleotide transhydrogenase. I. Indirect evidence for the reaction and purification of the enzyme. *J. Biol. Chem.* **195,** 95.
70. Cooksey, K. E., and Rainbow, C. (1962). Metabolic patterns in acetic acid bacteria. *J. Gen. Microbiol.* **27,** 135.
71. Crane, R. K., and Sols, A. (1953). The association of hexokinase with particulate fractions of brain and other tissue homogenates. *J. Biol. Chem.* **203,** 273.
72. Crane, R. K. (1962). Hexokinases and pentokinases. *In* "The Enzymes" (P. D. Boyer, H. Lardy, and K. Myrbäck, eds.), 2nd ed., Vol. 6, p. 47. Academic Press, New York.
73. Craven, G. R., Steers, E., Jr., and Anfinsen, C. B. (1965). Purification, composition, and molecular weight of the β-galactosidase of *Escherichia coli* K-12. *J. Biol. Chem.* **240,** 2468.
74. Dahms, A. S., and Anderson, R. L. (1969). 2-Keto-3-deoxy-L-arabonate aldolase and its role in a new pathway of L-arabinose degradation. *Biochem. Biophys. Res. Commun.* **36,** 809.
75. Darnall, D. W., and Murray, L. V. (1972). Effects of ATP and 2,3-diphosphoglycerate on glyceraldehyde-3-phosphate dehydrogenase. *Biochem. Biophys. Res. Commun.* **46,** 1222.
76. Datta, A. G., and Katznelson, H. (1957). Oxidation of 2,5-diketogluconate by a cell-free extract from *Acetobacter melanogenum*. *Nature (London)* **179,** 153.
77. David, J., and Wiesmeyer, H. (1970). Regulation of ribose metabolism in *Escherichia coli*. I. The ribose catabolic pathway. *Biochim. Biophys. Acta* **208,** 45.
78. David, J., and Wiesmeyer, H. (1970). Regulation of ribose metabolism in *Escherichia coli*. II. Evidence for two ribose-5-phosphate isomerase activities. *Biochim. Biophys. Acta* **208,** 56.
79. David, J., and Wiesmeyer, H. (1970). Regulation of ribose metabolism in *Escherichia coli*. III. Regulation of ribose utilization *in vivo*. *Biochim. Biophys. Acta* **208,** 68.

80. Davis, B. D. (1961). The teleonomic significance of biosynthetic control mechanisms. *Cold Spring Harbor Symp. Quant. Biol.* **26,** 1.
81. DeLey, J. (1954). Phospho-2-keto-D-gluconate, an intermediate in the carbohydrate metabolism of *Aerobacter cloacae*. *Biochim. Biophys. Acta* **13,** 302.
82. DeLey, J., and van Damme, J. (1955). The metabolism of sodium 2-keto-D-gluconate by microorganisms. *J. Gen. Microbiol.* **12,** 162.
83. DeLey, J., and Stouthamer, A. H. (1959). The mechanism and localization of hexonate metabolism in *Acetobacter suboxydans* and *Acetobacter melanogenum*. *Biochim. Biophys. Acta* **34,** 171.
84. DeLey, J. (1960). Comparative carbohydrate metabolism and localization of enzymes in *Pseudomonas* and related microorganisms. *J. Appl. Bacteriol.* **23,** 400.
85. DeLey, J. (1961). Comparative carbohydrate metabolism and a proposal for a phylogenetic relationship of the acetic acid bacteria. *J. Gen. Microbiol.* **24,** 31.
86. DeLey, J. (1964). *Pseudomonas* and related genera. *Annu. Rev. Microbiol.* **18,** 17.
87. DeLey, J., and Kersters, K. (1964). Oxidation of aliphatic glycols by acetic acid bacteria. *Bacteriol. Rev.* **28,** 164.
88. DeMoss, R. D., Bard, R. C., and Gunsalus, I. C. (1951). The mechanism of the heterolactic fermentation: A new route of ethanol formation. *J. Bacteriol.* **62,** 499.
89. DeMoss, R. D. (1953). Routes of ethanol formation in bacteria. *J. Cell. Comp. Physiol.* **41,** Suppl. 1, 207.
90. DeMoss, R. D., Gunsalus, I. C., and Bard, R. C. (1953). A glucose-6-phosphate dehydrogenase in *Leuconostoc mesenteroides*. *J. Bacteriol.* **66,** 10.
91. DeMoss, R. D. (1954). Oxidation of 6-phosphogluconate by *Leuconostoc mesenteroides*. *Bacteriol. Proc.* p. 109.
92. DeVries, W., and Stouthamer, A. H. (1967). Pathway of glucose fermentation in relation to the taxonomy of Bifidobacteria. *J. Bacteriol.* **93,** 574.
93. Dietz, G. W., and Heppel, L. A. (1971). Studies on the uptake of hexose phosphates. II. The induction of the glucose 6-phosphate transport system by exogenous but not by endogenously formed glucose 6-phosphate. *J. Biol. Chem.* **246,** 2885.
94. Dixon, M., and Webb, E. C. (1964). "Enzymes," 2nd ed. Longmans, Green, New York.
95. Dobrogosz, W. J., and Demos, R. D. (1963). The regulation of ribosephosphate isomerase activity in *Pediococcus pentosaceus*. *Biochim. Biophys. Acta* **77,** 629
96. Doelle, H. W., and Manderson, G. J. (1971). Comparative studies of fructose-1,6-diphosphate aldolase from *Escherichia coli* 518 and *Lactobacillus casei* var. *rhamnosus* ATCC 7469. *Antonie van Leeuwenhoek; J. Microbiol. Serol.* **37,** 21.
97. Doelle, H. W. (1972). Kinetic characteristics of phosphofructokinase from *Lactobacillus casei* var. *rhamnosus* ATCC 7469 and *Lactobacillus plantarum* ATCC 14917. *Biochim. Biophys. Acta* **258,** 404.
98. Doelle, H. W. (1974). Microbial metabolism. *In* "Benchmark Papers in Microbiology" (W. W. Umbreit, ed.). Vol. 5, p. 107. Dowden, Hutchinson, and Ross, Inc., Stroudsburg, Pennsylvania.
98a. Doelle, H. W., Hollywood, N., and Westwood, A. W. (1974). Effect of glucose concentration on a number of enzymes involved in the aerobic and anaerobic utilization of glucose in turbidostat-cultures of *Escherichia coli*. *Microbios* **9,** 221.
99. Doi, R., Halvorson, H., and Church, B. D. (1959). Intermediate metabolism of aerobic spores. III. The mechanism of glucose and hexose phosphate oxidation in extracts of *Bacillus cereus* spores. *J. Bacteriol.* **77,** 43.
100. Domagk, G. F., and Horecker, B. L. (1965). Fructose and erythrose metabolism

in *Alcaligenes faecalis. Arch. Biochem. Biophys.* **109,** 342; *Chem. Abstr.* **62,** 8145 (1965).

101. Doudoroff, M., Palleroni, N. J., MacGee, J., and Ohara, M. (1956). Metabolism of carbohydrate by *Pseudomonas saccharophila.* I. Oxidation of fructose by intact cells and crude cell free preparations. *J. Bacteriol.* **71,** 196.

102. DuToit, P. J., and Kotze, J. P. (1970). The isolation and characterization of sorbitol-6-phosphate dehydrogenase from *Clostridium pasteurianum. Biochim. Biophys. Acta* **206,** 333.

103. Dyson, J. E. D., and D'Orazio, R. E. (1971). 6-Phosphogluconate dehydrogenase from sheep liver: Inhibition of the catalytic activity by fructose 1,6-diphosphate. *Biochem. Biophys. Res. Commun.* **43,** 183.

104. Eagon, R. G., and Wang, C. H. (1961). Dissimilation of glucose and gluconic acid by *Pseudomonas natriegens. J. Bacteriol.* **83,** 879.

105. Eagon, R. G. (1963). Rate-limiting effects of pyridine nucleotides on carbohydrate catabolic pathways of microorganisms. *Biochem. Biophys. Res. Commun.* **12,** 274.

106. Eagon, R. G., and Wang, C. H. (1963). Dissimilation of glucose and gluconic acid by *Pseudomonas natriegens. J. Bacteriol.* **85,** 879.

107. Easterling, S. B., Johnson, E. M., Wohlhieter, J. A., and Baron, L. S. (1969). Nature of lactose-fermenting *Salmonella* strains obtained from clinical sources. *J. Bacteriol.* **100,** 35.

108. Egan, J. B., and Morse, M. L. (1966). Carbohydrate transport in *Staphylococcus aureus.* III. Studies of the transport process. *Biochim. Biophys. Acta* **112,** 63.

108a. Eggleston, L. V., and Krebs, H. A. (1974). Regulation of the pentose phosphate cycle. *Biochem. J.* **138,** 425.

109. Egyud, L. G., and Whelan, W. J. (1963). Substrate specificity of glucose-6-phosphate dehydrogenase from different sources. *Biochem. J.* **86,** 11.

110. Eisenberg, R. C., and Dobrogosz, W. J. (1967). Gluconate metabolism in *Escherichia coli. J. Bacteriol.* **93,** 941.

111. Englard, S., and Avigad, G. (1965). 5-Keto-D-fructose. II. Patterns of formation and of associated dehydrogenase activities in *Gluconobacter cerinus. J. Biol. Chem.* **240,** 2297.

112. Englard, S., Avigad, G., and Prosky, L. (1965). 5-Keto-D-fructose. III. Proof of structure based on stereospecific patterns of enzymatic reduction. *J. Biol. Chem.* **240,** 2302.

113. Entner, N., and Stanier, R. Y. (1951). Studies on the oxidation of glucose by *Pseudomonas fluorescens. J. Bacteriol.* **62,** 181.

114. Entner, N., and Doudoroff, M. (1952). Glucose and gluconic acid oxidation by *Pseudomonas saccharophila. J. Biol. Chem.* **196,** 853.

115. Farmer, J. J., III, and Eagon, R. G. (1969). Aldohexuronic acid catabolism by a soil *Aeromonas. J. Bacteriol.* **97,** 97.

116. Ferdinand, W. (1966). The interpretation of nonhyperbolic rate curves for two substrate enzymes. A possible mechanism for phosphofructokinase. *Biochem. J.* **98,** 278.

117. Ferdinandus, J., and Clark, J. B. (1969). The phosphofructokinase of *Arthrobacter crystallopoietes. Biochem. J.* **113,** 735.

118. Fleet, G. H. (1969). Fructose diphosphate aldolase—A parameter in microbiology. M.Sc. Thesis, University of Queensland, Brisbane, Australia.

119. Fossit, D., Mortlock, R. P., Anderson, R. L., and Wood, W. A (1964). Pathways of L-arabitol and xylitol metabolism in *Aerobacter aerogenes. J. Biol. Chem.* **239,** 2110.

120. Fraenkel, D. G., and Horecker, B. L. (1964). Pathways of D-glucose metabolism in *Salmonella typhimurium.* A study of a mutant lacking phosphoglucose isomerase. *J. Biol. Chem.* **239,** 2765.
121. Fraenkel, D. G., and Horecker, B. L. (1965). Fructose-1,6-diphosphatase and acid hexose phosphatase of *Escherichia coli. J. Bacteriol.* **90,** 837.
122. Fraenkel, D. G., Pontremoli, S., and Horecker, B. L. (1966). The specific fructose diphosphatase of *Escherichia coli.* Properties and partial purification. *Arch. Biochem. Biophys.* **114,** 4; *Chem. Abstr.* **64,** 17926 (1966).
123. Fraenkel, D. G., and Levisohn, S. R. (1967). Glucose and gluconate metabolism in an *Escherichia coli* mutant lacking phosphoglucose isomerase. *J. Bacteriol.* **93,** 1571.
124. Fraenkel, D. G. (1968). The accumulation of glucose 6-phosphate from glucose and its effect in an *Escherichia coli* mutant lacking phosphoglucose isomerase and glucose-6-phosphate dehydrogenase. *J. Biol. Chem.* **243,** 6451.
125. Fraenkel, D. G. (1968). The phosphoenolpyruvate-initiated pathway of fructose metabolism in *Escherichia coli. J. Biol. Chem.* **243,** 6458.
126. Frampton, E. W., and Wood, W. A. (1961). Carbohydrate oxidation by *Pseudomonas fluorescens.* VI. Conversion of 2-keto-6-phosphogluconate to pyruvate. *J. Biol. Chem.* **235,** 2165.
127. Francis, S. H., Meriwether, B. P., and Park, J. H. (1971). Interaction between adenine nucleotides and 3-phosphoglyceraldehyde dehydrogenase. I. Inhibition of the hydrolysis of *S*-acetyl–enzyme intermediates in the esterase activity. *J. Biol. Chem.* **246,** 5427.
128. Francis, S. H., Meriwether, B. P., and Park, J. H. (1971). Interaction between adenine nucleotides and 3-phosphoglyceraldehyde dehydrogenase. II. A study of the mechanism of catalysis and metabolic control of the multifunctional enzyme. *J. Biol. Chem.* **246,** 5433.
129. Frateur, J. (1950). Essai sur la systematique des *Acetobacters. Cellule* **53,** 297 [as cited by DeLey (85)].
130. Freeze, H., and Brock, T. D. (1970). Thermostable aldolase from *Thermus aquaticus. J. Bacteriol.* **101,** 541.
131. Fromm, H. J. (1958). Ribitol dehydrogenase. I. Purification and properties of the enzyme. *J. Biol. Chem.* **233,** 1049.
132. Fromm, H. J., and Zewe, V. (1962). Kinetic studies of brain hexokinase reaction. *J. Biol. Chem.* **237,** 1661.
133. Fruton, J. S., and Simmons, S. (1959). Fermentation and glycolysis. *In* "General Biochemistry," 2nd ed., p. 468. Wiley, New York.
134. Fuller, R. C., Smillie, R. M., Sisler, E. C., and Kornberg, H. L. (1961). Carbon metabolism in *Chromatium. J. Biol. Chem.* **236,** 2140.
135. Fulton, J. C., and Smith, P. J. C. (1960). Carbohydrate metabolism in *Spirochaeta recurrentis.* I. The metabolism of spirochaetes in vivo and vitro. *Biochem. J.* **76,** 491.
136. Fulton, J. C., and Smith, P. J. C. (1960). Carbohydrate metabolism in *Spirochaeta recurrentis.* II. Enzymes associated with disintegrated cells and extracts of spirochaetes. *Biochem. J.* **76,** 500.
137. Furfine, C. S., and Velick, S. F. (1965). The acyl–enzyme intermediate and the kinetic mechanism of the glyceraldehyde-3-phosphate dehydrogenase reaction. *J. Biol. Chem.* **240,** 844.
138. Galante, E., Scalaffa, P., and Lanzani, G. A. (1963). Attivita Enzymatiche di *Acetobacter suboxydans.* I. Glucosiodeidrogenasi. *Enzymologia* **26,** 23.

139. Gary, N. D., Kupferberg, L. L., and Graf, L. H. (1955). Demonstration of an iron-activated aldolase in some extracts of *Brucella suis*. *J. Bacteriol.* **69,** 478.
140. Geizer, E. (1967). *Shigella sonnei* mutants rapidly fermenting lactose. I. Formation and properties. *J. Hyg., Epidemiol., Microbiol., Immunol.* **11,** 443.
141. Geizer, E. (1968). *Shigella sonnei* mutants rapidly fermenting lactose. II. Influence on fermentation of lactose and salicin. *J. Hyg., Epidemiol., Microbiol., Immunol.* **12,** 335.
142. Ghuretti, F., and Barron, E. S. G. (1954). Der Gang der Glukose-Oxydation in *Corynebacterium creatinovorum*. *Biochim. Biophys. Acta* **15,** 445.
143. Gibbins, L. N., and Simpson, F. J. (1963). The purification and properties of D-allose-6-kinase from *Aerobacter aerogenes*. *Can. J. Microbiol.* **9,** 769.
144. Gibbins, L. N., and Simpson, F. J. (1964). The incorporation of D-allose into the glycolytic pathway by *Aerobacter aerogenes*. *Can. J. Microbiol.* **10,** 829.
145. Gibbs, M., and DeMoss, R. D. (1954). Anaerobic dissimilation of ^{14}C-labeled glucose and fructose by *Pseudomonas lindneri*. *J. Biol. Chem.* **207,** 689.
146. Gibbs, M., Sokatch, J. T., and Gunsalus, I. C. (1955). Product labeling of [1-^{14}C]-glucose fermentation by homofermentative and heterofermentative lactic acid bacteria. *J. Bacteriol.* **70,** 572.
147. Goldman, M., and Blumenthal, H. J. (1960). Pathway of glucose catabolism in intact heat-activated spores of *Bacillus cereus*. *Biochem. Biophys. Res. Commun.* **3,** 164.
148. Goldman, M., and Blumenthal, H. J. (1963). Pathways of glucose catabolism in *Bacillus subtilis*. *J. Bacteriol.* **86,** 303.
149. Goldman, M., and Blumenthal, H. J. (1964). Pathways of glucose catabolism in *Bacillus cereus*. *J. Bacteriol.* **87,** 377.
150. Goldstone, J. M., and Magasanik, B. (1954). Inositol dehydrogenase and keto-inositol dehydrase from inositol-adapted *Aerobacter aerogenes*. *Fed. Proc., Fed. Amer. Soc. Exp. Biol.* **13,** 218.
151. Gracy, R. W., and Noltman, E. A. (1968). Studies on phosphomannose isomerase. III. A mechanism for catalysis and for the role of zinc in the enzymatic and the nonenzymatic isomerization. *J. Biol. Chem.* **243,** 5410.
152. Grazi, E., Cheng, T., and Horecker, B. L. (1962). The formation of a stable aldolase–DHAP complex. *Biochem. Biophys. Res. Commun.* **7,** 250.
153. Grazi, E., Rowley, P. T., Cheng, T., Tchola, O., and Horecker, B. L. (1962). The mechanism of action of aldolases. III. Schiff base formation with lysine. *Biochem. Biophys. Res. Commun.* **9,** 38.
154. Grazi, E., Meloche, H., Martinez, G., Wood, W. A., and Horecker, B. L. (1963). Evidence for Schiff base formation in enzymatic aldol condensation. *Biochem. Biophys. Res. Commun.* **10,** 4.
154a. Gregory, E. M., and Fridovich, I. (1974). Oxygen metabolism in *Lactobacillus plantarum*. *J. Bacteriol.* **117,** 166.
155. Gromet, Z., Schramm, M., and Hestrin, S. (1957). Synthesis of cellulose by *Acetobacter xylinum*. IV. Enzyme system present in a crude extract of glucose-grown cells. *Biochem. J.* **67,** 679.
156. Groves, D. J., and Gronlund, A. F. (1969). Carbohydrate transport in *Clostridium perfringens* Type A. *J. Bacteriol.* **100,** 1256.
157. Groves, D. J., and Gronlund, A. F. (1969). Glucose degradation in *Clostridium perfringens* Type A. *J. Bacteriol.* **100,** 1420.
158. Halvorson, H., and Church, B. D. (1957). Intermediate metabolism of aerobic spores. II. The relationship between oxidative metabolism and germination. *J. Appl. Bacteriol.* **20,** 359.

159. Hammerstedt, R. H., Mohler, H., Decker, K. A., and Wood, W. A. (1971). Structure of 2-keto-3-deoxy-6-phosphogluconate aldolase. I. Physical evidence for a three-subunit molecule. *J. Biol. Chem.* **246,** 2069.
160. Hanson, T. E., and Anderson, R. L. (1966). D-Fructose-1-phosphate kinase, a new enzyme instrumental in the metabolism of D-fructose. *J. Biol. Chem.* **241,** 1644.
161. Hanson, T. E., and Anderson, R. L. (1968). Phosphoenolpyruvate-dependent formation of D-fructose 1-phosphate by a four-component phosphotransferase system. *Proc. Nat. Acad. Sci. U.S.* **61,** 269.
162. Hauge, J. G., King, T. E., and Cheldelin, V. H. (1954). Alternate pathways of glycerol oxidation in *Acetobacter suboxydans*. *Nature (London)* **174,** 1104.
163. Hauge, J. G., King, T. E., and Cheldelin, V. H. (1955). Alternate conversions of glycerol to dihydroxyacetone in *Acetobacter suboxydans*. *J. Biol. Chem.* **214,** 1.
164. Hauge, J. G., King, T. E., and Cheldelin, V. H. (1955). Oxidation of dihydroxyacetone via the pentose cycle in *Acetobacter suboxydans*. *J. Biol. Chem.* **214,** 11.
165. Hauge, J. G. (1964). Glucose dehydrogenase of *Bacterium anitratum* and enzyme with a novel prosthetic group. *J. Biol. Chem.* **239,** 3630.
166. Hayaishi, S., and Lin, E. C. C. (1967). Purification and properties of glycerol kinase from *Escherichia coli*. *J. Biol. Chem.* **242,** 1030.
167. Heath, E. C., Hurwitz, J., and Horecker, B. L. (1956). Acetyl phosphate formation in the phosphorolytic cleavage of pentose phosphate. *J. Amer. Chem. Soc.* **78,** 5449.
168. Heath, E. C., Hurwitz, J., Horecker, B. L., and Ginsburg, A. (1958). Pentose fermentation by *Lactobacillus plantarum*. I. The cleavage of xylulose 5-phosphate by phosphoketolase. *J. Biol. Chem.* **231,** 1009.
169. Heath, E. C., Horecker, B. L., Smyrniotis, P. Z., and Takagi, Y. (1958). Pentose fermentation by *Lactobacillus plantarum*. II. L-Arabinose isomerase. *J. Biol. Chem.* **231,** 1031.
170. Hempfling, W. P., Hofer, M., Harris, E. J., and Pressman, B. C. (1967). Correlation between changes in metabolite concentrations and rate of ion transport following glucose addition to *Escherichia coli* B. *Biochim. Biophys. Acta* **141,** 391.
171. Hengstenberg, W., Eagon, J. B., and Morse, M. L. (1967). Carbohydrate transport in *Staphylococcus aureus*. V. The accumulation of phosphorylated carbohydrate derivatives and evidence for a new enzyme splitting lactose phosphate. *Proc. Nat. Acad. Sci. U.S.* **58,** 274.
172. Hengstenberg, W. (1968). Carbohydrate transport in *Staphylococcus aureus*. *J. Biol. Chem.* **243,** 1881.
173. Hengstenberg, W., Peuberthy, W. K., Hill, K. L., and Morse, M. L. (1969). Phosphotransferase system of *Staphylococcus aureus*. Its requirement for the accumulation and metabolism of galactosides. *J. Bacteriol.* **99,** 383.
174. Herbert, D., Gordon, V., Subrahmanyan, V., and Green, D. E. (1940). Zymohexase. *Biochem. J.* **34,** 1108.
175. Hespell, R. B., and Canale-Parola, E. (1970). Carbohydrate metabolism in *Spirochaeta stenostrepta*. *J. Bacteriol.* **103,** 216.
176. Hey-Ferguson, A., and Elbein, A. D. (1970). Purification of a D-mannose isomerase from *Mycobacterium smegmatis*. *J. Bacteriol.* **101,** 777.
177. Hill, R. L., and Mills, R. C. (1954). The anaerobic glucose metabolism of *Bacterium tularense*. *Arch. Biochem. Biophys.* **53,** 174.
178. Hofer, H. W. (1970). Changing the catalytic properties of phosphofructokinase by chemical modification of functionally important amino acids. *Hoppe-Seyler's Z. Physiol. Chem.* **357,** 649.

179. Holzer, H., and Schroter, W. (1962). Zum Wirkungsmechanismus der Phosphoketolase. I. Oxydation verschiedener Substrate mit Ferricyanid zu Glycolsäure. *Biochim. Biophys. Acta* **65,** 271.

180. Hommes, F. A. (1966). Effect of glucose on the level of glycolytic enzymes in yeast. *Arch. Biochem. Biophys.* **114,** 231.

181. Hooft Visser't, F. (1925). Biochemische onderzoekingen over het geslacht *Acetobacter*. Ph.D. Thesis, Technische Universiteit, Delft [as cited by DeLey and Kersters (87)].

182. Horne, R. N., Anderson, W. B., and Nordlie, R. C. (1970). Glucose dehydrogenase activity of yeast glucose-6-phosphate dehydrogenase. Inhibition by adenosine 5′-triphosphate and other nucleoside 5′-triphosphates and diphosphates. *Biochemistry* **9,** 610.

183. Horwitz, S. B., and Kaplan, N. O. (1964). Hexitol dehydrogenase of *Bacillus subtilis. J. Biol. Chem.* **239,** 830.

184. Hromatka, O., and Polesovsky, W. (1962). Untersuchungen über die Essigsäuregärung. VII. Über die Oxidation verschiedener primärer Alkohole und Glykole. *Enzymologia* **24,** 372.

185. Hulley, S. B., Joergensen, S. B., and Lin, E. C. C. (1963). Ribitol dehydrogenase in *Aerobacter aerogenes* 1033. *Biochim. Biophys. Acta* **67,** 219 [as cited by Mortlock and Wood (285)].

186. Hullin, R. P., and Hassall, H. (1962). Butane-2,3-diol metabolism in *Pseudomonas. Biochem. J.* **83,** 298.

187. Izui, K., Yoshinaga, T., Morikawa, M., and Katsuki, H. (1970). Activation of phosphoenolpyruvate carboxylase of *Escherichia coli* by free fatty acids or their coenzyme A derivatives. *Biochem. Biophys. Res. Commun.* **40,** 949.

188. Janota-Bassalik, L., Popowski, J., Kukukka, E., and Matz, L. (1969). Sucrose and lactose degradation by thermophilic *Bacillus coagulans. Acta Microbiol. Pol., Ser. B* **1,** 73.

189. Johnson, P. A., and Quayle, J. R. (1965). Microbial growth on C_1 compounds. Synthesis of cell constituents by methane- and methanol-grown *Pseudomonas methanica. Biochem. J.* **95,** 859.

190. Juni, E., and Heym, G. A. (1956). A cyclic pathway for the bacterial dissimilation of 2,3-butanediol, acetoin, and diacetyl. I. General aspects of the 2,3-butanediol cycle. *J. Bacteriol.* **71,** 425.

191. Juni, E., and Heym, G. A. (1956). Cyclic pathway for the bacterial dissimilation of 2,3-butanediol, acetylmethyl carbinol, and diacetyl. III. A comparative study of 2,3-butanediol dehydrogenase from various microorganisms. *J. Bacteriol.* **74,** 757.

192. Kaback, H. R. (1969). Regulation of sugar transport in isolated bacterial membrane preparations from *Escherichia coli. Proc. Nat. Acad. Sci. U.S.* **63,** 724.

193. Kaback, H. R. (1970). Transport. *Annu. Rev. Biochem.* **39,** 561.

193a. Kaback, H. R., and Hong, J. (1973). Membranes and transport. *CRC Critical Rev. Microbiol.* **2,** 333.

194. Kafkewitz, D., and Delwiche, E. A. (1969). Utilization of D-ribose by *Veillonella. J. Bacteriol.* **98,** 903.

195. Kafkewitz, D., and Delwiche, E. A. (1972). Ribose utilization by *Veillonella alcalescens. J. Bacteriol.* **109,** 1144.

195a. Kaklij, G. S., and Nadkarni, G. B. (1974). Fructose-1,6-diphosphate aldolase from *Lactobacillus casei*. I. Functional similarities with the rabbit muscle aldolase. *Archs. Biochem. Biophys.* **160,** 47.

195b. Kaklij, G. S., and Nadkarni, G. B. (1974). Fructose-1,6-diphosphate aldolase from *Lactobacillus casei*. II. Multiple forms and their characteristics. *Arch. Biochem. Biophys.* **160,** 52.
196. Kamel, M. Y., and Anderson, R. L. (1966). Metabolism of D-mannose in *Aerobacter aerogenes*. Evidence for a cyclic pathway. *J. Bacteriol.* **92,** 1689.
197. Kamel, M. Y., Allison, D. P., and Anderson, R. L. (1966). Stereospecific D-glucokinase of *Aerobacter aerogenes*. Purification and properties. *J. Biol. Chem.* **241,** 690.
198. Katz, J., and Wood, H. G. (1960). The use of glucose-^{14}C for the evaluation of the pathways of glucose metabolism. *J. Biol. Chem.* **235,** 2165.
199. Katz, J., and Wood, H. G. (1963). The use of $^{14}CO_2$ yields from glucose-1 and glucose-6-^{14}C for the evaluation of pathways of glucose metabolism. *J. Biol. Chem.* **238,** 517.
200. Katz, J., and Rogustadt, R. (1967). The labelling of pentose phosphate from glucose-^{14}C and estimation of the rates of transaldolase, transketolase, the contribution of the pentose cycle, and ribose phosphate synthesis. *Biochemistry* **6,** 2227.
201. Katznelson, H., Tanenbaum, S. W., and Tatum, E. L. (1953). Glucose, gluconate, and 2-ketogluconate oxidation by *Acetobacter melanogenum*. *J. Biol. Chem.* **204,** 43.
202. Katznelson, H. (1958). Hexose phosphate metabolism by *Acetobacter melanogenum*. *Can. J. Microbiol.* **4,** 25.
203. Keele, B. B., Jr., Hamilton, P. B., and Elkan, G. H. (1969). Glucose catabolism in *Rhizobium japonicum*. *J. Bacteriol.* **97,** 1184.
204. Kelker, N. E., Hanson, T. E., and Anderson, R. L. (1970). Alternate pathways of D-fructose metabolism in *Aerobacter aerogenes*. *J. Biol. Chem.* **245,** 2060.
205. Kelker, N. E., and Anderson, R. L. (1971). Sorbitol metabolism in *Aerobacter aerogenes*. *J. Bacteriol.* **105,** 160.
206. Kemp, M. B., and Quayle, J. R. (1966). Microbial growth on C_1 compounds. Incorporation of C_1 units into allulose phosphate by extracts of *Pseudomonas methanica*. *Biochem. J.* **99,** 41.
207. Kemp, M. B., and Quayle, J. R. (1967). Microbial growth on C_1 compounds. Uptake of [^{14}C]formaldehyde and [^{14}C]formate by methane-grown *Pseudomonas methanica* and determination of the hexose labeling pattern after brief incubation with [^{14}C]methanol. *Biochem. J.* **102,** 94.
208. Kennedy, E. P., and Scarborough, G. A. (1967). Mechanism of hydrolysis of *o*-nitrophenyl-β-galactosides in *Staphylococcus aureus* and its significans for theories of sugar transport. *Proc. Nat. Acad. Sci. U.S.* **58,** 225.
209. Kersters, K., and DeLey, J. (1963). The oxidation of glycols by acetic acid bacteria. *Biochim. Biophys. Acta* **71,** 311.
210. Kersters, K., Wood, W. A., and DeLey, J. (1965). Polyol dehydrogenases of *Gluconobacter oxydans*. *J. Biol. Chem.* **240,** 965.
211. Kersters, K., and DeLey, J. (1968). The occurrence of the Entner-Doudoroff pathway in bacteria. *Antonie van Leeuwenhoek; J. Microbiol. Serol.* **34,** 393.
212. King, T. E., and Cheldelin, V. H. (1956). Oxidation of acetaldehyde by *Acetobacter suboxydans*. *J. Biol. Chem.* **220,** 177.
213. Kistler, W. S., Hirsch, C. A., Cozzarelli, N. R., and Lin, E. C. C. (1969). Second pyridine nucleotide-independent L-α-glycerophosphate dehydrogenase in *Escherichia coli* K-12. *J. Bacteriol.* **100,** 1133.
214. Kitos, P., Wang, C. H., Mohler, B. A., King, T. E., and Cheldelin, V. H. (1958). Glucose and gluconate dissimilation in *Acetobacter suboxydans*. *J. Biol. Chem.* **233,** 1295.

215. Koepsell, H. J. (1950). Gluconate oxidation by *Pseudomonas fluorescens*. *J. Biol. Chem.* **186,** 743.

216. Kono, N., and Uyeda, K. (1971). Cold labile phosphofructokinase. *Biochem. Biophys. Res. Commun.* **42,** 1095.

217. Kornberg, H. L., and Smith, J. (1970). Role of phosphofructokinase in the utilization of glucose by *Escherichia coli*. *Nature* (*London*) **227,** 44.

217a. Kornberg, H. L., and Soutar, A. K. (1973). Utilization of gluconate by *Escherichia coli*. Induction of gluconate kinase and 6-phosphogluconate dehydratase activities. *Biochem. J.* **134,** 489.

218. Kovachevich, R., and Wood, W. A. (1955). Carbohydrate metabolism of *Pseudomonas fluorescens*. III. Purification and properties of a 6-phosphogluconate dehydrase. *J. Biol. Chem.* **213,** 745.

219. Kovachevich, R., and Wood, W. A. (1955). Carbohydrate metabolism of *Pseudomonas fluorescens*. IV. Purification and properties of 2-keto-3-deoxy-6-phosphogluconate aldolase. *J. Biol. Chem.* **213,** 757.

220. Kundig, W. F., Kundig, D., Anderson, B., and Roseman, S. (1966). Restoration of active transport of glycoside in *E. coli* by a component of the phosphotransferase system. *J. Biol. Chem.* **241,** 3243.

221. Kundig, W. F., and Roseman, S. (1971). Sugar transport. I. Isolation of a phostransferase system from *Escherichia coli*. *J. Biol. Chem.* **246,** 1393.

222. Kwiek, S., Gabrys, A., and Witecki, J. (1970). The effect of glucose and galactose on catalase activity of *Salmonella typhimurium* in aerobic and anaerobic cultures. *Acta Microbiol. Pol., Ser. B* **2,** 115.

223. Kwon, T. W., and Brown, W. D. (1966). Augmentation of aldolase activity by high concentration of fructose 1,6-diphosphate. *J. Biol. Chem.* **241,** 1509.

224. Lamborg, M., and Kaplan, N. O. (1960). A comparison of some vic glycol dehydrogenase systems found in *Aerobacter aerogenes*. *Biochim. Biophys. Acta* **38,** 272.

225. Lamborg, M., and Kaplan, N. O. (1960). Adaptive formation of a vic glycol dehydrogenase in *Aerobacter aerogenes*. *Biochim. Biophys. Acta* **38,** 284.

226. Lardy, H. A., and Parks, R. E., Jr. (1956). Influence of ATP concentration on rates of some phosphorylation reactions. *In* "Enzymes: Units of Biological Structure and Function" (O. H. Gaebler, ed.), p. 584. Academic Press, New York.

227. Lascelles, J. (1964). Oxygen and the evolution of biochemical pathways. *Proc. Symp. Oxygen Animal Organism, 1964* p. 657.

228. Laue, P. (1968). Studies on the relation of thiomethyl-β-D-galactoside accumulation to thio-β-D-galactoside phosphorylation in *Staphylococcus aureus* HS 1159. *Biochim. Biophys. Acta* **165,** 410.

229. Laue, P., and MacDonald, R. E. (1968). Identification of thio-β-D-galactoside 6-phosphate accumulated by *Staphylococcus aureus*. *J. Biol. Chem.* **243,** 680.

230. LeBlanc, D. J., and Mortlock, R. P. (1971). Metabolism of D-arabinose: Origin of a D-ribulokinase activity in *Escherichia coli*. *J. Bacteriol.* **106,** 82.

231. LeBlanc, D. J., and Mortlock, R. P. (1971). Metabolism of D-arabinose: A new pathway in *Escherichia coli*. *J. Bacteriol.* **106,** 90.

232. LeCam, M., Madec, Y., Bernard, S., and Fradine, R. (1970). Intermediary metabolism of mycobacteria. I. Comparison of two metabolic pathways in *Mycobacterium phlei* and *Mycobacterium tuberculosis*, strain $H_{37}RA$. *Ann. Inst. Pasteur, Paris* **118,** 158.

233. Lee, C. K., and Dobrogosz, W. J. (1965). Oxidative metabolism in *Pediococcus pentosaceus*. III. Glucose dehydrogenase system. *J. Bacteriol.* **90,** 653.

234. Lee, C. K., and Ordal, Z. J. (1967). Regulatory effect of pyruvate on the glucose metabolism of *Clostridium thermosaccharolyticum*. *J. Bacteriol.* **94,** 530.
235. Lee, N., and Bendet, I. (1967). Crystalline L-ribulokinase from *Escherichia coli*. *J. Biol. Chem.* **242,** 2043.
236. Lee, N., Patrick, J. W., and Masson, M. (1968). Crystalline L-ribulose-5-phosphate-4-epimerase from *Escherichia coli*. *J. Biol. Chem.* **243,** 4700.
237. Leifson, E. (1954). The flagellation and taxonomy of species of *Acetobacter*. *Antonie van Leeuwenhoek; J. Microbiol. Serol.* **20,** 102.
238. Lessie, T., and Neidhardt, F. C. (1967). Adenosine triphosphate linked control of *Pseudomonas aeruginosa* glucose-6-phosphate dehydrogenase. *J. Bacteriol.* **93,** 1337.
239. Lienhard, G. E., and Rose, I. A. (1964). The mechanism of action of 6-phosphogluconate dehydrogenase. *Biochemistry* **3,** 190.
240. Lihkosherstov, L. M., Martynova, M. D., and Derevitskaya, V. A. (1968). Specificity of the glucosidases of an enzyme preparation made from *Clostridium perfringens* (Type A). *Biokhimiya* **33,** 1135.
241. Lin, E. C. C., Levin, A. P., and Magasanik, B. (1960). The effect of aerobic metabolism on the inducible glycerol dehydrogenase of *Aerobacter aerogenes*. *J. Biol. Chem.* **235,** 1824.
242. Lin, E. C. C. (1961). An inducible D-arabitol dehydrogenase from *Aerobacter aerogenes*. *J. Biol. Chem.* **236,** 31.
243. Linke, H. A. B. (1969). Der Fruktose-Stoffwechsel von *Clostridium aceticum*. *Zentralbl. Bakteriol.* (*Naturw.*) **123,** 369.
244. Liss, M., Horwitz, S. B., and Kaplan, N. O. (1962). D-Mannitol-1-phosphate dehydrogenase and D-sorbitol-6-phosphate dehydrogenase in *Aerobacter aerogenes*. *J. Biol. Chem.* **237,** 1342.
245. Lloyd, A. G., Large, P. F., Tudball, N., and Dodgson, K. S. (1965). Kinetic studies on the oxidation of D-glucose 6-*O*-sulfate by glucose 6-phosphate dehydrogenase. *Biochem. J.* **97,** 42.
245a. London, J , and Kline, K. (1973). Aldolase of lactic acid bacteria: A case history in the use of an enzyme as an evolutionary marker. *Bacteriol. Rev.* **37,** 453.
246. Low, I. E., Eaton, M. D., and Proctor, P. (1968). Relation of catalase to substrate utilization by *Mycoplasma pneumoniae*. *J. Bacteriol.* **95,** 1425.
247. Luppis, B., Traniello, S., Wood, W. A., and Pontremoli, S. (1964). Evidence for two forms of fructose diphosphatase. *Biochem. Biophys. Res. Commun.* **15,** 458.
248. McComb, B. B., and Yushok, W. D. (1959). Properties of particulate hexokinase of the Krebs-2-Ascites tumor. *Biochim. Biophys. Acta* **34,** 515.
249. McDonald, T. L., and Mallavia, L. (1971). Biochemistry of *Coxiella burnetii:* Embden-Meyerhof pathway. *J. Bacteriol.* **107,** 864.
250. McGill, D. J., and Dawes, E. A. (1971). Glucose and fructose metabolism in *Zymomonas anaerobia*. *Biochem. J.* **125,** 1059.
251. McGinnis, J. F., and Paigen, K. (1969). Catabolite inhibition: a general phenomenon in the control of carbohydrate utilization. *J. Bacteriol.* **100,** 902.
251a. McGinnis, J. F., and Paigen, K. (1973). Site of catabolite inhibition of carbohydrate metabolism. *J. Bacteriol.* **114,** 885.
252. McLean, P., and Gumaa, K. K. (1969). *Contr. Mech. Intermediary Metab., Symp.*, Univ. Miami Florida.
253. Maeba, P., and Sanwal, B. D. (1968). The regulation of pyruvate kinase of *Escherichia coli* by fructose diphosphate and adenylic acid. *J. Biol. Chem.* **243,** 448.
254. Maeba, P., and Sanwal, B. D. (1969). Phosphoenolpyruvate carboxylase of *Salmonella*. Some chemical and allosteric properties. *J. Biol. Chem.* **244,** 2549.

255. Mamkaeva, K. A. (1966). Formation of 2-ketogluconic acid by *Pseudomonas*. *Tr. Petergof. Biol. Inst., Leningrad. Gos. Univ.* **19,** 104; *Chem. Abstr.* **65,** 19267 (1966).
256. Manderson, G. J., and Doelle, H. W. (1971). Comparative studies of fructose-1,6-diphosphate aldolase from *Escherichia coli* 518 and *Lactobacillus casei* var. *rhamnosus* ATCC 7469. *Antonie van Leeuwenhoek; J. Microbiol. Serol.* **37,** 21.
257. Manderson, G. J., and Doelle, H. W. (1972). The effect of oxygen and pH on the glucose metabolism of *Lactobacillus casei* var. *rhamnosus* ATCC 7469. *Antonie van Leeuwenhoek; J. Microbiol. Serol.* **38,** 223.
258. Mansour, T. E. (1965). Studies on heart phosphofructokinase: Active and inactive forms of the enzyme. *J. Biol. Chem.* **240,** 2165.
259. Mansour, T. E. (1966). Phosphofructokinase. II. Heart muscle. *In* "Methods in Enzymology" (W. A. Wood, ed.), Vol. 9, p. 340, Academic Press, New York.
260. Marcus, L., and Marr, A. G. (1961). Polyol dehydrogenase of *Azotobacter agilis*. *J. Bacteriol.* **82,** 224.
261. Marmur, J., and Hotchkiss, R. D. (1955). Mannitol metabolism, a transferable property of *Pneumococcus*. *J. Biol. Chem.* **214,** 383.
262. Marr, A. G., Ingraham, J. L., and Squires, C. L. (1964). Effect of the temperature of growth of *Escherichia coli* on the formation of β-galactosidase. *J. Bacteriol.* **87,** 356.
263. Martinez, R. J., and Rittenberg, S. C. (1959). Glucose dissimilation by *Clostridium tetani*. *J. Bacteriol.* **77,** 156.
264. Martinez-deDrets, G., and Arias, A. (1972). Enzymatic basis for differentiation of *Rhizobium* into fast- and slow-growing groups. *J. Bacteriol.* **109,** 467.
265. Marus, A., and Bell, E. J. (1966). Carbohydrate catabolism of *Mima polymorpha*. II. Abortive catabolism of glucose. *J. Bacteriol.* **91,** 2229.
266. Mastroni, P., and Contadini, V. (1965). Incubation temperature and metabolism of glucose in *Serratia marcescens*. *Riv. Ist. Sieroter. Ital.* **40,** 90; *Chem. Abstr.* **63,** 15252 (1965).
267. Matsushima, K., and Simpson, F. J. (1965). The ribosephosphate and glucosephosphate isomerases of *Aerobacter aerogenes*. *Can. J. Microbiol.* **11,** 967.
268. Matsushima, K., and Simpson, F. J. (1966). The purification and properties of D-allosephosphate isomerase of *Aerobacter aerogenes*. *Can. J. Microbiol.* **12,** 313.
269. Mayo, J. W., and Anderson, R. L. (1958). Pathway of L-mannose degradation in *Aerobacter aerogenes*. *J. Biol. Chem.* **243,** 6330.
270. Mayo, J. W., and Anderson, R. L. (1969). Basis for the mutational acquisition of the ability of *Aerobacter aerogenes* to grow on L-mannose. *J. Bacteriol.* **100,** 948.
271. Melchior, J. B. (1965). The role of metal ions in the pyruvic kinase reaction. *Biochemistry* **4,** 1518.
272. Meloche, H. P., and Wood, W. A. (1964). The mechanism of 6-phosphogluconic dehydrase. *J. Biol. Chem.* **239,** 3505.
273. Meloche, H. P., and Wood, W. A. (1964). The mechanism of 2-keto-3-deoxy-6-phosphogluconate aldolase. *J. Biol. Chem.* **239,** 3511.
274. Metzger, R. P., Parson, R. L., and Wick, A. N. (1971). D-Xylose oxidation catalyzed by rat liver glucose-6-phosphate dehydrogenase. *Fed. Proc., Fed. Amer. Soc. Exp. Biol.* **30,** 1222.
275. Meyerhof, O., and Lohmann, K. (1934). Über die enzymatische Gleichgewichtsreaktion zwischen Hexosediphosphorsäure und Dioxyacetonphosphorsäure. *Biochem. Z.* **271,** 89 [as cited by Fruton and Simmons (133)].
276. Meyerhof, O., and Beck, L. V. (1944). Triose phosphate isomerase. *J. Biol. Chem.* **156,** 109.

277. Meyerhof, O. (1951). Aldolase and isomerase. *In* "The Enzymes. Chemistry and Mechanism of Action" (J. B. Sumner and K. Myrback, eds.), Vol. 2, Part 1, p. 162. Academic Press, New York.
278. Mickelson, M. N. (1967). Aerobic metabolism of *Streptococcus agalactiae*. *J. Bacteriol.* **94,** 184.
279. Mitchell, P. (1954). Transport of phosphate through osmotic barrier. *Symp. Soc. Exp. Biol.* **8,** 254 [as cited by Marr *et al.* (262)].
280. Mitchell, P. D. (1959). Biochemical cytology of microorganism. *Annu. Rev. Microbiol.* **13,** 407.
281. Mitchell, P. D. (1967). Translocation through natural membranes. *Advan. Enzymol.* **29,** 33.
282. Model, P., and Rittenberg, D. (1967). Measurement of the activity of the hexose monophosphate pathway of glucose metabolism with the use of [^{18}O]glucose. Variations in its activity in *E. coli*. *Biochemistry* **6,** 69.
283. Monod, J., Changeux, L. P., and Jacob, F. (1963). Allosteric proteins and cellular control systems. *J. Mol. Biol.* **6,** 306.
284. Morris, J. G. (1960). Studies on the metabolism of *Arthrobacter globiformis*. *J. Gen. Microbiol.* **22,** 564.
285. Mortlock, R. P., and Wood, W. A. (1964). Metabolism of pentoses and pentitols by *Aerobacter aerogenes*. I. Demonstration of pentose isomerase, pentololokinase, and pentitol dehydrogenase enzyme families. *J. Bacteriol.* **88,** 838.
286. Mortlock, R. P., and Wood, W. A. (1964). Metabolism of pentoses and pentitols by *Aerobacter aerogenes*. II. Mechanism of acquisition of kinase, isomerase, and dehydrogenase activity. *J. Bacteriol.* **88,** 845.
287. Moss, V., and Sharp, P. B. (1970). Adenosine 3′5′-cyclic monophosphate and catabolite repression in *Escherichia coli*. *Biochem. J.* **118,** 481.
288. Mukkada, A. J., and Bell, E. J. (1969). Fructose-1,6-diphosphatase of *Acinetobacter:* Inhibition by ATP and citrate. *Biochem. Biophys. Res. Commun.* **37,** 340.
289. Murphy, W. H., and Rosenblum, E. D. (1964). Mannitol catabolism by *Staphylococcus aureus*. *Arch. Biochem. Biophys.* **107,** 292.
289a. Nakazawa, T., and Yokota, T. (1973). Requirement of adenosine-3′,5′-cyclic monophosphate for L-arabinose isomerase synthesis in *Escherichia coli*. *J. Bacteriol.* **113,** 1412.
290. Narrod, S. A., and Wood, W. A. (1954). Gluconate and 2-ketogluconate phosphorylation by extracts of *Pseudomonas fluorescens*. *Bacteriol. Proc.* p. 108.
291. Narrod, S. A., and Wood, W. A. (1956). Carbohydrate oxidation by *Pseudomonas fluorescens:* Evidence for glucokinase and 2-ketoglucokinase. *J. Biol. Chem.* **220,** 45.
292. Neal, D. L., and Kindel, P. K. (1970). D-Apiose reductase from *Aerobacter aerogenes*. *J. Bacteriol.* **101,** 910.
293. Norris, F. C., and Campbell, J. J. R. (1949). The intermediate metabolism of *Pseudomonas aeruginosa*. III. The application of paper chromatography to the identification of gluconic and 2-ketogluconic acids, intermediates in glucose oxidation. *Can. J. Res., Sect. C* **27,** 253.
294. Novotny, C. P., and Englesberg, E. (1966). The L-arabinose permease system in *Escherichia coli* B/r. *Biochim. Biophys. Acta* **117,** 217.
295. Okamoto, K. (1963). Enzymatic studies on the formation of 5-ketogluconic acid. I. Glucose dehydrogenase. *J. Biochem.* (*Tokyo*) **53,** 348.
296. Okamoto, K. (1963). Enzymatic studies on the formation of 5-ketogluconic acid. II. 5-Ketogluconate reductase. *J. Biochem.* (*Tokyo*) **53,** 448.

297. Oliver, E. J., and Mortlock, R. P. (1971). Metabolism of D-arabinose by *Aerobacter aerogenes:* Purification of the isomerase. *J. Bacteriol.* **108,** 293.

298. Osmond, C. B., and Rees, T. A. P. (1969). Control of the pentose phosphate pathway in yeast. *Biochim. Biophys. Acta* **184,** 35.

299. Paigen, K. (1966). Role of the galactose pathway in the regulation of β-galactosidase. *J. Bacteriol.* **92,** 1394.

300. Palumbo, S. A., and Witter, L. D. (1969). The influence of temperature on the pathways of glucose catabolism in *Pseudomonas fluorescens. Can. J. Microbiol.* **15,** 995.

301. Palumbo, S. A., and Witter, L. D. (1969). Influence of temperature on glucose utilization by *Pseudomonas fluorescens. Appl. Microbiol.* **18,** 137.

302. Parlasova, E., and Harold, F. M. (1969). Energy coupling in the transport of β-galactosides by *Escherichia coli:* Effect of proton conductors. *J. Bacteriol.* **98,** 189.

303. Passonneau, J. V., and Lowry, O. H. (1962). Phosphofructokinase and the Pasteur effect. *Biochem. Biophys. Res. Commun.* **7,** 10.

304. Passoneau, J. V., and Lowry, O. H. (1963). Phosphofructokinase and the control of the citric acid cycle. *Biochem. Biophys. Res. Commun.* **13,** 372.

305. Pastan, I., and Perlman, R. L. (1969). Repression of β-galactosidase synthesis by glucose in phosphotransferase mutants of *Escherichia coli. J. Biol. Chem.* **244,** 5836.

306. Patrick, J. W., and Lee, N. (1968). Purification and properties of an L-arabinose isomerase from *Escherichia coli. J. Biol. Chem.* **243,** 4312.

307. Patui, N. J., and Alexander, J. K. (1971). Utilization of glucose by *Clostridium thermocellum.* Presence of glucokinase and other glycolytic enzymes in cell extracts. *J. Bacteriol.* **105,** 220.

308. Patui, N. J., and Alexander, J. K. (1971). Catabolism of fructose and mannitol in *Clostridium thermocellum:* Presence of phosphoenolpyruvate:fructose phosphotransferase, fructose-1-phosphate kinase, phosphoenolpyruvate:mannitol phosphotransferase, and mannitol-1-phosphate dehydrogenase in cell extracts. *J. Bacteriol.* **105,** 226.

309. Pepper, R. E., and Costilow, R. N. (1964). Glucose catabolism by *Bacillus popilliae* and *Bacillus lentimorbus. J. Bacteriol.* **87,** 303.

310. Peterson, M. H., Friedland, W. C., Denison, F. W., Jr , and Sylvester, J. C. (1956). The conversion of mannitol to fructose by *Acetobacter suboxydans. Appl. Microbiol.* **4,** 316.

311. Pine, L., and Howell, A., Jr. (1956). Comparison of physiological and biochemical characters of *Actinomyces* spp. with those of *Lactobacillus bifidus. J. Gen. Microbiol.* **15,** 428.

312. Platt, T. B., and Foster, E. M. (1957). Products of glucose metabolism by homofermentative streptococci under anaerobic conditions. *J. Bacteriol.* **75,** 453.

313. Pogson, C. I., and Randle, P. J. (1966). The control of rat heart phosphofructokinase by citrate and other regulators. *Biochem. J.* **100,** 683.

314. Polesovsky, W. (1951). Untersuchungen über die Bildung von Carbonsäuren durch submerse Vergärung primärer Alkohole. Ph.D. Thesis, University of Wien, Vienna, Austria [as cited by DeLey and Kersters (87)].

315. Racker, E. (1965). "Mechanisms of Bioenergetics," p. 207. Academic Press, New York.

316. Ragland, T. E., Kawasaki, T., and Lowenstein, J. M. (1966). Comparative aspects of some bacterial dehydrogenases and transhydrogenases. *J. Bacteriol.* **91,** 236.

317. Rainbow, C. (1961). The biochemistry of the acetobacters. *Progr. Ind. Microbiol.* **3,** 43.
318. Rainbow, C. (1966). Nutrition and metabolism of acetic acid bacteria. *Wallerstein Lab. Commun.* **94,** 615.
319. Raj, H. D. (1967). Radiorespirometric studies of *Leucothrix mucor*. *J. Bacteriol.* **94,** 615.
320. Rajkumar, T. V., Penhoet, E., and Rutter, W. J. (1966). Subunits and multiple forms of fructose diphosphate aldolase. *Fed. Proc., Fed. Amer. Soc. Exp. Biol.* **25,** 523.
321. Ramakrishnan, T., Indiva, M., and Maller, R. K. (1962). Evaluation of the route of glucose utilization in virulent and avirulent strains of *Mycobacterium tuberculosis*. *Biochim. Biophys. Acta* **59,** 529.
322. Ramakrishnan, T., Murthy, P. S., and Gopinathan, K. P. (1972). Intermediary metabolism of *Mycobacteria*. *Bacteriol. Rev.* **36,** 65.
323. Reeves, R. E. (1971). Pyruvate-phosphate dikinase from *Bacteroides symbiosus*. *Biochem. J.* **125,** 531.
324. Richards, O. C., and Rutter, W. J. (1961). Comparative properties of yeast and muscle aldolase. *J. Biol. Chem.* **236,** 3185.
325. Robertson, D. C., Hammerstedt, R. H., and Wood, W. A. (1971). Structure of 2-keto-3-deoxy-6-phosphogluconate aldolase. II. Chemical evidence for a three-subunit molecule. *J. Biol. Chem.* **246,** 2075.
326. Rogosa, M., Krichevsky, M. I., and Bishop, F. S. (1965). Truncated glycolytic system in *Veillonella alkalescens*. *J. Bacteriol.* **90,** 164.
327. Rohlfing, S. R., and Crawford, I. P. (1966). Purification and characterization of the β-galactosidase of *Aeromonas formicans*. *J. Bacteriol.* **91,** 1085.
328. Romano, A. H., Eberhard, S. J., Dingle, S. L., and McDowell, T. D. (1970). Distribution of the phosphoenolpyruvate:glucose phosphotransferase system in bacteria. *J. Bacteriol.* **104,** 808.
329. Roodyne, D. B. (1957). The binding of aldolase to isolated nuclei. *Biochim. Biophys. Acta* **25,** 128.
330. Rose, A. H. (1967). "Chemical Microbiology," 2nd ed., Butterworth, London.
331. Rose, I. A., Warms, J. V. B., and O'Connell, E. L. (1964). Role of inorganic phosphate in stimulation of glucose utilization of human red blood cells. *Biochem. Biophys. Res. Commun.* **15,** 33.
332. Rose, I. A., O'Connell, E. L., and Mehler, A. H. (1965). Mechanism of the aldolase reaction. *J. Biol. Chem.* **240,** 1758.
333. Rose, S. P., and Fox, C. F. (1971). The β-glucoside system of *Escherichia coli*. II. Kinetic evidence for a phosphorylenzyme II intermediate. *Biochem. Biophys. Res. Commun.* **45,** 376.
334. Roseman, S. (1969). The transport of carbohydrates by a bacterial phosphotransferase system. *J. Gen. Physiol.* **54,** 138s–148s.
335. Ruiz-Amil, M., Aparicio, M. L., and Canovas, J. L. (1969). Regulation of the synthesis of glyceraldehyde-3-phosphate dehydrogenase in *Pseudomonas putida*. *FEBS (Fed. Eur. Biochem. Soc.) Lett.* **3,** 65.
336. Rutter, W. J. (1964). Evolution of aldolase. *Fed. Proc., Fed. Amer. Soc. Exp. Biol.* **23,** 1248.
337. Sadoff, H. L., Hitchins, A. D., and Celikkol, E. (1969). Properties of fructose-1,6-diphosphate aldolase from spores and vegetative cells of *Bacillus cereus*. *J. Bacteriol.* **98,** 1208.
338. Saier, M. H., Jr., Young, W. S., III, and Roseman, S. (1971). Utilization and

transport of hexose by mutant strains of *Salmonella typhimurium* lacking enzyme I of the phosphoenolpyruvate-dependent phosphotransferase system. *J. Biol. Chem.* **246,** 5838.

339. Saier, M. H., Jr., and Roseman, S. (1972). Inducer exclusion and repression of enzyme synthesis in mutants of *Salmonella typhimurium* defective in enzyme I of the phosphoenolpyruvate:sugar phosphotransferase system. *J. Biol. Chem.* **247,** 972.

340. Saito, N. (1965). Contribution of glucose phosphorylation to glucose metabolism in *Brevibacterium fuscum. Agr. Biol. Chem.* **29,** 621.

341. Sanwal, B. D. (1970). Allosteric controls of amphibolic pathways in bacteria. *Bacteriol. Rev.* **34,** 20.

342. Sanwal, B. D. (1970). Regulatory mechanisms involving nicotinamide adenine nucleotides as allosteric effectors. III. Control of glucose-6-phosphate dehydrogenase. *J. Biol. Chem.* **245,** 1626.

343. Sapico, V., and Anderson, R. L. (1969). D-Fructose-1-phosphate kinase and D-fructose-6-phosphate kinase from *Aerobacter aerogenes*. A competitive study of regulatory properties. *J. Biol. Chem.* **244,** 6280.

344. Sapico, V., and Anderson, R. L. (1970). Regulation of D-fructose-1-phosphate kinase by potassium ion. *J. Biol. Chem.* **245,** 3252.

345. Scardovi, V., and Trovatelli, L. D. (1965). Fructose-6-phosphate shunt as peculiar pattern of hexose degradation in the genus *Bifidobacterium. Ann. Microbiol. Enzimol.* **15,** 19.

346. Schaeffer, S., Malamy, A., and Green, I. (1969). Phospho-β-glucosidases and β-glucoside permeases in *Streptococcus, Bacillus,* and *Staphylococcus. J. Bacteriol.* **99,** 434.

347. Schindler, J., and Schlegel, H. G. (1969). Regulation der Glucose 6-phosphat Dehydrogenase aus verschiedenen Bakterienarten durch ATP. *Arch. Mikrobiol.* **66,** 69.

348. Schramm, M., Gromet, Z., and Hestrin, S. (1957). Role of hexose phosphate in synthesis of cellulose by *Acetobacter xylinum. Nature (London)* **179,** 28.

349. Schramm, M., and Racker, E. (1957). Formation of erythrose 4-phosphate and acetyl phosphate by a phosphoroclastic cleavage of fructose 6-phosphate. *Nature (London)* **179,** 1349.

350. Schramm, M., Klybas, V., and Racker, E. (1958). Phosphorolytic cleavage of fructose 6-phosphate by fructose-6-phosphate phosphoketolase from *Acetobacter xylinum. J. Biol. Chem.* **233,** 1283.

350a. Schray, K. J., Benkovic, C. J., Benkovic, P. A., and Rose, I. A. (1973). Catalytic reactions of phosphoglucose isomerase with cyclic forms of glucose 6-phosphate and fructose 6-phosphate. *J. Biol. Chem.* **248,** 2219.

351. Scolnick, E. M., and Linn, E. C. C. (1962). Parallel induction of D-arabitol and D-sorbitol dehydrogenases. *J. Bacteriol.* **84,** 631.

352. Scopes, A. W. (1962) The infrared spectra of some acetic acid bacteria. *J. Gen. Microbiol.* **28,** 69.

353. Sebek, O. K., and Randles, C. I. (1952). The oxidative dissimilation of mannitol and sorbitol by *Pseudomonas fluorescens. J. Bacteriol.* **63,** 693.

354. Shaw, D. R. D. (1956). Polyol dehydrogenases. 3. Galactitol dehydrogenase and D-iditol dehydrogenase. *Biochem. J.* **64,** 394.

355. Shaw, W. N., and Stadie, W. C. (1959). Two identical Embden-Meyerhof enzyme systems in normal rat diaphagm differing in cytological location and response to insulin. *J. Biol. Chem.* **234,** 2491.

356. Shchalkunova, S. A. (1966). The effect of phosphates on the growth of acetic acid bacteria and the oxidation of glycerol by them. *Tr. Petergof. Biol. Inst., Leningrad. Gos. Univ.* **19,** 57; *Chem. Abstr.* **65,** 19024 (1966).

357. Shimwell, J. L., and Carr, J. G. (1960). Support for differentiation of *Acetobacter* and *Acetomonas. Antonie van Leeuwenhoek; J. Microbiol. Serol.* **26,** 430.

358. Shimwell, J. L. (1958). Flagellation and taxonomy of *Acetobacter* and *Acetomonas. Antonie van Leeuwenhoek; J. Microbiol. Serol.* **24,** 187.

359. Shuster, C. W., and Doudoroff, M. (1967). Purification of 2-keto-3-deoxy-6-phosphogluconate aldolase of *Pseudomonas saccharophila. Arch. Mikrobiol.* **59,** 279.

360. Simmons, R. J., and Costilow, R. N. (1962). Enzymes of glucose and pyruvate catabolism in cells and spores and germinated spores of *Clostridium botulinum. J. Bacteriol.* **84,** 1274.

361. Simon, R. D., Smith, M. F., and Roseman, S. (1968). Resolution of a staphylococcal phosphotransferase system into four protein components and its relation to sugar transport. *Biochem. Biophys. Res. Commun.* **31,** 804.

362. Simpson, F. J., Wolin, M. J., and Wood, W. A. (1958). Degradation of L-arabinose by *Aerobacter aerogenes.* I. A pathway involving phosphorylated intermediates. *J. Biol. Chem.* **230,** 457.

363. Simpson, F. J., and Wood, W. A. (1958). Degradation of L-arabinose by *Aerobacter aerogenes.* II. Purification and properties of L-ribulokinase. *J. Biol. Chem.* **270,** 473.

364. Singleton, R., Jr., Kimmel, J. R., and Amelunxen, R. E. (1969). The amino acid composition and other properties of thermostable glyceraldehyde-3-phosphate dehydrogenase from *Bacillus stearothermophilus. J. Biol. Chem.* **244,** 1623.

365. Slaughter, J. C., and Davies, D. D. (1968). Inhibition of 3-phosphoglycerate dehydrogenase by L-serine. *Biochem. J.* **109,** 749.

366. Sly, L. I., and Doelle, H. W. (1968). Glucose-6-phosphate dehydrogenase in cell free extracts of *Zymomonas mobilis. Arch. Mikrobiol.* **63,** 197.

367. Sly, L. I., and Doelle, H. W. (1968). 6-Phosphogluconate dehydrogenase in cell-free extracts of *Escherichia coli* K-12. *Arch. Mikrobiol.* **63,** 214.

368. Smith, P. J. C. (1960). Carbohydrate metabolism in *Spirochaeta recurrentis.* III. Properties of aldolase in spirochaetes. *Biochem. J.* **76,** 508.

369. Sokatch, J. T., and Gunsalus, I. C. (1957). Aldonic acid metabolism. I. Pathway of carbon in an inducible gluconate fermentation by *Streptococcus faecalis. J. Bacteriol.* **73,** 452.

370. Spring, T. G., and Wold, F. (1971). The purification and characterization of *Escherichia coli* enolase. *J. Biol. Chem.* **246,** 6797.

371. Stadtman, E. R. (1966). Allosteric regulation of enzyme activity. *Advan. Enzymol.* **28,** 41.

372. Stein, W. D. (1967). "The Movements of Molecules Across Cell Membranes." Academic Press, New York.

373. Stern, I. J., Wang, C. H., and Gilmour, C. M. (1960). Comparative catabolism of carbohydrates in *Pseudomonas* species. *J. Bacteriol.* **79,** 601.

374. Stewart, D. J. (1959). Production of 5-ketogluconic acid by a species of *Pseudomonas. Nature (London)* **183,** 1133.

375. Stouthamer, A. H. (1959). Oxidative possibilities in the catalase-positive *Acetobacter* species. *Antonie van Leeuwenhoek; J. Microbiol. Serol.* **25,** 242.

375a. Stribling, D., and Perham, R. N. (1973). Purification and characterization of two fructose-diphosphate aldolases from *Escherichia coli* (Crookes' strain). *Biochem. J.* **131,** 833.

376. Tanaka, S., Lerner, S. A., and Lin, E. C. (1967). Replacement of a phosphoenol pyruvate-dependent phosphotransferase by a nicotinamide adenine dinucleotide-linked dehydrogenase for the utilization of mannitol. *J. Bacteriol.* **93,** 642.
377. Thomas, A. D., Doelle, H. W., Westwood, A. W., and Gordon, G. L. (1972). The effect of oxygen on a number of enzymes involved in the aerobic and anaerobic utilization of glucose in *Escherichia coli. J. Bacteriol.* **112,** 1099.
378. Tiwari, N. P., and Campbell, J. J. R. (1969). Enzymatic control of the metabolic activity of *Pseudomonas aeruginosa* grown in glucose or succinate media. *Biochim. Biophys. Acta* **192,** 395.
379. Trentham, D. R. (1968). Aspects of the chemistry of D-glyceraldehyde-3-phosphate dehydrogenase. *Biochem. J.* **109,** 603.
380. Tsumura, N., and Sato, T. (1965). Enzymic conversion of D-glucose to D-fructose. V. Partial purification and properties of the enzyme from *Aerobacter aerogenes. Agr. Biol. Chem.* **29,** 1123; *Chem. Abstr.* **64,** 8551 (1966).
381. Tyler, B., and Magasanik, B. (1970). Physiological basis of transient repression of catabolic enzymes in *Escherichia coli. J. Bacteriol.* **102,** 411.
382. Vandemark, P. J., and Wood, W. A. (1956). The pathways of glucose dissimilation by *Microbacterium lacticum. J. Bacteriol.* **71,** 385.
383. Van der Wyk, J., and Lessie, T. (1969). Preferential inhibition by ATP of NAD reduction by glucose-6-phosphate dehydrogenase of *Pseudomonas multivorans. Bacteriol. Proc.* p. 118.
384. Van Thiedemann, H., and Born, J. (1959). Versuche zum Mechanismus der Pasteur-Reaktion. Der Einfluss von Phosphationen auf die Aktivität der strukturgebundenen Hexokinase. *Z. Naturforsch. B* **14,** 447.
385. Veste, J., and Reino, M. L. (1963). Hepatic glucokinase in a direct effect of insulin. *Science* **142,** 590.
386. Vinuela, E., Salas, M. L., Salas, M., and Sols, A. (1964). Two interconvertible forms of phosphofructokinase with different sensitivity to end product inhibition. *Biochem. Biophys. Res. Commun.* **5,** 243.
387. Votaw, R. G., and Krampitz, L. O. (1966). Mechanism of action of phosphoketolase. *Fed. Proc., Fed. Amer. Soc. Exp. Biol.* **25,** 342.
387a. Walter, Jr., R. W., and Anderson, R. L. (1973). Evidence that the inducible phosphoenolpyruvate:D-fructose 1-phosphotransferase system of *Aerobacter aerogenes* does not require "HPr." *Biochem. Biophys. Res. Commun.* **52,** 93.
388. Wang, C. H., Stern, I., Gilmour, C. M., Klungsoyr, S., Reed, D. J., Bialy, J. J., Christensen, B. E., and Cheldelin, V. H. (1958). Comparative studies of glucose catabolism by the radiorespirometric method. *J. Bacteriol.* **76,** 207.
389. Warburg, O., and Christian, W. (1933). Über das gelbe Oxydationsferment. *Biochem. Z.* **257,** 492.
390. Warburg, O., and Christian, W. (1943). Isolierung und Kristallisierung des Gärungsfermentes Zymohexase. *Biochem. Z.* **314,** 149.
391. Warburton, R. H., Eagles, B. A., and Campbell, J. J. R. (1951). The intermediate metabolism of *Pseudomonas aeruginosa.* V. The identification of pyruvate as an intermediate in glucose oxidation. *Can. J. Bot.* **29,** 143.
392. Watson, B. F., and Dworkin, M. (1968). Comparative intermediary metabolism of vegetative cells and microcysts of *Myxococcus xanthus. J. Bacteriol.* **96,** 1465.

393. Watson, H. C., and Banaszak, L. J. (1964). Structure of glyceraldehyde-3-phosphate dehydrogenase. *Nature* (*London*) **204,** 918

393a. Westwood, A. W., and Doelle, H. W. (1974). Glucose 6-phosphate and 6-phosphogluconate dehydrogenases and their control mechanisms in *Escherichia coli* K-12. *Microbios* **9,** 143.

393b. Weinhouse, H., and Benziman, U. (1974). Regulation of hexose phosphate metabolism in *Acetobacter xylinum*. *Biochem. J.* **138,** 537.

394. White, D. C. (1966). The obligatory involvement of the electron transport system in the catabolic metabolism of *Haemophilus parainfluenzae*. *Antonie van Leeuwenhoek; J. Microbiol. Serol.* **32,** 139.

395. Widmer, C., King, T. E., and Cheldelin, V. H. (1956). Particulate oxidase systems in *Acetobacter suboxydans*. *J. Bacteriol.* **71,** 737.

396. Willard, J. M., Schulman, M., and Gibbs, H. (1965). Aldolase in *Anacystis nidulans* and *Rhodopseudomonas spheroides*. *Nature* (*London*) **206,** 195.

397. Williams, P. J. deB., and Rainbow, C. (1964). Enzymes of the tricarboxylic acid cycle in acetic acid bacteria. *J. Gen. Microbiol.* **35,** 237.

398. Wittenberger, C. L., and Angelo, N. (1970). Purification and properties of fructose 1,6-diphosphate-activated lactate dehydrogenase from *Streptococcus faecalis*. *J. Bacteriol.* **101,** 717.

399. Wolff, J. B., and Kaplan, N. O. (1956). Hexitol metabolism in *Escherichia coli*. *J. Bacteriol.* **71,** 557.

400. Wolff, J. B., and Kaplan, N. O. (1956). D-Mannitol-1-phosphate dehydrogenase from *Escherichia coli*. *J. Biol. Chem.* **218,** 849.

401. Wolin, M. J. (1964). Fructose 1,6-diphosphate requirement of streptococcal lactic dehydrogenases. *Science* **146,** 775.

402. Wood, H. G. (1952). Fermentation of [3,4-^{14}C]- and [1-^{14}C]-labeled glucose by *Clostridium thermoaceticum*. *J. Biol. Chem.* **199,** 579.

403. Wood, W. A., and Schwerdt, R. F. (1952). Evidence for alternate routes of carbohydrate oxidation in *Pseudomonas fluorescens*. *Bacteriol. Proc.* p. 138.

404. Wood, W. A., and Schwerdt, R. F. (1953). Carbohydrate oxidation by *Pseudomonas fluorescens*. I. The mechanism of glucose and gluconate oxidation. *J. Biol. Chem.* **201,** 501.

405. Wood, W. A., and Schwerdt, R. F. (1954). Carbohydrate oxidation by *Pseudomonas fluorescens*. II. Mechanism of hexose phosphate oxidation. *J. Biol. Chem.* **206,** 625.

406. Wood, W. A., McDonough, M. J., and Jacobs, B. L. (1961). Ribitol and D-arabitol utilization by *Aerobacter aerogenes*. *J. Biol. Chem.* **236,** 2190.

407. Wu, R. (1966). Further analysis of the mode of inhibition and activation of Novikoff Ascites tumor phosphofructokinase. *J. Biol. Chem.* **241,** 4680.

408. Yoshida, M., Oshima, T., and Imahori, K. (1971). The thermostable allosteric enzyme: Phosphofructokinase from an extreme thermophile. *Biochem. Biophys. Res. Commun.* **43,** 36.

409. Zablotny, R., and Fraenkel, D. G. (1967). Glucose and gluconate metabolism in a mutant of *Escherichia coli* lacking gluconate-6-phosphate dehydrase. *J. Bacteriol.* **93,** 1579.

410. Zagallo, A. C., and Wang, C. H. (1962). Comparative carbohydrate catabolism in *Arthrobacter*. *J. Gen. Microbiol.* **29,** 389.

411. Zhdan-Pushkina, S. M., and Kreneva, R. A. (1963). Sorbitol oxidation with

reference to intensive and delayed reproduction of *Acetobacter suboxydans. Mikrobiologiya* **32,** 711.
412. Zipser, D. (1963). Studies on the ribosome-bound β-galactosidase of *Escherichia coli. J. Mol. Biol.* **7,** 739.
413. Zwaig, N., Kistler, W. S., and Lin, E. C. (1970). Glycerol kinase, the pacemaker for the dissimilation of glycerol in *Escherichia coli. J. Bacteriol.* **102,** 753.

Questions

1. Give the overall reactions of the three major pathways (EMP, HMP, ED) of carbohydrate metabolism.
2. What are the basic reasons for calling hexokinase, phosphofructokinase, FDP-aldolase, and glyceraldehyde-3-phosphate dehydrogenase regulatory enzymes of the EMP pathway?
3. Discuss ATP production in the EMP and HMP pathways and explain why some microorganisms use both pathways for their glucose utilization.
4. Give the key reactions that make it possible to determine whether a microorganism is using the EMP, the HMP, or the ED pathway during glucose utilization.
5. "In glucose utilization the combination EMP/HMP pathway is much more common than the EMP/ED or HMP/ED pathways." Discuss this statement.
6. Name and characterize the four major processes of glucose transport across the cell membrane.
7. Discuss the different modes of action of β-galactosidase and the PEP–phosphotransferase system.
8. Discuss the importance of the Bertrand-Hudson rule for the action of the dehydrogenases in acetic acid bacteria.
9. Discuss the classification of acetic acid bacteria according to their oxidative versatility.
10. What is meant by catabolite repression, transient repression, catabolite inhibition, and allosteric enzyme control?
11. Which enzymes are thought to play a major role in the regulation of the three major pathways of glucose utilization?
12. Discuss the different control mechanisms of glucose-6-phosphate dehydrogenase in *Escherichia coli* and pseudomonads.
13. Explain the cyclic and noncyclic mechanism of mannose metabolism by *Aerobacter aerogenes*.
14. Discuss the different gluconate metabolisms of *Acetomonas suboxydans* and *Acetobacter melanogenum*.
15. What are the differences in gluconate metabolism between acetic acid bacteria and pseudomonads?
16. "Bacterial species can be divided into two groups with respect to their mode of catabolism of mannitol." Discuss this statement.
17. What is the main difference in pentose and pentitol metabolism between the Enterobacteriaceae and the Lactobacillaceae?
18. Compare the D-arabinose metabolisms of *Aerobacter aerogenes*, *Escherichia coli*, and the pseudomonads.
19. What is the main difference between pentose and pentitol metabolisms?
20. Discuss the significance of glycol oxidation by acetic acid bacteria.

6

Aerobic Respiration—Chemolithotrophic Bacteria

Aerobic or oxidative respiration involves a considerably greater number of processes than fermentation or anaerobic respiration. It is the enzymatic oxidation of fuel molecules by molecular oxygen. A great number of books and reviews on this subject deal only with the tricarboxylic acid (TCA) cycle as the epitome of aerobic respiration. In microbiology, however, we have a complex group of organisms that are not able to use the TCA cycle but use molecular oxygen as their final electron acceptor. These microorganisms are the chemolithotrophic bacteria, which are mainly autotrophs and derive their energy by oxidizing inorganic compounds. They incorporate CO_2 into the Calvin cycle for cellular biosynthesis (30, 35, 50, 166, 236). In contrast to this group of microorganisms, stands the vast majority, which are called "chemoorganotrophs" and use organic compounds, such as carbohydrates, as electron donors.

There are only a few bacterial groups that are able to oxidize an inorganic compound for the production of energy: (*a*) the "nitroso" group of genera—*Nitrosomonas*, *Nitrosococcus*, *Nitrosocystis*, *Nitrosoglea*, and *Nitrosospira*—which oxidize ammonia; (*b*) the "nitro" group of genera—*Nitrobacter* and *Nitrocystis*—which oxidize nitrite; (*c*) the genus *Hydrogenomonas* or hydrogen bacteria (*Knallgas* bacteria)—which oxidize hydrogen; (*d*) the ferrous iron-oxidizing bacteria—*Ferrobacillus* and *Thiobacillus ferrooxidans;* and (*e*) the sulfur-oxidizing bacteria—*Thiobacillus*.

The "Nitroso" Group of Genera

The bacterial species that are involved in the oxidation of ammonia and that have been studied most extensively belong to the genus *Nitrosomonas* and *Nitrosocystis*. These organisms oxidize ammonia to nitrite

$$NH_4^+ + \tfrac{3}{2}O_2 + H_2O \rightarrow NO_2^- + 2H_3O^+$$

which reaction involves a net transfer of six electrons, causing a valence change of the nitrogen atom from 3− to 3+. Assuming that the normal electron transfer consists of two electrons, a series of three two-electron steps could be suggested (135) in which energy and reducing power released would be efficiently used by the cell (see also 199). If the reaction is exposed to hydrazine, the oxidation of ammonia does not proceed to nitrite, but hydroxylamine accumulates instead (130)

$$NH_4^+ + \tfrac{1}{2}O_2 \rightarrow NH_2OH + H^+$$

This accumulation of hydroxylamine, as well as the oxidation of this compound without a lag period, makes this compound an almost certain intermediate in the ammonia to nitrite oxidation (279).

The free energy change (ΔF) at pH 7.0 was calculated (22) by reducing the standard free energy change in kilocalories by 9.7 kcal (= 7 pH units × 1.38 kcal) for each H^+ ion appearing on the right-hand side of the equation:

$$NH_4^+(-19.0) + \tfrac{1}{2}O_2(0.0) = NH_2OH(-5.6) + H^+(0.0)$$

$$\Delta F = +13.4 \text{ kcal}$$

$$NH_2OH(-5.6) + O_2(0.0) = NO_2^-(-8.25) + H_2O(-56.7) + H^+$$

$$\Delta F = -59.4 \text{ kcal}$$

$$N_2H_4(30.6) + O_2(0.0) = N_2(0.0) + 2H_2O(-113.4)$$

$$\Delta F = -144.0 \text{ kcal}$$

At pH 7.0, the oxidations of hydroxylamine and hydrazine are exergonic, whereas the oxidation of ammonia to hydroxylamine requires energy.

Investigations into the thermodynamics of this reaction revealed (186) that one of the oxygen atoms in nitrite produced from the oxidation of ammonia is derived from molecular oxygen. This observation, together with the finding of a cytochrome P-450-like pigment (185), poses the question of whether this reaction is a monooxygenase or a hydroxylase reaction. However, it is still possible that the additional oxygen is derived from water if the oxidation of ammonia to nitrite involves the ultimate transfer of electrons to atmospheric oxygen via a cytochrome oxidase type of system exclusively. However, if one or more steps in the oxidation of ammonia to nitrite is mediated by an oxygenase type of enzymatic system, then at least one of the oxygen atoms of nitrite could be derived from atmospheric oxygen (213). Although such an oxygenase has not been isolated as yet, the standard redox potential of +0.899 V for NH_4^+/NH_2OH makes it impossible for ammonia to be oxidized by any of the electron transport carriers of the electron transport chain.

In support of the existence of an oxidizing system came the report that cell-free extracts of *Nitrosomonas europaea* actively oxidized ammonia when the system was protected by the addition of bovine serum albumin, Mg^{2+}, or such polyamines as spermine (239), and inhibited by compounds such as SKF 525, which interacts with cytochrome P-450 of mammalian microsomes (92). As the addition of ATP to the system was without any effect, the activation theory of ammonia oxidation by ATP (272) is losing ground. There is also not enough evidence for the suggestion that H_2O_2 is involved in the ammonia oxidation, although a peroxidase has been purified from *Nitrosomonas europaea* (24). So far, the mechanism of ammonia oxidation remains obscure, although the evidence very strongly favors an oxygenase-type reaction.

The oxidations of hydroxylamine and hydrazine are exergonic, which suggests that these reactions must be connected with a respiratory chain system (165, 277). However, the oxidation of hydroxylamine to nitrite is still a four-electron step. As a result of his search for an intermediate with an oxidation state of 1+, Lees (135) suggested that nitroxyl (NOH) could be the necessary intermediate; it could be formed by a dehydrogenation of hydroxylamine. However, this is an unstable and, as yet, unidentified intermediate. It could serve as a branch point for the formation of nitrite under aerobic conditions and of N_2O and molecular nitrogen under anaerobic conditions. This suggested that the aerobic metabolism of hydroxylamine to nitrite must occur in at least two steps: (*1*) a dehydrogenation of hydroxylamine with the formation of an intermediate in the same oxidation state as nitroxyl and (*2*) a conversion of nitroxyl to nitrite by an enzyme requiring oxygen. Subsequently, Aleem and Lees (6) concluded that the oxidation of NH_2OH to NOH and of the latter to NO_2^- might be a semicyclic process, involving an initial oxidative condensation between the two products of ammonium oxidation, i.e., NH_2OH and NO_2^-. Nitrohydroxylamine would be the "unknown" intermediate, as indicated by the occurrence of the following reactions catalyzed by the *Nitrosomonas* cell-free extracts

$$NH_2OH + 2\ \text{cyt. c}\ Fe^{3+} \longrightarrow (NOH) + 2\ \text{cyt. c}\ Fe^{2+} + H^+$$

$$(NOH) + HNO_2 \longrightarrow NO_2 \cdot NHOH$$

$$NO_2 \cdot NHOH + \tfrac{1}{2} O_2 \longrightarrow 2\ HNO_2$$

$$2\ H^+ + 2\ \text{cyt. c}\ Fe^{2+} + \tfrac{1}{2} O_2 \longrightarrow 2\ \text{cyt. c}\ Fe^{3+} + H_2O$$

Although NH_2OH oxidation involved the mediation of cytochrome systems, the oxidation of $NO_2 \cdot NHOH$ proceeded stoichiometrically without the involvement of cytochromes. It is probable that this reaction is catalyzed by a mixed function oxidase or, more likely, an oxygenase (17) or a "direct" oxidase (16). Anderson (21) observed that under anaerobic conditions the metabolism of hydroxylamine was accompanied by the formation of nitrogenous gas and not nitrite. This finding is still under dispute (194).

In the light of all these observations by Aleem and Lees (6), Anderson (21), and Nicholas and Jones (165), the reactions involved in hydroxylamine oxidation appear to be as indicated in Fig. 6.1. It is important to realize that the high standard redox potential of the NO_2^-/NH_2OH system, +0.066 V, does not allow the coupling of the hydroxylamine dehydrogenase to the NAD level of the respiratory system. The entrance at the flavoprotein

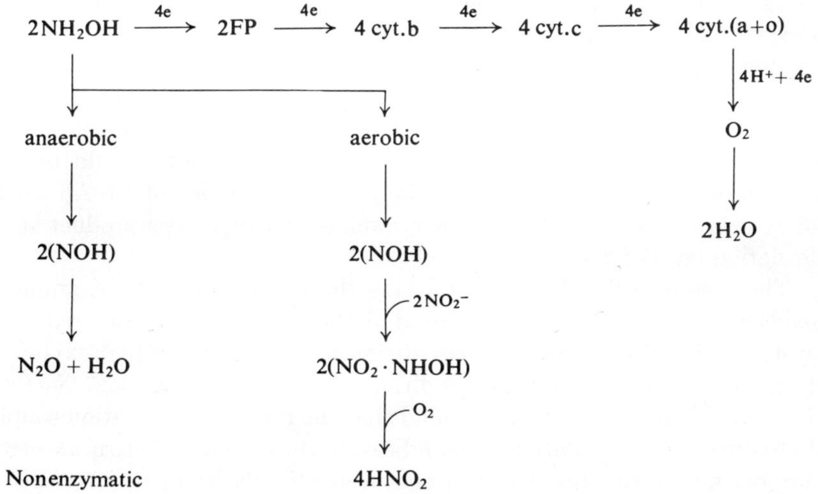
2NH₂OH
4e
2FP
4e
4 cyt.b
4e
4 cyt.c
4e
4 cyt.(a+o)
4H⁺+ 4e
O₂
2H₂O
anaerobic
aerobic
2(NOH)
2(NOH)
2NO₂⁻
N₂O + H₂O
2(NO₂ · NHOH)
O₂
Nonenzymatic
4HNO₂

Fig. 6.1. Hydroxylamine oxidation by *Nitrosomonas*.

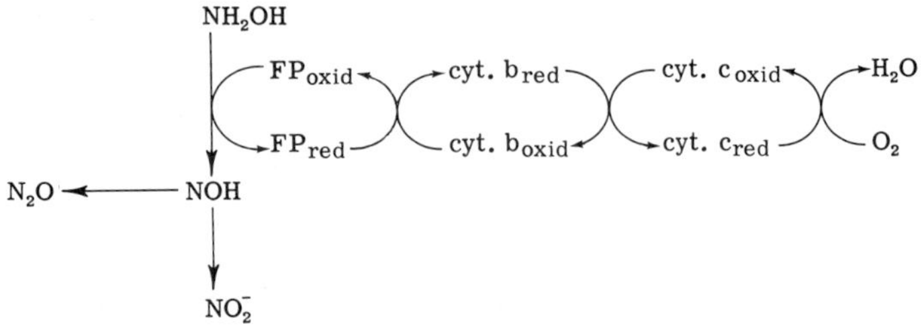
NH₂OH
FPoxid
FPred
cyt. bred
cyt. boxid
cyt. coxid
cyt. cred
H₂O
O₂
N₂O
NOH
NO₂⁻

Fig. 6.2. Proposed scheme of hydroxylamine oxidation and its connection to the electron transport system (reprinted with permission of *The Biochemical Journal*, London).

level, however, should involve at least two energy-conservation steps, mediated by the flavoprotein and cytochrome systems of the electron transport chain (Fig. 6.2). The terminal oxidase in *Nitrosomonas* appears to be a cytochrome of the a (187, 188) and o type (185). The formation of nitric oxide from hydroxylamine would involve the production of three equivalents of reducing power, which can be channeled through the respiratory chain, and probably most of the energy released by ammonia oxidation. In order for this to occur, it must be coupled with a system that generates ATP. There is no reduced NAD formation, which is thought to be necessary for the reduction of carbon dioxide.

No definite identification has been made of the intermediates between NOH and NO_2^-. As mentioned earlier, $NO_2 \cdot NHOH$ is claimed as the most likely intermediate; it requires NO_2^- for its formation. Another claim is (23) that NO is a intermediate between NOH and NO_2^-. The two different relationships between the intermediates of ammonia oxidation to nitrite would therefore be as outlined in Fig. 6.3, with the formation of N_2O being a nonenzymatic process (23). However, nitrite formation has not as yet been demonstrated from NO (194). A further possibility would be N_2O as an intermediate. This is very unlikely, as the reduction of 2.1–2.6 μmoles of cytochrome by 1 μmole of hydroxylamine (90) suggests a product at the oxidation level of 2+.

The present difficulty in establishing the last member of the ammonia oxidation to nitrite is probably caused by the presence of a nitrate reductase system (91). This enzyme reduces nitrite in the presence of hydroxylamine to ammonia under anaerobic conditions. During the reduction, N_2O and NO were identified. It was also found that the rate of O_2 utilization coupled to hydroxylamine oxidation was inhibited in the presence of nitrite, whereas aerobic hydroxylamine disappearance was stimulated in the presence of nitrite. The nitrite reductase system appears to be very similar to that in *Pseudomonas aeruginosa*. It therefore appears that *Nitrosomonas europaea*

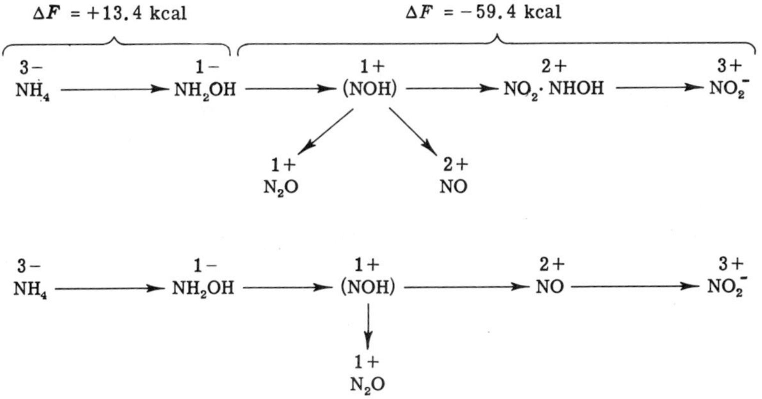
ΔF = +13.4 kcal
ΔF = −59.4 kcal
3− NH4
1− NH2OH
1+ (NOH)
2+ NO2 · NHOH
3+ NO2−
1+ N2O
2+ NO
3− NH4
1− NH2OH
1+ (NOH)
2+ NO
3+ NO2−
1+ N2O

Fig. 6.3. Relationship between the intermediates of ammonia oxidation (reprinted with permission of *The Biochemical Journal*, London).

is not only a nitrifier under aerobic conditions but also a denitrifying bacterium under anaerobic conditions (270). The reductase required FAD or FMN and also produced ammonia, which could be further metabolized by an NADPH-specific glutamate dehydrogenase (271). These reactions make it clear that the reductase is an assimilatory and not a dissimilatory enzyme. The presence of this enzyme could well obscure the step after the action of hydroxylamine–cytochrome c oxidoreductase on hydroxylamine. The partial purification of a hydroxylamine oxidase (hydroxylamine:oxygen oxidoreductase, EC 1.7.3.4) (195) indicates a direct conversion of hydroxylamine to nitrite without intermediates:

$$NH_2OH + O_2 \rightarrow NO_2^- + H_2O$$

The intermediates could be a result of nitrite reductase (nitric oxide: (acceptor) oxidoreductase, EC 1.7.99.3), that is not involved in the oxidation of ammonia but present in *Nitrosomonas.* The present knowledge on the function of this oxidase, however, does not allow further conclusions as to its real function.

Obligate autotrophs utilize carbon dioxide as their only carbon source for growth. The fixation of carbon dioxide and the formation of cell material occurs via the autotrophic CO_2 mechanism, or Calvin cycle (see p. 358). As this mechanism is the reverse of respiration, and as CO_2 is at a much higher oxidation state than organic compounds, reducing power in the form of $NADH + H^+$ is required for the reduction and CO_2 fixation. The thermodynamics of a direct reduction of NAD^+ or $NADP^+$, however, are unfavorable; the redox potential of $NAD^+/NADH + H^+$ is -0.32 V, compared with one for NH_2OH/NH_4^+ of +0.899 V or for NO_2^-/NH_2OH of +0.066 V. In the latter two cases, electrons cannot be transferred from the oxidation of ammonia or hydroxylamine to NAD^+. The reduction of NAD^+ must therefore be an energy-dependent process similar to that underlying the energy-dependent flow from succinate to NAD^+ in animal mitochondria (51).

This problem seems to be characteristic for all chemoautotrophic bacteria (139). It is therefore thought that a reversed electron flow from ferrocytochrome c to NAD^+ or $NADP^+$, with the possible mediation of flavoprotein, may occur. This system would require an input of energy (5, 11–13), however. The required energy could be supplied either by the oxidation of hydroxylamine or by exogenous ATP (13). The observed oxidation of 2 moles of reduced cytochrome c to cause the reduction of 1 mole of NAD^+ by utilizing the energy of 5 moles of ATP indicates a 40% efficiency of the energy-linked reaction (13). The fact that a 5-minute lag occurs in the ATP-dependent reduction of $NAD(P)^+$ indicates the possible formation of a high-energy compound effective in this reduction:

$$ATP + X \rightarrow {\sim}X + ADP + P_i$$

$$AH_2 + {\sim}X + NAD(P)^+ \rightarrow A + NAD(P)H + H^+ + X$$

where AH_2 may be either hydroxylamine or succinate or even reduced cytochrome c.

Because cytochrome appears to be an intermediate between the electron transfer chain and the formation of a high-energy intermediate utilized by energy-requiring processes, rather than a component of an intermediate that functions between the breakdown of ATP and the formation of the high-energy intermediate (151), the electron transport would then be as outlined in Fig. 6.4. The downward direction of the arrows indicates the energy transfer involving the above-mentioned high-energy intermediates (∼) generated either by the electron transport system or from externally supplied ATP. The electrons donated by hydroxylamine are ultimately accepted by oxygen. This path is certainly mediated by cytochromes of the b, c, and oxidase type (a or o) (6). From cytochrome c branches the pathway of energy transfer (∼) involving hypothetical high-energy intermediates. This possible formation of an high-energy intermediate in some ways supports the observation of strong phosphate (46) and polyphosphate (247) activity, which results in a net loss of ^{32}Pi into organic fractions, including ATP and ADP. However, no such high-energy intermediate has yet been found in *Nitrosomonas*.

As both oxidative phosphorylation and the energy-linked reversal of electron flow systems are present in the *Nitrosomonas* particles, these

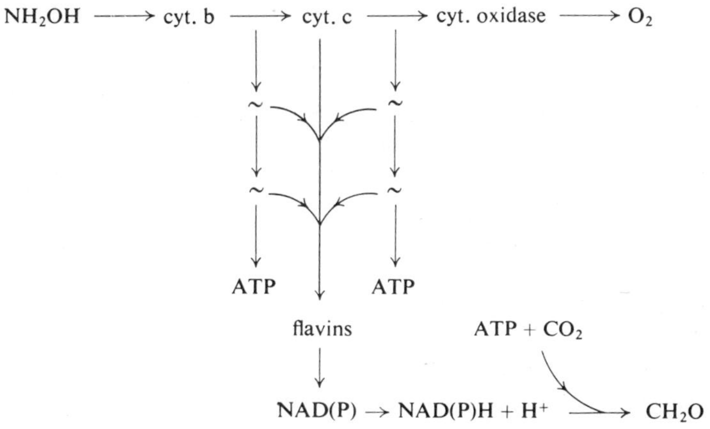
NH2OH
cyt. b
cyt. c
cyt. oxidase
O2
~
~
~
~
ATP
ATP
flavins
ATP + CO2
NAD(P)
NAD(P)H + H+
CH2O

Fig. 6.4. Reversed electron transport system in *Nitrosomonas* (13) (reprinted with permission of Elsevier Publ. Co., Amsterdam).

organisms appear to be well suited to the "Mitchell hypothesis" of ATP synthesis. This hypothesis involves a change in the membrane potential caused by an acid–base transition in the cell (161). If this phenomenon is applicable for the chemoautotrophic organisms, it should drive the reverse electron flow from cytochrome c to NAD^+ and the ATP obtained from oxidative phosphorylation could be used in the CO_2 reduction.

The "nitroso" group belongs to the obligate chemolithotrophs (196) or obligate autotrophs, as they are unable to utilize any organic compounds (184). Recent investigations with *Nitrosolobus multiformis* (274), which showed a slight heterotrophic potential in the absence of an inorganic energy source, indicate a trend toward facultative autotrophy. So do studies with *Nitrosocystis oceanus* (126, 273, 276), which exhibits all Embden-Meyerhof-Parnas pathway (Chapter 5) enzymes, except phosphofructokinase, as well as the TCA cycle enzymes, an $NADH_2$ oxidase, and the phosphoenolpyruvate carbon dioxide-fixing system (see Chapter 5).

The "Nitro" Group of Genera

All bacterial species that are involved in the oxidation of nitrite to nitrate belong to this group of genera. The organism most used for the study on the metabolic events in this group is *Nitrobacter agilis,* which has been characterized as a mixotroph (196) or facultative autotroph (222). The genus *Nitrobacter* is generally found in association with *Nitrosomonas*. Whereas *Nitrosomonas* requires intermediate steps for its oxidation of ammonia,

Nitrobacter oxidizes nitrite directly to nitrate. During this oxidative step, the nitrogen atom undergoes a change in valency from 3+ to 5+. This enzymatic conversion is stoichiometric when molecular oxygen serves as the terminal electron acceptor (16). As cytochrome c and cytochrome oxidase-like components were reduced by *Nitrobacter* cells (3, 4, 136), the following electron transport chain would function:

$$NO_2^- \rightarrow \text{cyt.c} \rightarrow \text{cyt.oxidase} \rightarrow O_2$$

Cytochrome c from *Nitrobacter*, however, has a redox potential of +0.265 V (123) compared with +0.42 V of the NO_3^-/NO_2^- system (5). In order for the electrons to enter at the cytochrome c level, an activation energy of 15 kcal would be necessary (264). The use of oxidative phosphorylation uncouplers (15) and evidence that the reduction of cytochrome c by nitrite was energy-dependent (217) suggested that the electrons from nitrite oxidation enter not at the cytochrome c, but at the cytochrome a_1 level. In the latter case, the reduction of cytochrome c involves an energy-dependent reversal of electron transfer from cytochrome a_1-like components. However, this thermodynamic consideration was again disputed (169) when cytochrome c reduction by nitrite was found to be more rapid than the nitrite oxidase reaction.

Using oxygen-18 and $H_2{}^{18}O$ in *Nitrobacter*, Aleem and co-workers (7, 9) were able to demonstrate for the first time that water and not molecular oxygen participated in the oxidation of nitrite to nitrate and that the hydrogen donor for the concomitant reduction of NAD^+ was also water. Nitrite first donates electrons to the cytochrome–electron transport chain

$$NO_2^- + n\text{ADP} + n\text{P}_i + H_2O \longrightarrow NO_3^- + n\text{ATP} + 2\,H^+ \quad (1)$$

$$NAD^+ + 2\,H^+ + 2e + \text{energy} \longrightarrow NADH + H^+ \quad (2)$$

$$NADH + H^+ + ADP + P_i + \tfrac{1}{2}\,O_2 \longrightarrow NAD^+ + ATP + H_2O \quad (3)$$

(reaction 1) that are ultimately accepted by molecular oxygen (reaction 3). This reaction is coupled with ATP formation. One part of this energy is required for reaction (2), which is the reduction of NAD^+. This reduction involves the participation of electrons from a reduced component, such as a cytochrome of the respiratory chain, and protons donated by water. Reaction (3) gives the energy required to fulfill the requirement of the reduction. Water as source of a reductant would make the need for other types of high-energy intermediates unnecessary in the case of *Nitrosomonas*. With nitrite oxidation mediated by the terminal part of the electron trans-

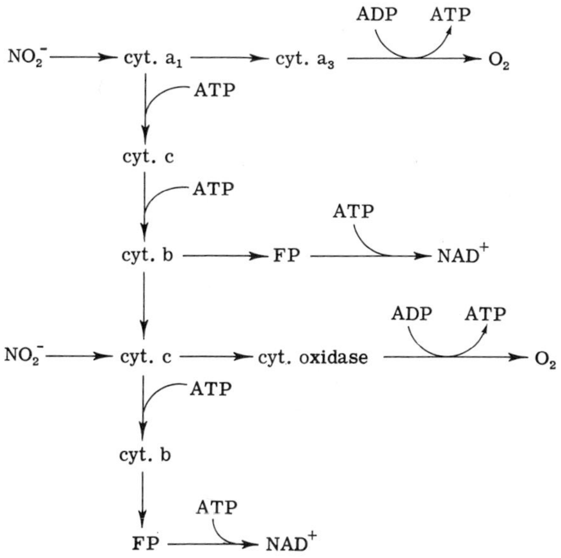
ADP ATP
NO_2^- cyt. a_1 cyt. a_3 O_2
ATP
cyt. c
ATP
ATP
cyt. b FP NAD^+
ADP ATP
NO_2^- cyt. c cyt. oxidase O_2
ATP
cyt. b
ATP
FP NAD^+

Scheme 6.1. Electron transport systems of *Nitrobacter*.

port system, the terminal electron acceptor must be reduced to water. It therefore follows (16) that the nitrite molecule should be hydrated before its electrons are removed and that the oxidation of the nitrite anion must be achieved at the expense of an oxygenation from water.

From the electron transport system in *Nitrobacter* it can be seen that the problem of forming reducing power is similar to that in *Nitrosomonas*. The reduction of NAD^+ by nitrite could be driven either at the expense of added ATP or by the energy generated from nitrite oxidation. The reduction of each mole of NAD^+ by nitrite required the utilization of 4–5 moles of ATP. This is in harmony with the calculated energetics of the overall reverse electron flow process, which involves a free energy gap of some 34 kcal. Depending on whether or not electrons from nitrite oxidation are able to enter the electron transport system at the cytochrome c level, two systems are proposed (Scheme 6.1) (16, 169).

Both suggestions, different as they may be, center around the question of how the thermodynamic gap between nitrite and cytochrome c is bridged. The presence of a highly active cytochrome c reductase (169) indicates substrate activation (84). Such an activation could overcome the relatively small thermodynamic gap between NO_2^-/NO_3^- and cytochrome c. Instead of the expenditure of ATP, and in the definite absence of an oxygenase, the substrate could form a complex with the cytochrome c reductase, resulting in a more negative potential than the substrate. Apart from this complex formation, the necessity of a hydrated form of nitrite or the involvement of an energy-rich phosphorylated compound (16) could well bridge the thermodynamic gap.

Whichever way the electrons enter the transport chain, *Nitrobacter* has only one oxidative phosphorylation site at the cytochrome oxidase or cytochrome a_3 level. This liberated energy is utilized in part to drive the energonic reduction and assimilation of carbon dioxide and in part to produce reduced NAD(P). During the autotrophic oxidation of 1 gram-mole of nitrite, approximately 17.5 kcal of energy becomes available. It is therefore quite reasonable to assume that at least part of this energy is dissipated in other reactions, such as ATP formation. Consideration of the overall yield suggests that 2 moles ATP might be expected (154), which supports the findings by Aleem and Nason (4). The presence of ATPase might be important for the regulation of the ATP requirement. The occurrence of polyphosphates in *Nitrobacter winogradskyi* (40, 67) could also be responsible for the ATP/ADP level in these organisms. An accumulation of this reserve material may be regarded as a metabolic buffer (85) to cope with excess ATP and supply of inorganic phosphate. The overall efficiency of nitrite oxidation has been calculated to be 43–50% (61) and is therefore very similar to the hydroxylamine oxidation of *Nitrosomonas*.

The energy metabolism of *Nitrobacter*, in comparison, has some similarities with photosynthesis and seems to stand between the energy metabolism of photosynthesis and that in *Nitrosomonas*.

The presence of an energy-linked transhydrogenase reaction in *Nitrobacter* warrants special interest (16). However, its importance for *Nitrobacter* has not been elucidated as yet.

In 1965, Schön (214) discovered that the inhibition of the nitrite oxidation at normal oxygen concentration did not result from NO_2^- inhibiting its own oxidation or from the influence of oxygen on the nitrite oxidation but resulted from the effect of oxygen on the carbon dioxide assimilation. The presence of 95% (v/v) oxygen inhibited the carbon dioxide assimilation completely but did not affect the nitrite oxidation of *Nitrobacter winogradskyi*. These findings provide firm evidence that the energy metabolism is independent of the carbon dioxide assimilation. During the same investigations, it was shown that a decrease in soluble oxygen to 2 mg of O_2 per liter (approximately 27% saturation) was harmful for the growth of *Nitrobacter*, whereas the limiting concentration for *Nitrosomonas* was 0.9 mg O_2 per liter. *Nitrosocystis oceanus* (83) was found to react similarly to *Nitrobacter winogradskyi*. Measurements of dissolved oxygen and inorganic nitrogen change indicate (275) that for efficient growth the oxygen–nitrogen ratios should be 3.22:1 in the case of ammonia to nitrite and 1.11:1 in the case of nitrite to nitrate oxidation.

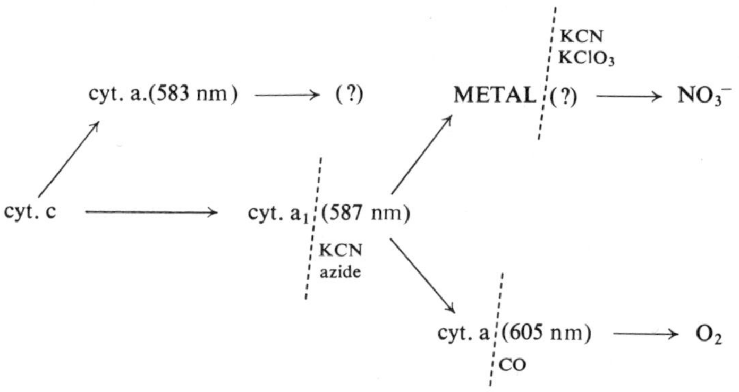
KCN
KClO3
cyt. a.(583 nm)
(?)
METAL
(?)
NO3⁻
cyt. c
cyt. a1
(587 nm)
KCN
azide
cyt. a
(605 nm)
O2
CO

Fig. 6.5. Electron transport scheme of *Nitrobacter* (232) (reprinted with permission of the American Society of Biological Chemists).

This separation of energy metabolism from the carbon dioxide assimilation was supported by the finding of cytochromes c, b, and a and two components of cytochrome a_1 (232), which suggests that *Nitrobacter* must have a respiratory chain that functions independently from the reducing power production and subsequent carbon dioxide assimilation.

Further studies on the electron transport system came from studies of nitrate reduction by *Nitrobacter*. Spectral studies indicate that the pathway of this nitrate reduction involves cytochrome c and one of the a_1 components, whereas the pathway for cytochrome oxidase activity (occurring in nitrite oxidation) involves the same cytochromes and also cytochrome a at 605 nm (232). The electron transport scheme outlined in Fig. 6.5 includes these new discoveries. According to this scheme, electrons are transferred from bacterial cytochrome c to both cytochrome a_1 components, namely, those which absorb at 583 and at 587 nm. The function of the former is still unknown, whereas the latter is definitely involved in electron transport. Because nitrate, as well as oxygen, oxidizes bacterial cytochrome c and a_1 at 587 nm, these two components are placed in a position common to both nitrate reductase and cytochrome oxidase. From cytochrome a_1 (587 nm) electrons may be donated either directly to nitrate or through cytochrome a (605 nm) to oxygen. The requirement of molybdenum for growth of *Nitrobacter* seems to be well established (70). Molybdenum is required in nitrate reduction.

Of particular interest is the fact that oxygen is not inhibitory to nitrate reduction in preparations of *Nitrobacter agilis* (232). This is in contrast to its effect on all other respiratory nitrate reductases. In fact, the addition or accumulation of nitrate prevents or inhibits the oxygen uptake in *N. agilis*.

The chemoautotrophic *Nitrobacter* is a unique example of metabolism

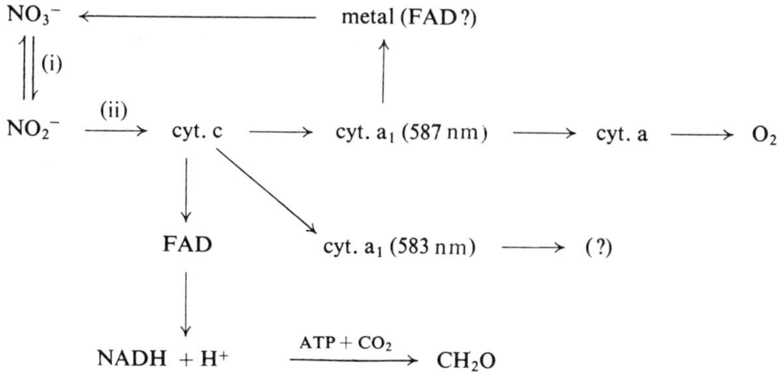

NO_3^-
metal (FAD?)
(i)
(ii)
NO_2^-
cyt. c
cyt. a_1 (587 nm)
cyt. a
O_2
FAD
cyt. a_1 (583 nm)
(?)
NADH + H^+
ATP + CO_2
CH_2O

Fig. 6.6. Electron transport system of *Nitrobacter* during nitrite oxidation.

in which growth depends on the oxidation of nitrite to nitrate, with oxygen as terminal electron acceptor, and the possession of an enzyme capable of reducing nitrate to nitrite. Any accumulation of nitrate inhibits the further production of nitrate, as the oxygen uptake is inhibited. In this case *Nitrobacter* may be able to reduce nitrate to nitrite, and this could mean a recycling process of the nitrite used for the nitrite oxidation. Should this newly discovered process prove to be functioning in *Nitrobacter* it would indicate a better regulatory mechanism than that mentioned above (49). However, there exists insufficient evidence for either of the two mechanisms (Fig. 6.6), although *Nitrobacter* seems to have a high cytochrome concentration (263).

Nitrobacter agilis not only oxidizes nitrite but also contains nitrate, nitrite and hydroxylamine reductases (270) under anaerobic conditions. $NADPH_2$ or $NADH_2$ can serve as electron donors, whereas ATP did not inhibit nitrate reductase. NADH oxidase activity was very small (69). The conversion of nitrate to ammonia under anaerobic conditions does not make the organism a denitrifyer, for no molecular nitrogen is found or formed (see p. 172). However, it supports the presence of assimilatory nitrate reduction. The same system appears to exist also in *Nitrosomonas*. As these enzymes are active only under anaerobic conditions, a relationship may exist between the oxygen sensitivity and the presence of these reductases that could control the intracellular level of nitrite compounds.

Apart from the assimilatory nitrate reduction, *Nitrobacter* and *Nitrosomonas* possess an assimilatory sulfate reduction system. Both genera reduce sulfate via APS and PAPS (see p. 159) and contain the enzymes sulfate adenyltransferase, adenylsulfate kinase, and PAPS reductase (266–268) as is the case in *Escherichia coli* and yeasts (266).

To summarize the present state of knowledge in the chemolithotrophic

Nitrobacter metabolism, the electron transport chain, together with the formation of ATP and reduced NAD, is outlined in Fig. 6.6, where (ii) is catalyzed by a cytochrome c reductase and (i) by nitrate reductase. The nitrate reductase may be of the ferrocytochrome:nitrate oxidoreductase (EC 1.9.6.1) type, although Straat and Nason (232) do not specify any NAD- or flavoprotein-linked reaction. Because there is a metal-containing flavoprotein as prosthetic group, it is anticipated that the nitrate reductase could also be reduced-NAD:nitrate oxidoreductase (EC 1.6.6.1). The electron transfer from nitrite to cytochrome c is almost certainly catalyzed by reduced-NAD:(acceptor) oxidoreductase (EC 1.6.99.3), because cytochrome c may act as acceptor instead of NAD if the preparation has been subjected to certain treatments.

Now that the nature and function of the electron transport chain and the important part played by water as the reductant have been established, the question of whether *Nitrobacter* really has two phosphorylation sites as has *Nitrosomonas* will be discussed. When substrate quantities of cytochrome c were used, no phosphorylation could be obtained under anaerobic conditions with NO_2^- as electron donor (17). However, the site between cytochrome c and molecular oxygen was operative. Therefore, with NO_2^- as the electron donor, ATP may be generated in the cytochrome oxidase region of the electron transport chain. Preliminary evidence, based on spectrophotometry and use of uncouplers, indicated that nitrite enters first at the level of cytochrome a_1 and the reduction of cytochrome c involves an energy-dependent reversal of electron transfer from cytochrome a_1-like components (15). Therefore, only one phosphorylation site is so far in evidence.

It was mentioned above that *Nitrobacter* can also metabolize a large variety of organic compounds (60). Acetate is one of the organic compounds that have been found to penetrate the cell membrane (94) and that can therefore be used as a carbon and energy source (223). Acetate itself had no effect on the rate of nitrite oxidation or exponential growth and contributed 33–39% of newly synthesized cell carbon. The assimilation of acetate into all of the major cell constituents occurred mainly in the absence of carbon dioxide or bicarbonate. When the supply of nitrite was completely exhausted, acetate was mainly assimilated to poly-β-hydroxybutyric acid (265). The addition of nitrite increased the acetate assimilation significantly. In the absence of nitrite, *Nitrobacter agilis* slowly metabolized acetate also to carbon dioxide and water; all the TCA cycle enzymes are present in this organism. Isocitratase was the only enzyme that increased on addition of acetate. The addition of nitrite suppressed the carbon dioxide production from acetate and increased the assimilation into cell material instead. With acetate and casein hydrolyzate, *Nitrobacter agilis* could be grown heterotrophically after adaptation but never lost the ability to grow autotrophically on addition of nitrite and carbon dioxide. The mean guanine plus cytosine content of their DNA was 61.2 ± 1%, whether the organism was grown autotrophically or heterotrophically.

This investigation demonstrates that *Nitrobacter agilis* can oxidize nitrite with carbon dioxide or acetate as carbon source and indicates again the entirely different control systems of energy and biosynthetic metabolism. In the absence of nitrite and carbon dioxide, the same organism can grow heterotrophically but appears to require casein hydrolyzates in addition to acetate. Cell-free extracts of *Nitrobacter agilis* also contained a particulate formate oxidase, which oxidizes formate to carbon dioxide plus water (170). However, any growth experiment with formate as sole carbon source has been unsuccessful.

Hydrogenomonas or *Knallgas* Bacteria

Members of this genus use the reaction between molecular hydrogen and molecular oxygen to obtain energy. Because this reaction is also called "the *Knallgas* reaction" these bacteria are quite commonly referred to as "*Knallgas* bacteria." The word *Knallgas* is German and means something like "explosive gas." Species of *Hydrogenomonas* are able to live autotrophically or heterotrophically and are strict aerobes. As the organisms are able to operate both systems at the same time, when provided with a gaseous environment of hydrogen, oxygen, and carbon dioxide (6:2:1) in addition to lactate, and as they also increase their heterotrophic yield compared with growth under strictly autotrophic conditions (59, 197), *Hydrogenomonas* can be regarded as mixotrophic.

In order to grow autotrophically, *Hydrogenomonas* cultures must be well aerated and must have a certain $H_2/CO_2/O_2$ ratio, for the carbon dioxide fixation depends very much on the oxygen contents in the gaseous mixture (280). During growth, hydrogenomonads consume hydrogen, oxygen, and carbon dioxide at a ratio somewhere between 5.7:1.7:1 (209) and 8:3:1 (20). Under these conditions, the organism can withstand a relatively high partial pressure of oxygen and also does not accumulate the reserve material, poly-β-hydroxybutyrate. It can therefore be assumed that these conditions are optimal. With oxygen as the limiting factor, the cells accumulate poly-β-hydroxybutyrate up to 23% of the dry weight (215), whereas the H_2/CO_2 ratio changes to 4.6. Under hydrogen limitation, the buildup of the reserve product did not occur, but the H_2/CO_2 ratio was 9.1. The only essential metal requirement for the growth appears to be nickel; the yield of protein formed was found to be proportional to the amount of nickel added (32). It is suggested that nickel participates in some carbon dioxide-fixation reactions.

In common with all bacteria that utilize molecular hydrogen in one way or another, *Hydrogenomonas* species possess hydrogenase activity for the activation of molecular hydrogen (211). Cells reduce nitrate and Methylene Blue at the expense of hydrogen and can assimilate 1 mole of carbon dioxide for every 2 moles of oxygen reduced in the *Knallgas* reaction (28). This CO_2 fixation is very rapid (35) and occurs via the Calvin cycle.

$$6\,H_2 + 2\,O_2 + CO_2 \rightarrow (CH_2O) + 5\,H_2O$$

The reducing power, reduced NAD, as well as ATP necessary for the CO_2 fixation is obtained via an oxidative phosphorylation.

Cell-free extract studies on *Hydrogenomonas* H 16 revealed that the hydrogenase activity can be separated into a soluble and a particulate fraction (66). The soluble fraction catalyzes the reduction of NAD^+ with molecular hydrogen (180) and does not require cofactors, although such a requirement may exist in *Hydrogenomonas eutropha* (191). This enzyme is only very slightly affected by the presence of ATP or $NADH + H^+$ and does not react with oxygen, $NADP^+$, FMN, FAD or Methylene Blue. The inhibition of the soluble hydrogenase by its reaction product $NADH_2$ is of an allosteric character (2) and could be responsible for a control of hydrogen transport similar to the above-described AMP inhibition of adenylylsulfate reductase (see p. 161). The particulate hydrogenase reduces Methylene Blue and oxygen as a physiological hydrogen acceptor. The soluble hydrogenase would be sufficient to provide the cell with reduced NAD and ATP for carbon dioxide reduction and Eberhardt (66) suggested that the second hydrogenase may be the first member of the electron transport chain, which is coupled with oxidative phosphorylation. There is also some evidence for the presence of a reduced NAD oxidase, which could oxidize $NADH + H^+$ with oxygen. This enzyme could be reduced-NAD:(acceptor) oxidoreductase (EC 1.6.99.3), which is synonymous

$$NADH + H^+ + O_2 \rightarrow NAD^+ + H_2O_2$$
$$\text{or}\quad 2\,NADH + 2\,H^+ + O_2 \rightarrow 2\,NAD^+ + 2\,H_2O$$

with reduced NAD dehydrogenase and cytochrome reductase. It is a flavoprotein and after purification can be coupled with cytochrome c as an acceptor. There is no ATP formation in this transfer. The formation of H_2O_2 presents no problem, for *Hydrogenomonas* H 16 exhibits a strong catalase activity.

Hydrogenomonas H 16 therefore has two hydrogenases, which have different functions. These functions may be similar to those of the two hydrogenases of the photosynthetic bacteria. Because one enzyme is obligately coupled with the phosphorylation of ADP to ATP (31), its function is controlled by the relative concentrations of ADP and ATP, being retarded by a deficiency of the former or an excess of the latter. Excess ATP may accumulate when there is a deficiency of carbon dioxide. The second hydrogenase transfers the electrons to oxygen via reduced NAD and is not necessarily affected by excess ATP, although it possibly is affected by reduced NAD (2). This control mechanism would be similar to the one of adenylylsulfate reductase in *Desulfovibrio desulfuricans*, where the reaction product AMP carries this regulatory function. The overall scheme has been described by Packer and Vishniac (173) and by Repaske (189).

Very little information is available in regard to the oxidation of reduced NAD by hydrogenomonads. Investigations with *Hydrogenomonas eutropha* revealed two possible pathways (190), one which has a menadione-dependent reductase and the other a menadione-independent cytochrome c reductase. As quinones are found in a number of microorganisms (108, 142), the scheme outlined in Fig. 6.7 has been proposed to exist in cell-free extracts of hydrogen bacteria (41).

The oxidative phosphorylation of hydrogen is thought to be limited to the sequence between hydrogen and cytochrome b. Three distinct c-type cytochromes were found present, which unlike normal cytochrome c did not bind with either CO or cyanide (68).

Very little work has been done to date on the energy metabolism of these bacteria. With carbon dioxide as the sole carbon source and hydrogen as the sole electron donor, *Hydrogenomonas ruhlandii* (173) performs in the following way:

$$CO_2 + H^+ \rightarrow (CH_2O) + H_2O \qquad \Delta F = +8.2 \text{ kcal}$$

$$H_2 + O_2 \rightarrow H_2O \qquad \Delta F = -56.5 \text{ kcal}$$

$$\overline{CO_2 + 2H_2 + O_2 \rightarrow (CH_2O) + 2H_2O \qquad \Delta F = -48.3 \text{ kcal}}$$

Similar results have been obtained with *Hydrogenomonas facilis* (181). The proportionality found (43) between the rate of H_2 oxidation and the rate of CO_2 fixation suggests that energy supply regulates the (maximum) rate of growth. Energy yield measurements revealed that 1 mole H_2 yields the equivalent of 2 moles ATP for *Hydrogenomonas eutropha*, and at least 5 moles ATP are required for the conversion of 1 mole CO_2 into cellular material. Because of the involvement of molecular hydrogen no reverse electron flow system is required, as there is no voltage span to the NAD^+/$NADH + H^+$ system (84).

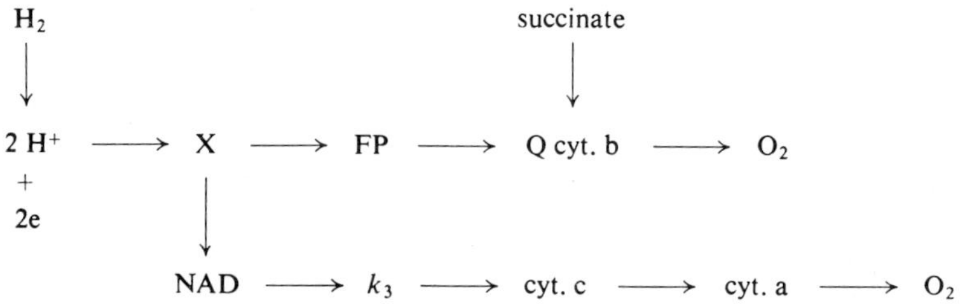
H2
succinate
2 H+
+
2e
X
FP
Q cyt. b
O2
NAD
k3
cyt. c
cyt. a
O2

Fig. 6.7. Electron transport system of *Hydrogenomonas eutropha* grown under autotrophic conditions (reprinted with permission of American Society of Microbiology).

Cells of the *Hydrogenomonas* species are able to synthesize poly-β-hydroxybutyric acid, not only heterotrophically from pyruvate and acetate, but also autotrophically from carbon dioxide (77, 79, 206). Under autotrophic conditions, acetate must be formed from the Calvin cycle. Schlegel and Gottschalk (208) postulated two pathways:

1. Acetyl-CoA could be formed from 3-phosphoglycerate via pyruvate and a decarboxylation reaction
2. Acetyl-CoA could be formed from fructose 6-phosphate or xylulose 5-phosphate via a phosphoketolase reaction (see p. 247).

As there was no phosphoketolase reaction detectable in *Hydrogenomonas* H 16 (204), the first reaction from 3-phosphoglycerate via pyruvate is the only possible mechanism. Acetyl-CoA is formed from pyruvate by a catalyzed reaction that requires pyruvate dehydrogenase [pyruvate:lipoate oxidoreductase (acceptor-acetylating), EC 1.2.4.1], CoA, NAD^+, and thiamine pyrophosphate

$$\text{pyruvate} + \text{CoA-SH} + NAD^+ \longrightarrow \text{acetyl-CoA} + CO_2 + NADH + H^+$$

The details of this reaction will be dealt with in the section on the TCA cycle in Chapter 7.

Under heterotrophic conditions, *Hydrogenomonas* species utilize the TCA cycle as their source of ATP generation. The withdrawal of acetyl-CoA from this cycle for poly-β-hydroxybutyrate synthesis would certainly jeopardize the ATP generation, for the TCA cycle would not function optimally. To replenish the TCA cycle with intermediates, *Hydrogenomonas*, like *Escherichia coli*, is able to form oxalacetate from phosphoenolpyruvate because of the presence of phosphoenolpyruvate carboxykinase (EC 4.1.1.32), PEP-carboxylase (EC 4.1.1.31), and PEP-carboxykinase(pyrophosphate) (EC 4.1.1.38). The PEP-carboxylase functions as an anaplerotic carbon dioxide-fixing enzyme (72, see also p. 422). *Hydrogenomonas* is thus able to form oxalacetate via a carboxylating reaction with pyruvate or phosphoenolpyruvate as substrate. With pyruvate or lactate as substrate, phosphoenolpyruvate can be formed via a PEP-synthetase reaction (73), wherein AMP has been found as a second product. The presence of reduced NAD-dependent malate dehydrogenases (EC 1.1.1.38, 39, and 40) also guarantees pyruvate synthesis if TCA cycle acids serve as carbon sources. Molar growth yield experiments with succinate and fumarate as carbon sources revealed higher yields from succinate-grown cells (42) owing to the additional mole of ATP generated. As these anaplerotic sequences are very similar to those in the other chemoorganotrophic bacteria, they will be dealt with in detail in Chapter 7.

Acetyl-CoA obtained via autotrophic or heterotrophic growth may be carboxylated to malonyl-CoA, which requires as cofactor biotin, or may condense in forming poly-β-hydroxybutyrate (see p. 119). The formation of malonate is assumed to also be the introductory step to lipid biosynthesis. The speed of poly-β-hydroxybutyrate formation is far greater under heterotrophic than under autotrophic conditions. The polymer is normally utilized again in the absence of a carbon source and can be used for protein synthesis in the presence of a nitrogen source (206).

Apart from metabolic intermediates (203), *Hydrogenomonas facilis* and H 16 can utilize certain carbohydrates, such as glucose (58), fructose (78), or ribose (150), whereas *H. eutropha* utilizes fructose and gluconic acid (56). The fructose metabolism follows the Entner-Doudoroff pathway (see p. 230), although the presence of 6-phosphogluconate dehydrogenase in some hydrogenomonads (129) may indicate some hexose monophosphate pathway activity. *Hydrogenomonas* H 16 exhibits a strong fructose-1,6-diphosphatase and only a weak phosphofructokinase, which makes the reverse reaction of the following sequence

$$\text{triose phosphate} \rightarrow \text{FDP} \rightarrow \text{F 6-P}$$

virtually impossible. The reaction

$$\text{6-phosphogluconate} \rightarrow \text{2-keto-3-deoxy-6-phosphogluconate}$$

thus becomes irreversible and the subsequent metabolic steps are to pyruvate and 3-phosphoglyceraldehyde.

Molecular hydrogen suppresses the induction of the enzymes of the Entner-Doudoroff pathway (37, 183), which results in a decrease of the rates of growth or function of poly-β-hydroxybutyrate by 80%. This phenomenon is called the "hydrogen effect." However, when *Hydrogenomonas facilis* is grown heterotrophically, the hydrogenase, although present, is not functional (143). Whenever these heterotrophically grown

Hydrogenomonas spp. are grown under hydrogen atmosphere (95% H_2 and 5% air), they exhibit a rapid and linear oxygen uptake (29) and a higher hydrogenase activity, which means that they can grow almost immediately under autotrophic conditions again.

In addition to poly-β-hydroxybutyrate, *Hydrogenomonas* spp. are also able to form polyphosphate as a second storage product. It is formed rapidly under aerated conditions from orthophosphate (101), but not under anaerobic conditions. The role of these polyphosphates appears to be similar to that of poly-β-hydroxybutyrate, that is, to supply the cell with phosphorus in the case of orthophosphate exhaustion (102, 207). It is almost certainly not used in the energy metabolism of *Hydrogenomonas*.

Purine Metabolism

Autotrophically and heterotrophically grown cells of *Hydrogenomonas* are able to use uric acid and allantoin as their sole nitrogen and carbon source for growth. Under autotrophic conditions these compounds are used solely as nitrogen source, whereas under heterotrophic conditions they can be used as both a carbon and nitrogen source (19, 104) (Fig. 6.8).

Uric acid is attacked by the enzyme urate oxidase (urate:oxygen oxidoreductase, EC 1.7.3.3), which has been found in a number of bacterial cell-free systems (27, 103, 167). The conversion of uric acid by this enzyme occurs by oxidative and hydrolytic mechanisms via a labile intermediate (153) to allantoin. The byproduct hydrogen peroxide is destroyed by a catalase. During the reaction, carbon dioxide is released. This degradation of uric acid was slightly stimulated by fructose, whereas ammonia addition did not effect the reaction rate (107). The conversion of allantoin via allantoic acid to glyoxylic acid and urea is assumed to be identical to the one carried out by other bacteria (280) (see Chapter 7). A tartronate-semialdehyde synthase [glyoxylate carboxy-lyase (dimerizing), EC 4.1.1.47] and reduced NAD- and NADP-linked D-glycerate 3-dehydrogenase (EC 1.1.1.29) (107) convert glyoxylate via tartronic semialdehyde to glycerate. The metabolism of the latter compound is then identical to that in *Pseudomonas* sp., whereby glycerate is converted via pyruvate and the TCA cycle (see p. 408). The second product of allantoin breakdown, urea, is converted to ammonia and CO_2 with the catalytic aid of urate oxidase (124, 125). An NADP-linked glutamate dehydrogenase is the main enzyme responsible for the incorporation of ammonia into amino acids (100) in *Hydrogenomonas eutropha*.

Although most of the pathway appears to be identical to the allantoate metabolism in *Pseudomonas* (see Chapter 7), the synthesis of the responsible enzymes is different. In the case of the tartronate-semialdehyde syn-

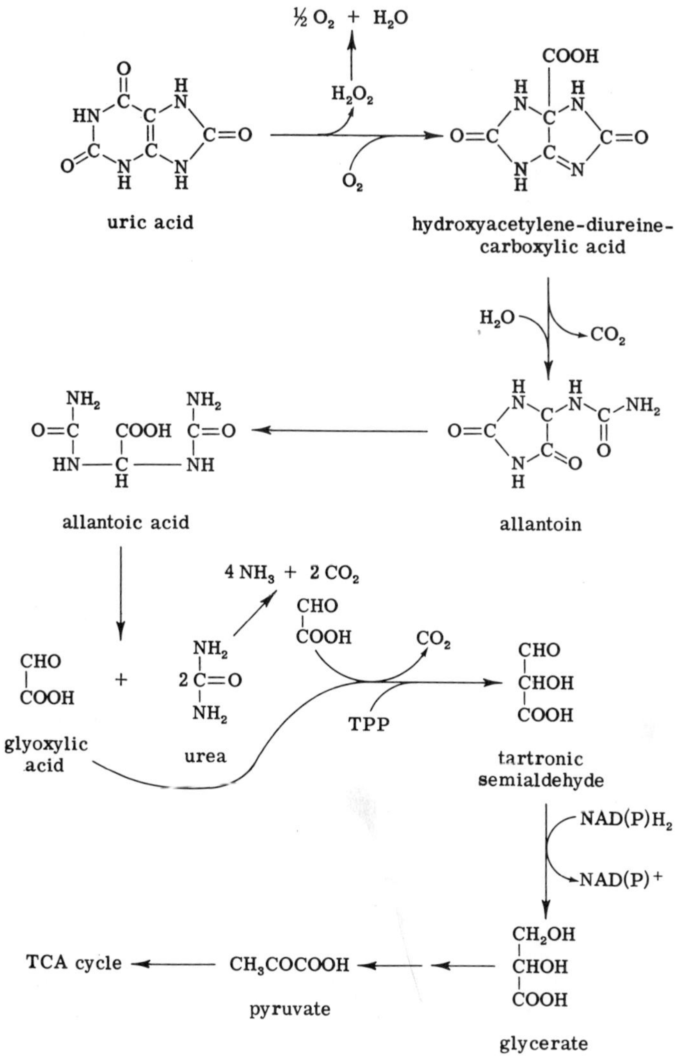
½ O2 + H2O
H2O2
O2
uric acid
COOH
hydroxyacetylene-diureine-carboxylic acid
H2O
CO2
allantoin
allantoic acid
4 NH3 + 2 CO2
CHO
COOH
CO2
glyoxylic acid
urea
TPP
tartronic semialdehyde
NAD(P)H2
NAD(P)+
TCA cycle
CH3COCOOH
pyruvate
glycerate

Fig. 6.8. Metabolism of uric acid by hydrogenomonads.

thase [glyoxylate carboxy-lyase (dimerizing), EC 4.1.1.47], the addition of fructose suppressed the synthesis only when glyoxylate was the inducing substrate but not when uric acid was (106). A completely different regulatory mechanism also appears to exist in the case of urate oxidase synthesis (62, 124, 127, 152, 230), as urea does not exhibit any influence on induction of this enzyme. It is considered that urate oxidase formation, at least in *Hydrogenomonas* H 16, is controlled by repression only and that ammonium ions serve as the repressing substrate (107) via catabolite repression.

This type of regulation is connected with the fact that uric acid is a relatively poor energy source for growing cells, whereas glyoxylate is a very good energy source. In the presence of fructose, the growth of cells on uric acid would have an additional energy source to overcome the initial problem. Fructose, of course, could easily be replaced by molecular hydrogen (19).

Thymine, cytosine, and uracil can also serve as nitrogen sources (105), as could amino acids (71), if the molecular hydrogen supplies the energy. Cytosine is completely utilized by *Hydrogenomonas facilis*, but only deaminated to uracil by *Hydrogenomonas* H 16. The detailed reactions await further investigations.

Deoxyribonucleic acid homology studies further support the close relationship between *Hydrogenomonas* and the Pseudomonadaceae, which lead to the suggestion to abandon the genus *Hydrogenomonas* altogether and to classify them into the genus *Pseudomonas* (182). This suggestion has so far not been recognized internationally.

The Iron-Oxidizing Bacteria

The microbial transformation of iron is brought about in two ways (48): (*1*) by specific organisms, which use the oxidation of ferrous ions to ferric ions as a source of energy for growth (some blue-green algae and the "true iron bacteria," e.g., *Ferrobacillus ferrooxidans* and *Thiobacillus ferrooxidans*); and (*2*) by nonspecific organisms. These may act by metabolizing the organic substances that chelate with iron and hold it in solution at pH 7.0. Removal of the chelate releases the Fe^{3+}, which precipitates as $Fe(OH)_3$ or chelates with the substances of the bacterial cell. This results in iron encrustation on cell sheaths or capsules.

Only the specific organisms, which use iron oxidation as a source of energy, shall be dealt with, for the second group transforms iron only as result of metabolism and does not require iron specifically for energy metabolism (133).

Ferrobacillus ferrooxidans

Information concerning the metabolism of this bacterium is very limited, probably because of the lack of suitable methods for obtaining mass cell growth. Observed yields of only 1–2 gm of cells per 18 liters of medium

(39) do not encourage metabolic studies. *Ferrobacillus ferrooxidans* derives its energy from the following reaction:

$$4\,FeCO_3 + O_2 + 6\,H_2O \rightarrow 4\,Fe(OH)_3 + 4\,CO_2 \qquad \Delta F = -40 \text{ kcal}$$

Although 92% of the theoretical amount of oxygen required for the oxidation of iron has been accounted for (138), the efficiency during the oxidation of 50 μmoles of iron was $20.5 \pm 4.3\%$.

Despite the higher efficiency, *Ferrobacillus* has the same problems with regard to the formation of ATP and reducing power as was shown above in the other chemolithotrophs. Preliminary findings by Aleem *et al.* (5) did, in fact, reflect the energy-dependent reversal of electron transport transfer from ferrocytochrome c to NAD^+ in cell-free extracts of this chemoautotroph. It contains cytochromes of the b, c, and a types and the E_m of Fe^{2+}/Fe^{3+} is 0.77 V. It was observed (39), however, that this potential could be reduced when iron was complexed with an ion, for example, iron oxalate. The potential of iron oxalate is approximately zero at pH 4–7. It is also possible that the enzymatic reduction of cytochrome c by ferrous ions under these conditions will prove to be an energy-yielding reaction, provided the potential of Fe^{2+}/Fe^{3+} is sufficiently lowered. This was found to be the case with iron–cytochrome c reductase (ferrocytochrome c:iron oxidoreductase, EC 1.9.99.1) (278) which was found capable of transferring electrons from ferrous ion to cytochrome c. The oxidation of $Fe^{2+} \rightarrow Fe^{3+}$ is accompanied by the production of an acid, which seems to be always H_2SO_4 (65). In formulating any scheme for iron oxidation, this requirement for SO_4^{2-} (212) must be considered, as must the fact that the ferrous ion must either enter the cell or be attached to a binding site at the surface of the cell in order to couple its oxidation with the carbon reduction mechanism within the cell. This has led to the proposal presented in Fig. 6.9 (65).

The oxidation of iron is linked to an energy source where an iron–oxygen complex is formed. The iron in this complex is oxygenated but not oxidized, for no electron flow has taken place. The complex can be formed either in solution or on the cell surface, where it reacts with iron oxidase (or oxygenase). This reaction releases an electron, which is transported into the cell via sulfate or flavoprotein (FP). In the cell, this electron is assumed to go via an ubiquinone to cytochrome c, then to cytochrome a, and finally to oxygen as final acceptor. It is presumed that the electron transfer to cytochrome c is coupled to a phosphorylation step as in *Nitrobacter* and *Nitrosomonas*.

The source of reducing power is also sought in a reverse electron flow, because it seems impossible to form an iron complex that would lower the potential to the E_m (−0.32 V) of the $NAD^+/NADH + H^+$ system. This

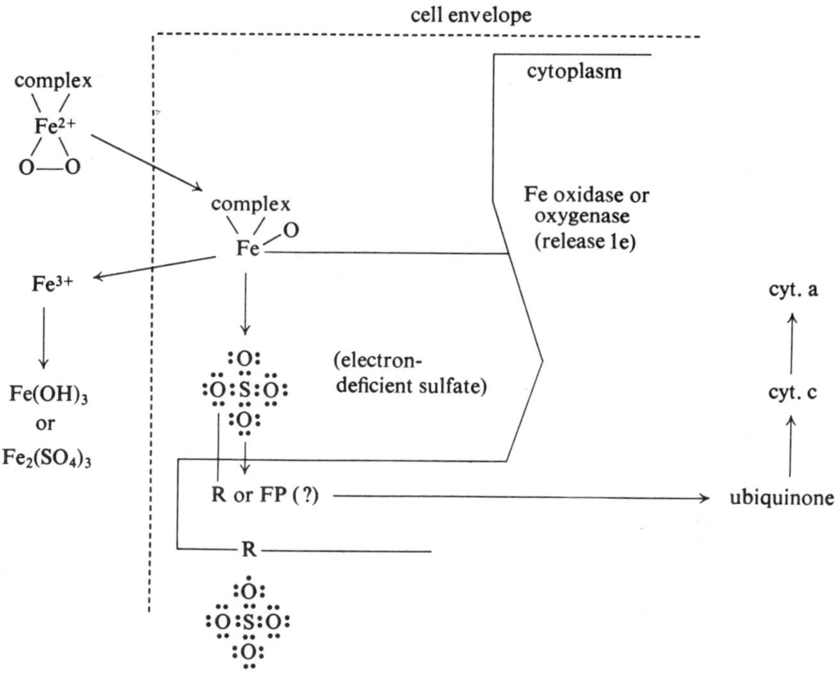
cell envelope
cytoplasm
complex
Fe2+
O—O
complex
Fe
O
Fe oxidase or
oxygenase
(release 1e)
Fe3+
Fe(OH)3
or
Fe2(SO4)3
(electron-
deficient sulfate)
cyt. a
cyt. c
R or FP (?)
ubiquinone
R

Fig. 6.9. The oxidation of ferrous ion by *Ferrobacillus ferrooxidans.* Sulfate may be bound to an R group or else electrons may flow directly to flavoproteins; R may be a flavoprotein (reprinted with permission of the American Society of Microbiology).

would mean the presence of the same electron transport chain outlined in *Nitrosomonas,* which would support the findings of Dugan and Lundgren (65). *Ferrobacillus ferrooxidans* is also able to grow on elemental sulfur (156) and to oxidize pyrite (234). The oxidation of iron was as rapid by sulfur-grown cells as by iron-grown cells. It is suggested that the different oxidative capacities are owing to the heterogenous mixture of cell types in the culture population.

Recent investigations indicate that *Ferrobacillus ferrooxidans* has also lost its obligate autotrophy and must be categorized as a facultative autotroph. Following a brief adaptation period to glucose, *Ferrobacillus ferrooxidans* is able to grow not only on glucose, but also on mannitol, several other sugars, and a few amino acids in the absence of an oxidizable iron source (218). This strict aerobe metabolizes glucose via the Embden-Meyerhof-Parnas pathway, as no glucose 6-phosphate and 6-phosphogluconate dehydrogenase could be detected (25), as well as via the tricarboxylic acid cycle. This catabolic system also changes the electron transport, as the electrons can now enter the respiratory chain at the $NAD^+/NADH + H^+$ level. In order to replenish dicarboxylic acids used for biosynthetic purposes, *Ferrobacillus ferrooxidans* also possesses a phosphoenolpyruvate carboxylase, which functions as an anaplerotic enzyme in oxalacetate formation (64). Neither pyruvate nor pyruvate plus ATP was able to replace phosphoenolpyruvate, which excludes the possible presence of a pyruvate carboxylase or phosphoenolpyruvate carboxykinase. Phosphoenolpyruvate carboxylase is activated by acetyl-CoA and inhibited by aspartate. This activation and inhibition mechanism is identical to the one found in enteric bacteria and will therefore be dealt with at a later stage (see p. 422). All indications are that no mixotrophy exists, as the organism grows either autotrophically or heterotrophically. After adaptation to glucose, *Ferrobacillus ferrooxidans* becomes obligately organotrophic (218).

Thiobacillus ferrooxidans

The metabolism of *Thiobacillus ferrooxidans* has not been studied as extensivcly as that of *Ferrobacillus ferrooxidans.* The sulfate requirement (134), the effect of phosphate (34), and the rate of iron oxidation (131) give the impression that the autotrophic metabolism of *Thiobacillus ferrooxidans* must be very similar to, or even the same as, that observed in *Ferrobacillus ferrooxidans.*

As its name indicates, *Thiobacillus ferrooxidans* is also able to oxidize reduced sulfur compounds. This oxidation, of course, is at a far slower rate than the oxidation of iron (132). Expressed on a molar basis, the organism can oxidize approximately 180 moles of ferrous ion in the same time as that required for the oxidation of 1 mole of sulfur. Both oxidations have been shown to occur simultaneously (132). Differences exist, however, in the heterotrophic growth and metabolism. Whereas *Ferrobacillus ferrooxidans* utilizes glucose via the Embden-Meyerhof pathway and possibly cannot grow mixotrophically, *Thiobacillus ferrooxidans* catabolizes glucose similarly to the hydrogenomonads, that is, via the Entner-Doudoroff pathway, and is also able to grow mixotrophically.

The addition of glucose to an iron-containing medium repressed the ability of the organism to oxidize this substrate for autotrophic growth (241). The uptake of glucose parallels the increase in glucose-6-phosphate dehydrogenase and poly-β-hydroxybutyric acid formation, neither of which have been observed as yet in *Ferrobacillus ferrooxidans.* Experiments with [^{14}C]glucose, together with enzymatic determinations, revealed that the enzymes of the Entner-Doudoroff pathway are induced if *Thiobacillus ferrooxidans* is grown on an iron–glucose supplemented medium or on

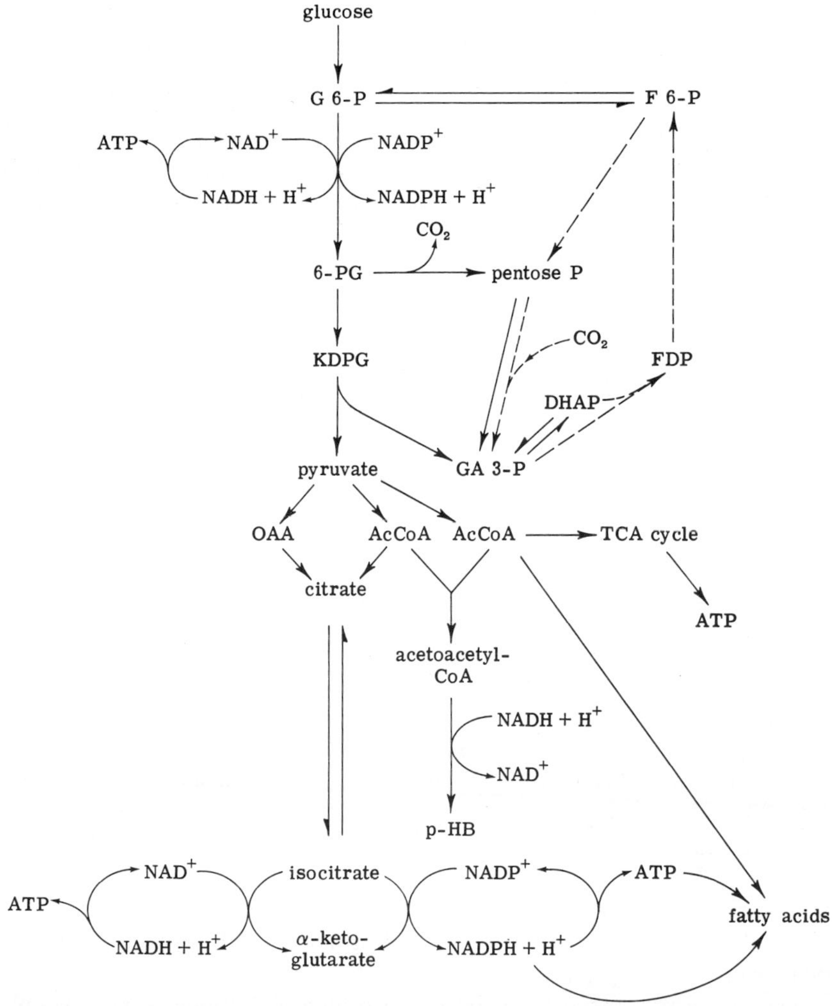
glucose
G 6-P
F 6-P
ATP
NAD+
NADP+
NADH + H+
NADPH + H+
CO2
6-PG
pentose P
CO2
KDPG
FDP
DHAP
pyruvate
GA 3-P
OAA
AcCoA
AcCoA
TCA cycle
citrate
ATP
acetoacetyl-CoA
NADH + H+
NAD+
p-HB
NAD+
isocitrate
NADP+
ATP
ATP
fatty acids
NADH + H+
α-keto-glutarate
NADPH + H+

Fig. 6.10. Mixotrophic (autotrophic, – – –; heterotrophic, ——) growth and metabolism of *Thiobacillus ferrooxidans*. G 6-P: glucose 6-phosphate; F 6-P: fructose 6-phosphate; 6-PG: 6-phosphogluconate; KDPG: 2-keto-3-deoxy-6-phosphogluconate; DHAP: dihydroxyacetone P; OAA: oxalacetate; p-HB: poly-β-hydroxybutyric acid; GA 3-P: glyceraldehyde 3-P; FDP: fructose diphosphate (reprinted with permission of the authors and the American Society of Microbiology).

glucose alone. Whereas heterotrophically grown cells possess the complete TCA cycle, autotrophically grown cells lack the enzymes α-ketoglutarate dehydrogenase and NADH oxidase (242).

The existence of two isocitrate dehydrogenases is of interest, as the NAD-linked enzyme is constitutive and the NADP-linked enzyme adaptive for heterotrophic growth. The NAD^+-linked enzyme is certainly the electron transport-linked enzyme, whereas the $NADP^+$-linked enzyme could be used for biosynthetic purposes. A pyruvate carboxylase replenishes those dicarboxylic acids which are required for amino acid or fatty acid biosynthesis. The low levels of fructose-diphosphate aldolase also indicate that the Embden-Meyerhof pathway is only used in reverse for biosynthetic purposes (gluconeogenesis). On all these accounts, the metabolic chart given in Fig. 6.10 could be drawn (242) as representing the mixotrophic growth and metabolism of *Thiobacillus ferrooxidans.*

The presence of the transketolase–transaldolase reaction also indicates that the HMP pathway is functioning. The other interesting enzyme, apart from isocitrate dehydrogenase, is glucose-6-phosphate dehydrogenase, which also is present as an NAD- and an NADP-linked enzyme (243). This key enzyme for the hexose monophosphate and Entner-Doudoroff pathway is inhibited by reduced NAD as well as by ATP. In contrast to the hydrogenomonads, however, no allostery could be observed. Whether or not this allostery has something to do with the fact that both pathways are functioning in *Thiobacillus ferrooxidans* and only the Entner-Doudoroff pathway in *Hydrogenomonas* spp. is unknown. The action of ATP on glucose-6-phosphate dehydrogenase can certainly be considered as a product inhibition. It is also not known as yet what regulates autotrophic and heterotrophic growth. As glucose suppresses iron oxidation and as hydrogen is not involved in autotrophic growth, the regulatory mechanism must be different from the one observed in *Hydrogenomonas.* It is possible, however, that glucose-6-phosphate dehydrogenase, and possibly isocitrate dehydrogenase, may play a vital role.

The electron transport under heterotrophic conditions is very simple:

$$NAD \rightarrow cyt.c \rightarrow cyt.a \rightarrow O_2$$

and generates all the necessary ATP.

Other Iron Bacteria

A number of bacteria have been described in the literature that can also oxidize ferrous and even manganous compounds (36) and that deposit these compounds in the form of a colloidal sheath. However, the taxonomy of these organisms is very complex and under dispute at this moment. A number of species have been found to be cultural variants of two species, *Sphaerotilus discophorus* (*Leptothrix discophorus*) and *Sphaerotilus natans*, which is reflected in their almost identical DNA base ratio compositions, 69.5 ± 0.5% and 70.0 ± 0.5% guanine plus cytosine (155), respectively. These organisms are also able to accumulate poly-β-hydroxybutyric acid on addition of glucose to the medium (231). The rate of formation of this polymer is stimulated by manganese and magnesium ions. The requirement for calcium has been assigned to sheath formation (63).

A very interesting genus is *Gallionella*, which deposits ferric hydroxide on the surface of a stalk. The absence of pure cultures of species of such genera as *Gallionella*, *Siderocapsa*, *Siderophaera*, *Siderobacter*, *Sideromonas*, and *Ferribacterium* has delayed biochemical investigation of these groups. For a review of iron bacteria in general, see Schlegel (205).

The Sulfur-Oxidizing Bacteria

Members of the genus *Thiobacillus*, generally called the "nonphotosynthetic sulfur bacteria," are known to oxidize thiosulfate to sulfate via various pathways. Their members can be autotrophic or heterotrophic. Rittenberg (196) catagorized *Thiobacillus thiooxidans*, *T. thioparus*, and *T. neapolitanus* under the obligate chemolithotrophs and the species *Thiobacillus intermedius* and *T. novellus* under the mixotrophs. Only one species, *Thiobacillus perometabolis*, is placed under the chemolithotrophic heterotrophs. This great versatility makes it difficult to map a general metabolic pathway for these microorganisms, particularly as the autotrophs oxidize not only thiosulfate but a variety of other sulfur compounds (see Table 6.1). The permeability problem, discussed in the section on iron-oxidizing bacteria (see p. 335), is also an important factor in the metabolism of *Thiobacillus* species.

The following four general pathways are recognized at the present moment by taxonomists (175):

1. Thiosulfate is oxidized first to the tetrathionate, which in turn is further oxidized to sulfate. However, the organisms that apparently perform this reaction are not able to oxidize tetrathionate alone: *Thiobacillus concretivorus* and *Thiobacillus thiooxidans*. *Thiobacillus neapolitanus* is the only bacterium in this group that oxidizes tetrathionate:

$$6\,Na_2S_2O_3 + 5\,O_2 \rightarrow 4\,Na_2SO_4 + 2\,Na_2S_4O_6$$

$$2\,Na_2S_4O_6 + 6\,H_2O + 7\,O_2 \rightarrow 2\,Na_2SO_4 + 6\,H_2SO_4$$

TABLE 6.1
The Sulfur Bacteria

	$S_2O_3^{2-}$ oxidized to:					Oxidation of:			pH (1% $S_2O_3^{2-}$)		Strict auto-troph	Utilizes		
Thiobacillus	$S_4O_6^{2-}$	$S_5O_6^{2-}$	$S_3O_6^{2-}$	SO_4^{2-}	S^0	$S_4O_6^{2-}$	S^0	H_2S	Initial	Final		NH_3	NO_3^-	Motility
concretivorus	+			+	+	−	+ (SO_4^{2-})	+ SO_4^{2-}	4.4	2.3	+	+	+	+
thiooxidans	+			+	+	−	+ SO_4^{2-}	+ SO_4^{2-}	4.4	2.2	+	+	−	+
thioparus	−			+	+	−	+ SO_4^{2-}	−	6.6	4.5	+	+	+	±
X	+	+		+	+	+	+ SO_4^{2-}	+ SO_4^{2-}	6.6	3.3 ↓ 3.0	+	+	+ +NO_2	−
novellus	−	−		+		−	−	−	7.8	5.8	−			−
M_{20}, M_{77}, M_{79}	+	+	+	+	+	−	+ SO_4^{2-}	−	6.6	7.8 → 7.0	−			
T (Trautwein)	+	+	+	+		−	−	+ $S_2O_3^{2-}$ SO_4^{2-}	6.6	Slight rise then slight fall	−			
B (Waksman)	+	+	+	+		−	−	−	6.6	Slight rise then slight fall				
K (Trautwein)	+	+	+	+		−	−	−						
ferrooxidans	−			+	+		Poor		5.8	2.0	+	+	+	+
caprolytiens				+	+		+ SO_4^{2-}		7.6	6.1	−			+
denitrificans				+			+ (SO_4^{2-})	+ SO_4^{2-}			?			+

2. *Thiobacillus thioparus* oxidizes thiosulfate first to sulfur, which in turn is oxidized to sulfate:

$$5\,Na_2S_2O_3 + H_2O + 4\,O_2 \rightarrow 5\,Na_2SO_4 + H_2SO_4 + 4\,S$$

$$2S + 3O_2 + 2H_2O \rightarrow 2H_2SO_4$$

3. The facultative autotroph *Thiobacillus novellus* is able to oxidize thiosulfate directly to sulfate (18):

$$Na_2S_2O_3 + 2O_2 + H_2O \rightarrow Na_2SO_4 + H_2SO_4$$

4. Some facultative autotrophs are able to perform this reaction, which is coupled with a rise in pH that is sometimes followed by a return to the original value:

$$2\,Na_2S_2O_3 + H_2O + \tfrac{1}{2}O_2 \rightarrow Na_2S_4O_6 + 2\,NaOH$$

Although some of the thiobacilli are able to form elemental sulfur, they never seem to store it. Instead, they oxidize it further or excrete it (226).

The early work with *Thiobacillus* therefore revealed thiosulfate as playing a key role in sulfur metabolism. Studies on the reduction of elemental sulfur S^0 to S^{2-} with a reduced glutathione requirement (235, 255, 269) suggested a possible permeability barrier. This function, although probably non-enzymatic, appears very similar to the oxidation of iron by *Ferrobacillus*, which requires sulfate for its electron transport into the cell. A path of sulfur oxidation was therefore postulated (269) as presented in Fig. 6.11.

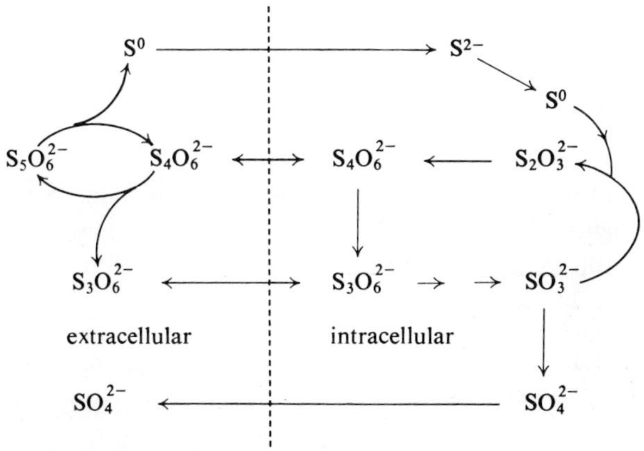
S0
S2−
S0
S5O6 2−
S4O6 2−
S4O6 2−
S2O3 2−
S3O6 2−
S3O6 2−
SO3 2−
extracellular
intracellular
SO4 2−
SO4 2−

Fig. 6.11. Sulfur oxidation by thiobacilli (reprinted with permission of the American Society of Microbiology).

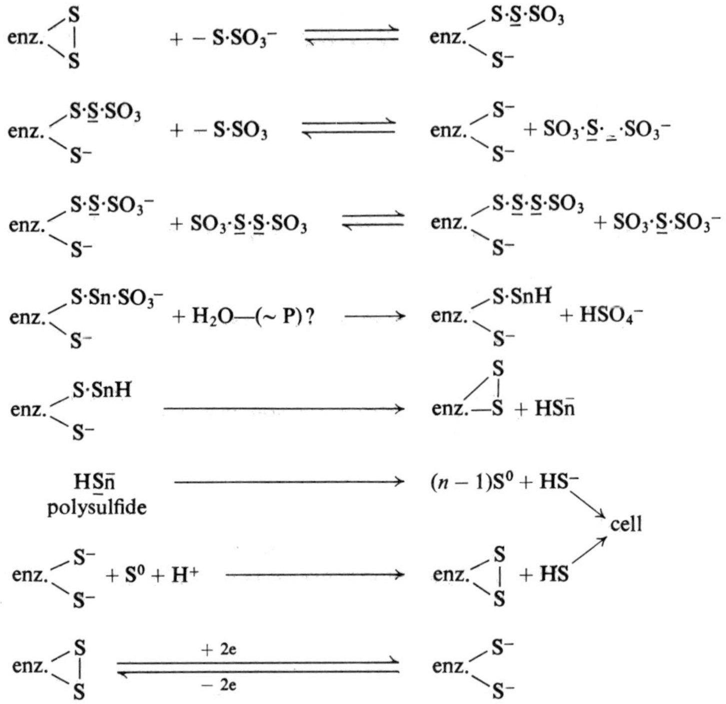

$$\text{enz.}\diagup_{S}^{S} + {-}S\cdot SO_3^- \rightleftarrows \text{enz.}\diagup_{S-}^{S\cdot S\cdot SO_3}$$

$$\text{enz.}\diagup_{S-}^{S\cdot \underline{S}\cdot SO_3} + {-}S\cdot SO_3 \rightleftarrows \text{enz.}\diagup_{S-}^{S-} + SO_3\cdot \underline{S}\cdot\underline{\ }\cdot SO_3^-$$

$$\text{enz.}\diagup_{S-}^{S\cdot \underline{S}\cdot SO_3^-} + SO_3\cdot \underline{S}\cdot\underline{S}\cdot SO_3 \rightleftarrows \text{enz.}\diagup_{S-}^{S\cdot \underline{S}\cdot \underline{S}\cdot SO_3} + SO_3\cdot \underline{S}\cdot SO_3^-$$

$$\text{enz.}\diagup_{S-}^{S\cdot Sn\cdot SO_3^-} + H_2O\text{---}(\sim P)? \longrightarrow \text{enz.}\diagup_{S-}^{S\cdot SnH} + HSO_4^-$$

$$\text{enz.}\diagup_{S-}^{S\cdot SnH} \longrightarrow \text{enz.---}S\diagup_{|}^{S} + HS\bar{n}$$

$$\underset{\text{polysulfide}}{HS\bar{n}} \longrightarrow (n-1)S^0 + HS^-$$

$$\text{enz.}\diagup_{S-}^{S-} + S^0 + H^+ \longrightarrow \text{enz.}\diagup_{S}^{S}{\Big|} + HS \searrow_{\text{cell}}\nearrow$$

$$\text{enz.}\diagup_{S}^{S}{\Big|} \underset{-2e}{\overset{+2e}{\rightleftarrows}} \text{enz.}\diagup_{S-}^{S-}$$

Fig. 6.12. Thiosulfate oxidation in relation to the problem of transporting the sulfur molecule across the cell membrane (reprinted with permission of *Annual Reviews*, Inc.).

The extracellular circular reactions are thiosulfate dependent. The discovery (220) that only the outer S atom of thiosulfate is metabolized to form SO_3^-, does not agree with the postulated scheme. It was therefore suggested that only S^{2-} can enter the cell, whereas all other reactions occur on the cell surface (137, 138), as presented in Fig. 6.12. This mechanism would also explain the requirement for reduced glutathione in sulfur oxidation and would be very similar to, although probably more complex than, the requirement for reduced glutathione in the iron bacteria. If these oxidations occur on the cell surface, the reactions within the cell would be reduced to that of oxidation of S^{2-} to sulfate.

This mechanism corresponds with the findings of tracer element work (112) on *Thiobacillus* strain C (Fig. 6.13); [^{35}S]thiosulfate was completely oxidized to sulfate, and [^{35}S]sulfate was formed more rapidly from $^{35}S\cdot SO_3^{2-}$ as substrate than with $S\cdot ^{35}SO_3^{2-}$. The elemental sulfur produced could then penetrate the cell membrane.

This permeability barrier theory has received more support (256, 257)

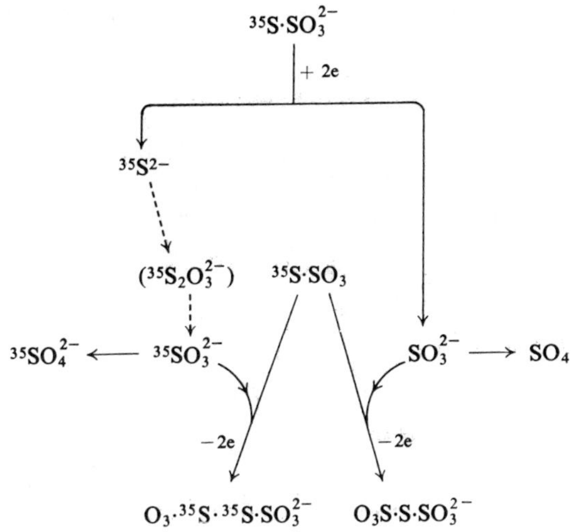

$^{35}S{\cdot}SO_3^{2-}$
+ 2e
$^{35}S^{2-}$
$(^{35}S_2O_3^{2-})$
$^{35}S{\cdot}SO_3$
$^{35}SO_4^{2-}$
$^{35}SO_3^{2-}$
SO_3^{2-}
SO_4
−2e
−2e
$O_3{\cdot}^{35}S{\cdot}^{35}S{\cdot}SO_3^{2-}$
$O_3S{\cdot}S{\cdot}SO_3^{2-}$

Fig. 6.13. Oxidation of [^{35}S] thiosulfate by *Thiobacillus* strain C (112) (reprinted with permission of *The Biochemical Journal*, London).

since it was found that SO_3^{2-} inhibits thiosulfate oxidation in cell-free extracts. Work with thiol-binding reagents indicated the existence of thiol groups on the membrane. Sulfite reacts readily with disulfide groups to form sulphenyl sulfites. It is therefore quite possible that SO_3^{2-} can displace $S_2O_3^{2-}$ from a sulphenyl thiosulfate and thus inhibit thiosulfate oxidation. It is therefore assumed that $S_2O_3^{2-}$ is carried into the cell with its S—S bond intact. The competition between $S_2O_3^{2-}$ and SO_3^{2-} for the binding site on the membrane (254) therefore occurs only outside the bacterial cell. However, this opens the question of S^0 oxidation to $S_2O_3^{2-}$, which was thought to occur with the help of reduced glutathione on the cell membrane as it was also shown that $S_2O_3^{2-}$ is an intermediate of S^{2-} formation. Adair (1) has provided evidence that the oxidation of S^0 to $S_2O_3^{2-}$, which may involve a glutathione polysulfide as intermediate (237), occurs only in the presence of reduced glutathione and may be linked to the cell wall–membrane complex.

The initial product of sulfur oxidation by *Thiobacillus thioparus* and *Thiobacillus thiooxidans* was identified as sulfite (238). The enzyme catalyzing this reaction (sulfur-oxidizing enzyme) has been purified and proposed to be an iron-containing oxygenase that requires reduced glutathione as cofactor

$$S + O_2 + H_2O \rightarrow H_2SO_3$$

whereas thiosulfate is formed through a secondary, nonenzymatic reaction

$$H_2SO_3 + S \rightarrow H_2S_2O_3$$

This purified sulfur-oxidizing enzyme is a nonheme iron oxygenase that possibly contains labile sulfide (EC 1.13.11.18). If it is assumed that the initial scission of thiosulfate by reduced glutathione leads to sulfur and sulfite instead of sulfide and sulfite, then this oxygenase could play an important role in the oxidation of the other sulfur atom of thiosulfate:

$$\begin{aligned} S \cdot SO_3^{2-} + GS^- &\rightarrow GSS^- + SO_3^{2-} \\ GSS^- + O_2 + H_2O &\rightarrow GS^- + SO_3^{2-} + 2\,H^+ \\ 2\,SO_3^{2-} + 2\,H_2O &\rightarrow 2\,SO_4^{2-} + 4e + 4\,H^+ \\ 4e + O_2 + 4\,H^+ &\rightarrow 2\,H_2O \\ \hline S \cdot SO_3^{2-} + 2\,O_2 + H_2O &\rightarrow 2\,SO_4^{2-} + 2\,H^+ \end{aligned}$$

In *Thiobacillus neapolitanus* a soluble, free system also catalyzed the oxidation of elemental sulfur to sulfate with a concomitant uptake of oxygen (244). This system did not require reduced glutathione. The inhibitory effect of sodium azide indicated that this oxidation could be linked to a heavy-metal pathway of electron transport to oxygen, which could involve cytochromes. It is therefore not clear whether the sulfur oxidation requires an oxygenase, which would not be connected to oxidative phosphorylation, or a dehydrogenase-type enzyme, which would be connected to oxidative phosphorylation, or even the help of wetting agents (202) excreted by the organism (229). This last would explain the attachment of *Thiobacillus thiooxidans* to the sulfur crystals (201) for its oxidation. It is also not clear whether or not the wetting agents replace the requirements for reduced glutathione. The permeability barrier theory requires that elemental sulfur (S^0) be converted to $S_2O_3^{2-}$ with the help of reduced glutathione probably via a glutathione polysulfide, wherefore reduced glutathione, because of its disulfide bond, plays an important role in the transport of thiosulfate across the membrane.

Within the cell, thiosulfate can undergo a great variety of oxidations. It can form polythionates and S^{2-} but the key intermediate to sulfate seems to be sulfite (Fig. 6.14).

The initial attack on thiosulfate is the reductive cleavage to sulfite and sulfide catalyzed by the enzyme thiosulfate sulfurtransferase, which may be thiosulfate:cyanide sulfurtransferase (EC 2.8.1.1). The involvement of this enzyme would mean that thiosulfate is first reduced in a manner

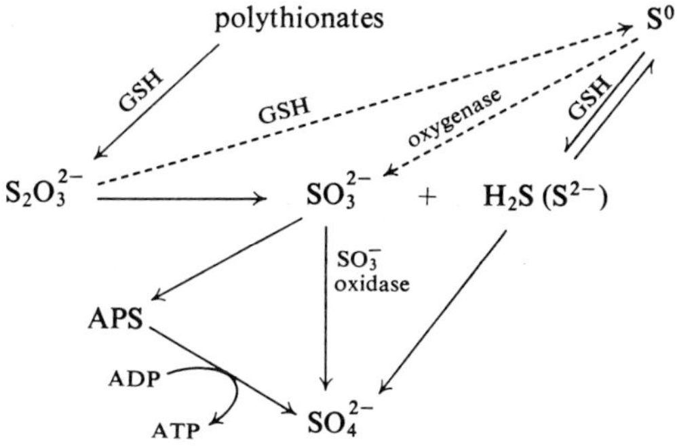
polythionates
S0
GSH
GSH
oxygenase
GSH
S2O3 2−
SO3 2−
+
H2S (S2−)
SO3−
oxidase
APS
ADP
ATP
SO4 2−

Fig. 6.14. Thiosulfate oxidation by thiobacilli.

similar to that in the system described for *Thiobacillus denitrificans* in anaerobic respiration. There exists a number of reports, however, that claim that an oxidation of thiosulfate does occur, which would result in the formation of tetrathionate and sulfate as an intermediate. Whether thiosulfate is oxidized or reduced appears to depend on the pathway involved. Let us first consider the pathway in which thiosulfate is reduced first to sulfite and subsequently oxidized to sulfate:

$$4H^+ + 4e + 2S_2O_3^{2-} \rightarrow 2SO_3^{2-} + 2H_2S \quad (a)$$

$$2H_2S + O_2 \rightarrow 2S^0 + 2H_2O \quad (b)$$

$$2SO_3^{2-} + 2AMP \rightleftharpoons 2APS + 4e \quad (c)$$

$$2APS + 2P_i \rightleftharpoons 2ADP + 2SO_4^{2-} \quad (d)$$

$$2ADP \rightleftharpoons AMP + ATP \quad (e)$$

$$2S_2O_3^{2-} + AMP + O_2 + 2P_i + 4H^+ \rightarrow 2S^0 + 2SO_4^{2-} + ATP + H_2O$$

After the reduction of thiosulfate to sulfite and hydrogen sulfide, a sulfide oxidase oxidizes H_2S to elemental sulfur (162–164). Sulfite combines with AMP in a reversible reaction catalyzed by adenylylsulfate reductase (EC 1.8.99.2) and forms APS. In reaction (e), the enzyme ADP:sulfate adenylyltransferase (EC 2.7.7.5) catalyzes the formation of sulfate and ADP before ATP is produced by the action of ATP:AMP phosphotransferase (EC 2.7.4.3), or adenylate kinase. In the presence of reduced glutathione reaction (b) becomes obsolete, for thiosulfate is always reductively cleaved to elemental sulfur and sulfite. The oxidation of sulfite to sulfate produces a high-energy sulfate at the substrate level and 1 mole ATP is formed per mole sulfate formed. Reduced glutathione can also be responsible for the reduction of elemental sulfur to H_2S before further oxidation to sulfate takes place. *Thiobacillus* strain C appears to possess an iron-containing oxygenase (EC 1.13.11.18) (238) whereby elemental sulfur is oxidized to sulfite.

The second pathway for sulfite oxidation is the direct conversion of sulfite by a sulfite oxidase (EC 1.8.3.1) (52, 89, 164). In *Thiobacillus neapolitanus* this enzyme exhibits an allosteric interaction with AMP. The AMP-dependent rate stimulation is prevented by the addition of reduced glutathione (89). In *Thiobacillus concretivorus* it was observed (164) that cytochrome c is partially reduced by sulfite, which suggests not only that this oxidation is connected with a respiratory chain system, but that the electrons must enter this system prior to cytochrome c. This direct oxidation step is certainly connected with oxidative phosphorylation

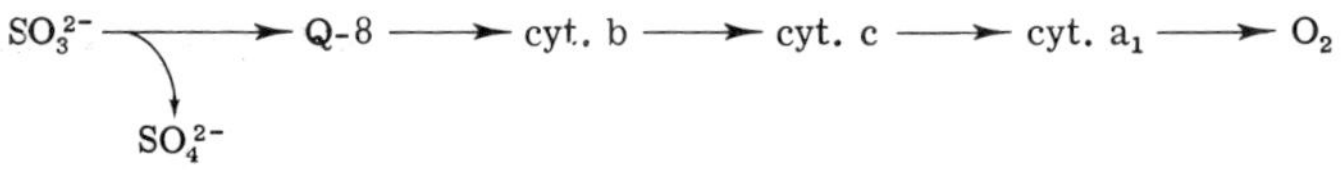
SO_3^{2-}
Q-8
cyt. b
cyt. c
cyt. a_1
O_2
SO_4^{2-}

All those thiobacilli which possess the ability to oxidize sulfite to sulfate and sulfide to elemental sulfur by an oxidase catalyzed reaction therefore possess an electron transport system coupled with oxidative phosphorylation, whereby the electrons enter the system at the flavin or ubiquinone level.

Another pathway of thiosulfate oxidation has been described (219) wherein 2 moles of thiosulfate combine to tetrathionate, which in turn is oxidized to trithionate and finally to thiosulfate, releasing 2 moles of sulfate during its cyclic mechanism (reprinted with permission of the authors and the publishers of *Can. J. Microbiology*):

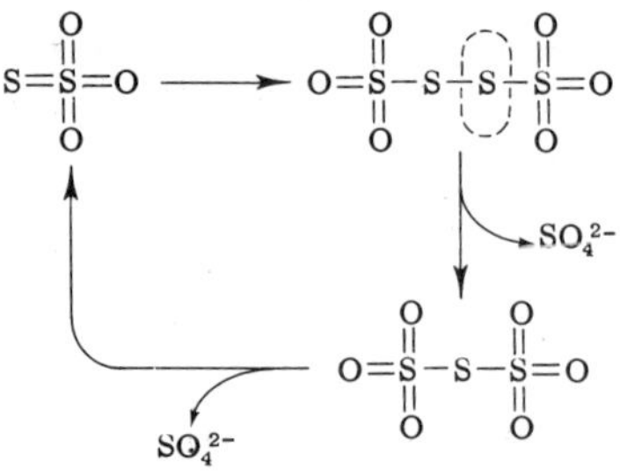
SO_4^{2-}
SO_4^{2-}

This pathway would explain other observations (145, 251, 252) on the formation of polythionates as well as the hypothesis that the inner sulfur atom is being used for the formation of sulfate. This pathway could start with the oxidation of elemental sulfur to thiosulfate, which occurs in the presence of reduced glutathione. The cyclic mechanism would also be very similar to the reaction sequence described by London and Rittenberg (145):

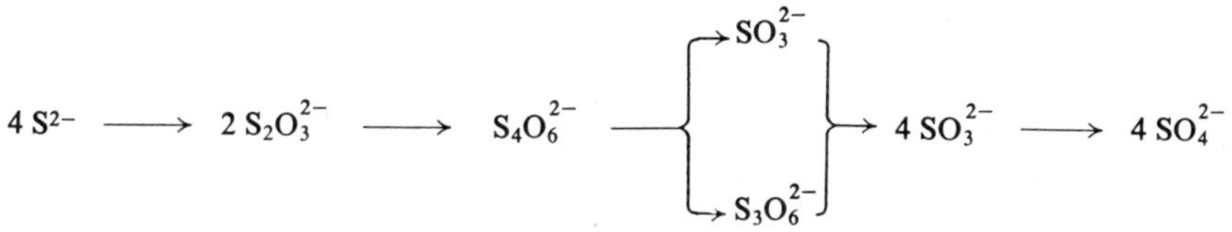
$4\,S^{2-}$
$2\,S_2O_3^{2-}$
$S_4O_6^{2-}$
SO_3^{2-}
$S_3O_6^{2-}$
$4\,SO_3^{2-}$
$4\,SO_4^{2-}$

with the exception that in the latter scheme sulfite is produced, which is not the case in the first thiosulfate oxidation system. It also would agree with the observations that under low aeration tetrathionate accumulates because of incomplete thiosulfate oxidation (253) and that thiosulfate can function as an intermediate in the oxidation of S^0 by *Thiobacillus* X (238).

Tetrathionate cannot function as a substrate for ATP formation unless reduced glutathione is present (171, 172), when it can be rapidly reduced nonenzymatically to thiosulfate. At high cell concentrations, however, tetrathionate oxidation could take place after the addition of a trace amount of thiosulfate in *Thiobacillus* X and *Thiobacillus thioparus* (250). This oxidation did not take place at low cell concentration and was also inhibited by 100% oxygen. Decreased oxygen concentrations allowed tetrathionate oxidation to proceed even without the presence of thiosulfate. Tetrathionate can be metabolized best anaerobically. It is therefore suggested that under aerobic conditions tetrathionate can be oxidized depending on age, strain, and cell concentration.

If the whole sulfur cycle known in nature is combined, including dissimilatory sulfate reducers (*Desulfovibrio desulfuricans*), assimilatory sulfur reduction, and sulfur oxidation (thiobacilli, mammalian, and plants), the scheme shown in Fig. 6.15 emerges (176, 177). In this general metabolic scheme of the sulfur cycle in nature, reactions 1, 5, and 6 are used by dissimilatory sulfate reducers, 2, 3, and 4 are used by assimilatory sulfate reducers, 7–12 and 14 by thiobacilli, and 13 by most microorganisms. The individual reactions are catalyzed as follows: (1) sulfate adenylyltransferase (EC 2.7.7.5); (2) adenylylsulfate kinase (EC 2.7.1.25); (3) aryl sulfotransferase (EC 2.8.2.1); (4) PAPS-reductase; (5) adenylylsulfate kinase (EC 1.7.1.25); (6) sulfite reductase (EC 1.8.99.1); (7) sulfide oxidase; (8) thiosulfate reductase; (9) adenylylsulfate reductase (EC 1.8.99.2); (10) adenylylsulfatase (EC 3.6.2.1); (11) sulfite oxidase (EC 1.8.3.1); (12)

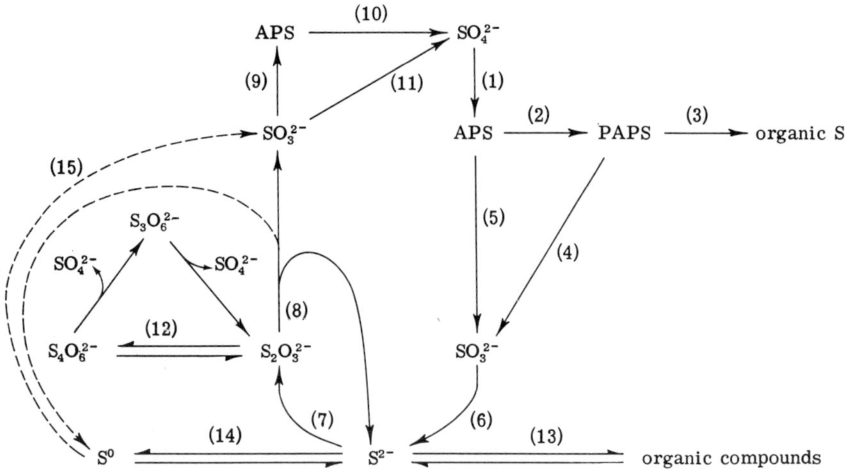
APS
(10)
SO_4^{2-}
(9)
(11)
(1)
SO_3^{2-}
APS
(2)
PAPS
(3)
organic S
(15)
$S_3O_6^{2-}$
(5)
SO_4^{2-}
SO_4^{2-}
(4)
(8)
(12)
$S_4O_6^{2-}$
$S_2O_3^{2-}$
SO_3^{2-}
(7)
(6)
(14)
(13)
S^0
S^{2-}
organic compounds

Fig. 6.15. The oxidation (*Thiobacillus* spp.) and reduction (*Desulfovibrio*) of sulfur compounds in nature.

tetrathionase, GSH requirement; (13) cystathionine β-synthase (EC 4.2.1.22); (14) nonenzymatic, but GSH required; and (15) sulfur dioxygenase (EC 1.13.11.18).

This variety in the reduction and oxidation of sulfur compounds by microorganisms in nature could be of the greatest importance in the geochemistry of sulfide deposits (149).

Autotrophic carbon dioxide fixation, which occurs in the thiobacilli, requires reduced NAD or NADP and ATP (10, 109). The discovery of cytochromes of the c type (144, 248) and of a thiosulfate-oxidizing system (249) connected the $S_2O_3^{2-}$ transfer across the cell membrane, with its subsequent oxidation, together with an electron transfer. The thiosulfate-oxidizing system is a thiosulfate-activating enzyme called "thiosulfate–cytochrome c reductase" (12), which is not involved in an oxidative phosphorylation process (88). However, the transfer of electrons to molecular oxygen is mediated by cytochromes b, c, o, and a cytochrome a type (17). This process is coupled to ATP formation (200) and yields P/O ratios close to 1.0 (15, 200). The energy generated could be utilized to drive the reverse electron flow for the reduction of NAD(P) (11, 14):

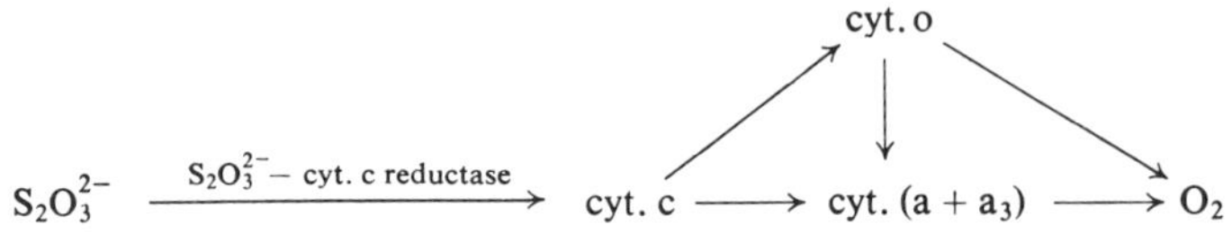
cyt. o
$S_2O_3^{2-}$
$S_2O_3^{2-}$ – cyt. c reductase
cyt. c
cyt. (a + a_3)
O_2

The direct reduction of pyridine nucleotides by inorganic sulfur compounds is not thermodynamically feasible and involves highly endergonic reactions (76):

$$(S^{2-} + HS^{+} + H_2S) + 3\,H_2O + 3\,NAD(P) \rightarrow SO_3^{2-} + 3\,NAD(P)H + 5\,H^{+} \qquad \Delta F = +186 \text{ kcal/mole}$$

$$SO_3^{2-} + H_2O + NAD(P) \rightarrow SO_4^{2-} + NAD(P)H + H^{+} \qquad \Delta F = +33.7 \text{ kcal/mole}$$

The mode of forming the reducing power necessary for carbon dioxide fixation is therefore the same as in other autotrophs (11, 111) (Fig. 6.16). We therefore have an electron pathway toward oxygen and an energy-dependent reverse electron flow system that forms reduced NAD(P) necessary for carbon dioxide fixation.

During the sulfite oxidation there is a coupled phosphorylation with the formation of AMP and ATP. The ATP- and NADH-dependent carbon dioxide assimilation is markedly inhibited by AMP (96) and ADP (160). A molar ratio of 5:1 of AMP or ADP/ATP reduces the specific activity from 0.22 to 0.12, whereas NADH + H^{+} did not stimulate carbon dioxide fixation.

The oxidase-type reactions, e.g., those catalyzed by sulfide oxidase and sulfite oxidase, enter their electrons at the flavin level (164), which would shorten the length of the reversed electron flow system. *Thiobacillus neapolitanus* and *Thiobacillus* strain C are also reported able to introduce their electrons at the NAD level (259). As this phenomenon has been noted in only a single report, and not universally, further work should clarify whether chemolithotrophic autotrophs are able to produce ATP as well as reduced NAD without at least a partly reversed electron flow system. Such a combined electron transport system has been described for *Thiobacillus denitrificans* (179) (Fig. 6.17).

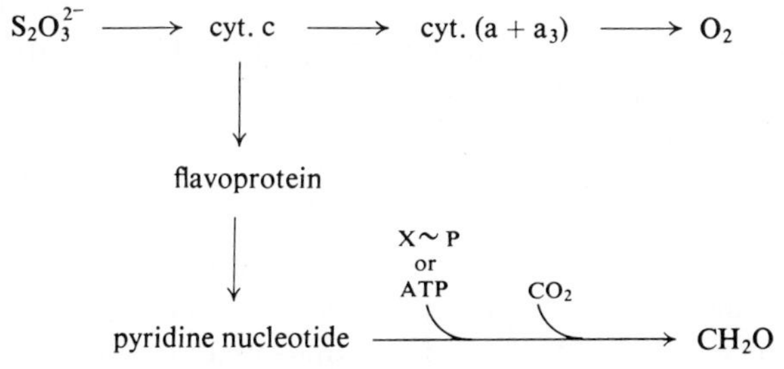
$S_2O_3^{2-}$
cyt. c
cyt. (a + a_3)
O_2
flavoprotein
X~P
or
ATP
CO_2
pyridine nucleotide
CH_2O

Fig. 6.16. Electron transport in the thiosulfate oxidation of thiobacilli (reprinted with permission of the American Society of Microbiology).

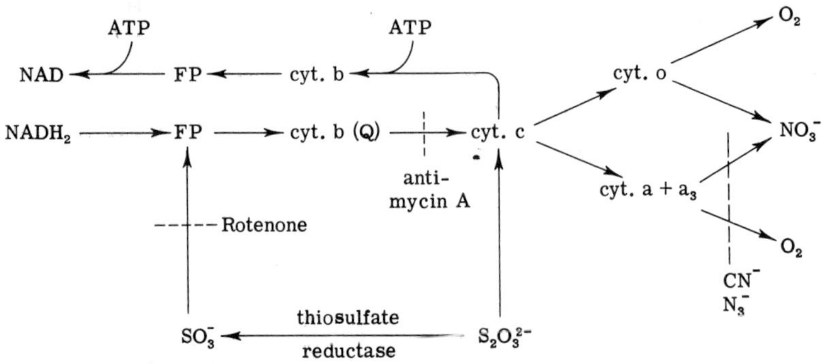
ATP
ATP
NAD
FP
cyt. b
NADH$_2$
FP
cyt. b (Q)
cyt. c
anti-
mycin A
Rotenone
cyt. o
cyt. a + a$_3$
O$_2$
NO$_3^-$
O$_2$
CN$^-$
N$_3^-$
SO$_3^-$
thiosulfate
reductase
S$_2$O$_3^{2-}$

Fig. 6.17. Electron transport system of *Thiobacillus denitrificans* (reprinted with permission of the authors and the publisher of *Arch. Mikrobiologie*).

It was mentioned above that AMP markedly inhibits the ATP- and $NADH_2$-dependent carbon dioxide fixation. AMP also inhibits the carbon dioxide fixation when cell-free extracts are primed with ribose 5-phosphate and ATP. The observations that carbon dioxide fixation with ribulose diphosphate as acceptor and the phosphoenolpyruvate carboxylation are hardly affected by AMP indicate that AMP causes a competitive inhibition in systems in which ATP is involved (15, 75). This competitive inhibition could possibly represent a basic control mechanism. There appears to exist an AMP regulation of carbon dioxide fixation that could possibly represent a basic control mechanism. There is a complete lack of research in the regulation of carbon dioxide fixation in thiobacilli. The only reports are on the biosynthesis of aromatic amino acids (118–120), which is similar to that described for a number of bacteria; that is, the end product acts on the initial DHAP synthetase [7-phospho-2-oxo-3-deoxy-D-*arabino*-heptonate D-erythrose-4-phosphate-lyase (pyruvate-phosphorylating), EC 4.1.2.15]. This could explain the reports on growth inhibition by amino acids (95).

Some species of *Thiobacillus* are able to utilize such organic material as aspartic acid (47), yeast extract, glucose (44), glutamate, acetate (113), and other organic materials (146), but no growth occurred in the absence of thiosulfate. In all cases ribulose-1,5-diphosphate carboxylase synthesis is repressed. The assimilation of these organic compounds (98, 113) provided the evidence for the presence of TCA cycle enzymes in *Thiobacillus thioparus* (57). Thiobacilli appear to require certain organic compounds for their assimilation (110, 245) despite the functioning of the Calvin cycle. Provided that energy is available from thiosulfate oxidation, *Thiobacillus neapolitanus* strain C (114) is able to activate exogenous amino acids, to incorporate them together with carbon dioxide into protein, and also to

TABLE 6.2
Free Energy Efficiency of Chemolithotrophic Bacteria[a]

Organism	Reaction	$-\Delta F^\circ$ 298°C (kcal)	Free energy efficiency (%)
Hydrogenomonas sp.	$H_2 + \frac{1}{2} O_2 = H_2O$	57.4	30
Thiobacillus thiooxidans	$S^0 + \frac{3}{2} O_2 = H_2SO_4$	118	max. 50
Thiobacillus denitrificans	$5\ Na_2S_2O_3 + 8\ KNO_3 + 2\ NaHCO_3 = 6\ Na_2SO_4 + 2\ K_2SO_4 + 4\ N_2 + 2\ CO_2 + H_2O$	893	max. 25
Thiobacillus thiocyanooxidans	$NH_4CNS + 2\ O_2 + 2\ H_2O = (NH_4)_2SO_4 + CO_2$	40	25
Thiobacillus ferrooxidans	$Fe^{2+} = Fe^{3+} + e$	11.3	3
Nitrosomonas sp.	$NH_4^+ + \frac{3}{2} O_2 = NO_2^- + H_2O + 2\ H^+$	66.5	max. 20

[a] From Senez (216). Reprinted with permission of the American Society of Microbiology.

synthesize proline and arginine from glutamate and adenine and guanine from glycine. Biosynthesis in this organism depends on mechanisms almost like those of heterotrophs (18) but requires a chemolithotrophic energy supply (see Table 6.2).

The best studied system is that of acetate incorporation by the chemoautotroph *Thiobacillus neapolitanus* strain C (116). This organism cannot grow on any organic substrate unless CO_2 and energy from thiosulfate oxidation are also available. *Thiobacillus neapolitanus* can use acetate as a major source for lipid and amino synthesis, however, acetate must be activated to acetyl-CoA. The condensation of acetyl-CoA with oxalacetate, formed presumably from pyruvate and the CO_2-fixation cycle, forms citrate. As fluoroacetate inhibits citrate synthase, the inhibition of growth (117) could be explained by the reaction sequence outlined in Fig. 6.18. Growth inhibition by fluoroacetate also indicates that acetate incorporation is necessary for autotrophic growth. A comparative study with a number of organic acids (121) such as formate, acetate, propionate, butyrate, valerate, and pyruvate, gave further evidence for the use of these compounds for the buildup of cell material. There is no catabolism involved (99).

In contrast to the strictly autotrophic thiobacilli are those which can convert to heterotrophic growth and do not require a chemolithotrophic energy supply (14). The membrane of *Thiobacillus novellus* has been shown to undergo structural changes (262) during the organism's change from

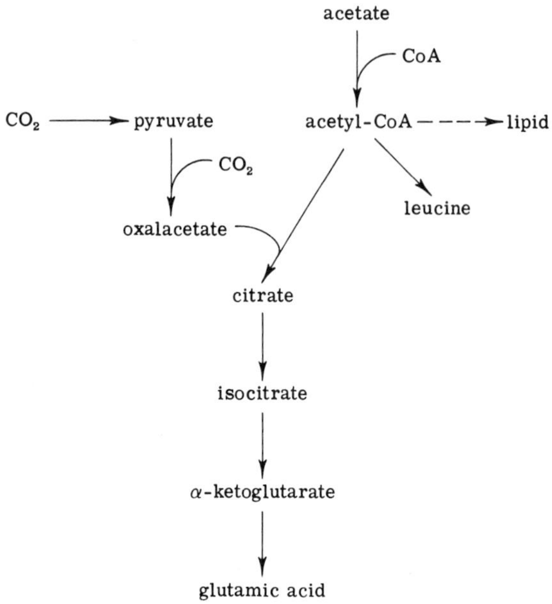
acetate
CoA
CO_2
pyruvate
acetyl-CoA
lipid
CO_2
leucine
oxalacetate
citrate
isocitrate
α-ketoglutarate
glutamic acid

Fig. 6.18. Acetate metabolism of *Thiobacillus neapolitanus* strain C (reprinted with permission of the author and the publisher of *Arch. Mikrobiologie*).

autotrophy to heterotrophy. Under heterotrophic conditions *Thiobacillus novellus* possesses all the enzymes of the TCA and glyoxylate cycles (see Chapter 7) irrespective of the growth substrate (53). The possession of fully operative TCA and glyoxylate cycles may distinctly distinguish the heterotrophs from the autotrophs. Although the enzymes are present, their activities change with alteration of the substrate. This organism has also an NAD- and NADP-specific glutamate dehydrogenase (5, 14). The NAD-specific enzymes exhibit allosteric properties toward ATP and could be assumed to play a regulatory role. The increase in affinity is shown only in the direction of glutamate synthesis. AMP, which altered the kinetic parameter and which is formed during biosynthesis via ATP-utilizing reactions, continuously regulates the equilibrium between glutamate catabolism and glutamate synthesis. *Thiobacillus novellus* returns to autotrophic growth as soon as carbon dioxide and organic acids are in the medium together with thiosulfate. Thiosulfate oxidation appears to be suppressed by carbon sources that can be metabolized both fermentatively and aerobically (140), e.g., glucose and glycerol. The reasons and mechanism of this catabolite repression is not known. It is likely, however, that

α-ketoglutarate inhibits the enzyme citrate synthase (246) and thus converts the TCA cycle from a catabolic to a biosynthetic sequence.

Thiobacillus intermedius is, like *Thiobacillus novellus*, a mixotrophic chemolithotroph, in which glucose metabolism has been investigated more closely (157, 158). As in *T. novellus*, glucose interferes with thiosulfate metabolism. Even though most of the glucose remains unused during growth, cell yields increase in direct proportion to the glucose concentration in the medium (157). It is assumed that *Thiobacillus intermedius* has an inefficient glucose transport system. Glucose itself is metabolized mainly via the Entner-Doudoroff pathway (158). The synthesis of the Entner-Doudoroff pathway enzymes is responsive to glucose concentration, which makes this pathway the major energy generator. The regulation of autotrophic or heterotrophic metabolism occurs at the glucose-6-phosphate dehydrogenase, NADP-linked isocitrate dehydrogenase, and α-ketoglutarate dehydrogenase levels. In an autotrophic environment, thiosulfate oxidation proceeds at a maximum rate, suppresses the synthesis of the glucose metabolizing enzymes, and reduces the TCA cycle activity to a biosynthetic role. In a mixotrophic environment, however, in which the presence of organic substrates should reduce the energy demand for growth, the thiosulfate-oxidizing system is repressed and the glucose metabolizing enzymes are induced. In the absence of thiosulfate, the Entner-Doudoroff pathway and TCA cycle become maximally induced. By controlling glucose-6-phosphate dehydrogenase, the organism can regulate the use of one or the other pathway. This regulation is carried out by ATP, which inhibits this enzyme. This control mechanism has certainly a close resemblance to the one described for *Hydrogenomonas* (see p. 330). It therefore appears that mixotrophic chemolithotrophs possess similar, if not identical, control mechanisms.

Thiobacillus parometabolis is a chemolithotrophic heterotroph (159) and is very similar to *Thiobacillus intermedius* in its glucose metabolism. Glucose is utilized via the Entner-Doudoroff pathway followed by the TCA cycle. One observation, however, may become of great importance for further studies. Whereas the obligate chemolithotrophs possess mainly an NAD-linked isocitrate dehydrogenase, the mixotrophic and obligately heterotrophic thiobacilli possess a predominantly NADP-linked isocitrate dehydrogenase. It therefore appears that the NAD-linked enzyme is involved in biosynthesis and the NADP-linked enzyme in catabolism.

The energy metabolism of the heterotrophic thiobacilli is not as yet clear. Although an electron transport chain whereby the electrons enter at the NAD level could be assumed in the chemolithotrophic heterotrophs, the mixotrophs appear to possess the reversed electron flow rate system (14).

In these cases it is assumed that organic (acids) oxidations yield high-energy cellular intermediates, such as ATP or precursors of it ($\sim$X), which then reverse the normal electron flow in the respiratory chain so that NAD^+ is reduced by cytochrome c as demonstrated in the other chemolithotrophs:

$$\text{succinate} + \text{FP} \rightarrow \text{fumarate} + \text{FPH} + \text{H}^+$$

$$\text{FPH} + \text{H}^+ + 2\ \text{cyt. Fe}^{3+} + (\text{X}) \rightarrow (\sim\text{X}) + \text{FP} + 2\ \text{cyt. Fe}^{2+} + 2\,\text{H}^+$$

$$(\sim\text{X}) + 2\ \text{cyt. Fe}^{2+} + \text{NAD}^+ + 2\,\text{H}^+ \rightarrow (\text{X}) + 2\ \text{cyt. Fe}^{3+} + \text{NADH} + \text{H}^+$$

On the other side of the chain, cytochrome **c** is oxidized by molecular oxygen under the production of a high-energy compound. The ($\sim$X) generated

$$2\ \text{cyt. Fe}^{2+} + 2\,\text{H}^+ + \tfrac{1}{2}\text{O}_2 + (\text{X}) \rightarrow 2\ \text{cyt. Fe}^{3+} + \text{H}_2\text{O} + (\sim\text{X})$$

at the terminal or cytochrome oxidase site can also be used for the reversal of electron transfer to reduce $NAD(P)^+$ (14). Therefore, the microorganisms appear in principle not to change their respiratory chains when inorganic compounds are replaced by organic compounds as electron donors. There are no reports of poly-β-hydroxybutyric acid formation. The thiobacilli do not require this polymer because they are to use organic sulfur compounds or polythionates as storage products.

For the sake of completeness, it should be mentioned that for many years not only autotrophic but also heterotrophic bacteria [which can oxidize inorganic sulfur compounds (80–82)] have been known to exist. Some workers even consider that heterotrophs may play the dominant role in the oxidation of sulfur compounds in soil. *Pseudomonas fluorescens*, *P. aeruginosa*, and *Achromobacter stutzeri* represent some of the identified organisms that can oxidize thiosulfate to tetrathionate (227, 228). However, there is as yet little information available on their requirements and metabolic activities (258).

Autotrophy and Heterotrophy

Most phototrophs and chemolithotrophs are strict autotrophs and perform a total synthesis of cell material from inorganic nutrients, with CO_2 serving as the sole carbon source. There exists, however, a significant number of bacteria in both categories that are also able to grow under heterotrophic conditions. In so-called "borderline" cases, bacteria are able to utilize both their autotrophic and their heterotrophic systems. These observations brought confusion as to what these bacteria should be called. The question, of course, is: "what is an autotroph?" or "what are the criteria for autotrophy?" The definition, as such, is quite clear and states that all microorganisms that can utilize carbon dioxide as their sole carbon source are autotrophs. However, what about those organisms which grow better on the addition of organic substrates? *Desulfovibrio*, for example, was once called an autotroph, but as this bacterium grows better on addition of yeast extract, it is now being referred to as a heterotroph. The composition of autotrophic cells is similar to that of heterotrophs (18). They possess similar enzymes and require similar cofactors and vitamins. They also have a phosphorylation system for ATP formation. Autotrophs, however, have a number of additional enzymes that enables them to oxidize inorganic salts and to reduce carbon dioxide for the synthesis of cell material. Although they may differ with regard to the former, they all fix carbon dioxide via the Calvin cycle (see p. 358). Why, then, are strict autotrophs unable to grow heterotrophically? It will be shown below that they can, on occasion, utilize glucose, but not as an energy source.

Umbreit (261) attempts to answer this question by stating that "the strict autotroph is adapted to life in what is essentially a toxic environment and that it has therefore so changed its permeability properties that all but a few essential materials are excluded." This is a hypothesis and not an answer, and it ignores the question of which came first—the autotrophs or the heterotrophs. It also fails to explain why the heterotrophs have not found the environment toxic. [For more information on potentials and permeability, see Cirillo (54)].

In any search for an explanation of autotrophy it should first be accepted that there may be no single explanation for all autotrophs. Kelly (115, 122) states six factors that may be responsible, either individually or in combination. These factors are as follows.

1. Limited permeability to organic nutrients, which has been mentioned above but which appears to be losing its significance on evidence that will be presented below.

2. Inability to oxidize organic nutrients or to obtain energy from their oxidation, thus making chemo- or photosynthetic energy indispensable. This may be the most attractive explanation, because the membrane of the organism could be permeable to organic nutrients and yet the organism be unable to obtain sufficient energy for its oxidation.

3. A limited ability to synthesize all the compounds necessary for growth from all or most carbon sources other than carbon dioxide. This may be owing to a lack of a number of enzymes (184) that are required for the synthesis of essential metabolites for biosynthetic reactions. This has recently been explained (221) in some blue-green algae and thiobacilli on the basis of a defective metabolism caused by the absence of $NADH_2$

oxidase and α-ketoglutarate dehydrogenase and unusually low levels of malic and succinic dehydrogenase enzymes. Such an explanation, however, is not in harmony with the observation that the electron transport particles from *Nitrobacter* actively oxidize $NADH_2$, involving normal carriers of the electron transport chain. Moreover, this process is coupled to the generation of adenosine triphosphate (15). The missing metabolites, however, could be formed in the presence of carbon dioxide via the Calvin cycle. In spite of all this, the specificity toward an inorganic energy-yielding substrate for autotrophic growth and metabolism remains a mystery the solution to which might perhaps lie in the specific nature of the cellular protein. Perhaps the sequence of amino acids in protein fractions catalyzing inorganic oxidations reflects the unusual characteristics of an autotroph which might otherwise be different from those catalyzing organic oxidations in chemoorganotrophs (17).

4. Some biochemical block of growth on excess external organic nutrients, related in some way to the carbon dioxide-based metabolism.

5. Self-inhibition by products of the metabolism of organic compounds (148). It has been shown above that pyruvic acid (174), a metabolite of glucose breakdown, inhibits the growth of *Thiobacillus* quite strongly.

6. Some special dependence either on an intermediate of the inorganic respiratory processes in the chemoautotroph, or on light in the photoautotroph, for some specific reactions in the cells.

Whatever factor(s) is involved, no conclusion can be drawn at the present state of knowledge and more study of the regulatory mechanisms of the carbon dioxide-based metabolism of autotrophs is necessary.

Various attempts have been made to overcome this problem (178, 196). The major aim was to find a terminology to characterize the appropriate organism and its metabolic versatility. Taking the presence of the carboxydismutase activity (see p. 359) under all conditions of growth as criterion, Peck (178) suggested the three groups obligate autotroph, facultative autotroph, and assimilative autotroph. His definitions were as follows.

Obligate autotroph—an autotroph that can obtain energy only by the oxidation of inorganic substrates and that cannot assimilate simple organic substrates, such as acetate, in the same way as carbon dioxide.

Facultative autotroph—an autotroph equally capable of utilizing either inorganic substrate or organic substrate as the energy source for growth.

Assimilatory autotroph—an autotroph that can obtain energy only by the oxidation of inorganic substrate but that is capable of incorporating a simple organic substrate in the same manner as carbon dioxide.

The extreme difficulties outlined above in determining what is an autotroph lead Rittenberg (196) to suggest a different terminology. He suggests that it would be better to call "obligate autotrophs" and "facultative autotrophs" "obligate chemolithotrophs" and "mixotrophs," respectively. The "assimilative autotroph" would thus become a "chemolithotrophic heterotroph."

Whichever way the trend goes, it is this author's opinion that chemolithotrophy is easier to observe than autotrophy. For example, *Nitrosomonas europaea* and *Thiobacillus thiooxidans* are obligate chemolithotrophs or obligate autotrophs; *Hydrogenomonas*, *Nitrobacter*, and *Thiobacillus intermedius* are mixotrophs or facultative autotrophs; and *Desulfovibrio* would be a typical chemolithotrophic heterotroph. In order to avoid any confusion, both terminologies will be used here, wherever applicable. It is hoped, however, that eventually an international agreement on the terminology will be obtained.

The position of the phototrophs does not clarify the situation. It was mentioned above (Chapter 3) that green bacteria and many purple sulfur bacteria share with plants the use of CO_2 as sole carbon source, but because they are obligate anaerobes and lack the responsible enzyme, they cannot use water as a reductant. They consequently depend on other inorganic reductants. They are also unable to use fermentative mechanisms to obtain the energy needed (15, 22). This group would, in principle, follow the general scheme of plant photosynthesis and thus be autotrophs (33). The other group of purple bacteria, which are categorized as photoorganotrophs, can make use of simple organic compounds as principal carbon sources and sources of reducing power. In such cases, the exogenous carbon source is similar in oxidation state to cell material, and the requirement for an exogenous inorganic reductant disappears.

A strict, overall oxidation–reduction balance must be maintained. When the organic substrate is more oxidized than cell material, this balance is achieved by anaerobic oxidation of part of the substrate to CO_2, and the energy released provides reducing power for the synthesis of cell material from CO_2 and other substrate molecules. If, in contrast, the organic material is more reduced than cell material, this balance is achieved by partial oxidation of the substrate, coupled with reduction and assimilation of carbon dioxide. In other words, there are two exogenous carbon sources, and the assimilation of the organic substrate is mandatorily linked with CO_2 assimilation. Table 6.3 summarizes the differences between green plants, green bacteria, and purple bacteria. The general principle of photochemical ATP generation and reducing power was dealt with in Chapter 3. The actual mechanisms depend entirely on the enzymatic outfit and the environmental conditions required for the individual bacteria.

These considerations of autotrophy and heterotrophy in phototrophs and chemolithotrophs suggest that this problem is very similar, if not identical,

TABLE 6.3

Differences among Green Plants, Green Bacteria, and Purple Bacteria

	Green plants	Green bacteria	Purple bacteria
Source of reducing power	H_2O	H_2S; other inorganic compounds	As green; organic compounds
Photosynthetic O_2 evolution	Yes	No	No
Principal C source	CO_2	CO_2	CO_2; organic compounds
Relation to oxygen when photosynthesizing	Aerobic	Anaerobic	Anaerobic

in anaerobic and aerobic bacteria. Both have autotrophy in common when reducing power can be obtained from inorganic compounds and heterotrophy when this reducing power is provided by an organic compound. In the aerobic bacteria, both systems can be operative together. The respective microorganisms must therefore have some sort of regulatory mechanism that determines autotrophic and/or heterotrophic growth.

The reduction of carbon dioxide to cell material is done in a cycling se-

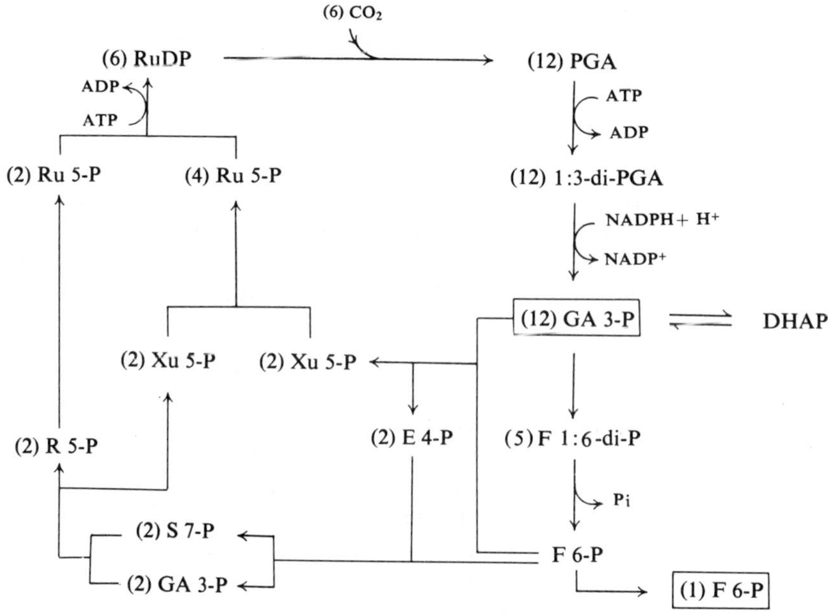
(6) CO_2
(6) RuDP
(12) PGA
ADP
ATP
ATP
ADP
(2) Ru 5-P
(4) Ru 5-P
(12) 1:3-di-PGA
NADPH+ H+
NADP+
(12) GA 3-P
DHAP
(2) Xu 5-P
(2) Xu 5-P
(2) R 5-P
(2) E 4-P
(5) F 1:6-di-P
Pi
(2) S 7-P
(2) GA 3-P
F 6-P
(1) F 6-P

Fig. 6.19. The cyclic mechanism of CO_2 fixation, or Calvin cycle, in autotrophs.

quence (see Fig. 6.19) the sum of which is

$$6\,CO_2 + 18\,ATP + 12\,(NADPH + H^+) \rightarrow 1\,F\,6\text{-}P + 18\,ADP + 12\,NADP^+$$

Carbon dioxide is incorporated with ribulose 1,5-diphosphate by the enzyme ribulosediphosphate carboxylase [3-phospho-D-glycerate carboxy-lyase (dimerizing), EC 4.1.1.39] in the formation of 2 moles of 3-phosphoglycerate. The first energy-requiring step follows as 3-phosphoglycerate is

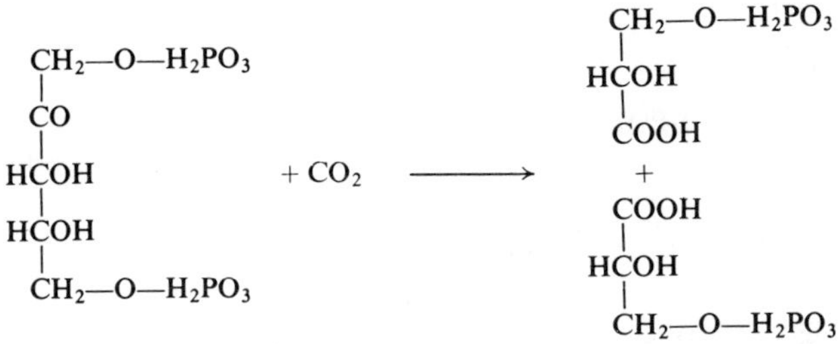

D-ribulose 1,5-diphosphate (RuDP) (2) 3-phosphoglycerate (PGA)

converted to a diphosphoglycerate in a reaction catalyzed by phosphoglycerate kinase (ATP:3-phospho-D-glycerate 1-phosphotransferase, EC

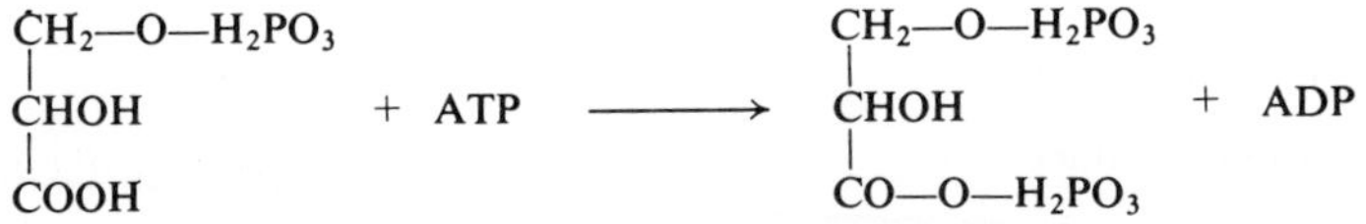

3-phospho-D-glycerate 1,3-diphospho-D-glycerate (1,3-diPGA)

2.7.2.3).

A further oxidation with a NADP-linked dehydrogenase [D-glyceraldehyde-3-phosphate:NADP oxidoreductase (phosphorylating), EC 1.2.1.13] produces glyceraldehyde 3-phosphate, which is held in an equilibrium with dihydroxyacetone phosphate by the triosephosphate isomerase (D-glycer-

1,3-diphospho-D-glycerate D-glyceraldehyde 3-phosphate (GA-3-P)

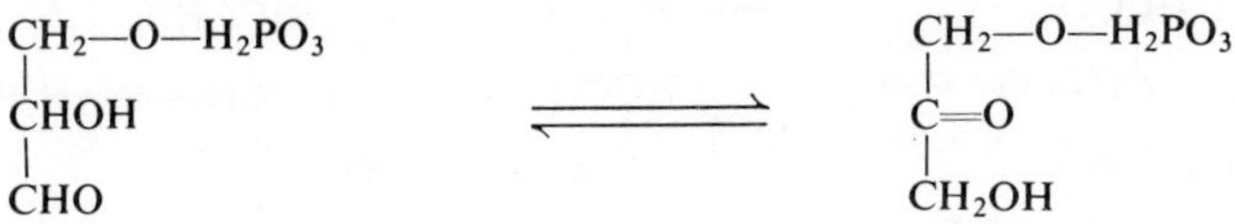

D-glyceraldehyde 3-phosphate dihydroxyacetonephosphate (DHAP)

aldehyde-3-phosphate keto-isomerase, EC 5.3.1.1). Fructose-diphosphate aldolase (EC 4.1.2.13) joins both triose phosphates with the formation of fructose 1,6-diphosphate. The following step is a dephosphorylation step

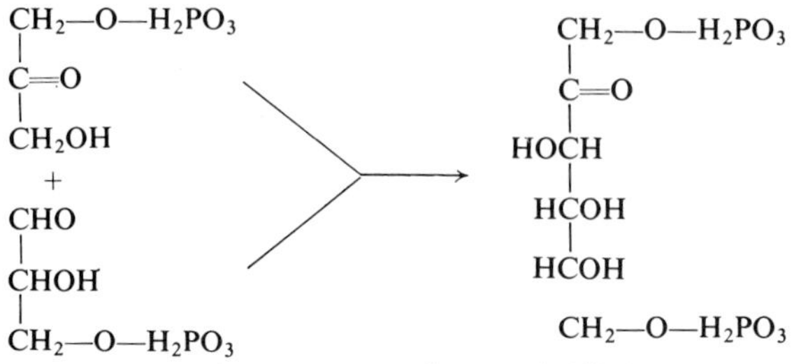

fructose 1,6-diphosphate (F 1,6-diP)

that leads to fructose 6-phosphate and that is catalyzed by hexosediphos-

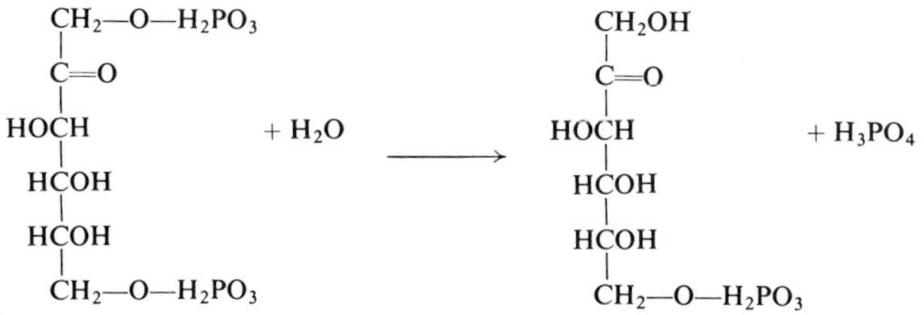

fructose 1,6-diphosphate   D-fructose 6-phosphate (F 6-P)

phatase (D-fructose-1,6-diphosphate 1-phosphohydrolase, EC 3.1.3.11). Fructose 6-phosphate can now be converted in two different ways. It can produce either sucrose or pentoses. The latter is the case in autotrophic microorganisms, for pentoses provide the organisms with vital precursors for the formation of such cell material as RNA and DNA and also close the cyclic reaction sequence by again forming ribulose 1,5-diphosphate.

In order to obtain such pentoses, transketolase (EC 2.2.1.1) first cleaves

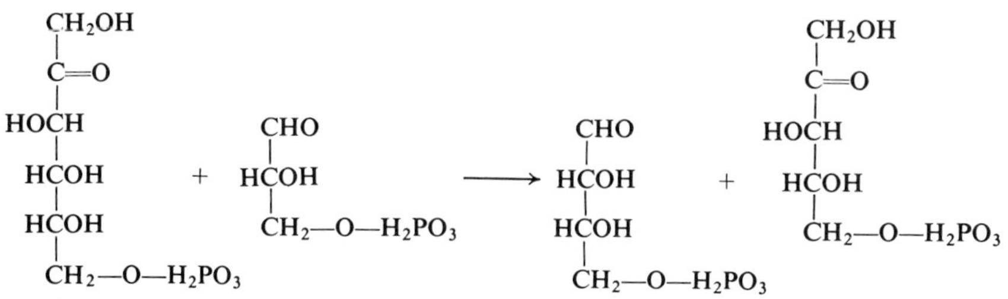

fructose 6-P   GA 3-P   erythrose 4-P (E 4-P)   xylulose 5-P (Xu 5-P)

D-glyceraldehyde 3-phosphate and D-fructose 6-phosphate, with the formation of D-erythrose 4-phosphate and D-xylulose 5-phosphate. The reaction sequence continues as in the hexose monophosphate pathway (see p. 221), with a transaldolase reaction producing sedoheptulose 7-phosphate and glyceraldehyde 3-phosphate. The reaction is followed by a second transketolase (EC 2.2.1.1) cleavage of sedoheptulose 7-phosphate and glycer-

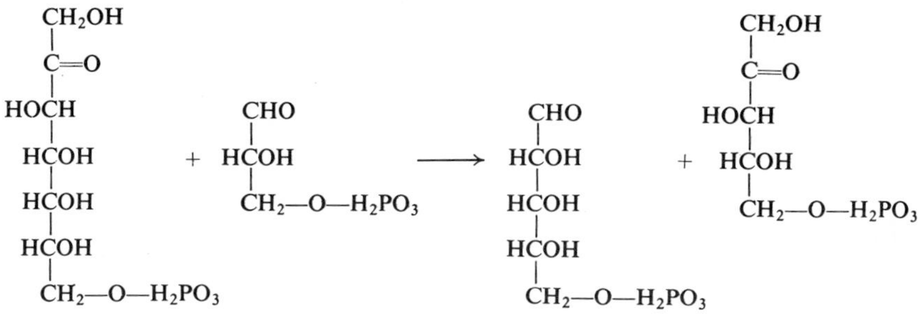

sedoheptulose 7-P GA 3-P ribose 5-P (R 5-P) Xu 5-P

aldehyde 3-phosphate, which forms ribose 5-phosphate and xylulose 5-phosphate. These products are both converted, in two separate reactions, to ribulose 5-phosphate. In the first instance, ribose-5-phosphate ketolisomerase (EC 5.3.1.6) and in the latter ribulose-5-phosphate 3-epimerase

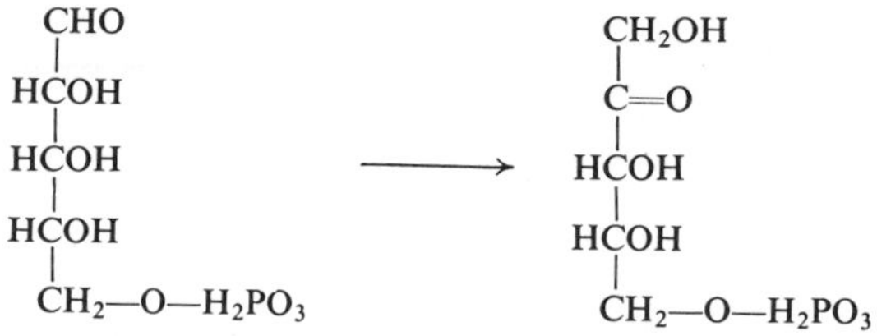

R 5-P ribulose 5-P (Ru 5-P)

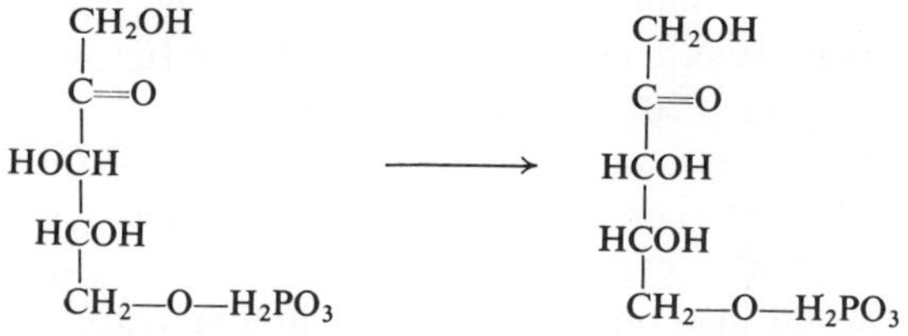

Xu 5-P Ru 5-P

(EC 5.1.3.1) are the responsible enzymes.

The second unique enzyme of the Calvin cycle closes the cyclic mechanism by phosphorylating ribulose 5-phosphate to ribulose 1,5-diphosphate, which is required for autotrophic carbon dioxide fixation. This enzyme is known to be phosphoribulokinase (ATP:D-ribulose-5-phosphate 1-phos-

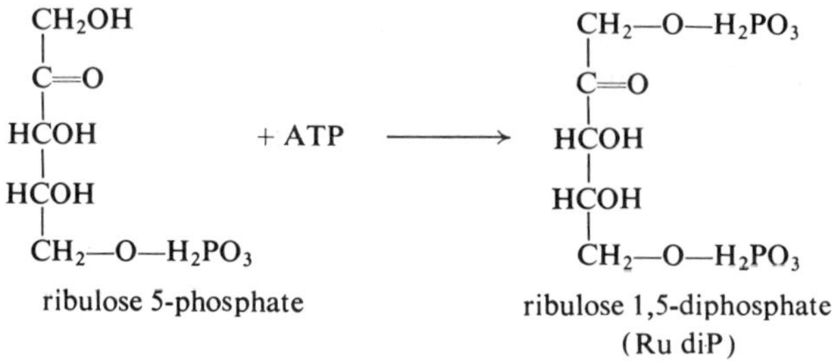

photransferase, EC 2.7.1.19). The presence or absence of this autotrophic carbon dioxide-fixation mechanism of the Calvin cycle is quite commonly determined by testing the activity of ribulosediphosphate carboxylase (EC 4.1.1.39). In the older literature this enzyme is referred to as "carboxydismutase."

With this mechanism working in most autotrophic microorganisms, they are also able to draw on each of the intermediates for further synthetic reactions of all types of cell material. The claim that these organisms are able to form hexoses from carbon dioxide could be misleading, for they do not accumulate hexoses. They draw on the intermediates instead of forming sugars.

Not all autotrophic microorganisms, however, possess this Calvin cycle as described above. There are a number of known microorganisms that follow a slightly altered mechanism, depending on whether they require two carbon dioxide-fixation steps or only one.

Radioautographs of reaction products in *Chromatium* (74) indicated that phosphoglycerate and aspartic acid were the earliest stable products of carbon dioxide fixation. The high initial incorporation of carbon dioxide into aspartic acid was found to be owing to a second carboxylation step at the phosphoenolpyruvate level (74, 147), which in turn was formed from phosphoglycerate (Fig. 6.20). Oxalacetate was produced in the presence of phosphoenolpyruvate carboxylase and aspartate, probably formed via transamination. It has therefore been postulated that *Chromatium* uses this pathway of "double carbon dioxide" fixation as a rapid method for incorporating carbon into organic acids and amino acids to produce new cellular material. It may also explain the function of the TCA cycle in autotrophs. It is not yet resolved whether part of the phosphoglycerate produces ribulose 5-phosphate via glyceraldehyde 3-phosphate or through the intact Calvin cycle.

A similar two-cycle system has also recently been found in sugarcane

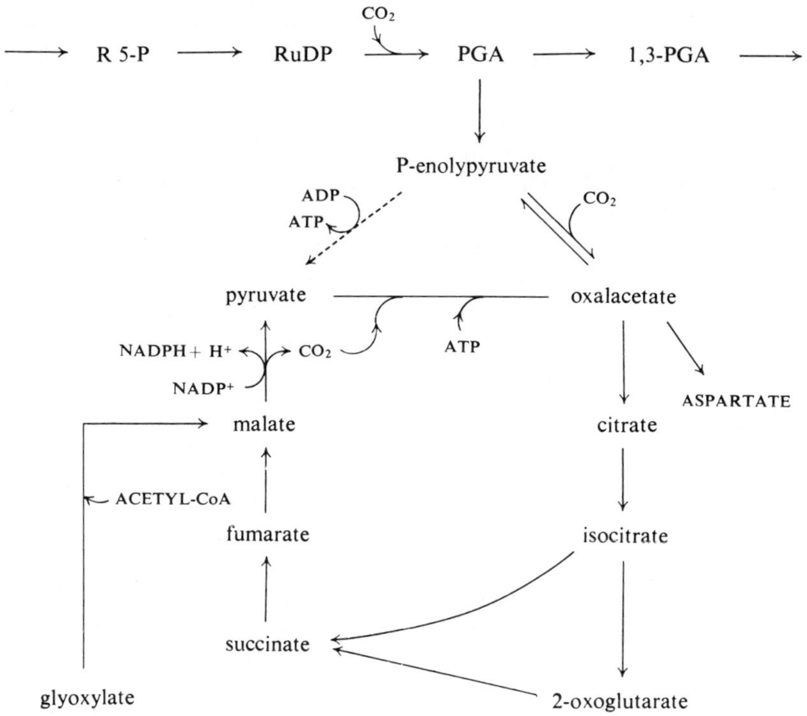
CO2
R 5-P
RuDP
PGA
1,3-PGA
P-enolypyruvate
ADP
ATP
CO2
pyruvate
oxalacetate
NADPH + H+
CO2
ATP
NADP+
ASPARTATE
malate
citrate
ACETYL-CoA
fumarate
isocitrate
succinate
glyoxylate
2-oxoglutarate

Fig. 6.20. "Double carbon dioxide" fixation in *Chromatium*.

leaves (86, 87). In these leaves, the first product of the major carboxylation reaction is either malate or oxalacetate, and aspartate is formed by a side reaction. This carboxylation reaction is considered to proceed via either malate dehydrogenase (EC 1.1.1.40) or phosphoenolpyruvate carboxylase (EC 4.1.1.31). The additional cycle seems to operate in regenerating the carboxyl acceptor for a transcarboxylation reaction.

The presence of fructose-biphosphate aldolase is still in dispute, as its activity was found to be very low (260). This observation coincides with those made in *Hydrogenomonas* H 16 (77), *Anacystis nidulans* (192), and *Rhodopseudomonas spheroides* (240). Experiments with the key enzyme of the autotrophic CO_2-fixation mechanism, ribulosediphosphate carboxylase, indicate that high oxygen tension during the growth of *Hydrogenomonas eutropha* on glucose repressed the synthesis of this enzyme (128). During mixtrophic growth, the addition of organic substrates appears also to affect the enzyme activity (233). The various levels of ribulosediphosphate carboxylase, however, appear to be only indirect effects of the change in growth conditions (38). The hydrogen effect therefore seems to have nothing to do with the activity of this key enzyme.

As a working hypothesis it was then postulated (210) that repression of catabolic enzymes depends on a compound that is in equilibrium with the redox system of the electron transport system and that in its reduced form can act as a repressor of the induced enzyme formation. Therefore, this control system must also be in equilibrium with the hydrogenase, the reduced pyridine nucleotides, or other redox catalysts. Determinations of the $NADH_2/NAD^+$ ratio revealed that under strict aerobic conditions with low hydrogen gas in the atmosphere the NAD system was predominantly in the oxidized state, with only 20% reduced, whereas in the presence of hydrogen the degree of reduction is always higher than 50% and can even reach 80%. The presence of carbon dioxide lowers the degree of reduction even further. This observation indicated a regulatory function of the NAD system.

It was now necessary to find the enzyme that would be affected by the reduced NAD system. Glucose-6-phosphate dehydrogenase exhibits an allosteric behavior toward ATP (38) and was also found to be effected by $NADH + H^+$ (45). The addition of hydrogen therefore reduces the NAD system and by doing so produces excess reduced NAD. This, in turn, inhibits glucose-6-phosphate dehydrogenase, which is a key enzyme in the catabolism of fructose via the Entner-Doudoroff pathway. It therefore appears that this enzyme plays the same role in the "hydrogen effect" as does phosphofructokinase in the "Pasteur effect," although ATP is the main effector in the latter case. However, it is not as yet certain whether or not the regulation is more complicated, for glucose-6-phosphate dehydrogenase is an allosteric enzyme with ATP as the negative allosteric effector. This brings the parallel to the "Pasteur effect" closer. Nothing is known as yet, however, of what role ATP plays in regulating autotrophic and heterotrophic growth. There is a possibility that the inhibition by reduced NAD controls autotrophic and heterotrophic growth, whereas the ATP allostery could have some regulatory control in the catabolism. The inhibition of glucose-6-phosphate dehydrogenase also explains why autotrophic carbon dioxide fixation and heterotrophic growth can occur at the same time, which would be the best explanation of the term "mixotrophy."

Factors common to both phototrophs and chemolithotrophs are that glucose induces glucose-6-phosphate dehydrogenase in *Hydrogenomonas*, to allow heterotrophic utilization via the Entner-Doudoroff pathway (see p. 330), and in *Rhodopseudomonas spheroides* and *Euglena gracilis* (168), to allow heterotrophic utilization via the hexose monophosphate pathway (26, 93, 97, 193, 198, 224). *Thiobacillus ferrooxidans* induces glucose-6-phosphate dehydrogenase, allowing heterotrophic utilization via both the Entner-Doudoroff and the hexose monophosphate pathways. Glucose-6-phosphate dehydrogenase is present as an NAD- and NADP-linked enzyme (243) and exhibits no allostery toward ATP. It can be seen from these investigations that the basic regulatory mechanisms for autotrophy–heterotrophy transition are very similar, yet it is not known why one, or the other, or both heterotrophic pathways are chosen. The role of glucose-6-phosphate dehydrogenase is very similar to the role of phosphofructokinase in the Pasteur effect (see Chapter 8).

References

1. Adair, F. W. (1966). Membrane-associated oxidation by the autotroph *Thiobacillus thiooxidans*. *J. Bacteriol.* **92,** 899.
2. Ahrens, J., and Schlegel, H. G. (1966). Zur Regulation der NAD-abhängigen Hydrogenase-Aktivität. *Arch. Mikrobiol.* **55,** 257.
3. Aleem, M. I. H., and Nason, A. (1959). Nitrite oxidase, a particulate cytochrome electron transport system from *Nitrobacter*. *Biochem. Biophys. Res. Commun.* **1,** 323.
4. Aleem, M. I. H., and Nason, A. (1960). Phosphorylation coupled to nitrite oxidation by particles from the chemoautotroph, *Nitrobacter agilis*. *Proc. Nat. Acad. Sci. U.S.* **46,** 763.
5. Aleem, M. I. H., and Lees, H. (1963). ATP-dependent reduction of NAD by ferrocytochrome c in chemoautotrophic bacteria. *Nature (London)* **200,** 759.
6. Aleem, M. I. H., and Lees, H. (1963). Autotrophic enzyme systems. I. Electron transport systems concerned with hydroxylamine oxidation in *Nitrosomonas*. *Can. J. Biochem. Physiol.* **41,** 763.
7. Aleem, M. I. H. (1965). Path of carbon and assimilatory power in chemosynthetic bacteria. I. *Nitrobacter agilis*. *Biochim. Biophys. Acta* **107,** 14.
8. Aleem, M. I. H. (1965). Thiosulfate oxidation and electron transport in *Thiobacillus novellus*. *J. Bacteriol.* **90,** 95.
9. Aleem, M. I. H., Hock, G. E., and Vanner, J. E. (1965). Water as the source of oxidant and reductant in bacterial chemosynthesis. *Proc. Nat. Acad. Sci. U.S.* **54,** 869.
10. Aleem, M. I. H., and Huang, E. (1965). CO_2 fixation and carboxydismutase in *Thiobacillus novellus*. *Biochem. Biophys. Res. Commun.* **20,** 515.
11. Aleem, M. I. H. (1966). Generation of reducing power in chemosynthesis. III. Energy-linked reduction of pyridine nucleotides in *Thiobacillus novellus*. *J. Bacteriol.* **91,** 729.
12. Aleem, M. I. H. (1966). Electron transfer pathways in *Thiobacillus novellus*. *J. Bacteriol.* **91,** 735.
13. Aleem, M. I. H. (1966). Generation of reducing power in chemosynthesis. II. Energy-linked reduction of pyridine nucleotides in the chemoautotroph *Nitrosomonas europaea*. *Biochim. Biophys. Acta* **113,** 216.
14. Aleem, M. I. H. (1966). Generation of reducing power in chemosynthesis. IV. Energy-linked reduction of pyridine nucleotides by succinate in *Thiobacillus novellus*. *Biochim. Biophys. Acta* **128,** 1.
15. Aleem, M. I. H. (1967). Energy conversions in the chemoautotroph *Nitrobacter agilis*. *Bacteriol. Proc.* p. 112.
16. Aleem, M. I. H. (1970). Oxidation of inorganic nitrogen compounds. *Annu. Rev. Plant Physiol.* **21,** 67.

17. Aleem, M. I. H. (1970). Personal communication.
18. Amemiya, K., and Umbreit, W. W. (1974). Heterotrophic nature of the cell-free protein-synthesizing system from the strict chemolithotroph *Thiobacillus thiooxidans*. *J. Bacteriol.* **117,** 834.
19. Amman, E. C. B., and Reed, L. L. (1967). Metabolism of nitrogen compounds by *Hydrogenomonas eutropha*. I. Utilization of uric acid, allantoin, hippuric acid, and creatinin. *Biochim. Biophys. Acta* **141,** 135.
20. Amman, E. C. B., Reed, L. J., and Durichek, J. E., Jr. (1968). Gas consumption and growth rate of *Hydrogenomonas eutropha* in continuous culture. *Appl. Microbiol.* **16,** 822.
21. Anderson, J. H. (1964). The metabolism of hydroxylamine to nitrite by *Nitrosomonas*. *Biochem. J.* **91,** 8.
22. Anderson, J. H. (1964). Studies on the oxidation of ammonia to hydroxylamine by *Nitrosomonas*. *Biochem. J.* **92,** 16.
23. Anderson, J. H. (1965). Studies on the oxidation of ammonia by *Nitrosomonas*. *Biochem. J.* **95,** 688.
24. Anderson, J. R., Strumeyer, D. J., and Prumer, D. (1968). Purification and properties of peroxidase from *Nitrosomonas europaea*. *J. Bacteriol.* **96,** 93.
25. Anderson, K. J., and Lundgren, D. G. (1969). Enzymatic studies of the iron-oxidizing bacterium, *Ferrobacillus ferrooxidans*, evidence for a glycolytic pathway and Krebs cycle. *Can. J. Microbiol.* **15,** 73.
26. Anderson, L. E., and Fuller, R. C. (1969). Photosynthesis in *Rhodospirillum rubrum*. IV. Isolation and characterization of ribulose-1,5-diphosphate carboxylase. *J. Biol. Chem.* **244,** 3105.
27. Arima, K., and Nose, K. (1968). Studies on bacterial urate:oxygen oxidoreductase. I. Purification and properties of the enzyme. *Biochim. Biophys. Acta* **151,** 54.
28. Atkinson, D. E., and McFadden, B. A. (1954). The biochemistry of *Hydrogenomonas*. I. The hydrogenase of *Hydrogenomonas facilis* in cell-free preparations. *J. Biol. Chem.* **210,** 885.
29. Atkinson, D. E. (1955). The biochemistry of *Hydrogenomonas*. II. The adaptive oxidation of organic substrates. *J. Bacteriol.* **69,** 310.
30. Aubert, J. P., Milhaud, G., and Millet, J. (1957). L'assimilation de l'anhydride carbonique par les bactéries chimiautotrophes. *Ann. Inst. Pasteur, Paris* **92,** 515.
31. Barthe, R. (1962). *Arch. Mikrobiol.* **41,** 313 [as cited by Eberhardt (65)].
32. Barthe, R., and Ordal, E. J. (1965). Nickel-dependent chemolithotrophic growth of two *Hydrogenomonas* strains. *J. Bacteriol.* **89,** 1015.
33. Basham, J. A., and Krause, G. H. (1959). Free energy changes and metabolic regulation in steady-state photosynthetic carbon reduction. *Biochim. Biophys. Acta* **189,** 207.
34. Beck, J. V., and Hafia, F. M. (1964). Effect of phosphate ions and 2,4-dinitrophenol on the activity of intact cells of *Thiobacillus ferrooxidans*. *J. Bacteriol.* **88,** 850.
35. Bergmann, F. H., Towne, J. C., and Burris, R. H. (1958). Assimilation of CO_2 by hydrogen bacteria. *J. Biol. Chem.* **230,** 13.
36. Bisset, K. A., and Grace, J. B. (1954). Iron and manganese oxidizing bacteria. *Symp. Soc. Gen. Microbiol.* **4,** 44.
37. Blackkolb, F., and Schlegel, H. G. (1968). Katabolische Repression und Enzymhemmung durch molekularen Wasserstoff bei *Hydrogenomonas*. *Arch. Mikrobiol.* **62,** 129.

38. Blackkolb, F., and Schlegel, H. G. (1968). Regulation der Glukose-6-phosphat Dehydrogenase aus *Hydrogenomonas* H 16 durch ATP und $NADH_2$. *Arch. Mikrobiol.* **63,** 177.
39. Blaycock, B. A., and Nason, A. (1963). Electron transport systems of the chemoautotroph *Ferrobacillus ferrooxidans*. I. Cytochrome c-containing iron oxidase. *J. Biol. Chem.* **238,** 3453.
40. Bock, E. (1968). pH-Metrische Untersuchungen über den Phosphatstoffwechsel lebender Zellen von *Nitrobacter winogradskyi* Buch. *Arch. Mikrobiol.* **63,** 70.
41. Bongers, L. (1967). Phosphorylation in hydrogen bacteria. *J. Bacteriol.* **93,** 1615.
42. Bongers, L. (1970). Yields of *Hydrogenomonas eutropha* from growth on succinate and fumarate. *J. Bacteriol.* **102,** 598.
43. Bongers, L. (1970). Energy generation and utilization in hydrogen bacteria. *J. Bacteriol.* **104,** 145.
44. Borischewski, R. M., and Umbreit, W. W. (1966). Growth of *Thiobacillus thiooxydans* on glucose. *Arch. Biochem. Biophys.* **116,** 97; *Chem. Abstr.* **65,** 15816 (1966).
45. Bowien, B., Cook, A. M., and Schlegel, H. G. (1974). Evidence for the *in vivo* regulation of glucose-6-phosphate dehydrogenase activity in *Hydrogenomonas eutropha* H 16 from measurements of the intracellular concentrations of metabolic intermediates. *Arch. Mikrobiol.* **97,** 273.
46. Burge, W. D., Malavolta, E., and Delwiche, C. C. (1963). Phosphorylation by extracts of *Nitrosomonas europaea*. *J. Bacteriol.* **85,** 106.
47. Butler, R. G., and Umbreit, W. W. (1966). Absorption and utilization of organic matter by the strict autotroph *Thiobacillus thiooxidans*, with special reference to aspartic acid. *J. Bacteriol.* **91,** 661.
48. Butlin, K. R., and Postgate, J. R. (1954). Microbial transformation of inorganic iron. *Symp. Soc. Gen. Microbiol.* **4,** 294.
49. Butt, W. D., and Lees, H. (1960). The biochemistry of the nitrifying organisms. 6. The effect of oxygen on nitrite oxidation in the presence of different inorganic ions. *Biochem. J.* **76,** 425.
50. Campbell, A. E., Hellebust, J. A., and Watson, S. W. (1966). Reductive pentose phosphate cycle in *Nitrosocystis oceanus*. *J. Bacteriol.* **91,** 1178.
51. Chance, B., and Hollunger, G. (1961). The interaction of energy and electron transfer reactions in mitochondria. I. General properties and nature of the products of succinate-linked reduction of pyridine nucleotide. *J. Biol. Chem.* **236,** 1534.
52. Charles, A. M., and Suzuki, I. (1965). Sulfite oxidase of the facultative autotroph *Thiobacillus novellus*. *Biochem. Biophys. Res. Commun.* **19,** 686.
53. Charles, A. M. (1971). Effect of growth substrate on enzymes of the citric and glyoxylic acid cycles in *Thiobacillus novellus*. *Can. J. Microbiol.* **17,** 617.
54. Cirillo, V. P. (1966). Symposium on bioelectrochemistry of microorganisms. 1. Membrane potentials and permeability. *Bacteriol. Rev.* **30,** 68.
55. Cole, J. S., III, and Aleem, M. I. H. (1970). Oxidative phosphorylation in *Thiobacillus novellus*. *Biochem. Biophys. Res. Commun.* **38,** 736.
56. Cook, D. W., Tischer, R. G., and Brown, C. R. (1967). Carbohydrate metabolism in *Hydrogenomonas eutropha*. *Can. J. Microbiol.* **13,** 701.
57. Cooper, R. C. (1964). Evidence for presence of certain TCA cycle enzymes in *Thiobacillus thioparus*. *J. Bacteriol.* **88,** 624.
58. Crouch, D. J., and Ramsay, H. H. (1962). Oxidation of glucose by *Hydrogenomonas facilis*. *J. Bacteriol.* **84,** 1340.

59. DeCicco, B. T., and Stukus, P. E. (1968). Autotrophic and heterotrophic metabolism of *Hydrogenomonas*. I. Growth yields and patterns under dual substrate conditions. *J. Bacteriol.* **95,** 1469.
60. Delwiche, C. C., and Finstein, M. S. (1965). Carbon and energy sources for the nitrifying autotroph *Nitrobacter*. *J. Bacteriol.* **90,** 102.
61. Dessers, A., Chiang, C., and Laudelout, H. (1970). Calorimetric determination of free energy efficiency in *Nitrobacter winogradskyi*. *J. Gen. Microbiol.* **64,** 71.
62. DeTurk, W. E. (1955). The adaptive formation of urease by washed suspensions of *Pseudomonas aeruginosa*. *J. Bacteriol.* **70,** 187.
63. Dias, F. F., Okrend, H., and Doudero, N. C. (1968). Calcium nutrition of *Sphaerotilus* growing in a continuous-flow apparatus. *Appl. Microbiol.* **16,** 1364.
64. Din, G. A., Suzuki, I., and Lees, H. (1967). Carbon dioxide fixation and phosphoenolpyruvate carboxylase in *Ferrobacillus ferrooxidans*. *Can. J. Microbiol.* **13,** 1413.
65. Dugan, P., and Lundgren, D. G. (1965). Energy supply for the chemoautotroph *Ferrobacillus ferrooxidans*. *J. Bacteriol.* **89,** 825.
66. Eberhardt, W. (1966). Über das Wasserstoff aktivierende System von *Hydrogenomonas* H 16. I. Verteilung der Hydrogenase-Aktivität auf zwei Zellfraktionen. *Arch. Mikrobiol.* **53,** 288.
67. Eigener, V., and Bock, E. (1972). Auf- und Abbau der Polyphosphatfraktion in Zellen von *Nitrobacter winogradskyi* Buch. *Arch. Mikrobiol.* **81,** 367.
68. Fang, F. S., and Burris, R. H. (1968). Cytochrome c in *Hydrogenomonas eutropha*. *J. Bacteriol.* **96,** 298.
69. Faull, K. F., Wallace, W., and Nicholas, D. J. D. (1969). Nitrite oxidase and nitrite reductase in *Nitrobacter agilis*. *Biochem. J.* **113,** 449.
70. Finstein, M. S., and Delwiche, C. C. (1965). Molybdenum as a micronutrient for *Nitrobacter*. *J. Bacteriol.* **89,** 123.
71. Fraser-Smith, E. C. B., Austin, M. A., and Reed, L. L. (1969). Utilization of amino acids as a source of nitrogen by *Hydrogenomonas eutropha*. *J. Bacteriol.* **97,** 457.
72. Frings, W., and Schlegel, H. G. (1971). Zur Synthese von C_4-Dicarbonsäuren aus Pyruvate durch *Hydrogenomonas eutropha* Stamm H 16. *Arch. Mikrobiol.* **79,** 204.
73. Frings, W., and Schlegel, H. G. (1971). Synthese von Phosphoenolpyruvat aus Pyruvat durch Extrakte aus *Hydrogenomonas eutropha* Stamm H 16. *Arch. Mikrobiol.* **79,** 220.
74. Fuller, R. C., Smillie, R. M., Sisler, E. C., and Kornberg, H. L. (1961). Carbon metabolism in *Chromatium*. *J. Biol. Chem.* **236,** 2140.
75. Gale, N. L., and Beck, J. V. (1967). Evidence for the Calvin cycle and hexose monophosphate pathway in *Thiobacillus ferrooxidans*. *J. Bacteriol.* **94,** 1052.
76. Gibbs, M., and Schiff, J. A. (1960). Chemosynthesis: The energy of chemoautotrophic organisms. *Plant Physiol.* **1B,** 27.
77. Gottschalk, G. (1964). Die Biosynthese der Poly-β-hydroxybuttersäure durch Knallgasbakterien. III. Synthese aus CO_2. *Arch. Mikrobiol.* **47,** 236.
78. Gottschalk, G., Eberhardt, U., and Schlegel, H. G. (1964). Verwertung von Fructose durch *Hydrogenomonas* H 16 (1.). *Arch. Mikrobiol.* **48,** 95.
79. Gottschalk, G., and Schlegel, H. G. (1965). Preparation of ^{14}C-D(−)-β-hydroxybutyric acid from $^{14}CO_2$ using "Knallgas" bacteria (*Hydrogenomonas*) *Nature (London)* **205,** 308.
80. Guittonneau, G. (1925). Sur la transformation du soufre en sulfate per voie d'association microbiènne. *C. R. Acad. Sci.* **181,** 261 [as cited by Trudinger (253)].

81. Guittonneau, G. (1927). Sur l'oxydation microbiènne du soufre an cours de l'ammonisation. *C. R. Acad. Sci.* **184,** 45 [as cited by Trudinger (253)].
82. Guittonneau, G., and Keilling, J. (1932). L'évolution et la solubilisation du soufre élementaire dans la terre avable. *Ann. Agron.* (N.S.) **2,** 690 [as cited by Trudinger (253)].
83. Gundersen, K. (1966). The growth and respiration of *Nitrosocystis oceanus* at different partial pressures of oxygen. *J. Gen. Microbiol.* **42,** 387.
84. Gundersen, K. (1968). The formation and utilization of reducing power in aerobic chemoautotrophic bacteria. *Z. Allg. Mikrobiol.* **8,** 445.
85. Harold, F. M. (1966). Inorganic polyphosphates in biology: Structure, metabolism, and function. *Bacteriol. Rev.* **30,** 772.
86. Hatch, M. D., and Slack, R. C. (1966). Photosynthesis by sugarcane leaves. A new carboxylation reaction and the pathway of sugar formation. *Biochem. J.* **101,** 103.
87. Hatch, M. D., Slack, C. R., and Johnson, H. S. (1967). Further studies on a new pathway of photosynthetic CO_2 fixation in sugarcane and its occurrence in other plant species. *Biochem. J.* **102,** 417.
88. Hempfling, W. P., and Vishniac, W. (1963). Oxidative phosphorylation in extracts of *Thiobacillus* X. *Biochem. Z.* **342,** 272; *Chem. Abstr.* **63,** 12020 (1965).
89. Hempfling, W. P., and Trudinger, P. A. (1967). Purification and some properties of sulfite oxidase from *Thiobacillus neapolitanus. Arch. Mikrobiol.* **59,** 149.
90. Hooper, A. B., and Nason, A. (1965). Characterization of hydroxylamine–cytochrome c reductase from the chemoautotrophs *Nitrosomonas europaea* and *Nitrosocystis oceanus. J. Biol. Chem.* **240,** 4044.
91. Hooper, A. B. (1968). A nitrite-reducing enzyme from *Nitrosomonas europaea.* Preliminary characterization with hydroxylamine as electron donor. *Biochim. Biophys. Acta* **162,** 49.
92. Hooper, A. B., and Terry, K. R. (1973). Specific inhibitors of ammonia oxidation in *Nitrosomonas. J. Bacteriol.* **115,** 480.
93. Hurlbert, R. E., and Lascelles, J. (1963). Ribulosediphosphate carboxylase in Thiorhodaceae. *J. Gen. Microbiol.* **33,** 445.
94. Ida, S., and Alexander, M. (1965). Permeability of *Nitrobacter agilis* to organic compounds. *J. Bacteriol.* **90,** 151.
95. Johnson, C. L., and Vishniac, W. (1970). Growth inhibition in *Thiobacillus neapolitanus* by histidine, methionine, phenylalanine, and threonine. *J. Bacteriol.* **104,** 1145.
96. Johnson, E. J., and Peck, H. D., Jr. (1965). Coupling of phosphorylation and CO_2 fixation in extracts of *Thiobacillus thioparus. J. Bacteriol.* **89,** 1041.
97. Johnson, E. J. (1966). Occurrence of the adenosine monophosphate inhibition of carbon dioxide fixation in photosynthetic and chemosynthetic autotrophs. *Arch. Biochem. Biophys.* **114,** 178.
98. Johnson, E. J., and Abraham, S. (1969). Assimilation and metabolism of exogenous organic compounds by the strict autotrophs *Thiobacillus thioparus* and *Thiobacillus neapolitanus. J. Bacteriol.* **97,** 1198.
99. Johnson, E. J., and Abraham, S. (1969). Enzymes of intermediary carbohydrate metabolism in the obligate autotrophs *Thiobacillus thioparus* and *Thiobacillus neapolitanus. J. Bacteriol.* **100,** 962.
100. Joseph, A. A., and Wixom, R. L. (1970). Ammonia incorporation in *Hydrogenomonas eutropha. Biochim. Biophys. Acta* **201,** 295.

101. Kaltwasser, H., and Schlegel, H. G. (1959). Nachweis und quantitative Bestimmung der Polyphosphate in wasserstoff-oxydierenden Bakterien. *Arch. Mikrobiol.* **34,** 76.
102. Kaltwasser, H. (1962). Die Rolle der Polyphosphate im Phosphatstoffwechsel eines Knallgasbakteriums (*Hydrogenomonas* Stamm 20). *Arch. Mikrobiol.* **41,** 282.
103. Kaltwasser, H. (1967). Bildung partikelgebundener Uricase bei *Hydrogenomonas* H 16 und anderen aeroben Bakterien. *Zentralbl. Baktriol., Parasitenk., Infektionskr. Hyg., Abt. 1: Orig.* **205,** 87.
104. Kaltwasser, H. (1968). Induktive Bildung partikelgebundener Uricase bei *Hydrogenomonas* H 16 und anderen aeroben Bakterien. *Arch. Mikrobiol.* **60,** 160.
105. Kaltwasser, H., and Kramer, J. (1968). Verwertung von Cytosin und Uracil durch *Hydrogenomonas facilis* und *Hydrogenomonas* H 16. *Arch. Mikrobiol.* **60,** 172.
106. Kaltwasser, H. (1968). Harnsäureabbau und Biosynthese der Enzyme Uricase, Glyoxylatcarboligase, und Urease bei *Hydrogenomonas* H 16. I. Bildung von Glyoxylatcarboligase und D-Glycerat-3-Dehydrogenase. *Arch. Mikrobiol.* **64,** 71.
107. Kaltwasser, H. (1969). Harnsäureabbau und Biosynthese der Enzyme Uricase, Glyoxylatcarboligase, und Urease bei *Hydrogenomonas* H 16. II. Einfluss von Harnsäure, Fructose, und Stickstoffmangel. *Arch. Mikrobiol.* **65,** 288
108. Kashket, E. R., and Brodie, A. F. (1963). Oxidative phosphorylation in fractionated bacterial systems. X. Different roles for the natural quinones of *Escherichia coli* W in oxidative metabolism. *J. Biol. Chem.* **238,** 2564.
109. Kelly, D. P., and Syrett, P. J. (1964). Inhibition of formation of ATP in *Thiobacillus thioparus* by 2,4-dinitrophenol. *Nature (London)* **202,** 597.
110. Kelly, D. P. (1965). Assimilation of organic compounds by a strictly chemoautotrophic *Thiobacillus*. *J. Gen. Microbiol.* **41,** v.
111. Kelly, D. P., and Syrett, P. J. (1966). Energy coupling during sulfur compound oxidation by *Thiobacillus* sp. strain C. *J. Gen. Microbiol.* **43,** 109.
112. Kelly, D. P., and Syrett, P. J. (1966). [^{35}S]Thiosulfate oxidation by *Thiobacillus* strain C. *Biochem. J.* **98,** 537.
113. Kelly, D. P. (1966). Influence of organic compounds on *Thiobacillus*. *Biochem. J.* **100,** 9P.
114. Kelly, D. P. (1967). Influence of amino acids and organic antimetabolites on growth and biosynthesis of the chemoautotroph *Thiobacillus neapolitanus* strain C. *Arch. Mikrobiol.* **56,** 91.
115. Kelly, D. P. (1967). Problems of the autotrophic microorganisms. *Sci. Progr. (London)* **55,** 35.
116. Kelly, D. P. (1967). The incorporation of acetate by the chemoautotroph *Thiobacillus neapolitanus* strain C. *Arch. Mikrobiol.* **58,** 99.
117. Kelly, D. P. (1968). Fluoroacetate toxicity in *Thiobacillus neapolitanus* and its relevance to the problem of obligate chemoautotrophy. *Arch. Mikrobiol.* **61,** 59.
118. Kelly, D. P. (1969). Regulation of chemoautotrophic metabolism. I. Toxicity of phenylalanine to thiobacilli. *Arch. Mikrobiol.* **69,** 330.
119. Kelly, D. P. (1969). Regulation of chemoautotrophic metabolism. II. Competition between amino acids for incorporation into *Thiobacillus*. *Arch. Mikrobiol.* **69,** 343.
120. Kelly, D. P. (1969). Regulation of chemoautotrophic metabolism. III. DHAP-synthetase in *Thiobacillus neapolitanus*. *Arch. Mikrobiol.* **69,** 360.
121. Kelly, D. P. (1970). Metabolism of organic acids by *Thiobacillus neapolitanus*. *Arch. Mikrobiol.* **73,** 177.
122. Kelly, D. P. (1971). Autotrophy: Concepts of lithotrophic bacteria and their organic metabolism. *Ann. Rev. Microbiol.* **25,** 177.

123. Ketchum, P. A., Sanders, H. K., Gryder, J. W., and Nason, A. (1969). Characterization of cyt. c from *Nitrobacter agilis*. *Biochim. Biophys. Acta* **189,** 360.
124. König, C., Kaltwasser, H., and Schlegel, H. G. (1966). Die Bildung von Urease nach Verbrauch der äusseren *N*-Quellen bei *Hydrogenomonas* H 16. *Arch. Mikrobiol.* **53,** 231.
125. König, C., and Schlegel, H. G. (1967). Oscillationen der Ureaseaktivität von *Hydrogenomonas* H 16 in statischer Kultur. *Biochim Biophys. Acta* **139,** 182.
126. Koops, H. P. (1969). Der Nutzeffekt der NH_4^+-Oxydation durch *Nitrosocystis oceanus* Watson. *Arch. Mikrobiol.* **65,** 115.
127. Kramer, J., Kaltwasser, H., and Schlegel, H. G. (1967). Die Bedeutung der Ureaserepression für die taxonomische Klassifizierung von Bakterien. *Zentralbl. Bakteriol.* (*Naturw.*) **121,** 414.
128. Kuehn, G. D., and McFadden, B. A. (1968). Factors affecting the synthesis and degradation of ribulose-1,5-diphosphate carboxylase in *Hydrogenomonas facilis* and *Hydrogenomonas eutropha*. *J. Bacteriol.* **95,** 937.
129. Kuehn, G. D., and McFadden, B. A. (1968). Enzymes of the Entner-Doudoroff path in fructose-grown *Hydrogenomonas eutropha*. *Can. J. Microbiol.* **14,** 1259.
130. Lamanna, C., and Mallette, M. F. (1965). "Basic Bacteriology," 3rd ed., p. 842. Williams & Wilkins, Baltimore, Maryland.
131. Landesman, J., Duncan, D. W., and Walden, C. C. (1966). Iron oxidation by washed cell suspensions of the chemoautotroph *Thiobacillus ferrooxidans*. *Can. J. Microbiol.* **12,** 25.
132. Landesman, J., Duncan, D. W., and Walden, C. C. (1966). Oxidation of inorganic sulfur compounds by washed cell suspensions of *Thiobacillus ferrooxidans*. *Can. J. Microbiol.* **12,** 957.
133. Lankford, C. E. (1973). Bacterial assimilation of iron. *CRC Critical Rev. Microbiol.* **2,** 273.
134. Lazaroff, N. (1963). Sulfate requirement for iron oxidation by *Thiobacillus ferrooxidans*. *J. Bacteriol.* **85,** 78.
135. Lees, H. (1954). The biochemistry of the nitrifying bacteria. *Symp. Soc. Gen. Microbiol.* **4,** 84.
136. Lees, H., and Simpson, J. R. (1957). The biochemistry of the nitrifying organisms. 5. Nitrite oxidation by *Nitrobacter*. *Biochem. J.* **65,** 297.
137. Lees, H. (1960). Energy metabolism of thiobacilli. *Annu. Rev. Microbiol.* **14,** 83.
138. Lees, H. (1960). Energy metabolism in chemolithotrophic bacteria. *Annu. Rev. Microbiol.* **14,** 91.
139. Lees, H. (1962). Some thoughts on the energetics of chemosynthesis. *Bacteriol. Rev.* **26,** 165.
140. LeJohn, H. B., Van Caeseele, L., and Lees, H. (1967). Catabolite repression in the facultative chemoautotroph *Thiobacillus novellus*. *J. Bacteriol.* **94,** 1484.
141. LeJohn, H. B., Suzuki, I., and Wright, J. A. (1968). Glutamate dehydrogenase of *Thiobacillus novellus*. Kinetic properties and a possible control mechanism. *J. Biol. Chem.* **243,** 118.
142. Lester, R. L., and Crane, F. L. (1959). The natural occurrence of coenzyme Q and related compounds. *J. Biol. Chem.* **234,** 2169.
143. Linday, E. M., and Syrett, P. J. (1958). The induced synthesis of hydrogenase by *Hydrogenomonas facilis*. *J. Gen. Microbiol.* **19,** 223.
144. London, J. (1963). Cytochrome in *Thiobacillus thiooxidans*. *Science* **140,** 409.
145. London, J., and Rittenberg, R. C. (1964). Path of sulfur in sulfide and thiosulfate oxidation by thiobacilli. *Proc. Nat. Acad. Sci. U.S.* **52,** 1183.

146. London, J., and Rittenberg, R. C. (1966). Effects of organic matter on growth of *Thiobacillus intermedians*. *J. Bacteriol.* **91,** 1062.
147. Losada, M., Trebst, A. V., Ogata, S., and Arnon, D. I. (1960). Equivalence of light and ATP in bacterial photosynthesis. *Nature* (*London*) **186,** 753.
148. Lu, M. C., Matin, A., and Rittenberg, R. C. (1971). Inhibition of growth of obligately chemolithotrophic thiobacilli by amino acids. *Arch. Mikrobiol.* **79,** 354.
149. Lyalikova, N. N. (1967). Oxidation of antimonite by a new culture of Thione bacteria. *Dokl. Akad. Nauk SSSR* **176,** 1432.
150. McFadden, B. A., and Homann, H. R. (1965). Characteristics and intermediates of short-term $^{14}CO_2$ incorporation during ribose oxidation by *Hydrogenomonas facilis*. *J. Bacteriol.* **89,** 839.
151. MacLennan, D. G., Lenaz, G., and Szarkowska, L. (1966). Studies on the mechanism of oxidative phosphorylation. IV. Effect of cytochrome c on energy-linked processes. *J. Biol. Chem.* **241,** 5251.
152. Magnana-Plaza, I., and Ruiz-Herrera, J. (1967). Mechanism of regulation of urease biosynthesis in *Proteus rettgeri*. *J. Bacteriol.* **93,** 1294.
153. Mahler, H. R., Baum, H. M., and Hübscher, G. (1956). Enzymatic oxidation of urate. *Science* **124,** 705.
154. Malavolta, E., Delwiche, C. C., and Burge, W. D. (1960). CO_2 fixation and phosphorylation by *Nitrobacter agilis*. *Biochem. Biophys. Res. Commun.* **2,** 445.
155. Mandel, M., Johnson, A., and Stokes, J. L. (1966). Deoxyribonucleic acid base composition of *Sphaerotilus natans* and *Sphaerotilus discophorus*. *J. Bacteriol.* **91,** 1656.
156. Margalith, P., Silver, M., and Lundgre, D. G. (1966). Sulfur oxidation by the iron bacterium *Ferrobacillus ferrooxidans*. *J. Bacteriol.* **92,** 1706.
157. Matin, A., and Rittenberg, S. C. (1970). Utilization of glucose in heterotrophic media by *Thiobacillus intermedius*. *J. Bacteriol.* **104,** 234.
158. Matin, A., and Rittenberg, S. C. (1970). Regulation of glucose metabolism in *Thiobacillus intermedius*. *J. Bacteriol.* **104,** 239.
159. Matin, A., and Rittenberg, S. C. (1971). Enzymes of carbohydrate metabolism in *Thiobacillus* species. *J. Bacteriol.* **107,** 179.
160. Mayeux, J. V., and Johnson, E. J. (1967). Effect of adenosine monophosphate, adenosine diphosphate, and reduced nicotinamide adenine dinucleotide on adenosine triphosphate-dependent carbon dioxide fixation in the autotroph *Thiobacillus neapolitanus*. *J. Bacteriol.* **94,** 409.
161. Mitchell, P., and Moyle, J. (1967). Proton transport phosphorylation: Some experimental tests. *In* "Biochemistry of Mitochondria" (E. C. Slater, A. Kaniuga, and L. Wojtczak, eds.), p. 53. Academic Press, New York.
162. Moriarty, D. J. W., and Nicholas, D. J. D. (1969). Enzymic sulfide oxidation by *Thiobacillus concretivorus*. *Biochim. Biophys. Acta* **184,** 114.
163. Moriarty, D. J. W., and Nicholas, D. J. D. (1970). Products of sulfide oxidation in extracts of *Thiobacillus concretivorus*. *Biochim. Biophys. Acta* **197,** 143.
164. Moriarty, D. J. W., and Nicholas, D. J. D. (1970). Electron transfer during sulfide and sulfite oxidation by *Thiobacillus concretivorus*. *Biochim. Biophys. Acta* **216,** 130.
165. Nicholas, D. J. D., and Jones, O. T. G. (1960). Oxidation of hydroxylamine in cell-free extracts of *Nitrosomonas europaea*. *Nature* (*London*) **185,** 512.
166. Nicholas, D. J. D., and Rao, P. S. (1964). The incorporation of labeled CO_2 into cells and extracts of *Nitrosomonas europaea*. *Biochim. Biophys. Acta* **82,** 394.
167. Nose, K., and Arima, K. (1968). Studies on bacterial urate:oxygen oxidoreductase.

II. Observations concerning the properties and components of the active site. *Biochim. Biophys. Acta* **151,** 63.

168. Ohlmann, E., Rindt, K. P., and Borris, R. (1969). Glukose-6-phosphat Dehydrogenase in autotrophen Mikroorganismen. I. Die Regulation der Synthese der Glukose-6-phosphat Dehydrogenase in *Euglena gracilis* und *Rhodopseudomonas spheroides* in Abhängigkeit von den Kulturbedingungen. *Z. Allg. Mikrobiol.* **9,** 557.
169. O'Kelly, J. C., Becker, G. E., and Nason, A. (1970). Characterization of the particulate nitrite oxidase and its component activities from the chemoautotroph *Nitrobacter agilis. Biochim. Biophys. Acta* **205,** 409.
170. O'Kelly, J. C., and Nason, A. (1970). Particulate formate oxidase from *Nitrobacter agilis. Biochim. Biophys. Acta* **205,** 426.
171. Okuzumi, M. (1966). Biochemistry of thiobacilli. VIII. Dismutation of tetrathionate by *Thiobacillus thiooxydans. Agr. Biol. Chem.* **30,** 313; *Chem. Abstr.* **65,** 1068 (1966).
172. Okuzumi, M. (1966). Biochemistry of thiobacilli. IX. Reduction of trithionate by *Thiobacillus thiooxydans. Agr. Biol. Chem.* **30,** 713; *Chem. Abstr.* **65,** 1068 (1966).
173. Packer, L., and Vishniac, W. (1955). Chemosynthetic fixation of CO_2 and characteristics of hydrogenase in resting cell suspensions of *Hydrogenomonas ruhlandii* nov. spec. *J. Bacteriol.* **70,** 216.
174. Pan, P., and Umbreit, W. W. (1972). Growth of obligate autotrophic bacteria on glucose in a continuous flow-through apparatus. *J. Bacteriol.* **109,** 1149.
175. Parker, C. D., and Prisk, J. (1953). The oxidation of inorganic compounds of sulfur by various sulfur bacteria. *J. Gen. Microbiol.* **8,** 344.
176. Peck, H. D., Jr. (1962). Comparative metabolism of inorganic sulfur compounds in microorganisms. *Bacteriol. Rev.* **26,** 67.
177. Peck, H. D., Jr. (1962). The oxidation of reduced sulfur compounds. *Bacteriol. Rev.* **26,** 83.
178. Peck, H. D., Jr. (1968). Energy-coupling mechanisms in chemolithotrophic bacteria. *Annu. Rev. Microbiol.* **22,** 489.
179. Peeters, T., and Aleem, M. I. H. (1970). Oxidation of sulfur compounds and electron transport in *Thiobacillus denitrificans. Arch. Mikrobiol.* **71,** 319.
180. Pfitzner, J., Linke, H. A. B., and Schlegel, H. G. (1970). Eigenschaften der NAD-spezifischen Hydrogenase aus *Hydrogenomonas* H 16. *Arch. Mikrobiol.* **71,** 67.
181. Pugh, L. H., and Umbreit, W. W. (1966). Anaerobic CO_2 fixation by autotrophic bacteria, *Hydrogenomonas* and *Ferrobacillus. Arch. Biochem. Biophys.* **115,** 122; *Chem. Abstr.* **65,** 5901 (1966).
182. Ralston, E., Palleroni, N. J., and Doudoroff, M. (1972). Deoxyribonucleic acid homologies of some so-called "*Hydrogenomonas*" species. *J. Bacteriol.* **109,** 465.
183. Ramsay, H. H. (1968). Autotrophic and heterotrophic metabolism in *Hydrogenomonas facilis. Antonie van Leeuwenhoek; J. Microbiol. Serol.* **34,** 71.
184. Rao, P. S., and Nicholas, D. J. D. (1966). Studies on the incorporation of CO_2 by cells and cell-free extracts of *Nitrosomonas europaea. Biochim. Biophys. Acta* **124,** 221.
185. Rees, M., and Nason, A. (1965). A P-450-like cytochrome and a soluble terminal oxidase identified as cytochrome o from *Nitrosomonas europaea. Biochem. Biophys. Res. Commun.* **21,** 248.
186. Rees, M., and Nason, A. (1966). Incorporation of atmospheric oxygen into nitrite formed during ammonia oxidation by *Nitrosomonas europaea. Biochim. Biophys. Acta* **113,** 398.

187. Rees, M. (1968). Studies of the hydroxylamine metabolism of *Nitrosomonas europaea*. I. Purification of hydroxylamine oxidase. *Biochemistry* **7,** 353.

188. Rees, M. (1968). Studies of the hydroxylamine metabolism in *Nitrosomonas europaea*. II. Molecular properties of the electron transport particle, hydroxylamine oxidase. *Biochemistry* **7,** 366.

189. Repaske, R. (1962). The electron transport system of *Hydrogenomonas eutropha*. I. Diphosphopyridine nucleotide reduction by hydrogen. *J. Biol. Chem.* **237,** 1351.

190. Repaske, R., and Lizotte, C. L. (1965). The electron transport system of *Hydrogenomonas eutropha*. II. Reduced nicotinamide adenine dinucleotide-menadione reductase. *J. Biol. Chem.* **240,** 4774.

191. Repaske, R., and Dans, C. L. (1968). A factor for coupling NAD to hydrogenase in *Hydrogenomonas eutropha. Biochem. Biophys. Res. Commun.* **30,** 136.

192. Richter, G. (1959). Comparison of sugar metabolism in two photosynthetic algae: *Anacystis nidulans* and *Chlorella pyrenoidosa. Nature* (*London*) **46,** 604.

193. Rindt, K. P., and Ohlmann, E. (1969). NADH and AMP as allosteric effectors of ribulose-5-phosphate kinase in *Rhodopseudomonas spheroides. Biochem. Biophys. Res. Commun.* **36,** 357.

194. Ritchie, G. A. F., and Nicholas, D. J. D. (1972). Identification of the sources of nitrous oxide produced by oxidative and reductive processes in *Nitrosomonas europaea. Biochem. J.* **126,** 1181.

195. Ritchie, G. A. F., and Nicholas, D. J. D. (1974). The partial characterization of purified nitrite reductase and hydroxylamine oxidase from *Nitrosomonas europaea. Biochem. J.* **138,** 471.

196. Rittenberg, S. C. (1969). The roles of exogenous organic matter in the physiology of chemolithotrophic bacteria. *Advan. Microbial Physiol.* **3,** 159.

197. Rittenberg, S. C., and Goodman, N. S. (1969). Mixotrophic growth of *Hydrogenomonas eutropha. J. Bacteriol.* **98,** 617.

198. Rolls, J. P., and Lindstrom, E. S. (1966). Coupling of thiosulfate oxidation in *Rhodopseudomonas palustris. Fed. Proc., Fed. Amer. Soc. Exp. Biol.* **25,** 739.

199. Rose, A. H. (1965). "Chemical Microbiology." Butterworth, London.

200. Ross, A. J., Schoenhoff, R. L., and Aleem, M. I. H. (1968). Electron transport and coupled phosphorylation in the chemoautotroph *Thiobacillus neapolitanus. Biochem. Biophys. Res. Commun.* **32,** 301.

201. Schaeffer, W. I., Holbert, P. E., and Umbreit, W. W. (1963). Attachment of *Thiobacillus thiooxidans* to sulfur crystals. *J. Bacteriol.* **85,** 137.

202. Schaeffer, W. I., and Umbreit, W. W. (1963). Phosphatidyl inositol as a wetting agent in sulfur oxidation by *Thiobacillus thiooxidans. J. Bacteriol.* **85,** 492.

203. Schatz, A., and Borell, C., Jr. (1952). Growth and hydrogenase activity of a new bacterium, *Hydrogenomonas facilis. J. Bacteriol.* **63,** 87.

204. Schindler, J. (1964). Die Synthese von Poly-β-hydroxybuttersäure durch *Hydrogenomonas* H 16: Die zu β-Hydroxybutyryl-CoA führenden Reaktionsschritte. *Arch. Mikrobiol.* **49,** 236.

205. Schlegel, H. G. (1960). *In* "Handbuch der Pflanzenphysiologie" (W. Ruhland, ed.), Vol. 5, Part 2, p. 649. Springer-Verlag, Berlin and New York [as cited by Lees (133)].

206. Schlegel, H. G., Gottschalk, G., and von Barthe, R. (1961). Formation and utilization of poly-β-hydroxybutyric acid by Knallgas bacteria (*Hydrogenomonas*). *Nature* (*London*) **191,** 463.

207. Schlegel, H. G., and Kaltwasser, H. (1961). Veränderungen des Polyphosphat-Gehaltes während des Wachstums von Knallgasbakterien unter Phosphatmangel. *Flora (Jena)* **1510,** 259 [as cited by Schlegel (205)].
208. Schlegel, H. G., and Gottschalk, G. (1962). Poly-β-hydroxybuttersäure, ihre Verbreitung, Funktion, und Biosynthese. *Angew. Chem.* **74,** 342.
209. Schlegel, H. G., and Lafferty, R. (1965). Growth of "Knallgas" bacteria (*Hydrogenomonas*) using direct electrolysis of culture medium. *Nature (London)* **205,** 308.
210. Schlegel, H. G. (1966). Physiology and biochemistry of Knallgas bacteria. *Advan. Comp. Physiol. Biochem.* **2,** 185.
211. Schlegel, H. G. (1969). "Allgemeine Mikrobiologie." Thieme, Stuttgart.
212. Schnaitman, C. A., Korczynski, M. S., and Lundgren, D. G. (1969). Kinetic studies of iron oxidation by whole cells of *Ferrobacillus ferrooxidans. J. Bacteriol.* **99,** 552.
213. Schöberl, G., and Engel, H. (1964). Das Verhalten der nitrifizierenden Bakterien gegenüber gelöstem Sauerstoff. *Arch. Mikrobiol.* **48,** 393.
214. Schön, G. (1965). Untersuchungen über den Nutzeffekt von *Nitrobacter winogradskyi* Buch. *Arch. Mikrobiol.* **50,** 111.
215. Schuster, E., and Schlegel, H. G. (1967). Chemolithotrophes Wachstum von *Hydrogenomonas* H 16 im Chemostaten mit elektrolytischer Knallgaserzeugung. *Arch. Mikrobiol.* **58,** 380.
216. Senez, J. C. (1962). Some considerations on the energetics of bacterial growth. *Bacteriol. Rev.* **26,** 96.
217. Sewell, D. L., and Aleem, M. I. H. (1969). Generation of reducing power in chemosynthesis. V. The mechanism of pyridine nucleotide reduction by nitrite in the chemoautotroph *Nitrobacter agilis. Biochim. Biophys. Acta* **172,** 467.
218. Shafia, F., and Wilkinson, R. F., Jr. (1969). Growth of *Ferrobacillus ferrooxidans* on organic matter. *J. Bacteriol.* **97,** 256.
219. Sinha, D. B., and Walden, C. C. (1966). Formation of polythionates and their relationships during oxidation of thiosulfate by *Thiobacillus ferrooxidans. Can. J. Microbiol.* **12,** 1041.
220. Skarzynski, B., Ostrowski, W., and Krawczyk, A. (1957). *Bull. Acad. Pol. Sci., Ser. Sci. Biol.* **5,** 159 [as cited by Lees (131)].
221. Smith, A. J., London, J., and Stanier, R. Y. (1967). Biochemical basis of obligate autotrophy in blue-green algae and thiobacilli. *J. Bacteriol.* **94,** 972.
222. Smith, A. J., and Hoare, D. S. (1968). *Nitrobacter agilis*—An obligate or a facultative autotroph? *Biochem. J.* **106,** 40P.
223. Smith, A. J., and Hoare, D. S. (1968). Acetate assimilation by *Nitrobacter agilis* in relation to its "obligate autotrophy." *J. Bacteriol.* **95,** 844.
224. Stachow, C. S., and Springgate, C. F. (1970). Guanosine 5′-triphosphate as the allosteric effector of fructose-1,6-diphosphatase in *Rhodopseudomonas palustris. Biochem. Biophys. Res. Commun.* **39,** 637.
225. Stanier, R. Y. (1961). Photosynthetic mechanisms in bacteria and plants. Development of a unitary concept. *Bacteriol. Rev.* **25,** 1.
226. Stanier, R. Y., Doudoroff, M., and Adelberg, E. (1961). "General Microbiology." Macmillan, New York.
227. Starkey, R. L. (1934). Isolation of some bacteria which oxidize thiosulfate. *Soil Sci.* **39,** 197.
228. Starkey, R. L. (1934). The production of polythionate from thiosulfate by microorganisms. *J. Bacteriol.* **28,** 387.

229. Starkey, R. L., Jones, G. E., and Frederich, L. R. (1956). Effects of medium agitation and wetting agents on oxidation of sulfur by *Thiobacillus thiooxidans*. *J. Gen. Microbiol.* **15,** 329.
230. Stewart, D. J. (1965). The urease activity of fluorescent pseudomonads. *J. Gen. Microbiol.* **41,** 169.
231. Stokes, J. L., and Powers, M. T. (1967). Stimulation of polyβ-hydroxybutyrate oxidation in *Sphaerotilus discophorus* by manganese and magnesium. *Arch. Mikrobiol.* **59,** 295.
232. Straat, P. A., and Nason, A. (1965). Characterization of a nitrate reductase from the chemoautotroph *Nitrobacter agilis*. *J. Biol. Chem.* **240,** 1412.
233. Stukus, P. E., and DeCicco, B. T. (1970). Autotrophic and heterotrophic metabolism of *Hydrogenomonas:* Regulation of autotrophic growth by organic substrates. *J. Bacteriol.* **101,** 339.
234. Stumm-Zollinger, E. (1972). Die bakterielle Oxydation von Pyrit. *Arch. Mikrobiol.* **83,** 110.
235. Suzuki, I., and Werkman, C. H. (1958). Glutathione and sulfur oxidation by *Thiobacillus thiooxidans*. *Proc. Nat. Acad. Sci. U.S.* **45,** 239.
236. Suzuki, I., and Werkman, C. H. (1958). Chemoautotrophic carbon dioxide fixation by extracts of *Thiobacillus thiooxidans*. II. Formation of phosphoglyceric acid. *Arch. Biochem. Biophys.* **77,** 112.
237. Suzuki, I. (1965). Oxidation of elemental sulfur by an enzyme system of *Thiobacillus thiooxidans*. *Biochim. Biophys. Acta* **104,** 359.
238. Suzuki, I., and Silver, M. (1966). The initial product and properties of the sulfur-oxidizing enzyme of thiobacilli. *Biochim. Biophys. Acta* **122,** 22.
239. Suzuki, I., and Kwok, S. (1970). Cell-free ammonia oxidation by *Nitrosomonas europaea* extracts: Effects of polyamines, Mg^{2+}, and albumin. *Biochem. Biophys. Res. Commun.* **39,** 950.
240. Szymona, M., and Doudoroff, M. (1960). Carbohydrate metabolism in *Rhodopseudomonas spheroides*. *J. Gen. Microbiol.* **22,** 167.
241. Tabita, R., and Lundgren, D. G. (1971). Utilization of glucose and the effect of organic compounds on the chemolithotrophic *Thiobacillus ferrooxidans*. *J. Bacteriol.* **108,** 328.
242. Tabita, R., and Lundgren, D. G. (1971). Heterotrophic metabolism of the chemolithtrophic *Thiobacillus ferrooxidans*. *J. Bacteriol.* **108,** 334.
243. Tabita, R., and Lundgren, D G. (1971). Glucose-6-phosphate dehydrogenase from the chemolithotrophic *Thiobacillus ferrooxidans*. *J. Bacteriol.* **108,** 343.
244. Taylor, B. F. (1968). Oxidation of elemental sulfur by an enzyme system from *Thiobacillus neapolitanus*. *Biochim. Biophys. Acta* **170,** 112.
245. Taylor, B. F., and Hoare, D. S. (1969). New facultative *Thiobacillus* and a reevaluation of the heterotrophic potential of *Thiobacillus novellus*. *J. Bacteriol.* **100,** 487.
246. Taylor, B. F. (1970). Regulation of citrate synthase activity in strict and facultatively autotrophic thiobacilli. *Biochem. Biophys. Res. Commun.* **40,** 957.
247. Terry, K. R., and Hooper, A. B. (1970). Polyphosphate and orthophosphate content of *Nitrosomonas europaea* as function of growth. *J. Bacteriol.* **103,** 199.
248. Trudinger, P. A. (1961). Thiosulfate oxidation and cytochromes in *Thiobacillus* X (*neapolitanus*). *Biochem. J.* **78,** 673.
249. Trudinger, P. A. (1961). Thiosulfate oxidation and cytochromes in *Thiobacillus* X. 2. Thiosulfate-oxidizing enzyme. *Biochem. J.* **78,** 680.

250. Trudinger, P. A. (1964). The effects of thiosulfate and oxygen concentration on tetrathionate oxidation by *Thiobacillus* X and *Thiobacillus thioparus*. *Biochem. J.* **90,** 640.
251. Trudinger, P. A. (1964). The metabolism of trithionate by *Thiobacillus* X. *Aust. J. Biol. Sci.* **17,** 459.
252. Trudinger, P. A. (1964). Evidence for a four-sulfur intermediate in $S_2O_3^{2-}$ oxidation by *Thiobacillus* X. *Aust. J. Biol. Sci.* **17,** 577.
253. Trudinger, P. A. (1964). Oxidation of thiosulfate by intact cells of *Thiobacillus* X. Effects of some experimental conditions. *Aust. J. Biol. Sci.* **17,** 738.
254. Trudinger, P. A. (1965). Effect of thiol-binding reagents on metabolism of thiosulfate and tetrathionate by *Thiobacillus neapolitanus*. *J. Bacteriol.* **89,** 617.
255. Trudinger, P. A. (1965). Effect of thiol-binding reagents on the metabolism of thiosulfate and tetrathionate by *Thiobacillus neapolitanus*. *J. Bacteriol.* **89,** 622.
256. Trudinger, P. A. (1965). Permeability of *Thiobacillus neapolitanus* ("*Thiobacillus* X") to thiosulfate. *Aust. J. Biol. Sci.* **18,** 563.
257. Trudinger, P. A. (1967). The metabolism of inorganic sulfur compounds by thiobacilli. *Rev. Pure Appl. Chem.* **17,** 1.
258. Trudinger, P. A. (1967). Metabolism of thiosulfate and tetrathionate by heterotrophic bacteria from soil. *J. Bacteriol.* **93,** 550.
259. Trudinger, P. A., and Kelly, D. P. (1968). Reduced nicotinamide adenine dinucleotide oxidation by *Thiobacillus neapolitanus* and *Thiobacillus* strain C. *J. Bacteriol.* **95,** 1962.
260. Trüper, H. G. (1964). CO_2-fixierung und Intermediärstoffwechsel bei *Chromatium okenii* Perty. *Arch. Mikrobiol.* **49,** 23.
261. Umbreit, W. W. (1962). Comparative physiology of autotrophic bacteria. *Bacteriol. Rev.* **26,** 145.
262. Van Caeseele, L., and Lees, H. (1969). The ultrastructure of autotrophically and heterotrophically grown *Thiobacillus novellus*. *Can. J. Microbiol.* **15,** 651.
263. Van Gool, A., and Laudelout, H. (1966). The mechanism of nitrite oxidation by *Nitrobacter winogradskyi*. *Biochim. Biophys. Acta* **113,** 41.
264. Van Gool, A., and Laudelout, H. (1967). Spectrophotometric and kinetic study of nitrite and formate oxidation in *Nitrobacter winogradskyi*. *J. Bacteriol.* **93,** 215.
265. Van Gool, A., Tobback, P. P., and Fischer, I. (1971). Autotrophic growth and synthesis of reserve polymers in *Nitrobacter winogradskyi*. *Arch. Mikrobiol.* **76,** 252.
266. Varma, A. K., and Nicholas, D. J. D. (1970). Studies on the incorporation of labeled sulfate into cells and cell-free extracts of *Nitrosomonas europaea*. *Arch. Mikrobiol.* **73,** 293.
267. Varma, A. K., and Nicholas, D. J. D. (1971). Metabolism of ^{35}S-sulfate and properties of APS kinase and PAPS reductase in *Nitrobacter agilis*. *Arch. Mikrobiol.* **78,** 99.
268. Varma, A. K., and Nicholas, D. J. D. (1971). Purification and properties of ATP sulfurylase from *Nitrobacter agilis*. *Biochim. Biophys. Acta* **227,** 373.
269. Vishniac, W., and Santer, M. (1957). The Thiobacilli. *Bacteriol. Rev.* **21,** 195.
270. Wallace, W., and Nicholas, D. J. D. (1968). Properties of some reductase enzymes in the nitrifying bacteria and their relationship to the oxidase systems. *Biochem. J.* **109,** 763.
271. Wallace, W., and Nicholas, D. J. D. (1969). Glutamate dehydrogenase in *Nitrosomonas europaea* and the effect of hydroxylamine, oximes, and related compounds on its activity. *Biochim. Biophys. Acta* **171,** 229.

272. Wallace, W., and Nicholas, D. J. D. (1969). The biochemistry of nitrifying bacteria. *Biol. Rev. Cambridge Phil. Soc.* **44,** 359.
273. Watson, S. W., Asbell, M. A., and Valois, F. W. (1970). Ammonia oxidation by cell-free extracts of *Nitrosocystis oceanus. Biochem. Biophys. Res. Commun.* **38,** 1113.
274. Watson, S. W., Graham, L. B., Remsen, C. C., and Valois, F. C. (1971). A lobular, ammonia-oxidizing bacterium, *Nitrosolobus multiformis* nov. gen. nov. sp. *Arch. Mikrobiol.* **76,** 183.
275. Wezernak, C. T., and Gannon, J. J. (1968). Oxygen–nitrogen relationships in autotrophic nitrification. *Appl. Microbiol.* **15,** 1211.
276. Williams, P. J. L., and Watson, S. W. (1968). Autotrophy in *Nitrosocystis oceanus. J. Bacteriol.* **96,** 1640.
277. Yamanaka, T., and Shinra, M. (1974). Cytochrome c-552 and cytochrome c-554 derived from *Nitrosomonas europaea. J. Biochem.* **75,** 1265.
278. Yates, M. G., and Nason, A. (1966). Electron transport systems of the chemoautotroph *Ferrobacillus ferrooxidans. J. Biol. Chem.* **241,** 4872.
279. Yoshida, T., and Alexander, M. (1964). Hydroxylamine formation by *Nitrosomonas europaea. Can. J. Microbiol.* **10,** 923.
280. Zolotukhin, N. V. (1970). Carbon dioxide fixation by a developing population of *Hydrogenomonas. Izv. Akad. Nauk SSSR, Ser. Biol.* p. 58.

Questions

1. Name the five different groups of chemolithotrophic genera that carry out aerobic respiration.

2. The oxidation of ammonia to nitrite involves a net transfer of 6 electrons. Name the proposed intermediates in the sequence and justify your proposals.

3. The first step in ammonia oxidation exhibits a positive free energy change. Discuss the energy requirement for this step.

4. "The oxidations of hydroxylamine and hydrazine are exergonic, which suggests that these reactions must be connected with a respiratory chain system." Discuss this statement.

5. Explain how chemolithotrophic bacteria obtain their energy (ATP) and reducing power ($NADH + H^+$), taking into consideration that the redox potential of $NH_2OH/NH_4^+ = +0.899$ V or of $NO_2^-/NH_2OH = +0.066$ V, compared with -0.32 of the $NAD^+/NADH + H^+$ system.

6. What are the main differences between *Nitrosomonas* and *Nitrobacter*?

7. What is the *Knallgas* reaction?

8. *Hydrogenomonas* species can be grown autotrophically and heterotrophically:
   a. Which is the key enzyme for autotrophic growth?
   b. Under what conditions does *Hydrogenomonas* form poly-β-hydroxybutyrate?
   c. What is the "hydrogen effect"?

9. State the metabolic differences between the two iron-oxidizing bacteria *Ferrobacillus ferrooxidans* and *Thiobacillus ferrooxidans.*

10. Compare the thiosulfate and sulfate oxidations of thiobacilli with the sulfate and thiosulfate reductions of *Desulfovibrio desulfuricans.*

11. Discuss the respiratory chain system of thiobacilli.

12. What would you regard as criteria for calling a microorganism an autotroph?

13. Explain the terms:
   a. Obligate autotroph
   b. Facultative autotroph
   c. Assimilatory autotroph
   d. Chemolithotrophic heterotroph
   e. Mixotroph
14. What is meant by "double carbon dioxide" fixation and where does it occur?
15. Discuss the regulatory mechanism of mixotrophic metabolism.
16. Discuss the thermodynamics of chemolithotrophic bacteria.
17. What are the reasons for grouping these five bacterial families together under the name "aerobic chemolithotrophs"?
18. A mixotrophic microorganism that can oxidize either an inorganic compound (e.g., sulfate) or an organic compound (e.g. glucose) under aerobic conditions possesses a complex electron transport system. Draw and justify such a respiratory chain.

7

Aerobic Respiration—Chemoorganotrophic Bacteria

A great number of microorganisms and other living cells meet their energy needs with the oxidation of organic compounds by molecular oxygen and liberation of free energy. Appropriate groups of enzymes catalyze a series of consecutive transformations, including dehydrogenations, of these substrates, resulting in their complete oxidation to carbon dioxide and water. The electrons removed from the substrates during these oxidations flow through an organized arrangement of electron carriers from the lowest to the highest potential and finally to oxygen. In the course of this energy flow, ATP is generated and becomes available for biosynthesis. Processes that lead to pyruvate from hexoses, pentoses, and trioses or polysaccharides have been considered in the previous chapter. The complete oxidation of pyruvate, and other carboxylic acids, and amino acids will be considered in this chapter.

Tricarboxylic Acid (TCA) Cycle

The conversion of pyruvate into water and carbon dioxide occurs by its oxidative decarboxylation, followed by a series of reactions called either the "tricarboxylic acid cycle," the "Krebs cycle," or because citrate holds a key position among the intermediates the "citric acid cycle."

In addition to its role in terminal respiration, the tricarboxylic acid cycle also plays an important role in the synthesis of cell material (108, 323, 382).

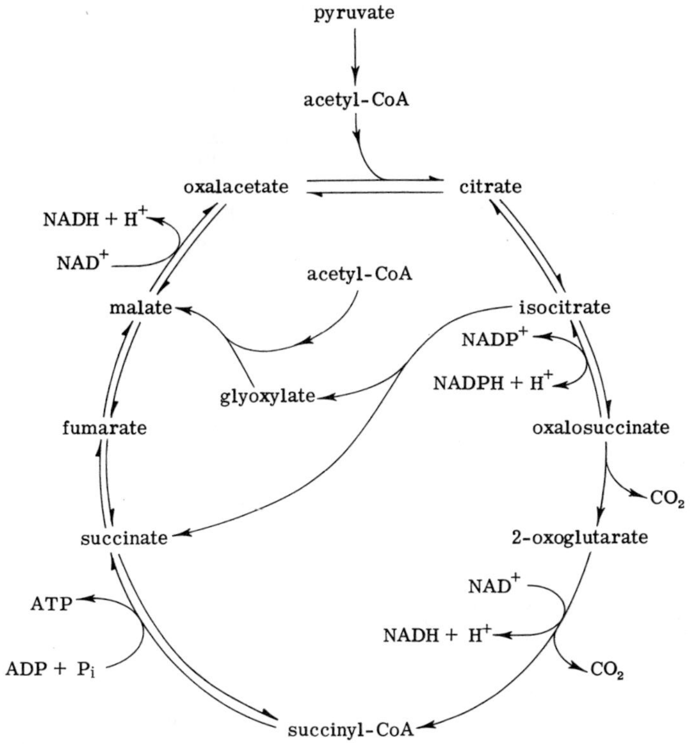
pyruvate
acetyl-CoA
oxalacetate
citrate
NADH + H+
NAD+
acetyl-CoA
malate
isocitrate
NADP+
NADPH + H+
glyoxylate
fumarate
oxalosuccinate
CO2
succinate
2-oxoglutarate
NAD+
ATP
NADH + H+
ADP + Pi
CO2
succinyl-CoA

Fig. 7.1. Tricarboxylic acid cycle and glyoxylate bypass in aerobic respiration.

It provides 2-oxoglutarate, which is the precursor of glutamic acid, a key compound in amino acid and protein biosynthesis, or oxalacetate which supplies aspartate and other amino acids (211). Because these processes occur simultaneously, there is a continual tapping off of both 2-oxoglutarate and all those four-carbon dicarboxylic acids that serve as important precursors for biosynthetic processes. Such a draining off of intermediates will only continue if the resulting deficit of four-carbon dicarboxylic acids is made up. The organisms therefore use a second system by which they make use of six of the eight reactions of the tricarboxylic acid cycle and bypass those steps which result in the evolution of carbon dioxide. They replace these with reactions the net effect of which is an increase in the organic acid carbon of the system (210). This new reaction sequence is called the "glyoxylate bypass" of the tricarboxylic acid cycle (Fig. 7.1), or simply the "glyoxylate" or "dicarboxylic acid cycle" (203–205), and operates mainly if the carbon source is a short-chain fatty acid (see p. 403). When grown on carbohydrates or pentoses, the organisms employ so-called

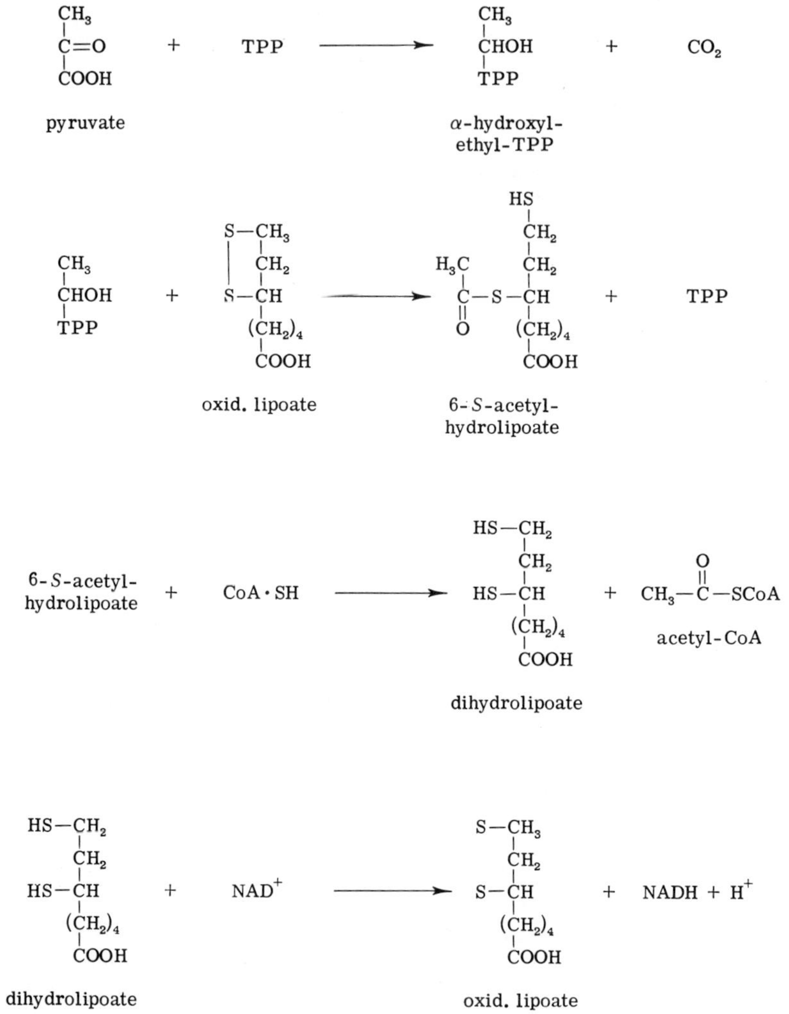
CH3
C=O
COOH
+
TPP
CH3
CHOH
TPP
+
CO2
pyruvate
α-hydroxyl-
ethyl-TPP
CH3
CHOH
TPP
+
S—CH3
CH2
S—CH
(CH2)4
COOH
HS
CH2
CH2
H3C
C—S—CH
O
(CH2)4
COOH
+
TPP
oxid. lipoate
6-S-acetyl-
hydrolipoate
6-S-acetyl-
hydrolipoate
+
CoA·SH
HS—CH2
CH2
HS—CH
(CH2)4
COOH
+
O
CH3—C—SCoA
acetyl-CoA
dihydrolipoate
HS—CH2
CH2
HS—CH
(CH2)4
COOH
+
NAD+
S—CH3
CH2
S—CH
(CH2)4
COOH
+
NADH + H+
dihydrolipoate
oxid. lipoate

Scheme 7.1. The formation of acetyl-CoA from pyruvate by the pyruvate dehydrogenase system.

"anaplerotic sequence" reactions (see p. 422) for the replenishment of dicarboxylic acids.

The tricarboxylic acid cycle itself, however, accepts only acetic acid in the form of an activated derivative. Pyruvate must therefore be broken down to yield a two-carbon acid derivative, acetyl-CoA or, as in *Leptospira* (28), acetyl-CoA must be formed from fatty acid degradation. This conversion is achieved enzymatically with the help of a multienzyme system

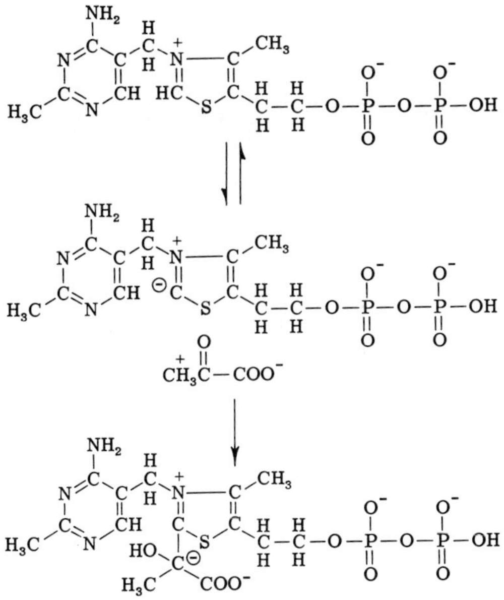

Scheme 7.2. First step in pyruvate utilization, where pyruvate is attached to thiamine pyrophosphate.

(109) consisting of not less than three different enzymes as well as the cofactors, thiamine pyrophosphate (TPP), lipoic acid, and NAD^+. The reaction sequence is shown in Scheme 7.1. The three enzymes involved in this multienzyme system are (*a*) pyruvate dehydrogenase [pyruvate:lipoate oxidoreductase (decarboxylating and acceptor-acetylating), EC 1.2.4.1]; (*b*) lipoate acetyltransferase (acetyl-CoA:dihydrolipoate *S*-acetyltransferase, EC 2.3.1.12); and (*c*) lipoamide dehydrogenase (reduced-NAD:lipoamide oxidoreductase, EC 1.6.4.3).

The first step in the multienzyme system is the reaction between pyruvate and thiamine pyrophosphate, whereby a proton is given off from the 2′ position of thiamine pyrophosphate. This enables pyruvate to be attached at this position (see Scheme 7.2). After it is attached to the carbanion of

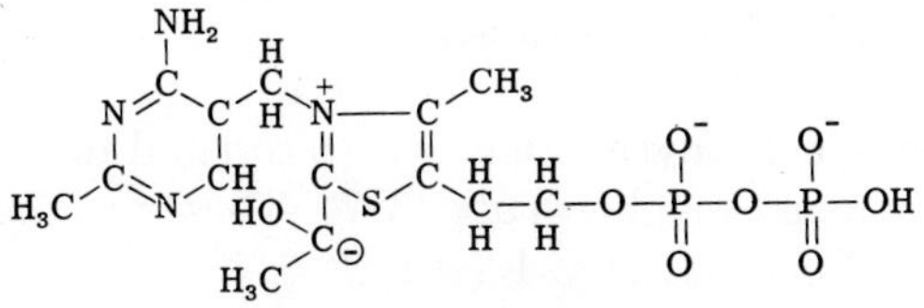

(–)-2-α-hydroxyethyl-
thiamine pyrophosphate

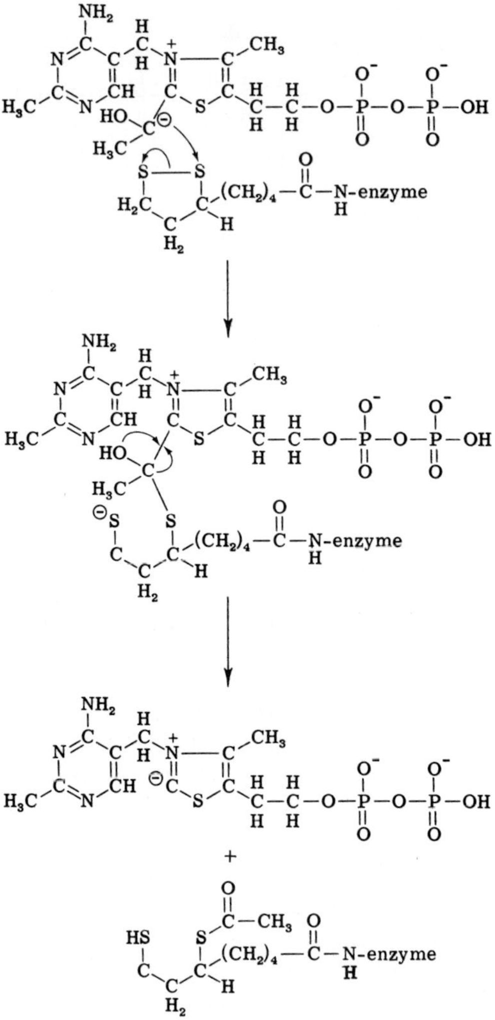

6-*S*-acetylhydrolipoate

Scheme 7.3. Second step in pyruvate utilization, where acetyl-CoA is formed under the catalytic action of lipoate acetyltransferase.

thiamine pyrophosphate, pyruvate is decarboxylated to CO_2 and (−)-2-α-hydroxyethylthiamine pyrophosphate (387). These reactions are catalyzed by pyruvate dehydrogenase (EC 1.2.4.1).

The next step in the multienzyme system is a transfer reaction (343), whereby the hydroxyethyl group from (−)-2-α-hydroxyethylthiamine pyrophosphate is transferred to lipoic acid. During this transfer, an oxidation of the hydroxyethyl group to an acetyl group occurs and as this acetyl group is attached to the 6 position on the lipoic acid molecule, the compound 6-*S*-acetylhydrolipoate is formed. This reaction also involves the protonation of the TPP-carbanion back to thiamine pyrophosphate (Scheme 7.3). The acyl generation can thus be visualized as a reductive acylation of protein-bound lipoic acid. It is the "active acetaldehyde" which attacks the disulfide linkage of lipoic acid in a nucleophilic displacement reaction (311). As far as the enzyme lipoate acetyltransferase (EC 2.3.1.12) is concerned, the carboxyl group of lipoic acid is in an amide linkage with an ε-amino group of a lysine residue. The intermediate 6-*S*-acetylhydrolipoate has not been isolated as yet and there are doubts as to whether this compound occurs as a free intermediate (247).

The second part of the reaction is the acyl transfer of the acetyl group from the thiol group of 6-*S*-acetylhydrolipoate to the thiol group of coenzyme A:

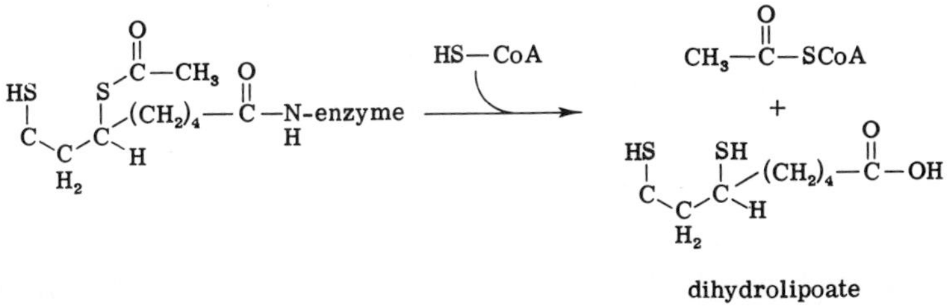
HS—CoA
CH3—C—SCoA
dihydrolipoate

The third and last step of the pyruvate dehydrogenase multienzyme system is the oxidation of dihydrolipoate to lipoate, with the concomitant reduction of NAD^+ by the enzyme lipoamide dehydrogenase (EC 1.6.4.3). The enzyme contains FAD (179) and its functions are well established (391, 391a). There is good evidence that lipoamide dehydrogenase contains a reactive disulfide group that participates in the catalytic cycle of the enzyme. Staub's diaphorase appears to be identical with the lipoamide dehydrogenase (179).

The aerobic breakdown of pyruvate carried out by this multienzyme system is inducible and pyruvate appears to be the metabolite causing this induction (82). However, short-chain acyl-CoA esters inhibit the overall reaction, as they compete for the CoA·SH site (95).

Reduced NAD, which in itself is an inhibitor of pyruvate dehydrogenase (122), is reoxidized by transfer of its hydrogens to the electron transport chain.

The role of coenzyme A has been discussed in Chapter 1; it should be regarded as a carrier of acetyl groups (25), just as ATP is a carrier of phosphate groups and NAD is a carrier of electrons in the cell. The free energy of acetyl-CoA hydrolysis is about $\Delta F = -8800$ cal/mole, which is somewhat higher than the free energy of ATP hydrolysis (238). The acetyl group of acetyl-CoA is therefore the immediate fuel for the TCA cycle and

$$CH_3CO{-}S{-}CoA + H_2O \rightarrow CH_3COO^- + H^+ + CoA{-}SH$$

is made available to the cycle by an enzymatic transfer reaction.

In the first step of the tricarboxylic acid cycle, acetyl-CoA donates the acetyl group to the four-carbon dicarboxylic acid oxalacetate to form citric acid, a six-carbon tricarboxylic acid.

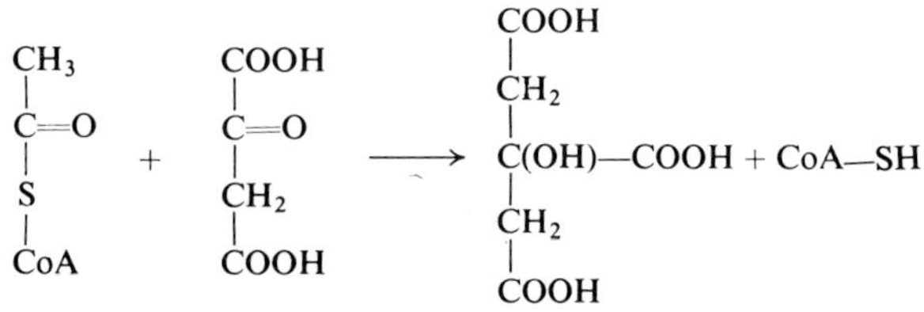

acetyl-CoA oxalacetic acid citric acid

During this reaction, catalyzed by the enzyme citrate synthase [citrate oxaloacetate-lyase (CoA-acetylating), EC 4.1.3.7] free CoA is generated and can reenter in the formation of acetyl-CoA. Citrate synthase is a condensing enzyme and has also been known as "citrate condensing enzyme," "citrogenase," or "oxalacetate transacetase" (156). In this reaction, a binary complex is formed between the enzyme and oxalacetate. This complex formation is accompanied by a change in protein conformation (91, 359).

This step, whereby a carbon enters the cycle, is very important and it is not surprising that citrate synthase underlies certain control mechanisms. One of the "end products" of the tricarboxylic acid cycle is NADH + H^+, which was found to have an allosteric effect on citrate synthase (360, 396, 397). This inhibition would therefore constitute a feedback control inhibition, as the enzyme is insensitive toward adenylate control (398). Such a control mechanism does not exist in mammals (162, 348) or yeast (128); it appears to be unique to microorganisms. Further investigations (400), however, revealed that it was only the enzymes from gram-negative bacteria that were insensitive to NADH inhibition with the exception of halophilic bacteria (56a). Enzyme purification (73a), together with Sephadex column chromatography, showed the existence of two types of citrate synthases, one with a large molecular size and the second with a small molecular size. The "large" citrate synthase is sensitive to NADH + H⁺ regulation. Citrate synthase is also inhibited by 2-oxoglutarate (416), which appears to serve as an allosteric inhibitor similarly to reduced NAD.

The next two steps are disputable at the present time, for one single enzyme forms two other tricarboxylic acids. First, *cis*-aconitate and then isocitrate are formed, with the enzyme aconitate hydratase [citrate (iso-

$$\begin{array}{c} CH_2-COOH \\ | \\ C(OH)-COOH \\ | \\ CH_2-COOH \end{array} \longrightarrow \begin{array}{c} CH_2-COOH \\ | \\ C-COOH \\ \| \\ CH-COOH \end{array} + H_2O$$

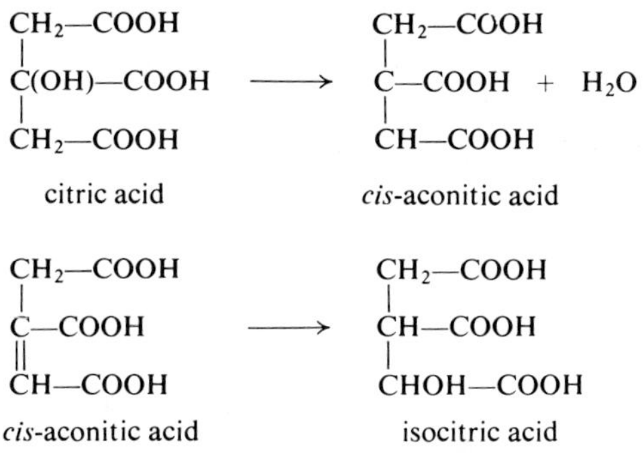

citric acid   *cis*-aconitic acid

$$\begin{array}{c} CH_2-COOH \\ | \\ C-COOH \\ \| \\ CH-COOH \end{array} \longrightarrow \begin{array}{c} CH_2-COOH \\ | \\ CH-COOH \\ | \\ CHOH-COOH \end{array}$$

cis-aconitic acid   isocitric acid

citrate) hydro-lyase, EC 4.2.1.3] catalyzing both reactions. Investigations by Speyer and Dickman (357) and Englard and Colowick (93), however, showed that *cis*-aconitic acid cannot be an intermediate of the tricarboxylic acid cycle, although it is in equilibrium with both citrate and isocitrate. The single intermediate common to all three substrates in the aconitase reaction is now believed (181) to be a carbonium ion with an intramolecular hydrogen transfer system. The actual reaction therefore would be:

$$\begin{array}{c} OH \\ | \\ H_2C-C-CH_2 \\ | \quad | \quad | \\ COOH \ COOH \ COOH \end{array} \rightleftharpoons \left[\begin{array}{c} \quad\quad\quad H \\ \quad\quad\quad \cdot \\ H_2C-C-C \\ | \quad | \quad + | \\ COO^- \ COO^- \ COO^- \end{array} \right]$$

$$\begin{array}{c} H_2C-C=CH \\ | \quad | \quad | \\ COOH \ COOH \ COOH \end{array} \quad\quad \begin{array}{c} CH_2-CH-CHOH \\ | \quad | \quad | \\ COOH \ COOH \ COOH \end{array}$$

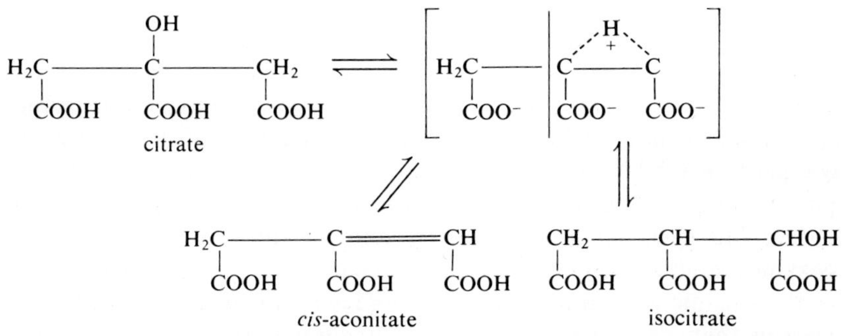

cis-aconitate   isocitrate

Speyer and Dickman (357) regard this carbonium ion as being different from the classical carbonium ion, because it occurs as a constituent of an enzyme–metal–substrate complex and it is formed and directed under the influence of an enzyme. This intermediate could never occur freely, therefore, but only as part of a complex.

All three acids, citric acid, *cis*-aconitic acid, and isocitric acid, form an equilibrium mixture.

Isocitric acid is converted into oxalosuccinic acid by an NADP-dependent isocitrate dehydrogenase [*threo*-D_s-isocitrate:NADP oxidoreductase (de-

$$\begin{array}{c} CH_2\text{---}COOH \\ | \\ CH\text{---}COOH \\ | \\ CHOH\text{---}COOH \end{array} + NADP \longrightarrow \begin{array}{c} CH_2\text{---}COOH \\ | \\ CH\text{---}COOH \\ | \\ CO\text{---}COOH \end{array} + NADPH + H^+$$

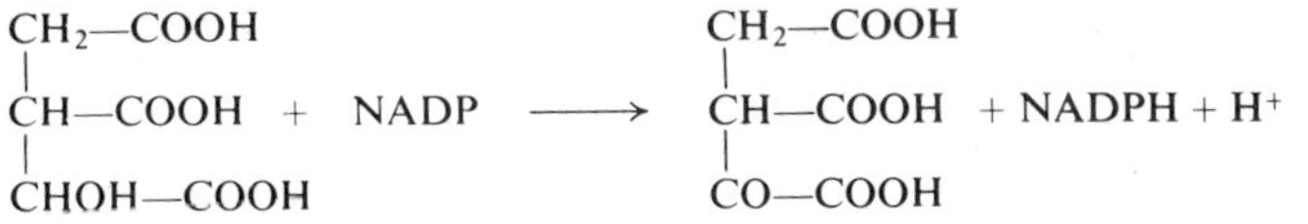

isocitric acid   oxalosuccinic acid

carboxylating), EC 1.1.1.42]. Consequently, a reduced-NADP:(acceptor) oxidoreductase (EC 1.6.99.1) and a cytochrome oxidase (EC 1.9.3.1) react in the oxidation of NADPH + H^+. The same isocitrate dehydrogenase also catalyzes the subsequent conversion of oxalosuccinate to 2-oxoglutaric acid

$$\begin{array}{c} CH_2\text{---}COOH \\ | \\ CH\text{---}COOH \\ | \\ CO\text{---}COOH \end{array} \longrightarrow \begin{array}{c} CH_2\text{---}COOH \\ | \\ CH_2 \\ | \\ CO\text{---}COOH \end{array} + CO_2$$

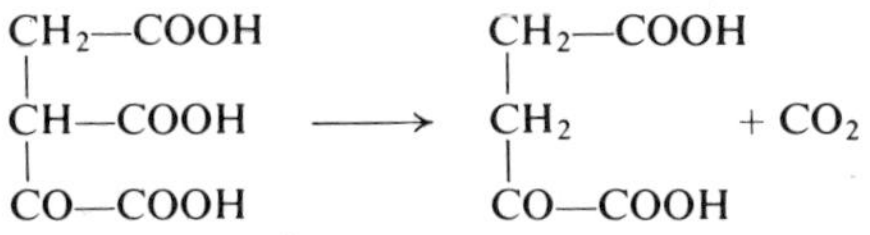

oxalosuccinic acid   2-oxoglutaric acid

and evolves one molecule of carbon dioxide. This molecule of CO_2 is the first of the two that arise from the two-carbon acetyl-CoA fed into the cycle.

Isocitrate dehydrogenase is another enzyme holding a key position in the tricarboxylic acid cycle, as isocitrate is the branching point not only of the TCA cycle but also the glyoxylate cycle. It therefore underlies regulatory control functions (295b). In contrast to the citrate synthase, ATP and GTP can inhibit isocitrate dehydrogenase, whereby these nucleotides appear to have a dual action (254). At low concentrations they are competitive inhibitors of $NADP^+$, and at higher concentrations they are primarily competitive inhibitors of the substrate isocitrate. The enzyme has been purified from a great number of microorganisms (116, 148, 150, 320, 328, 295a). The electrophoretic mobilities and the differences in the observed enzyme stabilities (148) indicate that the physical and chemical properties of the enzyme may differ widely among bacterial species. It may be noteworthy that microorganisms possess predominantly the $NADP^+$-specific enzyme, whereas fungi (332) and yeasts (57) possess the NAD^+-specific isocitrate dehydrogenase.

The second molecule of carbon dioxide arises in the following step, which again involves a multienzyme complex system, similar to the pyruvate dehydrogenase system (200, 211, 272, 312, 313). The conversion of 2-oxoglutarate to succinate by the 2-oxoglutarate dehydrogenase complex requires the participation of thiamine pyrophosphate, α-lipoic acid, coenzyme A, NAD^+, and Mg^{2+}. NAD^+ is the oxidant in this reaction, and the resulting NADH + H^+ is oxidized by the electron transport system.

The early work on the role of lipoic acid (lip-*S*) (331) and the lipoamide dehydrogenase (201, 202) led to the following overall sequence (120):

$$\text{2-Oxoglutarate} + \text{TPP} \rightleftharpoons \text{(succinic semialdehyde)–TPP} + CO_2$$

$$\text{(succinic semialdehyde)–TPP} + \text{lip-}S_{ox} \rightleftharpoons S^6\text{-succinylhydrolipoate} + \text{TPP}$$

$$S^6\text{-succinylhydrolipoate} + \text{CoA-SH} \rightleftharpoons \text{succinyl-}S\text{CoA} + \text{lip-}S_{red}$$

$$\text{lip-}S_{red} + NAD^+ \rightleftharpoons \text{lip-}S_{ox} + NADH + H^+$$

The first reaction is catalyzed by 2-oxoglutarate dehydrogenase (2-oxoglutarate:lipoate oxidoreductase (decarboxylating and acceptor-succinylating), EC 1.2.4.2) and leads to S^6-succinylhydrolipoate and CO_2 via an "active succinic semialdehyde" as intermediate. This enzyme catalysis requires TPP and lipoate and is possibly a system and not a single enzyme (295a). The transfer of the succinyl group from the complex to CoA and the concomitant formation of succinyl-CoA is the work of the second, not yet well established enzyme in this multienzyme system. The name given to this enzyme is "lipoate transsuccinylase (succinyl-CoA:dihydrolipoate *S*-succinyltransferase)." Since this reaction is very similar to the one described in the pyruvate dehydrogenase multienzyme system (see p. 382), the most likely name would be lipoate succinyltransferase. The final reaction is carried out by lipoamide dehydrogenase (EC 1.6.4.3), which links the oxidation of the hydrogen acceptors in the complex (lipoic acid and FAD) to the external hydrogen acceptor (NAD^+).

The 2-oxoglutarate dehydrogenase system plays an important regulatory role in facultative anaerobic bacteria. It is very sensitive to oxygen deficiency (122, 174, 218, 377) and is also subject to glucose repression (7, 8, 113). This enzyme complex also appears to play a vital role in the control and regulation of the concurrent operation of the TCA cycle and the glyoxylate bypass (363). All these observations led to the suggestion (8) that the tricarboxylic acid cycle proceeds in a cyclic manner under aerobic conditions, but under anaerobic conditions it is modified to provide a branched, noncyclic pathway. This pathway has an oxidative branch to 2-oxoglutarate, serving a purely biosynthetic role, whereas a reductive branch leads to succinate and serves an amphibolic role (75).

The enzyme succinyl-coenzyme A synthetase [succinate:CoA ligase (ADP), EC 6.2.1.5] is responsible for the conversion of succinyl-CoA to succinic acid in *Escherichia coli* (117, 138, 267a). With the release of CoA,

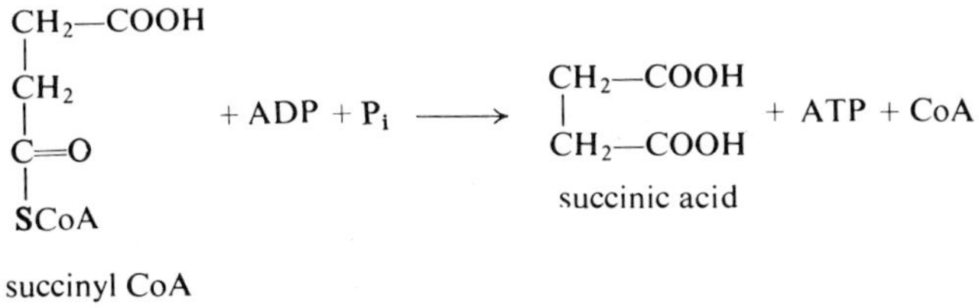

CH₂—COOH
CH₂
C=O
SCoA
succinyl CoA
+ ADP + Pᵢ
CH₂—COOH
CH₂—COOH
succinic acid
+ ATP + CoA

the organism is able to form an additional mole of ATP. Evidence is accumulating that suggests a "high-energy" compound, a phosphorylated form of the enzyme, as intermediate (271, 308). The reaction therefore has the following, more detailed mechanism (308):

$$\text{Succinyl-CoA} + \text{enzyme} \rightleftharpoons \text{enzyme-CoA} + \text{succinate}$$

$$\text{Enzyme-CoA} + \text{Pi} \rightleftharpoons \text{enzyme-phosphate} + \text{CoA}$$

$$\text{Enzyme-phosphate} + \text{ADP} \rightleftharpoons \text{enzyme} + \text{ATP}$$

$$\text{Succinyl-CoA} + \text{Pi} + \text{ADP} \rightleftharpoons \text{succinate} + \text{CoA} + \text{ATP}$$

This detailed mechanism, however, is somewhat in contradiction to the suggestion of succinyl phosphate as intermediate (138). If parallels are drawn to similar reactions, the formation of an enzyme–CoA and enzyme–phosphate exchange would emerge as the preferred mechanism, as the one for the succinyl-CoA:3-oxoacid CoA-transferase (EC 2.8.3.5) reaction transfers its CoA via an enzyme–CoA intermediate (31).

An alternative pathway for this reaction is the involvement of succinyl-CoA hydrolase (EC 3.1.2.3), whereby no energy is gained from the thioester bond

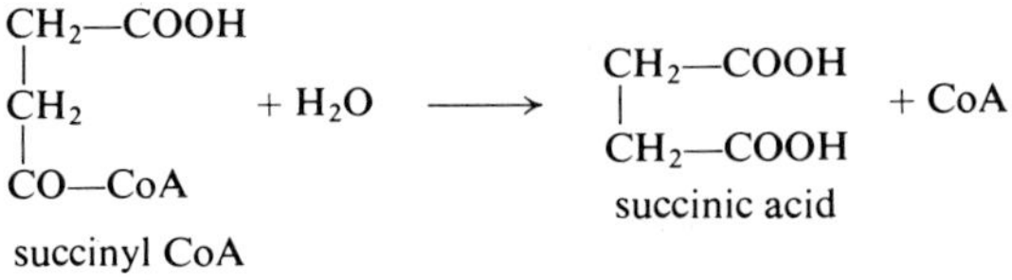

CH₂—COOH
CH₂
CO—CoA
succinyl CoA
+ H₂O
CH₂—COOH
CH₂—COOH
succinic acid
+ CoA

The next step in the oxidation via the tricarboxylic acid cycle is the dehydrogenation of succinate. Succinate is oxidized to fumarate by the enzyme succinate dehydrogenase [succinate: (acceptor) oxidoreductase, EC

succinic acid fumaric acid

1.3.99.1]. The active group of this enzyme accepts the hydrogen atoms and becomes reduced. It is an established fact that succinic dehydrogenase is closely linked to the electron transport chain and enters this system at the flavoprotein level. As small NAD^+ reductions were observed in *Escherichia coli* particulate fractions, it was not surprising to find that a certain proportion of reverse electron flow can occur (368), which supplies the organism with $NADH + H^+$. Because of the existence in *Escherichia coli* of such a reverse electron flow system (see p. 400) apart from the normal electron transport via the cytochromes, it has been suggested that the reaction occurs as follows:

$$\text{Succinate} + NAD^+ + ATP \rightleftharpoons \text{fumarate} + NADH + H^+ + ADP + Pi$$

The ATP is required to obtain reduced NAD. The necessity of this reaction, however, is not quite clear as yet. *Escherichia coli* is also able to carry out an ATP-dependent reduction of $NADP^+$ by $NADH + H^+$ (369) with an energy-linked transhydrogenase, whereby 1 mole ATP is hydrolyzed per mole $NADPH + H^+$ formed. *Escherichia coli* also possesses a transhydrogenase that is not energy linked and that appears to be a different enzyme or a different form of the same enzyme. Reduced $NADP^+$ is required for a great number of anabolic or biosynthetic reactions in the cell. The occurrence of these reactions shows the great importance of the succinic dehydrogenase for the generation of ATP via the electron transport chain and the formation of reduced $NADP^+$ for biosynthesis of cell material.

Succinate dehydrogenase is inhibited by oxalacetate (410, 429). This inhibition is thought to be a tight bond between the enzyme and oxalacetate. Alternatively, the latter could promote a reversible conformation change leading from an active to an inactive form of the enzyme.

As the succinate dehydrogenase reaction has an equilibrium strongly favoring fumarate formation, *Escherichia coli*, as a facultative anaerobe, has in addition a fumarate-reducing activity catalyzed by fumarate reductase (EC 1.3.1.6), which is distinctly different from succinic dehydrogenase (140).

Fumarate itself is now hydrated at the double bond (307) to form malic acid by the action of fumarate hydratase (L-malate hydro-lyase, EC 4.2.1.2 , also known under the name "fumarase") Whether or not this

$$\begin{array}{c} \text{CH}-\text{COOH} \\ \| \\ \text{CH}-\text{COOH} \end{array} + H_2O \longrightarrow \begin{array}{c} \text{CH}_2-\text{COOH} \\ | \\ \text{CHOH}-\text{COOH} \end{array}$$

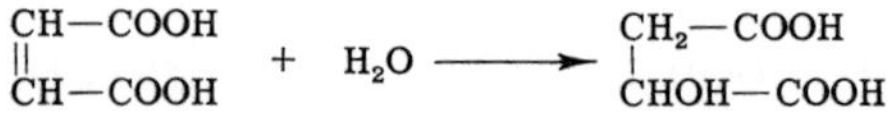

fumaric acid   malic acid

enzyme is inhibited by ATP in bacteria is not certain, although yeast and pig heart fumarases (296) show sigmoidal kinetics.

The final reaction in the tricarboxylic acid cycle is the dehydrogenation of malate to oxalacetate, catalyzed by malate dehydrogenase (L-malate:

$$\begin{array}{c} \text{CH}_2-\text{COOH} \\ | \\ \text{CHOH}-\text{COOH} \end{array} + NAD^+ \longrightarrow \begin{array}{c} \text{CH}_2-\text{COOH} \\ | \\ \text{CO}-\text{COOH} \end{array} + NADH + H^+$$

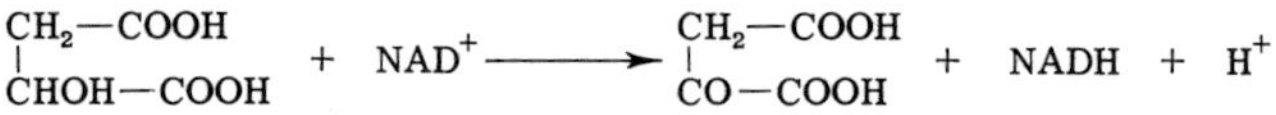

malic acid   oxalacetic acid

NAD oxidoreductase, EC 1.1.1.37). It appears that the enzyme from *Bacillus subtilis* and *B. stearothermophilus* is composed of four subunits and has a molecular weight of 117,000, whereas the *Escherichia coli* enzyme has only two subunits and a molecular weight of 60,000 (274). *Escherichia coli*, however, possesses a second malic enzyme, which is NADP+ linked (EC 1.1.1.40). Whereas the NAD-linked enzyme is claimed to be constitutive (373, 424), the NADP-linked enzyme appears to play a regulatory role in the C_4-dicarboxylic acid metabolism. It catalyzes the reaction

$$\begin{array}{c} \text{CH}_2-\text{COOH} \\ | \\ \text{CHOH}-\text{COOH} \end{array} + NADP^+ \longrightarrow \begin{array}{c} \text{CH}_3 \\ | \\ \text{C}=O \\ | \\ \text{COOH} \end{array} + CO_2 + NADPH + H^+$$

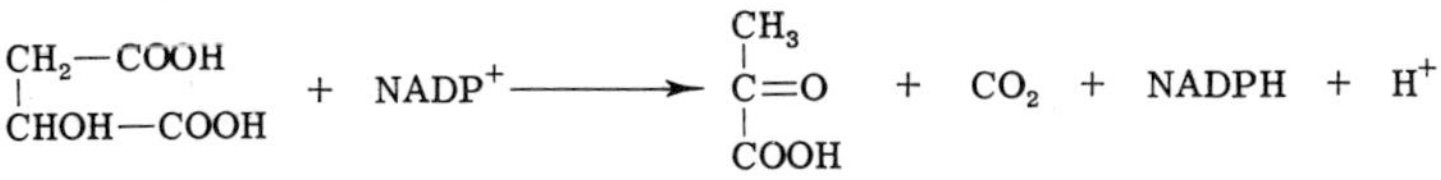

malic acid   pyruvic acid

The activities of both enzymes vary with the composition of the growth medium (183). The NADP-linked malic enzyme is inhibited strongly by oxalacetate, whereas the NAD-linked malic enzyme is only slightly affected. Kinetic investigations into the NADP-linked malic enzyme revealed that it is an allosteric protein (337), the activity of which is controlled by a negative-feedback inhibition by acetyl-CoA and oxalacetate (334). The discovery that cyclic 3',5'-AMP also affects the NADP+-linked malic enzyme (338) as an allosteric inhibitor certainly indicates a major control function for this enzyme in the facultative anaerobe *Escherichia coli*. The diversion from malate to pyruvate could be beneficial for the cell in two ways. It furnishes the cell with additional acetyl-CoA for the tricarboxylic acid cycle and fatty acid biosynthesis and at the same time provides the fatty acid biosynthesis with NADPH + H^+. If, however, the carbon source becomes limited, i.e., in starvation conditions, the cell would economize and use only the NAD^+-linked enzyme. This indicates that the NAD^+-linked malic enzyme (EC 1.1.1.37) is a constitutive enzyme playing a catabolic role, whereas the $NADP^+$-linked malic enzyme (EC 1.1.1.40) is an inducible enzyme playing an anabolic or biosynthetic role (273). In *Micrococcus* sp. (261), the $NADP^+$-linked malic enzyme is the only malic enzyme present and plays an important role in carbon dioxide fixation into pyruvate—a clear indication of the biosynthetic role of this enzyme.

Some bacteria, such as *Serratia* (129), have a malate dehydrogenase (D-malate:NAD oxidoreductase) that oxidizes malate to pyruvate. This enzyme, as are the other two, is also known under the name "malic enzyme" and could be either EC 1.1.1.38 or 1.1.1.39, depending on whether or not it also decarboxylates oxalacetate.

The various uses for malate make the oxidation of malate to oxalacetate a key step in the tricarboxylic acid cycle, but this step can be carried out in different ways by different microorganisms.

1. The most common situation is the presence of the soluble NAD-linked dehydrogenase (EC 1.1.1.37) as it occurs in *Escherichia coli.*

2. In addition to the soluble NAD-linked dehydrogenase, some bacteria possess a particulate oxidation system as it occurs in *Azotobacter agilis* (4) and *Micrococcus lysodeikticus* (68).

3. The soluble NAD-linked dehydrogenase can be absent and the oxidation of malate is coupled to oxygen and brought about by a particulate system as in *Serratia marcescens* (129), *Pseudomonas fluorescens*, or *Pseudomonas ovalis* (101). These microorganisms also require cytochrome c, which may be similar to *Pseudomonas* cytochrome c-511 (9, 10, 114) because nicotinamides and adenine nucleotides did not show any influence on it. This enzyme, known under the name "malic enzyme" was also found in a number of the Lactobacillaceae (305, 306).

4. The fourth possible mode of malate conversion to oxalacetate is achieved by a sequential oxidative decarboxylation of malate to pyruvate with a subsequent recarboxylation of pyruvate to oxalacetate. This mechanism has only been found so far in *Chromatium,* which lacks both the soluble and the particulate malate dehydrogenases (104).

The cycling reactions of the tricarboxylic acid cycle will now be evaluated. One molecule of acetyl-CoA and one molecule of oxalacetate are fed into the cycle. The reaction yields two molecules of carbon dioxide and one molecule of oxalacetate. It is therefore possible to write the overall equation:

$$\text{acetate} + \text{oxalacetate} \rightarrow 2\,CO_2 + \text{oxalacetate}$$

As one molecule of oxalacetate was supplied and received, the formula could be simplified to

$$\text{acetate} \rightarrow 2\,CO_2$$

In addition there were four dehydrogenation steps in the tricarboxylic acid cycle which extract four pairs of electrons enzymatically from the intermediates of the cycle, to which reference will shortly be made.

Electron Transport in Aerobic Microorganisms

In considering the oxidation of pyruvate via the tricarboxylic acid cycle, it was mentioned that pyruvate is completely broken down to carbon dioxide and water. So far only the formation of carbon dioxide has been evaluated and attention will now be given to the system that produces the water. When molecular oxygen is available as final electron acceptor, aerobic respiration takes place and the oxygen is reduced to water. Water is therefore the end product of the electron transport chain. The physiological function of this electron transport chain has been treated in Chapter 2. It is therefore necessary only to demonstrate how the different aerobic microorganisms are able to form the end products water and ATP, indicating the diversity which does exist. In connection with the treatment of this diversity the author wishes to draw attention also to the excellent review by White and Sinclair (402) on branched electron transport systems in bacteria.

The tricarboxylic acid cycle primarily concerns the fate of the carbon skeleton of acetic acid and is not involved in energy conservation. Of the four dehydrogenation steps each revolution of the cycle, three are connected to NAD^+, forming three molecules of $NADH + H^+$. In the fourth step, the pair of electrons removed from succinate is accepted by the active group of succinic dehydrogenase, a flavoprotein. These flavoproteins are a class of oxidizing enzymes that contain FAD as electron acceptors similar to NAD^+ in its action. All four molecules donate their electrons to another series of enzymes, which constitute the respiratory chain (115) (see Fig. 7.2). Electrons can enter the chain of cytochromes only via one of the two flavoproteins. One of these accepts electrons from $NADH + H^+$ and the other from succinate. The connection of the tricarboxylic acid cycle to the respiratory chain can be seen in Fig. 7.2.

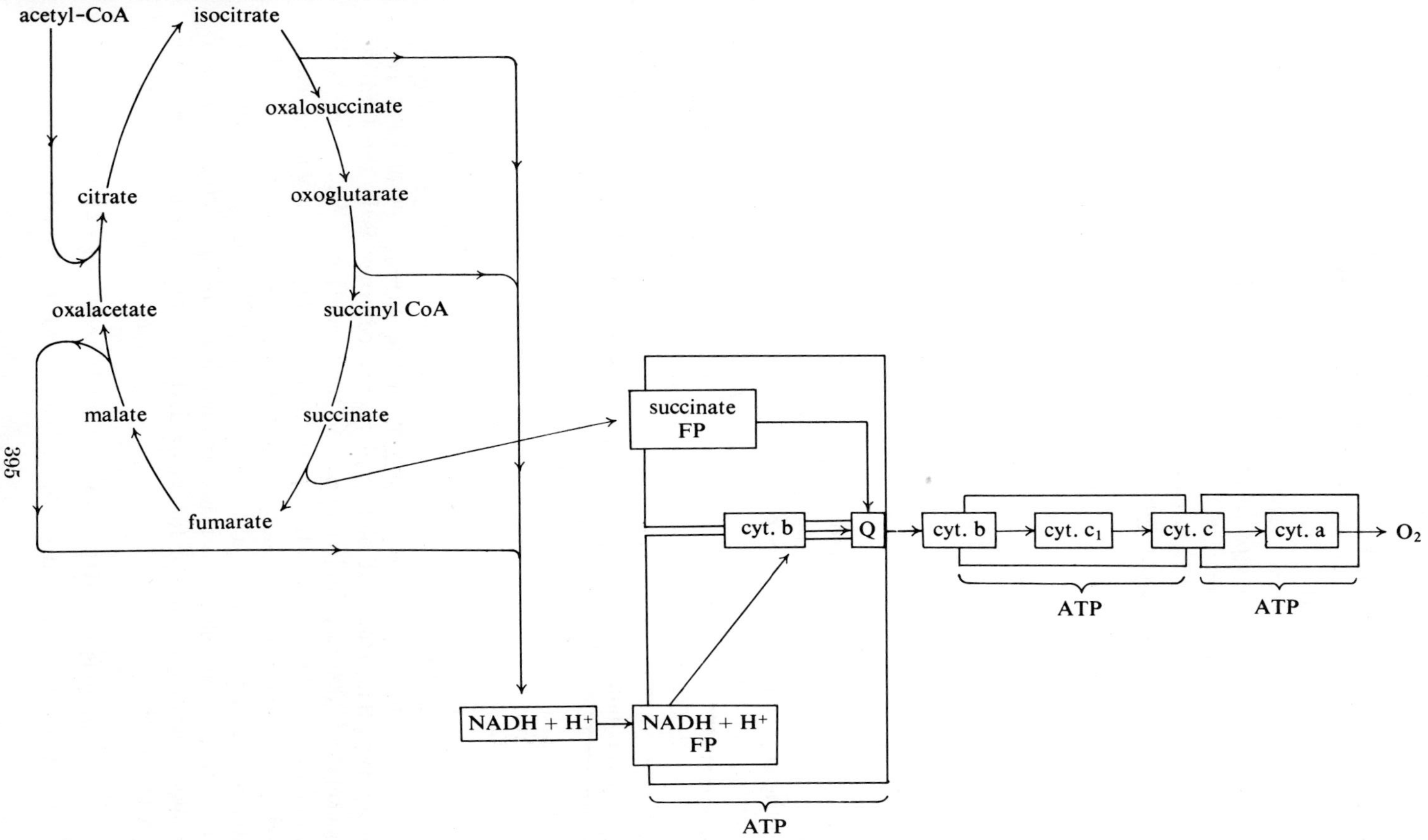
acetyl-CoA
isocitrate
oxalosuccinate
citrate
oxoglutarate
oxalacetate
succinyl CoA
malate
succinate
fumarate
succinate
FP
cyt. b
Q
cyt. b
cyt. c1
cyt. c
cyt. a
O2
ATP
ATP
NADH + H+
NADH + H+
FP
ATP

Fig. 7.2. Electron transport system of aerobic chemoorganotrophs (reprinted with permission of Business Publications, Ltd.) (115).

The two carriers in the chain, coenzyme Q and cytochrome c, are thought to be relatively free moving molecules and can be transferred from one protein molecule to another. This multiplicity of catalysts provides a device for tapping the energy of the system. It also takes into account the great variety of cytochrome pathways in microorganisms (see Chapter 2).

A molecule of cytochrome can carry one electron at a time, whereas those of NAD and flavoprotein can carry two at a time. Therefore, each cytochrome must react twice during the reoxidation of reduced NAD or FAD:

$$NADH + H^+ + FAD^+ \rightarrow NAD^+ + FADH + H^+$$

$$FADH + H^+ + 2\ \text{cyt. } b_{ox} \rightarrow FAD^+ + 2\ \text{cyt. } b_{red}$$

$$2\ \text{cyt. } b_{red} + 2\ \text{cyt. } c_{ox} \rightarrow 2\ \text{cyt. } b_{ox} + 2\ \text{cyt. } c_{red}$$

$$2\ \text{cyt. } c_{red} + 2\ \text{cyt. } a_{ox} \rightarrow 2\ \text{cyt. } c_{ox} + 2\ \text{cyt. } a_{red}$$

$$2\ \text{cyt. } a_{red} + \tfrac{1}{2}O_2 \rightarrow 2\ \text{cyt. } a_{ox} + H_2O$$

When electrons are transferred from one compound to another, an oxidation–reduction reaction takes place; the electron donor is the reducing agent and the electron acceptor the oxidizing agent.

During complete oxidation of 1 mole of glucose, 12 pairs of electrons pass down the respiratory chain with each pair giving a free energy of $\Delta F' = -52{,}000$ cal/mole. This means a total of $\Delta F' = -624{,}000$ cal/mole of glucose oxidized against a free energy of combustion of glucose of $\Delta F' =$ 686,000 cal/mole. The phosphorylation of 1 mole ADP to 1 mole of ATP requires an input of at least 7000 cal. Because it is assumed that 3 moles of ATP are formed, and because the oxidation of $NADH + H^+$ delivers 52,000 cal, we can deduce that the oxidative phosphorylation of 3 moles of ADP conserves $3(7000/52{,}000) \times 100 = 41\%$ of the total energy yield when 1 mole of $NADH + H^+$ is oxidized by oxygen (172). This high delivery of energy by reduced NAD oxidation makes it clear why a respiratory chain must have so many carriers acting in sequence and not just one or two. The standard biological energy currency is in the form of "packets" of 7000 cal, the free energy of the formation of ATP from ADP and phosphate. The electron transport chain is therefore really a molecular device for delivering the 52,000 cal of energy in a series of small packets, three of which are energetically equivalent to ATP. This is only possible if the energy of the electrons is lowered in a series of small steps. It is now possible to write the equation for the oxidation of two molecules of pyruvate to acetyl-CoA:

$$2\ \text{pyruvate} + 2NAD^+ + CoA \rightarrow 2\ \text{acetyl-CoA} + 2CO_2 + 2NADH + H^+$$

$$2\ NADH + H^+ + 6P_i + 6ADP + O_2 \rightarrow 2NAD^+ + 8H_2O + 6ATP$$

and also for the coupled oxidation of 2 moles of acetate via two operations of the tricarboxylic acid cycle

$$2\ \text{acetate} + 24P_i + 24ADP + 4O_2 \rightarrow 4CO_2 + 28H_2O + 24ATP$$

The complete oxidation of 2 moles of pyruvate to carbon dioxide and water forms 30 moles of ATP. This figure looks very impressive, for it conserves 42% of the energy produced. It is very likely, however, that this figure is only minimal, because cells are open systems and the efficiency is more likely to be around 60%.

Open systems in general exist in a dynamic steady state. It has been recognized that many enzyme systems in the cell have self-adjusting and self-regulating features; thus, the rate of the overall process being catalyzed is geared to the needs of the cell for the products of the system. The ADP: ATP ratio, for example, is very critical for the rate of respiration. If this ratio is low, the rate of respiration is low because it is limited by the supply of ADP. The amount of ADP available therefore represents the dynamic balance of the system if there are no restrictions on acetate or oxygen supplies.

The actual coupling mechanism that conserves energy by converting ADP to ATP is still obscure in microorganisms (224). It is assumed that it is the same as that studied in mitochondria. There is some evidence (42, 324) that in bacteria the chain is shorter, with fewer catalysts, for P/O ratios of 1.0 and below are usually obtained with bacterial cell-free extracts. This small ratio suggests that there is only one phosphorylation site in the electron transport chain. It is possible, however, that the chain operates only in whole cells. It is also possible that bacterial cytochromes have different oxidation–reduction potentials because their absorption bands often differ from those of animal cells. This probably is one of the reasons the cytochrome oxidase reaction is not in all circumstances

$$O_2 + 4H^+ \rightarrow 2H_2O$$

Some cytochrome oxidases transfer only one pair of hydrogen atoms to molecular oxygen, thus forming hydrogen peroxide

$$O_2 + 2H^+ \rightarrow H_2O_2$$

Hydrogen peroxide, however, is extremely toxic to bacteria and must be rapidly removed. This can be accomplished in the presence of two iron porphyrin enzymes, catalase (hydrogen-peroxide:hydrogen-peroxide oxidoreductase, EC 1.11.1.6) and either (*a*) peroxidase (donor:hydrogen-peroxide oxidoreductase, EC 1.11.1.7), or (*b*) reduced-NAD:hydrogen-peroxide oxidoreductase, EC 1.11.1.1; (*c*) reduced-NADP:hydrogen-peroxide oxidoreductase, EC 1.11.1.2; (*d*) ferrocytochrome c:hydrogen-peroxide oxidoreductase, EC 1.11.1.5; or (*e*) glutathione:hydrogen-peroxide oxidoreductase, EC 1.11.1.9. The difference in the reactions is

$$\text{catalase} \quad 2H_2O_2 \rightarrow 2H_2O + O_2$$

$$\text{peroxidase} \quad H_2A + H_2O_2 \rightarrow A + 2H_2O$$

where A may be an electron donor, cytochrome c, reduced NAD, reduced NADP, or glutathione. Catalase is found in most cytochrome-containing aerobic microorganisms, whereas most anaerobic microorganisms seem to possess peroxidase or substrate peroxidase. This generalization, however, is still disputed, because it was found that the catalase test carried out on microorganisms is not specific but could interfere with peroxidases present. Electrochemical analysis of enzymatic reactions (118, 240) could help solve these difficulties. A new approach to studying electron transfer and metabolism, called the "coulokinetic" technique, should give better insight into the present problems in this field (5).

Mycobacterium phlei

The respiratory components of *Mycobacterium phlei* contain bound NAD^+, flavins, a naphthoquinone (K_9H), and cytochromes b, c, c_1, a, and a_3. Three distinct respiratory chains were revealed (18); namely, one with succinic oxidase, one with malate, and one with an NAD^+-linked mechanism, wherein X represents a light-sensitive component (360 nm) and Y a nonheme iron (223). The malate respiratory chain (Fig. 7.3) consists of both particulate and soluble fractions (16, 17, 276). The soluble fraction

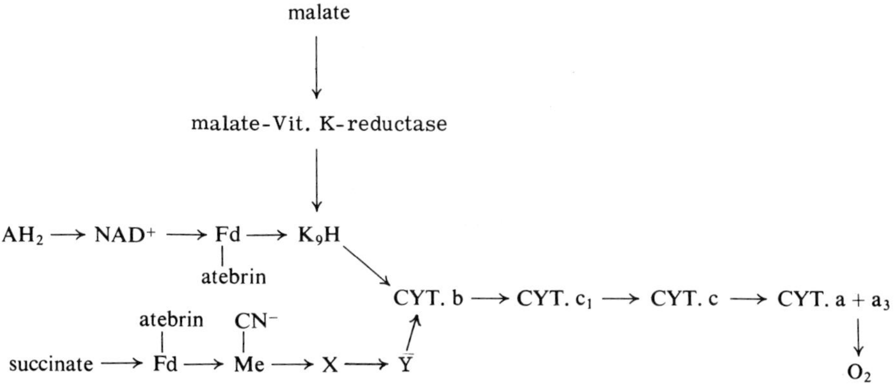
malate
malate-Vit. K-reductase
AH₂ → NAD⁺ → Fd → K₉H
atebrin
CYT. b → CYT. c₁ → CYT. c → CYT. a + a₃
atebrin
CN⁻
succinate → Fd → Me → X → Y
O₂

Fig. 7.3. Branched electron transport system in *Mycobacterium phlei.*

could be resolved into a reductase fraction and a phosphorylating factor free of reductase. The addition of the reductase fraction to the particulate fraction stimulated malate oxidation. However, for phosphorylation to occur the addition of the phosphorylating factor fraction was necessary (19, 134, 276, 425, 426). The reductase was inhibited by AMP and ATP, which could be reversed by FAD.

It is now almost certain that *Mycobacterium phlei* possesses two pathways for NADH + H^+ utilization. One of these involves the oxidation with molecular oxygen via the respiratory chain, located in the cytoplasmic membrane, whereas the other appears to be a nonphosphorylative pathway in the soluble fraction (426). There is some doubt as to whether reduced-NAD:menadione reductase (possibly EC 1.6.99.2) (19) or reduced-NAD: dichlorophenolindophenol reductase (425) is involved in the nonphosphorylated pathway.

A particulate transhydrogenase was responsible for oxidizing reduced NADP, thus associating phosphorylation with the NAD^+-linked pathway (275). The oxidation of reduced NADP resulted in a reduction of NAD^+ and supplied the NAD^+-linked transport system with NADPH + H^+. However, the nonphosphorylative bypass enzymes are also capable of oxidizing NADPH + H^+ and may thus play a regulatory function.

Bacillus

In this genus there appear to exist two slightly different electron transport chains (Scheme 7.4). The electron transport of *Bacillus subtilis* (266, 267) has some similarities to that of *Mycobacterium phlei*, as the succinate oxidation does not involve a menaquinone or vitamin K. Even during the reduced NAD oxidation (20, 21, 346, 378) it is reported possible to transfer electrons from the flavoprotein level to the cytochrome c-554 or cytochrome c-550 level without the requirement of vitamin K_2.

Investigations into the electron transport system of *Bacillus megaterium*

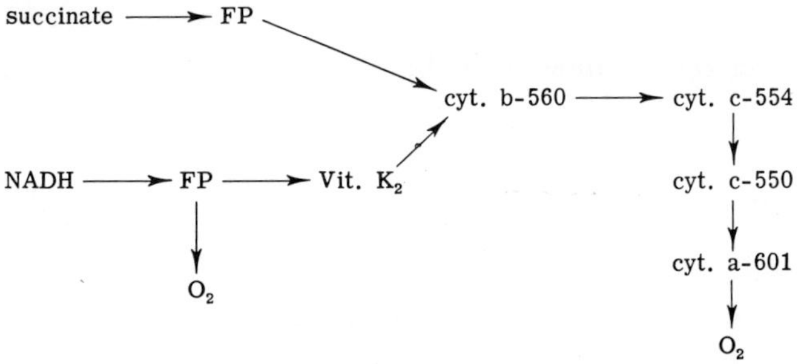
succinate
FP
cyt. b-560
cyt. c-554
NADH
FP
Vit. K2
cyt. c-550
cyt. a-601
O2
O2

Scheme 7.4. Electron transport system in *Bacillus subtilis*.

(221) demonstrated that menaquinones play an important role in the electron transport system of this organism. Apart from menaquinone, cytochromes b, c, a, and o were found to participate, making the electron transport chain very similar to that in the gram-negative bacteria, *Escherichia coli*, etc. The function of the menaquinone was very similar to that of ubiquinone. The position of menaquinone was decided to be between the

$$\text{NADH} + \text{H}^+ \rightarrow \text{MK} \rightarrow \text{cyt.b}_1 \rightarrow \text{cyt.c} \rightarrow \text{cyt.a,o} \rightarrow \text{O}_2$$

NADH dehydrogenase and cytochrome b_1.

Enterobacteriaceae

The tendency of investigations clearly indicates that quinones in general could be considered as an integrated part of the electron transport system leading from reduced NAD to the formation of water (43, 96, 221). As the members of the Enterobacteriaceae are facultative anaerobes, studies into the effect of aerobic or anaerobic conditions on the quinones indicate that ubiquinones are predominant under aerobic and menaquinones under anaerobic conditions (298, 401). *Escherichia coli* (72, 76) and *Aerobacter aerogenes* (192) possess a heme-containing cytochrome b_1 as well as ubiquinone-8, and cytochrome oxidases. The concentration of ubiquinone is required to be 27 times that of cyt.b_1 in order to obtain maximal oxidase activity (281). These studies also revealed that the electron transport system with nitrate and/or oxygen as terminal electron acceptor utilizes the same carriers, cytochrome b_1, and ubiquinone-8 (192). However, for *Escherichia coli* a scheme has been proposed whereby the ubiquinone is complexed to an electron carrier in at least two positions in the electron transport sequence (72). A further difference between *Escherichia coli* and

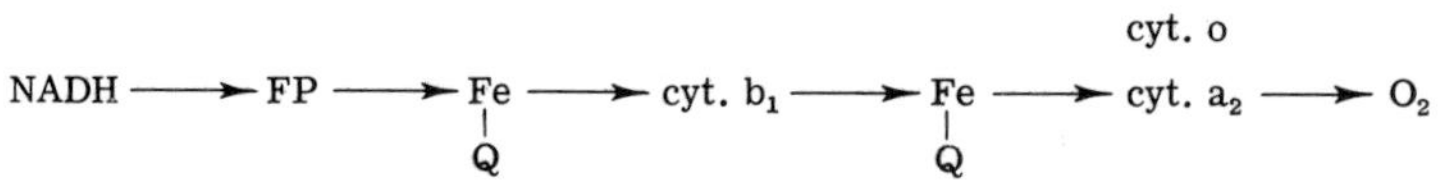
NADH
FP
Fe
Q
cyt. b1
Fe
Q
cyt. o
cyt. a2
O2

Aerobacter aerogenes appears to be the additional cyt.a_1 in *Aerobacter aerogenes*.

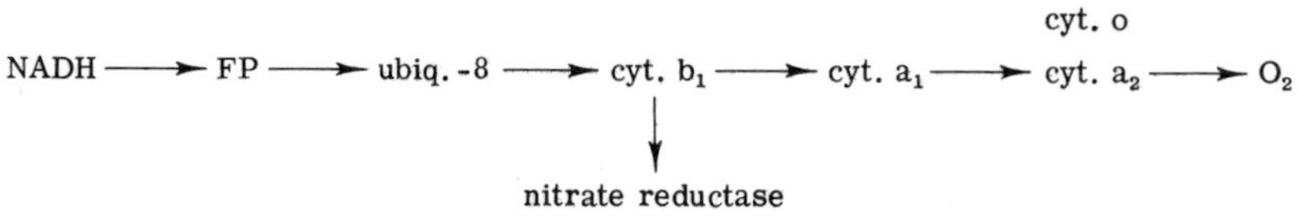
NADH
FP
ubiq.-8
cyt. b1
cyt. a1
cyt. o
cyt. a2
O2
nitrate reductase

The branched pathway of this organism is very similar, although not identical, to the one of *Escherichia coli*.

Salmonella typhimurium exhibits almost identical carrier components with ubiquinone-8 (84) and cytochromes a_1, a_2, and b_1. The possibility of a cytochrome c has been raised (85). It can therefore be assumed that this microorganism exhibits a similar electron transport sequence to *Aerobacter aerogenes.*

The electron sequence in *Azotobacter* has been mainly studied with *Azotobacter vinelandii.* It was found not only to contain ubiquinone-8 and the terminal cytochromes a_1, a_2, and o (175, 193, 367), but also cytochromes b_1, c_4, and c_5 (172).

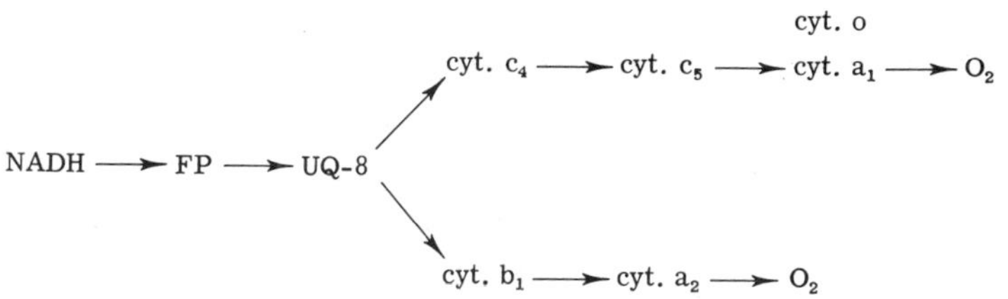
cyt. o
cyt. c4 → cyt. c5 → cyt. a1 → O2
NADH → FP → UQ-8
cyt. b1 → cyt. a2 → O2

Sensitivity studies with KCN revealed the functioning of a branched electron transport sequence, for cytochromes a_1 and o were highly sensitive, whereas cytochrome a_2 exhibited a low sensitivity toward KCN (172). The purified cytochrome c components of the particulate fraction (366) were found to react directly with the terminal oxidases. It is not yet known, however, whether or not ubiquinone-8 plays an important role at the branching point. Oxidative phosphorylation studies indicate (92, 173) that the pathway via cytochrome b_1 is phosphorylating, for the P/NADH ratio is not influenced in the presence of 100 μM cyanide. The very high oxidation rate and low phosphorylation yield could indicate the presence of a phosphorylating and a nonphosphorylating electron transport sequence. *Azotobacter vinelandii* also possesses a pyridine nucleotide transhydrogenase (66).

Halophilic Bacteria

Because of their extreme salt requirement for growth, the cell wall and the pattern of the membrane-bound electron transport of halophilic bacteria were the subject of special investigation. The organism itself requires about 4 M salt for maintaining its cell integrity (229). Despite the high salt concentrations, broken cell preparations rapidly oxidize NADH + H^+, α-glycerophosphate, and succinate. These preparations also exhibited at least two b-type and two c-type cytochromes (214). Of the terminal oxidases, a variation of between 1 and 3 has been reported, indicating a possible complex electron transport system. The NADH + H^+ pathway in *Halobacterium cutirubrum* was found (227) to contain a flavoprotein that was reduced by NADH + H^+. The succinate pathway was regarded as

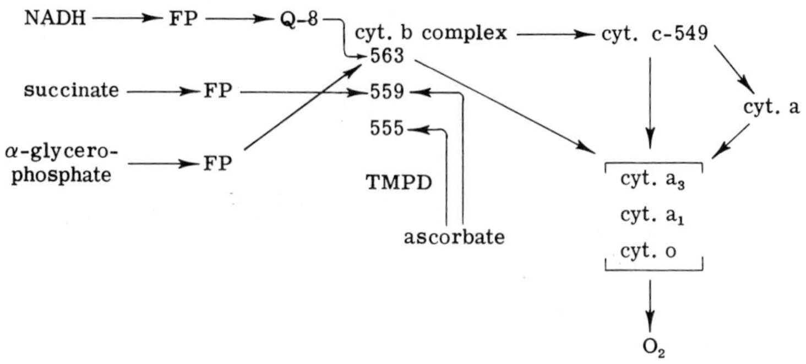
NADH
FP
Q-8
cyt. b complex
cyt. c-549
563
succinate
FP
559
cyt. a
555
α-glycero-
phosphate
FP
TMPD
cyt. a3
cyt. a1
ascorbate
cyt. o
O2

Scheme 7.5. Electron transport system in *Halobacterium.*

a separate branch consisting of a flavoprotein and cytochrome b-559. Both branches are thought to meet at the cytochrome c level. A second cytochrome b was also reduced by NADH + H^+ but was considered to lie outside the main branch of the pathway. In a search for quinone participation (228), a very strict dependence of menadione reductase (EC 1.6.99.2) on the salt concentration was observed but the enzyme was not thought to be an intermediate carrier in any of the two branches. As far as terminal oxidases are concerned, two functional oxidases, cytochrome a_3 and cytochrome o were suggested to be present (61). These two oxidases were also found in *Halobacterium halobium* and *H. salinarum* (62). These oxidases require up to 5.0 *M* NaCl for maximal activity (243) and stability. It is suggested that hydrophobic forces predominate in giving the stability at high salt concentrations. The main difference between *H. cutirubrum* and *H. salinarum* appears to be that the terminal oxidases are cytochromes o and a_1 in the latter instead of cytochromes o and a_3. *Halobacterium salinarum* appears to have the same system as *H. cutirubrum* (63). If all the obtainable data on the electron transport particle studies of halophilic bacteria are combined, the complex system shown in Scheme 7.5 would arise. Halophilic bacteria therefore appear to have three different branched pathways, with ascorbate being reduced by at least two b-type cytochromes (64). The incorporation of ubiquinone-8 is questionable as it has only been reported once (253).

Carboxylic Acid Metabolism

Glyoxylate or Dicarboxylic Acid Cycle

The discovery of the enzyme isocitrate lyase (*threo*-D_s-isocitrate glyoxylate-lyase, EC 4.1.3.1) (352) and malate synthase [L-malate glyoxylate-

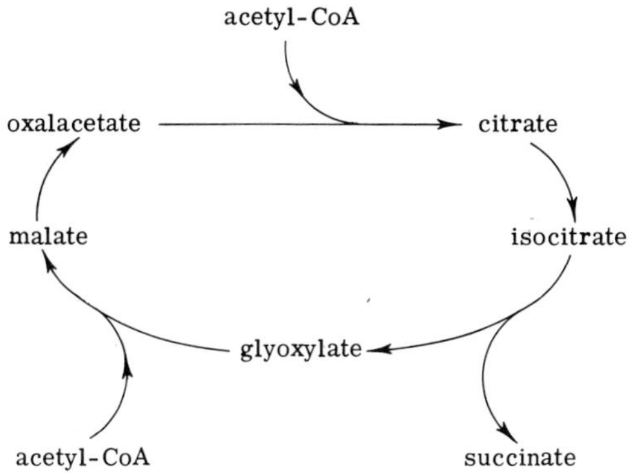
acetyl-CoA
oxalacetate
citrate
malate
isocitrate
glyoxylate
acetyl-CoA
succinate

Fig. 7.4. The glyoxylate cycle in microorganisms.

lyase (CoA acetylating), EC 4.1.3.2] (414) led to the proposition of the existence of a cyclic mechanism for replenishing C_4 acids from the tricarboxylic acid cycle for cellular biosynthesis (Fig. 7.4). This cyclic mechanism plays an important role in the metabolism of short-chain fatty acids (395) and is called the "glyoxylate cycle" (203–205).

As outlined above (see p. 394), the tricarboxylic acid cycle was responsible for terminal respiration and was also the only one for the complete oxidation of acetate (207). The cycle is therefore unable to produce compounds more oxidized than acetate. In addition to its role in terminal respiration, the tricarboxylic acid cycle also plays an important role in cell material synthesis (108, 323). It provides 2-oxoglutarate, which is the precursor of glutamic acid, one of the key compounds in amino acid and protein biosynthesis, as well as oxalacetate, which supplies aspartate and other amino acids (122).

If the cycle operates only for the oxidation of acetate, then all that is required is a supply of acetate and a catalytic amount of oxalacetate, which is regenerated in the cycle. If, however, biosynthetic processes are going on at the same time, there is a continual tapping off of both 2-oxoglutarate and of the four-carbon dicarboxylic acids which also serve as important precursors for these processes. Such a draining off of intermediates will only continue to operate if the resulting deficit of four-carbon dicarboxylic acids is made up. This problem is even greater when microorganisms use acetate as both a carbon and an energy source. The cell must be able to synthesize the four-carbon dicarboxylic acids from acetate. The organisms therefore use a second system, by which they make use of six of the eight reactions of the tricarboxylic acid cycle and bypass those steps which result in the evolution of carbon dioxide (isocitrate → 2-oxoglutarate → succinate). They replace these with reactions the net effect of which is an increase in the organic acid carbon of the system (210).

The formation of L-isocitrate from acetyl-CoA and oxalacetate is the same as in the tricarboxylic acid cycle, involving the two enzymes citrate synthase (EC 4.1.3.7) and aconitate hydratase (EC 4.2.1.3). Instead of being metabolized to oxalosuccinate, isocitrate undergoes an aldo cleavage to succinate and glyoxylate (289). This reaction is catalyzed by isocitrate

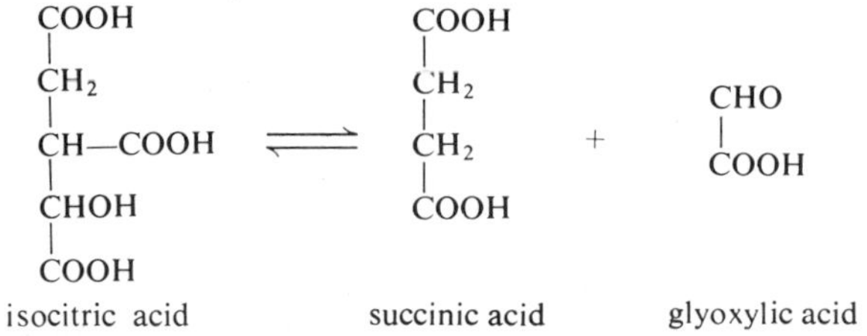
COOH
CH2
CH—COOH
CHOH
COOH
isocitric acid
COOH
CH2
CH2
COOH
succinic acid
+
CHO
COOH
glyoxylic acid

lyase (48, 342, 358) (*threo*-D-isocitrate glyoxylate-lyase, EC 4.1.3.1).

The second step of this bypass involves the condensation of acetyl-CoA with glyoxylate to form malate, with malate synthase [L-malate glyoxylate-lyase (CoA acetylating), EC 4.1.3.2] as catalyst. There is no combustion

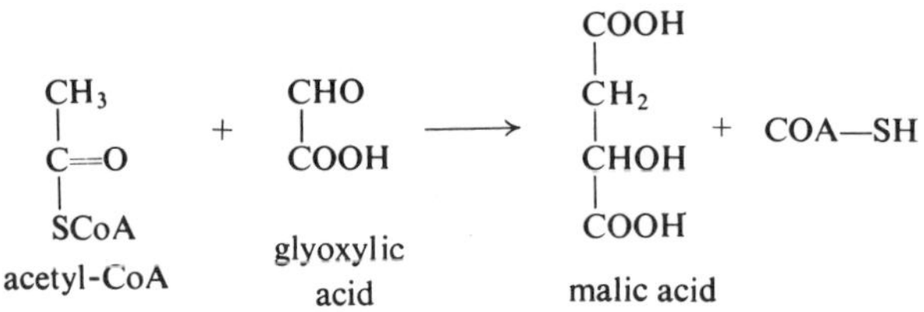
CH3
C=O
SCoA
acetyl-CoA
+
CHO
COOH
glyoxylic acid
COOH
CH2
CHOH
COOH
malic acid
+ COA—SH

of acetate with energy release but there is a net provision of four-carbon dicarboxylic acids. The subsequent metabolism of these acids therefore provides the precursors of most cell constituents and allows the organism to grow on acetate as the sole carbon source.

The functioning of the glyoxylate cycle is controlled by the first enzyme of this bypass, isocitrate lyase (209, 214, 215, 314). The mechanism of this control will be dealt with later in this chapter.

Citrate Metabolism

Like all the other dicarboxylic acids, citrate can only be metabolized by those microorganisms with a special permease system. It is suggested that the bacterial citrate transport system (234) must have an energy-dependent permease system similar to that described above (see Chapter 6).

Citrate can be metabolized aerobically and fermentatively by members of the Enterobacteriaceae (287, 288), *Halobacterium* (3), and *Pseudomonas* (379).

Citrate is catalyzed via two different pathways depending on the presence or absence of sodium. According to its end product formation, one is called the "aerobic" and the second the "fermentative" pathway.

The aerobic pathway utilizes citrate via the tricarboxylic acid cycle to oxalacetate. This compound is decarboxylated to pyruvate and carbon dioxide by oxalacetate decarboxylase (oxalacetate carboxy-lyase, EC 4.1.1.3). Pyruvate is further metabolized via the gluconeogenic pathway (reversed EMP pathway) and acetate or acetyl-CoA to higher fatty acids, each of which is a biosynthetic pathway to forming cellular material. This pathway also operates in *Halobacterium salinarum* (3) and *Pseudomonas aeruginosa* (379).

The fermentative pathway, which can occur both aerobically and anaerobically, starts with a cleavage reaction of citrate to acetate and oxalacetate that is catalyzed by citrate lyase (citrate oxaloacetate-lyase, EC 4.1.3.6).

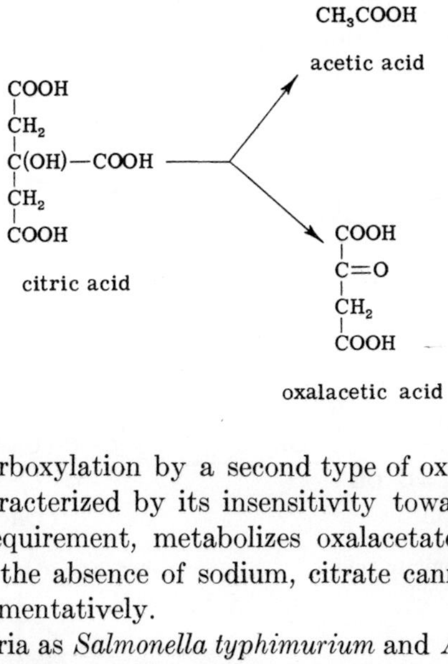
CH3COOH
acetic acid
COOH
CH2
C(OH)—COOH
CH2
COOH
citric acid
COOH
C=O
CH2
COOH
oxalacetic acid

The following decarboxylation by a second type of oxalacetate decarboxylase, which is characterized by its insensitivity toward EDTA and its absolute sodium requirement, metabolizes oxalacetate to pyruvate and carbon dioxide. In the absence of sodium, citrate cannot be metabolized anaerobically or fermentatively.

Such enterobacteria as *Salmonella typhimurium* and *Aerobacter aerogenes* therefore use the aerobic citrate pathway in sodium-free medium and the fermentative pathway in the presence of 5–10 m*M* sodium.

It has recently been discovered (329), with regard to the citrate transport into the cell, that oxalacetate decarboxylase also functions as a carrier protein by a novel mechanism. This sodium-activated oxalacetate decarboxylase is induced by citrate. This role of oxalacetate decarboxylase as a member of the membrane transport system may also be the reason for the higher citrate utilization under aerobic conditions (288). The main difference between the aerobic and anaerobic metabolism is therefore the citrate cleavage reaction, which is necessary because 2-oxoglutarate dehydrogenase is absent under anaerobic conditions (286). No glyoxylate cycle was observed during aerobic or anaerobic citrate metabolism.

Malate Metabolism

Aerobic malate metabolism has been reported from *Azotobacter vinelandii* (177) and pseudomonads (147). Whereas *A. vinelandii* oxidizes L-malate only with an FAD-dependent malate dehydrogenase, pseudomonads use an NADP-dependent L-malic enzyme [L-malate:NADP oxidoreductase (decarboxylating), EC 1.1.1.40] to decarboxylate L-malate to oxalacetate. An inducible NAD-dependent D-malic enzyme appears also to be present in the fluorescent species of *Pseudomonas* (147), which is able to oxidize D-malate.

Acetobacter xylinum was found to accumulate oxalacetate if grown on malate. The malate oxidation must be accomplished by a mechanism other than the one using the NAD-linked malic dehydrogenase [L-malate:NAD

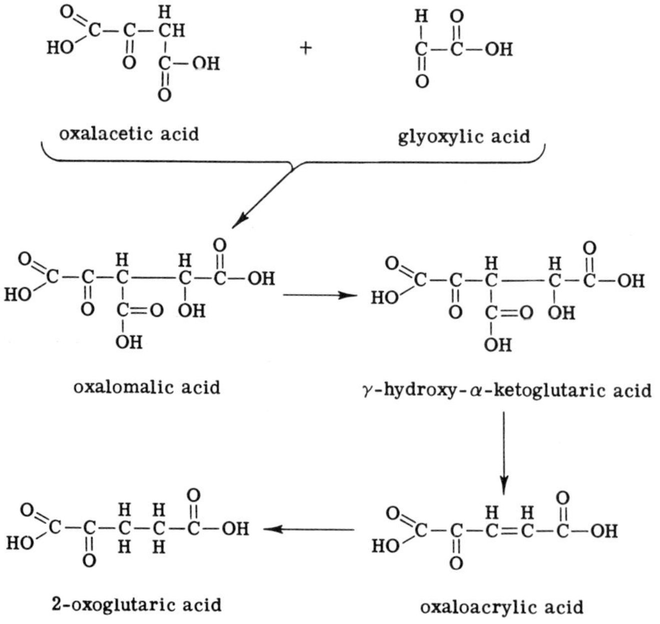
oxalacetic acid
glyoxylic acid
oxalomalic acid
γ-hydroxy-α-ketoglutaric acid
2-oxoglutaric acid
oxaloacrylic acid

Scheme 7.6. Formation of 2-oxoglutarate by *Acetomonas suboxydans*.

oxidoreductase (decarboxylating), EC 1.1.1.38], because whenever this enzyme is present, oxalacetate cannot accumulate (309). It appears that in *Acetobacter xylinum* malate dehydrogenase (L-malate:NAD oxidoreductase, EC 1.1.1.37) (34) functions to form oxalacetate from malate with the reduction of NAD^+ and the coupled formation of ATP. Kornberg and Phizackerley (212) first noted this reaction with *Pseudomonas*, in which the malate-oxidizing system oxidizes malate with oxygen or dichlorophenol-indophenol (DPI) as the acceptor. Investigations on cell-free extracts of *Acetobacter xylinum* revealed that oxalacetate is decarboxylated to pyruvate and carbon dioxide (33). It therefore appears that *Acetobacter xylinum* does not oxidize malate in one step but requires two steps with malate dehydrogenase (EC 1.1.1.37) and oxalacetate decarboxylase (oxalacetate carboxy-lyase, EC 4.1.1.3) as catalyzing enzymes. ATP, ADP, GTP, GDP, and inorganic phosphate do not promote the decarboxylation.

Although *Acetomonas suboxydans* is known not to possess the tricarboxylic acid cycle, the formation of 2-oxoglutarate is still a puzzle. It can be formed either from 2-ketogluconate or 2,5-diketogluconate or from a second pathway found with cell homogenates. It is not known as yet how the precursors glyoxylate and oxalacetate are formed (15) (see Scheme 7.6). This pathway could well operate in all other groups of *Acetobacter*, as they possess the glyoxylate cycle. 2-Oxoglutarate is used for amino acid biosynthesis.

Glycolate Metabolism

The growth of microorganisms on C_2 compounds as the sole source of carbon necessitates reactions whereby the C_2 substrates provide both metabolic energy and the carbon skeleton for cellular constituents. In growth on acetate both these functions are fulfilled with the operation of the tricarboxylic acid cycle and glyoxylate cycle. These cycles, however, are not sufficient for C_2 compounds at "higher oxidation levels than acetate" (207). Glycolate therefore must undergo some preliminary reactions to form a key intermediate that could connect with the tricarboxylic acid or glyoxylate cycle and that also could be able to build cell constituents.

The sequence in Fig. 7.5 has therefore been postulated (211). Glycolate is oxidized to glyoxylate by a still unidentified enzyme, which could be similar to the flavoprotein glycolic acid oxidase found in plants (67, 428) (glycolate:oxygen oxidoreductase EC 1.1.3.1). This reaction requires molecular oxygen and is exergonic, with $\Delta F = -41$ kcal/mole at pH 8.0

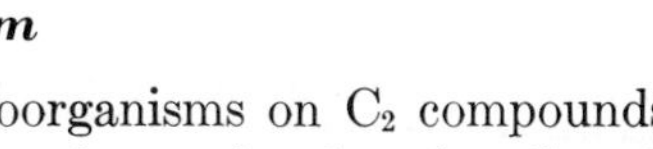
CH2OH
COOH
CHO
COOH

glycolic acid   glyoxylic acid

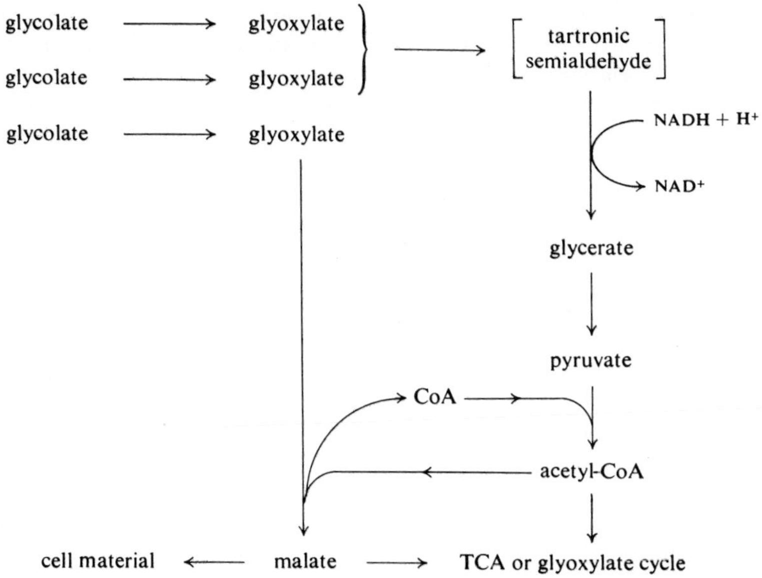
glycolate
glycolate
glycolate
glyoxylate
glyoxylate
glyoxylate
tartronic semialdehyde
NADH + H+
NAD+
glycerate
pyruvate
CoA
acetyl-CoA
cell material
malate
TCA or glyoxylate cycle

Fig. 7.5. Postulated route for the formation of cell constituents from glycolate by *Pseudomonas* (B_2 aba) (reprinted with the permission of *The Biochemical Journal*, London) (211).

(51). Under the influence of a carboligase [glyoxylate carboxy-lyase(dimerizing), EC 4.1.1.47] (212), two glyoxylate molecules then condense to an intermediate called "tartronic semialdehyde." Tartronic semialdehyde

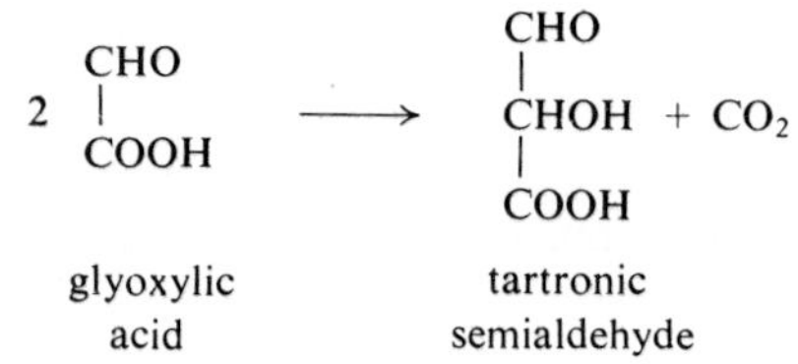
2
CHO
COOH
CHO
CHOH
COOH
+ CO2
glyoxylic acid
tartronic semialdehyde

reductase [D-glycerate:NAD(P) oxidoreductase, EC 1.1.1.60] reduces this semialdehyde to glycerate in an NAD^+-linked reaction. It is assumed

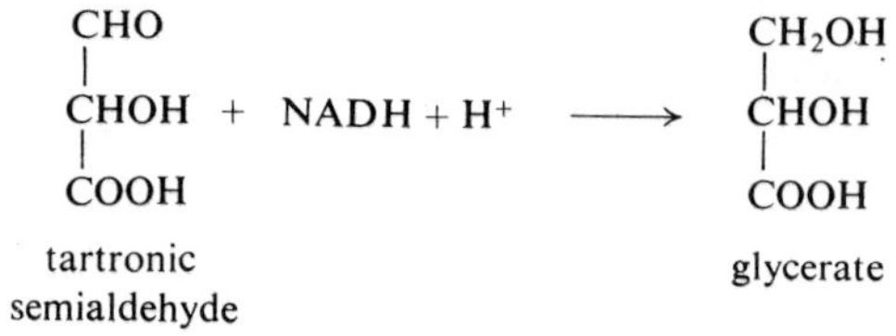
CHO
CHOH
COOH
+ NADH + H+
CH2OH
CHOH
COOH
tartronic semialdehyde
glycerate

that glycerate itself is then phosphorylated in a kinase reaction that almost certainly would involve glycerate kinase (ATP:glycerate 3-phosphotransferase, EC 2.7.1.31) and the formation of 3-phosphoglycerate. The latter product is further metabolized to pyruvate via phosphoenolpyruvate. High activities of isocitrate lyase (EC 4.1.3.1.) and malate synthase (EC 4.1.3.2) suggest that acetyl-CoA and malate are the two key intermediates for the complete oxidation of glycolate to carbon dioxide and for cell material. The presence of malate synthase, however, also suggests a direct conversion of glyoxylate to malate with the formation of acetyl-CoA, as occurs in *Azotobacter chroococcum* (223a).

Some *Pseudomonas* species are also able to reduce glyoxylate to glycolate, which is catalyzed by glyoxylate reductase (glycolate: NAD oxidoreductase, EC 1.1.1.26) (26).

Tartrate Metabolism

The breakdown of tartrate is marked by its stereospecific behavior. Stereospecific dehydrases attack (+)-, (−)-, or *meso*-tartrate and form

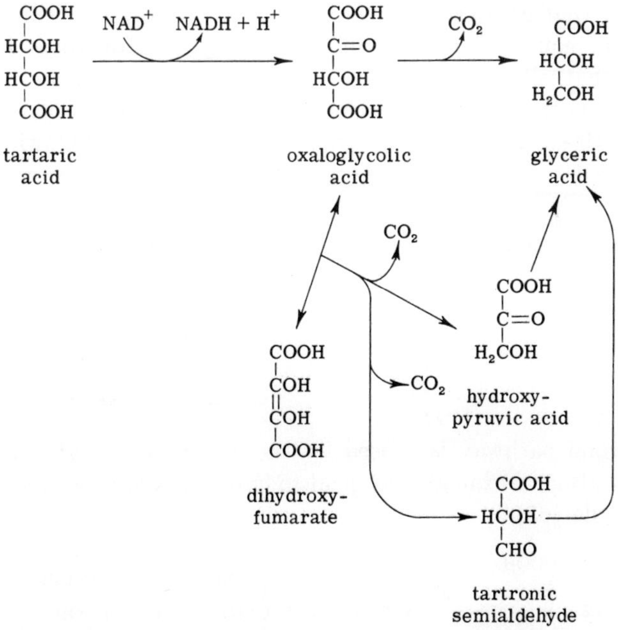

NAD+
NADH + H+
CO2
tartaric acid
oxaloglycolic acid
glyceric acid
dihydroxy-fumarate
hydroxy-pyruvic acid
tartronic semialdehyde

Scheme 7.7. Metabolism of tartaric acid by *Pseudomonas putida* and *Pseudomonas acidovorans* (reprinted with permission of the authors and the American Society of Biological Chemists).

oxalacetic acid. These reactions depend also on the availability of tartrate permeases, which bring the product to be metabolized across the cell membrane.

Pseudomonas putida and *P. acidovorans*, however, form glyceric acid from either L(+)- or *meso*-tartaric acid (73) (Scheme 7.7). This pathway contains a dismutation reaction (194) by which the tartaric acids are first oxidized by NAD^+ to oxaloglycolic acid.

The first step of tartaric acid metabolism is a NAD^+-linked oxidation by tartrate dehydrogenase (EC 1.1.1.93) (196) that converts tartaric acid to oxaloglycolic acid. This enzyme is specific for tartrate and does not attack L(−)- or D(+)-malic acid. *Pseudomonas acidovorans*, however, possesses an L-malic dehydrogenase (195) that is also NAD^+ linked and carries out both reactions, the conversion of L-malate to oxalacetate as well as of *meso*-tartrate to oxaloglycolic acid.

Oxaloglycolic acid, the α-keto acid intermediate in this pathway, is in equilibrium with dihydroxyfumarate, with the equilibrium constant favoring the formation of dihydroxyfumarate. This existing equilibrium is the reason for the two different pathways to glyceric acid. A reductive oxaloglycolic acid reductase (decarboxylating) (EC 1.1.1.92) has been crystallized (197) that catalyzes the conversion of oxaloglycolic acid to glyceric acid. As the equilibrium constant favors dihydroxyfumarate, it has been suggested that, although oxaloglycolic acid is the actual substrate, dihydroxyfumarate is the convertible substrate. As a possible mechanism for the $NADH + H^+$- or $NADPH + H^+$-requiring reaction, the following sequence was proposed

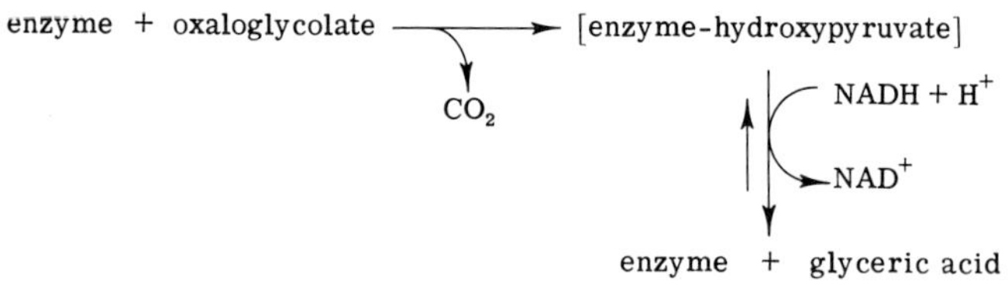
enzyme + oxaloglycolate
[enzyme-hydroxypyruvate]
CO2
NADH + H+
NAD+
enzyme + glyceric acid

The second pathway is caused by the cation catalyzed, nonenzymatic decarboxylation of oxaloglycolic acid to hydroxypyruvic acid and tartronic semialdehyde (65)

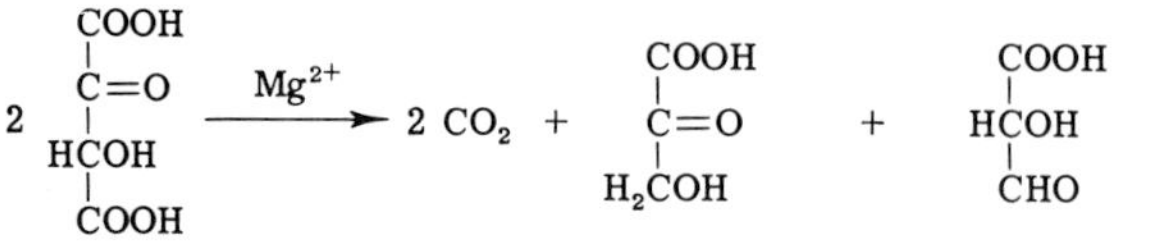
COOH
C=O
HCOH
COOH
2
Mg2+
2 CO2 +
COOH
C=O
H2COH
+
COOH
HCOH
CHO

oxaloglycolic acid | hydroxy-pyruvic acid | tartronic semialdehyde

Either of the two products are now converted to glyceric acid by either a NAD(P)H + H^+-linked hydroxypyruvate reductase (D-glycerate dehydrogenase) (EC 1.1.1.29) (198) or a NADH + H^+-linked tartronic semialdehyde reductase (EC 1.1.1.60) (199). The latter enzyme from *Pseudomonas putida* appears to be different from that of *Pseudomonas ovalis*.

The end product of the tartaric acid metabolism, glyceric acid, is probably phosphorylated to 3-phosphoglycerate in a fashion similar to that in the metabolism of oxalic acid. From this 3-phosphoglycerate level, metabolism could proceed via the EMP pathway into the tricarboxylic acid cycle.

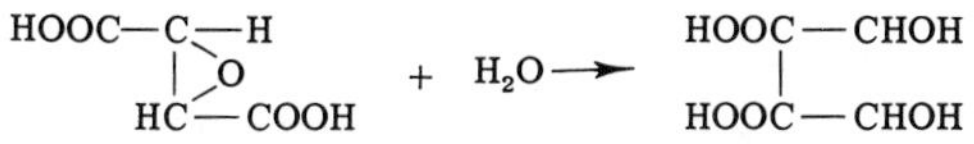

Pseudomonas putida possesses in tartrate epoxidase (6) an additional enzyme, which converts quantitatively both optical isomers of *trans*-epoxysuccinate to *meso*-tartrate.

Glucarate Metabolism

Glucarate can be metabolized as the sole source of carbon and energy by enterobacteria and *Pseudomonas* spp. (383) as well as by *Agrobacterium tumefaciens* (58). Whereas the pseudomonads and *Agrobacterium tumefaciens* use the same pathway, enterobacteria possess a different one (Fig. 7.6).

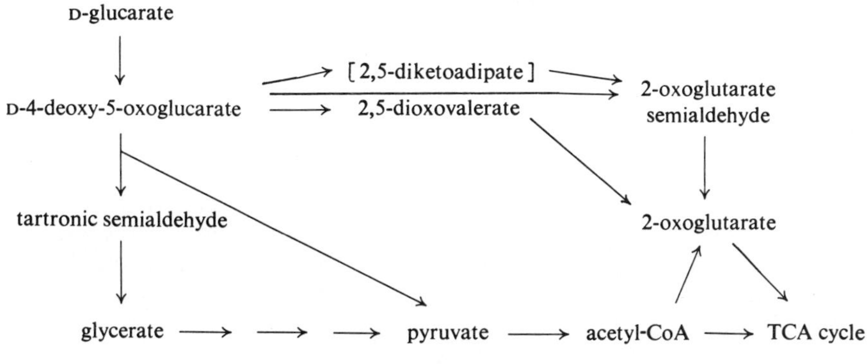
D-glucarate
[2,5-diketoadipate]
D-4-deoxy-5-oxoglucarate
2,5-dioxovalerate
2-oxoglutarate
semialdehyde
tartronic semialdehyde
2-oxoglutarate
glycerate
pyruvate
acetyl-CoA
TCA cycle

Fig. 7.6. Glucarate metabolism in microorganisms.

The first step in the glucarate metabolism appears to be common for all microorganisms. D-Glucarate is acted upon by a glucarate dehydratase (EC 4.2.1.40) (165) that converts D-glucarate to D-4-deoxy-5-oxoglucarate. It was purified from *Klebsiella aerogenes* and *Pseudomonas acidovorans*

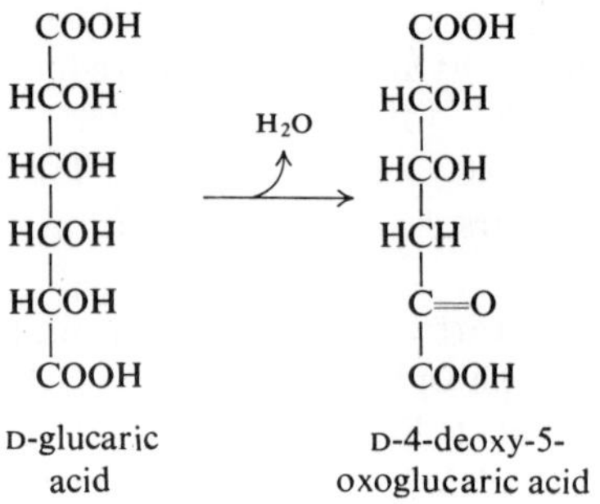
COOH
HCOH
HCOH
HCOH
HCOH
COOH
D-glucaric acid
H₂O
COOH
HCOH
HCOH
HCH
C=O
COOH
D-4-deoxy-5-oxoglucaric acid

(166a). The pathway taken by the *Pseudomonas* strains and *Agrobacterium tumefaciens* branches off at the D-4-deoxy-5-oxoglucarate level. An enzyme extracted from *Pseudomonas acidovorans* was purified (126, 166) and shown to catalyze the simultaneous dehydration and decarboxylation of D-4-deoxy-5-oxoglucarate by a Schiff base reaction with a lysine residue in the enzyme (see Scheme 7.8). The suggested name for the enzyme catalyzing this reaction is deoxyglucarate dehydratase [D-4-deoxy-5-oxoglucarate hydro-lyase (decarboxylating)].

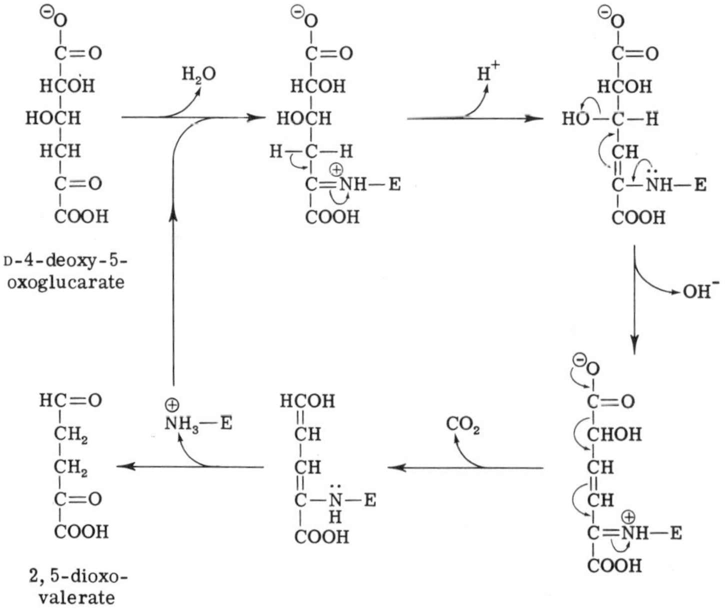
D-4-deoxy-5-oxoglucarate
H₂O
H⁺
OH⁻
CO₂
NH₃—E
NH—E
2, 5-dioxovalerate

Scheme 7.8. Schiff base reaction during D-4-deoxy-5-oxoglucarate conversion to 2,5-dioxovalerate by *Pseudomonas acidovorans* (reprinted with permission of the authors and the Biochemical Society, London).

2,5-Dioxovalerate is now oxidized by a NAD⁺-linked dehydrogenase (EC 1.2.1.26) and produces the key intermediate of the tricarboxylic acid

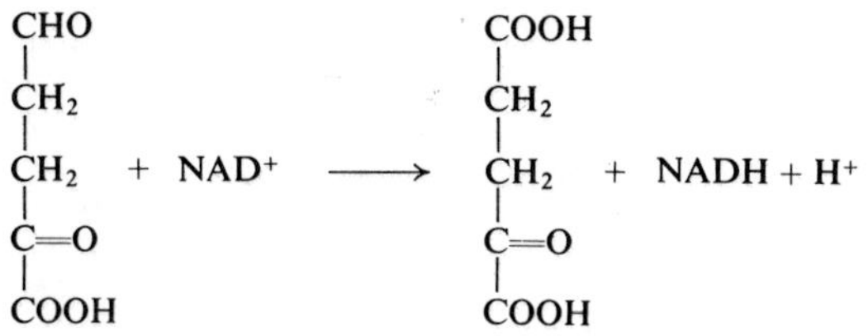
CHO
CH2
CH2
C=O
COOH
+ NAD+
COOH
CH2
CH2
C=O
COOH
+ NADH + H+

2,5-dioxovalerate 2-oxoglutarate

cycle, 2-oxoglutarate.

Variations of 4-deoxy-5-oxoglucarate metabolism have been reported (164) from an unidentified species of *Pseudomonas,* with 2-oxoglutarate semialdehyde as intermediate instead of 2,5-dioxovalerate.

Agrobacterium tumefaciens, however, appears to form 2,5-diketoadipate and 2-oxoglutaric semialdehyde as intermediates. Some of these intermediates have been found only by tracer element work.

The enterobacteria are not able to carry out the dehydration step from D-4-deoxy-5-oxoglucarate to 2,5-dioxovalerate but carry out a dismutation reaction step forming tartronic semialdehyde and pyruvate, which in turn are metabolized as described in the metabolism of tartrate (see p. 410). Whether or not this dismutation reaction is an aldolase cleavage carried out by the enzyme 2-keto-3-deoxy-D-glucarate aldolase (2-keto-3-deoxy-D-glucarate tartronate-semialdehyde-lyase, EC 4.1.2.20) is not known. This enzyme has, however, been established in *Pseudomonas acidovorans* (166a), where it was shown that the observed aldolase activity is associated with the glucarate dehydratase (glucarate hydro-lyase, EC 4.2.1.40) protein.

Oxalate and Formate Metabolism

Pseudomonas oxalaticus is able to grow on formate or oxalate as sole carbon source. In growth on formate, an oxidation with formate dehydrogenase (formate:ferricytochrome b_1 oxidoreductase, EC 1.2.2.1) produces CO_2, which is fixed by a mechanism similar to the Calvin cycle. Carbon dioxide enters at the level of phosphoglyceric acid or malic acid. Carbon dioxide, however, can also be fixed into pyruvate or phosphoenolpyruvate:

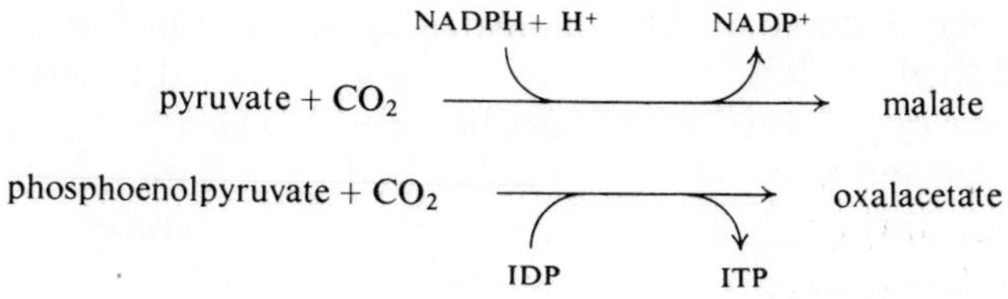
NADPH + H+
NADP+
pyruvate + CO2
malate
phosphoenolpyruvate + CO2
oxalacetate
IDP
ITP

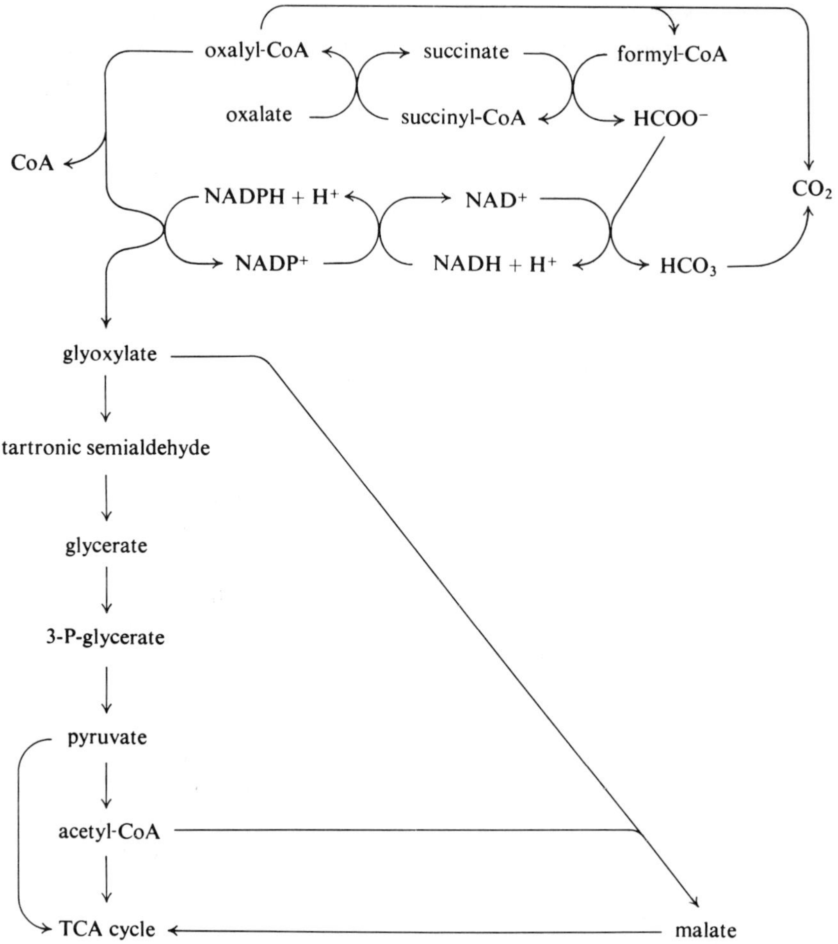

oxalyl-CoA
succinate
formyl-CoA
oxalate
succinyl-CoA
HCOO⁻
CoA
CO2
NADPH + H+
NAD+
NADP+
NADH + H+
HCO3
glyoxylate
tartronic semialdehyde
glycerate
3-P-glycerate
pyruvate
acetyl-CoA
TCA cycle
malate

Fig. 7.7. Oxalate and formate metabolism of *Pseudomonas oxalaticus*.

It is necessary for *Pseudomonas oxalaticus* to have a cosubstrate, such as pyruvate or glycine, present in order to metabolize formate (146).

If *P. oxalaticus* is grown on oxalate, ribulose diphosphate carboxylase [3-phospho-D-glycerate carboxy-lyase (dimerizing), EC 4.1.1.39] or carboxydismutase is absent (301–303) and the Calvin cycle is therefore inoperative. Kornberg suggested (206) the pathway outlined in Fig. 7.7. Oxalate is oxidized to glyoxylate. In order to derive a functional tricarboxylic acid or glyoxylate cycle, acetyl-CoA must be formed, which is done via several intermediates and pyruvate. Acetyl-CoA and glyoxylate are then able to form malate with the help of malate synthase [L-malate:glyoxylate-lyase (CoA-acetylating), EC 4.1.3.2]. Malate itself, as are acetyl-CoA and pyruvate, is able to carry out the oxidation of oxalate to completion. It is assumed that the glyoxylate cycle is mainly used to build up cell material of poly-β-hydroxybutyric acid, whereas the tricarboxylic acid cycle provides the energy.

With *Pseudomonas ovalis*, Kornberg (206) and Morris (269) clarified the pathway from glyoxylate to glycerate. They demonstrated that the conversion of glyoxylate to tartronic semialdehyde is carried out by a tartronate-semialdehyde synthase [glyoxylate carboxy-lyase (dimerizing), EC 4.1.1.47] with the tartronic semialdehyde reductase (383) [D-glycerate: NAD(P) oxidoreductase, EC 1.1.1.60] being responsible for the formation of glycerate. Glycerate kinase (ATP:D-glycerate 3-phosphotransferase, EC 2.7.1.31) phosphorylates glycerate and forms 3-phosphoglycerate. The continuation of the pathway is identical with the EMP pathway.

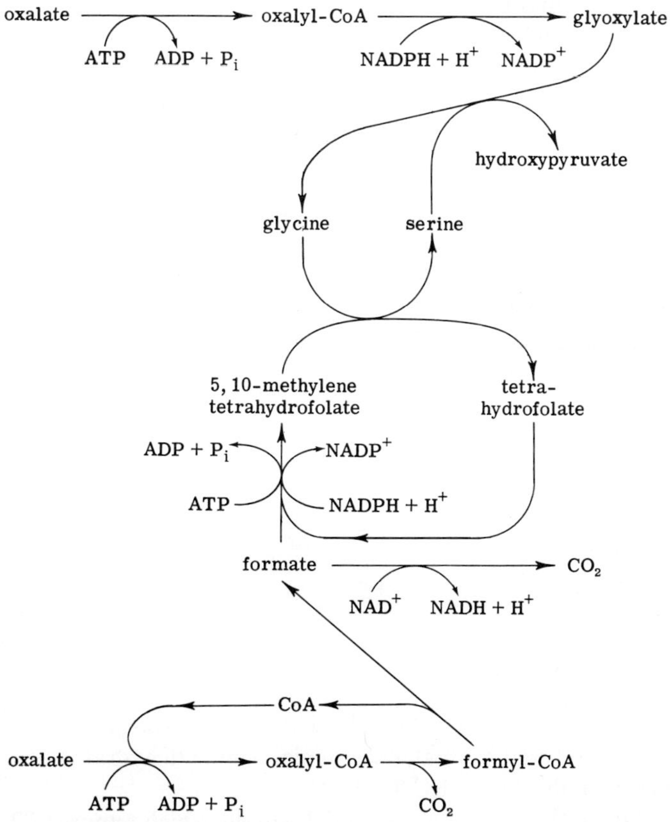
oxalate
oxalyl-CoA
glyoxylate
ATP
ADP + P$_i$
NADPH + H$^+$
NADP$^+$
hydroxypyruvate
glycine
serine
5, 10-methylene tetrahydrofolate
tetra-hydrofolate
ADP + P$_i$
NADP$^+$
ATP
NADPH + H$^+$
formate
CO$_2$
NAD$^+$
NADH + H$^+$
CoA
oxalate
oxalyl-CoA
formyl-CoA
ATP
ADP + P$_i$
CO$_2$

Fig. 7.8. Oxalate and formate metabolism by microorganisms using a serine–glyoxylate aminotransferase instead of tartronate semialdehyde synthase (reprinted with permission of the authors and The Biochemical Society, London) (38).

The decarboxylation of oxalyl-CoA to formyl-CoA and carbon dioxide is catalyzed by oxalyl-CoA decarboxylase (oxalyl-CoA carboxy-lyase, EC 4.1.1.8) (304). This reaction requires thiamine pyrophosphate as cofacter and is stimulated by Mg^{2+} or Mn^{2+} ions.

Microorganisms that do not possess the enzyme tartronate-semialdehyde synthase were found to use a different pathway, utilizing a serine–glyoxylate aminotransferase (EC 2.6.1.45), which produces hydroxypyruvate (39). The latter compound is then reduced to glycerate by D-glycerate dehydrogenase (EC 1.1.1.29). This pathway has great similarities to the pathway described above for the oxidation of formate by *Pseudomonas methanica* (see p. 441). The general metabolic sequence is outlined in Fig. 7.8.

Malonate Metabolism

The well-known inhibitor of succinic dehydrogenase (EC 1.3.99.1), malonate, can be oxidized by several strains of *Pseudomonas* nearly to completion. Malonate is first oxidized to malonate semialdehyde by a malonate semialdehyde dehydrogenase [malonate-semialdehyde:NAD(P) oxidoreductase, EC 1.2.1.15]. Malonic semialdehyde is also a hydration

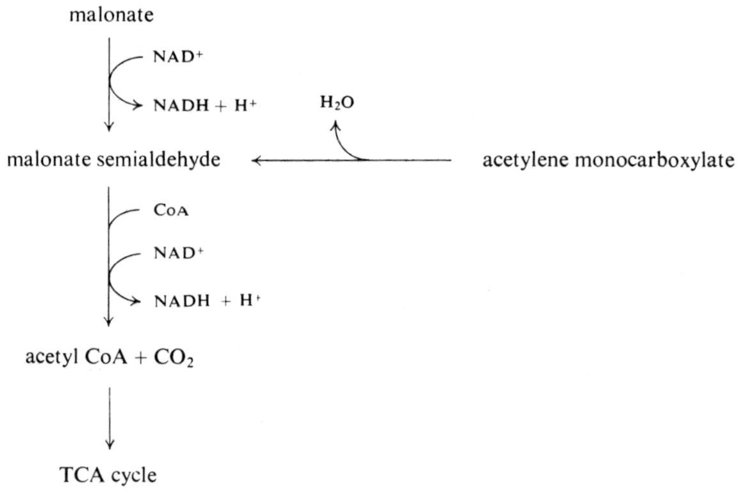
malonate
NAD⁺
NADH + H⁺
H₂O
malonate semialdehyde
acetylene monocarboxylate
CoA
NAD⁺
NADH + H⁺
acetyl CoA + CO₂
TCA cycle

product of acetylene monocarboxylic acid in *Pseudomonas fluorescens*. The hydration is catalyzed by malonate semialdehyde dehydratase (malonate-semialdehyde hydro-lyase, EC 4.2.1.27).

A decarboxylation of malonate semialdehyde, which requires ATP, CoA, and Mg^{2+} (411–413, 420, 421), produces acetyl-CoA, which can then enter the tricarboxylic acid cycle. The enzyme responsible for this last step is malonate semialdehyde dehydrogenase [malonate-semialdehyde:NAD(P) oxidoreductase (acylating CoA), EC 1.2.1.18] and could well be a dehydrogenase complex.

Itaconate Metabolism

Pseudomonas sp. "B_2s abo" and *P. fluorescens* are able to metabolize itaconate and *P. fluorescens,* in addition, methylsuccinate. Itaconate can be derived from the tricarboxylic acid cycle via aconitate. Aconitate decarboxylase (*cis*-aconitate carboxy-lyase, EC 4.1.1.6) is the catalyst for this reaction. The overall mechanism for itaconate metabolism has been elucidated by Cooper and Kornberg (69) and diagrammed by DeLey (81) as can be seen in Fig. 7.9. The enzymes involved in this pathway are (*a*) suc-

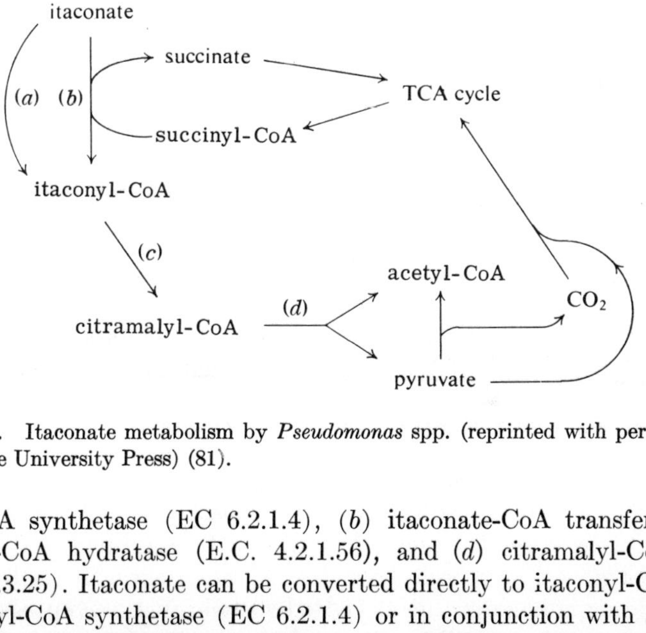
itaconate
succinate
TCA cycle
(a) (b)
succinyl-CoA
itaconyl-CoA
(c)
acetyl-CoA
(d)
CO2
citramalyl-CoA
pyruvate

Fig. 7.9. Itaconate metabolism by *Pseudomonas* spp. (reprinted with permission of Cambridge University Press) (81).

cinyl-CoA synthetase (EC 6.2.1.4), (*b*) itaconate-CoA transferase, (*c*) itaconyl-CoA hydratase (E.C. 4.2.1.56), and (*d*) citramalyl-CoA lyase (EC 4.1.3.25). Itaconate can be converted directly to itaconyl-CoA with a succinyl-CoA synthetase (EC 6.2.1.4) or in conjunction with succinyl-CoA and an itaconate-CoA transferase. In addition to itaconyl-CoA, the latter reaction produces succinic acid, which immediately reenters the

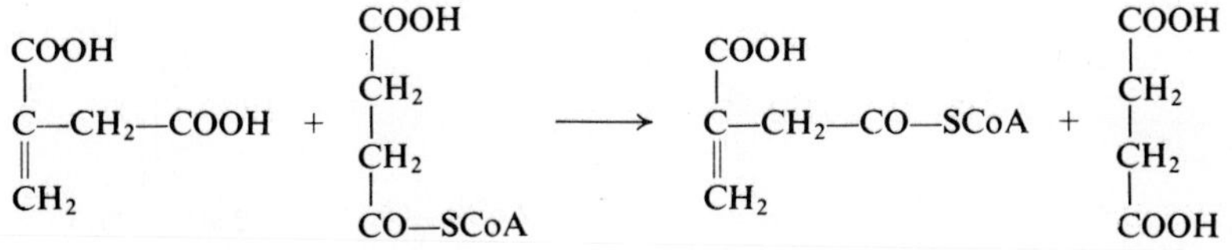

itaconate succinyl-CoA itaconyl-CoA succinate

tricarboxylic acid cycle. Itaconyl-CoA hydratase (citramalyl-CoA hydrolyase, EC 4.2.1.56) converts itaconyl-CoA to citramalyl-CoA:

$$\begin{array}{c} \text{COOH} \\ | \\ \text{C—CH}_2\text{—CO—SCoA} \\ || \\ \text{CH}_2 \end{array} \longrightarrow \begin{array}{c} \text{COOH} \\ | \\ \text{HOC—CH}_2\text{—CO—SCoA} \\ | \\ \text{CH}_3 \end{array}$$

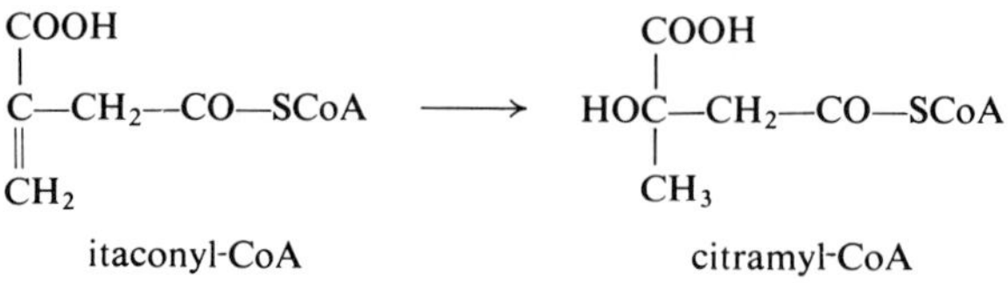

itaconyl-CoA → citramyl-CoA

The splitting of the carbon chain into pyruvate and acetyl-CoA is caused by citramalyl-CoA lyase (citramalyl-CoA pyruvate-lyase, EC 4.1.3.25):

$$\begin{array}{c} \text{COOH} \\ | \\ \text{HOC—CH}_2\text{—CO—SCoA} \\ | \\ \text{CH}_3 \end{array} \longrightarrow \begin{array}{c} \text{H}_3\text{C—CO—SCoA} \\ \text{acetyl-CoA} \\ + \\ \text{H}_3\text{C—CO—COOH} \\ \text{pyruvate} \end{array}$$

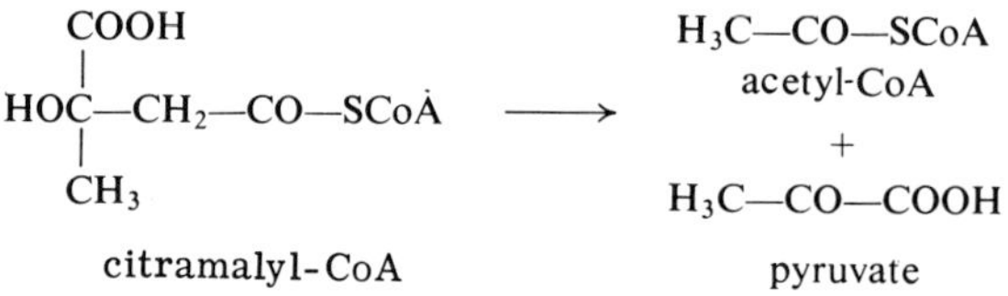

citramalyl-CoA

Both end products enter the tricarboxylic acid cycle for further metabolism. This itaconate degradation was also found to occur in *Micrococcus* sp. (70).

Glutarate Metabolism

The metabolism by *Pseudomonas fluorescens* of glutarate to carbon dioxide and acetyl-CoA (285) is similar to that in animal tissue (385) (Scheme 7.9). Glutarate must first be activated (264), or a CoA transferase reaction must take place, to form the CoA derivative glutaryl-CoA.

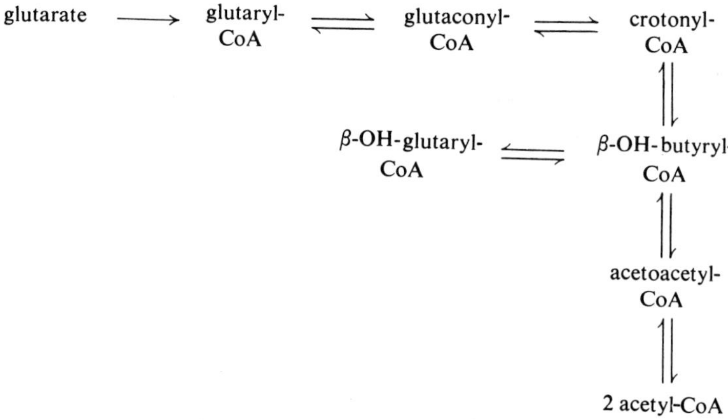

glutarate ⟶ glutaryl-CoA ⇌ glutaconyl-CoA ⇌ crotonyl-CoA

β-OH-glutaryl-CoA ⇌ β-OH-butyryl-CoA

acetoacetyl-CoA

2 acetyl-CoA

Scheme 7.9. Glutarate catabolism by *Pseudomonas fluorescens*.

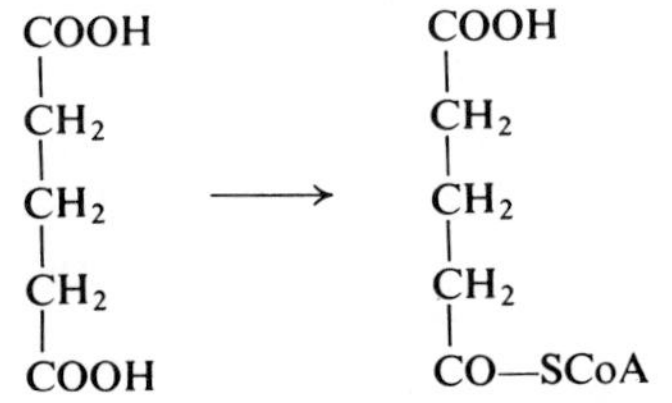

glutarate glutaryl-CoA

Glutarate cannot be metabolized as free acid (385, 386). This would parallel findings by Gholson and co-workers (107), who found that not glutarate, but glutaryl-CoA was formed from β-ketoadipate during the metabolism of tryptophan in animal tissue. The first product in the oxidation of glutaryl-CoA by a dehydrogenase was glutaconyl-CoA, which was immediately followed by a decarboxylation to crotonyl-CoA. The enzyme glutaryl-CoA dehydrogenase [glutaryl-CoA: (acceptor) oxidoreductase (decarboxylating), EC 1.3.99.7] catalyzes both reactions, the dehydrogenation as well

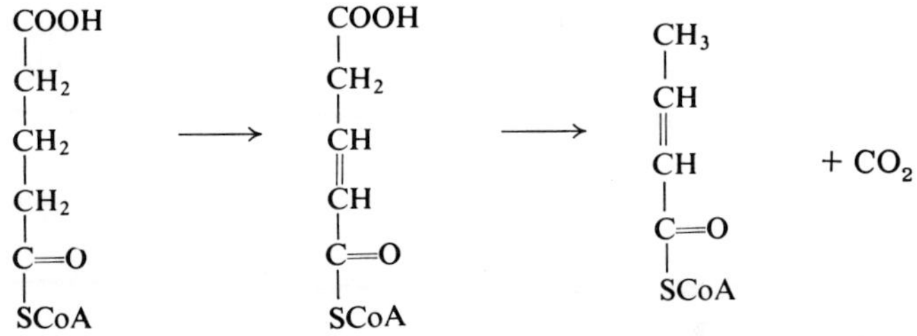

glutaryl-CoA glutaconyl-CoA crotonyl-CoA

as the decarboxylation, and is a flavoprotein.

Crotonyl-CoA can now undergo a hydration by the action of crotonase (L-3-hydroxyacyl-CoA hydro-lyase, EC 4.2.1.17) and forms β-hydroxybutyryl-CoA (284)

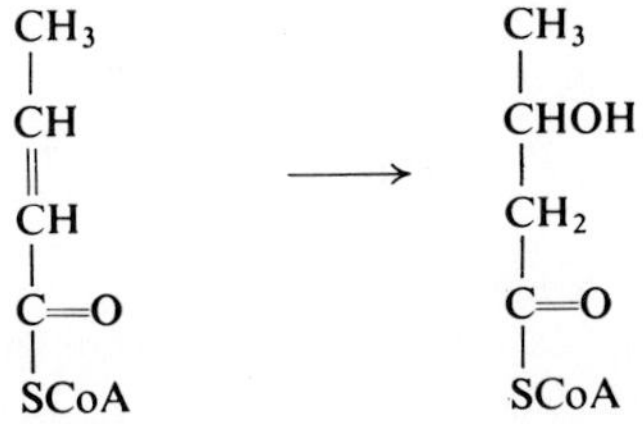

crotonyl-CoA β-hydroxybutyryl-CoA

Crotonyl-CoA itself can also enter the fatty acid oxidation pathway, where it is converted to acetoacetyl-CoA and to two molecules of acetyl-CoA. Whether or not *Pseudomonas* can take the alternate route via α-hydroxyglutarate to β-hydroxybutyryl-CoA or even metabolize α-hydroxyglutarate via acetone dicarboxylate → acetoacetate → acetate (141) is not confirmed.

Propionate Metabolism

Escherichia coli metabolizes propionate aerobically via the glyoxylate cycle to an as yet uncertain end product (Fig. 7.10). *Escherichia coli* metabolizes propionate first to its CoA derivative, which condenses with glyoxylate and a 2-hydroxyglutarate synthase [2-hydroxyglutarate glyoxylate-lyase (CoA-propionylating), EC 4.1.3.9] to α-hydroxyglutarate (315). A cleavage reaction catalyzed by α-hydroxyglutarate synthetase follows this α condensation and forms lactate plus acetate (316). Whereas acetate returns via acetyl-CoA into the tricarboxylic acid cycle and/or glyoxylate cycle, lactate is possibly transformed by a lactyl-CoA synthetase (263) to its CoA derivative. Lactyl-CoA is converted to pyruvyl-CoA and subsequently to hydroxypyruvic aldehyde (317). The further utilization of this compound is unknown. It is also not as yet known whether or not hydroxypyruvic aldehyde could be metabolized via glyceric acid, which would be similar to tartaric acid metabolism (see p. 410).

The conversion of acetate to acetyl-CoA could be catalyzed by an acetyl-CoA synthetase, similarly to that occurring in mammalian tissue (122).

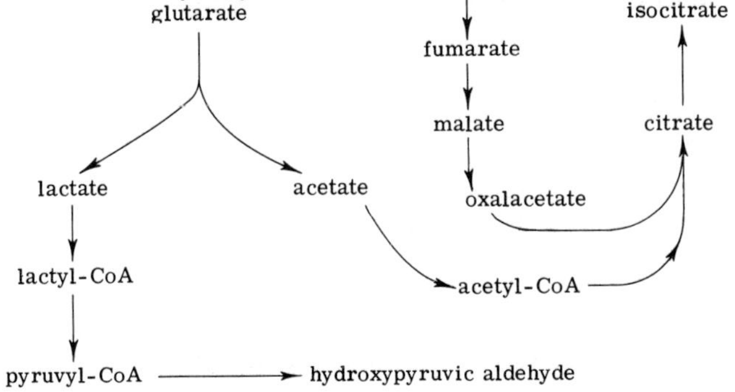
propionate
propionyl-CoA
glyoxylate
α-hydroxy-glutarate
succinate
fumarate
malate
isocitrate
citrate
lactate
acetate
oxalacetate
lactyl-CoA
acetyl-CoA
pyruvyl-CoA
hydroxypyruvic aldehyde

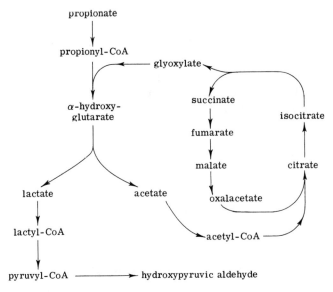

Fig. 7.10. Propionate metabolism in microorganisms.

Studies with various *Escherichia coli* mutants indicate that the substrate concentration at which *E. coli* is grown markedly influences the pathway by which propionate is metabolized (395). High substrate concentrations led to the conversion of propionate predominantly to acetate. In this case, isocitrate lyase activity increased and made it possible for the glyoxylate cycle to function. Low substrate concentrations resulted in the conversion of propionate via carboxylation reactions to other as yet not determined C_4 acids. It is possible that the lactate pathway is used in this instance.

Lactate Metabolism

The enzymatic mechanism for the aerobic lactate breakdown by growing cells is essentially identical in *Acetobacter* and *Acetomonas* (80). Both isomers of the lactate are metabolized by two sets of enzymes that do not require any cofactor and that are located in the particulate fraction of the cell. Resting cells of *Acetobacter peroxydans* are able to metabolize D-lactate about four times as fast as L-lactate, and the activity is located in the soluble fraction (78). In the particulate fraction, no NAD^+- and $NADP^+$-linked lactate dehydrogenase or lactate racemase was found; instead, a D-lactate oxidase [D-lactate:oxygen oxidoreductase (decarboxylating), EC 1.13.12 group] was found to be present:

$$\text{D-lactate} + O_2 \rightarrow \text{acetate} + CO_2 + H_2O$$

Acetobacter therefore can oxidize lactate to acetate similarly to *Azotobacter vinelandii* (176) and also lactate to pyruvate. *Mycobacterium smegmatis,* in contrast, possesses (364) the L-lactate oxidase [L-lactate:oxygen oxidoreductase (decarboxylating), EC 1.13.12.4] as a constitutive enzyme. It is believed that these enzymes are monooxygenases and that the substrate serves as an internal electron donor (130). Pyruvate itself is metabolized in three different ways by acetic acid bacteria:

1. A pyruvate decarboxylase (2-oxoacid carboxy-lyase, EC 4.1.1.1) that acts on pyruvate and forms acetaldehyde and carbon dioxide. This mechanism has been observed in *Acetomonas suboxydans* (188) and *Acetobacter peroxydans* (78).
2. A pyruvate oxidase system that was found to be a joint action of pyruvate decarboxylase (EC 4.1.1.1), which forms acetaldehyde and carbon dioxide, and an NAD^+- or $NADP^+$-linked aldehyde dehydrogenease (80). This system requires thiamine pyrophosphate and was found in *Acetobacter pasteurianum* (190) and *A. liquefaciens.* This system seems to be semiinducible. for no or doubtful activities were found in lactate-grown cells. The aldehyde dehydrogenase is not activated by CoA or glutathione

and could therefore be any of the three pyridine nucleotide-linked dehydrogenases (EC 1.2.1.3, EC 1.2.1.4, or EC 1.2.1.5),

3. A combination of a pyruvate decarboxylase (EC 4.1.1.1), a cytochrome system that does not require cofactors, and molecular oxygen as hydrogen acceptor (78). This combination appears to be similar to the function of pyruvate dehydrogenase (pyruvate:ferricytochrome b_1 oxidoreductase, EC 1.2.2.2), which requires thiamine pyrophosphate. In this pyruvate dehydrogenase system, the combination of pyruvate decarboxylase (EC 4.1.1.1), which is a thiamine pyrophosphate flavoprotein, and a cytochrome is certainly found. There is, however, no conclusive evidence for the latter statement available as yet. In either case, the product of this reaction is acetate.

The presence of a pyruvate decarboxylase indicates that *Acetomonas suboxydans* must have a yeast-type system of pyruvate decarboxylation, which leads directly to acetaldehyde. It is one of the few bacteria that can use this system, and this is certainly associated with this organism's failure to form acetyl phosphate or acetyl-CoA. It may also be why acetate is not further metabolized. It would be of interest to see whether this applies to all species of *Acetomonas*.

In *Acetobacter peroxydans*, in contrast, the further breakdown of acetate does not seem to be interrupted. The role of the marked peroxidase activity exhibited for lactate and alcohol metabolism (79) is still obscure.

The electron transport system in *A. peroxydans* (78) from the dehydrogenase to oxygen certainly involves cytochromes and cytochrome oxidase. This is in contrast to *Acetomonas suboxydans* where only a cytochrome a_1 (590 nm), which must function as an oxygen-transferring enzyme (158), was found. Lactate dehydrogenase is almost certainly involved with cytochrome c_1 (554 nm).

The metabolism of short-chain fatty acids exhibited undoubtedly the key role of glyoxylate in these catabolic events. However, there are many more reports on the use of glyoxylate as a condensing substrate with a number of other fatty acids (105, 262, 390, 394, 422, 423). As it is the aim to consider catabolic routes, rather than individual reactions, the reader is referred to the excellent review by Wegener *et al.* (395).

Regulatory Mechanisms of Carboxylic Acid Metabolism

Anaplerotic Sequences

The close interrelationship between catabolic control and anabolism makes it necessary to deal first with those anabolic sequences known as anaplerotic sequences.

Under aerobic conditions, it is the tricarboxylic acid cycle that serves as the central pathway between catabolic products with energy production and anabolic precursors. This cycle is therefore important for the respiration and biosynthesis of the cell. If intermediates of this cycle are tapped off for biosynthetic purposes, respiration can become vulnerable, as these intermediates would be lost for energy production. The organism must therefore have an ancilary system that takes care of the replenishment of these intermediates. The routes required for this replenishment are collectively called "anaplerotic sequences" (216). There is no doubt that the catabolic events must be controlled in conjunction with these anaplerotic sequences.

One anaplerotic sequence has already been mentioned in discussing the glyoxylate cycle. Isocitrate lyase (EC 4.1.3.1) can well be described as an anaplerotic enzyme for all those organisms utilizing citrate or acetate as their sole carbon source. The second anaplerotic enzyme would therefore be malate synthase (EC 4.1.3.2), which catalyzes the condensation of acetyl-CoA with glyoxylate to form malate. The combined action of both enzymes therefore leads to the net formation of a C_4-dicarboxylic acid. With these two systems, the tricarboxylic acid and glyoxylate cycles, in action, oxalacetate can be used as a precursor for cell component synthesis, which could go via gluconeogenesis.

The situation changes slightly when aerobic glucose utilization is considered. The oxidation of pyruvate results in a loss of one of the carbon atoms. As the intermediates 2-oxoglutarate, succinate, and oxalacetate are continuously withdrawn from the cycle, it is necessary to find other ways of replenishing this loss of carbon atoms in order to keep the tricarboxylic acid cycle intact. In contrast to the glyoxylate cycle, which replenishes C_4 compounds for the metabolism of C_2 compounds, such as acetate, it is now necessary to obtain C_4 compounds to replenish the loss of carbon compounds owing to the activity of the tricarboxylic acid cycle. However, this can only be done by CO_2 fixation (415). Altogether, the bacterial world has five enzymes at its disposal for such reactions:

1. A biotin-dependent pyruvate carboxylase [pyruvate: carbon-dioxide ligase (ADP), EC 6.4.1.1), which produces oxalacetic acid from pyruvate with the transformation of ATP to ADP plus inorganic phosphate (54, 55, 365), appears to be different from the mammalian enzyme (184, 344, 345, 388), which requires acetyl-CoA instead of biotin.

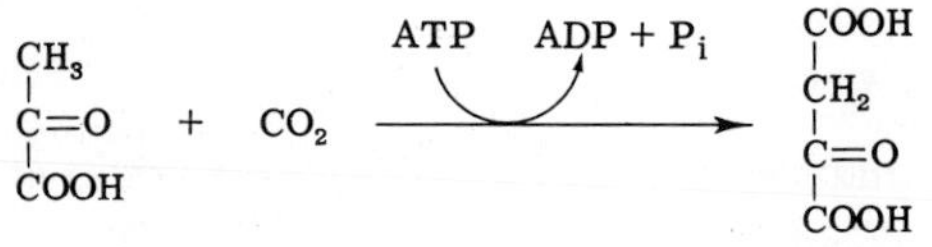
CH3
C=O
COOH
+
CO2
ATP
ADP + Pi
COOH
CH2
C=O
COOH

pyruvic acid oxalacetic acid

2. A phosphoenolpyruvate carboxylase [orthophosphate:oxaloacetate carboxy-lyase (phosphorylating), EC 4.1.1.31], which catalyzes the reaction of phosphoenolpyruvate to oxalacetate (49, 325). In this reaction, the inorganic phosphorus group makes the need for ATP redundant.

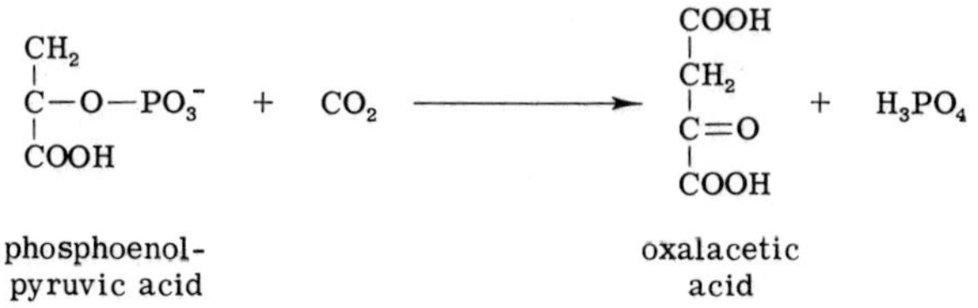
phosphoenol-
pyruvic acid
oxalacetic
acid

3. A phosphoenolpyruvate carboxykinase [ATP: oxaloacetate carboxy-lyase (transphosphorylating), EC 4.1.1.49] also forms oxalacetate from phosphoenolpyruvate but requires the help of ADP, which accepts the phosphorus group and forms ATP.

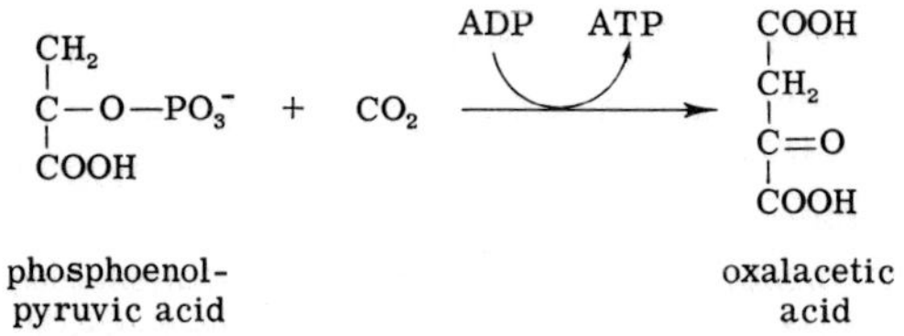
ADP
ATP
phosphoenol-
pyruvic acid
oxalacetic
acid

4. A phosphoenolpyruvate carboxytransphosphorylase [pyrophosphate:oxaloacetate carboxy-lyase (phosphorylating), EC 4.1.1.38] performs a reaction almost identical to that of phosphoenolpyruvate carboxykinase, requiring bicarbonate and inorganic phosphate, which is itself converted to pyrophosphate (244).

5. A malic enzyme [L-malate:NADP oxidoreductase (decarboxylating), EC 1.1.1.40], which catalyzes the reductive carboxylation of pyruvate to malate while oxidizing NADPH + H^+ to $NADP^+$.

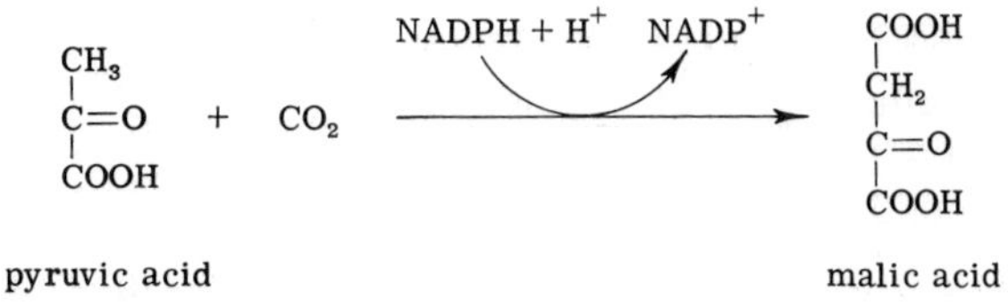
NADPH + H+
NADP+
pyruvic acid
malic acid

Two of these anaplerotic reactions have their origin at pyruvate and the other three at phosphoenolpyruvate level. There is no doubt (216) that not all of these enzymes are present in the same organism and also that

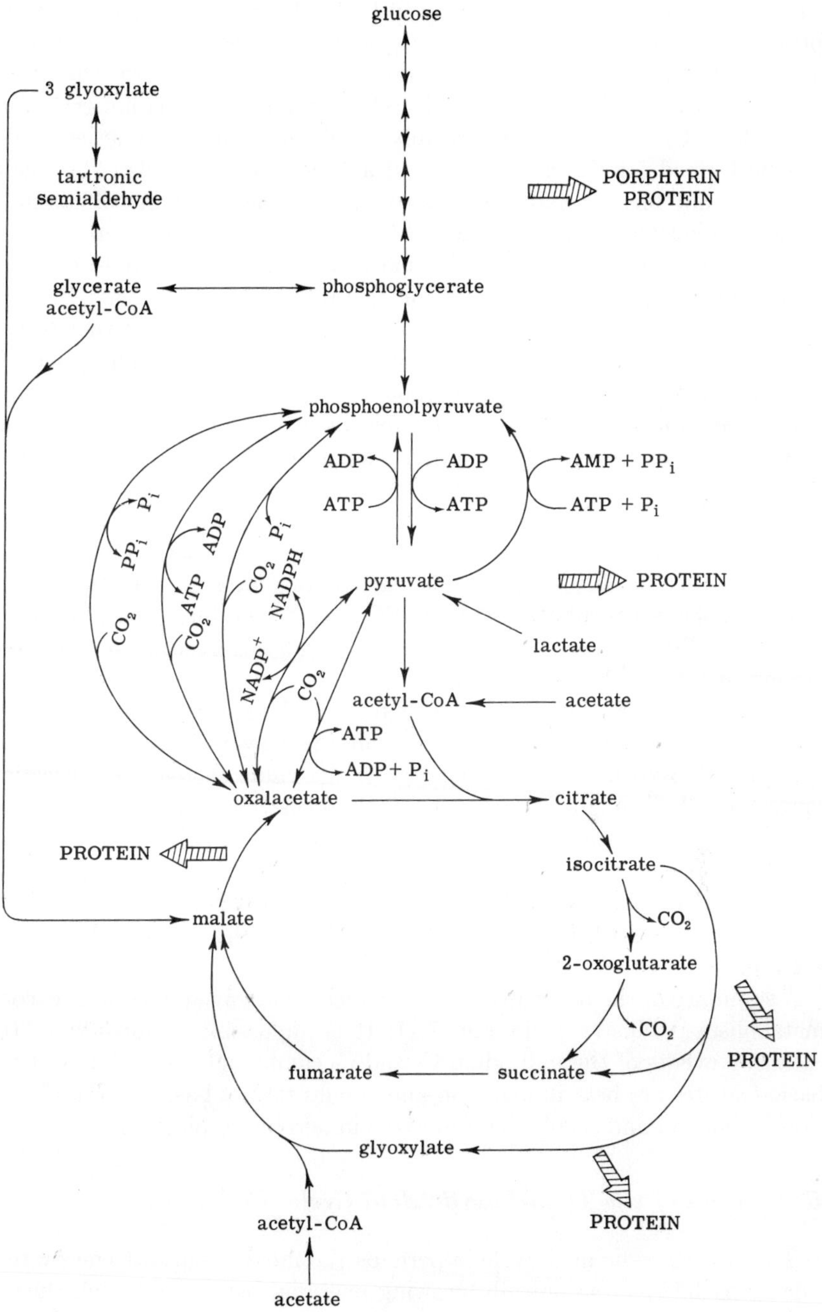
glucose
3 glyoxylate
tartronic semialdehyde
PORPHYRIN PROTEIN
glycerate acetyl-CoA
phosphoglycerate
phosphoenolpyruvate
ADP
ATP
ADP
ATP
AMP + PP$_i$
ATP + P$_i$
P$_i$
PP$_i$
ADP
ATP
P$_i$
CO_2
NADPH
pyruvate
PROTEIN
CO_2
CO_2
CO_2
NADP$^+$
lactate
CO_2
acetyl-CoA
acetate
ATP
ADP+ P$_i$
oxalacetate
citrate
PROTEIN
isocitrate
malate
CO_2
2-oxoglutarate
CO_2
PROTEIN
fumarate
succinate
glyoxylate
PROTEIN
acetyl-CoA
acetate

Fig. 7.11. Anaplerotic sequences in aerobic carbon metabolism.

some of these enzymes could very well be used for anabolic functions. It appears that phosphoenolpyruvate carboxylase is the main anaplerotic enzyme, at least in the Enterobacteriaceae (49). If this is so, the other enzymes must function mainly in decarboxylation reactions under conditions where C_4 compounds are readily available. Pentose moieties could thus be formed for the biosynthesis of aromatic amino acids. The malic enzyme could furnish additional pyruvate for the synthesis of alanine, valine, and leucine as well as of acetyl-CoA for the synthesis of fatty acids.

Because the pyruvate kinase reaction is an almost irreversible reaction, aerobic microorganisms that grow on lactate or do not possess the tricarboxylic acid cycle would be at a great disadvantage, for neither of the five anaplerotic enzymes mentioned could convert lactate into phosphoenolpyruvate. Two additional enzymes have been found for the conversion of pyruvate to phosphoenolpyruvate for these particular microorganisms. The first enzyme is a phosphoenolpyruvate synthase (37), which requires the input of 1 mole ATP per mole pyruvate. A second enzyme performing the same conversion of pyruvate to phosphoenolpyruvate has been found as the controlling enzyme for gluconeogenesis in *Acetobacter xylinum* (37). In the absence of glucose, pyruvate phosphate dikinase (ATP:pyruvate, orthophosphate phosphotransferase, EC 2.7.9.1) performs this reaction, requiring ATP and inorganic phosphate, which are converted to AMP and pyrophosphate. A kinetic competition appears to exist between the phosphoenolpyruvate synthetase and pyruvate dehydrogenase complex (65a). Whereas a high energy charge value favors the conversion to phosphoenolpyruvate (biosynthesis), a low energy charge value changes the direction toward catabolism and thus acetyl-CoA formation. This regulation in response to adenylate energy charge in addition to the differences in feedback inhibition certainly adapts the cell to the momentary metabolic needs. This competition appears to be absent in cells grown on glucose and very active in cells grown on lower molecular weight compounds such as acetate.

A summary of the main anaplerotic sequences involved in aerobic carbon metabolism is illustrated in Fig. 7.11. It is impossible to include all the catabolic events of the individual C_2 and C_3 compound utilization, but the basic features are held in common and should form a basis for the discussion on control and regulatory functions in aerobic respiration.

Regulation of the Tricarboxylic Acid Cycle

The tricarboxylic acid cycle is perhaps the most important energy producing pathway of aerobically growing bacteria, because for each turn of the cycle, 8 moles of ATP are produced. It is therefore most probable that this pathway is controlled to a large extent by the energy status of the cells, namely by adenylates (220) and/or reduced pyridine nucleotides (340). However, the tricarboxylic acid cycle also serves as an intermediate producer for the biosynthesis of various amino acids, which makes it necessary for the anaplerotic sequences to function. If the organism is growing on C_3 or C_4 compounds, the tricarboxylic acid or glyoxylate cycles serve in an amphibolic manner. It is therefore necessary to deal with the controls of both functions separately.

THE TRICARBOXYLIC ACID CYCLE AS ENERGY PRODUCER. If the TCA cycle is seen as an amphibolic pathway, it should be expected to follow the common rules of regulation as a biosynthetic pathway, e.g., feedback inhibition by end products, and as a catabolic pathway, e.g., feedback by such energy indicators as ATP. Although the tricarboxylic acid cycle appears to be a cycle without end products, closer consideration revealed that reduced NAD as well as ATP could be regarded as the respective end products (220). Some investigators go further and divide the TCA cycle into operational units (416), each unit being controlled independently of another. The first unit consists of the sequences leading from acetyl-CoA to 2-oxoglutarate. Following the rules of feedback inhibition, the first enzyme of the pathway, citrate synthase, certainly was found to exhibit an allosteric control by reduced NAD (98, 397, 398), but it was ATP insensitive. This type of control may be of great advantage, as biosynthesis requires the availability of ATP, and this buildup can occur without inhibition of the citrate synthase. Under semianaerobic conditions, when 2-oxoglutarate dehydrogenase activity is low or absent, 2-oxoglutarate itself functions as an inhibitor of citrate synthase.

In *Bacillus subtilis,* in contrast, citrate synthase is controlled by 2-oxoglutarate and ATP (98). ATP inhibition is competitive with acetyl-CoA but not with oxalacetate. It is assumed that the synthesis of citrate synthase is sensitive to both anabolic and catabolic signals. Once the enzyme is formed, however, its activity is regulated primarily by catabolic effectors. A survey of the inhibition of citrate synthase by reduced NAD was conducted among a large number of microorganisms (399) and two groups emerged. The enzyme from all the gram-negative organisms showed reduced NAD inhibition, whereas no such inhibition was observed in gram-positive organisms. The gram-negative organisms could be further subdivided according to whether or not AMP relieved the inhibition. This subdivision coincides with that between strict aerobes and facultative anaerobes. This phenomenon was explained in that facultative anaerobes possess already catabolic regulatory enzymes in their EMP pathway, whereas strict aerobes do not possess similar enzymes in the HMP or ED pathway. These results have since been confirmed (99). Purification and Sephadex column chromatography of citrate synthase (400) revealed a relationship between the allosteric regulation and molecular size of the enzyme. The "large" citrate synthases are all sensitive to reduced NAD regulation and are present in gram-negative bacteria, whereas the "small" ones are insensitive and are found in gram-positive bacteria.

Microorganisms grown on carbohydrates do not substitute the TCA cycle by the glyoxylate cycle. The reason for this is the synergistic inhibition of isocitrate dehydrogenase by glyoxylate and oxalacetate (94, 145, 300, 330, 349). This enzyme is also inhibited by ATP (99, 255–258). The discovery of isozymes (318, 319) has revealed the existence of multiple forms of $NADP^+$-specific isocitrate dehydrogenases in crude extracts of *E. coli*. The two separate $NADP^+$-specific isocitrate dehydrogenases from *Acinetobacter lwoffi* (295) differed in their reaction to adenosine nucleotides, but a similar resolution at *E. coli* isozymes were unsuccessful (347). It thus appears that the inhibition by ATP is used for reducing the overall TCA cycle activity when energy is abundant, whereas the concerted action of glyoxylate and oxalacetate could control whether the TCA or the glyoxylate cycle is used (121). It is therefore likely that once ATP inhibits isocitrate dehydrogenase, the glyoxylate cycle could function and the abundant supply of oxalacetate or glyoxylate could reverse the reaction sequence again. This role of the substrates on isocitrate dehydrogenase appears also to exist in isocitrate lyase synthesis (133), wherein the presence of acetate induces the synthesis of this enzyme and represses isocitrate dehydrogenase (22, 144, 208).

Of great importance for the functioning of the tricarboxylic acid cycle is its maintenance. Carbohydrate-utilizing bacteria possess the following anaplerotic enzymes which ensure the continuous supply of C_4 compounds to replenish those intermediates withdrawn from the TCA cycle (for reactions, see p. 424):

1. Phosphoenolpyruvate carboxylase is poorly active in the presence of only its substrate (217, 249, 353) but is activated strongly if acetyl-CoA, fatty acids, and their derivatives (160) are present. This indicates that pyruvate is metabolized via acetyl-CoA to either citrate and/or fatty acids. When the demand for fatty acid has been met or when an oversupply takes place, the metabolic flow will encourage the formation of oxalacetate from phosphoenolpyruvate. The enzyme is also stimulated by FDP (292, 333). Apart from the activations, malate and aspartate (241) act as inhibitors to make certain that no oversupply of C_4 acids occurs (159).

2. $NADP^+$-linked malic enzymes furnish the organism with additional

malate from pyruvate. This enzyme, like citrate synthase (418), is inhibited by reduced NAD (339). Additional inhibitors for this enzyme are acetyl-CoA, oxalacetate (49, 336), and cyclic 3′,5′-AMP (334); the latter inhibition occurs only under starvation conditions.

3. *Escherichia coli* also has an NAD^+-linked malic enzyme, which is stimulated by aspartate (372) and inhibited by ATP (341) and CoA in an allosteric manner. The activation by aspartate could only reverse the inhibition of coenzyme A but not of ATP.

4. Phosphoenolpyruvate carboxykinase is an ADP-requiring enzyme (417) and is inhibited allosterically by reduced NAD. Like PEP-carboxylase, it furnishes the cell with oxalacetate and ATP.

A summary of all the reported investigations on regulatory functions in the aerobic, carbohydrate-utilizing organisms is attempted in Fig. 7.12. The figure may look bewildering, but the organisms must not only obtain their energy from the TCA cycle but must also use intermediates for protein and porphyrin biosynthesis. This in turn makes it necessary to have provisions for the maintenance of the TCA cycle itself. The latter is accomplished by two separate malic enzymes, a PEP-carboxylase and a PEP-

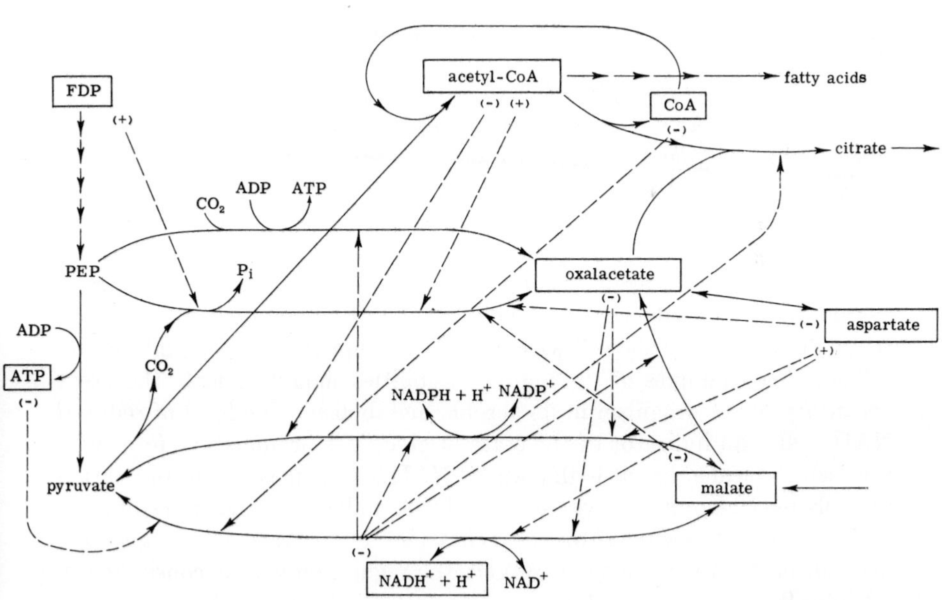

FDP
acetyl-CoA
fatty acids
CoA
citrate
(+)
(−) (+)
(−)
ADP ATP
CO_2
PEP
P_i
oxalacetate
aspartate
ADP
ATP
CO_2
NADPH + H^+ $NADP^+$
pyruvate
malate
$NADH^+$ + H^+
NAD^+

Fig. 7.12. Regulatory functions of aerobic carbohydrate-utilizing microorganisms. (+) - - → , activation; (−) - - → inhibition.

carboxykinase, which can furnish either malate or oxalacetate from phosphoenolpyruvate to pyruvate as the need arises.

Under normal aerobic conditions, pyruvate is metabolized via the TCA cycle with the production of CO_2, ATP, and H_2O or reduced NAD. With the concomitant biosynthesis, such intermediates as succinate, 2-oxoglutarate, and oxalacetate are withdrawn from the cycle. In order to maintain the cycle, PEP-carboxylase, being activated by FDP and acetyl-CoA, supplies the required oxalacetate for the condensation reaction to form citrate. Accumulation of acetyl-CoA would, of course, occur, if oxalacetate becomes a limiting factor. Acetyl-CoA itself or CoA inhibit the $NADP^+$- and NAD^+-linked malic enzymes, respectively, thus replacing the inhibitory effect of oxalacetate on both enzymes. An overproduction of aspartate, and subsequently malate, would inhibit PEP-carboxylase activity and stop additional oxalacetate supply.

Apart from regulating its carbon supply, the organism must also regulate its end product formation. The end product ATP inhibits pyruvate kinase, NAD^+-dependent malic enzyme, and the isocitrate dehydrogenase. The second end product of the tricarboxylic acid cycle, reduced NAD, inhibits a number of major enzymes, such as citrate synthase, PEP-carboxykinase, NADP-linked malic enzyme, and malate dehydrogenase. An overproduction of reduced NAD and ATP therefore facilitates fatty acid biosynthesis. Both end products inhibit differentially, in an allosteric manner. The enzymes inhibited by reduced NAD are not affected by ATP and vice versa. In the case of an ATP overproduction, only pyruvate kinase and the NAD-linked malic enzyme and isocitrate dehydrogenase are inhibited. The organism can thus utilize the functional glyoxylate cycle, avoiding the energy-producing steps in the TCA cycle and keeping up the supply of C_4-dicarboxylic acids. Only an excess of oxalacetate and glyoxylate will inhibit citrate synthase and again stimulate the fatty acid biosynthesis. During overproduction of reduced NAD and normal ATP production, the organism is still able to supply the cell with acetyl-CoA, malate, and/or oxalacetate.

These few examples of the many possibilities should indicate the great versatility of the organism in its aerobic metabolism. The level of reduced NAD is also an indicator of the state of glycolysis or anaerobic fermentation, which means that at high reduced NAD levels anaerobic fermentation prevails over respiration, whereas low levels indicate a higher respiratory activity. As this switch from fermentation to respiration is largely dependent on the availability of oxygen, the regulation will be considered in Chapter 9.

Organisms growing on lower molecular weight compounds, such as acetate and succinate, use the TCA cycle as their amphibolic route, the glyoxylate cycle as their anaplerotic route, and gluconeogenesis as their biosynthetic route. It is also in these metabolic events where the need for anaplerotic sequences is greatest.

Acetate is metabolized via the TCA cycle (Fig. 7.13) in a total oxidation to 2 units of carbon dioxide. The first step is the conversion of acetate to acetyl-CoA by acetyl-CoA synthetase (acetate:CoA ligase (AMP-forming), EC 6.2.1.1) (393), which then is able to condense with oxalacetate to form citrate. In order to replenish the C_4-dicarboxylic acids necessary to maintain the TCA cycle, microorganisms use mainly the glyoxylate cycle via isocitrate lyase (see p. 402). For biosynthesis of its components, the cell must operate the gluconeogenic pathway, which simulates the reverse of

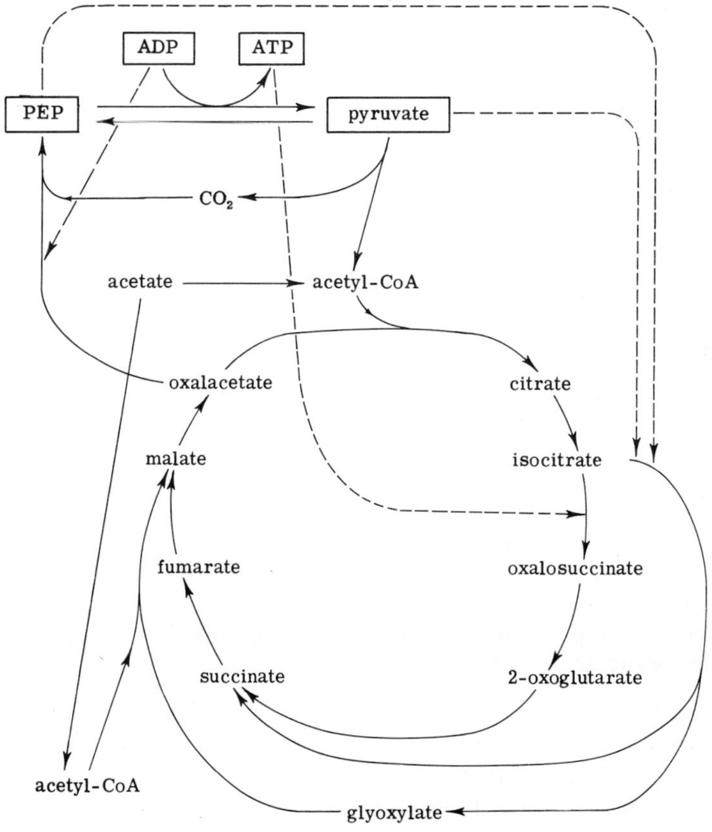
ADP
ATP
PEP
pyruvate
CO_2
acetate
acetyl-CoA
oxalacetate
citrate
malate
isocitrate
fumarate
oxalosuccinate
succinate
2-oxoglutarate
acetyl-CoA
glyoxylate

Fig. 7.13. Regulatory control of aerobic acetate metabolism in microorganisms. - - → , inhibition.

the EMP pathway. This way leads via pyruvate and phosphoenolpyruvate. The most important enzyme for this pathway is PEP-carboxylase, as it was found that without this enzyme no gluconeogenesis occurs (149, 322). It is also understandable that phosphoenolpyruvate (PEP) and pyruvate control the isocitrate lyase activity (217). However, ADP accumulation inhibits PEP-carboxylase competitively with its substrate (35). Such an increase would activate pyruvate kinase, channeling the carbons through pyruvate and the TCA cycle. This switching to catabolism would increase ATP formation (34), which in turn inhibits isocitrate dehydrogenase (144), opens the glyoxylate cycle, and relieves the inhibition of PEP-carboxykinase. No stimulation of PEP-carboxylase by acetyl-CoA was observed. It has been suggested that the physiological role of PEP-carboxylase is to effect the net formation of C_4 acids from C_3 acids, which are essential for the maintenance of the TCA cycle during growth on glucose.

Apart from phosphoenolpyruvate and pyruvate there exists a second type of control of the glyoxylate cycle. This control is exerted on the induction and repression of isocitrate lyase synthesis (215) and malate synthase. Not very much is known as yet about the detailed mechanism.

Succinate, malate, 2-oxoglutarate, and oxalacetate play a multiple role in aerobic metabolism. In order to be completely oxidized, these acids must first be converted to acetyl-CoA. The key enzyme for this pathway is the NAD^+-linked malic enzyme, which converts malic acid into pyruvate and carbon dioxide. This enzyme is repressed by acetate, an action that preserves dicarboxylic acids in the presence of other sources of acetyl-CoA (291, 421a). The stimulation of the activity of the enzymes that form dicarboxylic acids from C_3 compounds and carbon dioxide by acetyl-CoA has also been observed in several bacteria (49, 50). However, no inhibition of the malic enzyme has been found to be caused by acetyl-CoA (161). The TCA cycle plays the major catabolic role (380).

Glycolate and glyoxylate can also serve as sole carbon source in *Escherichia coli* (see p. 407). Both compounds are interconvertible by the organism utilizing either a FMN-dependent glycolate oxidase (427) or an NADPH-linked glyoxylate reductase (182) in the opposite direction. The metabolic pathway diverges at the glyoxylate level:

1. A malate synthase [L-malate:glyoxylate-lyase (CoA-acetylating), EC 4.1.3.2] condenses glyoxylate and acetyl-CoA to yield L-malate (414).
2. A glyoxylate carboxy-lyase (EC 4.1.1.47) catalyzes the conversion of glyoxylate to tartronic semialdehyde and CO_2 (219).

For a long time it was not certain which one of the two pathways played the major oxidative role and subsequently which one the anaplerotic role. One proposal (163) referred to the TCA cycle as the major oxidative or catabolic route, which meant that malate synthesis replenished the intermediates of the TCA cycle withdrawn for biosynthetic purposes (216). The second proposal (213) was exactly the reverse, with the TCA cycle being not essential. Mutant studies, however, indicated very convincingly (290) that the TCA cycle represents the catabolic route, whereas malate synthase is the anaplerotic enzyme for both glyoxylate- and acetate-grown cells.

All organisms grown on these low molecular weight compounds, such as acetate or succinate, have one characteristic in common. In order to synthesize cell material, they form pentoses, trioses, and hexoses via the gluconeogenic pathway. Gluconeogenesis is vital for the metabolism of these cells. Two enzymes are induced during growth on succinate or malate (335, 417), PEP-carboxykinase and the $NADP^+$-linked malic enzyme, which metabolize oxalacetate or malate, respectively. Microorganisms grown on lactate or pyruvate therefore require the inducible enzyme PEP-synthase (71). This additional enzyme becomes necessary as the pyruvate kinase reaction is virtually irreversible (216, 242). PEP-carboxykinase is inhibited by reduced NAD and malic enzyme by reduced NAD, acetyl-CoA, and cyclic 3′5′-AMP (335, 336). This differential inhibition indicates that the function of PEP-carboxykinase is to form phosphoenolpyruvate, whereas that of malic enzyme is to produce reduced NADP for biosynthesis together with acetyl-CoA via pyruvate. The allosteric control by reduced NAD of these two gluconeogenic enzymes can be considered as a mecha-

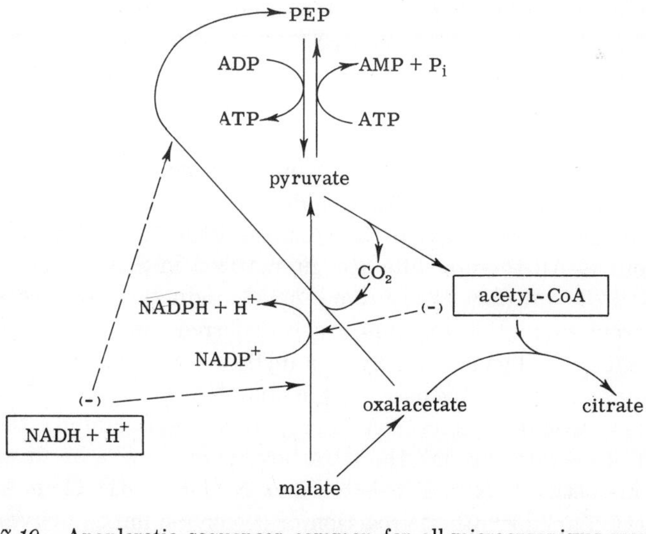
PEP
ADP
AMP + P_i
ATP
ATP
pyruvate
CO_2
acetyl-CoA
NADPH + H^+
(-)
$NADP^+$
(-)
oxalacetate
citrate
NADH + H^+
malate

Scheme 7.10. Anaplerotic sequences common for all microorganisms grown on low molecular weight compounds such as acetate or succinate.

nism that prevents the buildup of cellular components during vigorous growth on glucose. The levels of reduced NAD are therefore regarded as indicators of glucose utilization (Scheme 7.10).

Ethanol Metabolism

It is still questionable whether acetaldehyde is an intermediate in ethanol oxidation in *Acetobacter peroxydans* (23), for this species oxidizes ethanol, pyruvate, and a number of tricarboxylic acid cycle intermediates (32) to carbon dioxide and water but hardly oxidizes acetate at all. Crude cell-free extracts were found to only weakly oxidize ethanol, by means of an NAD^+-linked alcohol dehydrogenase, to acetate and intermediates of the tricarboxylic acid cycle beyond oxalacetate. This enzyme of *Acetobacter peroxydans* resembles the corresponding alcohol dehydrogenase of higher organisms (187).

Acetomonas suboxydans was shown to have a powerful aldehyde dehydrogenase, which reduces $NADP^+$ at four times the rate of NAD^+ reduction (189). It is not certain, however, whether the number of dehydrogenases involved is one or two. It is also possible that two alcohol–cytochrome reductases are present (278). The pH optimum of the alcohol dehydrogenase in the *Acetobacter oxydans* and *Acetobacter mesoxydans* groups was 3.7–4.7, whereas it was 5.7–6.2 in *Acetomonas*. Because the cytochromes a_1 (590 nm), b (565 nm), and c-553 were found, it has been postulated that the alcohol–cytochrome-553 reductase is one of the important components responsible for acetic acid production. In contrast, *Acetobacter* seemed to have cytochrome o as terminal oxidase (52, 53). *Acetomonas* therefore seems to have two different terminal oxidases, one for lactate and one for ethanol oxidation.

In *Acetobacter peroxydans* a NAD^+-linked ethanol dehydrogenase (EC 1.1.1.1) as well as a NAD^+-linked aldehyde dehydrogenase (EC 1.2.1.3 or 1.2.1.10) are present (24) if the organism is grown with ethanol as the carbon source. All these results are summarized in Scheme 7.11. Ethanol is oxidized by alcohol–cytochrome c-553 reductase to acetaldehyde. The electrons from this reaction are transferred via the electron transport system (15). The acetaldehyde is further metabolized by the NAD^+-linked aldehyde dehydrogenase. A possible $NADP^+$-linked aldehyde dehydrogenase would produce $NADPH + H^+$, which inhibits the further oxidation of acetate through the tricarboxylic acid cycle by upsetting the equilibrium between $NADPH + H^+$ and $NADH + H^+$. This favors the back reaction or biosynthetic reaction of coenzyme-linked dehydrogenases in the tricarboxylic acid cycle.

In *Pseudomonas aeruginosa*, a different aldehyde dehydrogenase appears

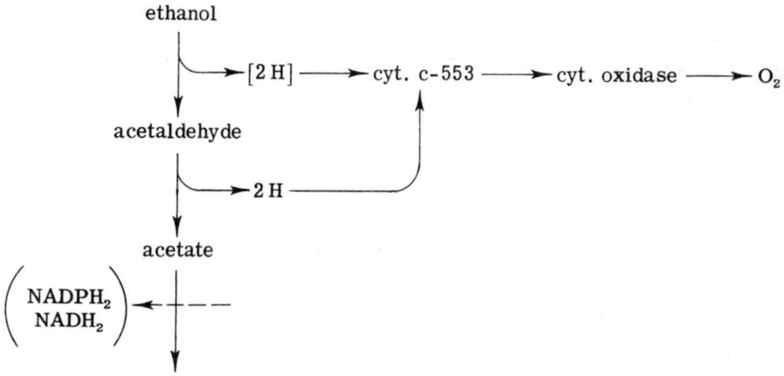
ethanol
[2 H]
cyt. c-553
cyt. oxidase
O_2
acetaldehyde
2 H
acetate
NADPH$_2$
NADH$_2$

Scheme 7.11. Ethanol oxidation by acetic acid bacteria.

to play the major role in ethanol metabolism (389). This enzyme had an ionic strength requirement and was activated by potassium or ammonium ions. This type of requirement is certainly different from the normally known pyridine nucleotide-dependent aldehyde dehydrogenases.

The presence of NAD^+-linked ethanol dehydrogenase (EC 1.1.1.1) in *Acetobacter* is therefore not well understood. Investigations into the substrate–energy relationships in *Acetomonas oxydans* (110), however, revealed that each of the oxidation steps from ethanol to acetate produces an equal amount of energy. Although *A. oxydans* is not able to grow with ethanol as the sole carbon source because it lacks a TCA cycle, it can use the energy produced during the oxidation from ethanol to acetate for its growth. Therefore, the ethanol in the glucose medium results in a higher ATP production. The P/O ratios obtainable were 0.28 (110), 0.3 (191), and 0.4 (362).

The presence of a pyruvate dikinase, which converts pyruvate, ATP, and orthophosphate (Pi) into phosphoenolpyruvate (36), AMP, and pyrophosphate assures gluconeogenesis and cell biosynthesis in acetic acid bacteria. The presence of a pyruvate kinase at the same time indicates that *Acetobacter xylinum* is also able to grow on glucose, using the latter enzyme for its catabolic route. Both enzymes are controlled by nucleotides; a high cellular ATP/ADP ratio favors gluconeogenesis, whereas a low ratio or high AMP concentration would favor the catabolic route. As AMP inhibits the dikinase reaction, a recycling of phosphoenolpyruvate is avoided.

Methane Oxidation

The main representative of the methane-oxidizing bacteria is *Methanomonas methanooxidans* (*Pseudomonas methanica*). There is a possibility that the name "*Methanomonas methanooxidans*" was given to the autotrophic strains and "*Pseudomonas methanica*" to the heterotrophic strains of the same species (168), for *Pseudomonas methanica* oxidizes methanol, methane, formaldehyde, and formate, as does *M. methanooxidans*, but has a growth factor requirement for calcium pantothenate (86, 236). Another suggestion, that methane-oxidizing bacteria using the autotrophic allulose pathway for cell material synthesis should be classified as "*Methylomonas*" instead of *Methanomonas* (100), indicates the difficulties in separating autotrophically from heterotrophically grown organisms. *Methylococcus capsulatus* was found (235) to be an autotrophic, methane-oxidizing bacterium, in which formate is incorporated into a pentose phosphate to form allulose phosphate. Difficulties with the different names, however, will certainly arise, because a methane-oxidizing bacterium can grow either way, or mixotrophically. Improved isolation and cultivation methods (47, 74) should uncover further information in this area in the near future.

Methane-oxidizing bacteria appear to occur quite commonly in the topmost layer of marine sedimentary materials and in soil, being particularly abundant where methane and free oxygen are present (154). Carbon dioxide is required for growth initiation and these bacteria are able to utilize ammonium chloride, potassium nitrate, peptone, or glutamate almost equally well as nitrogen sources (155). A most rapid utilization of methane occurred under an atmosphere of 10–40% oxygen, up to 70% methane, and 5–20% carbon dioxide.

Methane oxidation by *Pseudomonas methanica* (86, 170) and *Methanomonas methanooxidans* (45) proceeds as follows:

$$CH_4 \rightarrow CH_3OH \rightarrow HCHO \rightarrow HCOOH \rightarrow CO_2$$

This oxidation is carried out by both autotrophic and heterotrophic organisms. Whereas the autotrophs incorporate formate (HCHO) into ribose 5-phosphate to form allulose 6-phosphate (361), the heterotrophs incorporate formate into folate derivatives and fix CO_2 heterotrophically (see Fig. 7.14). In the strictest terminology, the so-called "autotrophic" methane oxidizers are not autotrophs. They do not grow on CO_2 as their sole carbon source and therefore do not incorporate carbon dioxide into the Calvin cycle. However, the biosynthesis of cell material follows the Calvin cycle very closely and they are therefore designated as autotrophs. As the autotrophic mechanism has been dealt with above (see p. 358), only the heterotrophic mechanism of methane oxidation will be described here. Most of the investigations have been carried out on *Pseudomonas methanica* or *Pseudomonas* AM 1.

The first step in methane utilization is the oxidation of methane. This

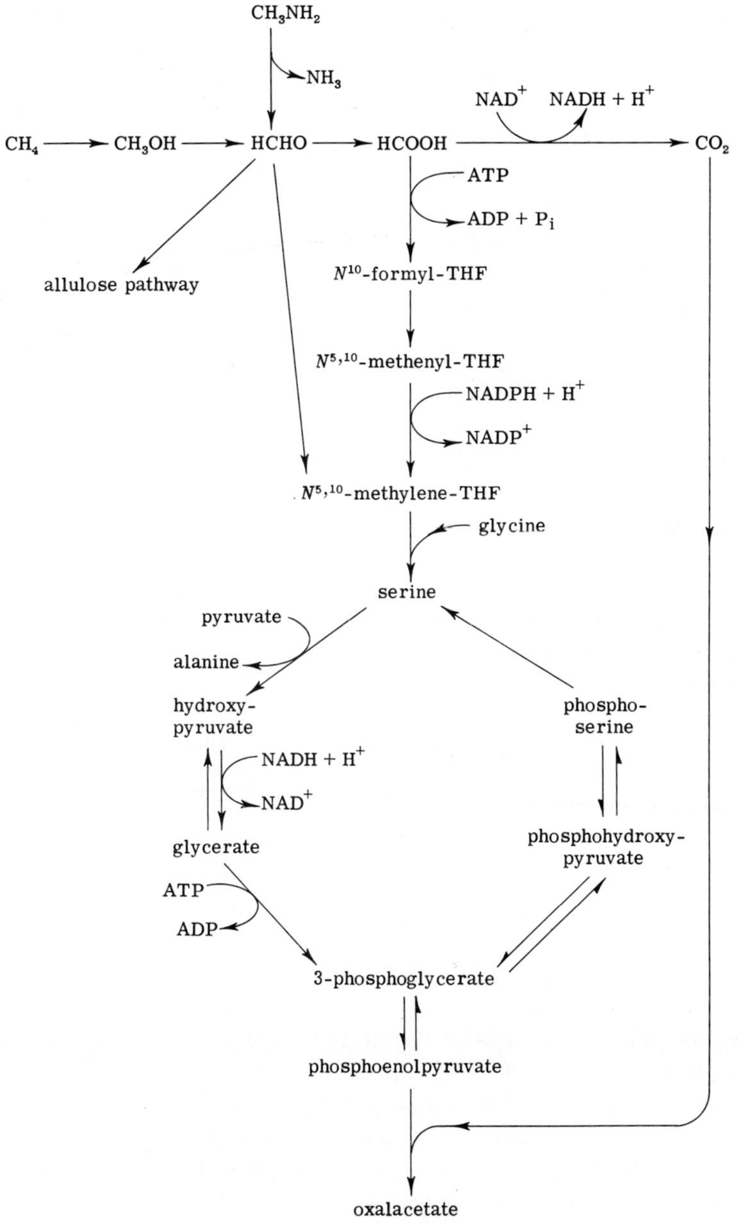

CH_3NH_2
NH_3
NAD^+
$NADH + H^+$
CH_4
CH_3OH
HCHO
HCOOH
CO_2
ATP
$ADP + P_i$
allulose pathway
N^{10}-formyl-THF
$N^{5,10}$-methenyl-THF
$NADPH + H^+$
$NADP^+$
$N^{5,10}$-methylene-THF
glycine
serine
pyruvate
alanine
hydroxy-pyruvate
phospho-serine
$NADH + H^+$
NAD^+
glycerate
phosphohydroxy-pyruvate
ATP
ADP
3-phosphoglycerate
phosphoenolpyruvate
oxalacetate

Fig. 7.14. Methane oxidation by *Pseudomonas methanica* and *Pseudomonas* AM 1.

reaction was under dispute for a long time. Using [^{18}O]-enriched oxygen gas and [^{18}O]water it was shown conclusively (137) that the oxygen in the reaction product methanol was derived from the oxygen gas and not from water:

$$2\ CH_4 + O_2 \rightarrow 2\ CH_3OH$$

There appears to be no doubt that an oxygenase system is involved, although the possibility of a mixed function oxidase or monooxygenase can not be discarded as yet:

$$CH_4 + O_2 + XH_2 \rightarrow CH_3OH + H_2O + X$$

where XH_2 is a reducing agent. These findings are certainly in contrast to the claim of the existence of a hydroxylase (405):

$$CH_4 + H_2O \rightarrow CH_3OH + 2\ [H]$$

which would involve water in its reaction system. Neither of the enzymes has been isolated as yet.

A suggestion that methanol is oxidized by *Pseudomonas methanica* via a catalase-linked peroxidase to formaldehyde, which in turn is oxidized by a substrate-specific aldehyde dehydrogenase requiring NAD and glutathione (125), was partly opposed by Anthony and Zatman (11) and Johnson and Quayle (169), who showed that methanol oxidation was independent of catalase. The alcohol dehydrogenase responsible for the oxidation of methanol to formaldehyde has been isolated and purified (12, 13) from *Pseudomonas* M 27. This enzyme also oxidized ethanol, was substrate

$$CH_3OH \rightarrow HCHO$$

specific for primary alcohols (356a) and unusual in possessing not a flavoprotein as prosthetic group but a pteridine derivative instead (14). This pteridine derivative could be a hydrogen-transfer coenzyme and C_1 carrier, for it was found that folate plays an important role in the further metabolic events (142, 143). It is therefore conceivable that the pteridine bound to the alcohol dehydrogenase is a folate (14).

There exist two opinions about the oxidation of formaldehyde; which is right depends on the need for ATP in autotrophs or the need for reducing power in heterotrophs. *Pseudomonas methanica* contains a high concentration of NAD^+-linked formaldehyde dehydrogenase (formaldehyde:NAD oxidoreductase, EC 1.2.1.1), which requires reduced glutathione (GSH).

The subsequent reaction could be catalyzed by formate:NAD oxidoreductase (EC 1.2.1.2), which produces carbon dioxide and a further molecule

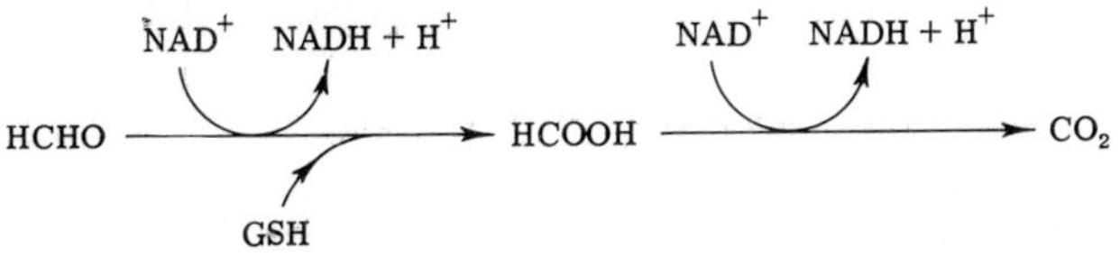
NAD$^+$
NADH + H$^+$
HCHO
HCOOH
CO$_2$
GSH

of NADH + H^+. Although two molecules of reduced NAD are formed, together with carbon dioxide for CO_2 fixation, the formation of ATP was missing. The alternative mechanism for formaldehyde formation (169) includes not only the production of ATP, but also the formation of reduced NADP:

$$\text{HCHO} + \text{tetrahydrofolate (THF)} \rightleftharpoons N^{5,10}\text{-methylene-THF}$$

$$N^{5,10}\text{-methylene-THF} + \text{NADP}^+ \rightleftharpoons N^{5,10}\text{-methenyl-THF} + \text{NADPH} + \text{H}^+$$

$$N^{5,10}\text{-methenyl-THF} + \text{H}_2\text{O} \rightleftharpoons N^{10}\text{-formyl-THF} + \text{H}^+$$

$$N^{10}\text{-formyl-THF} + \text{ADP} + \text{P}_i \rightleftharpoons \text{THF} + \text{HCOOH} + \text{ATP}$$

This proposed reaction sequence results in 1 mole of ATP and 1 mole of reduced NADP per mole of formaldehyde. It is not known, however, which scheme functions and whether the methane-oxidizing bacteria have a cytochrome system and another way of producing ATP.

Labeled isotope experiments with *Pseudomonas* AM 1 grown on [^{14}C]-methanol and [^{14}C]bicarbonate indicated the participation of serine and malate in the metabolism of these compounds (230, 231). As the hydroxymethyl group of serine is derived mainly from methanol and the carboxyl group of glycine mainly from carbon dioxide, a hydroxymethylation of glycine to give serine was anticipated.

If *Pseudomonas* AM 1 was grown on formate, the above-outlined reactions were found to be partially reversible. Formate is activated by ATP and is metabolized by a formyltetrahydrofolate synthetase [formate:tetra-

$$\text{HCOOH} + \text{ATP} + \text{THF} \rightleftharpoons N^{10}\text{-formyl-THF} + \text{ADP} + \text{Pi}$$

hydrofolate ligase (ADP), EC 6.3.4.3] (139) to N^{10}-formyltetrahydrofolate (403). This intermediate can now be attacked by an enzyme called meth-

$$N^{10}\text{-formyl-THF} + \text{H}^+ \rightleftharpoons N^{5,10}\text{-methenyl-THF} + \text{H}_2\text{O}$$

enyltetrahydrofolate cyclohydrolase (EC 3.5.4.9) to form $N^{5,10}$-methenyltetrahydrofolate. These reactions have also been found to occur in *Micrococcus aerogenes* (404). A reduced NADP-dependent reaction is necessary (127) to produce the substrate for the following hydroxymethylation reaction (40, 151):

$$N^{5,10}\text{-Methenyl-THF} + \text{NADPH} + H^+ \rightleftharpoons N^{5,10}\text{-methylene-THF} + \text{NADP}^+$$

$$N^{5,10}\text{-Methylene-THF} + \text{glycine} \rightleftharpoons \text{THF} + \text{serine}$$

In this reaction, which is catalyzed by serine hydroxymethyltransferase (5,10-methylenetetrahydrofolate:glycine hydroxymethyltransferase, EC 2.1.2.1), the hydroxy and methyl group is transferred to glycine, which results in the formation of serine (132, 233). It is assumed (124) that two serine hydroxymethyltransferases exist: one to supply the organism with glycine and C_1 units during growth on succinate (biosynthesis) and the second to carry out the reaction described above. *Pseudomonas* therefore has two metabolic routes; one is a nonphosphorylating pathway, which operates when *Pseudomonas* AM 1 is grown on C_1 compounds, whereas the second phosphorylating pathway could be regarded as a biosynthetic pathway acting when the organism is grown on succinate (132).

The first step in the nonphosphorylating pathway is a transamination

$$\text{Serine} + \text{pyruvate} \rightleftharpoons \text{hydroxypyruvate} + \text{alanine}$$

between pyruvate (233) or any other keto acid. The reduced NAD-depend-

$$\text{HO}\cdot\text{CH}_2\cdot\text{COCOOH} + \text{NADH} + H^+ \rightleftharpoons \text{HO}\cdot\text{CH}_2\cdot\text{CH(OH)COOH} + \text{NAD}^+$$

ent hydroxypyruvate reductase (D-glycerate:NAD oxidoreductase, EC 1.1.1.29) reduces hydroxypyruvate to glycerate. The ATP-dependent glycerate kinase (ATP:D-glycerate 3-phosphotransferase, EC 2.7.1.31) finally forms the member of the Embden-Meyerhof pathway, 3-phosphoglycerate.

The phosphorylating pathway that produces serine from 3-phosphoglycerate involves different enzymes. A phosphoglycerate dehydrogenase first converts 3-phosphoglycerate to phosphohydroxypyruvate. A transami-

$$\text{3-Phosphoglycerate} + \text{NAD}^+ \rightleftharpoons \text{phosphohydroxypyruvate} + \text{NADH} + H^+$$

nation between phosphohydroxypyruvate and α-ketoglutarate by a phosphoserine aminotransferase (*O*-phospho-L-serine:2-oxoglutarate aminotransferase, EC 2.6.1.52) forms phosphoserine, which itself is converted to

Phosphohydroxypyruvate + α-ketoglutarate ⇌ glutamate + phosphoserine

serine by a phosphoserine phosphohydrolase (phosphoserine phosphatase,

$$\text{Phosphoserine} + H_2O \rightarrow \text{serine} + \text{Pi}$$

EC 3.1.3.3).

Let us now leave the biosynthetic pathway and return to the further metabolism of 3-phosphoglycerate in methane oxidation. *Pseudomonas* AM 1 possesses the two enzymes of the EMP pathway to convert 3-phosphoglycerate via 2-phosphoglycerate to phosphoenolpyruvate. A carboxylation of phosphoenolpyruvate by phosphoenolpyruvate carboxylase [orthophosphate:oxaloacetate carboxy-lyase (phosphorylating), EC 4.1.1.31] utilizes thc CO_2 formed in the early stages of methane or methanol oxidation and forms the end product oxalacetate (232).

This pathway of methane or methanol oxidation could be subject to alteration, as it was found that glyoxylate and glycine are intermediates of the methanol oxidation pathway (123, 124). It is thought that the concerted action of serine hydroxymethyltransferase (EC 2.1.2.1) and serine–glyoxylate aminotransferase (EC 2.6.1.45) could constitute a cycle converting glyoxylate and methylene tetrahydrofolate into hydroxypyruvate. Two possible sources for glyoxylate have recently been found. The first source is due to the discovery of the enzyme malyl-CoA lyase (malyl-CoA glyoxylate-lyase, EC 4.1.3.24) in a number of C_1-unit-utilizing bacteria (330a). It catalyzes the reaction but the enzyme producing malyl-CoA from

ATP + malyl-CoA → acetyl-CoA + glyoxylate + ADP + Pi

malate has not yet been found in these organisms (see also 186a, 261a, 304a). The second source for glyoxylate could be acetate, which is rapidly metabolized to glycine during growth of *Pseudomonas* AM 1 on methanol (85c). Acetate, which is an intermediate in methanol utilization (see Fig. 7.14), can certainly serve as a precursor of glyoxylate (85a) with malate being an additional intermediate (85b) in this glyoxylate cycle pathway. If glyoxylate would be involved in the methanol oxidation and glycine

Serine + glyoxylate → hydroxypyruvate + glycine

becomes again available for serine formation.

Methylamine Metabolism

In addition to methane, methanol, and formate, *Pseudomonas* AM 1 is also able to utilize methylamine (87). The oxidation of methylamine to CO_2 plus water has formaldehyde and formate as intermediates. The enzyme responsible for the first reaction is methylamine dehydrogenase [amine: (acceptor) oxidoreductase (deaminating)] (89). It could be coupled to

$$CH_3\text{-}NH_2 \rightarrow HCHO + NH_3$$

an oxidative phosphorylation, which would be very important for the energy metabolism of the "autotrophic" methane oxidizer. The dehydrogenase does not react with oxygen otherwise one could expect either amine oxidase (flavin-containing) (EC 1.4.3.4) or amine oxidase (pyridoxal-containing) (EC 1.4.3.6) as catalyst in this reaction (41). The additional product H_2O_2 would necessitate a catalase to be present in these organisms. However, methylamine oxidation can also proceed via the glycine–serine pathway, as has been found to occur in a gram-negative diplococcus (237). *Pseudomonas* sp. MS can also utilize dimethylamine, trimethylamine, and trimethyl sulfonium salts (222). The mechanisms involved, however, are not clarified.

The energetics of the methane oxidation is not solved as yet (124a). The methanol-oxidizing step is probably the energy-yielding step and is coupled to a phosphorylation at cytochrome c level. The mechanism of CO_2 fixation or incorporation may be mainly concerned with the synthesis of C_4 compounds from C_3 compounds. In the "autotrophic" scheme, the C_1 incorporation appears to go via a modified pentose phosphate cycle (185), which involves the condensation of ribose 5-phosphate with formaldehyde to produce allulose 6-phosphate. This condensation is a hydroxymethylation reaction (186) of ribose 5-phosphate with formaldehyde (see p. 230) and appears to be the key reaction, as the serine–glycine hydroxymethylation is the key reaction for the heterotrophic mechanism. Formaldehyde appears to be the key compound at which the allulose pathway and the serine–glycine pathway branch. The regulation of the flow of carbon units between the two pathways awaits further investigation.

Pseudomonas methanica stores poly-β-hydroxybutyric acid, and 8% of the total lipid is in the form of monopalmitin (178). As in other bacteria, this polymer can serve as a substrate for endogenous metabolism in the absence of an exogenous carbon source (83).

Amino Acid Metabolism

Aerobic and facultative aerobic bacteria (e.g., Enterobacteriaceae, Pseudomonadaceae) are equally capable of oxidizing proteins and amino acids under aerobic conditions and can use these compounds as their sole source of carbon, nitrogen, and energy. Amino acid oxidases, which are flavoprotein enzymes with a redox potential of about −0.004 V (46), are mainly involved in the initial oxidation step, particularly in the D-amino acids. The amino acid is oxidized to an imino acid, followed by a hydrolyzation to the corresponding keto acid (419):

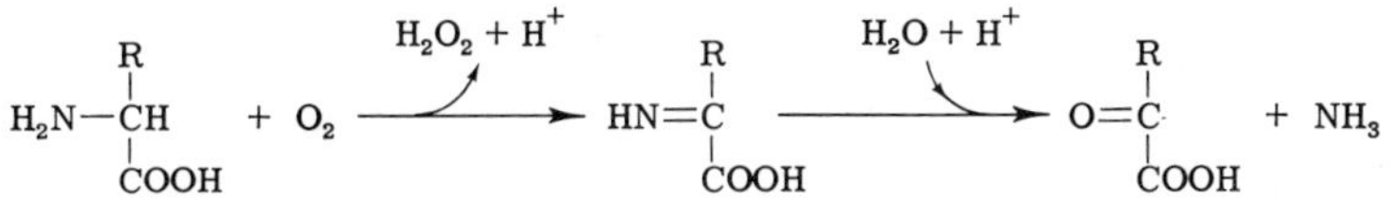
R
H_2N-CH
COOH
$+ O_2$
$H_2O_2 + H^+$
R
$HN=C$
COOH
$H_2O + H^+$
R
$O=C$
COOH
$+ NH_3$

These D-amino acid oxidases are frequently suppressed as soon as a second substrate, such as glucose or succinate, is present in the medium (259). L-Amino acids, in contrast, will undergo transamination reactions first, before further degradation takes place.

Spermidine Metabolism

A *Pseudomonas* species was found to metabolize the polyamine spermidine to β-alanine and succinate (Fig. 7.15). Spermidine is first cleaved to putres-

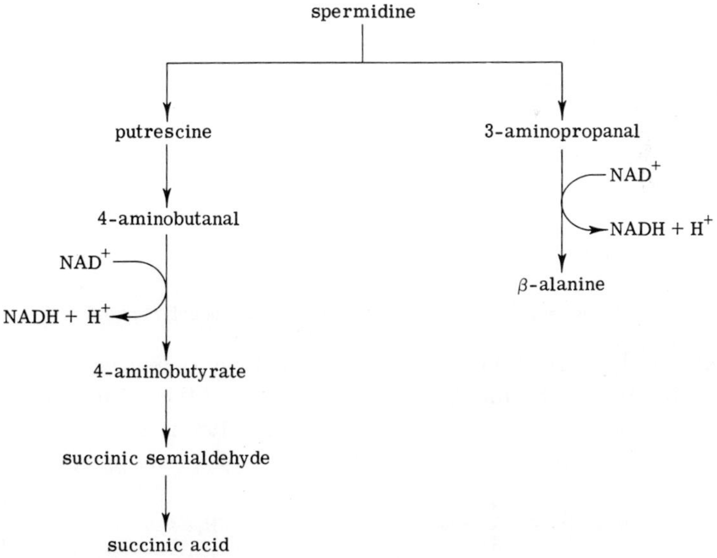
spermidine
putrescine
3-aminopropanal
NAD^+
$NADH + H^+$
4-aminobutanal
NAD^+
$NADH + H^+$
β-alanine
4-aminobutyrate
succinic semialdehyde
succinic acid

Fig. 7.15. Spermidine metabolism by pseudomonads (reprinted with permission of the authors and the American Society of Microbiology) (293).

cine and 3-aminopropanal (293). This is followed by an NAD-linked amino-

$$H_2N-\underset{H}{\overset{H}{C}}-\underset{H}{\overset{N}{C}}-\underset{H}{\overset{H}{C}}-\underset{NH_2}{\overset{H}{C}}-\underset{H}{\overset{H}{C}}-\underset{H}{\overset{H}{C}}-\underset{H}{\overset{H}{C}}-NH_2$$

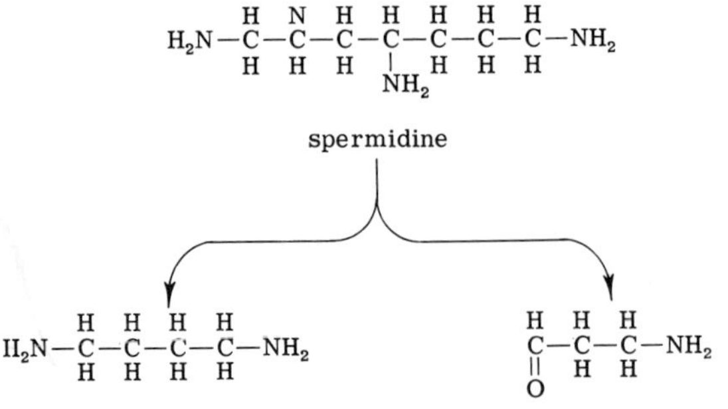

spermidine

$$H_2N-\underset{H}{\overset{H}{C}}-\underset{H}{\overset{H}{C}}-\underset{H}{\overset{H}{C}}-\underset{H}{\overset{H}{C}}-NH_2 \qquad \underset{O}{\overset{H}{C}}-\underset{H}{\overset{H}{C}}-\underset{H}{\overset{H}{C}}-NH_2$$

putrescine   3-aminopropanal

aldehyde dehydrogenase reaction, which oxidizes the aldehyde group of 3-aminopropanal and forms β-alanine. The other branch undergoes a trans-

$$\underset{O}{\overset{H}{C}}-\underset{H}{\overset{H}{C}}-\underset{H}{\overset{H}{C}}-NH_2 \longrightarrow HOOC-\underset{H}{\overset{H}{C}}-\underset{H}{\overset{H}{C}}-NH_2$$

3-aminopropanal   β-alanine

amination reaction with putrescine (43, 265), being converted to 4-amino-

$$H_2N-\underset{H}{\overset{H}{C}}-\underset{H}{\overset{H}{C}}-\underset{H}{\overset{H}{C}}-\underset{H}{\overset{H}{C}}-NH_2 \longrightarrow H_2N-\underset{H}{\overset{H}{C}}-\underset{H}{\overset{H}{C}}-\underset{H}{\overset{H}{C}}-\overset{H}{C}=O$$

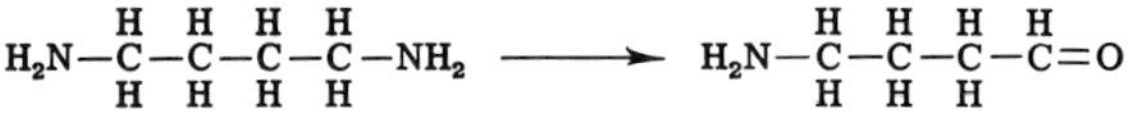

putrescine   4-aminobutanal

butanal. The enzyme catalyzing this reaction is the putrescine transaminase. A very similar reaction to the β-alanine formation occurs at the 4-amino-

$$H_2N-\underset{H}{\overset{H}{C}}-\underset{H}{\overset{H}{C}}-\underset{H}{\overset{H}{C}}-\overset{H}{C}=O \longrightarrow H_2N-\underset{H}{\overset{H}{C}}-\underset{H}{\overset{H}{C}}-\underset{H}{\overset{H}{C}}-COOH$$

4-aminobutanal   4-aminobutyric acid

butanal level, as an NAD-linked aminoaldehyde dehydrogenase converts 4-aminobutanal to 4-aminobutyric acid. A second transamination reaction (4-aminobutyrate aminotransferase, EC 2.6.1.19) is responsible for the formation of succinic semialdehyde. *Pseudomonas* species possess two differ-

$$H_2N-\underset{H}{\overset{H}{C}}-\underset{H}{\overset{H}{C}}-\underset{H}{\overset{H}{C}}-COOH \longrightarrow OHC-CH_2-CH_2-COOH$$

4-aminobutyric acid   succinic semialdehyde

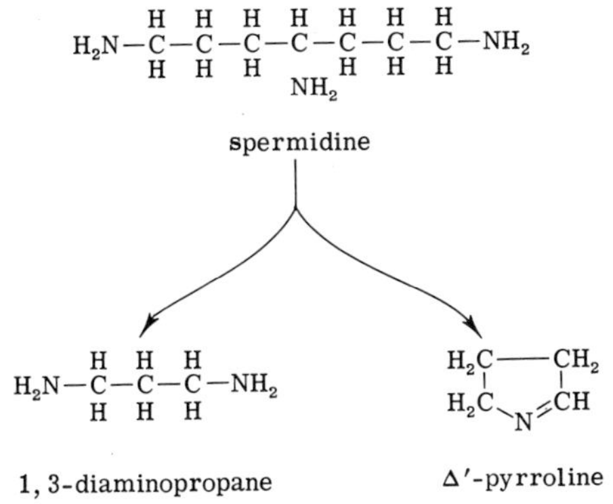
spermidine
1, 3-diaminopropane
Δ'-pyrroline

Scheme 7.12. Spermidine metabolism in *Serratia marcescens.*

ent enzymes for the utilization of succinic semialdehyde (277, 282, 294), a NAD-linked and a NADP-linked succinate semialdehyde dehydrogenase. It is thought that the NAD-linked enzyme (EC 1.2.1.24) is inducible and that the NADP-linked dehydrogenase (EC 1.2.1.16) is a constitutive enzyme.

Serratia marcescens utilizes spermidine in a different way (370). Spermidine dehydrogenase, a heme-FAD-enzyme protein, oxidizes spermidine to 1,3-diaminopropane and Δ′-pyrroline (see Scheme 7.12). Any further insight into the metabolism must await further developments.

Uric Acid and Allantoin Metabolism

The degradation of allantoin proceeds via three different pathways, which are distinguishable on the stereospecificity of the enzymes involved (392) (see Scheme 7.13). The most common pathway is the one used by *Pseudomonas aeruginosa* and *Pseudomonas fluorescens.* Uric acid is oxidized to (+)-allantoin with the help of the enzyme urate oxidase (urate:oxygen

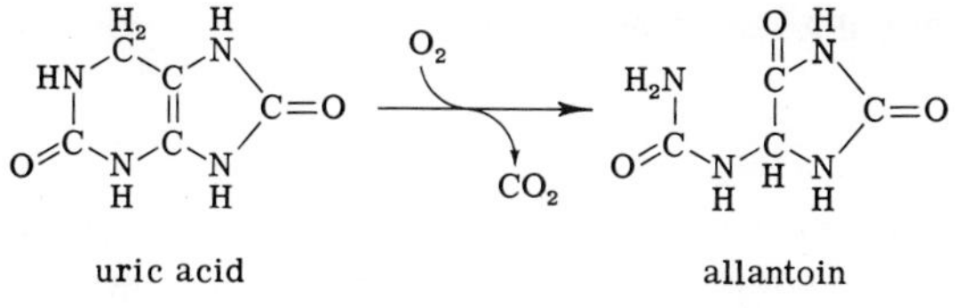
uric acid
allantoin

oxidoreductase, EC 1.7.3.3). Carbon dioxide is liberated in the same reaction. A hydration follows, catalyzed by allantoinase (allantoin amido-

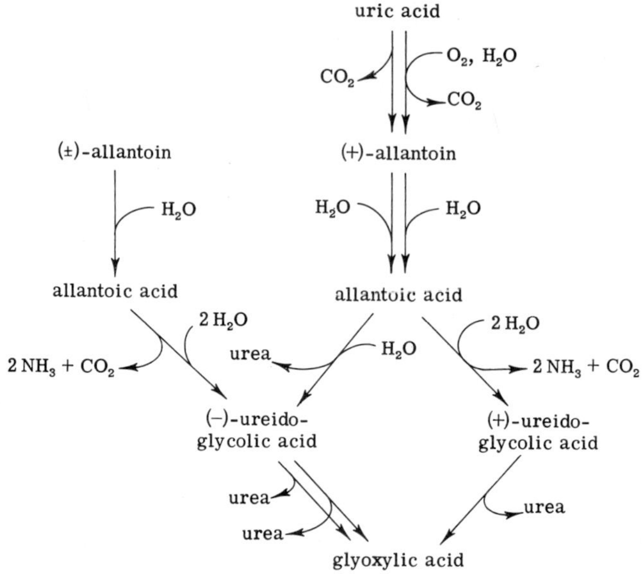
uric acid
O_2, H_2O
CO_2
CO_2
(±)-allantoin
(+)-allantoin
H_2O
H_2O
H_2O
allantoic acid
allantoic acid
2 H_2O
2 NH_3 + CO_2
urea
H_2O
2 H_2O
2 NH_3 + CO_2
(−)-ureido-
glycolic acid
(+)-ureido-
glycolic acid
urea
urea
urea
glyoxylic acid

Scheme 7.13. Metabolism of uric acid and allantoin (reprinted with permission of the author and the publisher of *Antonie van Leeuwenhoek J. Microbiol. Serol.*).

hydrolase, EC 3.5.2.5) (111), which produces allantoic acid. At this level,

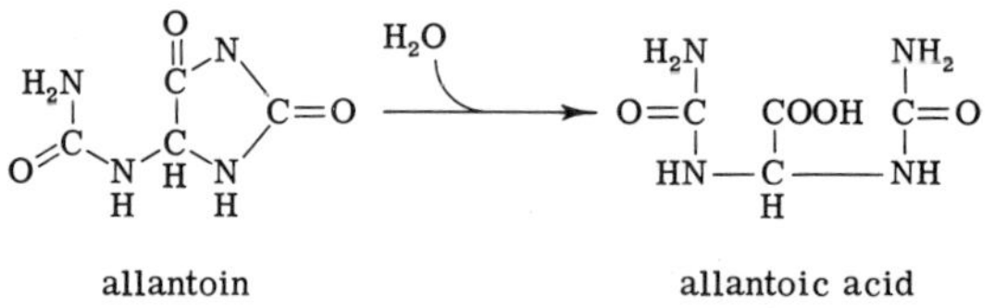
H_2O
allantoin
allantoic acid

the pathway separates. *Pseudomonas aeruginosa* and *Pseudomonas fluorescens* convert allantoic acid to (−)-ureidoglycolic acid and urea in a single step, with the aid of allantoicase (allantoate amidinohydrolase, EC 3.5.3.4)

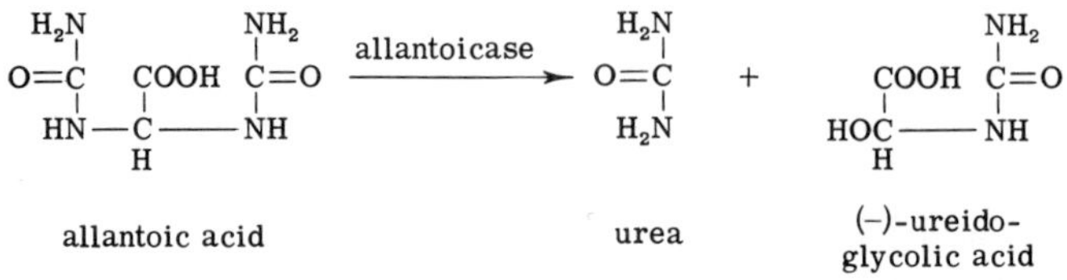
allantoicase
allantoic acid
urea
(−)-ureido-
glycolic acid

(226). This enzyme was purified from *Pseudomonas putida* (106) and found to contain a covalently bound α-ketobutyrate as prosthetic group. A second molecule of urea is liberated by the action of (−)-ureidoglycolate lyase

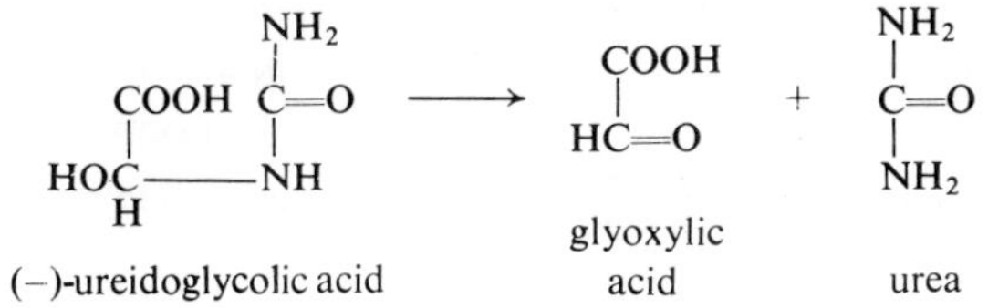
(−)-ureidoglycolic acid
glyoxylic acid
urea

(EC 4.3.2.3) on (−)-ureidoglycolic acid. The second end product of this reaction is glyoxylic acid.

Pseudomonas acidovorans follows a second pathway. This organism lacks the enzyme allantoicase (EC 3.5.3.4) but possesses allantoate amidohydrolase (112, 381), which transforms allantoic acid to (+)-ureidoglycolic acid,

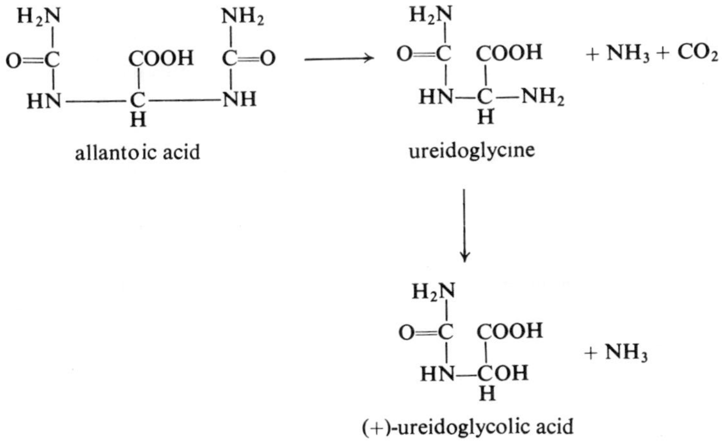
allantoic acid
ureidoglycine
(+)-ureidoglycolic acid

ammonia, and carbon dioxide. It is postulated that this conversion proceeds via ureidoglycine, as is the case in *Streptococcus allantoicus* (112). This enzyme is not identical to allantoin amidohydrolase. If ureidoglycine is an intermediate, the enzyme allantoate amidohydrolase would catalyze the first reaction and an additional enzyme, ureidoglycine aminohydrolase, would convert ureidoglycine to (+)-ureidoglycolate. These microorganisms substitute the enzyme allantoicase (allantoate amidinohydrolase) with two different enzymes, one an amidohydrolase and the second an aminohydrolase. (+)-Ureidoglycolic acid is then metabolized to glyoxylic acid and urea by (+)-ureidoglycolate lyase.

Ureidoglycine can also be transaminated to oxaluric acid in *Pseudomonas acidovorans*, whereby one amino group is transferred to glyoxylic acid form-

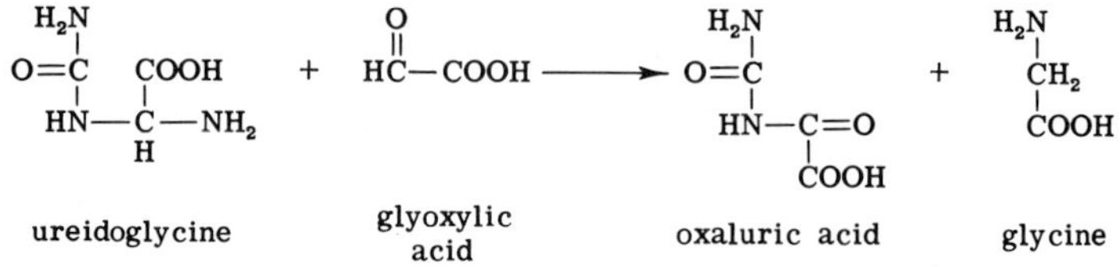
ureidoglycine
glyoxylic acid
oxaluric acid
glycine

ing glycine.

The third pathway is followed mainly by *Escherichia coli* and *Arthrobacter allantoicus*. Allantoic acid is transformed to (−)-ureidoglycolate, with the release of 2 moles of ammonia and 1 mole of carbon dioxide by the action of allantoate amidohydrolase. To keep biochemical unity, an intermediate, such as ureidoglycine, should also have been formed (11). (−)-Ureidoglycolic acid is metabolized to glyoxylic acid and urea by (−)-ureidoglycolate lyase.

The finding of the intermediate ureidoglycine between allantoic acid and (+)-ureidoglycolic acid led to the proposal that *Pseudomonas aeruginosa* may also have an intermediate, allanturate, which would lead to the following sequence:

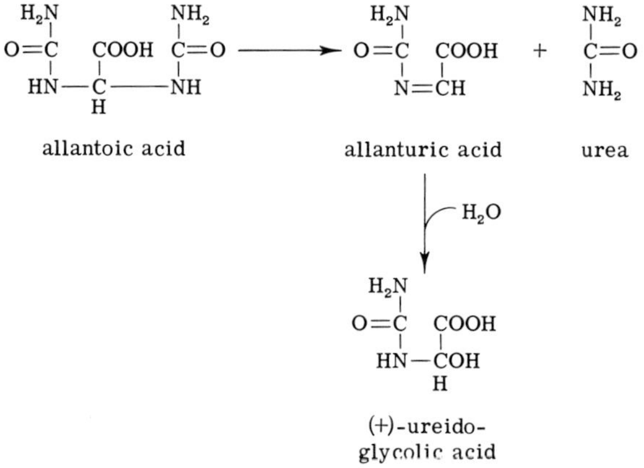
allantoic acid
allanturic acid
urea
H_2O
(+)-ureido-
glycolic acid

An additional enzyme has not been found as yet, but the presence of this intermediate would keep the biochemical unity.

Tryptophan Metabolism

Pseudomonads can be categorized according to their pathways of tryptophan metabolism:

1. The aromatic group, which degrades L-tryptophan via anthranilic acid (30)
2. The quinoline group, which degrades D- and L-tryptophan via kynurenic acid (29, 374)
3. The racemase–aromatic group, which degrades D- and L-tryptophan via anthranilic acid (30, 260)
4. The quinazoline group, which degrades D- and L-tryptophan via *O*-aminoacetophenone (252).

The overall degradation of D- and L-tryptophan is outlined in Fig. 7.16.

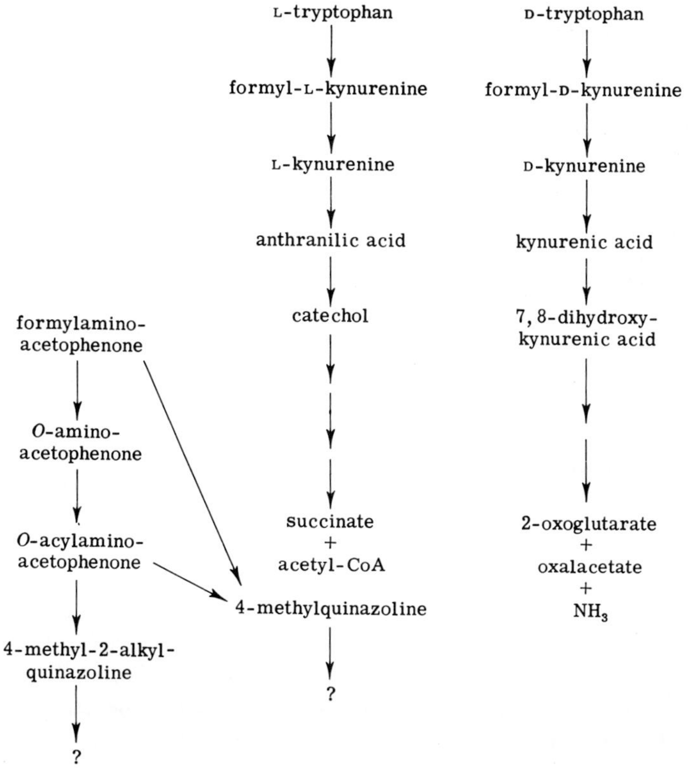
L-tryptophan
formyl-L-kynurenine
L-kynurenine
anthranilic acid
catechol
succinate + acetyl-CoA
D-tryptophan
formyl-D-kynurenine
D-kynurenine
kynurenic acid
7,8-dihydroxy-kynurenic acid
2-oxoglutarate + oxalacetate + NH3
formylamino-acetophenone
O-amino-acetophenone
O-acylamino-acetophenone
4-methyl-2-alkyl-quinazoline
4-methylquinazoline
?
?

Fig. 7.16. Tryptophan metabolism in pseudomonads.

The first group, which metabolizes L-tryptophan to catechol via anthranilic acid, has been known for a long time. Its main representative is *Pseudomonas fluorescens*. L-Tryptophan is attacked by a L-specific tryptophan 2,3-dioxygenase (157, 250) [L-tryptophan:oxygen 2,3-oxidoreductase (de-

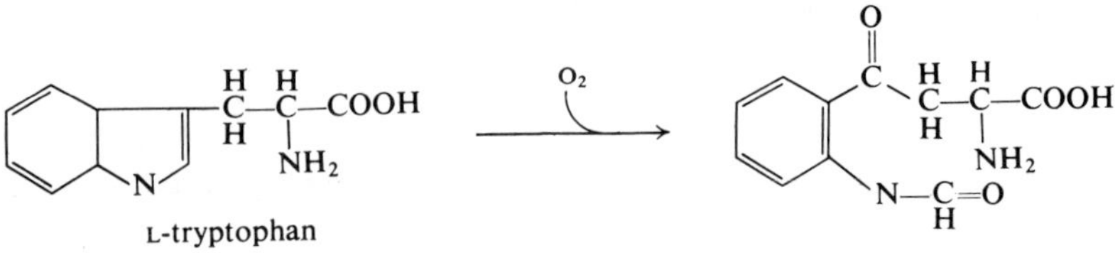
O2

L-tryptophan formyl-L-kynurenine

cyclizing), EC 1.13.11.11] and converts L-tryptophan to L-formylkynurenine. This enzyme is as yet the only dioxygenase reported that contains a hematin prosthetic group (251) and inserts the molecular oxygen into the pyrrole ring (297). A kynurenine formamidase (aryl-formylamine amido-

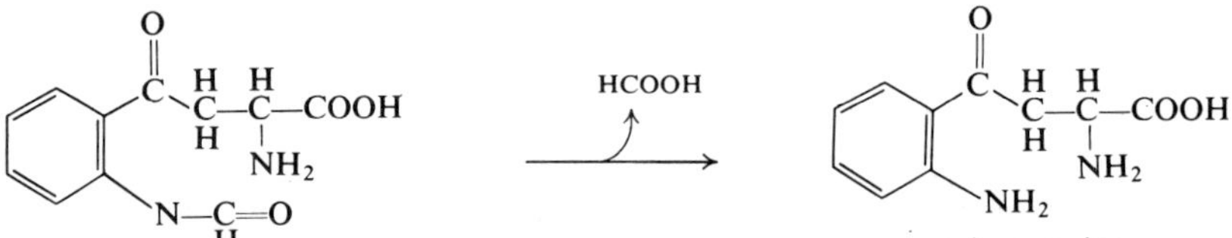
HCOOH

formyl-L-kynurenine L-kynurenine

hydrolase, EC 3.5.1.9) separates formate from formyl-L-kynurenine with the formation of L-kynurenine. The formation of anthranilic acid is brought

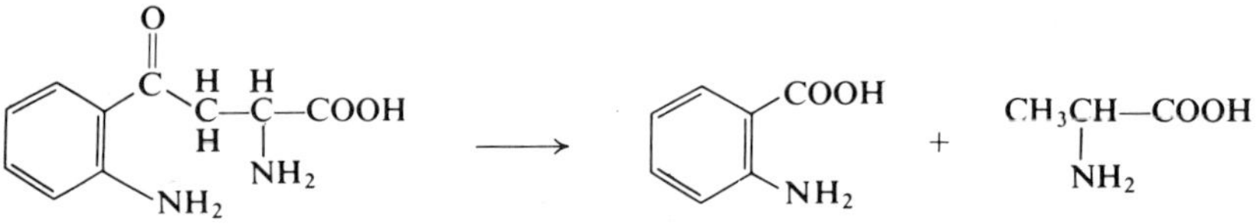

L-kynurenine anthranilic acid alanine

about by the action of kynureninase (L-kynurenine hydrolase, EC 3.7.1.3), which hydrolyzes the side chain of L-kynurenine to alanine. This enzyme has been crystallized from *Pseudomonas marginalis* (268). An oxygenase system that carries out an oxidative decarboxylation and a deamination and that incorporates oxygen molecules converts anthranilic acid to cate-

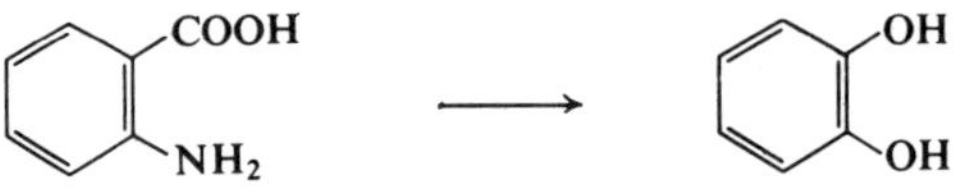

anthranilic acid catechol

chol, which undergoes an ortho fission and metabolizes via the β-ketoadipate pathway (see Chapter 8).

The second or quinoline group of pseudomonads converts D- and L-tryptophan via kynurenic acid. The main representative of this pathway is *Pseudomonas acidovorans*. Both D- and L-tryptophan are metabolized by their respective stereospecific oxygenase and kynurenine formamidase to D- or L-kynurenine. Instead of the kynureninase, these organisms possess an L- or D-kynurenine oxidase (374), which metabolizes L- or D-kynurenine

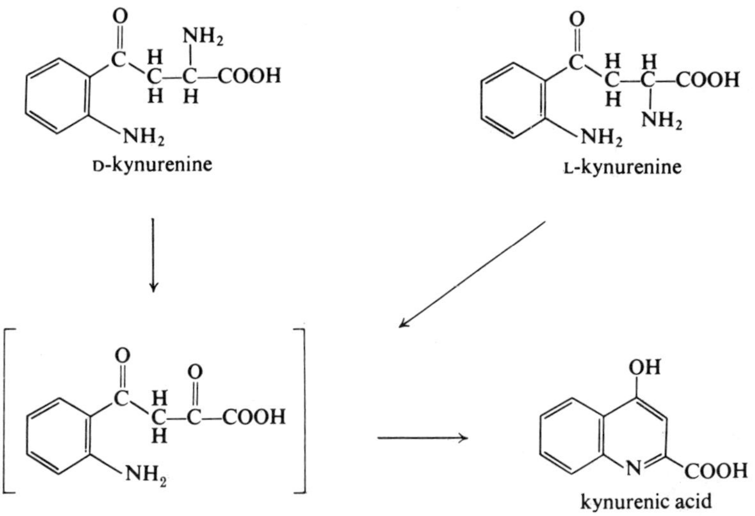
D-kynurenine
L-kynurenine
kynurenic acid

via an unknown intermediate to kynurenic acid. A hydroxylase, kynurenate 7,8-hydroxylase [kynurenate, hydrogen donor:oxygen oxidoreductase (hydroxylating), EC 1.14.99.2] converts kynurenic acid to 7,8-dihydroxy-

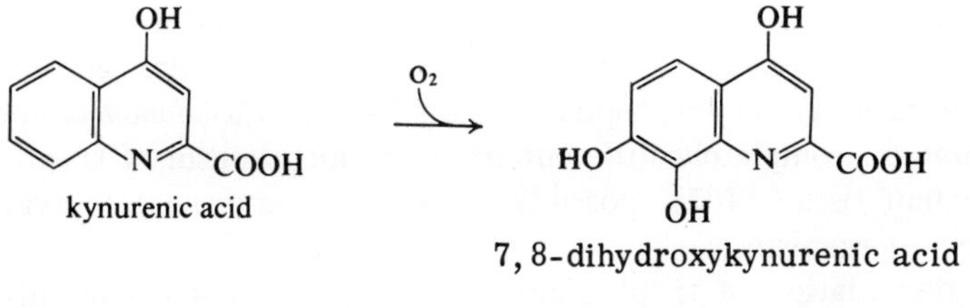
kynurenic acid
O_2
7,8-dihydroxykynurenic acid

kynurenic acid. The end products of this pathway are 2-oxoglutarate, oxalacetate, and ammonia, but the individual steps have not as yet been elucidated.

The third group, called the "racemase–aromatic" group, degrades D- and L-tryptophan via anthranilic acid. The only difference from the first pathway is the additional tryptophan racemase, which enables the organism to convert D-tryptophan to L-tryptophan.

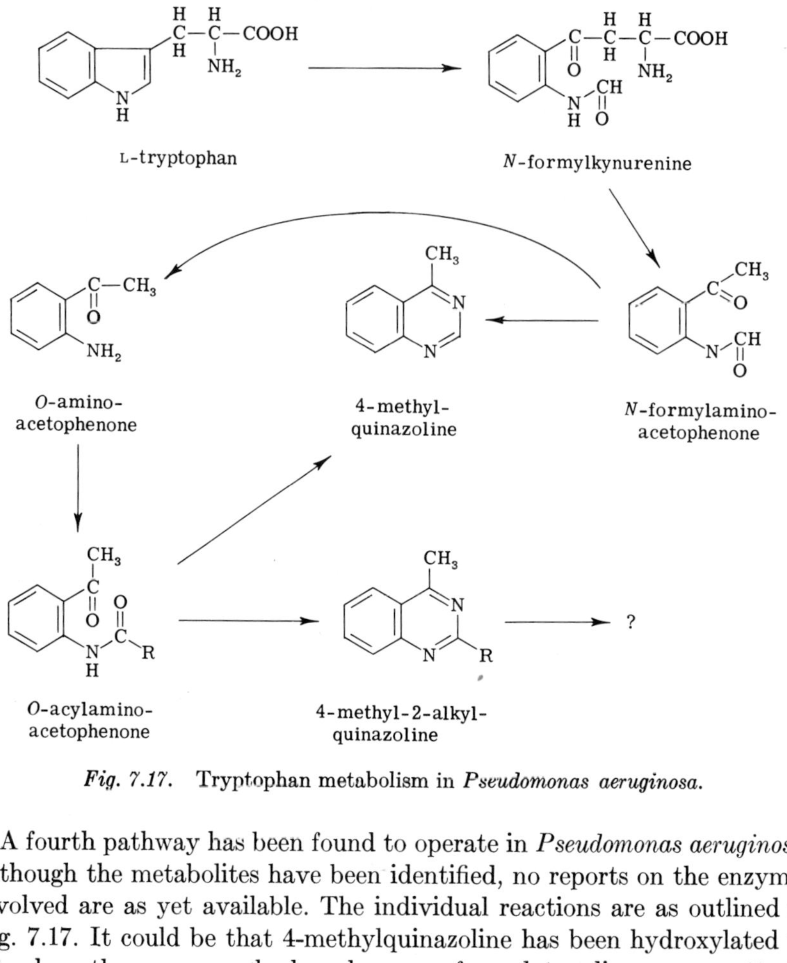
L-tryptophan
N-formylkynurenine
O-amino-acetophenone
4-methyl-quinazoline
N-formylamino-acetophenone
O-acylamino-acetophenone
4-methyl-2-alkyl-quinazoline
?

Fig. 7.17. Tryptophan metabolism in *Pseudomonas aeruginosa.*

A fourth pathway has been found to operate in *Pseudomonas aeruginosa.* Although the metabolites have been identified, no reports on the enzymes involved are as yet available. The individual reactions are as outlined in Fig. 7.17. It could be that 4-methylquinazoline has been hydroxylated to introduce the necessary hydroxyl groups for a later dioxygenase attack.

These pathways of tryptophan metabolism by *Pseudomonas* and *Flavobacterium* are completely different from the metabolism of tryptophan in mammalian tissue (107), possibly because an oxygenase is available in these microorganisms.

The degradation of tryptophan, however, does not occur only in the Pseudomonadaceae. *Streptomyces* (375) and *Bacillus cereus* (299) have been reported to metabolize D-tryptophan via anthranilic acid. This metabolism appears to play an important role during sporulation (299).

The first attempt at elucidating the regulation mechanism of these pathways has been carried out with *Pseudomonas acidovorans* (97, 327). These investigations revealed that L-tryptophan 2,3-dioxygenase is an allosteric enzyme, underlying an allosteric regulation by the substrate and molecular oxygen, whereas L-kynurenine induces the enzymes tryptophan 2,3-dioxygenase and kynurenine formamidase. Both enzymes, however, are under noncoordinate regulation (326), increasing the versatility of induction of either of the enzymes by other products or environmental conditions.

Threonine Metabolism

Under aerobic conditions, threonine can be metabolized via three different pathways, depending on the key enzyme available (Fig. 7.18). A number of pseudomonads are able to utilize L-threonine as sole carbon and energy

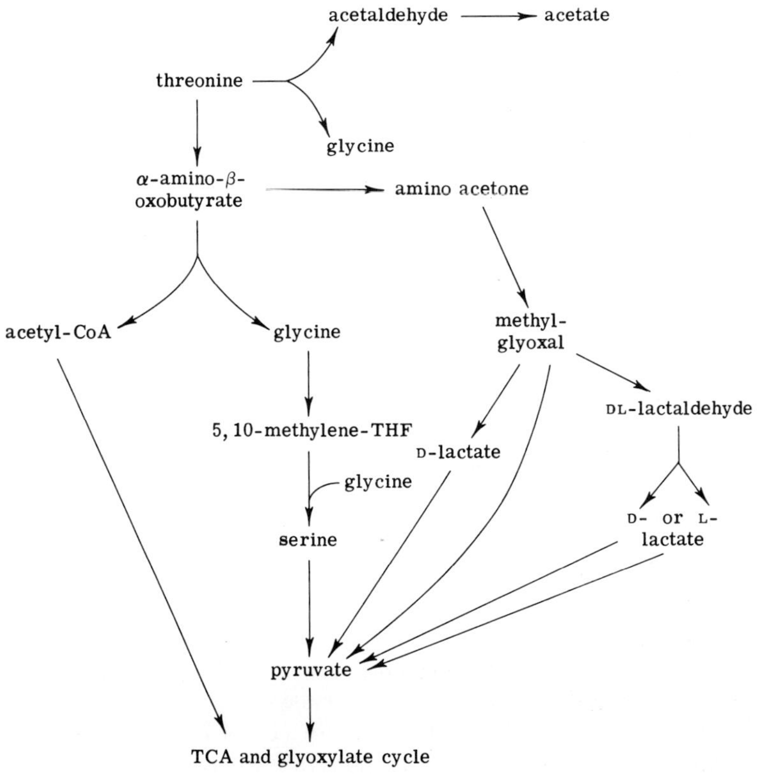
acetaldehyde
acetate
threonine
glycine
α-amino-β-oxobutyrate
amino acetone
acetyl-CoA
glycine
methyl-glyoxal
DL-lactaldehyde
5,10-methylene-THF
D-lactate
glycine
D- or L-lactate
serine
pyruvate
TCA and glyoxylate cycle

Fig. 7.18. Aerobic metabolism of threonine.

source. These organisms possess a very strong serine hydroxymethyltransferase (EC 2.1.2.1) activity (270) which cleaves threonine to acetaldehyde and glycine. A very active NAD^+-dependent aldehyde dehydrogenase converts acetaldehyde to acetate. These pseudomonads then follow the pathway of aerobic acetate metabolism (see p. 431), using mainly the glyoxylate cycle.

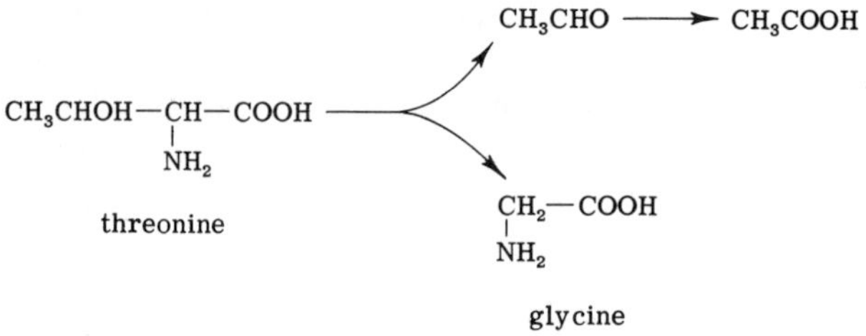
CH3CHO
CH3COOH
CH3CHOH—CH—COOH
NH2
threonine
CH2—COOH
NH2
glycine

A strain of *Arthrobacter* was isolated from soil that did not possess the serine hydroxymethyltransferase but instead possessed a L-threonine 3-dehydrogenase (EC 1.1.1.103) (245, 248). This enzyme oxidizes threonine

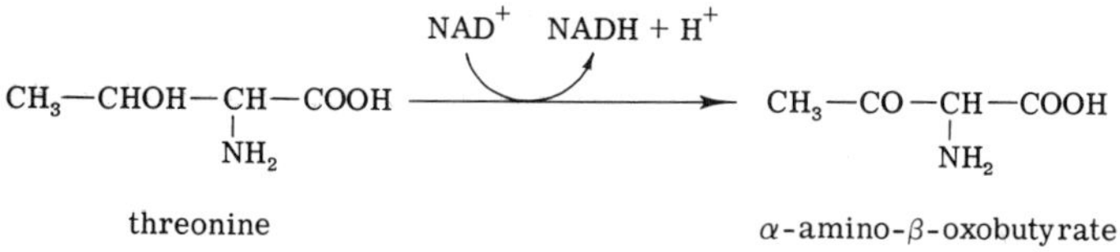
NAD+
NADH + H+
CH3—CHOH—CH—COOH
NH2
threonine
CH3—CO—CH—COOH
NH2
α-amino-β-oxobutyrate

to α-amino-β-oxobutyrate. A coenzyme A-dependent cleavage of α-amino-β-oxobutyrate catalyzed by an α-amino-β-oxobutyrate CoA ligase yields acetyl-CoA and glycine. Further metabolism follows the serine pathway,

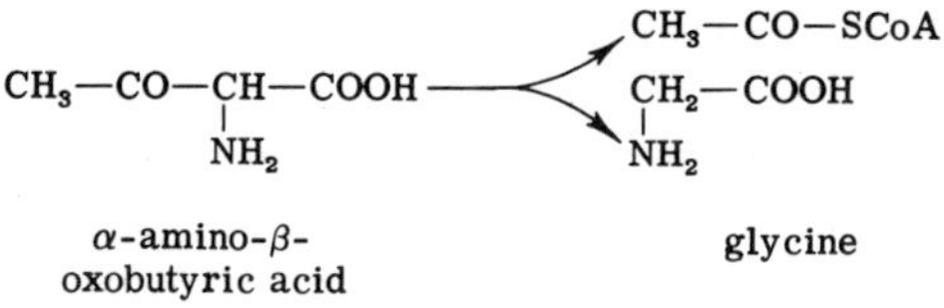
CH3—CO—SCoA
CH3—CO—CH—COOH
NH2
CH2—COOH
NH2
α-amino-β-
oxobutyric acid
glycine

involving tetrahydrofolate and a second molecule of glycine, as was described above (see p. 440). One of the glycine molecules supplies the methyl group for the tetrahydrofolate derivative, which in turn transfers this methyl group to a second molecule of glycine, forming serine. The metabolism ends in the TCA and glyoxylate cycle.

Pseudomonas oxalaticus was found to possess a glycine aminotransferase (EC 2.6.1.4) and glyoxylate carboxy-lyase (EC 4.1.1.47) activity (110), which indicated the operation of the glycerate pathway (see p. 415), used by this organism during growth on oxalate (38). The presence of a low isocitrate lyase activity, however, suggests that glycine is metabolized as an anaplerotic intermediate (39).

Bacillus subtilis (406–409) and *Pseudomonas* sp. NCIB 8858 (136) metabolize threonine and 1-aminopropan-2-ol via aminoacetone and methylglyoxal to pyruvate and the glyoxylate cycle. The branchoff occurs at the α-amino-β-oxobutyrate level, where an aminoacetone reductase (408) replaces the CoA-dependent ligase. The aminoketone is converted

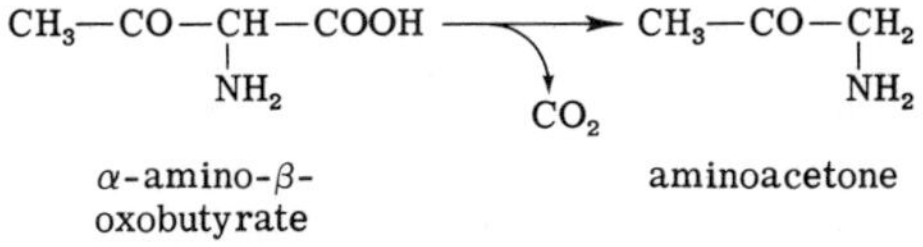
CH3—CO—CH—COOH
NH2
CO2
CH3—CO—CH2
NH2
α-amino-β-oxobutyrate
aminoacetone

to methylglyoxal, a reaction that is catalyzed by an aminoacetone amino-

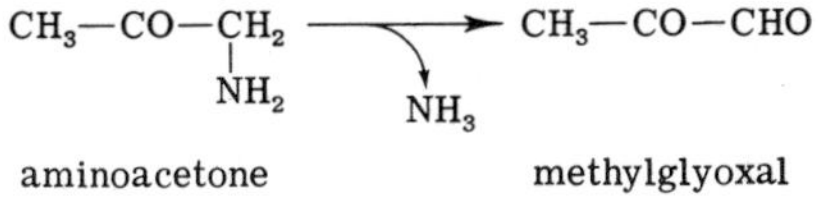
CH3—CO—CH2
NH2
NH3
CH3—CO—CHO
aminoacetone
methylglyoxal

transferase (135). Methylglyoxal can now follow three different routes to pyruvate. A NAD$^+$-linked methylglyoxal dehydrogenase converts

CH3—CO—CHO
CH3—CO—COOH
methylglyoxal
pyruvic acid

methylglyoxal directly to pyruvic acid. In the absence of this enzyme,

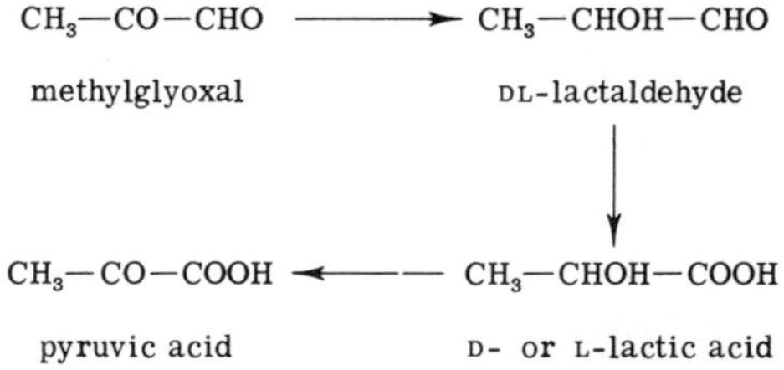
CH3—CO—CHO
CH3—CHOH—CHO
methylglyoxal
DL-lactaldehyde
CH3—CO—COOH
CH3—CHOH—COOH
pyruvic acid
D- or L-lactic acid

methylglyoxal can be metabolized by a NADPH-linked methylglyoxal reductase to DL-lactaldehyde, which is followed by NAD$^+$-linked lactaldehyde dehydrogenase reactions (EC 1.1.1.78; EC 1.2.1.22) forming D- or L-lactic acid. The involvement of stereospecific D- and L-lactate dehydrogenases finally produces pyruvic acid, which itself is metabolized via the TCA and glyoxylate cycle.

In the absence of the reduced NADP-linked methylglyoxal reductase, a methylglyoxalase system can convert methylglyoxal directly to D-lactic acid, which is oxidized to pyruvate by the NAD$^+$-linked D-lactate dehydrogenase.

These microorganisms are also able to utilize 1-aminopropan-2-ol instead of threonine. In this case, D- and L-aminopropan-2-ol:NAD$^+$ dehydrogenases (EC 1.1.1.74; EC 1.1.1.75) (409) convert 1-aminopropan-2-ol to aminoacetone and then follows the above-described route. Members of the Enterobacteriaceae also possess this enzyme (171, 246, 384) but require it for the formation of 1-aminopropan-2-ol from aminoacetone for the production of vitamin B_{12}.

Lysine Metabolism

Pseudomonas putida utilizes lysine via two distinct pathways (59, 321, 599). L-Lysine appears to be the substrate for the acyclic pathway and D-lysine for the cyclic pathway leading via pipecolate (Fig. 7.19).

Each of these pathways is selectively induced by L- or D-lysine and the

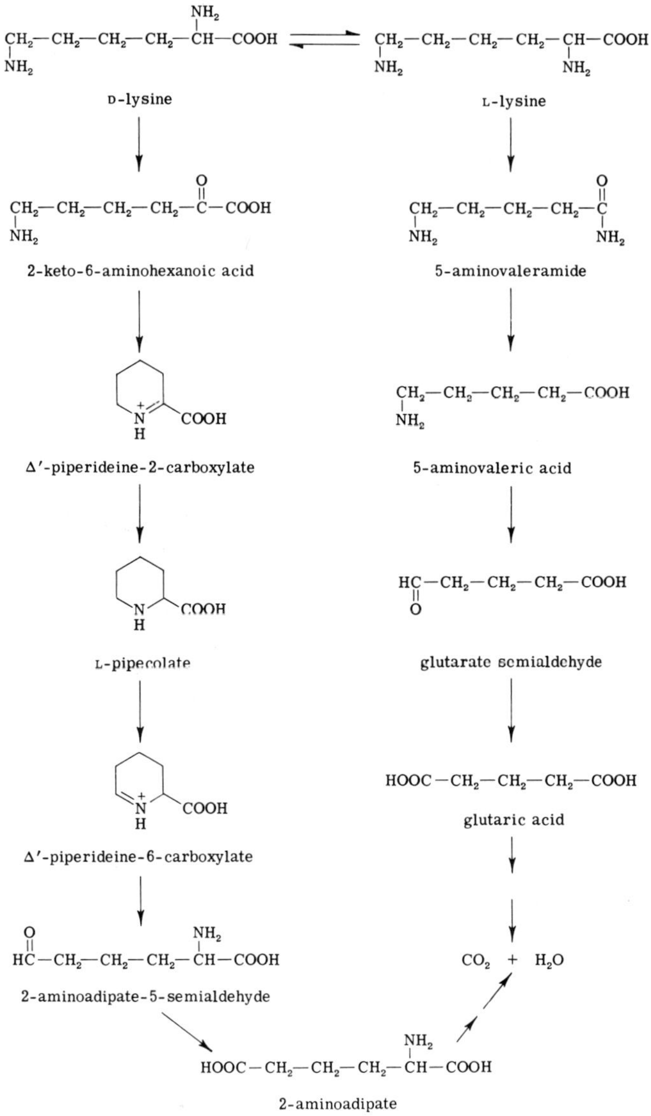
D-lysine
L-lysine
2-keto-6-aminohexanoic acid
5-aminovaleramide
Δ'-piperideine-2-carboxylate
5-aminovaleric acid
L-pipecolate
glutarate semialdehyde
glutaric acid
Δ'-piperideine-6-carboxylate
2-aminoadipate-5-semialdehyde
2-aminoadipate

Fig. 7.19. Lysine metabolism by *Pseudomonas putida* (reprinted with permission of the authors and the American Society of Biological Chemists) (267b).

appropriate intermediates. The enzyme that catalyzes the conversion of D-lysine to Δ′-piperideine-2-carboxylate is not known as yet. A Δ′-piperideine-2-carboxylate reductase (L-pipecolate: NADP 2-oxidoreductase; 59a), however, was found responsible for the formation of L-pipecolate. The further metabolism of L-pipecolate is identical to that described in Chapter 8.

Apart from the racemase (EC 5.1.1.5) (155a, 354), which can interconvert D- and L-lysine, an L-lysine 2-monooxygenase [L-lysine: oxygen 2-oxidoreductase (decarboxylating), EC 1.13.12.2] (120a, 131, 371, 371a) appears to be responsible for the conversion to γ-aminovaleramide. The release of the amide group is obtained by the action of δ-aminovaleramide amidase (321), which results in the formation of aminovalerate. The next amino group is transferred to α-ketoglutarate by a δ-aminovalerate transaminase (155b) producing glutamic acid and glutaric semialdehyde. The last intermediate, glutaric acid, is formed with the aid of an NAD^+-dependent glutaric semialdehyde dehydrogenase (glutarate-semialdehyde: NADP oxidoreductase, EC 1.2.1.20) (155c). Glutaric acid is further metabolized as described in Chapter 8.

Hydroxyproline Metabolism

Hydroxyproline can be metabolized by *Pseudomonas putida* (118), *Pseudomonas convexa*, and *Pseudomonas fluorescens* (376), forming 2-oxoglutarate as the key intermediate for the entry into the TCA cycle.

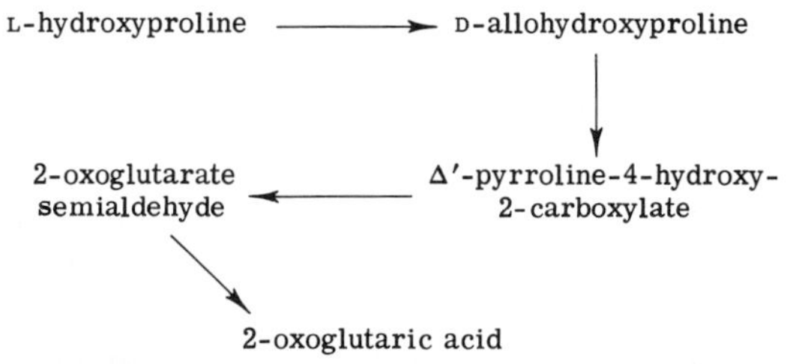
L-hydroxyproline
D-allohydroxyproline
Δ′-pyrroline-4-hydroxy-2-carboxylate
2-oxoglutarate semialdehyde
2-oxoglutaric acid

The first reaction, from hydroxyproline to D-allohydroxyproline, is a reversible reaction of a 2-epimerase (hydroxyproline, 2-epimerase, EC

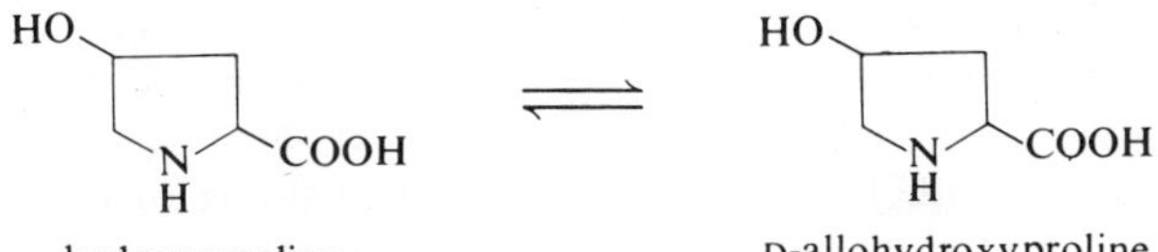
HO
N
H
COOH
HO
N
H
COOH

hydroxyproline D-allohydroxyproline

5.1.1.8). It is followed by an oxygenase reaction, with the formation of the

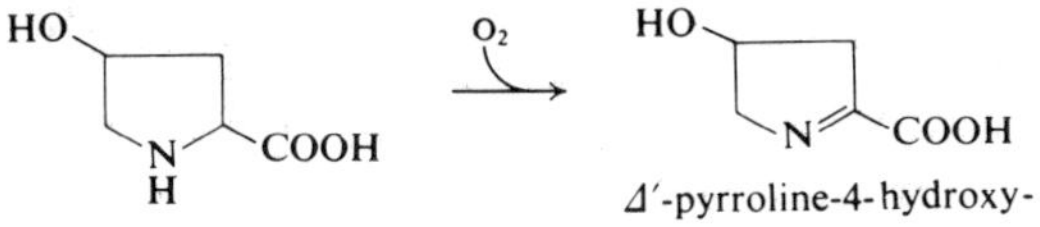

D-allohydroxyproline → Δ′-pyrroline-4-hydroxy-2-carboxylate

intermediate Δ′-pyrroline-4-hydroxy-2-carboxylate (1). This compound undergoes a deamination, which is catalyzed by the enzyme Δ′-pyrroline-4-

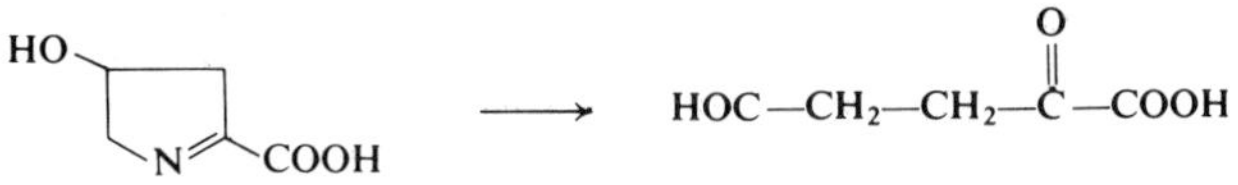

Δ′-pyrroline-4-hydroxy-2-carboxylate → 2-oxoglutaric semialdehyde

hydroxy-2-carboxylate deaminase (350) to form 2-oxoglutaric semialdehyde (351). An α-oxoglutarate semialdehyde dehydrogenase (NADP-linked) (2) converts the semialdehyde into 2-oxoglutaric acid (376).

The pathway is induced by either of the two epimers (119). The second and third enzyme are induced coordinately. This induction process appears to be insensitive to catabolite repression.

Valine Metabolism

Both isomers of valine are metabolized by *Pseudomonas aeruginosa* (283) (Fig. 7.20). Whereas L-valine is transaminated with 2-oxoglutarate, which forms glutamate and 2-ketoisovalerate, D-valine is oxidized with the help of D-valine dehydrogenase (259), releasing ammonia and producing the same intermediate. The glutamate from the transamination reaction is converted back to 2-oxoglutarate by glutamate dehydrogenase. A coenzyme A-dependent oxidative decarboxylation metabolizes 2-ketoisovalerate to isobutyryl-CoA and carbon dioxide. Apart from the CoA requirement, 2-ketoisovalerate dehydrogenase is also NAD^+ dependent [2-oxoisovalerate:NAD^+ oxidoreductase (CoA-isobutyrylating), EC 1.2.1.25] (279). An NAD^+-dependent dehydrogenase, together with two hydration steps, converts isobutyryl-CoA to 3-hydroxyisobutyric acid. These enzymes have not been well studied. 3-Hydroxyisobutyric acid is oxidized to methylmalonate semialdehyde with the aid of the NAD^+-dependent 3-hydroxyisobutyrate dehydrogenase (EC 1.1.1.31). The oxidation of methylmalonate semialdehyde to propionyl-CoA is catalyzed by a CoA- and NAD^+-depend-

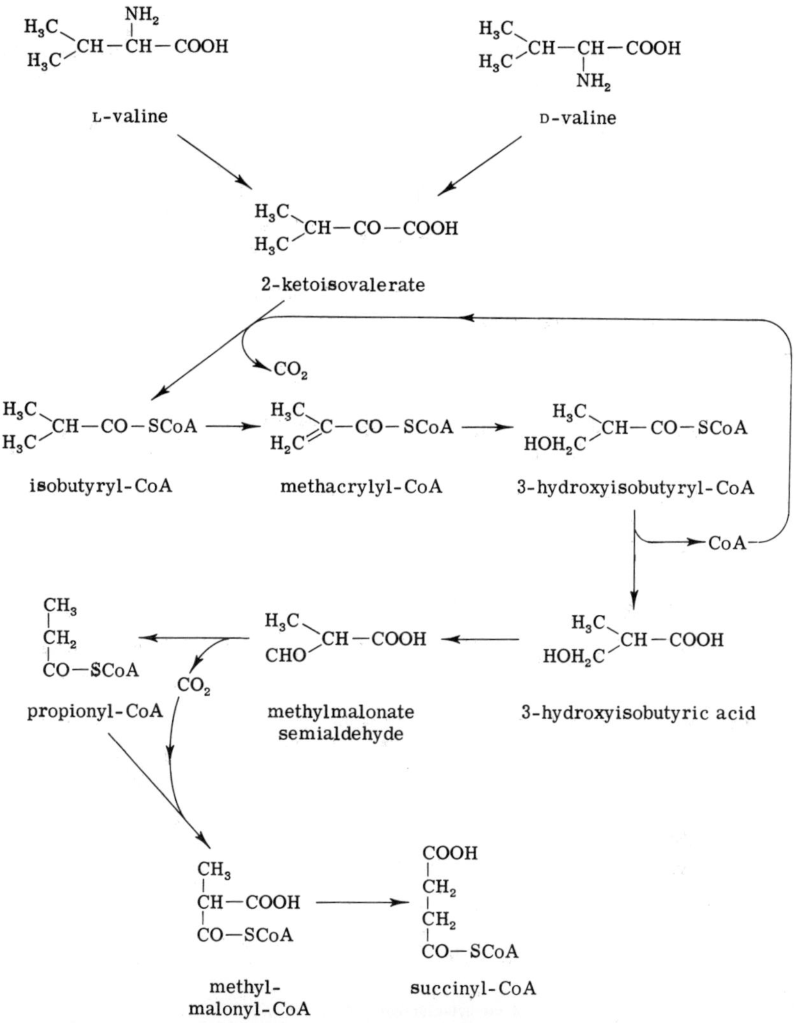
L-valine
D-valine
2-ketoisovalerate
CO_2
isobutyryl-CoA
methacrylyl-CoA
3-hydroxyisobutyryl-CoA
CoA
propionyl-CoA
CO_2
methylmalonate
semialdehyde
3-hydroxyisobutyric acid
methyl-
malonyl-CoA
succinyl-CoA

Fig. 7.20. Aerobic metabolism of valine. (reprinted with permission of the author and the American Society of Biological Chemists). (283).

ent dehydrogenase (27, 356), with the release of carbon dioxide. Propionyl-CoA can now be metabolized either via acrylyl-CoA to lactyl-CoA or via methylmalonyl-CoA and succinyl-CoA possibly to succinate and the TCA and glyoxylate cycle. *Pseudomonas aeruginosa* was also found to form alanine and aspartate during growth on valine, which would suggest the existence of further routes from propionyl-CoA (355).

Leucine and Isoleucine Catabolism

The newly proposed pathways for leucine (260c) and isoleucine (68a) catabolism are given in Figs. 7.21 and 7.22.

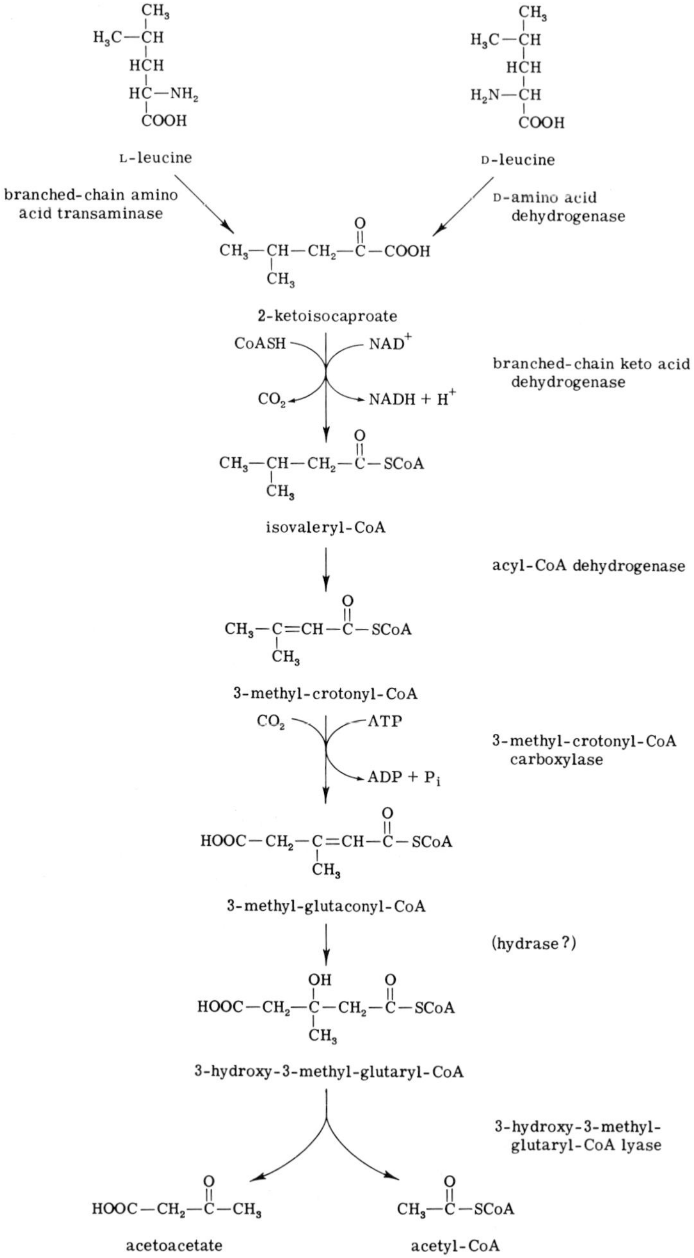

L-leucine
D-leucine
branched-chain amino acid transaminase
D-amino acid dehydrogenase
2-ketoisocaproate
CoASH
NAD+
CO2
NADH + H+
branched-chain keto acid dehydrogenase
isovaleryl-CoA
acyl-CoA dehydrogenase
3-methyl-crotonyl-CoA
CO2
ATP
ADP + Pi
3-methyl-crotonyl-CoA carboxylase
3-methyl-glutaconyl-CoA
(hydrase?)
3-hydroxy-3-methyl-glutaryl-CoA
3-hydroxy-3-methyl-glutaryl-CoA lyase
acetoacetate
acetyl-CoA

Fig. 7.21. D- and L-Leucine catabolism in *Pseudomonas putida* (260c).

Similar to the valine utilization, the D-isoleucine is oxidatively deaminated by a D-amino acid dehydrogenase, whereas the L-isomer is converted to the corresponding keto acid, L-2-keto-3-methylvaleric acid, by a branched-chain amino acid transaminase. The same mode of action also occurs in the case of D- and L-leucine catabolism. The so formed keto acid

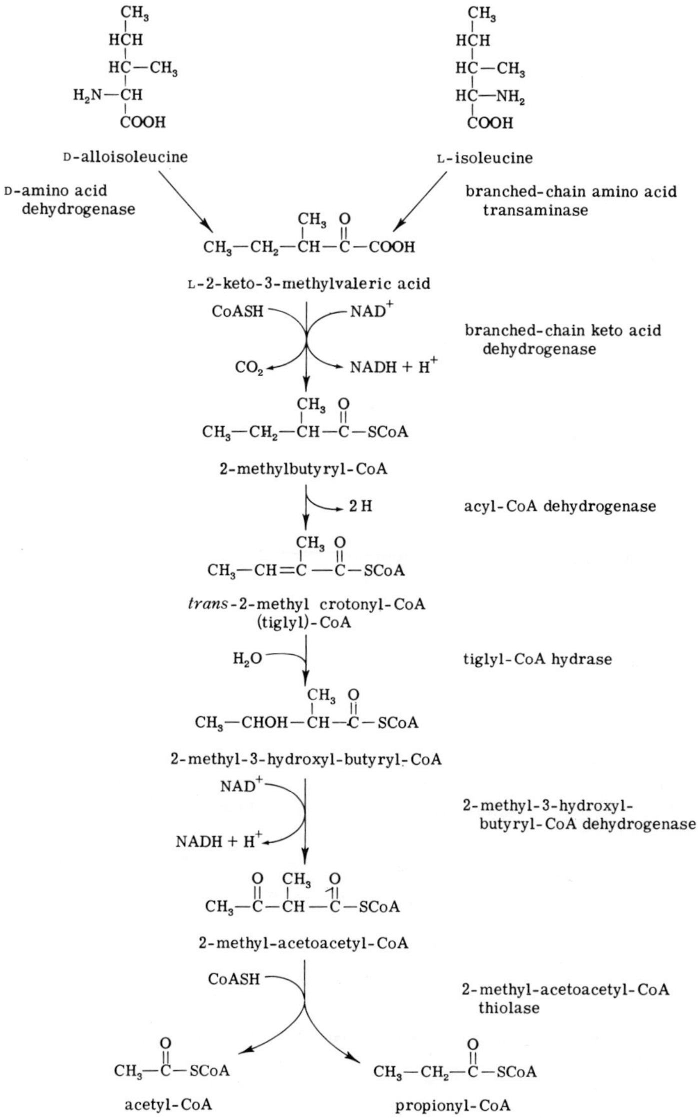
D-alloisoleucine
L-isoleucine
D-amino acid dehydrogenase
branched-chain amino acid transaminase
L-2-keto-3-methylvaleric acid
CoASH
NAD+
CO2
NADH + H+
branched-chain keto acid dehydrogenase
2-methylbutyryl-CoA
2 H
acyl-CoA dehydrogenase
trans-2-methyl crotonyl-CoA (tiglyl)-CoA
H2O
tiglyl-CoA hydrase
2-methyl-3-hydroxyl-butyryl-CoA
2-methyl-3-hydroxyl-butyryl-CoA dehydrogenase
2-methyl-acetoacetyl-CoA
2-methyl-acetoacetyl-CoA thiolase
acetyl-CoA
propionyl-CoA

Fig. 7.22. D- and L-Isoleucine catabolism in *Pseudomonas putida* (68a).

undergoes a coenzyme A-dependent oxidative decarboxylation with a branched-chain keto acid dehydrogenase, which leads to 2-methylbutyryl-CoA in the case of isoleucine and isovaleryl-CoA in the case of leucine degradation. The third common enzymatic step involves an acyl-CoA dehydrogenase, which forms methacrylyl-CoA, tiglyl-CoA, and 3-methylcrotonyl-CoA in the valine, isoleucine, and leucine pathways, respectively.

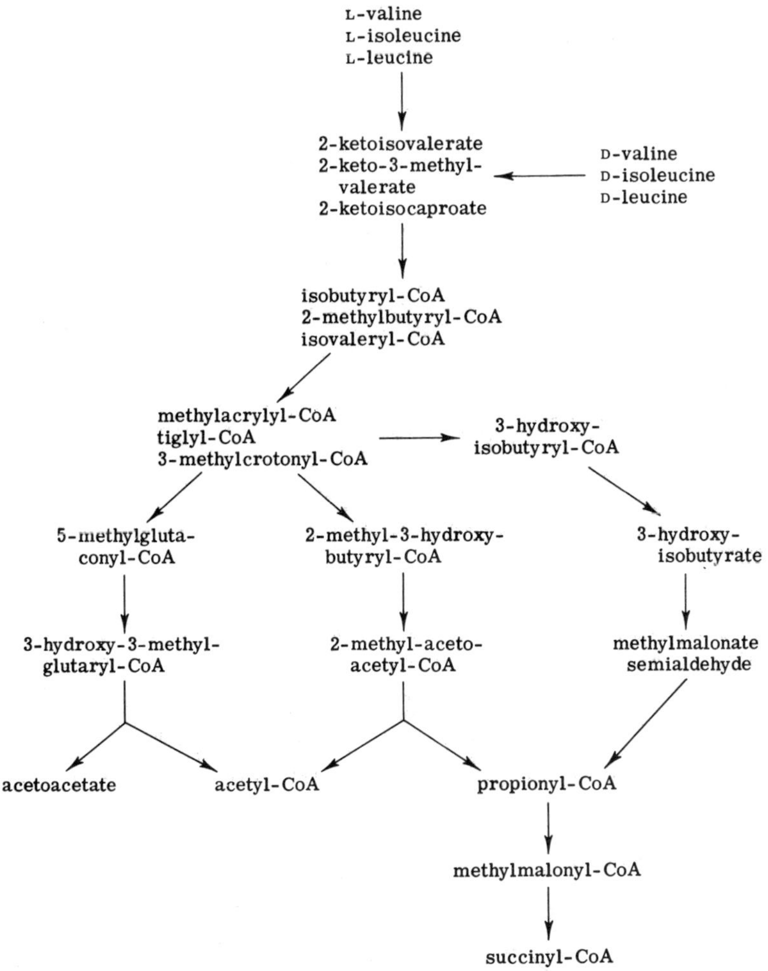
L-valine
L-isoleucine
L-leucine
2-ketoisovalerate
2-keto-3-methyl-valerate
2-ketoisocaproate
D-valine
D-isoleucine
D-leucine
isobutyryl-CoA
2-methylbutyryl-CoA
isovaleryl-CoA
methylacrylyl-CoA
tiglyl-CoA
3-methylcrotonyl-CoA
3-hydroxy-isobutyryl-CoA
5-methylglutaconyl-CoA
2-methyl-3-hydroxy-butyryl-CoA
3-hydroxy-isobutyrate
3-hydroxy-3-methyl-glutaryl-CoA
2-methyl-aceto-acetyl-CoA
methylmalonate semialdehyde
acetoacetate
acetyl-CoA
propionyl-CoA
methylmalonyl-CoA
succinyl-CoA

Fig. 7.23. Pathways for the metabolism of branched-chain amino acids in *Pseudomonas putida* (260b).

Now, each pathway follows its own way and uses pathway characteristic enzymes.

In the isoleucine pathway, two newly discovered enzymes (68a), tiglyl-CoA hydrase and 2-methyl-3-hydroxy-butyryl-CoA dehydrogenase lead to the formation of 2-methyl-acetoacetyl-CoA. The final end products, acetyl-CoA plus propionyl-CoA lead into the tricarboxylic acid cycle and are formed owing to the presence of 2-methyl-acetoacetyl-CoA thiolase.

The leucine pathway follows up with a carboxylation of 3-methyl-crotonyl-CoA to 3-methyl-glutaconyl-CoA. The introduction of an hydroxyl-group by an unknown enzyme (a hydrase ?) forms 3-hydroxy-3-methyl-glutaryl-CoA and the entry into the tricarboxylic acid cycle occurs after the formation of acetoacetate and acetyl-CoA by 3-hydroxy-3-methyl-glutaryl-CoA lyase (260c).

This close relationship in the branched-chain amino acid catabolism is therefore very similar to the biosynthesis of these amino acids, since it was established (260b) that the branched-chain amino acid transaminase and branched-chain keto acid dehydrogenases are common enzymes for all three pathways. If one combines all three pathways, the general scheme in Fig. 7.23 would eventuate.

With this common pathway for the deamination of the D- and L-isomers and the oxidative decarboxylation of the keto acids, regulation must be very similar (68a). It has been established for *Pseudomonas putida* (260a). that the branched-chain amino acid transaminase is a constitutive enzyme. One would therefore expect at least three separate inductive events, one for the D-amino acid dehydrogenase, one for the branched-chain keto acid dehydrogenase, and one for the induction of all those enzymes unique to the metabolism of the individual pathways.

Histidine Metabolism

The utilization of histidine is widespread among microorganisms. There are only a few differences between the anaerobic (see Chapter 9) and aerobic metabolisms of this amino acid (Fig. 7.24).

Aerobically, histidine is first deaminated by histidine ammonia lyase (EC 4.3.1.3) (102, 310) to urocanic acid. A urocanate hydratase (EC 4.2.1.49) that exhibits photoactivation (153) converts the unsaturated urocanic acid to imidazolone propionate. The ring fission occurs via imidazolone propionase (EC 3.5.2.7) forming formiminoglutamic acid. A second hydrolase splits the molecule into formamide and glutamic acid and was named "formiminoglutamase (EC 3.5.3.8)" (180). Glutamic acid can be further deaminated to 2-oxoglutaric acid, a member of the TCA cycle.

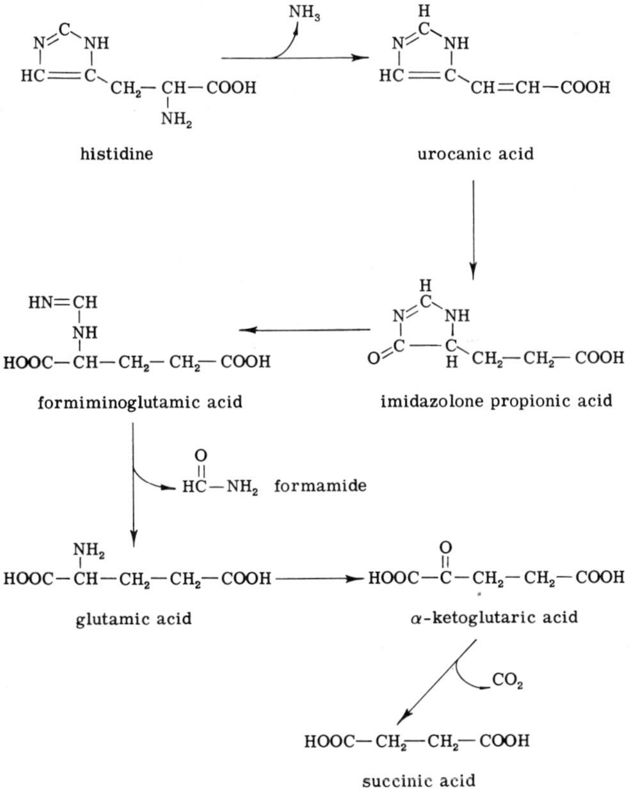
NH3
histidine
urocanic acid
formiminoglutamic acid
imidazolone propionic acid
formamide
glutamic acid
α-ketoglutaric acid
CO2
succinic acid

Fig. 7.24. Aerobic metabolism of histidine.

As far as regulation is concerned, there appear to be differences among the bacterial families. Pseudomonads use the first intermediate, urocanic acid, as their physiological inducer (152, 280), and it is thought that some sort of allosteric interaction with the histidine ammonia-lyase could exist (239). Succinate, however, appears to play the role of feedback inhibitor (152), which would suggest that it must be regarded as the end product of the histidine pathway and as the second intermediate of the TCA cycle pathway.

In the Enterobacteriaceae (*Aerobacter aerogenes, Salmonella typhimurium*) urocanic acid also plays the role of physiological inducer, whereas in *Bacillus subtilis* this role is taken by histidine (60). It appears that in the Enterobacteriaceae the synthesis of histidine ammonia-lyase can be influenced by environmental conditions, as evidenced by the close relationship between histidine catabolism and growth rate (167). Although urocanic acid is the inducer of this pathway, these facultative anaerobes seem to have an extensive catabolite control system, which has so far not been fully elucidated.

Arginine Metabolism

The catabolic pathway of L-arginine in *Streptococcus faecalis* joins the ornithine pathway (77, 103). Arginine deiminase (EC 3.5.3.6) converts

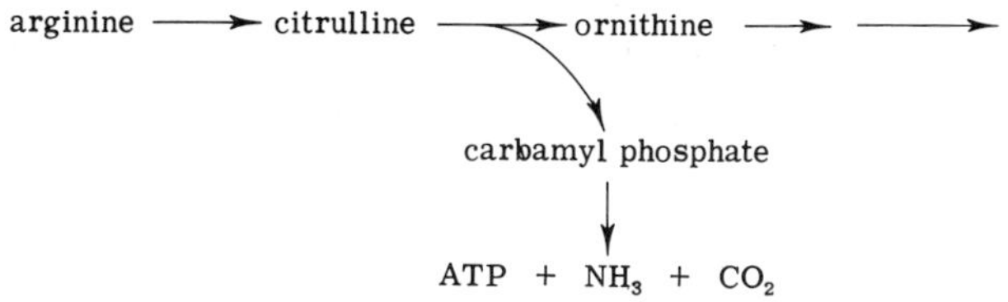
arginine
citrulline
ornithine
carbamyl phosphate
ATP + NH3 + CO2

arginine to citrulline, with the liberation of ammonia. Ornithine carbamoyltransferase (EC 2.1.3.3) cleaves citrulline to ornithine and carbamylphosphate. Whereas ornithine is further metabolized via putrescine carbamate kinase (EC 2.7.2.2) forms the end products ATP, ammonia, and carbon dioxide.

Metabolism of Various Amines

Pseudomonas aminovorans is able to oxidize a number of amines (41, 90). Dimethylamine, for example, is oxidized to methylamine and formaldehyde

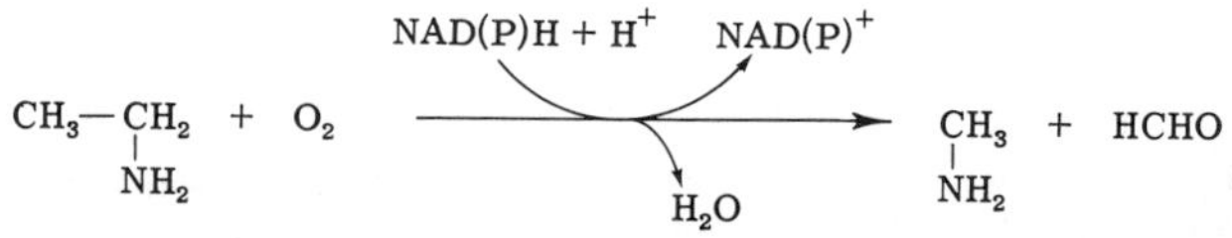
NAD(P)H + H+
NAD(P)+
CH3—CH2 + O2
NH2
H2O
CH3 + HCHO
NH2

by a reduced NAD(P)-dependent oxidative *N*-demethylation (88) by a hydroxylase. This enzyme was found to be a mixed function secondary amine oxidase system; it also attacks methylamine, trimethylamine, or trimethylamine *N*-oxide (90). A purification of the oxidase system revealed that it contains an enzymatically active cytochrome P-420-type hemoprotein.

Bacillus cereus degrades 2-aminoethyl phosphate to acetaldehyde and inorganic phosphate via an intermediate called "2-phosphonoacetaldehyde" (225). The first step in this metabolism is a transamination, whereby the

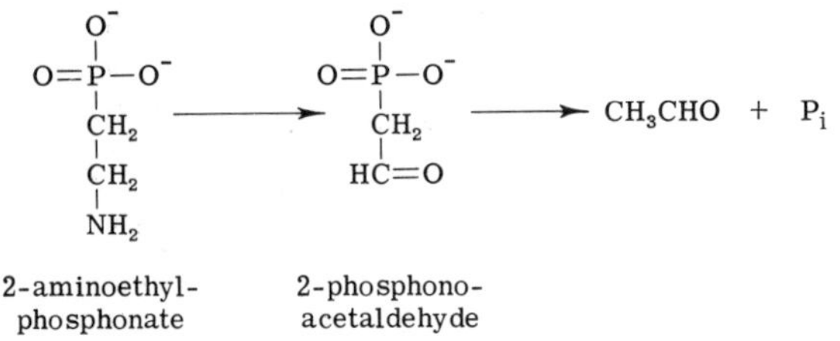
O⁻
O=P—O⁻
CH2
CH2
NH2
O⁻
O=P—O⁻
CH2
HC=O
CH3CHO + Pi
2-aminoethyl-phosphonate
2-phosphono-acetaldehyde

amino group is transferred to pyruvate with the help of pyridoxal phosphate and a transaminase. The second step requires an enzyme that cleaves the carbon–phosphorus bond of 2-phosphonoacetaldehyde. This reaction is carried out by an 2-phosphonoacetaldehyde phosphonohydrolase or phosphonatase (226).

References

1. Adams, E. (1959). Hydroxyproline metabolism. I. Conversion to α-ketoglutarate by extracts of *Pseudomonas*. *J. Biol. Chem.* **234,** 2073.
2. Adams, E., and Rose, G. (1967). α-Ketoglutaric semialdehyde dehydrogenase of *Pseudomonas*. *J. Biol. Chem.* **242,** 1802.
3. Aitken, D. M., and Brown, A. D. (1969). Citrate and glyoxylate cycle in the halophil, *Halobacterium salinarum*. *Biochim. Biophys. Acta* **177,** 351.
4. Alexander, M., and Wilson, P. W. (1956). Intracellular distribution of tricarboxylic acid cycle enzymes in *Azotobacter vinelandii*. *J. Bacteriol.* **71,** 252.
5. Allen, M. J. (1966). Symposium on bioelectrochemistry of microorganisms. II. Electrochemical aspects of metabolism. *Bacteriol. Rev.* **30,** 80.
6. Allen, R. H., and Jakoby, W. B. (1969). Tartaric acid metabolism. IX. Synthesis with tartrate epoxidase. *J. Biol. Chem.* **244,** 2078.
7. Amarasingham, C. R., and Davis, B. D. (1965). Regulation of malate and α-ketoglutarate dehydrogenases in *Escherichia coli*. *Fed. Proc., Fed. Amer. Soc. Exp. Biol.* **24,** 416.
8. Amarasingham, C. R., and Davis, B. D. (1965). Regulation of α-ketoglutarate dehydrogenase formation in *Escherichia coli*. *J. Biol. Chem.* **240,** 3664.
9. Ambler, R. P. (1963). The purification and amino acid composition of *Pseudomonas* cytochrome c-551. *Biol. J.* **89,** 341.
10. Ambler, R. P. (1963). The amino acid sequence of *Pseudomonas* cytochrome c-551. *Biochem. J.* **89,** 349.
11. Anthony, C., and Zatman, L. J. (1964). Microbial oxidation of methanol. The methanol-oxidizing enzyme of *Pseudomonas* M 27. *Biochem. J.* **92,** 614.
12. Anthony. C., and Zatman, L. J. (1965). Microbial oxidation of methanol. The alcohol dehydrogenase of *Pseudomonas* sp. M 27. *Biochem. J.* **96,** 808.
13. Anthony, C., and Zatman, L. J. (1967). The microbial oxidation of methanol. Purification and properties of the alcohol dehydrogenase of *Pseudomonas* sp. M 27. *Biochem. J.* **104,** 953.
14. Anthony, C., and Zatman, L. J. (1967). The microbial oxidation of methanol. The prosthetic group of the alcohol dehydrogenase of *Pseudomonas* sp. M 27: A new oxidoreductase prosthetic group. *Biochem. J.* **104,** 960.

15. Asai, T. (1968). "Acetic Acid Bacteria. Classification and Biochemical Activities." Univ. of Tokyo Press, Tokyo.
16. Asano, A., and Brodie, A. F. (1963). Oxidative phosphorylation in fractionated bacterial systems. XI. Separation of soluble factors necessary for oxidative phosphorylation. *Biochem. Biophys. Res. Commun.* **13,** 416.
17. Asano, A., and Brodie, A. F. (1963). Oxidative phosphorylation in fractionated bacterial systems. XII. The properties of malate-vitamin K reductase. *Biochem. Biophys. Res. Commun.* **13,** 423.
18. Asano, A., and Brodie, A. F. (1964). Oxidative phosphorylation in fractionated bacterial systems. XIV. Respiratory chain of *Mycobacterium phlei. J. Biol. Chem.* **239,** 4280.
19. Asano, A., and Brodie, A. F. (1965). The properties of the nonphosphorylative electron transport bypass enzymes of *Mycobacterium phlei. Biochem. Biophys. Res. Commun.* **19,** 131.
20. Ashton, D. H., and Blankenship, L. C. (1969). Soluble reduced nicotinamide adenine dinucleotide oxidase from *Bacillus cereus* T spores and vegetative cells. I. Purification. *Can. J. Microbiol.* **15,** 1309.
21. Ashton, D. H., and Blankenship, L. C. (1969). Soluble reduced nicotinamide adenine dinucleotide oxidase from *Bacillus cereus* T spores and vegetative cells. II. Properties. *Can. J. Microbiol.* **15,** 1313.
22. Ashworth, J. M., and Kornberg, H. L. (1964). The role of isocitrate lyase in *Escherichia coli. Biochim. Biophys. Acta* **89,** 383.
23. Atkinson, D. E. (1956). The oxidation of ethanol and tricarboxylic acid cycle intermediates by *Acetobacter peroxydans. J. Bacteriol.* **72,** 195.
24. Atkinson, D. E., and Serat, W. F. (1960). The cofactor specificity of ethanol dehydrogenase from *Acetobacter peroxydans. Biochim. Biophys. Acta* **39,** 154.
25. Baddiley, J. (1955). The structure of coenzyme A. *Advan. Enzymol.* **16,** 1.
26. Bailey, E., and Hullin, R. P. (1966). The metabolism of glyoxylate by cell-free extracts of *Pseudomonas* sp. *Biochem. J.* **101,** 755.
27. Bannerjee, D., Sanders, L. E., and Sokatch, J. R. (1970). Properties of purified methylmalonate semialdehyde dehydrogenase of *Pseudomonas aeruginosa. J. Biol. Chem.* **245,** 1828.
28. Baseman, J. B., and Cox, C. D. (1969). Intermediate energy metabolism of *Leptospira. J. Bacteriol.* **97,** 992.
29. Behrman, E. J. (1962). Tryptophan metabolism in *Pseudomonas. Nature (London)* **196,** 150.
30. Behrman, E. J., and Stella, E. J. (1963). Enrichment procedures for the isolation of tryptophan-oxidizing organisms. *J. Bacteriol.* **85,** 946.
31. Benson, R. W., and Boyer, P. D. (1969). The participation of an enzyme-bound group in a coenzyme A transferase reaction. *J. Biol. Chem.* **244,** 2366.
32. Benziman, M., and Abeliovitz, A. (1964). Metabolism of dicarboxylic acids in *Acetobacter xylinum. J. Bacteriol.* **87,** 270.
33. Benziman, M., and Heller, N. (1964). Oxalacetate decarboxylation and oxalacetate–CO_2 exchange in *Acetobacter xylinum. J. Bacteriol.* **88,** 1678.
34. Benziman, M., and Levy, L. (1966). Phosphorylation coupled to malate oxidation in *Acetobacter xylinum. Biochem. Biophys. Res. Commun.* **24,** 214.
35. Benziman, M. (1969). Role of phosphoenolpyruvate carboxylation in *Acetobacter xylinum. J. Bacteriol.* **98,** 1005.
36. Benziman, M., and Palgi, A. (1970). Characterization and properties of the pyruvate phosphorylation system of *Acetobacter xylinum. Biochem. J.* **112,** 631.

37. Benziman, M., and Eizen, N. (1971). Pyruvate-phosphate dikinase and the control of gluconeogenesis in *Acetobacter xylinum*. *J. Biol. Chem.* **246,** 57.
38. Blackmore, M. A., and Quayle, J. R. (1970). Microbial growth on oxalate by a route not involving glyoxylate carboligase. *Biochem. J.* **118,** 53.
39. Blackmore, M. A., and Turner, J. M. (1971). Threonine metabolism via two-carbon compounds by *Pseudomonas oxalaticus*. *J. Gen. Microbiol.* **67,** 243.
40. Blakley, R. L. (1955). The interconversion of serine and glycine: Participation of pyridoxal phosphate. *Biochem. J.* **61,** 315.
41. Boulton, C. A., Crabbe, M. J. C., and Large, P. J. (1974). Microbial oxidation of amines. Partial purification of a trimethylamine mono-oxygenase from *Pseudomonas aminovorans* and its role in growth on trimethylamine. *Biochem. J.* **140,** 253.
42. Brodie, A. F., and Adelson, J. (1965). Respiratory chains and sites of coupled phosphorylation. *Science* **149,** 265.
43. Brodie, A. F. (1969). Biological function of terpenoid quinones. *Biochem. J.* **113,** 25P.
44. Brohn, F., and Tchen, T. T. (1971). A single transaminase for 1,4-diaminobutanes and 4-aminobutyrate in a *Pseudomonas* species. *Biochem. Biophys. Res. Commun.* **45,** 578.
45. Brown, L. R., Strawinski, R. J., and McCleskey, C. S. (1964). The isolation and characterization of *Methanomonas methanooxidans* Brown and Strawinski. *Can. J. Microbiol.* **10,** 791.
46. Brunori, M., Rotilio, G. C., Antonini, E., Curti, B., Branzoli, U., and Massey, V. (1971). Oxidation–reduction potentials of D-amino acid oxidase. *J. Biol. Chem.* **246,** 3140.
47. Bryant, M. P., McBride, B. C., and Wolfe, R. S. (1968). Hydrogen-oxidizing bacteria. I. Cultivation and methanogenesis. *J. Bacteriol.* **95,** 1118.
48. Campbell, J. J., Smith, R. A., and Eagle, B. A. (1953). A deviation from the conventional tricarboxylic acid cycle in *Pseudomonas aeruginosa*. *Biochim. Biophys. Acta* **11,** 594.
49. Canovas, J. L., and Kornberg, H. L. (1965). Fine control of phosphopyruvate carboxylase activity in *Escherichia coli*. *Biochim. Biophys. Acta* **96,** 169.
50. Canovas, J. L., and Kornberg, H. L. (1966). Properties and regulation of PEP carboxylase activity in *E. coli*. *Proc. Roy. Soc., Ser. B* **165,** 189.
51. Cartwright, L. N., and Hullin, R. P. (1966). Purification and properties of two glyoxylate reductases from a species of *Pseudomonas*. *Biochem. J.* **101,** 781.
52. Castor, L. N., and Chance, B. (1955). Photochemical action spectra of carbon monoxide-inhibited respiration. *J. Biol. Chem.* **217,** 453.
53. Castor, L. N., and Chance, B. (1959). Photochemical determination of the oxidases of bacteria. *J. Biol. Chem.* **234,** 158.
54. Cazzulo, J. J., Sundaram, T. K., and Kornberg, H. L. (1969). Regulation of pyruvate carboxylase formation from the apoenzyme and biotin in a thermophilic bacillus. *Nature (London)* **223,** 1137.
55. Cazzulo, J. J., Sundaram, T. K., Dilks, S. N., and Kornberg, H. L. (1971). Synthesis of pyruvate carboxylase from its apoenzyme and (+)-biotin in *Bacillus stearothermophilus*. Purification and properties of the apoenzyme and the holoenzyme synthetase. *Biochem. J.* **122,** 653.
56. Cazzulo, J. J. (1973). On the regulatory properties of a halophilic citrate synthase. *FEBS Lett.* **30,** 339.

57. Cennamo, C., Razzoli, L., and Ferrari, F. (1970). Kinetic behavior of NAD^+-specific isocitrate dehydrogenase from baker's yeast and inhibition by AMP. *Ital. J. Biochem.* **19,** 100.
58. Chang, Y. F., and Feingold, D. S. (1970). D-Glucaric acid and galactaric acid catabolism by *Agrobacterium tumefaciens. J. Bacteriol.* **102,** 85.
59. Chang, Y. F., and Adams, E. (1971). Induction of separate catabolic pathways for L- and D-lysine in *Pseudomonas putida. Biochem. Biophys. Res. Commun.* **45,** 570.
59a. Chang, Y. F., and Adams, E. (1974). D-Lysine catabolic pathway in *Pseudomonas putida:* Interrelations with L-lysine catabolism. *J. Bacteriol.* **117,** 753.
60. Chasin, L. A., and Magasanik, B. (1968). Induction and repression of the histidine-degrading enzymes of *Bacillus subtilis. J. Biol. Chem.* **243,** 5165.
61. Cheah, K. S. (1969). Properties of electron transport particles from *Halobacterium cutirubrum.* The respiratory chain system. *Biochim. Biophys. Acta* **180,** 320.
62. Cheah, K. S. (1970). The membrane-bound carbon monoxide-reactive hemoproteins in the extreme halophiles. *Biochim. Biophys. Acta* **197,** 84.
63. Cheah, K. S. (1970). The membrane-bound ascorbate oxidase system of *Halobacterium halobium. Biochim. Biophys. Acta* **205,** 148.
64. Cheah, K. S. (1970). Properties of the membrane-bound respiratory chain system of *Halobacterium salinarum. Biochim. Biophys. Acta* **216,** 43.
65. Chow, C. T., and Vennesland, B. (1958). The nonenzymatic decarboxylation of diketosuccinate and oxaloglycolate (dihydroxyfumarate). *J, Biol. Chem.* **233,** 997.
65a. Chulavatnatol, M., and Atkinson, D. E. (1973). Kinetic competition *in vitro* between phosphoenolpyruvate synthetase and the pyruvate dehydrogenase complex from *Escherichia coli. J. Biol. Chem.* **248,** 2716.
66. Chung, A. E. (1970). Pyridine nucleotide transhydrogenase from *Azotobacter vinelandii. J. Bacteriol.* **102,** 438.
67. Clagett, C. O., Tolbert, N. E., and Burris, R. H. (1949). *J. Biol. Chem.* **178,** 977 [as cited by Cartwright and Hullin (51)].
68. Cohn, D. V. (1958). The enzymatic formation of oxalacetic acid by nonpyridine nucleotide malic dehydrogenase of *Micrococcus lysodeikticus. J. Biol. Chem.* **233,** 299.
68a. Conrad, R. S., Massey, L. K., and Sokatch, J. R. (1974). D- and L-Isoleucine metabolism and regulation of their pathways in *Pseudomonas putida. J. Bacteriol.* **118,** 103.
69. Cooper, R. A., and Kornberg, H. L. (1964). The utilization of itaconate by *Pseudomonas* sp. *Biochem. J.* **91,** 82.
70. Cooper, R. A., Itiaba, K., and Kornberg, H. L. (1965). The utilization of aconate and itaconate by *Micrococcus* sp. *Biochem. J.* **94,** 25.
71. Cooper, R. A., and Kornberg, H. L. (1967). The direct synthesis of phosphoenol pyruvate by *Escherichia coli. Proc. Roy. Soc., Ser. B* **168,** 263.
72. Cox, G. B., Newton, N. A., Gibson, F., Snoswell, A. M., and Hamilton, J. A. (1970). The function of ubiquinone in *Escherichia coli. Biochem. J.* **117,** 551.
73. Dagley, S., and Trudgill, P. W. (1963). The metabolism of tartaric acid by a *Pseudomonas.* A new pathway. *Biochem. J.* **89,** 22.
73a. Danson, M. J., and Weitzman, P. D. J. (1973). Functional groups in the activity and regulation of *Escherichia coli* citrate synthase. *Biochem. J.* **135,** 513.
74. Davies, S. L., and Whittenbury, R. (1970). Fine structure of methane- and other hydrocarbon-utilizing bacteria. *J. Gen. Microbiol.* **61,** 227.

75. Davis, B. D. (1961). The teleonomic significance of biosynthetic control mechanisms. *Cold Spring Harbor Symp. Quant. Biol.* **26,** 1.
76. Deeb, S. S., and Hager, L. P. (1964). Crystalline cytochrome b_1 from *Escherichia coli. J. Biol. Chem.* **239,** 1024.
77. Deibel, R. H. (1964). Utilization of arginine as an energy source for the growth of *Streptococcus faecalis. J. Bacteriol.* **87,** 988.
78. DeLey, J., and Schell, J. (1959). Studies on the metabolism of *Acetobacter peroxidans.* II. The enzymic mechanism of lactate metabolism. *Biochim. Biophys. Acta.* **35,** 154.
79. DeLey, J., and Vervloet, V. (1961). Studies on the metabolism of *Acetobacter peroxydans.* III. Some properties and localization of peroxidases. *Biochim. Biophys. Acta* **50,** 1.
80. DeLey, J., and Schell, J. (1962). Lactate and pyruvate catabolism in acetic acid bacteria. *J. Gen. Microbiol.* **29,** 589.
81. DeLey, J. (1964). *Pseudomonas* and related genera. *Annu. Rev. Microbiol.* **18,** 17.
82. Dietrich, J., and Henning, U. (1970). Regulation of pyruvate dehydrogenase complex synthesis in *Escherichia coli* K-12. *Eur. J. Biochem.* **14,** 258.
83. Doudoroff, M., and Stanier, R. Y. (1959). Role of poly-β-hydroxybutyric acid in assimilation of organic carbon by bacteria. *Nature (London)* **183,** 1440.
84. Drabikowska, A. K. (1969). Subcellular distribution and function of ubiquinone in *Salmonella typhimurium. Acta Biochem. Pol.* **16,** 135.
85. Drabikowska, A. K. (1970). Electron transport system of *Salmonella typhimurium* cells. *Acta Biochim. Pol.* **17,** 89.
85a. Dunstan, P. M., Anthony, C., and Drabble, W. T. (1972). Microbial metabolism of C_1 and C_2 compounds. The role of glyoxylate, glycollate and acetate in the growth of *Pseudomonas* AM 1 on ethanol and on C_1 compounds. *Biochem. J.* **128,** 107.
85b. Dunstan, P. M., Anthony, C., and Drabble, W. T. (1972). Microbial metabolism of C_1 and C_2 compounds. The involvement of glyoxylate in the metabolism of ethanol and acetate by *Pseudomonas* AM 1. *Biochem. J.* **128,** 99.
85c. Dunstan, P. M. and Anthony, C. (1973). Microbial metabolism of C_1 and C_2 compounds. The role of acetate during growth of *Pseudomonas* AM 1 on C_1 compounds, ethanol and β-hydroxybutyrate. *Biochem. J.* **132,** 797.
86. Dworkin, M., and Foster, J. W. (1956). Studies on *Pseudomonas methanica* (Sohngen) nov. comb. *J. Bacteriol.* **72,** 646.
87. Eady, R. R., and Large, P. J. (1968). Purification and properties of an amine dehydrogenase from *Pseudomonas* AM 1 and its role in growth on methylamine. *Biochem. J.* **106,** 245.
88. Eady, R. R., and Large, P. J. (1969). Bacterial oxidation of dimethylamine, a new monooxygenase reaction. *Biochem. J.* **111,** 37P.
89. Eady, R. R., and Large, P. L. (1971). Microbial oxidation of amines. Spectral and kinetic properties of the primary amine dehydrogenase of *Pseudomonas* AM 1. *Biochem. J.* **123,** 757.
90. Eady, R. R., Jarman, T. R., and Large, P. J. (1971). Microbial oxidation of amines. Partial purification of a mixed-function secondary amine oxidase system from *Pseudomonas aminovarans* that contains an enzymatically active cytochrome P-420-type hemoprotein. *Biochem. J.* **125,** 449.
91. Eggerer, H., Buckel, W., Lenz, H., Wunderwald, P., Cornforth, H. W., Dominger, C., Mallaby, R., Redmond, J. W., and Gottschalk, G. (1970). Stereochemistry of enzymic citrate synthesis and cleavage. *Nature (London)* **226,** 517.

92. Eilermann, L. J. M., Pandit-Hovenkamp, H. G., and Kolk, A. J. H. (1970). Oxidative phosphorylation in *Azotobacter vinelandii.* Phosphorylation sites and respiratory control. *Biochim. Biophys. Acta* **197,** 25.

93. Englard, S., and Colowick, S. P. (1957). On the mechanism of the aconitase and isocitrate dehydrogenase reaction. *J. Biol. Chem.* **226,** 1047.

94. Englesberg, E., and Levy, J. B. (1955). Induced synthesis of tricarboxylic acid cycle enzymes as correlated with the oxidation of acetate and glucose by *Pasteurella pestis, J. Bacteriol.* **69,** 418.

95. Erfle, J. D., and Sauer, F. (1969). The inhibitory effects of acyl-coenzyme A esters on the pyruvate and α-oxoglutarate dehydrogenase complex. *Biochim. Biophys. Acta* **178,** 441.

96. Erickson, S. K., and Parker, G. L. (1969). The electron transport system of *Micrococcus lutea* (*Sarcina lutea*). *Biochim. Biophys. Acta* **180,** 56.

97. Feigelson, P., Maeno, H., Poillon, W. N., and Rosenfeld, H. (1968). Regulation of tryptophan metabolism in Pseudomonas by enzyme and permease induction and the allosteric control of tryptophan oxygenase. *Med. J. Osaka Univ.* **19,** 35.

98. Flechtner, V. R., and Hanson, R. S. (1969). Coarse and fine control of citrate synthase from *Bacillus subtilis. Biochim. Biophys. Acta.* **184,** 252.

99. Flechtner, V. R., and Hanson, R. S. (1970). Regulation of the tricarboxylic acid cycle in bacteria. A comparison of citrate synthase from different bacteria. *Biochim. Biophys. Acta* **222,** 253.

100. Foster, J. W., and Davis, R. H. (1966). A methane-dependent coccus, with notes on classification and nomenclature of obligate, methane-utilizing bacteria. *J. Bacteriol.* **91,** 1924.

101. Francis, M. J. O., Hughes, D. E., and Phizackerley, P. J. R. (1963). The oxidation of L-malate by *Pseudomonas* sp. *Biochem. J.* **89,** 430.

102. Frankfater, A., and Fridovich, I. (1970). The purification and properties of oxidized derivatives of L-histidine ammonia-lyase. *Biochim. Biophys. Acta* **206,** 457.

103. Franks, N. E. (1971). Catabolism of L-arginine by entrapped cells of *Streptococcus faecalis* ATCC 8043. *Biochim. Biophys. Acta* **252,** 246.

104. Fuller, R. C., and Kornberg, H. L. (1961). A possible route for malate oxidation by *Chromatium. Biochem. J.* **79,** 8P.

105. Furmanski, P., Wegener, W. S., Reeves, H. C., and Ajl, S. J. (1967). Function of the glyoxylate-condensing enzymes. I. Growth of *Escherichia coli* on *n*-valeric acid. *J. Bacteriol.* **94,** 1075.

106. George, D. J., and Phillips, A. T. (1970). Identification of α-ketobutyrate as the prosthetic group of urocanase from *Pseudomonas putida. J. Biol. Chem.* **245,** 528.

107. Gholson, R. K., Nishizuka, Y., Ichiyama, A., Kawai, H., Nakamura, S., and Hayaishi, O. (1962). New intermediates in the catabolism of tryptophan in mammalian liver. *J. Biol. Chem.* **237,** PC2043.

108. Gilvarg, C., and Davis, B. D. (1956). The role of the tricarboxylic acid cycle in acetate oxidation in *Escherichia coli. J. Biol. Chem.* **222,** 307.

109. Ginsburg, A., and Stadtman, E. R. (1970). Multienzyme systems. *Annu. Rev. Biochem.* **39,** 429.

110. Glattli, H., and Ettlinger, L. (1970). Substrate–energy relationships in *Acetomonas oxydans. Arch. Mikrobiol.* **74,** 273.

111. 's-Gravenmade, E. J., and Vogels, G. D. (1969). Purification of allantoicase from *Pseudomonas aeruginosa. Antonie van Leeuwenhoek; J. Microbiol. Serol.* **35,** 463.

112. 's-Gravenmade, E. J., Vogels, G. D., and van der Drift, C. (1971). Hydrolysis,

racemization, and absolute configuration of ureidoglycolate, a substrate of allantoicase. *Biochim. Biophys. Acta* **198,** 569.

113. Gray, C. T., Wimpenny, J. W. T., and Mossman, M. R. (1966). Regulation of metabolism in facultative bacteria. II. Effects of aerobiosis, anaerobiosis, and nutrition on the formation of Kreb's cycle enzymes in *Escherichia coli. Biochim. Biophys. Acta* **117,** 33.
114. Gray, W. R., and Hartley, B. S. (1963). The structure of a chymotryptic peptide from *Pseudomonas* cytochrome c-551. *Biochem. J.* **89,** 379.
115. Green, D. E. (1962). Power from the mitochondria. *Discovery* **23,** 32.
116. Greenfield, S., and Claus, G. W. (1969). Isocitrate dehydrogenase and glutamate synthesis in *Acetobacter suboxydans. J. Bacteriol.* **100,** 1264.
117. Grinnell, F., and Nishimura, J. S. (1970). The inactivation and dissociation of *Escherichia coli* succinyl-CoA synthetase by sulfhydryl reagents. *Biochim. Biophys. Acta* **212,** 150.
118. Gryder, R. M., and Adams, E. (1969). Inducible degradation of hydroxyproline in *Pseudomonas putida:* Pathway regulation and hydroxyproline uptake. *J. Bacteriol.* **97,** 292.
119. Guilbault, G. G. (1966). Symposium on bioelectrochemistry of microorganisms. III. Electrochemical analysis of enzymatic reactions. *Bacteriol. Rev.* **30,** 94.
120. Hager, L. P., and Kornberg, H. L. (1961). On the mechanisms of 2-oxoglutarate oxidation in *Escherichia coli. Biochem. J.* **78,** 194.
120a. Hagihira, H., Hayashi, H., Ichihara, A., and Suda, M. (1960). Metabolism of L-lysine by bacterial enzymes. II. L-Lysine oxidase. *J. Biochem.* **48,** 267.
121. Hampton, M. L., and Hanson, R. S. (1969). Regulation of isocitrate dehydrogenase from *Thiobacillus thiooxidans* and *Pseudomonas fluorescens. Biochem. Biophys. Res. Commun.* **36,** 296.
122. Hansen, R. G., and Henning, U. (1966). Regulation of pyruvate dehydrogenase activity in *Escherichia coli* K-12. *Biochim. Biophys. Acta* **122,** 355.
123. Harder, W., and Quayle, J. R. (1971). The biosynthesis of serine and glycine in *Pseudomonas* AM 1 with special reference to growth on carbon sources other than C_1 compounds. *Biochem. J.* **121,** 753.
124. Harder, W., and Quayle, J. R. (1971). Aspects of glycine and serine biosynthesis during growth of *Pseudomonas* AM 1 on C_1 compounds. *Biochem. J.* **121,** 763.
124a. Harder, W. (1973). Microbial metabolism of organic C_1 and C_2 compounds. *Antonie van Leeuwenhoek J. Microbiol. Serol.* **39,** 650.
125. Harrington, A. A., and Kallio, R. E. (1960). Oxidation of methanol and formaldehyde by *Pseudomonas methanica. Can. J. Microbiol.* **6,** 1.
126. Hassall, H., Jeffcoat, R., and Dagley, S. (1969). The physical properties and mechanism of action of deoxyoxoglucarate dehydratase from *Pseudomonas acidovorans. Biochem. J.* **114,** 78P.
127. Hatefi, Y., Osborne, M. J., Kay, L. D., and Huennekens, F. M. (1957). Hydroxymethyl tetrahydrofolic dehydrogenase. *J. Biol. Chem.* **227,** 637.
128. Hathaway, J. A., and Atkinson, D. E. (1965). Kinetics of regulatory enzymes: Effect of adenosine triphosphate on yeast citrate synthase. *Biochem. Biophys. Res. Commun.* **20,** 661.
129. Hayaishi, M., Hayashi, M., and Unemoto, T. (1966). The presence of D-malate dehydrogenase (D-malate:NAD oxidoreductase) in *Serratia marcescens. Biochim. Biophys. Acta* **124,** 374.
130. Hayaishi, O., and Sutton, W. B. (1957). Enzymatic oxygen fixation into acetate concomitant with the enzymatic decarboxylation of L-lactate. *J. Amer. Chem. Soc.* **79,** 4809 [as cited by Sullivan (364)].

131. Hayaishi, O. (1966). Crystalline oxygenases of pseudomonads. *Bacteriol. Rev.* **30,** 720.

132. Heptinstall, J., and Quayle, J. R. (1970). Pathways leading to and from serine during growth of *Pseudomonas* AM 1 on C_1 compounds or succinate. *Biochem. J.* **117,** 563.

133. Herman, N. J., and Bell, E. J. (1970). Metabolic control in *Acinetobacter* sp. I. Effect of C_4 versus C_2 and C_3 substrates on isocitrate lyase synthesis. *Can. J. Microbiol.* **16,** 769.

134. Higashi, T., Bogin, E., and Brodie, A. F. (1969). Separation of a factor indispensable for coupled phosphorylation from the particulate fraction of *Mycobacterium phlei. J. Biol. Chem.* **244,** 500.

135. Higgins, I. J., Turner, J. M., and Willets, A. J. (1967). Enzyme mechanisms of aminoacetone metabolism by microorganisms. *Nature* (*London*) **215,** 887.

136. Higgins, I. J., and Turner, J. M. (1969). Enzymes of methylglyoxal metabolism in a pseudomonad which rapidly metabolizes aminoacetone. *Biochim. Biophys. Acta* **184,** 467.

137. Higgins, I. J., and Quayle, J. R. (1970). Oxygenation of methane by methane-grown *Pseudomonas methanica* and *Methanomonas methanooxidans. Biochem. J.* **118,** 201.

138. Hildebrand, J. G., and Spector, L. B. (1969). Succinylphosphate and the succinyl-coenzyme A synthetase reaction. *J. Biol. Chem.* **244,** 2606.

139. Himes, R. H., and Wilder, T. (1965). Formyltetrahydrofolate synthetase: Mechanism of cation activation. *Biochim. Biophys. Acta* **99,** 464.

140. Hirsch, C. A., Rasminsky, M., Davis, B. D., and Lin, E. C. C. (1973). A fumarate reductase in *Escherichia coli* distinct from succinate dehydrogenase. *J. Biol. Chem.* **238,** 3770.

141. Hobbs, D. C., and Koeppe, R. R. (1958). The metabolism of glutaric acid-3-^{14}C by the intact rat. *J. Biol. Chem.* **230,** 655.

142. Hollinshead, J. A. (1965). Microbial growth on C_1 compounds: The role of folate in the metabolism of *Pseudomonas* AM 1. *Biochem. J.* **96,** 49P

143. Hollinshead, J. A. (1966). Microbial growth of C_1 compounds: The role of folate in the metabolism of *Pseudomonas* AM 1. *Biochem. J.* **99,** 389.

144. Holms, W. H., and Bennett, P. M. (1971). Regulation of isocitrate dehydrogenase activity in *Escherichia coli* on adaptation to acetate. *J. Gen. Microbiol.* **65,** 57.

145. Hommes, F. A., and Stekhoven, F. (1964). Aperiodic changes of reduced NAD during anaerobic glycolysis in brewer's yeast. *Biochem. Biophys. Acta* **86,** 427.

146. Hopner, T., and Trautwein, A. (1971). *Pseudomonas oxalaticus:* Requirement of a cosubstrate for growth on formate. *Arch. Mikrobiol.* **77,** 26.

147. Hopper, D. J., Chapman, P. J., and Dagely, S. (1970). Metabolism of L-malate and D-malate by a species of *Pseudomonas. J. Bacteriol.* **104,** 1197.

148. Howard, R. L., and Becker, R. R. (1970). Isolation and some properties of the triphosphopyridine nucleotide isocitrate dehydrogenase from *Bacillus stearothermophilus. J. Biol. Chem.* **245,** 3186.

149. Hsie, A. W., and Rickenberg, H. V. (1966). A mutant of *Escherichia coli* deficient in phosphoenolpyruvate carboxykinase activity. *Biochem. Biophys. Res. Commun.* **25,** 676.

150. Hubbard, J. S., and Miller, A. B. (1969). Purification and reversible inactivation of the isocitrate dehydrogenase from an obligate halophile. *J. Bacteriol.* **99,** 161.

151. Huennekens, F. M., Osborne, M. J., and Whiteley, H. R. (1958). Folic acid enzymes. Metabolic reactions involving "active formate" and "active formaldehyde" are surveyed. *Science* **128,** 120.

152. Hug, D. H., Roth, D., and Hunter, J. (1968). Regulation of histidine catabolism by succinate in *Pseudomonas putida*. *J. Bacteriol.* **96,** 396.

153. Hug, D. H., and Hunter, J. K. (1970). Photoactivation of urocanase in *Pseudomonas putida*. Possible role in photoregulation of histidine metabolism. *J. Bacteriol.* **102,** 874.

154. Hutton, W. E., and ZoBell, C. E. (1949). The occurrence and characteristic of methane-oxidizing bacteria in marine sediments. *J. Bacteriol.* **58,** 463.

155. Hutton, W. E., and ZoBell, C. E. (1963). Production of nitrite from ammonia by methane-oxidizing bacteria. *J. Bacteriol.* **65,** 217.

155a. Ichihara, A., Furiya, S., and Suda, M. (1960). Metabolism of L-lysine by bacterial enzymes. III. Lysine racemase. *J. Biochem.* **48,** 277.

155b. Ichihara, A., Ichihara, E. A., and Suda, M. (1960). Metabolism of L-lysine by bacterial enzymes. IV. Aminovaleric acid-glutamic acid transaminase. *J. Biol. Chem.* **48,** 412.

155c. Ichihara, A., and Ichihara, E. A. (1961). Metabolism of L-lysine by bacterial enzymes. V. Glutamic semialdehyde dehydrogenase. *J. Biochem.* **49,** 154.

156. International Uninion of Biochemistry Commission. (1972). "Enzyme Nomenclature: Recommendations of the International Union of Biochemistry on the Nomenclature and Classification of Enzymes, Together with their Units and the Symbols of Enzyme Kinetics." Elsevier, Amsterdam.

157. Ishimura, Y., Nozaki, M., and Hayaishi, O. (1970). The oxygenated form of L-tryptophen-2,3-dioxygenase as reaction intermediate. *J. Biol. Chem.* **245,** 3593.

158. Iwasaki, Y. (1960). Components of the electron-transferring system in *Acetobacter suboxydans* and reconstruction of the lactate oxidation system. *J. Plant Cell Physiol.* **1,** 207.

159. Izui, K., Iwatani, A., Nishikido, T., Katsuki, H., and Tanaka, S. (1967). Regulation of phosphoenolpyruvate carboxylase activity in *Escherichia coli*. *Biochim. Biophys. Acta* **139,** 188.

160. Izui, K., Yoshinaga, T., Mosikawa, M., and Katsuki, H. (1970). Activation of phosphoenolpyruvate carboxylase of *Escherichia coli* by free fatty acids or their coenzyme A derivatives. *Biochem. Biophys. Res. Commun.* **40,** 949.

161. Jacobson, L. A., Bartholomans, R. C., and Gunsalus, I. C. (1966). Repression of malic enzyme by acetate in *Pseudomonas*. *Biochem. Biophys. Res. Commun.* **24,** 955.

162. Jangaard, N. O., Hathaway, J. A., and Atkinson, D. E. (1966). Effect of ATP on the kinetics of citrate synthase. *Fed. Proc., Fed. Amer. Soc. Exp. Biol.* **25,** 220.

163. Jansen, R. W., and Hayaishi, J. A. (1962). Glycolate metabolism in *Escherichia coli*. *J. Bacteriol.* **83,** 679.

164. Jeffcoat, R., Hassall, H., and Dagley, S. (1968). The metabolism of D-glucarate by a species of *Pseudomonas*. *Biochem. J.* **107,** 30P.

165. Jeffcoat, R., Hassall, H., and Dagley, S. (1969). The metabolism of D-glucarate by *Pseudomonas acidovorans*. *Biochem. J.* **115,** 969.

166. Jeffcoat, R., Hassall, H., and Dagley, S. (1969). Purification and properties of D-4-deoxy-5-oxoglucarate hydrolyase (decarboxylating). *Biochem. J.* **115,** 977.

166a. Jeffcoat, R. (1974). Studies on glucarate catabolism: The oxodeoxy-glucarate aldolase activity of glucarate hydro-lyase from *Pseudomonas acidovorans*. *Biochem. J.* **139,** 477.

167. Jensen, D. E., and Neidhardt, F. C. (1969). Effect of growth rate on histidine catabolism and histidase synthesis in *Aerobacter aerogenes*. *J. Bacteriol.* **98,** 131.

168. Johnson, J. L., and Temple, K. L. (1962). Some aspects of methane oxidation. *J. Bacteriol.* **84,** 456.

169. Johnson, P. A., and Quayle, J. R. (1964). Microbial growth on C_1 compounds. 6. Oxidation of methanol, formaldehyde, and formate by methanol-grown *Pseudomonas* AM 1. *Biochem. J.* **92,** 281.

170. Johnson, P. A., and Quayle, J. R. (1965). Microbial growth on C_1 compounds. Synthesis of cell constituents by methane and methanol-grown *Pseudomonas methanica. Biochem. J.* **95,** 859.

171. Jones, A., and Turner, J. M. (1971). Microbial metabolism of amino alcohols via aldehydes. *J. Gen. Microbiol.* **67,** 379.

172. Jones, C. W., and Redfearn, E. R. (1966). Electron transport in *Azotobacter vinelandii. Biochim. Biophys. Acta* **113,** 467.

173. Jones, C. W., Ackrell, B. A. C., and Erickson, S. K. (1971). Respiratory control in *Azotobacter vinelandii* membranes. *Biochim. Biophys. Acta* **245,** 54.

174. Jones, R. G. W., and Lascelles, J. (1967). The relationship of 4-hydroxybenzoic acid to lysine and methionine formation in *Escherichia coli. Biochem. J.* **103,** 709.

175. Jurtshuk, P., and Old, L. (1968). Cytochrome c oxidation by the electron transport fraction of *Azotobacter vinelandii. J. Bacteriol.* **95,** 1790.

176. Jurtshuk, P., and Harper, L. (1968). Oxidation of D(−)lactate by the electron transport fraction of *Azotobacter vinelandii. J. Bacteriol.* **96,** 678.

177. Jurtshuk, P., Bednarz, A. J., Zey, P. and Denton, C. H. (1969). L-Malate oxidation by the electron transport fraction of *Azotobacter vinelandii. J. Bacteriol.* **98,** 1120.

178. Kallio, R. E., and Harrington, A. A. (1960). Sudanophilic granules and lipid of *Pseudomonas methanica. J. Bacteriol.* **80,** 321.

179. Kalse, J. F., and Veeger, C. (1968). Relation between conformation and activities of lipoamide dehydrogenase. I. Relation between diaphorase and lipoamide dehydrogenase activities upon binding of FAD by the apoenzyme. *Biochim. Biophys. Acta* **159,** 244.

180. Kaminskas, E., Kimhi, Y., and Magasanik, B. (1970). Urocanase and *N*-formimino-L-glutamate formiminohydrolase of *Bacillus subtilis,* two enzymes of the histidine degradative pathway. *J. Biol. Chem.* **245,** 3536.

181. Karlson, P. (1965). "Introduction to Modern Biochemistry," 2nd ed. Academic Press, New York.

182. Katagiri, H., and Tochikura, T. (1960). Microbiological studies with *coli–aerogenes* bacteria. XIV. Competition between glyoxylic reductase and glutamic dehydrogenase. *Bull. Inst. Chem. Res., Kyoto Univ.* **38,** 379.

183. Katsuki, H., Takeo, K., Kameda, K., and Tanaka, S. (1967). Existence of two malic enzymes in *Escherichia coli. Biochem. Biophys. Res. Commun.* **27,** 331.

184. Keech, D. B., and Utter, M. F. (1963). Pyruvate carboxylase. II. Properties. *J. Biol. Chem.* **238,** 2609.

185. Kemp, M. B., and Quayle, J. R. (1966). Microbial growth on C_1 compounds. Incorporation of C-1 units into allulose phosphate by extracts of *Pseudomonas methanica. Biochem. J.* **99,** 41.

186. Kemp, M. B., and Quayle, J. R. (1967). Microbial growth on C_1 compounds. Uptake of [^{14}C]formaldehyde and [^{14}C]formate by methane-grown *Pseudomonas methanica* and determination of the hexose labeling pattern after brief incubation with [^{14}C]methanol. *Biochem. J.* **102,** 94.

186a. Kikuchi, G. (1973) The glycine cleavage system; composition, reaction mechanism and physiological significance. *Mol. Cell. Biochem.* **1,** 169.

187. King, T. E., and Cheldelin, V. H. (1954). Pyruvic carboxylase in *Acetobacter suboxydans. J. Biol. Chem.* **208,** 821.

188. King, T. E., and Cheldelin, V. H. (1954). Oxidations in *Acetobacter suboxydans*. *Biochim. Biophys. Acta* **14,** 108.
189. King, T. E., and Cheldelin, V. H. (1956). Oxidation of acetaldehyde by *Acetobacter suboxydans*. *J. Biol. Chem.* **220,** 177.
190. King, T. E., Kawasaki, E. H., and Cheldelin, V. H. (1956). Tricarboxylic acid cycle activity in *Acetobacter pasteurianum*. *J. Bacteriol.* **72,** 418.
191. Klungsoyr, L., King, T. E., and Cheldelin, V. H. (1957). Oxidative phosphorylation in *Acetobacter suboxydans*. *J. Biol. Chem.* **227,** 135.
192. Knook, D. L., and Planta, R. J. (1971). Function of ubiquinone in electron transport from reduced nicotinamide adenine dinucleotide to nitrate and oxygen in *Aerobacter aerogenes*. *J. Bacteriol.* **105,** 483.
193. Knowles, C. J., and Redfearn, E. R. (1966). Ubiquinone in the electron transport system of *Azotobacter vinelandii*. *Biochem. J.* **99,** 33P.
194. Kohn, L. D., and Jakoby, W. B. (1968). Tartaric acid metabolism. III. The formation of glyceric acid. *J. Biol. Chem.* **243,** 2465.
195. Kohn, L. D., and Jakoby, W. B. (1968). Tartaric acid metabolism. IV. Crystalline L-malic dehydrogenase from *Pseudomonas acidovorans*. *J. Biol. Chem.* **243,** 2472.
196. Kohn, L. D., Packman, P. M., Allen, R. H., and Jakoby, W. B. (1968). Tartaric acid metabolism. V. Crystalline tartrate dehydrogenase. *J. Biol. Chem.* **243,** 2479.
197. Kohn, L. D., and Jakoby, W. B. (1968). Tartaric acid metabolism. VI. Crystalline oxaloglycolate reductive decarboxylase. *J. Biol. Chem.* **243,** 2486.
198. Kohn, L. D., and Jakoby, W. B. (1968). Tartaric acid metabolism. VII. Crystalline hydroxypyruvate reductase (D-glycerate dehydrogenase). *J. Biol. Chem.* **243,** 2494.
199. Kohn, L. D. (1968). Tartaric acid metabolism. VIII. Crystalline tartronic semialdehyde reductase. *J. Biol. Chem.* **243,** 4426.
200. Koike, M., Reed, L. J., and Carroll, W. R. (1960). α-Keto acid dehydrogenation complexes. I. Purification and properties of pyruvate and α-ketoglutarate dehydrogenase complexes of *Escherichia coli*. *J. Biol. Chem.* **235,** 1924.
201. Koike, M., and Reed, L. J. (1960). α-Keto acid dehydrogenation complexes. II. The role of protein-bound lipoic acid and flavine adenine nucleotide. *J. Biol. Chem.* **235,** 1931.
202. Koike, M., Shah, P. C., and Reed, L. J. (1960). α-Keto acid dehydrogenation complexes. III. Purification and properties of dihydrolipoyl dehydrogenase of *Escherichia coli*. *J. Biol. Chem.* **235,** 1939.
203. Kornberg, H. L., and Krebs, H. A. (1957). Synthesis of cell constituents from C_2 units by a modified tricarboxylic acid cycle. *Nature* (*London*) **179,** 988.
204. Kornberg, H. L. (1958). The metabolism of C_2-compounds in microorganisms. 1. The incorporation of 2-^{14}C acetate by *Pseudomonas fluoresscens*, and by a Corynebacterium, grown on ammonium acetate. *Biochem. J.* **68,** 535.
205. Kornberg, H. L., and Madsen, N. B. (1958). The metabolism of C_2-compounds in microorganisms. 3. Synthesis of malate from acetate via the glyoxylate cycle. *Biochem. J.* **68,** 549.
206. Kornberg, H. L. (1959). Aspects of terminal respiration in microorganisms. *Ann. Rev. Microbiol.* **13,** 49.
207. Kornberg, H. L., and Sadler, J. R. (1960). Microbial oxidation of glycollate via a dicarboxylic acid cycle. *Nature* (*London*) **185,** 153.
208. Kornberg, H. L., Collins, J. F., and Bigley, D. (1960). The influence of growth substrate on metabolic pathways in *Micrococcus denitrificans*. *Biochim. Biophys.* Acta **39,** 9.

209. Kornberg, H. L. (1961). Selective utilization of metabolic routes by *Escherichia coli*. *Cold Spring Harbor Symp. Quant. Biol.* **26,** 257.
210. Kornberg, H. L., and Elsden, S. R. (1961). The metabolism of 2-carbon compounds by microorganisms. *Advan. Enzymol.* **23,** 401.
211. Kornberg, H. L., and Gotto, A. M. (1961). The metabolism of C_2 compounds in microorganisms. 6. Synthesis of cell constituents from glycolate by *Pseudomonas* sp. *Biochem. J.* **78,** 69.
212. Kornberg, H. L., and Phizackerley, P. J. R. (1961). Malate oxidation by *Pseudomonas* spp. *Biochem. J.* **79,** 10P.
213. Kornberg, H. L., and Sadler, J. R. (1961). The metabolism of C_2 compounds in microorganisms. 8. A dicarboxylic acid cycle as a route for the oxidation of glycolate by *Escherichia coli. Biochem. J.* **81,** 503.
214. Kornberg, H. L. (1965). The coordination of metabolic routes. *Symp. Soc. Gen. Microbiol.* **15,** 8.
215. Kornberg, H. L. (1966). The role and control of the glyoxylate cycle in *Escherichia coli. Biochem. J.* **99,** 1.
216. Kornberg, H. L. (1967). Anaplerotic sequences and their role in metabolism. *Essays Biochem.* **2,** 1.
217. Kornberg, H. L. (1970). The role and maintenance of the tricarboxylic acid cycle in *Escherichia coli. Biochem. Soc. Symp.* **30,** 155.
218. Korotiaev, A. L. (1966). Effect of chloramphenicol on α-ketoglutarate dehydrogenase activity of *Escherichia coli* and *Proteus vulgaris. Mikrobiologiya* **35,** 13.
219. Krakow, G. and Barkulis, S. (1956). Conversion of glyoxylate to hydroxypyruvate by extracts of *Escherichia coli. Biochim. Biophys. Acta* **21,** 593.
220. Krebs, H. A. (1970). Rate control of the tricarboxylic acid cycle. *Advan. Enzyme Regul.* **8,** 335.
221. Kroger, A., and Dadak, V. (1969). On the role of quinones in bacterial electron transport. The respiratory system of *Bacillus megaterium. Eur. J. Biochem.* **11,** 328.
222. Kung, H. F., and Wagner, C. (1970). Oxidation of C_1 compounds by *Pseudomonas* sp. MS. *Biochem. J.* **116,** 357.
223. Kurup, C. K. R., and Brodie, A. F. (1967) Oxidative phosphorylation in fractionated bacterial systems. XXIX. The involvement of nonheme iron in the respiratory pathways of *Mycobacterium phlei. J. Biol. Chem.* **242,** 5830.
223a. Kurz, W. G. W., and LaRue, T. A. G. (1973). Metabolism of glycolic acid by *Azotobacter chroococcum* PRL H 62. *Can. J. Microbiol.* **19,** 321.
224. Lamanna, C., and Mallette, M. F. (1965). "Basic Bacteriology," 3rd ed., p. 842. Williams & Wilkins, Baltimore Maryland.
225. LaNauze, J. M., and Rosenberg, H. (1968). The identification of 2-phosphonoacetaldehyde as an intermediate in the degradation of 2-aminoethyl phosphonate by *Bacillus cereus. Biochim. Biophys. Acta* **165,** 438.
226. LaNauze, J. M., Rosenberg, H., and Shaw, D. C. (1970). The enzymic cleavage of the carbon–phosphorus bond: Purification and properties of phosphonatase. *Biochim. Biophys. Acta* **212,** 332.
227. Lanyi, J. K. (1969). Studies on the electron transport chain of extremely halophilic bacteria. II. Salt dependence of reduced diphosphopyridine nucleotide oxidase. *J. Biol. Chem.* **244,** 2864.
228. Lanyi, J. K. (1969). Studies of the electron transport chain of extremely halophilic bacteria. III. Mechanisms of the effect of salt on menadione reductase. *J. Biol. Chem.* **244,** 4168.

229. Lanyi, J. K. (1971). Studies of the electron transport chain of extremely halophilic bacteria. VI. Salt-dependent dissolution of the cell envelope. *J. Biol. Chem.* **246,** 4552.
230. Large, P. J., Peel, D., and Quayle, J. R. (1961). Microbial growth on C_1 compounds. 2. Synthesis of cell constituents by methanol- and formate-grown *Pseudomonas* AM 1 and methanol-grown *Hyphomicrobium vulgare. Biochem. J.* **81,** 470.
231. Large, P. J., Peele, D., and Quayle, J. R. (1962). Microbial growth on C_1 compounds. C. Distribution of radioactivity in metabolites of methanol-grown *Pseudomonas* AM 1 after incubation with [^{14}C]methanol and [^{14}C]bicarbonate. *Biochem. J.* **82,** 483.
232. Large, P. J., Peel, D., and Quayle, J. R. (1962). Microbial growth on C_1 compounds. 4. Carboxylation of phosphoenol pyruvate in methanol-grown *Pseudomonas* AM 1. *Biochem. J.* **85,** 243.
233. Large, P. J., and Quayle, J. R. (1963). Microbial growth on C_1 compounds. 5. Enzyme activities in extracts of *Pseudomonas* AM 1. *Biochem. J.* **87,** 386.
234. Lawford, H. J., and Williams, G. R. (1971). The transport of citrate and other tricarboxylic acids in two species of *Pseudomonas. Biochem. J.* **123,** 571.
235. Lawrence, A. J., Kemp, M. B., and Quayle, J. R. (1970). Synthesis of cell constituents by methane-grown *Methylococcus capsulatus* and *Methanomonas methanooxidans. Biochem. J.* **116,** 631.
236. Leadbetter, E. R., and Foster, J. W. (1958). Studies on some methane-utilizing bacteria. *Arch. Mikrobiol.* **30,** 91.
237. Leadbetter, E. R., and Gottlieb, J. A. (1967). On methylamine assimilation in a bacterium. *Arch. Mikrobiol.* **59,** 211.
238. Lehninger, A. L. (1965). "Bioenergetics." Benjamin, New York.
239. Lessie, T. G., and Neidhardt, F. C. (1967). Formation and operation of the histidine-degrading pathway in *Pseudomonas aeruginosa. J. Bacteriol.* **93,** 1800.
240. Lewis, K. (1966). Symposium on bioelectrochemistry of microorganisms. IV. Biochemical fuel cells. *Bacteriol. Rev.* **30,** 101.
241. Liao, C. L., and Atkinson, D. E. (1971). Regulation at the phosphoenolpyruvate branch point in *Azotobacter vinelandii:* Phosphoenolpyruvate carboxylase. *J. Bacteriol.* **106,** 31.
242. Liao, C. L., and Atkinson, D. E. (1971). Regulation at the phosphoenolpyruvate branch point in *Azotobacter vinelandii:* Pyruvate kinase. *J. Bacteriol.* **106,** 37.
243. Lieberman, M. M., and Landyi, K. S. (1971). Studies of the electron transport chain of extremely halophilic bacteria. V. Mode of action of salts on cytochrome oxidase. *Biochim. Biophys. Acta* **245,** 21.
244. Lochmuller, H., Wood, H. G., and Davis, J. J. (1966). Phosphoenolpyruvate carboxytransphosphorylase. II. Crystallization and properties. *J. Biol. Chem.* **241,** 5678.
245. Lowe, C. R., and Dean, P. D. G. (1971). Affinity chromatography of enzymes on insolubilized cofactors. *FEBS* (*Fed. Eur. Biochem. Soc.*) *Lett.* **14,** 313.
246. Lowe, D. A., and Turner, J. M. (1970). Microbial metabolism of amino ketones: D-1-Aminopropan-2-ol and aminoacetone metabolism in *Escherichia coli. J. Gen. Microbiol.* **63,** 49.
247. Lowenstein, J. M. (1971). The pyruvate dehydrogenase complex and the citric acid cycle. *Compr. Biochem.* **18S,** 1.
248. McGivray, D., and Morris, J. G. (1969). Utilization of L-threonine by a species of *Arthrobacter.* A novel catabolic role for "aminoacetone synthase." *Biochem. J.* **112,** 657.

249. Maeba, P., and Sanwal, B. D. (1969). Phosphoenolpyruvate carboxylase of *Salmonella*. Some chemical and allosteric properties. J. *Biol. Chem.* **244,** 2549.

250. Maeno, H., and Feigelson, P. (1967). Spectral studies on the catalytic mechanism and activation of *Pseudomonas* tryptophan oxygenase (tryptophan pyrrolase). *J. Biol. Chem.* **241,** 596.

251. Maeno, H., and Feigelson, P. (1968). Studies on the interaction of carbon monoxide with tryptophan oxygenase of *Pseudomonas. J. Biol. Chem.* **243,** 301.

252. Mann, S. (1967). Besonderheiten im Tryptophanstoffwechsel von *Pseudomonas aeruginosa. Arch. Hyg.* **151,** 474.

253. Marquez, E. D., and Brodie, A. F. (1970). Electron transport in halophilic bacteria: Involvement of a menaquinone in the reduced nicotinamide adenine dinucleotide oxidative pathway. *J. Bacteriol.* **103,** 260.

254. Marr. J. J., and Weber, M. M. (1968). Studies on the mechanism of purine nucleotide inhibition of a triphosphopyridine nucleotide-specific isocitrate dehydrogenase. *J. Biol. Chem.* **243,** 4973.

255. Marr, J. J., and Weber, M. M. (1969). Multivalent inhibition of a soluble or NADP-specific isocitrate dehydrogenase—A regulatory enzyme. *Fed. Proc., Fed. Amer. Soc. Exp. Biol.* **28,** 342.

256. Marr. J. J., and Weber, M. M. (1969). NADP-specific isocitrate dehydrogenase and its role in metabolic regulation. *Bacteriol. Proc.* p. 118.

257. Marr. J. J., and Weber, M. M. (1969). Allosteric inhibition of a NADP-specific isocitrate dehydrogenase from *Crithidia fasciculata* by nucleotide phosphate. *J. Biol. Chem.* **244,** 2503.

258. Marr, J. J., and Weber, M. M. (1969). Feedback inhibition of an allosteric triphosphopyridine nucleotide-specific isocitrate dehydrogenase. *J. Biol. Chem.* **244,** 5709.

259. Marshall, V. P., and Sokatch, J. R. (1968). Oxidation of D-amino acids by a particulate enzyme from *Pseudomonas aeruginosa. J. Bacteriol.* **95,** 1419.

260. Martin, J. R., and Durham, N. N. (1966). Metabolism of D-tryptophan by a species of *Flavobacterium. Can. J. Microbiol.* **12,** 1269.

260a. Marshall, V. P., and Sokatch, J. R. (1972). Regulation of valine catabolism in *Pseudomonas putida. J. Bacteriol.* **110,** 1073.

260b. Martin, R. R., Marshall, V. D., Sokatch, J. R., and Unger, L. (1973). Common enzymes of branched-chain amino acid catabolism in *Pseudomonas putida. J. Bacteriol.* **115,** 198.

260c. Massey, L. K., Conrad, R. S., and Sokatch, J. R. (1974). Regulation of leucine catabolism in *Pseudomonas putida. J. Bacteriol.* **118,** 112.

261. Matula, T. I., McDonald, I. J., and Martin, S. M. (1969). CO_2 fixation by malic enzyme in a species of *Micrococcus. Biochem. Biophys. Res. Commun.* **34,** 795.

261a. Meedel, T. H., and Pizer, L. I. (1974). Regulation of one-carbon biosynthesis and utilization in *Escherichia coli. J. Bacteriol.* **118,** 905.

262. Megraw, R. E., and Beers, R. J. (1964). Glyoxylate metabolism in growth and sporulation of *Bacillus cereus. J. Bacteriol.* **87,** 1087.

263. Megraw, R. E., Reeves, H. C., and Ajl, S. J. (1965). Formation of lactyl-coenzyme A and pyruvyl-coenzyme A from lactic acid by *Escherichia coli. J. Bacteriol.* **90,** 984.

264. Menon, G. K. K., and Stern, J. R. (1959). Enzymic synthesis and metabolism of coenzyme A esters of dicarboxylic acids. *Fed. Proc., Fed. Amer. Soc. Exp. Biol.* **18,** 287.

265. Michaelis, R., and Tchen, T. T. (1971). Constitutive degradation of putrescine in a *Pseudomonas* species and its possible physiological significance. *Biochem. Biophys. Res. Commun.* **42,** 545.

266. Miki, K., and Okumuki, K. (1969). Cytochromes of *Bacillus subtilis.* II. Purification and spectral properties of cytochromes c-550 and c-554. *J. Biochem.* (*Tokyo*) **66,** 831.

267. Miki, K., and Okumuki, K. (1969). Cytochromes of *Bacillus subtilis.* III. Physicochemical and enzymatic properties of cytochromes c-550 and c-554. *J. Biochem.* (*Tokyo*) **66,** 845.

267a. Moffet, F. J., and Bridger, W. A. (1973). Succinyl-coenzyme A synthetase of *Escherichia coli:* Initial rate kinetics of succinyl-CoA cleavage and isotope exchange studies. *Can. J. Biochem.* **51,** 44.

267b. Miller, D. L., and Rodwell, V. W. (1971). Metabolism of basic amino acids in *Pseudomonas putida* catabolism of lysine by cyclic and acyclic intermediates. *J. Biol. Chem.* **246,** 2758.

268. Moriguchi, M., Yamamoto, T., and Soda, K. (1970). Crystalline kynureninase from *Pseudomonas marginalis. Biochem. Biophys. Res. Commun.* **44,** 752.

269. Morris, J. G. (1965). The assimilation of 2-C compounds other than acetate. *J. Gen. Microbiol.* **32,** 167.

270. Morris, G. (1969). Utilization of L-threonine by a pseudomonad: a catabolic role for L-threonine aldolase. *Biochem. J.* **115,** 603.

271. Moyer, R. W., Ramaley, R. F., Butler, L. G., and Boyer, P. D. (1967). The formation and reactions of a nonphosphorylated high energy form of succinyl-coenzyme A synthetase. *J. Biol. Chem.* **242,** 4299.

272. Mukherjee, B. B., Matthews, J., Horney, D., and Reed, L. J. (1965). Resolution and reconstitution of *Escherichia coli* α-ketoglutarate dehydrogenase complex. *J. Biol. Chem.* **240,** PC2268.

273. Murai, T., Tokushige, M., Nagai, J., and Katsuki, H. (1971). Physiological functions of NAD^+- and $NADP^+$-linked malic enzymes in *Escherichia coli. Biochem. Biophys. Res. Commun.* **43,** 875.

274. Murphy, W. H., Barnaby, C., Lin, F. C., and Kaplan, N. O. (1967). Malate dehydrogenases. II. Purification and properties of *Bacillus subtilis, Bacillus stearothermophilus,* and *Escherichia coli* malate dehydrogenase. *J. Biol. Chem.* **242,** 1548.

275. Murthy, P. S., and Brodie, A. F. (1964). Oxidative phosphorylation in fractional bacterial systems. XV. Reduced nicotinamide adenine phosphate-linked phosphorylation. *J. Biol. Chem.* **239,** 4292.

276. Murthy, P. S., Bogin, E., Higashi, T., and Brodie, A. F. (1969). Properties of the soluble malate-vitamin K reductase and associated phosphorylation. *J. Biol. Chem.* **244,** 3117.

277. Nakamura, K. (1960). Separation and properties of DPN- and TPN-linked succinic semialdehyde dehydrogenase from *Pseudomonas aeruginosa. Biochem. Biophys. Acta* **45,** 554.

278. Nakayama, T., and DeLey, J. (1965). Localization and distribution of alcohol-cytochrome 553 reductase in acetic acid bacteria. *Antonie van Leeuwenhoek; J. Microbiol. Serol.* **31,** 205.

279. Namba, Y., Yoshizawa, K., Ejima, A., Hayashi, T., and Kameda, T. (1969). Coenzyme A- and nicotinamide adenine dinucleotide-dependent branched chain α-keto acid dehydrogenase. I. Purification and properties of the enzyme from *Bacillus subtilis. J. Biol. Chem.* **244,** 4437.

280. Newell, C. P., and Lessie, T. G. (1970). Induction of histidine-degrading enzymes in *Pseudomonas aeruginosa. J. Bacteriol.* **104,** 596.

281. Newton, N. A., Cox, G. B., and Gibson, F. (1972). Function of ubiquinone in *Escherichia coli:* A mutant strain forming a low level of ubiquinone. *J. Bacteriol.* **109,** 69.

282. Nirenberg, M. W., and Jakoby, W. B. (1960). Enzymatic utilization of β-hydroxybutyric acid. *J. Biol. Chem.* **235,** 954.

283. Norton, J. E., and Sokatch, J. R. (1966). Oxidation of D- and L-valine by enzymes of *Pseudomonas aeruginosa. J. Bacteriol.* **92,** 116.

284. Numa, S., Ishimura, Y., Nishizuka, Y., and Hayaishi, O. (1961). β-Hydroxybutyryl-CoA, an intermediate in glutarate catabolism. *Biochem. Biophys. Res. Commun.* **6,** 38.

285. Numa, S., Ishimura, Y., Nakazawa, D., Okazaki, T., and Hayaishi, O. (1964). Enzymic studies on the metabolism of glutarate in *Pseudomonas. J. Biol. Chem.* **239,** 3915.

286. O'Brien, R. W., and Stern, J. R. (1969). Requirement for sodium in the anaerobic growth of *Aerobacter aerogenes* on citrate. *J. Bacteriol.* **98,** 388.

287. O'Brien, R. W., and Stern, J. R. (1969). Role of sodium in determining alternate pathways of aerobic citrate catabolism in *Aerobacter aerogenes. J. Bacteriol.* **99,** 389.

288. O'Brien, R. W., Frost, G. M., and Stern, J. R. (1969). Enzymatic analysis of the requirement for sodium in aerobic growth of *Salmonella typhimurium* on citrate. *J. Bacteriol.* **99,** 395.

289. Olson, J. A. (1954). The D-isocitrate lyase system: The formation of glyoxylic and succinic acids from D-isocitric acid. *Nature* (*London*) **174,** 695.

290. Ornston, L. N., and Ornston, M. K. (1969). Regulation of glyoxylate metabolism in *Escherichia coli* K-12. *J. Bacteriol.* **98,** 1098.

291. Ornston, L. N. (1971). Regulation of catabolic pathways in *Pseudomonas. Bacteriol. Rev.* **35,** 87.

292. Ozaki, H., and Shiio, I. (1969). Regulation of the TCA and glyoxylate cycle in *Brevibacterium flavum.* II. Regulation of phosphoenolpyruvate carboxylase and pyruvate kinase. *J. Biochem.* (*Tokyo*) **66,** 297.

293. Padmanabhan, R., and Kim, K. (1965). Oxidation of spermidine by a pseudomonad. *Biochem. Biophys. Res. Commun.* **19,** 1.

294. Padmanabhan, R., and Tchen, T. T. (1969). Nicotinamide adenine dinucleotide and nicotinamide adenine dinucleotide phosphate-linked succinic semialdehyde dehydrogenases in a *Pseudomonas* species. *J. Bacteriol.* **100,** 398.

295. Parker, M. G., and Weitzman, P. D. J. (1970). Regulation of NADP-specific isocitrate dehydrogenase activity in *Acinetobacter. FEBS* (*Fed. Eur. Biochem. Soc.*) *Lett.* **7,** 324.

295a. Parker, M. G., and Weitzman, P. D. J. (1973). The purification and regulatory properties of α-oxoglutarate dehydrogenase from *Acinetobacter lwoffi. Biochem. J.* **135,** 215.

295b. Pavada, J. L., and Ortega, M. V. (1972). Partial purification of the NADP-specific isocitrate dehydrogenase of *Salmonella typhimurium* and characterization of its inhibition by oxalacetate. *Rev. Lat. Amer. Microbiol.* **14,** 173.

296. Penner, P. E., and Cohen, L. H. (1969). Effects of adenosine triphosphate and magnesium ions on the fumarase reaction. *J. Biol. Chem.* **244,** 1070.

297. Poillon, W. N., Maeno, H., Koike, K., and Feigelson, P. (1969). Tryptophan oxygenase of *Pseudomonas acidovorans.* Purification, composition, and subunit structure. *J. Biol. Chem.* **244,** 3447.

298. Polglase, W. J., Run, W. T., and Withaar, J. (1966). Lipoquinones of *Escherichia coli. Biochim. Biophys. Acta* **118,** 42.

299. Prasad, C., and Srinivasan, V. R. (1970). Tryptophan metabolism during sporulation in *Bacillus cereus*. *Biochem. J.* **119,** 343.
300. Pye, E. K., and Eddy, A. A. (1965). The regulation of glycolysis in yeast. *Fed. Proc., Fed. Amer. Soc. Exp. Biol.* **24,** 537.
301. Quayle, J. R., and Keech, D. B. (1959). Carboxydismutase activity in formate- and oxalate-grown *Pseudomonas oxalaticus* (OX 1). *Biochim. Biophys. Acta* **31,** 587.
302. Quayle, J. R., and Keech, D. B. (1959). Carbon assimilation by *Pseudomonas oxalaticus* (OX 1). 1. Formate and carbon dioxide utilization during growth on formate. *Biochem. J.* **72,** 623.
303. Quayle, J. R., and Keech, D. B. (1959). Carbon assimilation by *Pseudomonas oxalaticus* (OX 1). 2. Formate and carbon dioxide utilization by cell-free extracts of the organisms grown on formate. *Biochem. J.* **72,** 631.
304. Quayle, J. R. (1963). Carbon assimilation by *Pseudomonas oxalaticus* (OX 1). 7. Decarboxylation of oxalyl-coenzyme A to formyl-coenzyme A. *Biochem. J.* **89,** 492.
304a. Quayle, J. R. (1972). The metabolism of one-carbon compounds by microorganisms. *Advan. Microbiol Physiol.* **7,** 119.
305. Radler, F. (1962). Über die Milchsäurebakterien des Weines und den biologischen Säureabbau. Übersicht. I. Systematik und chemische Grundlagen. *Vitis* **3,** 144.
306. Radler, F. (1962). Über die Milchsäurebakterien des Weines und den biologischen Säureabbau. Übersicht. II. Physiologie und Ökologie der Bakterien. *Vitis* **3,** 469.
307. Rajender, S., and McCollock, R. J. (1965). Infrared study of the fumarase reaction. *Anal. Biochem.* **13,** 469.
308. Ramaley, R. F., Bridger, W. A., Moyer, R. W., and Boyer, P. D. (1967). The preparation, properties and reactions of succinyl-coenzyme A synthetase and its phosphorylated form. *J. Biol. Chem.* **242,** 4287.
309. Ravel, D. N., and Wolfe, R. G. (1963). Malic dehydrogenase. V. Kinetic studies of substrate inhibition by oxalacetate. *Biochemistry* **2,** 220.
310. Rechler, M. M. (1969). The purification and characterization of L-histidine ammonia-lyase (*Pseudomonas*). *J. Biol. Chem.* **244,** 551.
311. Reed, L. J. (1966). Chemistry and function of lipoic acid. *Comp. Biochem.* **14,** 99.
312. Reed, L. J., and Cox, D. J. (1966). Macromolecular organization of enzyme systems. *Annu. Rev. Biochem.* **35,** 57.
313. Reed, L. J., and Oliver, R. M. (1968). The multienzyme α-ketoacid dehydrogenation complexes. *Brookhaven Symp. Biol.* **21,** 397.
314. Reeves, H. C., and Ajl, S. J. (1962). Function of malate synthetase and isocitratase in the growth of bacteria on two-carbon compounds. *J. Bacteriol.* **83,** 597.
315. Reeves, H. C., and Ajl, S. J. (1962). Alpha-hydroxyglutaric acid synthetase. *J. Bacteriol.* **84,** 186.
316. Reeves, H. C., and Ajl, S. J. (1963). Enzymatic formation of lactate and acetate from β-hydroxyglutarate. *Biochem. Biophys. Res. Commun.* **12,** 132.
317. Reeves, H. C., and Ajl. S. J. (1965). Enzymatic synthesis and metabolism of hydroxypyruvic aldehyde. *J. Biol. Chem.* **240,** 569.
318. Reeves, H. C., Brehmeyer, B. A., and Ajl, S. J. (1968). Multiple forms of isocitrate dehydrogenase in *Escherichia coli*. *Bacteriol. Proc.* p. 117.
319. Reeves, H. C., Brehmeyer, B. A., and Ajl, S. J. (1968). Multiple forms of bacterial NADP-specific isocitrate dehydrogenase. *Science* **162,** 359.
320. Reeves, H. C., Daumy, G. O., Lin, C. C., and Houston, M. (1972). $NADP^+$-specific isocitrate dehydrogenase of *Escherichia coli*. I. Purification and characterization. *Biochim. Biophys. Acta* **258,** 27.

321. Reitz, M. S., and Rodwell, V. W. (1970). α-Aminovaleramidase of *Pseudomonas putida*. *J. Biol. Chem.* **245,** 3091.

322. Renner, E. D., and Bernlohr, R. W. (1972). Characterization and regulation of pyruvate carboxylase of *Bacillus licheniformis*. *J. Bacteriol.* **109,** 764.

323. Roberts, R. B., Abelson, P. H., Cowie, D. B., Bolton, E. T., and Britton, R. J. (1955). Studies of biosynthesis in *Escherichia coli*. *Carnegie Inst. Wash. Publ.* **607** [as cited by Kornberg and Gotto (211)].

324. Rose, A. H. (1965). "Chemical Microbiology." Butterworth, London.

325. Rose, I. A., O'Connell, E. L., Noce, P., Utter, M. F., Wood, H. G., Willard, J. M., Cooper, T. G., and Benziman, M. (1969). Stereochemistry of the enzymatic carboxylation of phosphoenolpyruvate. *J. Biol. Chem.* **244,** 6130.

326. Rosenfeld, H., and Feigelson, P. (1969). Synergistic and product induction of the enzymes of tryptophan metabolism in *Pseudomonas acidovorans*. *J. Bacteriol.* **97,** 697.

327. Rosenfeld, H., and Feigelson, P. (1969). Product inhibition in *Pseudomonas acidovorans* of a permease system which transports L-tryptophan. *J. Bacteriol.* **97,** 705.

328. Rowe, J. J., and Reeves, H. C. (1971). Electrophoretic heterogeneity of bacterial nicotinamide adenine dinucleotide phosphate-specific isocitrate dehydrogenases. *J. Bacteriol.* **108,** 824.

329. Sacchan, D. S., and Stern, I. R. (1971). Studies of citrate transport in *Aerobacter aerogenes:* Binding of citrate by a membrane-bound oxalacetate decarboxylase. *Biochem. Biophys. Res. Commun.* **45,** 402.

330. Salas, M. L., Vinuela, E., Salas, M., and Sols, A. (1965). Citrate inhibition of PFK and the Pasteur effect. *Biochem. Biophys. Res. Commun.* **19,** 371.

330a. Salem, A. R., Hacking, A. J., and Quayle, J. R. (1973). Cleavage of malyl-coenzyme A into acetyl-coenzyme A and glyoxylate by *Pseudomonas* AM 1 and other C_1-unit-utilizing bacteria. *Biochem. J.* **136,** 89.

331. Sanadi, D. R., Langley, M., and White, F. (1959). α-Ketoglutarate dehydrogenase. VII. The role of thioctic acid. *J. Biol. Chem.* **234,** 183.

332. Sanwal, B. D., Zink, M. W., and Stachow, C. S. (1964). NAD-specific isocitrate dehydrogenase. A possible regulatory protein. *J. Biol. Chem.* **239,** 1587.

333. Sanwal, B. D., and Maeba, P. (1966). Regulation of the activity of phosphoenol-pyruvate carboxylase by fructose diphosphate. *Biochem. Biophys. Res. Commun.* **22,** 194.

334. Sanwal, B. D., Wright, J. A., and Smando, R. (1968). Allosteric control of the activity of malic enzyme in *Escherichia coli*. *Biochem. Biophys. Res. Commun.* **31,** 623.

335. Sanwal, B. D., and Smando, R. (1969). Regulatory role of cyclic 3′5′-AMP in bacteria: Control of malic enzyme of *Escherichia coli*. *Biochem. Biophys. Res. Commun.* **35,** 486.

336. Sanwal, B. D., and Smando, R. (1969). Malic enzyme of *Escherichia coli:* Diversity of effectors controlling enzyme activity. *J. Biol. Chem.* **244,** 1817.

337. Sanwal, B. D., and Smando, R. (1969). Malic enzyme of *Escherichia coli*. Possible mechanism for allosteric effects. *J. Biol. Chem.* **244,** 1824.

338. Sanwal, B. D., and Smando, R. (1969). Regulatory roles of cyclic 3′,5′-AMP in bacteria: Control of malic enzyme of *Escherichia coli*. *Biochem. Biophys. Res. Commun.* **35,** 486.

339. Sanwal, B. D. (1969). Regulatory mechanisms involving nicotinamide adenine nucleotides as allosteric effectors. I. Control characteristics of malate dehydrogenase. *J. Biol. Chem.* **244,** 1831.

340. Sanwal, B. D. (1970). Allosteric controls of amphibolic pathways in bacteria. *Bacteriol. Rev.* **34,** 20.
341. Sanwal, B. D. (1970). Regulatory characteristics of the diphosphopyridine nucleotide-specific malic enzyme of *Escherichia coli*. *J. Biol. Chem.* **245,** 1212.
342. Saz, H. J. (1954). The enzymatic formation of glyoxylate and succinate from tricarboxylic acids. *Biochem. J.* **58,** xx.
343. Schwartz, E. R., and Reed, L. J. (1969). α-Keto acid dehydrogenase complexes. XII. Effects of acetylation on the activity and structure of the dihydrolipoyl transacetylase of *Escherichia coli*. *J. Biol. Chem.* **244,** 6074.
344. Scrutton, M. C., and Utter, M. F. (1965). Pyruvate carboxylase. III. Some physical and chemical properties of the highly purified enzyme. *J. Biol. Chem.* **240,** 1.
345. Scrutton, M. C., Keech, D. B., and Utter, M. F. (1965). Pyruvate carboxylase. IV. Partial reactions and the locus of activation by acetyl-coenzyme A. *J. Biol. Chem.* **240,** 574.
346. Seddon, B., and Fynn, G. H. (1970). Terminal oxidations in *Bacillus brevis* ATCC 10068. 1. The measurement of the NADH oxidase activity of *Bacillus brevis* ATCC 10068. *Biochem. Biophys. Acta* **216,** 435.
347. Self, C. H., and Weitzman, P. D. J. (1970). Separation of isoenzymes by zonal centrifugation. *Nature* (*London*) **225,** 644.
348. Shepherd, D., and Garland, P. B. (1966). ATP controlled acetoacetate and citrate synthesis by rat liver mitochondria oxidizing palmitoyl carnitine, and the inhibition of citrate synthase by ATP. *Biochem. Biophys. Res. Commun.* **22,** 89.
349. Shiio, I., and Ozaki, H. (1968). Concerted inhibition of isocitrate dehydrogenase by glyoxylate and oxaloacetate. *J. Biochem.* (*Tokyo*) **64,** 45.
350. Singh, R. M. M., and Adams, E. (1965). Enzymatic deamination of Δ′-pyrroline-4-hydroxy-2-carboxylate to 2,5-dioxovalerate (α-ketoglutaric semialdehyde). *J. Biol. Chem.* **240,** 4344.
351. Singh, R. M. M., and Adams, E. (1965). Isolation and identification of 2,5-dioxovalerate, an intermediate in the bacterial oxidation of hydroxyproline. *J. Biol. Chem.* **240,** 4352.
352. Smith, R. A., and Gunsalus, I. C. (1954). Isocitratase: A new tricarboxylic acid cleavage system. *J. Amer. Chem. Soc.* **76,** 5002.
353. Smith, T. E. (1971). *Escherichia coli* phosphoenolpyruvate carboxylase. *J. Biol. Chem.* **246,** 4234.
354. Soda, K., and Osumi, T. (1969). Crystalline amino acid racemase with low substrate specificity. *Biochem. Biophys. Res. Commun.* **35,** 363.
355. Sokatch, J. R. (1966). Alanine and aspartate formation during growth on valine-^{14}C by *Pseudomonas aeruginosa*. *J. Bacteriol.* **92,** 72.
356. Sokatch, J. R., Sanders, L. E., and Marshall, V. P. (1968). Oxidation of methylmalonate semialdehyde to propionyl-CoA in *Pseudomonas aeruginosa* grown on valine. *J. Biol. Chem.* **243,** 2500.
356a. Sperl, G. T., Forrest, H. S., and Gibson, D. T. (1974). Substrate specificity of the purified primary alcohol dehydrogenases from methanol-oxidizing bacteria. *J. Bacteriol.* **118,** 541.
357. Speyer, J. F., and Dickman, S. R. (1956). On the mechanism of action of aconitase. *J. Biol. Chem.* **220,** 193.
358. Sprecher, M., Berger, R., and Sprinson, D. B. (1964). Stereochemical course of the isocitrate lyase reaction. *J. Biol. Chem.* **239,** 4268.
359. Srere, P. A. (1966). Citrate-condensing enzyme–oxalacetate binary complex. Studies on its physical and chemical properties. *J. Biol. Chem.* **241,** 2157.

360. Srere, P. A., and Whissen, N. (1967). Regulation of *E. coli* citrate synthase activity. *Fed. Proc., Fed. Amer. Soc. Exp. Biol.* **26,** 559.

361. Stieglitz, B., and Mateles, R. I. (1973). Methanol metabolism in *Pseudomonad* C. *J. Bacteriol.* **114,** 390.

362. Stouthamer, A. H. (1962). Energy production in *Gluconobacter liquefaciens. Biochim. Biophys. Acta* **56,** 19.

363. Sturm, R. N., Herman, N. J., and Bell, E. J. (1970). Metabolic control in *Acinetobacter species.* II. Effect of C_4 versus C_2 substrates on α-ketoglutarate dehydrogenase synthesis. *Can. J. Microbiol.* **16,** 817.

364. Sullivan, P. A. (1968). Crystallization and properties of L-lactate oxidase from *Mycobacterium smegmatis. Biochem. J.* **110,** 363.

365. Sundaram, T. K., Cazzulo, J. J., and Kornberg, H. L. (1971). Synthesis of pyruvate carboxylase from its apoenzyme and (+)-biotin in *Bacillus stearothermophilus. Biochem. J.* **122,** 663.

366. Swank, R. T., and Burris, R. H. (1969). Purification and properties of cytochrome c of *Azotobacter vinelandii. Biochim. Biophys. Acta* **180,** 473.

367. Swank, R. T., and Burris, R. H. (1969). Restoration by ubiquinone of *Azotobacter vinelandii*-reduced nicotinamide adenine dinucleotide oxidase activity. *J. Bacteriol.* **98,** 311.

368. Sweetman, A. J., and Griffiths, D. E. (1971). Studies on energy-linked reactions. Energy-linked reduction of oxidized nicotinamine adenine dinucleotide by succinate in *Escherichia coli. Biochem. J.* **121,** 117.

369. Sweetman, A. J., and Griffiths, D. E. (1971). Studies on energy-linked reactions. Energy-linked transhydrogenase reaction in *Escherichia coli. Biochem. J.* **121,** 125.

370. Tabor, C. W., and Kellogg, P. D. (1970). Identification of flavin adenine dinucleotide and heme in a homogeneous spermidine dehydrogenase from *Serratia marcescens. J. Biol. Chem.* **245,** 5425.

371. Takeda, H., and Hayaishi, O. (1966). Crystalline L-lysine oxygenase. *J. Biol. Chem.* **241,** 2733.

371a. Takeda, H., Yamamoto, S., Kojima, Y., and Hayaishi, O. (1969). Studies on monooxygenases. I. General properties of crystalline L-lysine monooxygenase. *J. Biol. Chem.* **244,** 2935.

372. Takeo, K., Murai, T., Nagai, J., and Katsuki, H. (1967). Allosteric activation of DPN-linked malic enzyme from *Escherichia coli* by aspartate. *Biochem. Biophys. Res. Commun.* **29,** 717.

373. Takeo, K. (1969). Existence and properties of two malic enzymes in *Escherichia coli*—especially of NAD-linked enzyme. *J. Biochem.* (*Tokyo*) **66,** 379.

374. Tashiro, M., Tsukada, M., Kobayashi, S., and Hayaishi, O. (1961). A new pathway of D-tryptophan metabolism: Enzymic formation of kynurenic acid via D-kynurenine. *Biochem. Biophys. Res. Commun.* **6,** 155.

375. Teuscher, G. (1967). Untersuchungen über den Tryptophan-stoffwechsel von Streptomyceten. I. Abbau von Tryptophan und entstehende Metaboliten. *Z. Allg. Mikrobiol.* **7,** 393.

376. Thacker, R. P. (1969). Conversion of L-hydroxyproline to glutamate by extracts of strains of *Pseudomonas convexa* and *Pseudomonas fluorescens. Arch Mikrobiol.* **64,** 235.

377. Thomas, A. D., Doelle, H. W., Westwood, A. W., and Gordon, G. L. (1972). The effect of oxygen on a number of enzymes involved in the aerobic and anaerobic utilization of glucose in *Escherichia coli. J. Bacteriol.* **112,** 1099.

378. Thompson, E. D., and Nakata, H. M. (1971). Reduction of activity of reduced

nicotinamide adenine dinucleotide oxidase by divalent cations in cell-free extracts of *Bacillus cereus* T. *J. Bacteriol.* **105,** 494.

379. Tiwari, N. P., and Campbell, J. J. R. (1969). Utilization of dicarboxylic acids by *Pseudomonas aeruginosa. Can. J. Microbiol.* **15,** 1095.

380. Tiwari, N. P., and Campbell, J. J. R. (1969). Enzymatic control of the metabolic activity of *Pseudomonas aeruginosa* grown in glucose or succinate media. *Biochim. Biophys. Acta* **192,** 395.

381. Trijbels, F., and Vogels, G. D. (1966). Allantoicase and ureidoglycolase in *Pseudomonas* and *Penicillium* species. *Biochim. Biophys. Acta* **118,** 387.

382. Trivett, T. L., and Meyer, E. A. (1971). Citrate cycle and related metabolism of *Listeria monocytogenes. J. Bacteriol.* **107,** 770.

383. Trudgill, P. W., and Widdus, R. (1966). D-Glucarate catabolism by Pseudomonadaceae and Enterobacteriaceae. *Nature (London)* **211,** 1097.

384. Turner, J. M. (1967). Microbial metabolism of aminoketones. L-1-Aminopropan-2-ol dehydrogenase and L-threonine dehydrogenase in *Escherichia coli. Biochem. J.* **104,** 112.

385. Tustanoff, E. R., and Stern, J. R. (1960). Enzymic carboxylation of crotonyl-CoA and the metabolism of glutaric acid. *Biochem. Biophys. Res. Commun.* **3,** 81.

386. Tustanoff, E. R., and Stern, J. R. (1961). Oxidation of glutaryl-coenzyme A to glutaconyl-coenzyme A. *Fed. Proc., Fed. Amer. Soc. Exp. Biol.* **20,** 272.

387. Ullrich, J., and Mannschreck, A. (1967). Studies on the properties of (−)-2-α-hydroxyethyl thiamine pyrophosphate ("active acetaldehyde"). *Eur. J. Biochem.* **1,** 110.

388. Utter, M. F., and Keech, D. B. (1963). Pyruvate carboxylase. I. The nature of the reaction. *J. Biol. Chem.* **238,** 2603.

389. Van Tigerstrom, R. G., and Razzell, W. E. (1968). Aldehyde dehydrogenase. I. Purification and properties of the enzyme from *Pseudomonas aeruginosa. J. Biol. Chem.* **243,** 2691.

390. Vestal, J. R., and Perry, J. J. (1969). Divergent metabolic pathways for propane and propionate utilization by a soil isolate. *J. Bacteriol.* **99,** 216.

391. Visser, J., and Veeger, C. (1968). Relation between conformations and activities of lipoamide dehydrogenase. III. Protein association–dissociation and the influence on catalytic properties. *Biochim. Biophys. Acta* **159,** 265.

391a. Vogel, O., and Henning, U. (1973). The subunit structure of the *Escherichia coli* K-12 pyruvate dehydrogenase complex. Dihydrolipoamide dehydrogenase component. *Eur. J. Biochem.* **35,** 307.

392. Vogels, G. D. (1969). Stereospecificity in the allantoin metabolism. *Antonie van Leeuwenhoek; J. Microbiol. Serol.* **35,** 236.

393. Webster, L. T., Jr., (1965). Studies of the acetyl-coenzyme A synthetase reaction. *J. Biol. Chem.* **240,** 4158.

394. Wegener, W. S., Reeves, H. C., and Ajl, S. J. (1965). Heterogeneity of the glyoxylate-condensing enzymes. *J. Bacteriol.* **90,** 594.

395. Wegener, W. S., Reeves, H. C., Rabin, R., and Ajl, S. J. (1968). Alternate pathways of metabolism of short-chain fatty acids. *Bacteriol. Rev.* **32,** 1.

396. Weitzman, P. D. J. (1966). Regulation of citrate synthase activity in *Escherichia coli. Biochim. Biophys. Acta* **128,** 213.

397. Weitzman, P. D. J. (1966). Reduced nicotinamide-adenine dinucleotide as an allosteric effector of citrate synthase activity in *Escherichia coli. Biochem. J.* **101,** 44c.

398. Weitzman, P. D. J. (1967). Allosteric fine control of citrate synthase in *Escherichia coli. Biochim. Biophys. Acta* **139,** 526.

399. Weitzman, P. D. J., and Jones, D. (1968). Regulation of citrate synthase and microbial taxonomy. *Nature (London)* **219,** 270.

400. Weitzman, P. D. J., and Dunmore, P. (1969). Citrate synthases: Allosteric regulation and molecular size. *Biochim. Biophys. Acta* **171,** 198.

401. Whistance, G. R., Dillon, J. F., and Threlfall, D. R. (1969). The nature, intergeneric distribution, and biosynthesis of isoprenoid quinones and phenols in gram-negative bacteria. *Biochem. J.* **111,** 461.

402. White, D. C., and Sinclair, P. R. (1971). Branched electron transport systems in bacteria. *Advan. Microbial Physiol.* **5,** 173.

403. Whitely, H. R., Osborne, M. J., and Huennekens, F. M. (1959). Purification and properties of the formate-activating enzyme from *Micrococcus aerogenes. J. Biol. Chem.* **234,** 1538.

404. Whitely, H. R. (1967). Induced synthesis of formyltetrahydrofolate synthetase in *Micrococcus aerogenes. Arch. Mikrobiol.* **59,** 315.

405. Whittenbury, R. (1969). *Process Biochem.* **4,** 51 [as cited by Higgins and Quayle (137)].

406. Willets, A. J., and Turner, J. M. (1970). Threonine metabolism in a strain of *Bacillus subtilis:* Enzymic oxidation of the intermediate DL-lactaldehyde. *Biochim. Biophys. Acta* **222,** 234.

407. Willets, A. J., and Turner, J. M. (1970). Threonine metabolism in a strain of *Bacillus subtilis:* Enzymes acting on methylglyoxal. *Biochim. Biophys. Acta* **222,** 668.

408. Willets, A. J., and Turner, J. M. (1970). Threonine metabolism in a strain of *Bacillus subtilis. Biochem. J.* **117,** 27P.

409. Willets, A. J., and Turner, J. M. (1971). Threonine metabolism in a strain of *Bacillus subtilis:* Enzymic oxidation of 1-aminopropan-2-ol and aminoacetone. *Biochim. Biophys. Acta* **252,** 98.

410. Wojtczak, L., Wojtczak, A. B., and Ernster, L. (1969). The inhibition of succinate dehydrogenase by oxaloacetate. *Biochim. Biophys. Acta.* **191,** 10.

411. Wolfe, J. B., Ivler, D., and Rittenberg, S. C. (1954). Malonate decarboxylation by *Pseudomonas fluorescens.* I. Observations with dry cells and cell-free extracts. *J. Biol. Chem.* **209,** 867.

412. Wolfe, J. B., Ivler, D., and Rittenberg, S. C. (1954). Malonate decarboxylation by *Pseudomonas fluorescens.* II. Mg^{2+} dependency and trapping of active intermediates. *J. Biol. Chem.* **209,** 875.

413. Wolfe, J. B., Ivler, D., and Rittenberg, S. C. (1954). Malonate decarboxylation by *Pseudomonas fluorescens.* III. Role of acetyl-CoA. *J. Biol. Chem.* **209,** 885.

414. Wong, D. T. O., and Ajl, S. J. (1956). Conversion of acetate and glyoxylate to malate. *J. Amer. Chem. Soc.* **78,** 3230.

415. Wood, H. G., and Utter, M. F. (1965). The role of CO_2 fixation in metabolism. *Essays Biochem.* **1,** 1.

416. Wright, J. A., Maeba, P., and Sanwal, B. D. (1967). Allosteric regulation of the activity of citrate synthetase of *Escherichia coli* by α-ketoglutarate. *Biochem. Biophys. Res. Commun.* **29,** 34.

417. Wright, J. A., and Sanwal, B. D. (1969). Regulatory mechanisms involving nicotinamide adenine nucleotides as allosteric effectors. II. Control of phosphoenolpyruvate carboxykinase. *J. Biol. Chem.* **244,** 1838.

418. Wright, J. A., and Sanwal, B. D. (1971). Regulatory mechanisms involving nicotinamide adenine nucleotides as allosteric effectors. IV. Physiochemical study and binding of ligands to citrate synthase. *J. Biol. Chem.* **246,** 1689.

419. Yagi, K., Nishikimi, M., Ohishi, N., and Takai, A. (1970). Release of α-imino acid as primary product in D-amino acid oxidase reaction. *Biochim. Biophys. Acta* **212,** 243.
420. Yamada, E. W., and Jacoby, W. B. (1959). Enzymatic utilization of acetylenic compounds. II. Acetylenemonocarboxylic acid hydrase. *J. Biol. Chem.* **234,** 941.
421. Yamada, E. W., and Jacoby, W. B. (1960). Enzymatic utilization of acetylenic compounds. V. Direct conversion of malonic semialdehyde to acetyl-coenzyme A. *J. Biol. Chem.* **235,** 589.
421a. Yamaguchi, M., Tokushige, M., and Katsuki, H. (1973). Studies on regulatory functions of malic enzymes. II. Purification and molecular properties of NAD-linked malic enzyme from *Escherichia coli. J. Biochem.* **73,** 169.
422. Yamasaki, H., and Moriyama, T. (1970). α-Ketoglutarate: glyoxylate carboligase activity in *Escherichia coli. Biochem. Biophys. Res. Commun.* **39,** 790.
423. Yamasaki, H., and Moriyama, T. (1971). Purification, general properties and two other catalytic activities of α-ketoglutarate:glyoxylate carboligase of *Mycobacterium phlei. Biochim. Biophys. Acta* **242,** 637.
424. Yoshida, A. (1965). Purification and chemical characterization of malate dehydrogenase of *Bacillis subtilis. J. Biol. Chem.* **240,** 1113.
425. Zagorski, W., and Kaniuga, Z. (1967). Isolation and some properties of reduced diphosphopyridine nucleotide: 2,6-dichlorophenol indophenol soluble reductase from *Mycobacterium phlei. Acta Microbiol. Pol.* **16,** 91.
426. Zagorski, W., Kaniuga, Z., Manteuffel-Cymborowska, M., and Stryzewska, E. (1969). Oxidation of reduced nicotinamide adenine dinucleotide by *Mycobacterium phlei. Acta Microbiol. Pol., Ser. A* **1,** 135.
427. Zelitch, I., and Ochoa, S. (1953). Oxidation of glycolic and glyoxylic acids in plants. I. Glycolic acid oxidase. *J. Biol. Chem.* **201,** 707.
428. Zelitch, I. (1953). Oxidation and reduction of glycolic and glyoxylic acids in plants. II. Glyoxylic acid reductase. *J. Biol. Chem.* **201,** 719 [as cited by Cartwright and Hullin (51)].
429. Zeylemaker, W. P., Klaasoe, A. D. M., and Slater, E. C. (1969). Studies on succinate dehydrogenase. V. Inhibition by oxaloacetate. *Biochim. Biophys. Acta* **191,** 229.

Questions

1. What are chemoorganotrophs as opposed to chemolithotrophs?
2. Give the main differences in the electron transport systems of aerobic chemoorganotrophs and aerobic chemolithotrophs.
3. Draw the tricarboxylic acid cycle and name the individual enzymes of the appropriate reactions.
4. Explain the detailed mechanism of pyruvate oxidation to acetyl-CoA.
5. Discuss the first step of the tricarboxylic acid cycle and its regulation.
6. Why can the oxidation of malate be regarded as a key step in the tricarboxylic acid cycle?
7. The tricarboxylic acid cycle is the main energy-producing cycle of aerobic chemoorganotrophs. How much energy in the form of ATP can the microorganism produce in one cycle?
8. What are the main differences between the electron transport system in *Mycobacterium phlei* and that in the Enterobacteriaceae?

9. What is the role of the glyoxylate cycle?

10. Name the key reaction responsible for the deviation from the tricarboxylic acid into the glyoxylate cycle.

11. What are anaplerotic sequences?

12. As the tricarboxylic acid cycle is regarded as the main energy-producing pathway, it must underly regulatory control. Name and discuss the major control points.

13. Describe the two different pathways of citrate metabolism.

14. Compare the aerobic metabolism of glycolate by *Pseudomonas* with the anaerobic metabolism of glycolate by *Micrococcus denitrificans.*

15. Compare the aerobic metabolism of tartrate by *Pseudomonas* with the tartrate metabolism suggested for the Lactobacillaceae.

16. What are the differences in glucarate metabolism of pseudomonads and enterobacteria?

17. Depending on the presence or absence of the enzyme tartronate-semialdehyde synthase, oxalate can be metabolized via two different pathways. Name these pathways and their specific differences.

18. Compare the aerobic and anaerobic metabolism of propionic acid.

19. Explain the regulation of aerobic acetate metabolism.

20. Describe the pathway for methane oxidation.

21. Aerobic and facultative aerobic bacteria are equally capable of oxidizing amino acids. What type of reactions do D- and L-amino acids undergo first?

22. Discuss briefly the four possible pathways of aerobic tryptophan metabolism.

23. Threonine can be metabolized via three different pathways, depending on the key enzyme available. Name this key enzyme and outline the pathways schematically.

8

Aerobic Respiration—Hydrocarbon Metabolism

Since the discovery that hydrocarbons can be metabolized by microorganisms (160), possible industrial applications have been considered (21). This is because three classes of industrially important products could be affected by microorganism metabolism (73): (*1*) a conversion to products that consist of oxygenated molecules with no change in the carbon skeleton; (*2*) oxygenated products with shorter chains, produced by oxidative degradation; and (*3*) biosynthetic products such as amino acids, vitamins, and lipids.

Any metabolism of hydrocarbons must, of course, be oxidative. A number of strains of *Pseudomonas* are able to adapt very quickly to using hydrocarbons as sole carbon sources and can revert to growth on nonhydrocarbon media as rapidly (75). Certain hydrocarbon compounds, however, are utilizable only as cosubstrates; that is, the organism uses a second compound for its energy source and at the same time oxidizes the hydrocarbon in question. The latter compound is not oxidized completely, but only for one or two steps. This type of mechanism is called "cooxidation" or "cometabolism" of hydrocarbons.

During hydrocarbon metabolism, fatty acids are formed and further metabolized. In order that the different ways by which fatty acids can be broken down be understood, the basic principles of these oxidations are considered first. In general, fatty acids may undergo alpha, beta, or omega oxidations.

ALPHA OXIDATION OF FATTY ACIDS. The alpha oxidation of long-chain fatty acids occurs at the second position of the chain. The necessary enzyme system catalyzes the 2-hydroxylation of the particular acid (76, 77). The result of this oxidative reaction, a decarboxylation of the α-hydroxy fatty acid, usually occurs, giving an odd-numbered fatty acid. The actual mechanism of this reaction is not known (249).

BETA OXIDATION OF FATTY ACIDS. This type of fatty acid oxidation is the best known and possibly the most frequently occurring one. The beta-oxidation mechanism results in the continual removal of acetate (C_2) units. This removal follows a sequence of reactions. The initial step in beta oxidation is the activation of a fatty acid by its transformation into the corresponding CoA-thioester, which is catalyzed by the appropriate acyl-CoA synthetases (EC 6.2.1.3).

$$\text{RCOOH} + \text{CoA-SH} + \text{ATP} \rightleftharpoons \text{RCOS-CoA} + \text{AMP} + \text{PPi}$$

The most commonly known reaction is that of acetate to acetyl-CoA, which requires the enzyme acetyl-CoA synthetase [acetate:CoA ligase (AMP), EC 6.2.1.1]. Another mechanism of obtaining fatty acid-CoA ester derivatives is carried out by a transferase such as propionate-CoA transferase (EC 2.8.3.1) and consists of a transacetylation whereby CoA is transferred to the appropriate fatty acid:

$$\text{Propionyl-CoA} + \text{RCOOH} \rightleftharpoons \text{propionate} + \text{RCOSCoA}$$

In the longer saturated fatty acid chains, the fatty acid is then converted to its unsaturated form by an acyl-CoA dehydrogenase [acyl-CoA: (acceptor)oxidoreductase, EC 1.3.99.3]

$$\text{RCH}_2\text{CH}_2\text{COSCoA} + \text{A} \rightarrow \text{RCH{=}CHCOSCoA} + \text{AH}_2$$

where A represents a hydrogen acceptor. The so-obtained α,β-unsaturated fatty acyl-CoA is hydrated by an CoA-dependent hydratase, forming the β-hydroxyacyl-CoA derivative of the particular fatty acid. This hydration

$$\text{RCH{=}CHCOSCoA} + \text{H}_2\text{O} \rightleftharpoons \text{RCHOHCH}_2\text{COSCoA}$$

is very often stereospecific, as the trans isomer is converted to the L(+) and the cis isomer to the D(−) isomer. β-Hydroxyacyl-CoA is now oxidized by an NAD-dependent L-3-hydroxyacyl-CoA:NAD oxidoreductase (EC 1.1.1.35) to the corresponding β-ketoacyl-CoA ester

$$\text{RCHOHCH}_2\text{COSCoA} + \text{NAD}^+ \rightleftharpoons \text{R}\underset{\underset{\text{O}}{\|}}{\text{C}}\text{CH}_2\text{COSCoA} + \text{NADH} + \text{H}^+$$

The NAD-linked dehydrogenase is very stereospecific but can oxidize various chain lengths. A thiolytic cleavage of the β-ketoacyl is the terminal step in beta oxidation, as it forms acetyl-CoA. A thiolase (acyl-CoA:acetyl-CoA *C*-acyltransferase, EC 2.3.1.16) catalyzes the reaction

$$RCCH_2COSCoA + CoASH \rightleftharpoons RCOSCoA + CH_3COSCoA$$

OMEGA OXIDATION OF FATTY ACIDS. Apart from the alpha and beta oxidations, a third pathway involves conversion of the acid to the ω-hydroxy compound, which may then be oxidized further to dicarboxylic acids. This system requires the participation of the electron transport system, possibly with cytochrome P-450 or thioredoxin and NADPH–cytochrome c reductase components. These oxidations predominantly occur as mixed function oxidase systems during hydrocarbon metabolism. For further reading on this subject, the excellent reviews by Wakil and Barnes (249) and Lennarz (151) should be consulted.

Oxidation of Alkanes and Alkenes

The oxidation of alkanes and alkenes involves the participation of oxygen, which is incorporated into the molecule (73, 143). The attack on the alkanes can be twofold: via a monoterminal or via a diterminal oxidation, whereby in both cases methyl groups are oxidized.

Monoterminal Oxidation

The most common type of oxidation for hydrocarbon chains is the monoterminal oxidation, followed by a β-oxidation. The β-oxidation of the fatty acids can lead to a great number of biosynthetic materials, which will not be considered here. The first product of this sequence is a primary alcohol, which is oxidized through to a fatty acid that via β-oxidation is able to form acetic acid (67, 223). Primary alcohols can induce NAD(P)-inde-

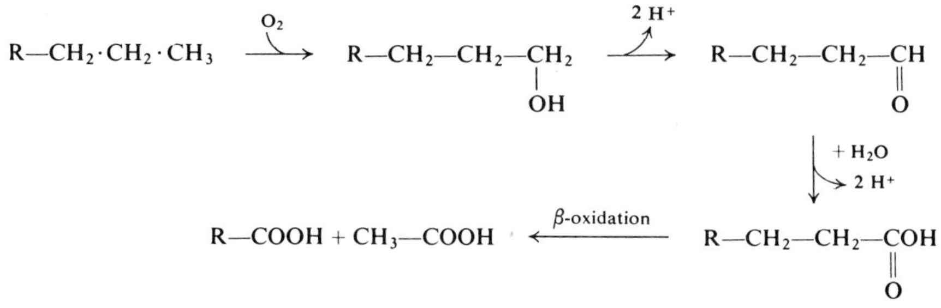
O2
R—CH2·CH2·CH3
R—CH2—CH2—CH2
OH
2 H+
R—CH2—CH2—CH
O
+ H2O
2 H+
β-oxidation
R—COOH + CH3—COOH
R—CH2—CH2—COH
O

pendent primary alcohol dehydrogenases (246), which are formal-bound to the cellular structure. In addition to these enzymes, *Pseudomonas aeruginosa* (strain 473) also possesses a soluble constitutive NADP-linked dehydrogenase. All of the available reports on *n*-alkane oxidation (6, 113, 150, 188, 248) emphasize the extreme susceptability of the C-1 position for oxidation.

It is, however, not only the terminal methyl group that can be susceptible to attack in a monoterminal attack. The methylene group next to the methyl group can also be oxidized. *Brevibacterium* sp. strain JOB 5 (247) was found to oxidize propane via acetone to acetic acid

$$CH_3CH_2CH_3 \longrightarrow CH_3CHOHCH_3 \longrightarrow CH_3COCH_3 \longrightarrow CH_3COCH_2OH \longrightarrow \cdots \longrightarrow CH_3COOH$$

Diterminal Oxidation

Most bacteria attack only one end of the chain, as was demonstrated above. *Pseudomonas aeruginosa* grows on 2-methylhexane and produces a mixture of acids (73).

$$\text{2-methylhexane} \rightarrow \begin{array}{l}\text{5-methylhexanoic acid}\\ \text{2-methylhexanoic acid}\end{array}$$

The formation of these two acids indicates that the organism must be able to attack the hydrocarbon chain from either end, in a so-called "diterminal" oxidation. It is still thought (73) that the bacterium attacks only one terminal methyl group at a time. Other acids that have been identified by similar oxidation steps are from decane or dodecane: 11-formylundecanoic acid, 12-hydroxydodecanoic acid, and 3-hydroxydodecanoic acid. Similar reactions also occur in *Corynebacterium* (141).

Whether a monoterminal or diterminal oxidation takes place, the initial attack is on one methyl group, which results in the formation of the corresponding alcohol, aldehyde (23), and fatty acid. The latter is then further

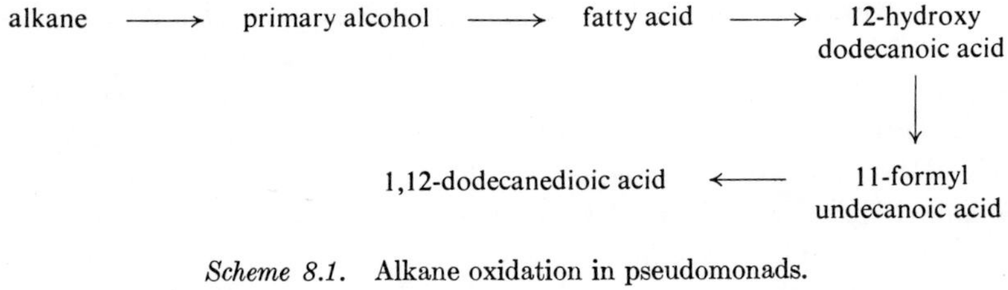
alkane
primary alcohol
fatty acid
12-hydroxy dodecanoic acid
11-formyl undecanoic acid
1,12-dodecanedioic acid

Scheme 8.1. Alkane oxidation in pseudomonads.

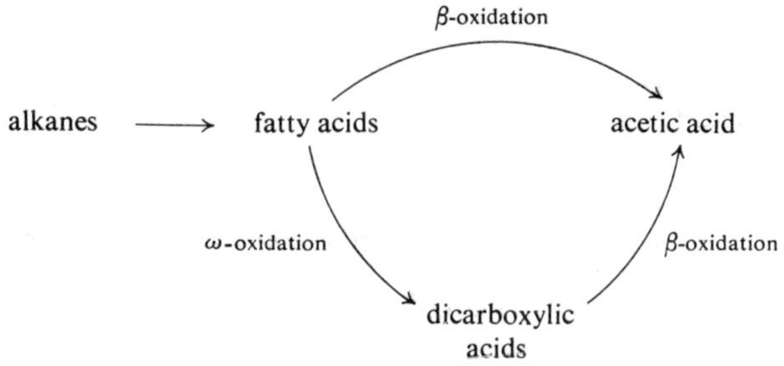
β-oxidation
alkanes
fatty acids
acetic acid
ω-oxidation
β-oxidation
dicarboxylic acids

Scheme 8.2. Alkane oxidation by β- and/or ω-oxidation.

degraded by β-oxidation. In addition to this β-oxidation an ω-oxidation can take place, which was originally thought to be present only in mammalian tissue (196) and may be more widespread than anticipated at the moment (147) (Scheme 8.1). The simultaneous use by *Pseudomonas* species of beta and omega oxidation connects alkanes to fatty acids and finally dicarboxylic acids. Where no β-oxidation occurs, higher molecular esters, such as wax, are formed (Scheme 8.2). Similar reaction sequences have been found to exist in mycobacteria (158).

From the reports available, it seems that aliphatic hydrocarbons with eight or more carbon atoms in the chain undergo exclusively an α,ω-oxidation first. This could be regarded as an alternative pathway to that of the formation of fatty acids. Asymmetric branched aliphatic hydrocarbons, such as 2-methylhexane, undergo oxidation on all terminal groups. As far as α-olefins are concerned, bacteria prefer to attack the saturated end of the molecule, but it was also demonstrated that the terminal double bond can be oxidized as well. This oxidation is carried out by enzymes that are associated with the oxidation of the saturated end of the molecules. These enzymes are therefore not very specific with respect to molecular configuration.

Methyl group hydroxylation itself has been, and still is, the subject of extensive research (9, 82, 181–183). The enzyme system responsible is a mixed function oxidase system. Whereas the mixed function oxidase itself is of a single-protein component type (107), most of the microorganisms require in addition a multicomponent electron transfer system. The best known mixed function oxidase system is undoubtedly the ω-hydroxylase system of *Pseudomonas oleovorans* (181, 185). Three protein components were found to be involved in this reaction (181); they have been purified and identified as rubredoxin (156, 183), rubredoxin–NAD reductase (NADH:rubredoxin oxidoreductase, EC 1.6.7.2) and an ω-hydroxylase (alkane, reduced-rubredoxin:oxygen 1-oxidoreductase, EC 1.14.15.3) (161) (Scheme 8.3). The inducible mixed function oxidase system is specific for the omega oxidation of fatty acids and hydrocarbons (146). The functional

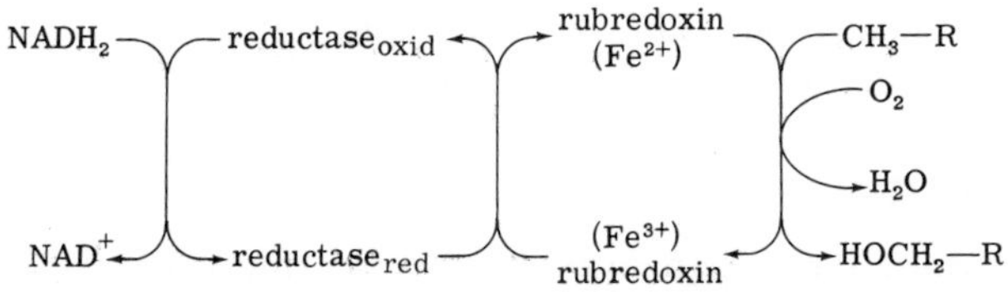
NADH$_2$
reductase$_{oxid}$
rubredoxin
(Fe^{2+})
CH$_3$—R
O$_2$
H$_2$O
NAD$^+$
reductase$_{red}$
(Fe^{3+})
rubredoxin
HOCH$_2$—R

Scheme 8.3. Mixed function oxidase system in *Pseudomonas oleovorans* (reprinted with permission of the author and Academic Press, Inc.).

group of this mixed function oxidase system does not contain cytochrome P-450 (184). The proper mechanism of this methyl group oxidation is still under dispute. McKenna and Kallio (160), in their excellent review, mention that a number of research groups propose a role for hydroperoxide formation as an intermediate in this reaction. Such an alkyl hydroperoxide formation, arising via a free radical mechanism in the oxidation of alkanes, could account for the formation of methylketones (150) and fatty acids of substrate chain length. The formation and metabolism of alkyl hydroperoxides are known to occur not only in bacteria (66, 222, 243) but also in mammalian tissue (19a, 42–44, 155). That such a mechanism also exists in *Pseudomonas oleovorans* has been clearly established (20). Although it was found that rubredoxin serves as an electron carrier for hydroperoxide reduction, the significance of such a reaction is not clear. The available evidence suggests, then, that the electron transfer components are rubredoxin, rubredoxin–NAD reductase (reduced-NAD:rubredoxin oxidoreductase, EC 1.6.7.2), and the ω-hydroxylase [alkane, rubredoxin:oxygen oxidoreductase (1-hydroxylating), EC 1.14.15.3]. The stoichiometry of the reaction corresponds to that of a monooxygenase (242a).

Methylene-group oxidation replaces the rubredoxin with a hemeprotein, cytochrome P-450, and the ω-hydroxylase with a methylene hydroxylase (136). *Pseudomonas putida* has also a NADH-dependent putidaredoxin reductase (flavoprotein) and putidaredoxin (nonheme-iron protein), whereby it is assumed that cytochrome P-450 plays the functional role (126). This cytochrome is thought to act as the oxygen-activated enzyme in numerous hydroxylations of steroids and of condensed aromatic, aliphatic, and cyclohydrocarbons (72, 115). Whether or not this system involves the formation of an oxygenated intermediate (56) in the form of a peroxide is not fully elucidated as yet. A very similar mixed function oxidase system appears to exist in *Corynebacterium* sp. strain 7 E1C (35, 36), whereby *n*-octane is oxidized to 1-octanol and octanoic acid. So far, a flavoprotein and cytochrome P-450 have been isolated.

It therefore appears that the methyl group oxidation is carried out by two mechanisms. In the Pseudomonadaceae, rubredoxin is the oxygen-activating enzyme, whereas in *Corynebacterium*, cytochrome P-450 takes over this particular role. The methylene group oxidation undoubtedly requires cytochrome P-450 for its function (56a, 92a).

Propane Oxidation

The main route for propane utilization is a terminal oxidation pathway, whereby propane is oxidized via propanol to propionic acid. Investigations

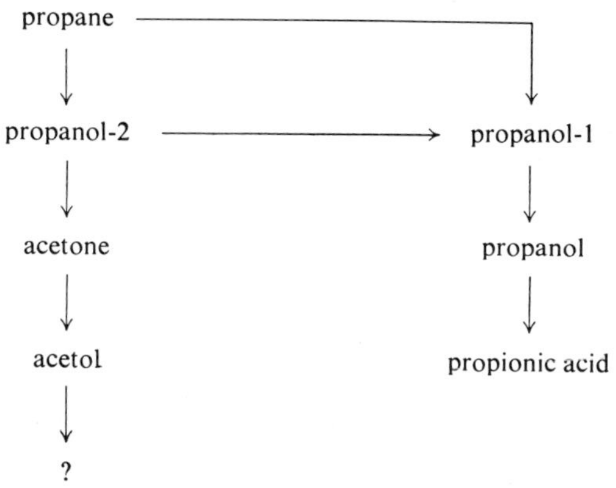
propane
propanol-2
propanol-1
acetone
propanol
acetol
propionic acid
?

with *Brevibacterium* sp. strain JOB 5 indicated that propionate can be metabolized further via methylmalonate to acetate (247). Induction experiments with isocitrate lyase revealed also that propane may not be oxidized via a terminal oxidation pathway, but rather via isopropanol and acetone, with the concomitant cleavage of acetol to acetate and carbon dioxide. Acetate would be the inducer of the isocitrate lyase. There is not enough evidence on whether or not the routes of propane metabolism are identical on pseudomonads and *Brevibacterium*, despite the close resemblance. The increase in isocitrate lyase on growth of the microorganisms on alkanes

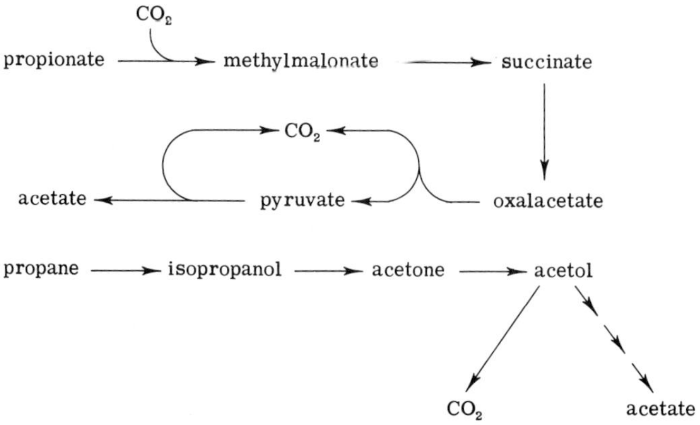
CO_2
propionate
methylmalonate
succinate
CO_2
acetate
pyruvate
oxalacetate
propane
isopropanol
acetone
acetol
CO_2
acetate

suggests the formation of acetyl-CoA (241) and a possible important role of the glyoxylate cycle.

n-Decane Oxidation

n-Decane is oxidized by *Mycobacterium rhodochrous* (74) and some pseudomonads to *n*-decanol, which is degraded further, under the influence of an alcohol and an aldehyde dehydrogenase, to *n*-decanoic acid. β-Oxidation is then responsible for the production of acetic acid (Fig. 8.1). The alternate pathway shows first α,ω-oxidation of *n*-decane, which is followed by β-oxidation after *n*-octane-1:8-dicarboxylic acid is formed. Both pathways can be linked together by omega oxidation of *n*-decanol, producing *n*-octane-1:8-dicarboxylic acid. The nature of the end product formed depends on whether the alkane has an odd or an even number of carbon atoms in the chain (29). β-Oxidation of an odd-numbered alkane after oxidation to acid would give acetic and propionic acid in a monoterminal attack and acetic and malonic acid in a diterminal attack. It therefore seems that the oxidation of straight-chain hydrocarbons follows a similar pattern. Attempts are being made to work out the kinetics (25) and thermodynamics of these processes (162).

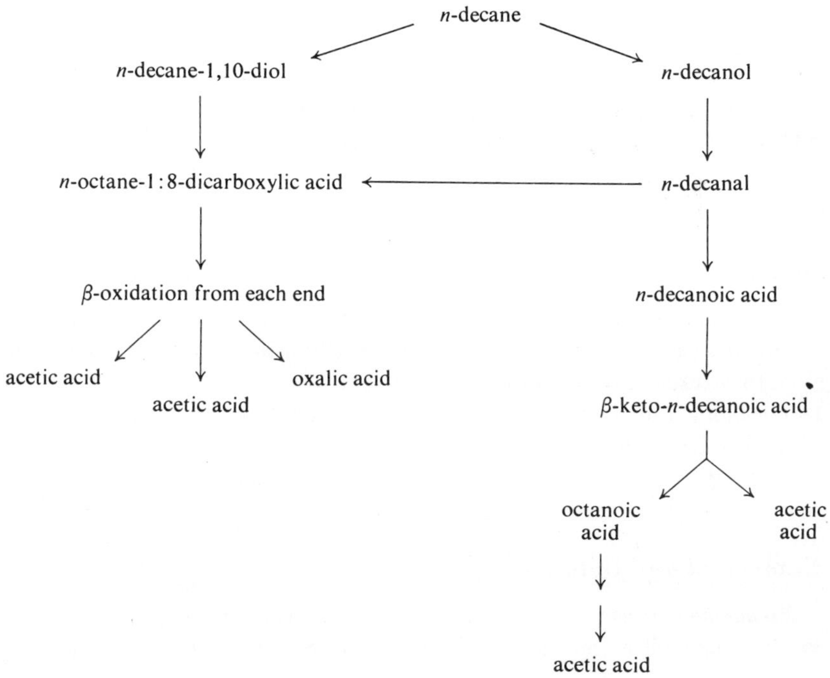
n-decane
n-decane-1,10-diol
n-decanol
n-octane-1:8-dicarboxylic acid
n-decanal
β-oxidation from each end
n-decanoic acid
acetic acid
acetic acid
oxalic acid
β-keto-n-decanoic acid
octanoic acid
acetic acid
acetic acid

Fig. 8.1. The oxidation of *n*-decane by microorganisms.

Some species of *Pseudomonas* are unable to form *n*-decanal. *Pseudomonas aeruginosa*, for example, metabolizes decane to 1-, 2-, 3-, 4-, and sometimes 5-decanol and the corresponding ketones (74). These products are further metabolized to decanoic, nonanoic, octanoic, and heptanoic acids, thus producing the acids from the corresponding alcohol.

Undecane Oxidation

An obligate halophilic, hydrocarbon-oxidizing *Pseudomonas* strain H that exhibited an oxidation spectrum from decane to octadecane was isolated from the North Sea (142). Undecane is attacked by a terminal oxidation that leads to undecanol, undecylic acid, and finally via a β-oxidation to

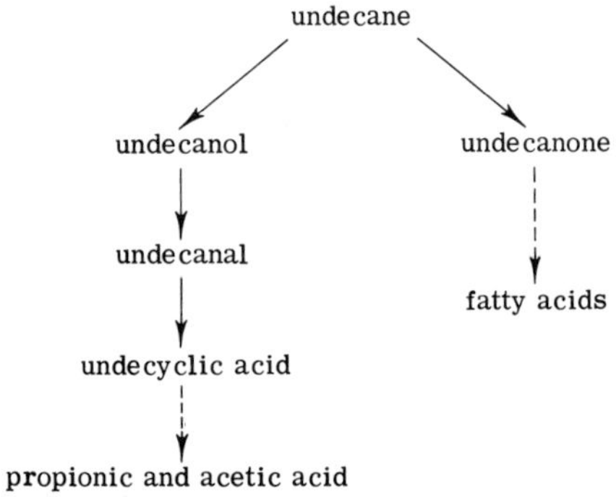
undecane
undecanol
undecanone
undecanal
fatty acids
undecyclic acid
propionic and acetic acid

propionic and acetic acids. Propionic acid itself is further metabolized via the methylmalonyl pathway, as described above (see p. 567). A subterminal oxidation is also possible and leads to undecanone. The metabolism of undecanone is not known.

Methylketone Oxidation

Methylketones are in general oxidized in a subterminal attack, possibly by monooxygenases (69). *Pseudomonas multivorans* and *P. aeruginosa* are able to oxidize 2-tridecanone to undecyl acetate, which is cleaved to 1-undecanol and acetate (70, 71). The latter product is thought to be catabolized via the glyoxylate cycle, furnishing the cell with the carbon and energy required for the formation of cellular material. The suggested sequence is as outlined in Fig. 8.2.

Hexanediol and Octanediol Oxidation

Pseudomonas aeruginosa is able to oxidize 1,6-hexanediol to succinate via beta as well as omega oxidation (245) (Scheme 8.4). Both pathways

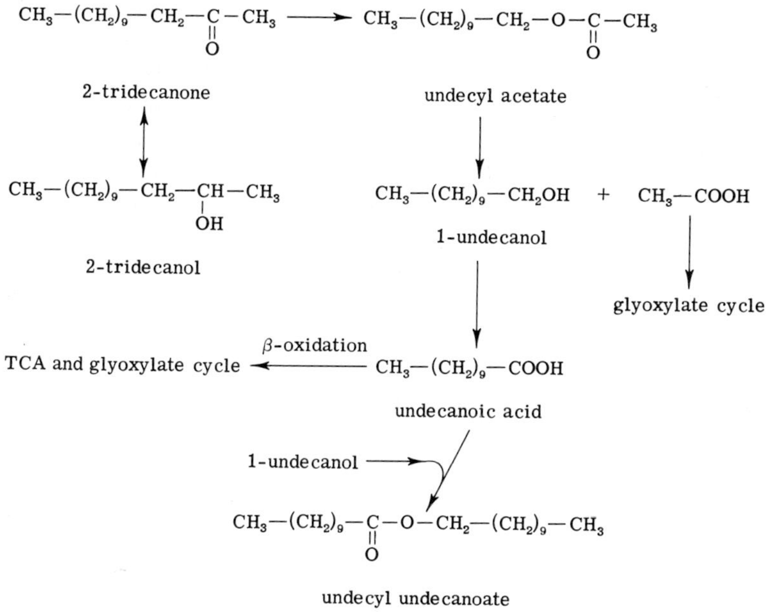
$CH_3-(CH_2)_9-CH_2-C(=O)-CH_3$
$CH_3-(CH_2)_9-CH_2-O-C(=O)-CH_3$
2-tridecanone
undecyl acetate
$CH_3-(CH_2)_9-CH_2-CH(OH)-CH_3$
$CH_3-(CH_2)_9-CH_2OH$ + CH_3-COOH
2-tridecanol
1-undecanol
glyoxylate cycle
β-oxidation
TCA and glyoxylate cycle
$CH_3-(CH_2)_9-COOH$
undecanoic acid
1-undecanol
$CH_3-(CH_2)_9-C(=O)-O-CH_2-(CH_2)_9-CH_3$
undecyl undecanoate

Fig. 8.2. Methylketone oxidation by *Pseudomonas multivorans* and *Pseudomonas aeruginosa* (reprinted with permission of the author and Academic Press, Inc.).

are regulated by the substrate. As long as the substrate is still available, omega oxidation is suppressed. Although β-oxidation of 1,6-hexanediol is operative, it appears to be a slow metabolizing pathway that leads to the accumulation of 6-hydroxyhexanoate. As soon as all the substrate has been oxidized to 6-hydroxyhexanoate, omega oxidation starts and becomes the preferentially used pathway. 6-Hydroxyhexanoate is metabolized through to succinate and possibly into the TCA or glyoxylate cycle. The β-oxidation of 6-hydroxyhexanoate leads to an intermediate, which is jointly shared with the 1,8-octanediol metabolism, and through to succinate. A constitutive $NADP^+$-linked α,ω-diol dehydrogenase is responsible for the metabolism of 1,6-hexanediol to 1,6-hydroxyhexanoate. The oxidation of 1,8-octanediol proceeds via a number of ω-hydroxy acids linked by β-oxidation.

Oxidation of Aromatic Hydrocarbons

The oxidation and degradation of aromatic hydrocarbons has received more and more attention during the past decades. The information on hydrocarbon utilization is increasing in general, but detailed mechanisms

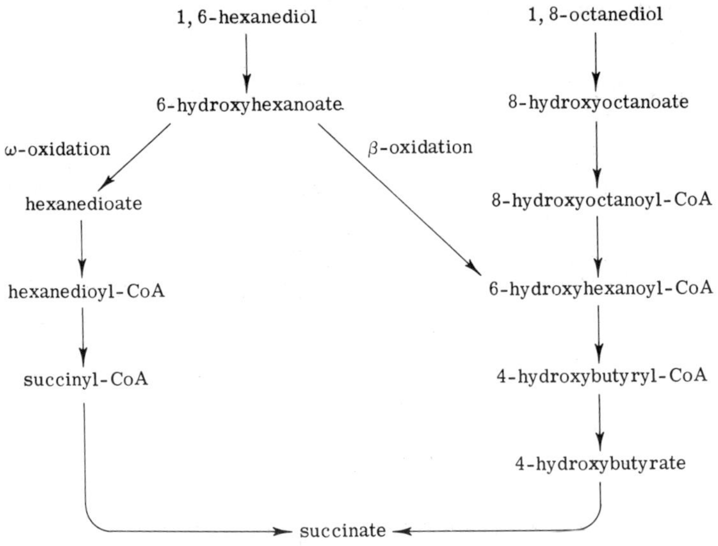
1, 6-hexanediol
6-hydroxyhexanoate
ω-oxidation
β-oxidation
hexanedioate
hexanedioyl-CoA
succinyl-CoA
1, 8-octanediol
8-hydroxyoctanoate
8-hydroxyoctanoyl-CoA
6-hydroxyhexanoyl-CoA
4-hydroxybutyryl-CoA
4-hydroxybutyrate
succinate

Scheme 8.4. Metabolism of 1,6-hexanediol and 1,8-octanediol by *Pseudomonas aeruginosa* (reprinted with permission of the author and the publisher of *Antonie van Leeuwenhoek J. Microbiol. Serol.*).

have been elucidated on a relatively small number of bacterial species of the genus *Pseudomonas* only and a still smaller number of other genera.

The great majority of hydrocarbons are utilized oxidatively and converge onto three major intermediates, which represent the last aromatic hydrocarbon before ring fission occurs. The compounds are catechol, protocatechuate, and gentisate

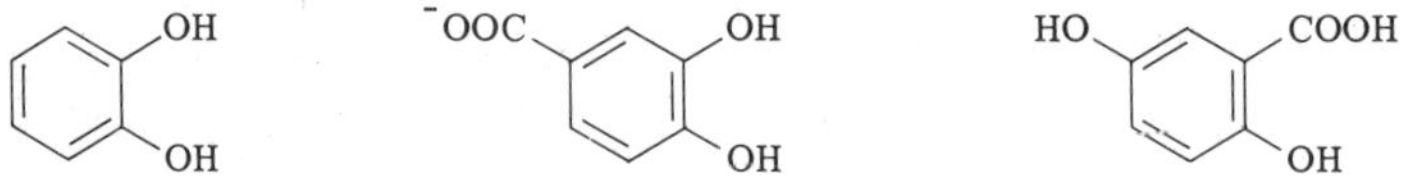
OH
OH
⁻OOC
OH
OH
HO
COOH
OH

catechol protocatechuate gentisate

According to their further utilization, catechol (Fig. 8.3) and protocatechuate follow the β-ketoadipate pathway (216a), whereas gentisate has its

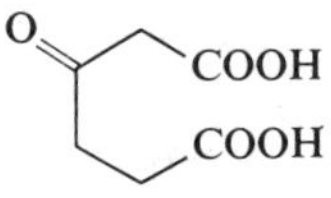
O
COOH
COOH

β-ketoadipate

own name given to the pathway. Whatever pathway is used, the aromatic hydrocarbons are utilized to form members of the tricarboxylic acid cycle.

ortho Ring Fission of the Benzene Ring

D-MANDELATE PATHWAY. One of the best studied pathways of hydrocarbon degradation is undoubtedly the mandelate pathway (108–110, 221) (Scheme 8.5). This pathway can be separated into two groups: (*a*) the mandelate group, which contains all the compounds from D-mandelate including benzoate, and (*b*) the catechol group, which contains the compounds from catechol to 3-oxoadipate. Both groups are joined by an oxygenase system.

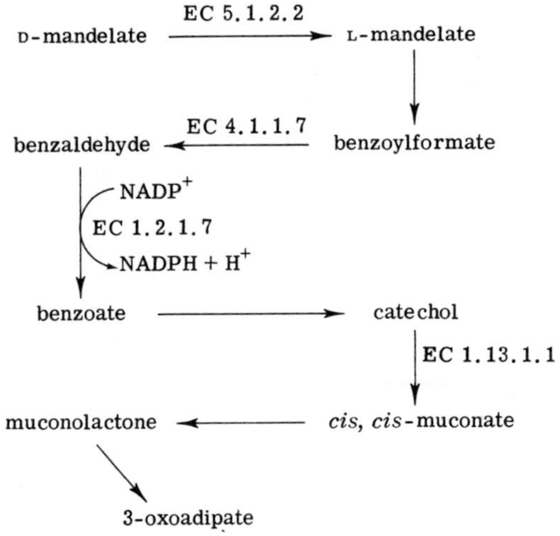
EC 5.1.2.2
D-mandelate
L-mandelate
EC 4.1.1.7
benzaldehyde
benzoylformate
NADP+
EC 1.2.1.7
NADPH + H+
benzoate
catechol
EC 1.13.1.1
muconolactone
cis, cis-muconate
3-oxoadipate

Scheme 8.5. General map of D-mandelate utilization to 3-oxoadipate.

D-Mandelate has first to be isomerized to its L isomer (254), which is carried out by a mandelate racemase (EC 5.1.2.2). This L isomer of man-

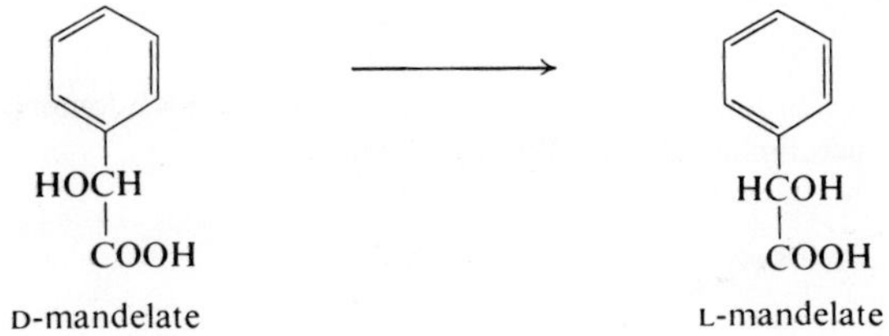
HOCH
COOH
D-mandelate
HCOH
COOH
L-mandelate

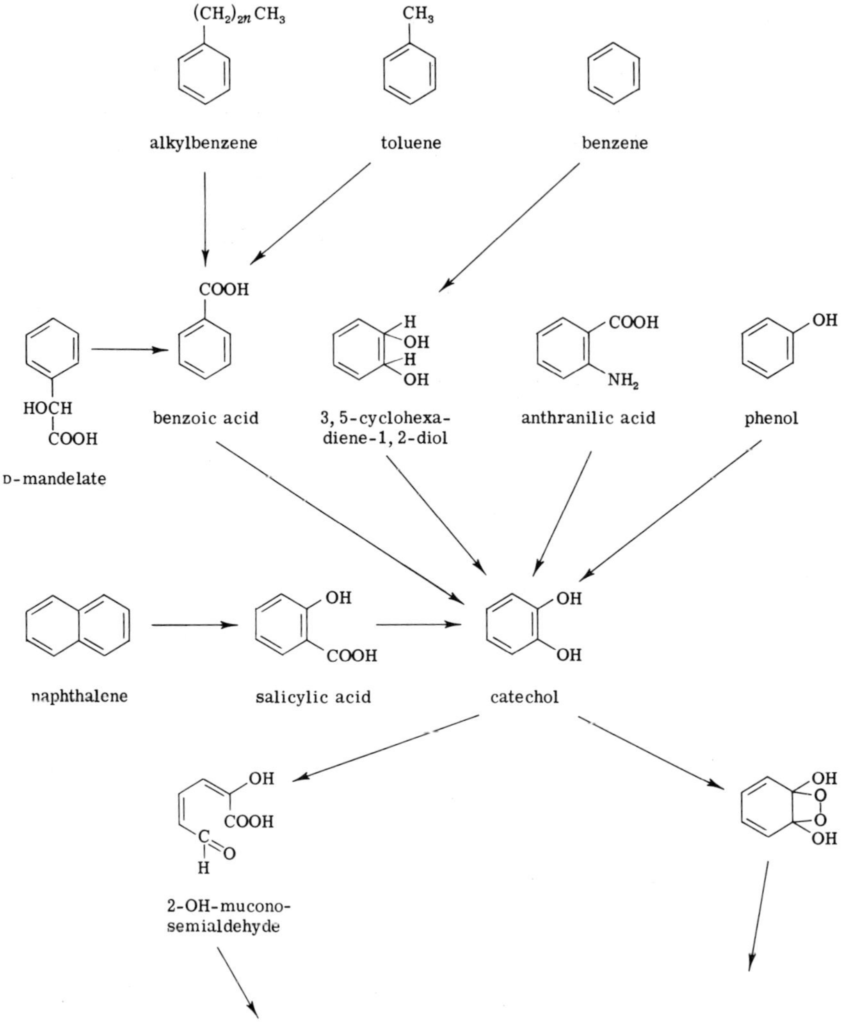

(CH2)2n CH3
CH3
alkylbenzene
toluene
benzene
COOH
HOCH
COOH
D-mandelate
benzoic acid
H
OH
H
OH
3, 5-cyclohexa-
diene-1, 2-diol
COOH
NH2
anthranilic acid
OH
phenol
naphthalene
OH
COOH
salicylic acid
OH
OH
catechol
OH
COOH
C
O
H
2-OH-mucono-
semialdehyde
OH
O
O
OH

Fig. 8.3. Degradation of aromatic hydrocarbon by pseudomonads via catechol (reprinted with permission of John Wiley and Sons).

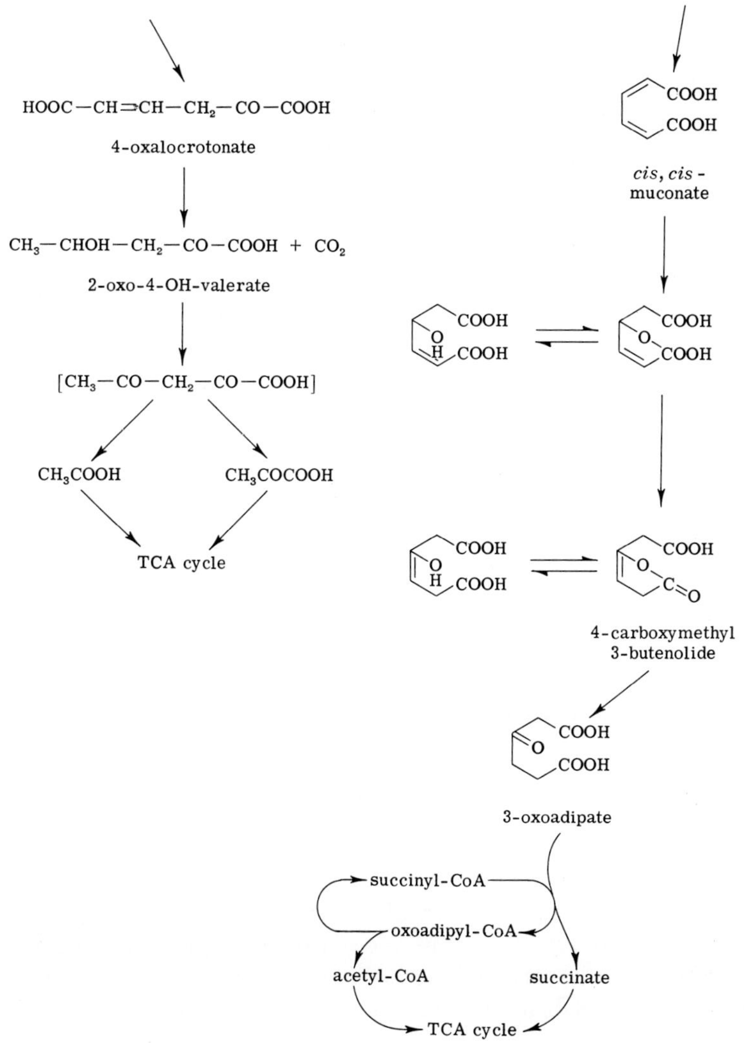
HOOC—CH=CH—CH2—CO—COOH
4-oxalocrotonate
CH3—CHOH—CH2—CO—COOH + CO2
2-oxo-4-OH-valerate
[CH3—CO—CH2—CO—COOH]
CH3COOH
CH3COCOOH
TCA cycle
COOH
COOH
cis, cis -
muconate
COOH
O
H
COOH
COOH
O
COOH
COOH
O
H
COOH
COOH
O
C=O
4-carboxymethyl
3-butenolide
COOH
O
COOH
3-oxoadipate
succinyl-CoA
oxoadipyl-CoA
acetyl-CoA
succinate
TCA cycle

Fig. 8.3. (Continued).

delate can now be oxidized by L-mandelate dehydrogenase (EC 1.1 group),

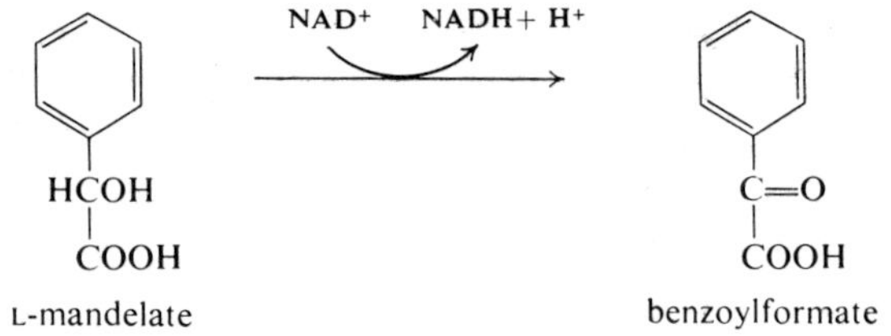
NAD+
NADH + H+
HCOH
COOH
L-mandelate
C=O
COOH
benzoylformate

which produces benzoylformate. A decarboxylation step liberates carbon

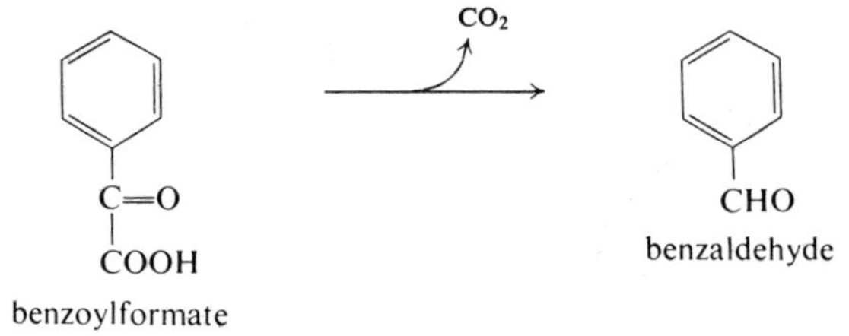
CO2
C=O
COOH
benzoylformate
CHO
benzaldehyde

dioxide and forms benzaldehyde. The enzyme responsible for this reaction is benzoylformate carboxy-lyase (EC 4.1.1.7). The following oxidation is a NAD^+-linked reaction,

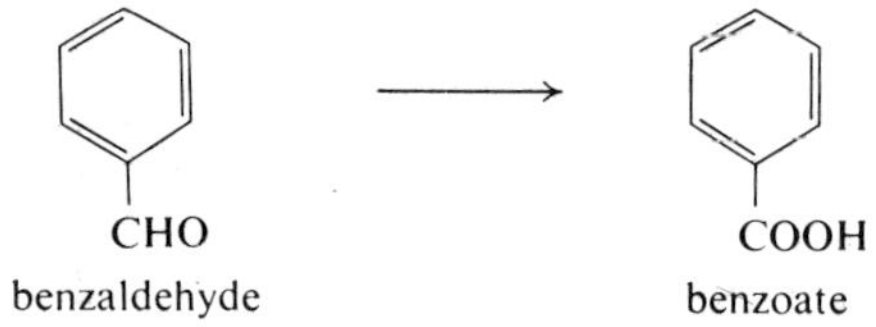
CHO
benzaldehyde
COOH
benzoate

catalyzed by benzaldehyde:$NADP^+$ oxidoreductase (EC 1.2.1.7), which forms benzoate, the last member of the mandelate group. Benzoate can also be formed if benzyl alcohol serves as carbon source (135). In general, benzoate is a very important intermediate, as it is also formed during the metabolism of alkylbenzene and toluene. A benzoate 1,2-dioxygenase

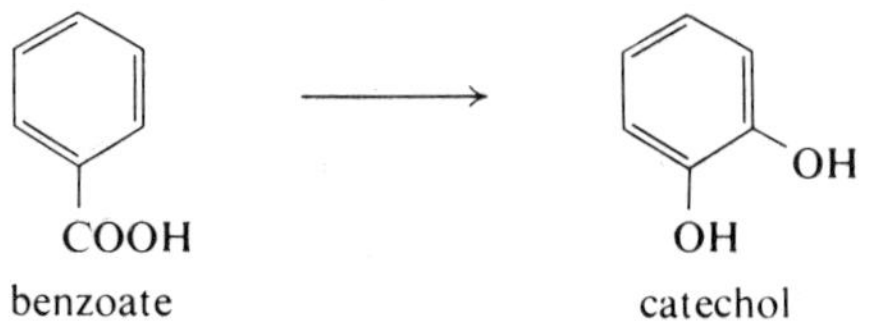
COOH
benzoate
OH
OH
catechol

(benzoate:oxygen oxidoreductase, EC 1.13.99.2) transforms benzoate into catechol, which is the last ring compound in this pathway. An ortho fission now occurs caused by the action of catechol 1,2-dioxygenase [catechol:oxy-

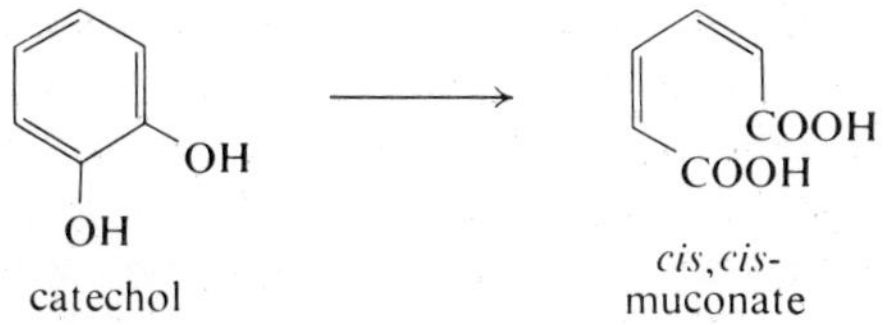
OH
OH
catechol
COOH
COOH
cis,cis-
muconate

gen 1,2-oxidoreductase, (decyclizing) EC 1.13.11.1], which splits the benzene ring of catechol between the two hydroxyl groups and forms *cis,cis*-muconate. The next enzymes involved in this degradation are a lactonizing

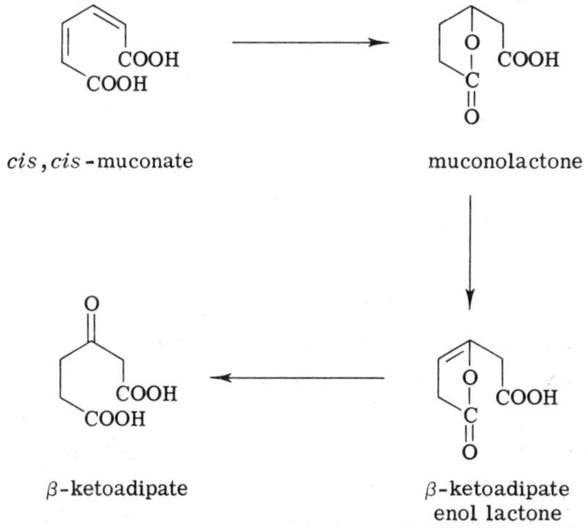
COOH
COOH
cis,cis-muconate
O
C
O
COOH
muconolactone
O
COOH
COOH
β-ketoadipate
O
C
O
COOH
β-ketoadipate
enol lactone

enzyme (167a), muconolactone isomerase (EC 5.3.3.4), and finally an 3-oxoadipate enol-lactonase (EC 3.1.1.24). This enzyme is an important inducible enzyme (137) as will be seen below (see p. 521). The formation of acetyl-CoA and succinyl-CoA from β-ketoadipate is carried out in two

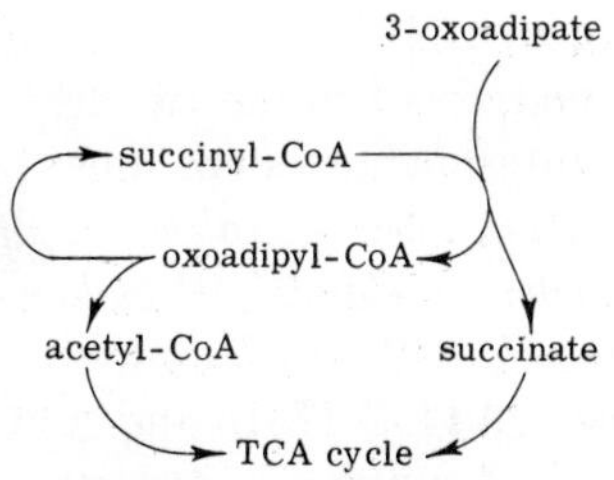
3-oxoadipate
succinyl-CoA
oxoadipyl-CoA
acetyl-CoA
succinate
TCA cycle

steps with the involvement of 3-oxoadipate-CoA transferase (succinyl-CoA:3-oxoadipate-CoA transferase, EC 2.8.3.6), which forms oxoadipyl-CoA, and β-ketoadipyl-CoA thiolase, which forms acetyl-CoA. This pathway appears to be functional also in a number of *Arthrobacter* species (220).

The pathways for 2-hydroxy, 4-hydroxy, 3,4-dihydroxy, and 4-hydroxy-3-methoxy derivatives of mandelate were investigated using the method of simultaneous adaptation (216). Whereas the 2-hydroxy-substituted compounds were oxidized to catechol by a parallel pathway, the 4-hydroxy, 3,4-dihydroxy, and 4-hydroxy-3-methoxy derivatives were all converted to 3,4-dihydroxybenzoate. This last product underwent ring cleavage, with the ultimate formation of β-ketoadipate. The detailed mechanism of the latter cleavage has not been elucidated.

METABOLISM OF PHENOL. Phenol can be metabolized by *Bacterium* NCIB 8250 (14) and *Pseudomonas aeruginosa* T 1 (58, 176). A phenol 2-mono-

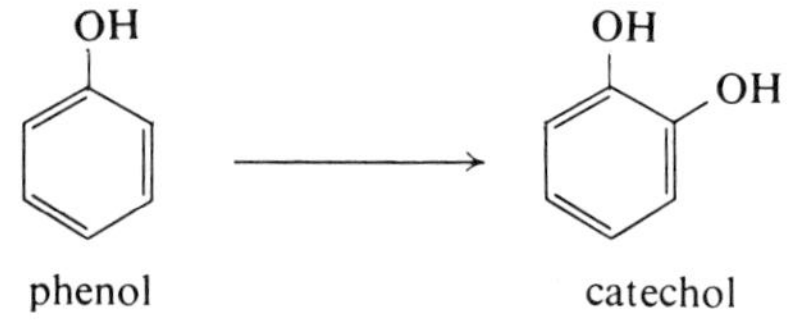
OH
OH
OH
phenol
catechol

oxygenase [phenol, NADPH:oxygen oxidoreductase(2-hydroxylating), EC 1.14.13.7] acts on phenol and converts this compound to catechol. The degradation of catechol follows the ortho-fission cleavage with the formation of β-ketoadipate and, finally, acetyl-CoA and succinate.

Of the monosubstituted phenols, only 3- and 4-fluorophenol, as well as 4-chlorophenol, were able to follow the same pathway (14) and to induce the corresponding enzymes. Monohydric phenols, in contrast, follow a meta-cleavage pathway (194) (see p. 512) and inhibit the utilization of phenol to catechol.

METABOLISM OF BENZOIC ACID AND BENZENE. During the degradation of mandelate to catechol, it was shown that the intermediate benzoic acid is directly converted to catechol by benzoate 1,2-dioxygenase (EC 1.13.99.2). This enzyme requires reduced NAD for its catalysis. As the COOH-group of benzoic acid must be converted to the two hydroxyl groups of catechol, it is proposed that some intermediate steps must be involved. Studies with cell-free extracts from *Acinetobacter calcoaceticus, Alcaligenes eutrophus* (ATCC 17697), *Azotobacter vinelandii, Pseudomonas aeruginosa* (ATCC 17504), *Pseudomonas cepacia* 249 and *Pseudomonas cepacia* 382 (formerly *Pseudomonas multivorans*) (ATCC 17616 and ATCC 17759), *Pseudomonas putida* ATCC 12633, and *Pseudomonas testosteroni* ATCC 15668 conclusively showed that [^{14}C]benzoic acid is converted to 3,5-cyclohexadiene-1,2-diol-1-carboxylic acid (1,2-dihydro-1,2-dihydroxybenzoic acid) first, before being decarboxylated to catechol (192). This intermediate makes this mechanism analogous to the benzene oxidation (84, 85) that forms the

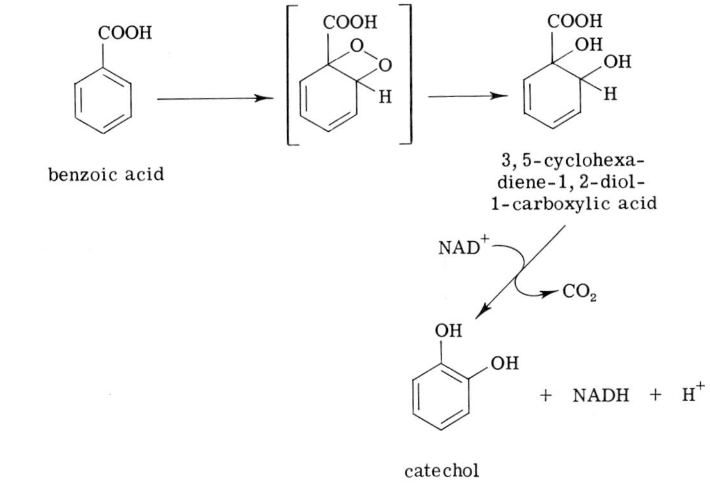
COOH
benzoic acid
COOH
O
O
H
COOH
OH
OH
H
3, 5-cyclohexa-diene-1, 2-diol-1-carboxylic acid
NAD+
CO2
OH
OH
+ NADH + H+
catechol

intermediate 3,5-cyclohexadiene-1,2-diol. Therefore the dioxygenase involved in the benzoic acid oxidation introduces the molecular oxygen first

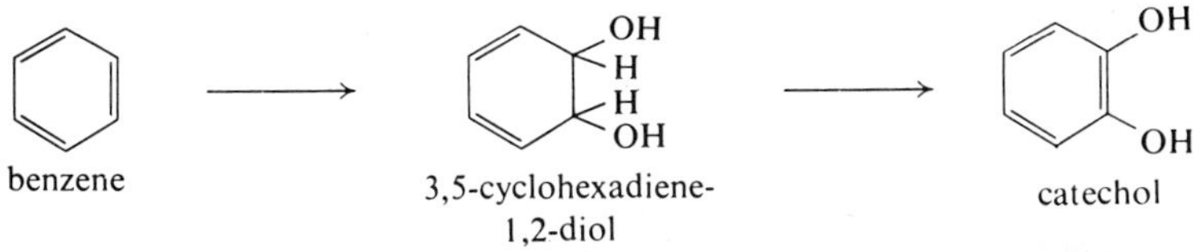
benzene
OH
H
H
OH
3,5-cyclohexadiene-1,2-diol
OH
OH
catechol

followed by a dehydrogenation and decarboxylation to give catechol. There are reports available (86, 129) that suggest a dioxetane in the form of benzene-1,2-oxide is the first intermediate in benzene oxidation. The mechanism of this reaction, however, is still unclear.

NAPHTHALENE METABOLISM. The oxidation of naphthalene occurs in a number of soil isolates (224, 235, 250) and obeys the criteria of sequential induction (see p. 521) (237). In the five *Pseudomonas* strains (236) investigated, as well as in *Pseudomonas fluorescens* (1), *Pseudomonas denitrificans,* and *Achromobacter* sp. (166), salicylic acid and catechol appeared as intermediates. However, two ring fissions are necessary (51), leading to the sequence outlined in Fig. 8.4. Naphthalene is oxidatively metabolized to 1,2-dihydro-1,2-dihydroxynaphthalene (130), in which reaction the formation of a dioxetane as intermediate is still obscure (86). This naphthalene diol is then dehydrogenated to 1,2-dihydroxynaphthalene. At this stage, the first ring fission takes place between the two hydroxyl groups during the formation of *cis-o*-hydroxybenzalpyruvic acid. An aldolase-type enzyme subsequently converts *cis-o*-hydroxybenzalpyruvic acid to salicylaldehyde and pyruvate. A NAD-specific dehydrogenase oxidizes salicyl-

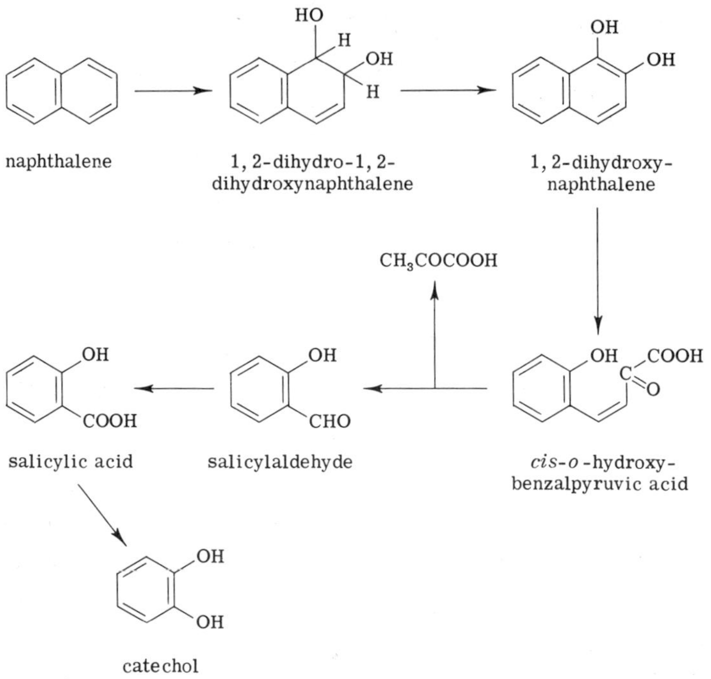
HO
H
OH
H
OH
OH
naphthalene
1, 2-dihydro-1, 2-dihydroxynaphthalene
1, 2-dihydroxy-naphthalene
CH3COCOOH
OH
COOH
OH
CHO
OH
COOH
C
O
salicylic acid
salicylaldehyde
cis-o-hydroxy-benzalpyruvic acid
OH
OH
catechol

Fig. 8.4. The oxidation of naphthalene by *Pseudomonas* strains (reprinted with permission of the author and Akademie-Verlag, Berlin).

aldehyde to salicylic acid, which itself is oxidatively decarboxylated to catechol. The enzyme salicylate 1-monooxygenase [salicylate, NADH: oxygen oxidoreductase (1-hydroxylating, 1-decarboxylating) EC 1.14.13.1] forms an enzyme–substrate complex during this reaction (226), converting salicylate to catechol with stoichiometric consumption of oxygen and reduced NAD. Catechol 1,2-dioxygenase (EC 1.13.11.1) in *Pseudomonas desmolytica* forms β-ketoadipate, proving the presence of the β-ketoadipate pathway.

PHENANTHRENE AND ANTHRACENE METABOLISM. The oxidation of phenanthrene and anthracene follows a pattern very similar to the oxidation of naphthalene (Fig. 8.5). *Flavobacterium* (65) metabolizes phenanthrene and anthracene. Both pathways go through very similar steps in their metabolism, breaking one ring after the other. After the formation of 1-hydroxy-2-naphthoic acid an oxidative decarboxylation could be seen as taking place, which would lead to 1,2-dihydroxynaphthalene. From this point the pathway is identical to the naphthalene pathway. The formation

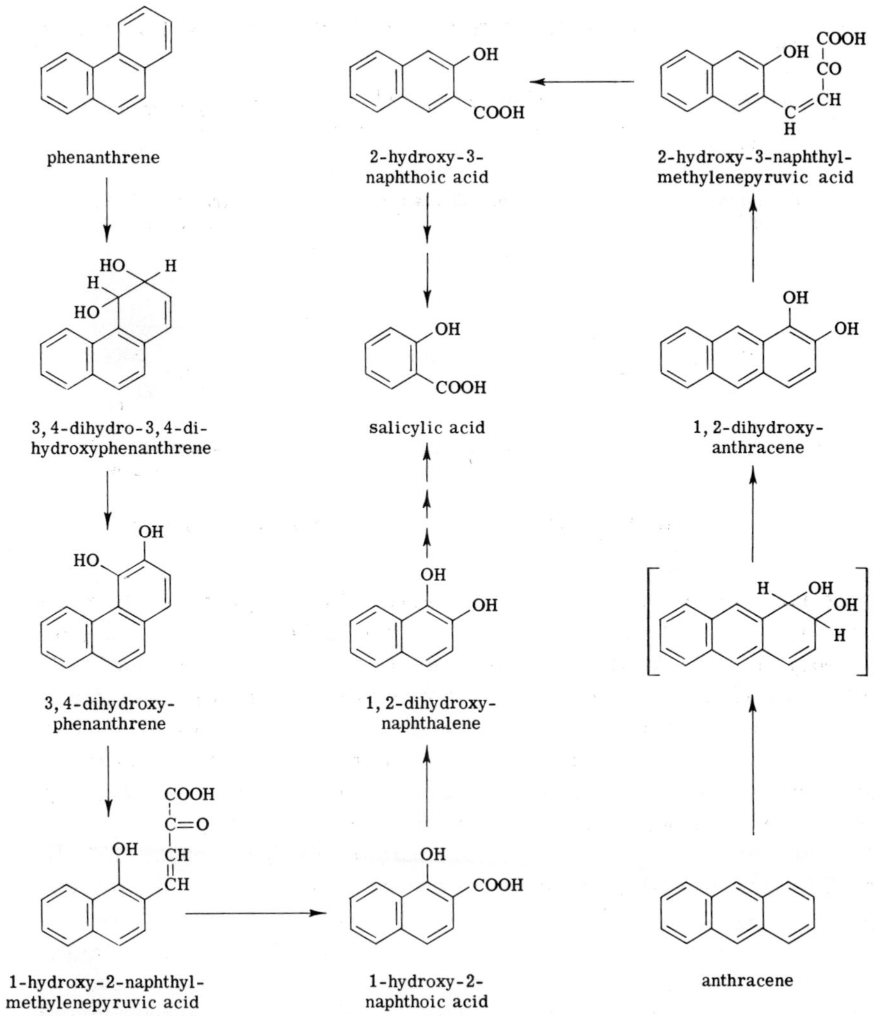
phenanthrene
2-hydroxy-3-
naphthoic acid
2-hydroxy-3-naphthyl-
methylenepyruvic acid
3,4-dihydro-3,4-di-
hydroxyphenanthrene
salicylic acid
1,2-dihydroxy-
anthracene
3,4-dihydroxy-
phenanthrene
1,2-dihydroxy-
naphthalene
1-hydroxy-2-naphthyl-
methylenepyruvic acid
1-hydroxy-2-
naphthoic acid
anthracene

Fig. 8.5. Metabolism of phenanthrene and anthracene by *Flavobacterium* sp. (reprinted with permission of the author and Akademie-Verlag, Berlin).

of 2-hydroxy-3-naphthoic acid makes further suggestion difficult, although both pathways will definitely go through salicylic acid (199, 200).

p-HYDROXYBENZOATE METABOLISM. A different type of hydrocarbon metabolism which also finally uses the β-ketoadipate pathway, is that using protocatechuate instead of catechol as key intermediate. The best studied metabolism in a number of bacteria (47, 94) is that of *p*-hydroxybenzoate (Fig. 8.6). The substrate *p*-hydroxybenzoate is initially attacked by the hydroxylation (170) in position 3 to yield protocatechuate (3,4-dihydroxy-

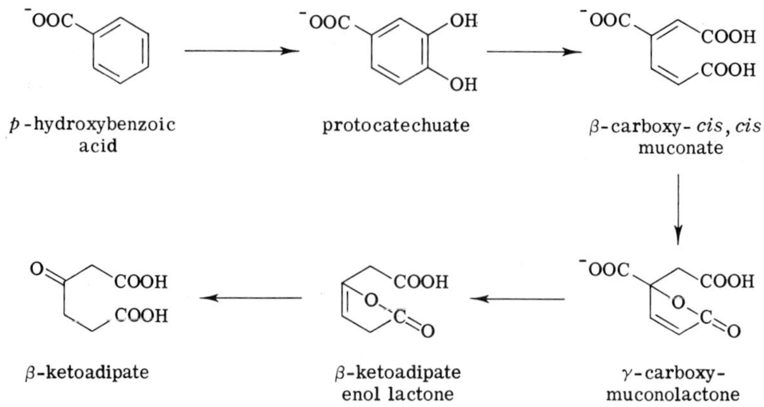

$^{-}$OOC
p-hydroxybenzoic acid
$^{-}$OOC
OH
OH
protocatechuate
$^{-}$OOC
COOH
COOH
β-carboxy-cis, cis-muconate
$^{-}$OOC
COOH
O
C
O
γ-carboxy-muconolactone
COOH
O
C
O
β-ketoadipate enol lactone
O
COOH
COOH
β-ketoadipate

Fig. 8.6. Metabolism of *p*-hydroxybenzoate in *Pseudomonas putida.*

benzoate), which is catalyzed by the enzyme 4-hydroxybenzoate 3-monooxygenase (EC 1.14.13.2) (123). This enzyme is FAD dependent and forms an enzyme–FAD–substrate complex for its oxygen incorporation (112, 230). In *Klebsiella aerogenes*, *p*-hydroxybenzoate is metabolized to catechol (180) and not to protocatechuate. This organism is also able to form phenol from *p*-hydroxybenzoate but cannot utilize this compound further; this may reflect an imperfection of metabolism. The conversion of *p*-hydroxybenzoate to catechol is a nonoxidative decarboxylation (91).

Protocatechuate is now oxidized to 3-carboxy-*cis,cis*-muconate (159) by protocatechuate 3,4-dioxygenase (protocatechuate:oxygen 3,4-oxidoreductase, EC 1.13.11.3). The product of this oxidation undergoes a lactonization whereby 4-carboxymuconolactone is formed. This reaction requires the γ-carboxy-*cis,cis*-muconate lactonizing enzyme (carboxy-*cis,cis*-muconate cycloisomerase, EC 5.5.1.2) (180a). A decarboxylation step converts γ-carboxymuconolactone into β-ketoadipate enol lactone with the help of γ-carboxymuconolactone decarboxylase (EC 4.1.1.44) (179a). This compound is identical to the one described in the metabolism of hydrocarbons via catechol to β-ketoadipate. Catechol and protocatechuate therefore serve the same function, as they are the last intermediate possessing an aromatic structure toward β-ketoadipate formation (211). β-Ketoadipate enol lactone (176) is identical with 4-carboxymethyl-3-butenolide (244).

It therefore seems that two completely separate pathways are involved in the metabolism of benzenoid substrates—the catechol and the protocatechuate pathways. *Pseudomonas putida* was found to possess the enzymes of both but exhibited preference for only one of them (177). Apart from pseudomonads, *Acinetobacter* (formerly *Moraxella*) *calcoaceticus* uses the same pathways in benzoate and *p*-hydroxybenzoate degradation, but the regulation of the enzymes appears to be distinctively different from that in pseudomonads (31; see p. 521).

2,4-XYLENOL METABOLISM. The oxidation of 2,4-xylenol was investigated in a fluorescent pseudomonad (41). The metabolism appears to be initiated by the conversion of the methyl group para to the hydroxyl group to a carboxyl group (Fig. 8.7), the intermediate being 4-hydroxy-3-methylbenzoic acid. It is assumed that this conversion occurs via the formation of an alcohol, and aldehyde group, which intermediates have not been identified as yet. The second part of the metabolism is thought to occur on the conversion of the second methyl group in a similar manner, which does lead to protocatechuate via 4-hydroxyisophthalic acid. Protocatechuate then undergoes an ortho fission followed by the above-described *β*-ketoadipate pathway.

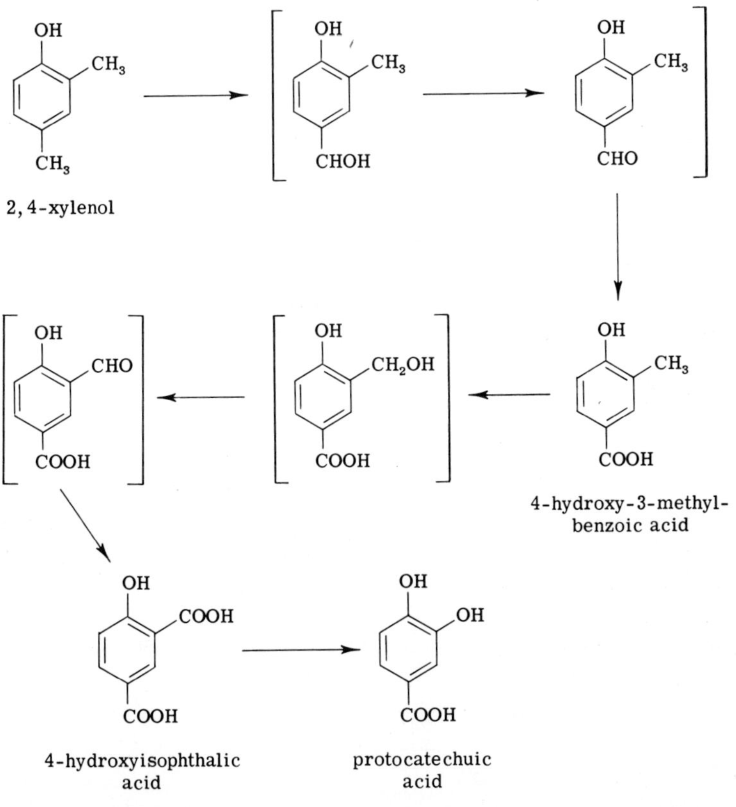
OH
CH_3
CH_3
2,4-xylenol
OH
CH_3
CHOH
OH
CH_3
CHO
OH
CH_3
COOH
4-hydroxy-3-methyl-benzoic acid
OH
CH_2OH
COOH
OH
CHO
COOH
OH
COOH
COOH
4-hydroxyisophthalic acid
OH
OH
COOH
protocatechuic acid

Fig. 8.7. Metabolism of 2,4-xylenol via protocatechuate by a species of *Pseudomonas* (reprinted with permission of the author and the Biochemical Society, London).

meta Fission of the Benzene Ring

The metabolism of the various aromatic compounds indicated that the β-ketoadipate pathway is the result of an ortho cleavage of catechol or protocatechuate, with the end products being acetyl-CoA and succinate. Both compounds can also undergo meta cleavage (12a). This meta cleavage is predominant in such substituted catechols as 3-methylcatechol or 4-methylcatechol. The end products of these pathways are pyruvate plus an aldehyde (205, 258) (Scheme 8.6). Catechol is cleaved to 2-hydroxy-

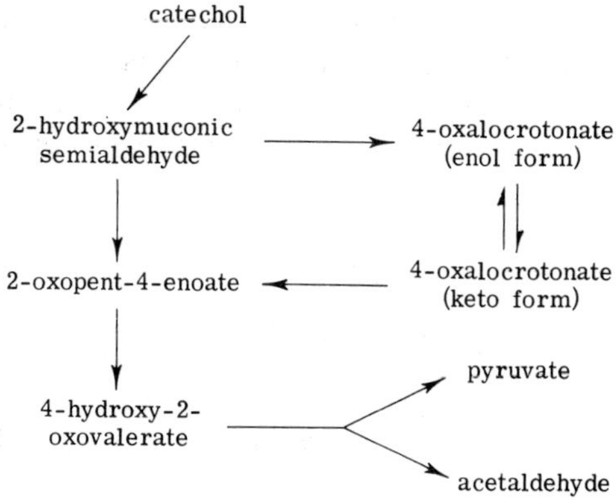
catechol
2-hydroxymuconic semialdehyde
4-oxalocrotonate (enol form)
2-oxopent-4-enoate
4-oxalocrotonate (keto form)
4-hydroxy-2-oxovalerate
pyruvate
acetaldehyde

Scheme 8.6. General scheme for catechol oxidation via meta fission.

muconic semialdehyde by a meta cleavage catalyzed by catechol 2,3-oxygenase (catechol:oxygen 2,3-oxidoreductase, EC 1.13.11.2) (45, 46,

OH
OH
H
CO
COOH
OH

catechol 2-hydroxymuconic semialdehyde

144, 173). Ring fission occurs not between but outside the hydroxyl groups. A hydrolase (2-hydroxymuconic semialdehyde hydrolyase) (171a) accepts

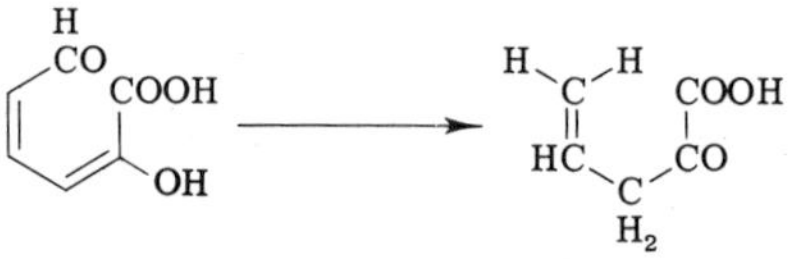
H
CO
COOH
OH
H
C
H
COOH
HC
CO
C
H2

2-hydroxymuconic semialdehyde 2-oxopent-4-enoate

a molecule of water and releases formate, with the formation of 2-oxopent-4-enoate. The latter intermediate is further hydrolyzed (2-oxopent-4-enoate hydratase) to 4-hydroxy-2-oxovalerate, which itself is cleaved by an

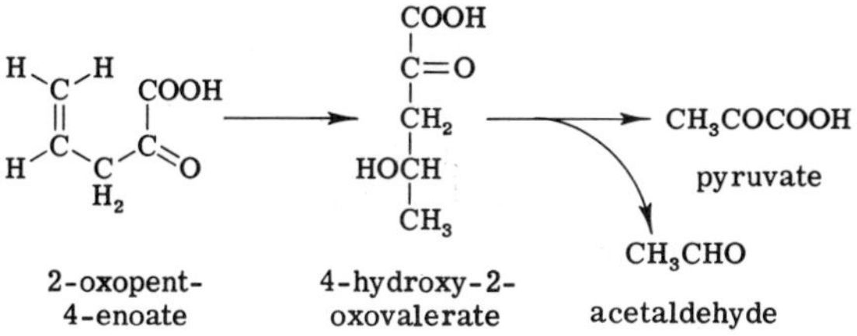
2-oxopent-4-enoate
4-hydroxy-2-oxovalerate
$CH_3COCOOH$
pyruvate
CH_3CHO
acetaldehyde

aldolase (4-hydroxy-2-oxovalerate aldolase) to pyruvate and acetaldehyde. A NAD-linked 2-hydroxymuconic semialdehyde dehydrogenase is responsible for the formation of 4-oxalocrotonate. 4-Oxalocrotonate tautomerase converts the enol form into the keto form and makes this compound liable to attack by 4-oxalocrotonate decarboxylase, whereby 2-oxopent-4-enoate is formed. This second pathway does not produce formate. Some microorganisms, such as *Pseudomonas putida* NCIB 10105 and naphthalene-grown pseudomonads, possess both (12a, 38), whereas others, such as *Pseudomonas acidovorans*, possess only the ortho-cleavage pathway (197). In *Azotobacter* and the naphthalene-grown pseudomonads (38, 204, 204a) it was shown that the oxidative pathway via 4-oxalocrotonate is the more important because of the extremely low levels of hydrolase. *Pseudomonas putida* (205a) and *Pseudomonas arvilla* (169a) are able to use both the hydrolytic and the oxidative pathway. It has been suggested (169a) that the 4-oxalocrotonate pathway may be preferred in the metabolism of catechol, 4-methylcatechol, and their precursors phenol and *p*-cresol, whereas the role of hydrolase may be limited to the metabolism of 3-methylcatechol and its precursors *o*- and *m*-cresol (12, 12a, 205, 205a).

There is also a report (238) suggesting that 3- and 4-methylcatechol are oxidized predominantly by meta cleavage in *Pseudomonas desmolyticum*, whereas catechol and protocatechuate mainly follow the ortho cleavage.

The oxidation of catechol, 3-methylcatechol, and 4-methylcatechol would therefore be as illustrated in Fig. 8.8 (37, 38, 83, 94, 204).

Very little is known about the meta-cleavage reactions of protocatechuate. *Pseudomonas desmolytica* (45, 239) and *Pseudomonas testosteroni* (49) have a protocatechuate 4,5-oxygenase (protocatechuate:oxygen 4,5-oxidoreductase, EC 1.13.11.8) (175) that forms α-carboxy-*cis,cis*-muconic semialdehyde and could follow a path similar to the one described for catechol. It appears that the side-chain substituents and the position of these in the benzene ring are important in relation to the cleavage of the benzene ring

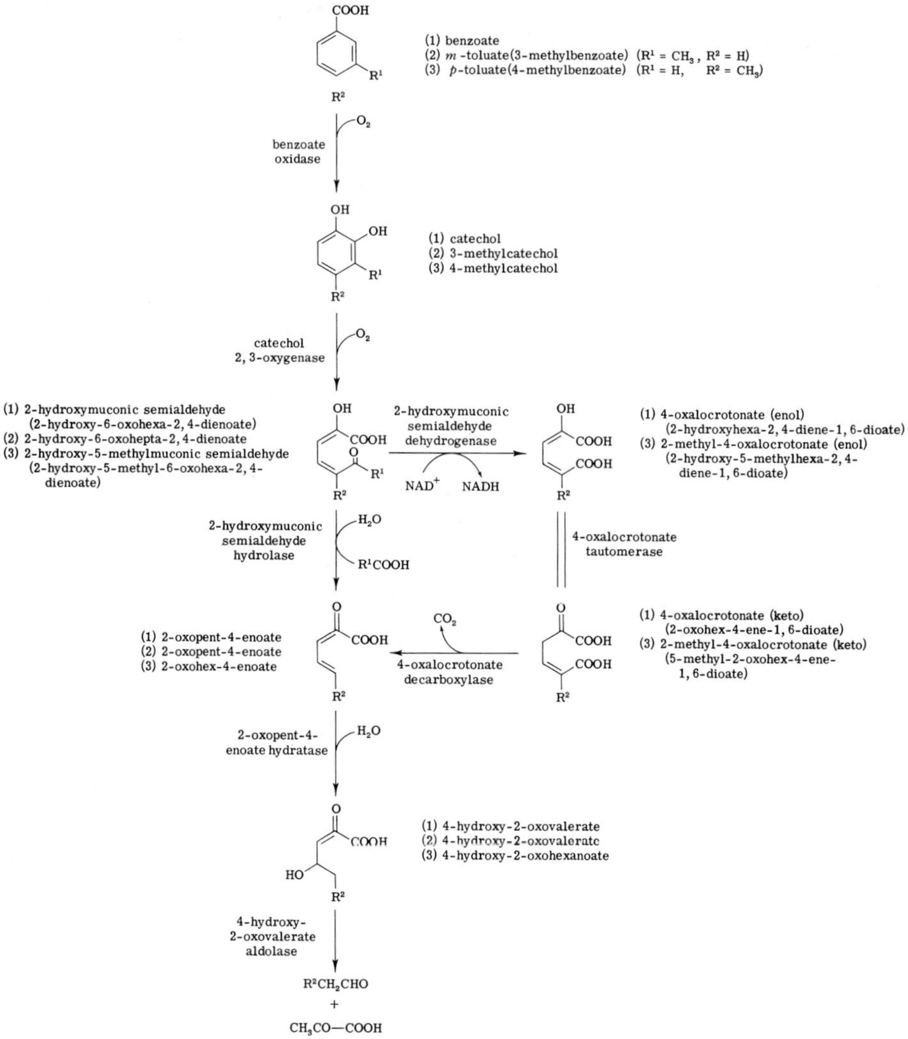
COOH
R¹
R²
(1) benzoate
(2) m-toluate(3-methylbenzoate) (R¹ = CH₃, R² = H)
(3) p-toluate(4-methylbenzoate) (R¹ = H, R² = CH₃)
O₂
benzoate oxidase
OH
OH
R¹
R²
(1) catechol
(2) 3-methylcatechol
(3) 4-methylcatechol
O₂
catechol 2, 3-oxygenase
(1) 2-hydroxymuconic semialdehyde (2-hydroxy-6-oxohexa-2, 4-dienoate)
(2) 2-hydroxy-6-oxohepta-2, 4-dienoate
(3) 2-hydroxy-5-methylmuconic semialdehyde (2-hydroxy-5-methyl-6-oxohexa-2, 4-dienoate)
OH
COOH
O
R¹
R²
2-hydroxymuconic semialdehyde dehydrogenase
NAD⁺
NADH
OH
COOH
COOH
R²
(1) 4-oxalocrotonate (enol) (2-hydroxyhexa-2, 4-diene-1, 6-dioate)
(3) 2-methyl-4-oxalocrotonate (enol) (2-hydroxy-5-methylhexa-2, 4-diene-1, 6-dioate)
2-hydroxymuconic semialdehyde hydrolase
H₂O
R¹COOH
4-oxalocrotonate tautomerase
(1) 2-oxopent-4-enoate
(2) 2-oxopent-4-enoate
(3) 2-oxohex-4-enoate
O
COOH
R²
CO₂
4-oxalocrotonate decarboxylase
O
COOH
COOH
R²
(1) 4-oxalocrotonate (keto) (2-oxohex-4-ene-1, 6-dioate)
(3) 2-methyl-4-oxalocrotonate (keto) (5-methyl-2-oxohex-4-ene-1, 6-dioate)
2-oxopent-4-enoate hydratase
H₂O
O
COOH
HO
R²
(1) 4-hydroxy-2-oxovalerate
(2) 4-hydroxy-2-oxovalerate
(3) 4-hydroxy-2-oxohexanoate
4-hydroxy-2-oxovalerate aldolase
R²CH₂CHO
+
CH₃CO—COOH

Fig. 8.8. Proposed pathway for metabolism of benzoate and *m*- and *p*-toluates by *P. arvilla*. Metabolites of benzoate are denoted (1), those of *m*-toluate (2), and those of *p*-toluate (3) (reprinted with permission of the authors and the *European J. of Biochemistry*).

(208). It also appears that this influence could vary from organism to organism.

It therefore emerges that all aromatic hydrocarbons metabolized via the β-ketoadipate pathway and via catechol must be unsubstituted, whereas substituted aromatic hydrocarbons, such as 4-methylnaphthalene, have the corresponding substituted catechol followed by meta cleavage.

GENTISATE METABOLISM. In addition to catechol and protocatechuate, gentisate is a third major intermediate in hydrocarbon metabolism. It is able to form the last aromatic ring compound intact before ring fission occurs. This pathway was found to occur whenever *Pseudomonas acidovorans* was grown on *m*-hydroxybenzoate. The latter is oxidized by a *m*-hydroxybenzoate 6-hydroxylase, which converts *m*-hydroxybenzoate to gentisate. This hydroxylase is specific for the 6 position (255); the hydroxylase for the 4 position would form protocatechuate, as occurs in *Pseudomonas testosteroni*. Gentisate is then oxidized by gentisate 1,2-dioxygenase (EC 1.13.11.4) to maleylpyruvate. The following isomerization to fumarylpyruvate required high concentrations of reduced glutathione (GSH). Because of this requirement, it is doubtful whether *Pseudomonas acidovorans* isomerizes maleylpyruvate. This organism could hydrolyze maleylpyruvate to maleate and pyruvate with subsequent hydration of maleate to malate, as was shown to occur in the case of *Pseudomonas* NCIB 9876 (116). There exist, however, *Pseudomonas* spp., that do isomerize maleylpyruvate to fumarylpyruvate, with subsequent hydrolysis to fumaric acid and pyruvic acid requiring only small amounts of reduced glutathione

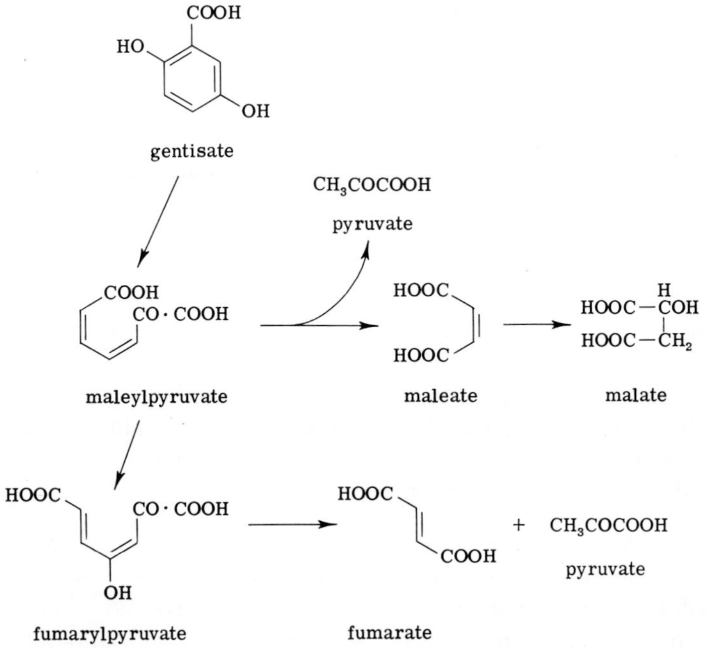
COOH
HO
OH
gentisate
$CH_3COCOOH$
pyruvate
COOH
CO · COOH
HOOC
HOOC
H
HOOC—COH
HOOC—CH_2
maleylpyruvate
maleate
malate
HOOC
CO · COOH
OH
HOOC
COOH
+ $CH_3COCOOH$
pyruvate
fumarylpyruvate
fumarate

Fig. 8.9. Metabolism of gentisic acid by species of *Pseudomonas* (reprinted with permission of the authors and Academic Press, Inc.).

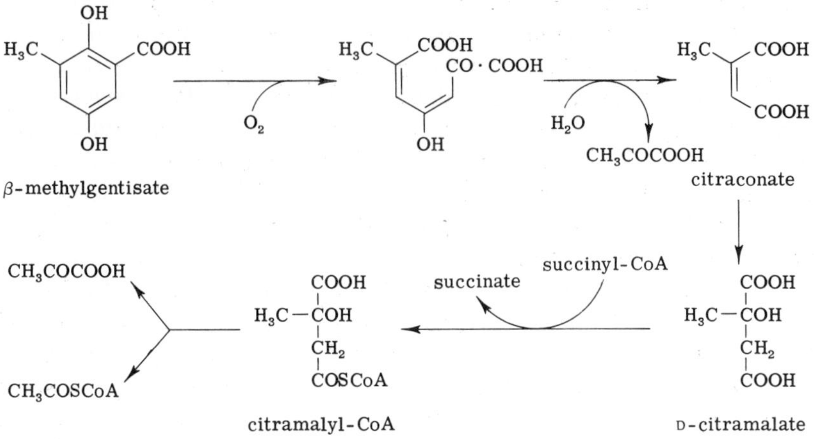
OH
H_3C
COOH
OH
β-methylgentisate
O_2
H_3C
COOH
CO · COOH
OH
H_2O
$CH_3COCOOH$
H_3C
COOH
COOH
citraconate
COOH
H_3C-COH
CH_2
COOH
D-citramalate
succinyl-CoA
succinate
COOH
H_3C-COH
CH_2
COSCoA
citramalyl-CoA
$CH_3COCOOH$
$CH_3COSCoA$

Fig. 8.10. Metabolism of 3-methylgentisate by a nonfluorescent pseudomonad (reprinted with permission of the authors and Academic Press, Inc.).

(148). In this case, both enzymes responsible for the isomerization and hydrolyzation have been separated (149, 191). *Pseudomonas testosteroni* uses the same mechanism and can therefore use the tricarboxylic acid cycle (Fig. 8.9).

A very similar mechanism exists for the metabolism of 3-methyl gentisate (Fig. 8.10), which is degraded via citraconate, followed by a hydration to D-citramalate to pyruvate and acetyl-CoA with the help of succinyl-CoA. Citramalyl-CoA is an intermediate and is cleaved by an aldolase-type reaction to form the end products acetyl-CoA and pyruvate (118).

m-Hydroxybenzoate can therefore be metabolized in two ways, depending on the type of hydroxylase present. This difference could possibly be used for taxonomic differentiation.

The gentisate pathway is used during the *m*-cresol oxidation (117) whereby *m*-cresol is converted to 3-hydroxybenzoic acid first before gentisic acid is formed. In a similar way, 2,5-xylenol is converted to 4-methylgentisate and 3,5-xylenol to 3-methylgentisate.

ANTHRANILIC ACID METABOLISM. Anthranilic acid, like *m*-hydroxybenzoic acid, can be metabolized via two different pathways (Fig. 8.11) (27). *Nocardia opaca* hydroxylates anthranilate in a reduced NAD-requiring reaction (125, 227) with catechol as its product. There appears to be no doubt that this route is the predominant one in the metabolism of anthranilic acid (26). *Nocardia,* however, has also an NADPH-linked anthran-

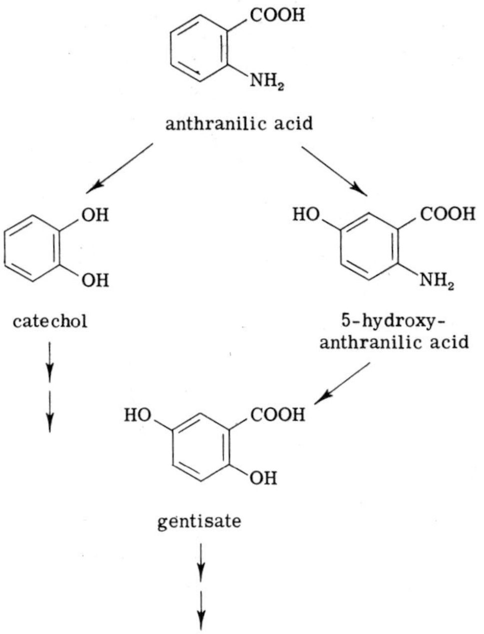
COOH
NH2
anthranilic acid
OH
OH
catechol
HO
COOH
NH2
5-hydroxy-
anthranilic acid
HO
COOH
OH
gentisate

Fig. 8.11. Metabolism of anthranilic acid by *Nocardia* (reprinted with permission of the author and the publisher of *Ant. v. Leeuwenh. J. Microbiol. and Serology*).

ilate-5-hydroxylase [anthranilate, reduced-NADP:oxygen oxidoreductase (5-hydroxylating)] (133) that converts anthranilic acid into 5-hydroxy-anthranilic acid and further to gentisate. This pathway is an inducible pathway and exhibits the ability of *Nocardia* to carry out a monohydroxylation of the anthranilate molecule.

THYMOL METABOLISM. *Pseudomonas putida* metabolizes thymol to isobutyrate, acetate, and 3-oxobutyrate via a meta-cleavage pathway (Fig. 8.12) (40). Thymol is hydroxylated twice to give 3-hydroxythymol-1,4-quinol. Soil pseudomonads can convert the quinol easily into the quinone form, which is recognizable as a deep purple color. The benzene ring of 3-hydroxythymol-1,4-quinol is cleaved by a dioxygenase and the final end products are isobutyrate, acetate, and 2-oxobutyrate.

p-XYLENE METABOLISM. The metabolism of *p*-xylene has many similarities to the oxidation of 2,4-xylenol (see p. 511). Species of *Pseudomonas* oxidize one of the two methyl groups to *p*-toluic acid (52). The latter intermediate is then decarboxylated and hydroxylated with the formation of 4-methyl

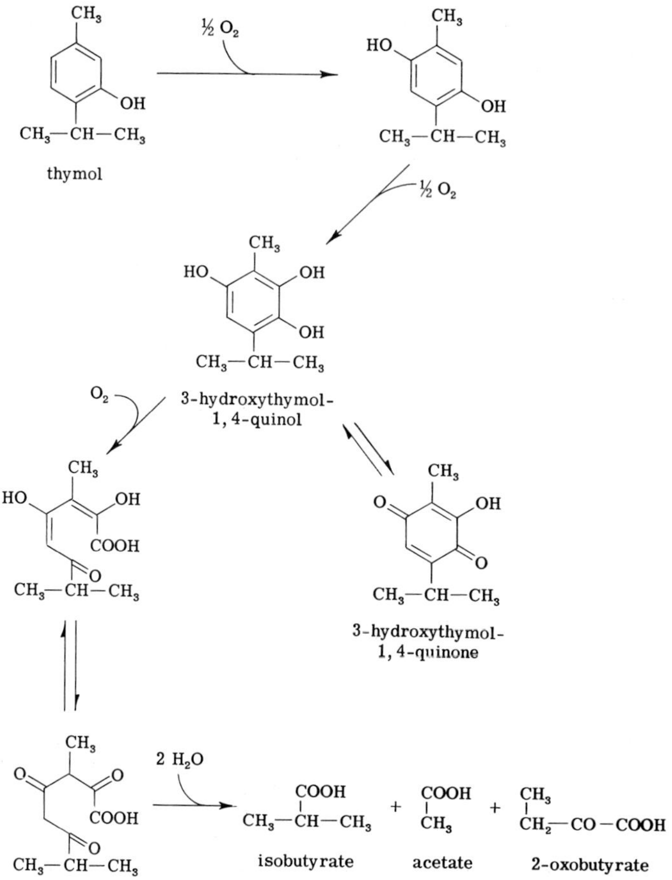

thymol
3-hydroxythymol-1,4-quinol
3-hydroxythymol-1,4-quinone
isobutyrate
acetate
2-oxobutyrate

Fig. 8.12. Metabolism of thymol by *Pseudomonas putida* (reprinted with permission of the author and Academic Press, Inc.).

catechol (Fig. 8.13). The second part of the pathway is different from the one of 2,4-xylenol metabolism. 4-Methyl catechol undergoes a meta cleavage to 2-hydroxy-5-methylmuconic semialdehyde, which is further metabolized via 2-oxopent-4-enoate to pyruvate and acetaldehyde as end products (see p. 513).

Metabolism of Halogenated Aromatic Hydrocarbons

Halogenated aromatic compounds are very toxic agents (212) and very little is known about their metabolism. *Nocardia erythropolis* is one of the very few organisms that have been investigated and cell-free extracts were

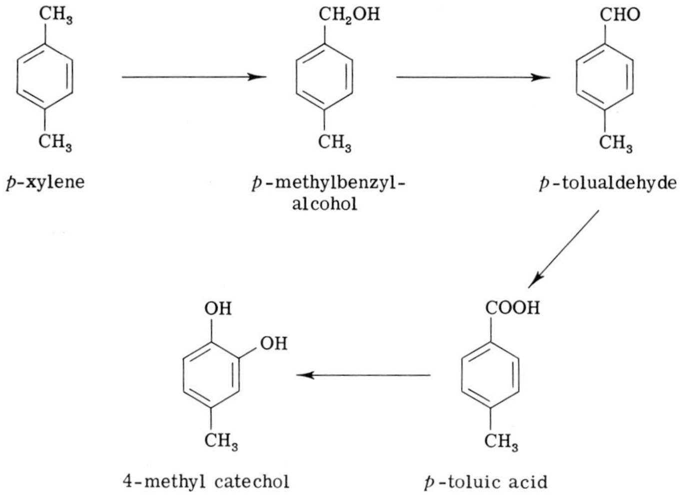
CH3
CH3
p-xylene
CH2OH
CH3
p-methylbenzyl-
alcohol
CHO
CH3
p-tolualdehyde
OH
OH
CH3
4-methyl catechol
COOH
CH3
p-toluic acid

Fig. 8.13. Metabolism of *p*-xylene by *Pseudomonas* species (reprinted with permission of the authors and the publishers of *Can. J. Microbiology*).

shown to be able to utilize *p*-nitrobenzoate, 2-fluoroprotocatechuate, 2-fluorobenzoate, and 4-fluorobenzoate (28, 89). It was observed that the toxic effect comes mainly from the formation of fluorocitrate, a potent inhibitor of the tricarboxylic acid cycle. The breakdown of *p*-nitrobenzoate and 2-fluoroprotocatechuate follows the protocatechuate pathway, with the end products being acetate plus succinate in the former and fluoroacetate plus succinate in the latter pathway.

2- and 4-fluorobenzoate, in contrast, follow the catechol pathway, whereby the former produces fluoroacetate plus succinate and the latter acetate plus fluorosuccinate as end products.

Investigations into the metabolism of fluorinated organic compounds by the Pseudomonadaceae were more successful, as these organisms can use these compounds as substrates for their growth (87, 89, 90, 169). With the aid of respiratory studies as well as intermediate identification, *p*-fluorobenzoate was found to be metabolized via two different pathways (99). Both branches differ in the step during which fluoride is liberated as HF (Fig. 8.14). The importance of any of these pathways is that all fluorine is cleaved before the tricarboxylic acid cycle intermediates are formed.

Different species of *Pseudomonas* (89, 169) were found to eliminate more than 80% of the fluorine during the oxygenase reaction with the formation of catechol. Oxidizing these fluoro compounds therefore depends on the ability of the organism to prevent the accumulation of fluoroacetate or

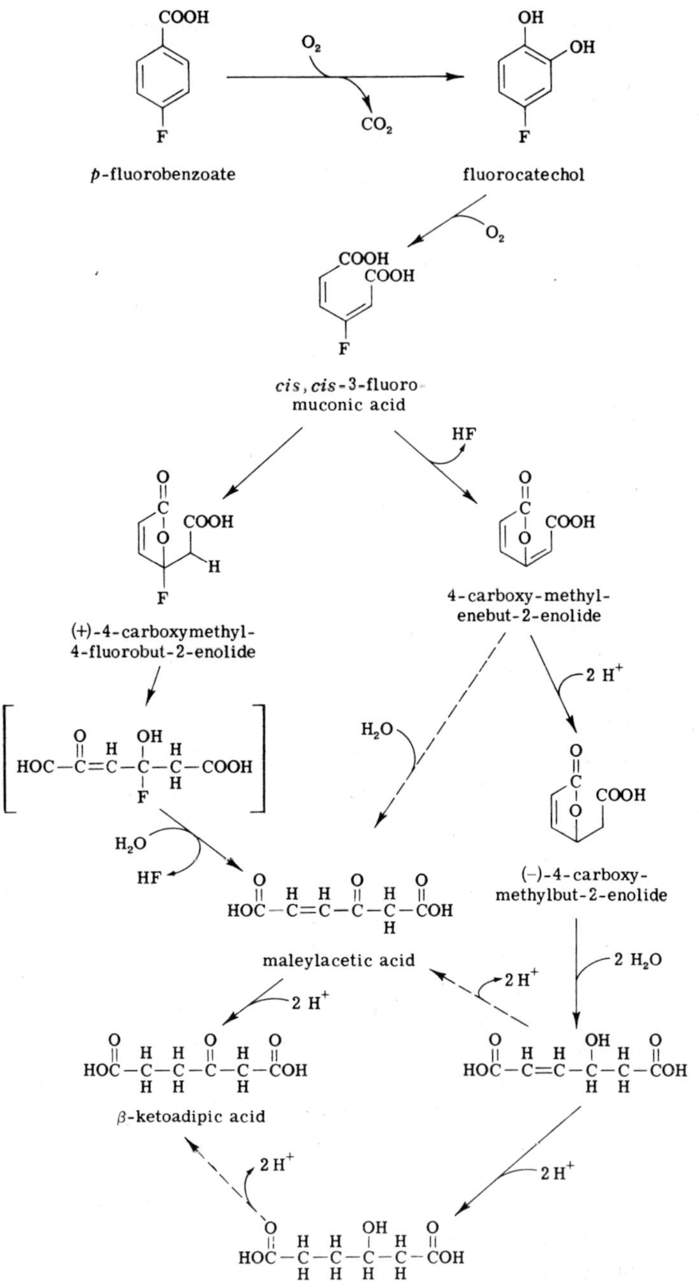

COOH
F
O_2
CO_2
OH
OH
F
p-fluorobenzoate
fluorocatechol
O_2
COOH
COOH
F
cis, cis-3-fluoro-
muconic acid
HF
COOH
H
F
COOH
(+)-4-carboxymethyl-
4-fluorobut-2-enolide
4-carboxy-methyl-
enebut-2-enolide
2 H+
H_2O
COOH
H_2O
HF
(−)-4-carboxy-
methylbut-2-enolide
maleylacetic acid
2 H_2O
2 H+
2 H+
β-ketoadipic acid
2 H+
2 H+

Fig. 8.14. Metabolism of *p*-fluorobenzoate by pseudomonads (reprinted with permission of the authors and the publishers of *Can. J. Microbiology*).

fluorocitrate by either defluorinating the latter (87, 88) or liberating fluorine well before β-ketoadipate is formed.

Regulation of the Aromatic Hydrocarbon Metabolism

The β-Ketoadipate Pathway

The best studied regulatory mechanism in hydrocarbon metabolism is that of the ketoadipate pathway (216a). As was outlined above, the keto-adipate pathway leads from *p*-hydroxybenzoate and benzoate to succinate and acetyl-CoA, via catechol or protocatechuate, respectively. These diphenols occur very readily as the last aromatic intermediates in the metabolism of a variety of benzenoid compounds. As has been indicated in this chapter, catechol is produced from salicylic acid, phenol, anthranilic acid, and benzene (46). The similarities of the reactions via catechol and protocatechuate are given in Fig. 8.15. Despite the similarities of the individual reactions, they are all catalyzed by physically separable enzymes of high specificity. The main difference is at the muconolactone level, where a decarboxylation or an isomerase functions to yield the enol lactone of β-ketoadipate.

As both pathways are inducible, it must be known whether when cells are induced to oxidize a certain compound they are also capable of oxidizing those metabolites which follow in the pathway system (216). If one of the metabolic intermediates is provided as a separate substrate, however, this does not necessarily mean that the organism obtains the ability to oxidize earlier compounds in the pathway sequence. This observation led to the suggestion of sequential induction, which means that the substrate and each intermediate in turn triggers the specific synthesis of the enzyme responsible for its conversion to the next intermediate (216). Two modifications have been added to this theory. First, enzymes could be synthesized or derepressed not only sequentially but also coordinately. This would mean that a group of enzymes in the same pathway could be derepressed together. The second modification takes into account that these enzymes could also be derepressed by the products of the pathway.

When *p*-hydroxybenzoate is given to *Pseudomonas putida* as growth substrate, it induces the synthesis of the hydroxylase and the so-formed protocatechuate initiates the synthesis of the oxygenase in a sequential induction step (124). The next steps are not inducers, but once β-keto-adipate has been formed it induces the carboxymuconate regulatory unit by a coordinate induction mechanism (179). This unit includes four enzymes that oxidize γ-carboxy-*cis,cis*-muconate to β-ketoadipyl-CoA. The striking

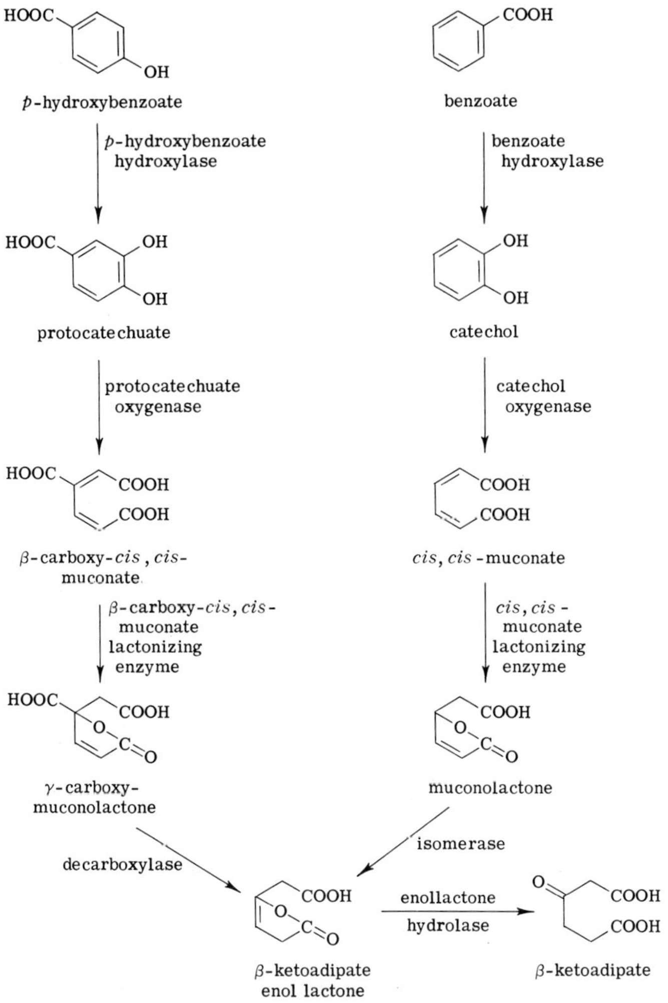
p-hydroxybenzoate
benzoate
p-hydroxybenzoate hydroxylase
benzoate hydroxylase
protocatechuate
catechol
protocatechuate oxygenase
catechol oxygenase
β-carboxy-cis, cis-muconate
cis, cis-muconate
β-carboxy-cis, cis-muconate lactonizing enzyme
cis, cis-muconate lactonizing enzyme
γ-carboxy-muconolactone
muconolactone
decarboxylase
isomerase
enollactone hydrolase
β-ketoadipate enol lactone
β-ketoadipate

Fig. 8.15. The β-ketoadipate pathway of hydrocarbon metabolism.

feature of this regulation is that the coordinate induction, not only of the last two enzymes of the protocatechuate branch but also of the following two enzymes, is common to both branches of the β-ketoadipate pathway. This means that two enzymes of the protocatechuate branch are induced even by cells grown on benzoate, which is metabolized via the catechol pathway.

Benzoate, like *p*-hydroxybenzoate, induces only the hydroxylase, which converts benzoate to catechol. It is *cis,cis*-muconate that derepresses the synthesis of the three enzymes of the catechol branch (15, 260). Genetic investigations with *Pseudomonas putida* (257, 261) revealed that *cis,cis*-muconate sequentially induces catechol oxygenase with *cat A* being the structural gene responsible, whereas the following two enzymes are coordinately induced by *cis,cis*-muconate with *cat B* and *cat C* as the responsible structural genes. The details of regulation at the molecular level are still unclear.

Acinetobacter calcoaceticus is able to metabolize *p*-hydroxybenzoate in the same manner as the pseudomonads do. The regulation of these pathways, however, was found to be entirely different (31–34, 50, 127). Protocatechuate is the only inducer and initiates the synthesis of all the enzymes, including the 3-oxoadipate-CoA transferase (EC 2.8.3.6) that converts β-ketoadipate to β-ketoadipyl-CoA. As a consequence of this type of induction by protocatechuate, *Acinetobacter* does not induce any of the protocatechuate enzymes during growth on benzoate.

When *Acinetobacter* is grown on benzoate, the growth substrate induces the hydroxylase enzyme and *cis,cis*-muconate induces the catechol 1,2-oxygenase. The synthesis of both of these enzymes is therefore sequentially derepressed, as in the pseudomonads. In *Acetobacter calcoaceticus, cis,cis*-muconate derepresses, similarly to protocatechuate, the synthesis of the total group of enzymes required for the conversion of *cis,cis*-muconate to β-ketoadipyl-CoA. These bacteria also synthesize two forms of 3-oxoadipate enol-lactonase (EC 3.1.1.24) (137) and 3-oxoadipate-CoA:transferase (34). The first is induced by protocatechuate and the second by *cis,cis*-muconate, coordinately with the enzyme operative in the respective branches of the pathway.

A third type of control system of the β-ketoadipate pathway exists in *Alcaligenes eutrophus* (131, 132). The control of the *p*-hydroxybenzoate branch of the pathway is very similar to that of *Acinetobacter*, although 3-oxoadipate-CoA:transferase appears not to be induced by protocatechuate. The control of the enzymes of the catechol branch, however, is much more complex. Catechol oxygenase cannot be induced by catechol, but only by its precursor benzoate or its 1,2-dihydro-1,2-dihydroxy derivative. The uniqueness of the regulation of the catechol branch is the dual regulation of the enzymes muconolactone isomerase (EC 5.3.3.4) and the enol hydrolase II by either benzoate or 1,2-dihydro-1,2-dihydroxybenzoate and also by either *cis,cis*-muconate or (+)-muconolactone. The last enzyme, 3-oxoadipate-CoA transferase, is induced by β-ketoadipate.

The available genetic data (139, 256) suggest that all coordinately synthesized enzymes are controlled by a single operon (178). However, loci-specifying enzymes sharing a common inducer were also clustered, which may indicate some degree of chromosomal specialization (139) and could possibly be helpful in determining evolutionary development. Investigations are underway to find out enzymatic responses to substrate mixtures. The indications are (114) that the response did not result in the metabolism of the better growth substrate but in that of the substrate requiring the synthesis of fewer enzymes.

The α-Keto Acid Pathway

This pathway represents all meta-cleavage mechanisms on catechol and protocatechuate that result in the formation of α-keto acids. Because some strains of *Pseudomonas* are able to form the catechol enzymes of both the α-ketoacid and the β-ketoadipate pathways (51), some regulatory mechanism must exist to bring about an ortho or meta fission of the aromatic ring. Whether ortho or meta cleavage occurs depends very much on the precursor of catechol (208). The regularity mechanism of this selective utilization has been studied mainly with *Pseudomonas putida* (63, 64). These investigations clearly indicate that every one of the enzymes involved in the metabolism of phenol via catechol and the α-keto acid pathway is, in fact, induced by the substrate itself. It is assumed that cresol could act in the same way. Catechol or methyl catechol can function as inducer of the meta-cleavage pathway (169b), which is demonstrated in the presence of the ortho-cleavage pathway only in catechol-grown cells. From these results it is deduced that the meta-cleavage pathway serves as a general mechanism for the catabolism of alkyl derivatives of catechol, whereas the ortho-cleavage pathway is more specific for catechol and its precursors.

The Mandelate Pathway

The metabolism of mandelate has been shown earlier to include four enzymatic reaction steps, with five enzymes being involved in *Pseudomonas putida*. Physiological studies with *Pseudomonas putida* (108, 110) revealed that the mandelate enzymes share a common regulatory gene. It has been suggested that the structural genes coding for the enzymes are manifested in one operon and that all five of these enzymes are coordinately regulated and can be induced by D(−)- or L(+)-mandelate or benzoylformate. Genetic analysis with phage pf 16 H_2 supported this view (39, 256). The intriguing part of the mandelate pathway, however, is the existence of three benzaldehyde dehydrogenases (92, 108, 221). Two of these are isofunctional, using either NAD^+ or $NADP^+$ as cofactor. The third benzaldehyde dehydrogenase is induced separately from the mandelate regulatory unit by benzaldehyde and *p*-hydroxybenzaldehyde, which would be sequential induction. All of these enzymes of the mandelate group can be specifically repressed by any metabolite of the catechol group of the pathway (165), justifying the above (p. 501) divisions of mandelate utilization.

Pseudomonas aeruginosa has a slightly different mode of regulation. The first enzyme, L(+)-mandelate dehydrogenase, is induced by its substrate (201). The following two enzymes, benzoylformate decarboxylase (EC 4.1.1.7) and benzaldehyde dehydrogenase, are both induced by benzoylformate. The linkage of the genes for the enzymes of benzoate and *p*-hydroxybenzoate (139) revealed that they form a cluster (202) and share a common inducer. The following enzymes, benzoate and anthranilate oxidase, are part of the catechol oxygenase, *cis,cis*-muconate lactonizing enzyme, and muconolactone isomerase cluster. The clustering of functionally related genes would give the organism a better chance to synthesize enzymes in a short period of time. As there also exist numerous examples of functionally related enzymes in other organisms by nonclustered genes (119, 206, 228), it is not as yet known whether or not clustering is of evolutionary (111) or populational importance (172, 215).

Metabolism of Phenoxyalkyl Carboxylic Acids

Pesticides have become more and more a subject of active research. Modern pest control research has been mainly concerned with the finding of potent organic compounds to control either a broad or narrow spectrum of pests, chiefly insects, weeds, and plant pathogens. The degradability of these pesticides is of great practical importance because the toxicity of these chemicals can endanger the life of man. Although pesticides may be destroyed or detoxified by various mechanisms in the soil, the major means of degradation is a consequence of biological activity. The most resistant pesticides are the chlorinated hydrocarbon insecticides, such as aldrin, chlordane, DDT, and dieldrin, and phenoxy herbicides, such as 2-(2,4-dichlorophenoxy)propionic acid (3). There appears to be a distinct influence of chemical structure on the persistence and rate of decomposition of phenoxyalkylcarboxylic acid herbicides. Compounds having a chlorine on the aromatic ring meta to the ether linkage and α-substituted phenoxyalkylcarboxylic acids are more resistant to biodegradation than those having no meta halogen on the aromatic ring and ω-substitution on the side chain. Most of the work has been done with the dichlorophenoxyacetic acid compounds, which are known to be degraded by soil microorganisms. For further information on this subject, the reader should consult the excellent reviews by Audus (5), Cripps (44a) and by Kearney and Kaufman (138).

The general mechanism of the initial oxidation of phenoxyalkylcarboxylic acids is still in question. Whereas some investigators (57, 217–219, 229, 252) suggest that the aliphatic moiety of these acids is metabolized by

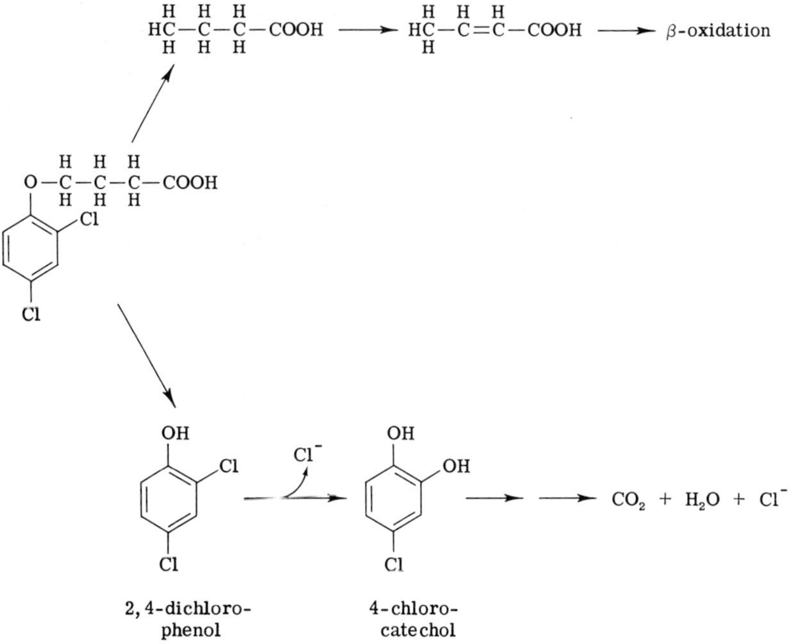
β-oxidation
Cl
Cl
OH
Cl
Cl⁻
OH
OH
Cl
CO2 + H2O + Cl⁻
Cl
2,4-dichloro-
phenol
4-chloro-
catechol

Scheme 8.7. General mechanism of the initial oxidation of phenoxyalkylcarboxylic acids (reprinted with permission of the authors and the Biochemical Society, London).

β-oxidation, studies with the decomposition of 4-(2,4-dichlorophenoxy)-butyric acid by *Flavobacterium* sp. give evidence for a degradation by cleavage of the ether linkage rather than by β-oxidation of fatty acid (164). This cleavage of the ether linkage forms 2,4-dichlorophenol and butyric acid. The 2,4-dichlorophenol is then dehalogenated at the ortho position, and the resultant 4-chlorocatechol is readily and completely degraded, possibly with the formation of γ-chloro-*cis,cis*-muconic acid (see Scheme 8.7).

It therefore appears that two pathways exist in the metabolism of phenoxyalkylcarboxylic acids (163). The first pathway is mainly used by *Nocardia*. It involves a β-oxidation of the aliphatic moiety (229, 252) and leads to the production of phytotoxic intermediates from the original substrate. The second pathway is indicated by a cleavage of the ether linkage, which leads to the immediate detoxication of the herbicidal activity.

Metabolism of 4-Chlorophenoxyacetic Acid

The hormone herbicide, 4-chlorophenoxyacetic acid, together with 2,4-dichlorophenoxyacetic acid and 4-chloro-2-methylphenoxyacetic acid has

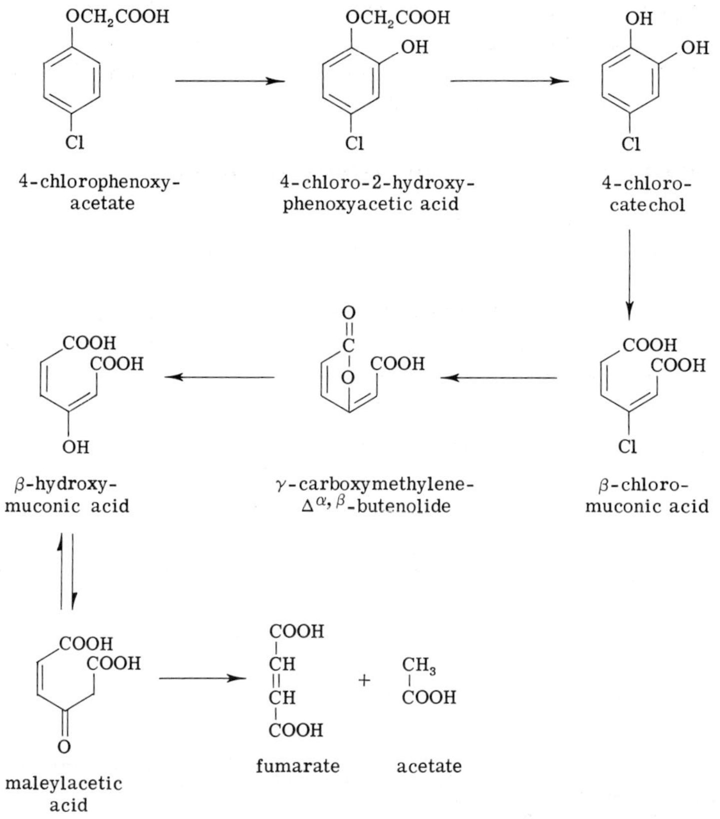

OCH2COOH
Cl
4-chlorophenoxy-acetate
OCH2COOH
OH
Cl
4-chloro-2-hydroxy-phenoxyacetic acid
OH
OH
Cl
4-chloro-catechol
COOH
COOH
OH
β-hydroxy-muconic acid
O
C
O
COOH
γ-carboxymethylene-Δα,β-butenolide
COOH
COOH
Cl
β-chloro-muconic acid
COOH
COOH
O
maleylacetic acid
COOH
CH
CH
COOH
fumarate
+
CH3
COOH
acetate

Fig. 8.16. Metabolism of 4-chlorophenoxyacetic acid by microorganisms (reprinted with permission of the authors and the Biochemical Society, London).

been very successful in crop protection and has become of worldwide importance. Although there are reports available that *Bacterium globiforme* (4), *Flavobacterium aquile, Nocardia coeliaca* (128), *Flavobacterium peregrinum* (17), *Pseudomonas* sp. (57), *Corynebacterium* sp. (198), *Mycoplana* (251), *Achromobacter* sp. (13, 48, 218), *Pseudomonas* NCIB 9340 (79), and *Arthrobacter* sp. (157) can attack some of these herbicides to varying degrees, only one soil pseudomonad (60) and *Arthrobacter* sp. (19) were found to completely metabolize 4-chlorophenoxyacetate to the tricarboxylic acid cycle intermediates fumarate and acetate (Fig. 8.16).

The attack on 4-chlorophenoxyacetic acid starts with a hydroxylation of the aromatic ring in the ortho position and 4-chloro-2-hydroxyphenoxyacetic acid is formed. The ether linkage is cleaved possibly with the formation of glyoxylate and 4-chlorocatechol, which itself undergoes an ortho fission to β-chloromuconic acid. This compound is transformed very rapidly to γ-carboxymethylene-$\Delta^{\alpha,\beta}$-butenolide, with release of the Cl^- ion. The latter lactone is very readily converted via maleylacetate to fumarate and acetate. The latter step could well go via fumarylacetate.

Metabolism of 4-Chloro-2-methylphenoxyacetate

One of the most effective hormone herbicides is probably 2-methyl-4-chlorophenoxyacetate, which is more persistent in the soil than 2,4-dichlorophenoxyacetate (5).

The first attack on 2-methyl-4-chlorophenoxyacetate is on the ether linkage (Fig. 8.17), which is substituted by a hydroxyl group (80) and forms 5-chloro-*o*-cresol and glyoxylate (78) by an oxidative cleavage reaction. This intermediate is then hydroxylated at the ortho position. This type of initial attack is different from the one described for 4-chlorophenoxyacetate. The hydroxylation leads to 5-chloro-3-methylcatechol, which itself undergoes ortho cleavage to form γ-chloro-α-methylmuconic acid. Three pathways open the way for the soil pseudomonad to obtain its next stable intermediate, γ-carboxymethylene-α-methyl-$\Delta^{\alpha,\beta}$-butenolide, depending on the release of the Cl^- ion (81). The lactonization could occur in two steps, one of which is a dehalogenation and the other the lactonization proper. However, none of these intermediates has been elucidated as yet. It is therefore not certain whether the lactonization step is a one- or two-step reaction. Whichever way the decision may go, the lactone is very rapidly metabolized via γ-hydroxy-α-methylmuconic acid and β-methylmaleylacetic acid to tricarboxylic acid intermediates.

Metabolism of 2,4-Dichlorophenoxyacetic Acid

The metabolism of 2,4-dichlorophenoxyacetic acid is very similar to that described for 4-chloro-2-methylphenoxyacetic acid, as the first attack is on the side chain and chlorine is not eliminated on the aromatic ring (61). This first step in the degradation of 2,4-dichlorophenoxyacetate seems to be the oxidation of the acetic acid moiety, together with the cleavage of the ether linkage (232), to give the corresponding phenol, 2,4-dichlorophenol (218), and glyoxylate. *Pseudomonas* NCIB 9340 and *Arthrobacter* sp (18) hydroxylate this phenol and 3,5-dichlorocatechol is produced. An ortho cleavage takes place to form α,γ-dichloromuconic acid (231) (see Scheme 8.8). There are reports that a meta cleavage occurs in *Achromobacter* sp. (121). *Flavobacterium peregrinum* and *Achromobacter* sp. are reported to convert 2,4-dichlorophenol to 4-chlorocatechol and further to *p*-chloromuconic acid (59). *Corynebacterium* is also able to metabolize 2,4-dichlorophenol (198). After the ring fission, α,γ-dichloromuconic acid is dehalogen-

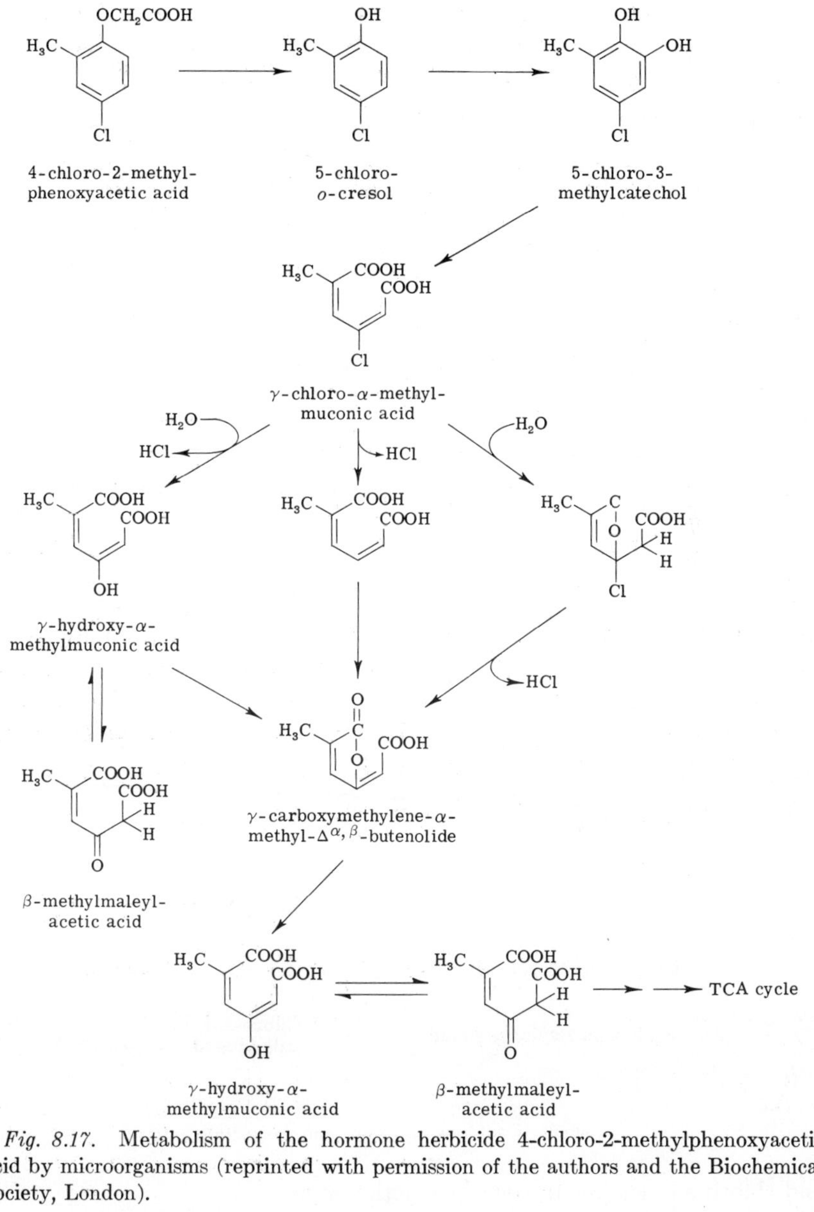

4-chloro-2-methyl-
phenoxyacetic acid
5-chloro-
o-cresol
5-chloro-3-
methylcatechol
γ-chloro-α-methyl-
muconic acid
γ-hydroxy-α-
methylmuconic acid
β-methylmaleyl-
acetic acid
γ-carboxymethylene-α-
methyl-Δα,β-butenolide
γ-hydroxy-α-
methylmuconic acid
β-methylmaleyl-
acetic acid
TCA cycle

Fig. 8.17. Metabolism of the hormone herbicide 4-chloro-2-methylphenoxyacetic acid by microorganisms (reprinted with permission of the authors and the Biochemical Society, London).

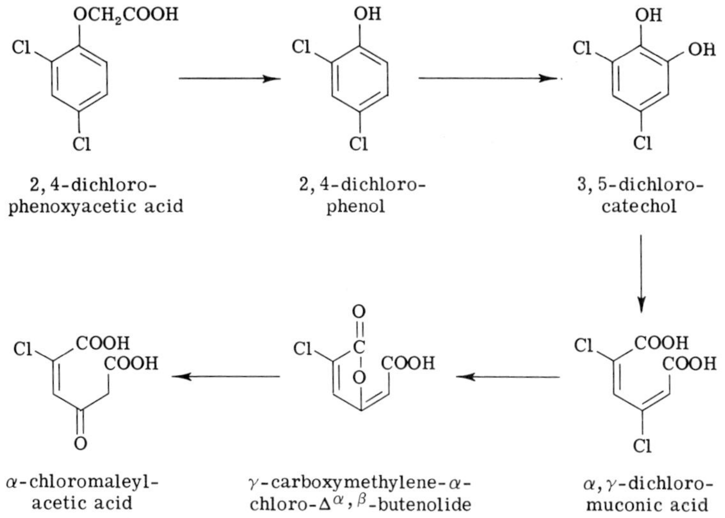

2, 4-dichloro-phenoxyacetic acid
2, 4-dichloro-phenol
3, 5-dichloro-catechol
α-chloromaleyl-acetic acid
γ-carboxymethylene-α-chloro-Δ^{α,β}-butenolide
α, γ-dichloro-muconic acid

Scheme 8.8. Metabolism of 2,4-dichlorophenoxyacetic acid (reprinted with permission of the authors and the Biochemical Society, London).

ated and lactonized to form γ-carboxymethylene-α-chloro-$\Delta^{\alpha,\beta}$-butenolide. β-Chloromaleylacetic acid is further catabolized to succinic acid (54). The intermediates in this reaction sequence depend on the stage of the removal of the remaining chlorine. If the chlorine is removed during the attack on β-chloromaleylacetic acid, the sequence is

β-chloromaleylacetic acid ⟶ 3-ketoadipic acid ⟶ succinic acid

If, however, the chlorine remains in the molecule, the sequence is

β-chloromaleylacetic acid ⟶ 2-chloro-4-keto-adipic acid ⟶ chlorosuccinic acid ⟶ succinic acid

and chlorine is set free in the last reaction step.

Metabolism of DDT, DDD, DDMS, and DDNU

A number of other well-known pesticides have been found to be attacked by microorganisms. However, their complete metabolic route has not been fully elucidated as yet.

The degradation of DDT [1,1,1-trichloro-2,2-bis(*p*-chlorophenyl)ethane] by *Aerobacter aerogenes* was found to have at least seven metabolites (253). The main attention is being focused on the chloride ions, which, if removed, would remove the potency of the pesticide. DDT is first converted to DDD [1,1-dichloro-2,2-bis(*p*-chlorophenyl)ethane] in releasing chloride ions and

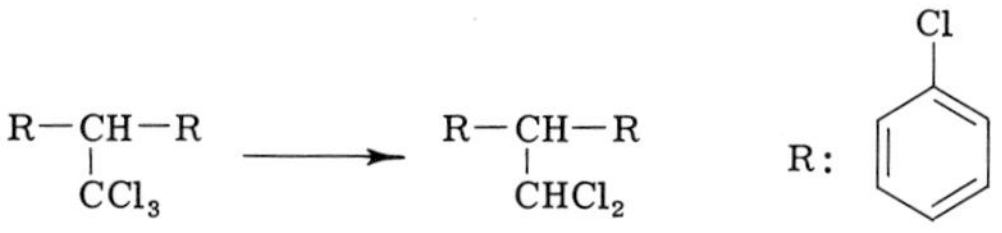

DDT DDD

replacing this ion by a hydrogen ion. A further chloride ion removal in

DDD DDMU

form of a HCl release produces DDMU [1-chloro-2,2-bis(*p*-chlorophenyl)-ethylene] from DDD. The reduction of DDMU gives DDMS [1-chloro-

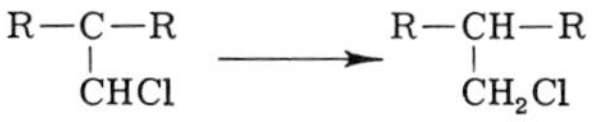

DDMU DDMS

2,2-bis(*p*-chlorophenyl)ethane]. The last chloride ion is removed now as HCl, and DDMS is converted to DDNU [unsym. bis(*p*-chlorophenyl)-

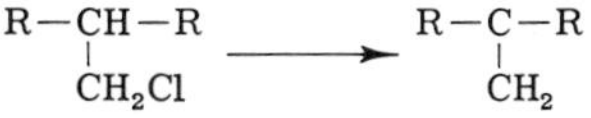

DDMS DDNU

ethylene]. A hydration together with an oxygenation metabolizes DDNU to DDA [2,2-bis(*p*-chlorophenyl)acetate]. A decarboxylation reaction

DDNU DDA DBP

results finally in the formation of DBP (4,4′-dichlorobenzophenone). The further metabolism is as yet not known.

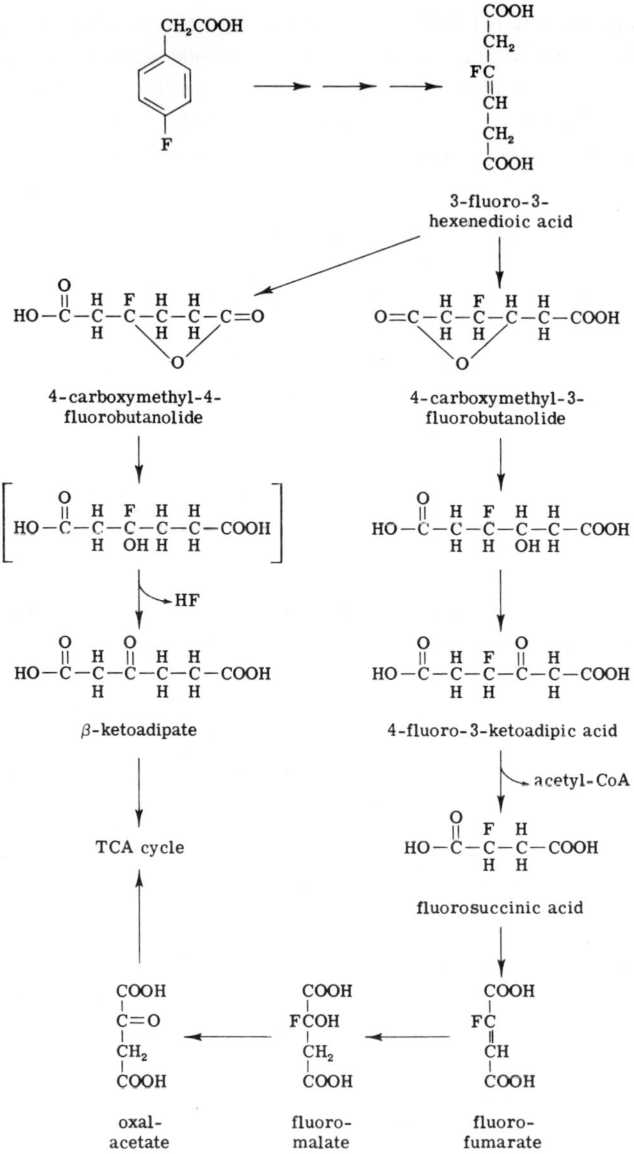
3-fluoro-3-
hexenedioic acid
4-carboxymethyl-4-
fluorobutanolide
4-carboxymethyl-3-
fluorobutanolide
HF
β-ketoadipate
4-fluoro-3-ketoadipic acid
acetyl-CoA
TCA cycle
fluorosuccinic acid
oxal-
acetate
fluoro-
malate
fluoro-
fumarate

Fig. 8.18. Degradation of *p*-fluorophenylacetic acid (reprinted with permission of the authors and the publishers of *Can. J. Microbiology*).

The degradation of Linuron and some other herbicides and fungicides by *Bacillus sphaericus* has just been reported but no pathways are suggested (55).

Metabolism of p-Fluorophenylacetic Acid

Very little is known about the initial attack on *p*-fluorophenylacetic acid (97, 98). A *Pseudomonas* sp. metabolizes this compound via unknown steps to 3-fluoro-3-hexenedioic acid, which can be metabolized in two different ways depending on the lactone formed (Fig. 8.18). 4-Carboxymethyl-4-fluorobutanolide can be hydrolyzed to give 3-hydroxy-3-fluoroadipic acid, which spontaneously eliminates fluoride to give β-ketoadipic acid. The second alternative is the lactonization to 4-carboxymethyl-3-fluorobutanolide, which is hydrolyzed to 3-keto-4-fluoroadipic acid. This intermediate does not liberate HF but keeps the fluoride on the molecule until fluoromalate is being produced. This unstable compound liberates HF and forms the tricarboxylic acid cycle intermediate, oxalacetate.

Metabolism of Riboflavin

Pseudomonas RF oxidizes riboflavin to butyric, pyruvic, propionic, and acetoacetic acid (Fig. 8.19). Riboflavin is hydroxylated to 1-ribityl-2,3-diketo-1,2,3,4-tetrahydro-6,7-dimethyl quinoxaline (96, 168). Oxygen is required and CO_2 and urea are liberated during the conversion. The ribose side chain is removed in the following oxidation step, with the formation of 6,7-dimethyl quinoxaline-2,3-diol (242), which itself can be metabolized further in two different ways (10). One of these pathways leads to the formation of 3,4-dimethyl-6-carboxy-α-pyrone (95, 203, 213), whereas the other leads to pyruvic, butyric, propionic, and acetoacetic acids. Whether the organism uses one or the other pathway appears to depend very much on the degree of aeration of the growth medium (10); high aeration induces the pathway leading to the smaller acid molecules, for high amounts of CO_2 could be measured. Isotope experiments also supported the suggestion as to the aliphatic acid formation, with butyric acid and pyruvate being the predominant end products.

Metabolism of Vitamin B_6 (Pyridoxine)

The metabolism of vitamin B_6 utilizes two different pathways, which depend on whether pyridoxine or pyridoxamine serves as the sole carbon and energy source (Fig. 8.20).

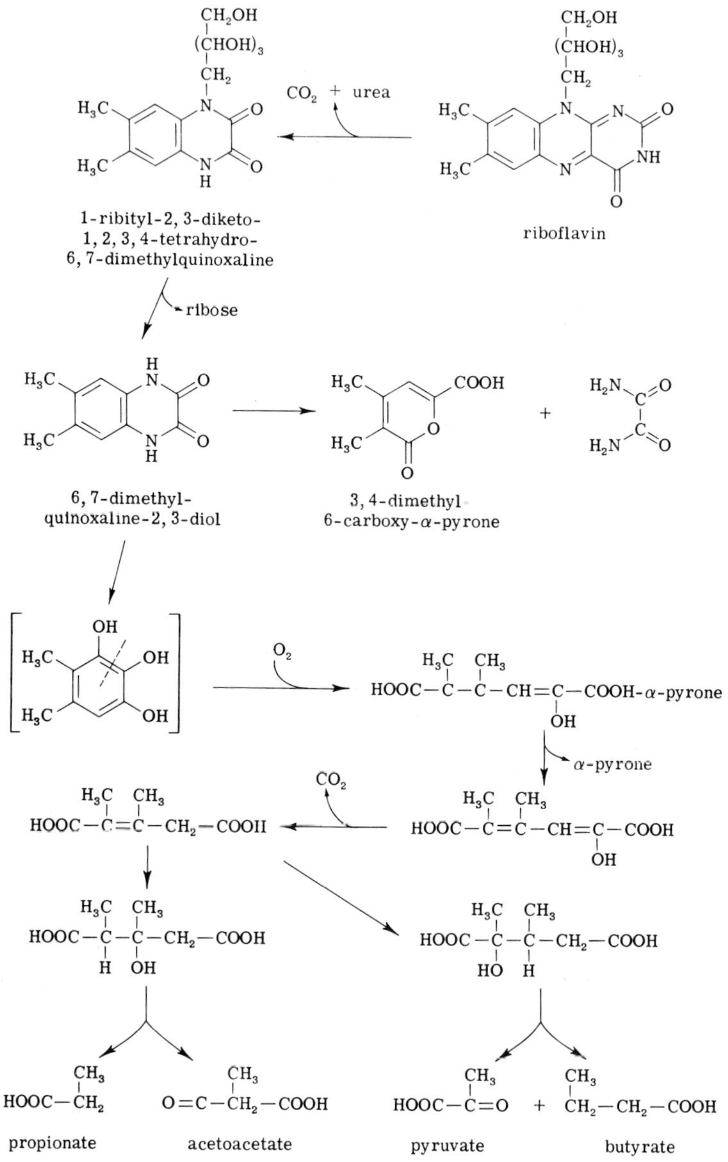
CO2 + urea
1-ribityl-2, 3-diketo-
1, 2, 3, 4-tetrahydro-
6, 7-dimethylquinoxaline
riboflavin
ribose
6, 7-dimethyl-
quinoxaline-2, 3-diol
3, 4-dimethyl
6-carboxy-α-pyrone
O2
HOOC—C—C—CH=C—COOH-α-pyrone
α-pyrone
CO2
HOOC—C=C—CH2—COOH
HOOC—C=C—CH=C—COOH
HOOC—C—C—CH2—COOH
propionate
acetoacetate
pyruvate
butyrate

Fig. 8.19. Degradation of riboflavin by *Pseudomonas* RF (reprinted with permission of the author and the publishers of *Arch. f. Mikrobiologie*).

The first degradative step in the pyridoxine metabolism of *Pseudomonas* 1A involves pyridoxine 5-dehydrogenase (EC 1.1.99.9), which is an FAD-dependent enzyme that uses 2,6-dichloroindophenol as electron acceptor (225). A completely different FAD-dependent enzyme, pyridoxine 4-oxidase (EC 1.1.3.12), was isolated from *Pseudomonas* MA-1, which catalyzes the conversion of pyridoxine to pyridoxal. This enzyme can use both oxygen and 2,6-dichloroindophenol as hydrogen acceptor. Isopyridoxal is further dehydrogenated to 5-pyridoxolactone, which itself is rapidly converted to 5-pyridoxic acid. There is still some doubt as to whether 5-pyridoxolactone is a true intermediate to 5-pyridoxic acid. At this stage of the pathway an oxygenase (5-pyridoxate dioxygenase, EC 1.14.12.5) opens the ring to form α-hydroxymethyl-α'(*N*-acetylaminomethylene)succinic acid.

The second pathway has been elucidated in much more detail. The first reaction in the metabolism of pyridoxamine is a transamination reaction, producing pyridoxal and converting at the same time pyruvate into alanine. An NAD-specific pyridoxal dehydrogenase forms the lactone of pyridoxal, 4-pyridoxolactone, which is cleaved to 4-pyridoxic acid by a 4-pyridoxolactonase (EC 3.1.1.27) (24). Unlike 5-pyridoxic acid, 4-pyridoxic acid cannot be attacked by an oxygenase. Instead, its alcohol group is first oxidized in two successive steps to a carboxyl group, and the formation of 2-methyl-3-hydroxypyridine 4,5-dicarboxylic acid is completed. A decarboxylation reaction produces 2-methyl-3-hydroxypyridine 5-carboxylic acid, which can now undergo an oxygenase cleavage (214). Oxygen-18 experiments indicated that the oxygenase is a dioxygenase that incorporates the oxygen predominantly into the acetyl and also into the newly formed carboxyl group of α-(*N*-acetylaminomethylene)succinic acid. It was observed, however, that some ^{18}O from $H_2{}^{18}O$ was also incorporated into the latter compound. The reaction also requires reduced NAD. These observations led to the proposal of two different reaction sequences (214), one with direct incorporation of two atoms from molecular oxygen, according to a dioxygenase reaction, and the second with the incorporation of one atom from molecular oxygen into the acetyl group, together with an oxygen atom from water into the carboxyl group. It is assumed that the oxygenase from the 5-pyridoxic acid conversion acts in a similar way.

The acyclic product from the oxygenase reaction is now hydrolyzed by a hydrolase that forms the end products acetate, CO_2, NH_3, and succinic semialdehyde (174).

Although both pathways of vitamin B_6 utilization are inducible, it is not as yet known whether the individual enzymes of these pathways appear by sequential induction or whether single inducers are involved. One pathway is used by *Pseudomonas* 1A, whereas the other is used by *Pseudomonas* MA-1.

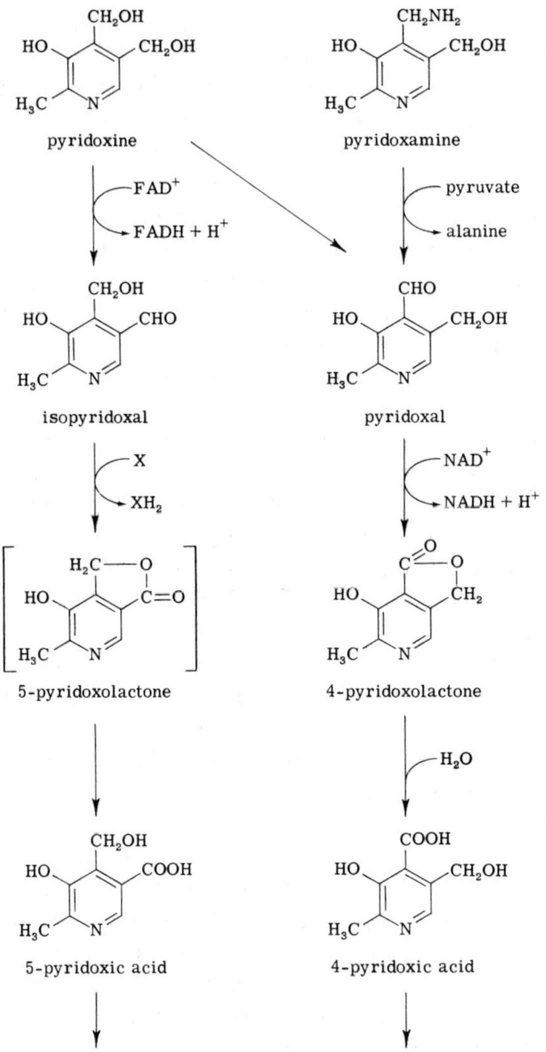
pyridoxine
pyridoxamine
FAD+
FADH + H+
pyruvate
alanine
isopyridoxal
pyridoxal
X
XH2
NAD+
NADH + H+
5-pyridoxolactone
4-pyridoxolactone
H2O
5-pyridoxic acid
4-pyridoxic acid

Fig. 8.20. Metabolism of vitamin B (pyridoxine) and pyridoxamine by microorganisms (reprinted with permission of the authors and the American Society of Biological Chemists).

Metabolism of Steroids

The possibility that microorganisms can transform a great number of steroids has gained in stature over the past years. These steroids can be attacked aerobically as well as anaerobically, but so far no clear picture has emerged from the various individual publications (2, 22, 53, 152, 189, 207,

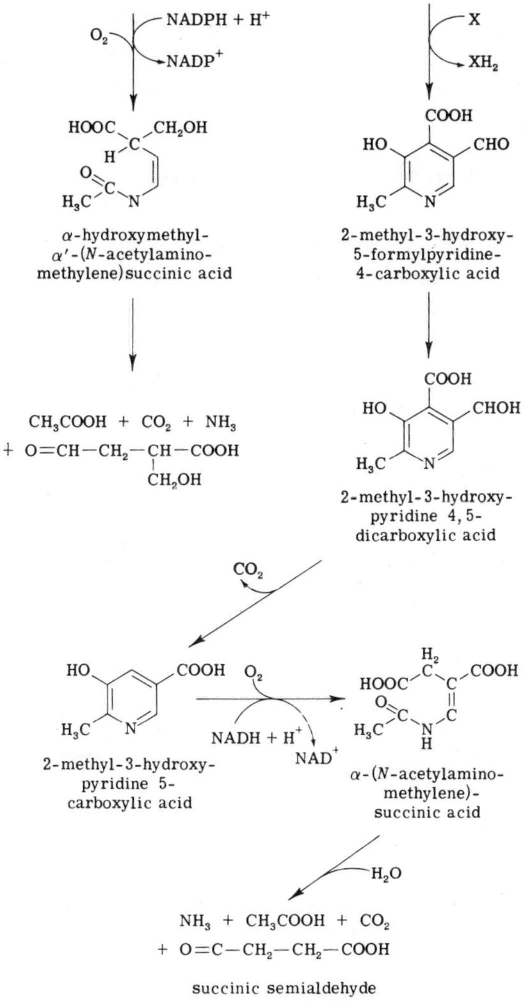
NADPH + H+
O2
NADP+
HOOC
CH2OH
H
O
H3C
N
α-hydroxymethyl-
α'-(N-acetylamino-
methylene)succinic acid
X
XH2
COOH
HO
CHO
H3C
N
2-methyl-3-hydroxy-
5-formylpyridine-
4-carboxylic acid
CH3COOH + CO2 + NH3
+ O=CH—CH2—CH—COOH
CH2OH
COOH
HO
CHOH
H3C
N
2-methyl-3-hydroxy-
pyridine 4,5-
dicarboxylic acid
CO2
HO
COOH
O2
H3C
N
NADH + H+
NAD+
2-methyl-3-hydroxy-
pyridine 5-
carboxylic acid
H2
HOOC
COOH
O
H3C
N
H
α-(N-acetylamino-
methylene)-
succinic acid
H2O
NH3 + CH3COOH + CO2
+ O=C—CH2—CH2—COOH
succinic semialdehyde

Fig. 8.20. (Continued).

209, 210, 233, 234, 259). The presently available knowledge has been summarized excellently in a recent review (120).

Metabolism of Aromatic Polycyclic Hydrocarbons

Microorganisms such as *Bacillus megaterium*, *Mycobacterium flavum*, and *M. rubrum* are reported to attack and utilize aromatic polycyclic hydrocarbons (62, 186, 187). Benz[*a*]pyrene appears to be the main target, although specific oxidative enzymes have also been shown to be induced on 9,10-dimethyl-1,2-benzanthracene, 1,2,5,6-dibenzanthracene, 1,2-benzanthracene, pyrene, 1,2-benzpyrene, 3,4-benzpyrene, 1,12-benzperylene, and perylene. So far no metabolic sequence has been elucidated.

Metabolism of *p*-Toluene Sulfonate

Members of the Pseudomonadaceae are also able to metabolize aromatic sulfonates, such as *p*-toluene sulfonate (68) (see Scheme 8.9). The sulfonate

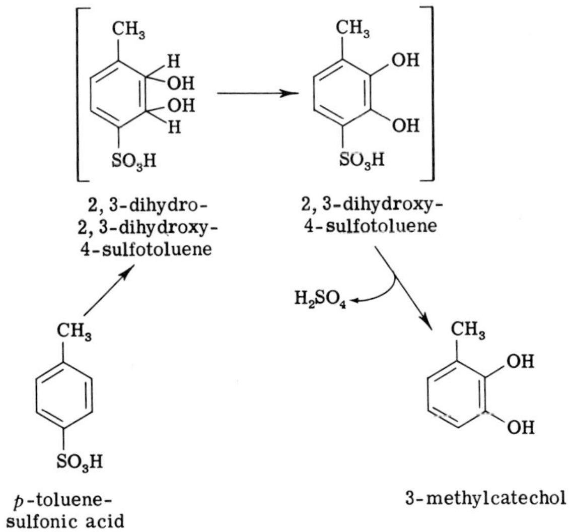

CH_3
H
OH
OH
H
SO_3H
CH_3
OH
OH
SO_3H
2, 3-dihydro-
2, 3-dihydroxy-
4-sulfotoluene
2, 3-dihydroxy-
4-sulfotoluene
H_2SO_4
CH_3
SO_3H
p-toluene-
sulfonic acid
CH_3
OH
OH
3-methylcatechol

Scheme 8.9. Metabolism of *p*-toluene sulfonate (reprinted with permission of the authors and the publishers of *Can. J. Microbiol.*).

group is liberated from the aromatic ring. 3-Methyl catechol is an established intermediate (29); it is metabolized by a meta-cleavage mechanism, as described above, producing pyruvate and acetate as tricarboxylic acid cycle intermediates. Bacteria of the genus *Alcaligenes* were also found to metabolize alkyl benzene sulfonates in the same way as described for *p*-toluenesulfonic acid (15a). The first attack occurs on the side chain and leads to the appropriate catechol derivative before fission occurs.

Benzene sulfonate can also be a good growth substrate for *Pseudomonas testosteroni* H-8 (195), whereas alkyl benzene sulfonates are first attacked on the side chain until benzene sulfonate or benzene is obtained (122). Benzene itself is oxidized as described above (see p. 506).

Metabolism of Coumarin

Arthrobacter metabolizes coumarin via a pathway that has not as yet been fully elucidated (Fig. 8.21). Because coumarin is in fact the lactone of *cis*-coumarinic acid, a cis–trans isomerization on the double bond in the side chain of the aromatic ring is assumed to occur, giving rise to an *o-trans*-coumaric acid (153). This isomerization is produced by enzymatic hydroly-

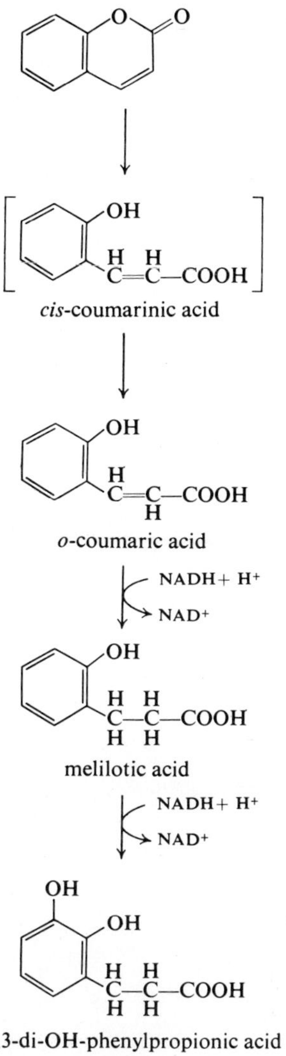
O
O
OH
H H
C=C—COOH
cis-coumarinic acid
OH
H
C=C—COOH
H
o-coumaric acid
NADH + H+
NAD+
OH
H H
C—C—COOH
H H
melilotic acid
NADH + H+
NAD+
OH
OH
H H
C—C—COOH
H H
2,3-di-OH-phenylpropionic acid

Fig. 8.21. Degradation of coumarin by *Arthrobacter* (reprinted with permission of the American Society of Biological Chemists).

sis between the oxygen and carbonyl carbon atom of the pyrone ring. *o*-Coumaric acid is reduced by the enzyme NADH:*o*-coumarate oxidoreductase to melilotic acid. In the presence of atmospheric oxygen and reduced NAD, a hydroxylation takes place that is mediated by an enzyme similar or identical to melilolate hydroxylase (154). This hydroxylation step produces 2,3-dihydroxyphenylpropionic acid, which could be metabolized further similarly to the metabolism of phenylacetic acid (16, 145). This aromatic hydrocarbon compound is converted via *p*-hydroxyphenylacetic acid and 3,4-dihydroxyphenylacetic acid to γ-carboxymethyl-α-hydroxymuconic semialdehyde. A conversion to catechol, however, would mean that a cleavage of the side chain occurs before ring fission.

Metabolism of Pipecolate

Pipecolic acid is metabolized by *Pseudomonas putida* (7). The oxidation of pipecolate (Fig. 8.22) to Δ′-piperideine-6-carboxylate is catalyzed by pipecolate dehydrogenase (EC 1.5.99.3) (8), also known as "*Pseudomonas* P 2 electron transport particle (P2-ETP)" (11), which is an inducible, membrane-bound dehydrogenase. A nonenzymatic reaction converts Δ′-piperideine-6-carboxylate to L-α-aminoadipate δ-semialdehyde. The enzyme L-α-aminoadipate-δ-semialdehyde:NAD(P)$^+$ oxidoreductase (EC

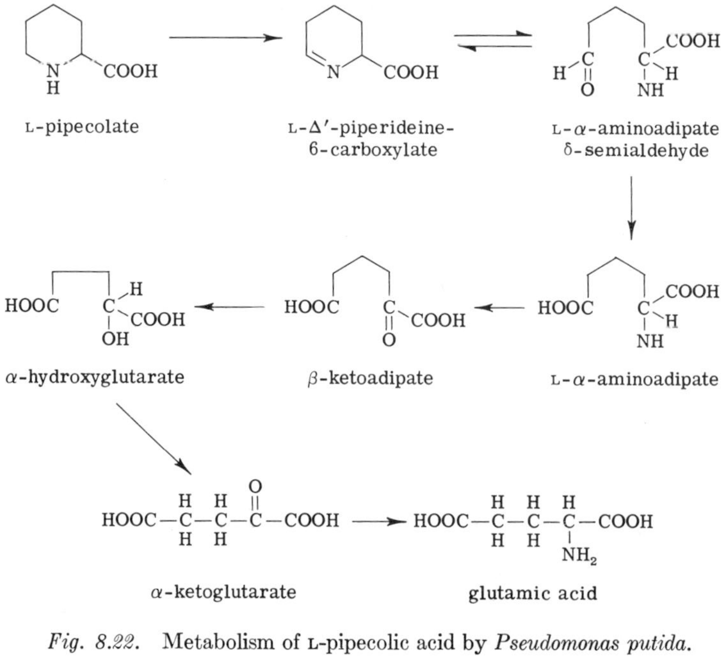

COOH
L-pipecolate
L-Δ′-piperideine-6-carboxylate
L-α-aminoadipate δ-semialdehyde
HOOC
α-hydroxyglutarate
β-ketoadipate
L-α-aminoadipate
HOOC—C—C—C—COOH
HOOC—C—C—C—COOH
NH2
α-ketoglutarate
glutamic acid

Fig. 8.22. Metabolism of L-pipecolic acid by *Pseudomonas putida*.

1.2.1.31) (30) oxidizes the semialdehyde to α-aminoadipate (190). A transamination reaction, together with a still uncharacterized enzyme, converts α-aminoadipate via β-ketoadipate to α-hydroxyglutarate. A second inducible, membrane-bound enzyme of the electron transport particle (ETP) of *Pseudomonas putida*, α-hydroxyglutarate oxidoreductase (EC 1.1.99.2), oxidizes α-hydroxyglutarate to α-ketoglutarate (193). The enzyme is specific for the L-isomer. As oxygen acts solely as a terminal electron acceptor, the electron carrier system is thought to be

$$\text{Substrate} \rightarrow \text{FP} \rightarrow \text{cyt.b} \rightarrow \text{cyt.c} \rightarrow O_2$$

The final conversion to glutamic acid involves a transaminase.

Metabolism of 2-Furoic Acid

The importance of this degradation pathway is that microorganisms are able to attack a furan ring, which is of interest to clinical research (240) (Fig. 8.23).

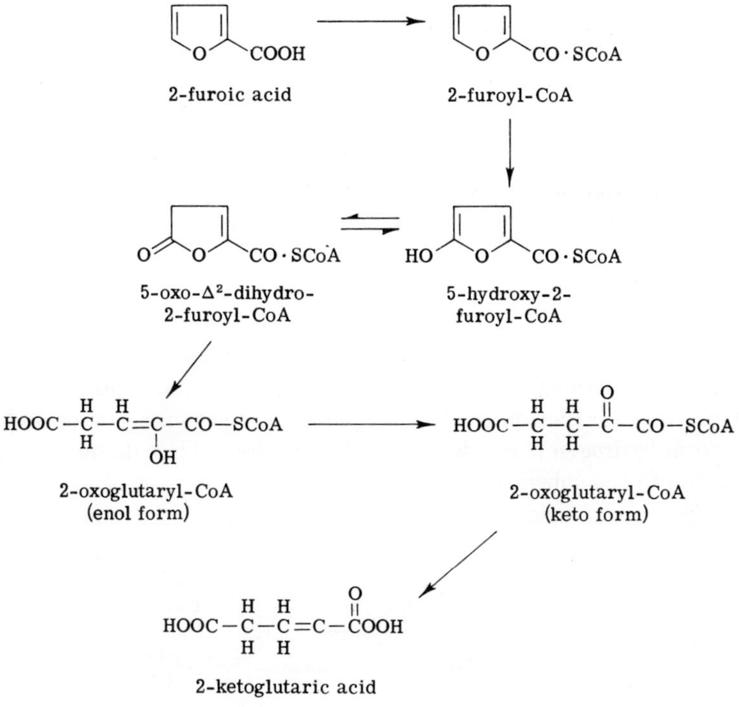
COOH
2-furoic acid
CO·SCoA
2-furoyl-CoA
CO·SCoA
5-oxo-Δ²-dihydro-
2-furoyl-CoA
HO
CO·SCoA
5-hydroxy-2-
furoyl-CoA
HOOC—C—C=C—CO—SCoA
OH
2-oxoglutaryl-CoA
(enol form)
HOOC—C—C—C—CO—SCoA
2-oxoglutaryl-CoA
(keto form)
HOOC—C—C=C—COOH
2-ketoglutaric acid

Fig. 8.23. Metabolism of 2-furoic acid (reprinted with permission of the author and the Biochemical Society, London).

2-Furoic acid is converted to its CoA ester via an energy-requiring reaction catalyzed by L-furoyl-CoA synthetase. The following hydroxylation system incorporates a hydroxyl group at the 5 position. A keto–enol tautomerism establishes an equilibrium between 5-hydroxy-2-furoyl-CoA and 5-oxo-Δ^2-dihydro-2-furoyl-CoA. The latter form is catalyzed by a lactone hydrolase that forms the enol form of 2-oxoglutaryl-CoA. This latter compound is converted to its keto form and by the action of a thiolester hydrolase into the end product 2-oxoglutarate. This oxidation of 2-furoate to 2-oxoglutarate was not inhibited by substrate analogs and no evidence was obtained for the occurrence of 2,5-dioxovalerate as an intermediate.

Oxygenases

Oxygenases are a group of enzymes that catalyze the incorporation of molecular oxygen into various organic compounds (93). They are recognized by the Enzyme Commission of the IUB as enzymes catalyzing the addition of molecular oxygen across a double bond between two carbon atoms. While doing so, they add both atoms of the oxygen molecule to the substrate (Fig. 8.24). It is important to note that both oxygen atoms are added to the substrate and that none of the added atoms arise from the solvent water. Hayaishi (105) uses the term "dioxygenase" for this type of enzyme.

In the same subgroup belong the enzymes called "hydroxylases." These are enzymes that also catalyze the addition of molecular oxygen to a substrate, but only one atom of the molecular oxygen is incorporated into the substrate with the second atom forming water (Fig. 8.25). In this reaction H_2X represents an electron donor. Hayaishi (105) uses the term "monooxygenase" for this type of enzyme, whereas Mason and co-workers (167) prefer to call these enzymes "mixed function oxidases."

An oxygenase should be distinctly separated from an oxidase. The latter are enzymes that use molecular oxygen as an electron acceptor and therefore form hydrogen peroxide or water as product. They do not incorporate oxygen into the substrate.

Oxygenases are widely distributed in nature. They are important in both

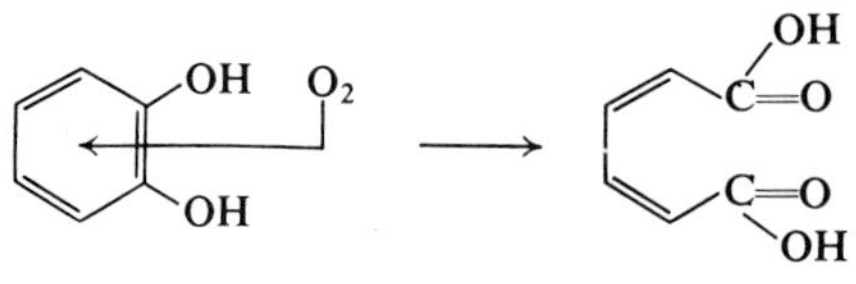
OH
O2
OH
OH
C=O
C=O
OH

Fig. 8.24. Mode of action of an oxygenase (dioxygenase).

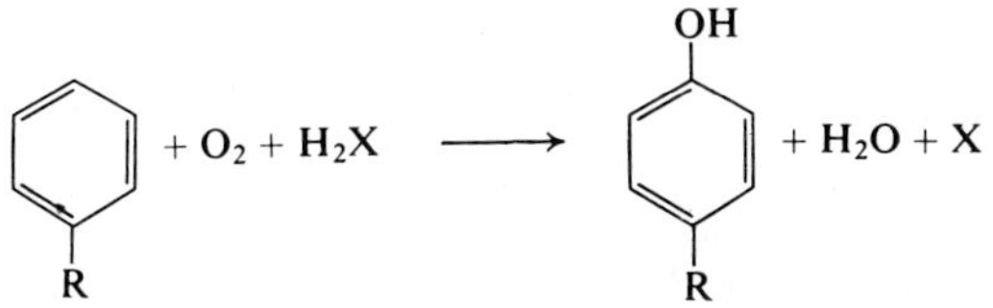

Fig. 8.25. Mode of action of a hydroxylase (monooxygenase).

catabolic and synthetic reactions, generally involving aromatic compounds. They may account in part for the wide versatility of the pseudomonads with respect to utilizable substrates. In these pathways they are often the first enzyme and transform generally nonreactive hydrocarbons, such as benzene and naphthalene, to ones that can be utilized by the common metabolic pathways. Oxygenases oxygenate *n*-alkanes to give the fatty acids, which are then metabolized by ω- or β-oxidation.

Further references on this subject in regard to the detailed mechanism of the oxygenases can be obtained in a number of excellent available reviews and papers (100–106, 134, 144, 171). There are, of course, still a great number of cyclic compounds for which the metabolism has not yet been clarified.

References

1. Abbott, B. J., and Gerhardt, G. (1970). Dialysis fermentation. I. Enhanced production of salicylic acid from naphthalene by *Pseudomonas fluorescens*. *Biotechnol. Bioeng.* **12,** 577.
2. Albrecht, K., Tomorkeny, E., and Szabo, A. (1970). Anaerobic transformation of steroids by *Mycobacterium phlei*. *Acta Microbiol.* **17,** 195.
3. Alexander, M. (1965). Biodegradation: Problems of molecular recalcitrance and microbial fallibility. *Appl. Microbiol.* **7,** 35.
4. Audus, L. J. (1950). Biological detoxication of 2,4-dichlorophenoxyacetic acid in soils: Isolation of an effective organism. *Nature (London)* **166,** 356.
5. Audus, L. J. (1964). "The Physiology and Biochemistry of Herbicides." Academic Press, New York.
6. Azoulay, E., Chouteau, J., and Davidovics, G. (1963). Isolement et caractérisation des enzymes responsibles de l'oxydation des hydrocarbures. *Biochim. Biophys. Acta* **77,** 554.
7. Baginsky, M. L., and Rodwell, V. W. (1966). Metabolism of pipecolic acid in a *Pseudomonas* species. IV. Electron transport particle of *Pseudomonas putida*. *J. Bacteriol.* **92,** 424.
8. Baginsky, M. L., and Rodwell, V. W. (1967). Metabolism of pipecolic acid in a *Pseudomonas* species. V. Pipecolate oxidase and dehydrogenase. *J. Bacteriol.* **94,** 1034.
9. Baptist, J. N., Gholson, R. K., and Minor, J. C. (1963). Hydrocarbon oxidation by a bacterial enzyme system. 1. Products of octane oxidation. *Biochim. Biophys. Acta* **69,** 40.

10. Barz, W., and Stadtman, E. R. (1969). Bacterial degradation of riboflavin. VII. Studies on the bacterial decomposition of 6,7-dimethylquinoxaline-2,3-diol. *Arch. Mikrobiol.* **67,** 128.
11. Basso, L. V., Rao, D. R., and Rodwell, V. W. (1962). Metabolism of pipecolic acid in *Pseudomonas* species. II. Δ'-Piperideine 6-carboxylic acid and α-aminoadipic acid-δ-semialdehyde. *J. Biol. Chem.* **237,** 2239.
12. Bayley, R. C., and Dagley, S. (1969). Oxoenoic acids as metabolites in the bacterial degradation of catechols. *Biochem. J.* **111,** 303.
12a. Bayly, R. C., and Wigmore, G. J. (1973). Metabolism of phenol and cresols by mutants of *Pseudomonas putida. J. Bacteriol.* **113,** 1112.
13. Bell, G. R. (1957). Some morphological and biochemical characteristics of a soil bacterium which decomposes 2,4-dichlorophenoxy acetic acid. *Can. J. Microbiol.* **3,** 821.
14. Beveridge, E. G., and Tall, D. (1969). The metabolic activity of phenol analogs to *Bacterium* NCIB 8250. *J. Appl. Bacteriol.* **32,** 304.
15. Bird, J. A., and Cain, R. B. (1968). *Cis,cis*-muconate, the product inducer of catechol 1,2-oxygenase in *Pseudomonas aeruginosa. Biochem. J.* **109,** 479.
15a. Bird, J. A., and Cain, R. B. (1974). Microbial degradation of alkylbenzenesulphonates. Metabolism of homologues of short alkyl-chain length by an *Alcaligenes* sp. *Biochem. J.* **140,** 121.
16. Blakley, E. R., Kurz, W., Halvorson, H., and Simpson, F. J. (1967). The metabolism of phenylacetic acid by a pseudomonad. *Can. J. Microbiol.* **13,** 147.
17. Bollag, J. M., Helling, C. S., and Alexander, M. (1967). Metabolism of 4-chloro-2-methylphenoxyacetic acid by soil bacteria. *Appl. Microbiol.* **15,** 1393.
18. Bollag, J. M., Helling, C. S., and Alexander, M. (1968). 2,4-D metabolism. Enzymatic hydroxylation of chlorinated phenols. *J. Agr. Food Chem.* **16,** 826.
19. Bollag, J. M., Briggs, G. G., Dawson, J. E., and Alexander, M. (1968). 2,4-D metabolism. Enzymatic degradation of chlorocatechols. *J. Agr. Food Chem.* **16,** 829.
19a. Boyd, G. S. (1972). Biological hydroxylation reactions. *In* "Biological Hydroxylation Mechanisms" (G. S. Boyd and R. M. S. Smellie, eds.), p. 1. Academic Press Inc., New York.
20. Boyer, R. F., Lode, E. T., and Coon, M. J. (1971). Reduction of alkyl hydroperoxides to alcohols: Role of rubredoxin, an electron carrier in the bacterial hydroxylation of hydrocarbons. *Biochem. Biophys. Res. Commun.* **44,** 925.
21. Brisbane, P. G., and Ladd, J. N. (1965). The role of microorganisms in petroleum exploration. *Annu. Rev. Microbiol.* **19,** 351.
22. Buki, K. G., Ambrus, G., and Szabo, A. (1969). Microbiological decomposition of 17-methyl-17-hydroxy steroids with androstane nucleus. *Acta Microbiol.* **16,** 253.
23. Büning-Pfaue, H., and Rehm, H.-J. (1972). Anreicherung der Aldehydzwischenstufe aus der wachsenden Kultur von *Pseudomonas aeruginosa* beim Abbau von Tetradecanol. *Arch. Mikrobiol.* **82,** 213.
24. Burg, R. W., and Snell, E. E. (1969). The bacterial oxidation of vitamin B_6. VI. Pyridoxal dehydrogenase and 4-pyridoxolactonase. *J. Biol. Chem.* **244,** 2585.
25. Caglar, M. A., Thompson, A. R., Houston, C. W., and Rose, V. C. (1969). Kinetics of microbial oxidation of *n*-heptane. I. Characterization of the process. *Biotechnol. Bioeng.* **11,** 417.
26. Cain, R. B. (1966). Utilization of anthranilic acid and nitrobenzoic acid by *Nocardia opaca* and a *Flavobacterium. J. Gen. Microbiol.* **42,** 219.

27. Cain, R. B. (1968). Anthranilic acid metabolism by microorganisms. Formation of 5-hydroxyanthranilate as an intermediate in anthranilate metabolism by *Nocardia opaca*. *Antonie van Leeuwenhoek; J. Microbiol. Serol.* **34,** 417.

28. Cain, R. B., Trauter, E. K., and Darrah, J. A. (1968). The utilization of some halogenated aromatic acids by *Nocardia*. Oxidation and metabolism. *Biochem. J.* **106,** 211.

29. Cain, R. B., and Farr, D. R. (1968). Metabolism of arylsulfonates by microorganisms. *Biochem. J.* **106,** 859.

30. Calvert, A. F., and Rodwell, V. W. (1966). Metabolism of pipecolic acid in a *Pseudomonas* species. III. L-α-aminoadipate-δ-semialdehyde:NAD oxidoreductase. *J. Biol. Chem.* **241,** 409.

31. Canovas, J. L., and Stanier, R. Y. (1967). Regulation of the enzymes of the β-ketoadipate pathway in *Moraxella calcoaceticus*. I. General aspects. *Eur. J. Biochem.* **1,** 28.

32. Canovas, J. L., Wheelis, M. L., and Stanier, R. Y. (1968). Regulation of the enzymes of the β-ketoadipate pathway in *Moraxella calcoaceticus*. 2. The role of protocatechuate as inducer. *Eur. J. Biochem.* **3,** 293.

33. Canovas, J. L., Johnson, B. F., and Wheelis, M. L. (1968). Regulation of the enzymes of the β-ketoadipate pathway in *Moraxella calcoaceticus*. 3. Effects of 3-hydroxy-4-methylbenzoate on the synthesis of enzymes of the protocatechuate branch. *Eur. J. Biochem.* **3,** 305.

34. Canovas, J. L., and Johnson, B. F. (1968). Regulation of the β-ketoadipate pathway in *Moraxella calcoacetica*. 4. Constitutive synthesis of β-ketoadipate succinyl-CoA transferase II and III. *Eur. J. Biochem.* **3,** 312.

35. Cardini, G., and Jurtshuk, P. (1968). Cytochrome P-450 involvement in the oxidation of *n*-octane by cell-free extracts of *Corynebacterium* sp. strain 7 E1C. *J. Biol. Chem.* **243,** 6070.

36. Cardini, G., and Jurtshuk, P. (1970). The enzymatic hydroxylation of *n*-octane by *Corynebacterium* sp. strain 7 E 1C. *J. Biol. Chem.* **245,** 2789.

37. Catelani, D., Fiecchi, A., and Galli, E. (1971). (+)-γ-Carboxymethyl-γ-methyl-Δ^{α}-butenolide. A 1,2-ring fission product of 4-methylcatechol by *Pseudomonas desmolyticum*. *Biochem. J.* **121,** 89.

38. Catterall, F. A., Sala-Trepat, J. M., and Williams, P. A. (1971). The coexistence of two pathways for the metabolism of 2-hydroxymuconic semialdehyde in a naphthalene-grown pseudomonad. *Biochem. Biophys. Res. Commun.* **43,** 463.

39. Chakrabarty, A. M., and Gunsalus, I. C. (1969). Autonomous regulation of a defective transducing phage in *Pseudomonas putida*. *Virology* **38,** 92.

40. Chamberlain, E. M., and Dagley, S. (1968). The metabolism of thymol by a *Pseudomonas*. *Biochem. J.* **110,** 755.

41. Chapman, P. J., and Hopper, D. J. (1968). The bacterial metabolism of 2,4-xylenol. *Biochem. J.* **110,** 491.

42. Chen, C., and Lin, C. C. (1968). Mechanism of aliphatic hydroxylation. Tetralin hydroperoxide as an intermediate in the hydroxylation of tetralin in rat-liver homogenates. *Biochim. Biophys. Acta* **170,** 366.

43. Chen, C., and Lin, C. C. (1969). [9-^{14}C]Fluorene hydroperoxide as a possible intermediate in the hydroxylation of [9-^{14}C]fluorene by rat-liver homogenates. *Biochim. Biophys. Acta* **184,** 634.

44. Christopherson, B. O. (1969). Reduction of linolenic acid hydroperoxide by a glutathione peroxidase. *Biochim. Biophys. Acta* **176,** 463.

44a. Cripps, R. E. (1971). The microbial breakdown of pesticides. *Soc. Appl. Bacteriol. Symp. Series* **1,** 255.
45. Dagley, S., and Stopher, D. A. (1959). A new mode of fission of the benzene nucleus by bacteria. *Biochem. J.* **73,** 16P.
46. Dagley, S., Evans, W. C., and Ribbons, D. W. (1960). New pathways in the oxidative metabolism of aromatic compounds by microorganisms. *Nature* (*London*) **188,** 560.
47. Dagley, S., Chapman, P. J., Gibson, D. T., and Wood, J. M. (1964). Degradation of the benzene nucleus by bacteria. *Nature* (*London*) **202,** 775.
48. Dagley, S., Chapman, P. J., and Gibson, D. T. (1965). The metabolism of β-phenylpropionic acid by an *Achromobacter*. *Biochem. J.* **97,** 643.
49. Dagley, S., Geary, P. J., and Wood, J. M. (1968). The metabolism of protocatechuate by *Pseudomonas testosteroni*. *Biochem. J.* **109,** 559.
50. Dagley, S. (1971). Catabolism of aromatic compounds by microorganisms. *Advan. Microbial Physiol.* **6,** 1.
51. Davies, J. I., and Evans, W. C. (1964). Oxidative metabolism of naphthalene by soil pseudomonads. The ring-fission mechanism. *Biochem. J.* **91,** 251.
52. Davis, R. S., Hossler, F. E., and Stone, R. W. (1968). Metabolism of *p*- and *m*-xylene by species of *Pseudomonas*. *Can. J. Microbiol.* **14,** 1005.
53. Druillet, R. E., Traxler, R. W., and Sobek, J. M. (1968). Bacterial utilization of cholesterol. *Antonie van Leeuwenhoek; J. Microbiol. Serol.* **34,** 315.
54. Duxbury, J. M., Tiedje, J. M., Alexander, M., and Dawson, J. E. (1970). 2,4-D metabolism: Enzymatic conversion of chloromaleylacetic acid to succinic acid. *J. Agr. Food Chem.* **18,** 199.
55. Engelhardt, G., Wallhofer, P. R., and Plapp, R. (1971). Degradation of Linuron and some other herbicides and fungicides by a linuron-inducible enzyme obtained from *Bacillus sphaericus*. *Appl. Microbiol.* **22,** 284.
56. Estabrook, R. W., Hildebrandt, A., and Ullrich, V. (1968). Oxygen interaction with reduced cytochrome P-450. *Hoppe-Seyler's Z. Physiol. Chem.* **349,** 1605.
56a. Estabrook, R. W., Baron, J., Peterson, J., and Ishimura, Y. (1972). Oxygenated cytochrome P-450 as an intermediate in hydroxylation reactions. *In* "Biological Hydroxylation Mechanisms" (G. S. Boyd and R. M. S. Smellie, eds.), p. 159. Academic Press Inc., New York.
57. Evans, W. C., and Smith, B. S. W. (1954). The photochemical inactivation and microbial metabolism of the chlorophenoxyacetic acid herbicides. *Biochem. J.* **57,** xxx.
58. Evans, W. C. (1957). Biochemistry of the oxidative metabolism of aromatic compounds by microorganisms. *Rep. Progr. Chem.* p. 279.
59. Evans, W. C., and Moss, P. (1957). The metabolism of the herbicide, *p*-chlorophenoxyacetic acid by a soil microorganism—The formation of a β-chloromuconic acid on ring fission. *Biochem. J.* **65,** 8P.
60. Evans, W. C., Smith, B. S. W., Moss, P., and Fernley, H. N. (1971). Bacterial metabolism of 4-chlorophenoxyacetate. *Biochem. J.* **122,** 509.
61. Evans, W. C., Smith, B. S. W., Fernley, H. N., and Davies, J. I. (1971). Bacterial metabolism of 2,4-dichlorophenoxyacetate. *Biochem. J.* **122,** 543.
62. Fedoseeva, G. E., Khesina, A. Y., Poglazova, M., Shabad, L. M., and Meisel, M. N. (1968). The oxidation of aromatic polycyclic hydrocarbons by microorganisms. *Dokl. Akad. Nauk SSSR* **183,** 208.
63. Feist, C. F., and Hegeman, G. D. (1969). Phenol and benzoate metabolism by *Pseudomonas putida:* Regulation of tangential pathways. *J. Bacteriol.* **100,** 869.

64. Feist, C. F., and Hegeman, G. D. (1969). Regulation of the meta cleavage pathway for benzoate oxidation by *Pseudomonas putida*. *J. Bacteriol.* **100,** 1121.
65. Fernley, H. N., Griffiths, E., and Evans, W. C. (1964). Oxidative metabolism of phenanthrene by soil bacteria: The initial ring fission step. *Biochem. J.* **91,** 15P.
66. Finnerty, W. R., Hawtrey, E., and Kallio, R. E. (1962). Alkane-oxidizing micrococci. *Z. Allg. Mikrobiol.* **2,** 169.
67. Finnerty, W. R., and Kallio, R. E. (1964). Origin of palmitic acid carbon in palmitates formed from hexadecane-1-^{14}C and tetradecane-1-^{14}C by *Micrococcus cerificans*. *J. Bacteriol.* **87,** 1261.
68. Focht, D. D., and Williams, F. D. (1970). The degradation of p-toluene sulfonate by a *Pseudomonas*. *Can. J. Microbiol.* **16,** 309.
69. Forney, F. W., Markovetz, A. J., and Kallio, R. E. (1967). Bacterial oxidation of 2-tridecanone to 1-undecanol. *J. Bacteriol.* **93,** 649.
70. Forney, F. W., and Markovetz, A. J. (1968). Oxidative degradation of methylketones. II. Chemical pathway for degradation of 2-tridecanone by *Pseudomonas multivorans* and *Pseudomonas aeruginosa*. *J. Bacteriol.* **96,** 1055.
71. Forney, F. W., and Markovetz, A. J. (1969). An enzyme system for aliphatic methylketone oxidation. *Biochem. Biophys. Res. Commun.* **37,** 31.
72. Forney, F. W., and Markovetz, A. J. (1970). Subterminal oxidation of aliphatic hydrocarbons. *J. Bacteriol.* **102,** 281.
73. Foster, J. W. (1962). Hydrocarbons as substrates for microorganisms. *Antonie van Leeuwenhoek; J. Microbiol. Serol.* **28,** 241.
74. Fredericks, K. M. (1967). Products of the oxidation of *n*-decane by *Pseudomonas aeruginosa* and *Mycobacterium rhodochrous*. *Antonie van Leeuwenhoek; J. Microbiol. Serol.* **33,** 41.
75. Fuhs, G. W. (1961). Der mikrobielle Abbau von Kohlenwasserstoffen. *Arch. Mikrobiol.* **39,** 394.
76. Fulco, A. J., and Mead, J. F. (1961). The biosynthesis of lignoceric cerebronic and nervonic acids. *J. Biol. Chem.* **236,** 2416.
77. Fulco, A. J. (1967). Chain elongation, 2-hydroxylation, and decarboxylation of long-chain fatty acids by yeast. *J. Biol. Chem.* **242,** 3608.
78. Gamar, Y., and Gaunt, J. K. (1971). Bacterial metabolism of 4-chloro-2-methylphenoxyacetate. Formation of glyoxylate by side-chain cleavage. *Biochem. J.* **122,** 527.
79. Gaunt, J. K., and Evans, W. C. (1961). Metabolism of 4-chloro-2-methylphenoxyacetic acid by a soil microorganism. *Biochem. J.* **79,** 25P.
80. Gaunt, J. K., and Evans, W. C. (1971). Metabolism of 4-chloro-2-methylphenoxyacetate by a soil pseudomonad. Preliminary evidence for the metabolic pathway. *Biochem. J.* **122,** 519.
81. Gaunt, J. K., and Evans, W. C. (1971). Metabolism of 4-chloro-2-methylphenoxyacetate by a soil pseudomonad. Ring fission, lactonizing, and delactonizing enzymes. *Biochem. J.* **122,** 533.
82. Gholson, R. K., Baptist, J. N., and Minor, J. C. (1963). Hydrocarbon oxidation by a bacterial enzyme system. II. Cofactor requirements for octanol formation from octane. *Biochem. Z.* **2,** 1155.
83. Gibson, D. T., Wood, J. M., Chapman, P. J., and Dagley, S. (1967). Bacterial degradation of aromatic compounds. *Biotechnol. Bioeng.* **9,** 33.
84. Gibson, D. T., Kocj, J. R., and Kallio, R. E. (1968). Oxidative degradation of aromatic hydrocarbons by microorganisms. I. Enzymatic formation of catechol from benzene. *Biochemistry* **7,** 2653.

85. Gibson, D. T., Cardini, G. E., Marseles, F. C., and Kallio, R. E. (1970). Incorporation of oxygen-18 into benzene by *Pseudomonas putida*. *Biochemistry* **9,** 1631.
86. Gibson, D. T. (1971). The microbial oxidation of aromatic hydrocarbons. *CRC Crit. Rev. Microbiol.* **1,** 199.
87. Goldman, P. (1965). The enzymatic cleavage of the carbon–fluorine bond on fluoroacetate. *J. Biol. Chem.* **240,** 3434.
88. Goldman, P., and Milne, G. W. A. (1966). Carbon–fluorine bond cleavage. II. Studies on the mechanism of the defluorination of fluoroacetate. *J. Biol. Chem.* **241,** 5557.
89. Goldman, P., Milne, G. W. A., and Pignatero, M. T. (1967). Fluorine-containing metabolites formed from 2-fluorobenzoic acid by *Pseudomonas* sp. *Arch. Biochem. Biophys.* **118,** 178.
90. Goldman, P., Milne, G. W. A., and Keister, D. B. (1968). Carbon–halogen bond cleavage. III. Studies on bacterial halido-hydrolases. *J. Biol. Chem.* **243,** 428.
91. Grant, D. J. W., and Patel, J. C. (1969). The nonoxidative decarboxylation of *p*-hydroxybenzoic acid, gentisic acid, protocatechuic acid, and gallic acid by *Klebsiella aerogenes* (*Aerobacter aerogenes*). *Antonie van Leeuwenhoek; J. Microbiol. Serol.* **35,** 325.
92. Gunsalus, C. F., Stanier, R. Y., and Gunsalus, I. C. (1953). The enzymatic conversion of mandelic acid to benzoic acid. III. Fractionation and properties of the soluble enzymes. *J. Bacteriol.* **66,** 548.
92a. Gunsalus, I. C., Lipscomb, J. D., Marshall, V., Frauenfelder, H., Greenbaum, E., and Munck, E. (1972). Structure and reactions of oxygenase active centers: Cytochrome P-450 and iron sulfur protein. *In* "Biological Hydroxylation Mechanisms" (G. S. Boyd and R. M. S. Smellie, eds.), p. 135. Academic Press Inc., New York.
93. Guroff, G., and Ho, T. (1963). Induced soluble phenylalanine hydroxylase from *Pseudomonas* grown on phenylalanine and tyrosine. *Biochim. Biophys. Acta* **77,** 157.
94. Hardisson, C., Sala-Trepat, J. M., and Stanier, R. Y. (1969). Pathways for the oxidation of aromatic compounds by *Azotobacter*. *J. Gen. Microbiol.* **59,** 1.
95. Harkness, D. R., Tsai, L., and Stadtman, E. R. (1964). Bacterial degradation of riboflavin. V. Stoichiometry of riboflavin degradation to oxamide and other products, oxidation of ^{14}C-labeled intermediates and isolation of the *Pseudomonas* effecting these transformations. *Arch. Biochem. Biophys.* **108,** 323.
96. Harkness, E. R., and Stadtman, E. R. (1965). Bacterial degradation of riboflavin. VI. Enzymatic conversion of riboflavin to 1-ribityl-2,3-diketo-1,2,3,4-tetrahydro-6,7-dimethyl quinoxaline, urea, and carbon dioxide. *J. Biol. Chem.* **240,** 4089.
97. Harper, D. B., and Blakley, E. R. (1971). The metabolism of *p*-fluorophenylacetic acid by a *Pseudomonas* sp. I. Isolation and identification of intermediates in degradation. *Can. J. Microbiol.* **17,** 635.
98. Harper, D. B., and Blakley, E. R. (1971). The metabolism of *p*-fluorophenylacetic acid by *Pseudomonas* sp. II. The degradative pathway. *Can. J. Microbiol.* **17,** 645.
99. Harper, D. B., and Blakley, E. R. (1971). The metabolism of *p*-fluorobenzoic acid by a *Pseudomonas* sp. *Can. J. Microbiol.* **17,** 1015.
100. Hayaishi, O., and Stanier, R. Y. (1951). Bacterial oxidation of tryptophan. III. Enzymatic activities of cell-free systems from bacteria employing the aromatic pathway. *J. Bacteriol.* **62,** 691.
101. Hayaishi, O., Katagiri, M., and Rothberg, S. (1955). Mechanism of the pyrocatechase reaction. *J. Amer. Chem. Soc.* **77,** 5450.

102. Hayaishi, O., Rothberg, S., Mehler, A. H., and Saito, Y. (1957). Studies on oxygenases, enzymatic formation of kynurenine from tryptophan. *J. Biol. Chem.* **229,** 889.

103. Hayaishi, O., Katagiri, M., and Rothberg, S. (1957). Studies on oxygenase-pyrocatechase. *J. Biol. Chem.* **229,** 905.

104. Hayaishi, O., ed. (1962). "Oxygenases." Academic Press, New York.

105. Hayaishi, O. (1966). Crystalline oxygenase of pseudomonads. *Bacteriol. Rev.* **30,** 720.

106. Hayaishi, O., and Nozaki, M. (1969). Nature and mechanisms of oxygenases. *Science* **164,** 389.

107. Hayaishi, O. (1969). Enzymic hydroxylation. *Annu. Rev. Biochem.* **38,** 21.

108. Hegeman, G. D. (1966). Synthesis of the enzymes of the mandelate pathway by *Pseudomonas putida*. I. Synthesis of enzymes by the wild type. *J. Bacteriol.* **91,** 1140.

109. Hegeman, G. D. (1966). Synthesis of the enzymes of the mandelate pathway by *Pseudomonas putida*. II. Isolation and properties of blocked mutants. *J. Bacteriol.* **91,** 1155.

110. Hegeman, G. D. (1966). Synthesis of the enzymes of the mandelate pathway of *Pseudomonas putida*. III. Isolation and properties of constitutive mutants. *J. Bacteriol.* **91,** 1161.

111. Hegeman, G. D., and Rosenberg, S. L. (1970). The evolution of bacterial enzyme systems. *Annu. Rev. Microbiol.* **24,** 429.

112. Hesp, B., Calvin, M., and Hosokawa, K. (1969). Studies on *p*-hydroxybenzoate hydroxylase from *Pseudomonas putida*. *J. Biol. Chem.* **244,** 5644.

113. Heydeman, M. T. (1960). Studies on a paraffin-utilizing pseudomonad. *Biochim. Biophys. Acta* **42,** 557.

114. Higgins, S. J., and Mandelstam, J. (1972). Regulation of pathways degrading aromatic substrates in *Pseudomonas putida*. Enzymic response to binary mixtures of substrates. *Biochem. J.* **126,** 901.

115. Hill, H. A. O., Roder, A., and Williams, R. J. P. (1970). Cytochrome P-450. Suggestions as to the structure and mechanisms of action. *Naturwissenschaften* **57,** 69.

116. Hopper, D. J., Chapman, P. J., and Dagley, S. (1968). Enzymic formation of D-malate. *Biochem. J.* **110,** 798.

117. Hopper, D. J., and Chapman, P. J. (1971). Gentisic acid and its 3- and 4-methyl-substituted homologs as intermediates in the bacterial degradation of *m*-cresol, 3,5-xylenol, and 2,5-xylenol. *Biochem. J.* **122,** 19.

118. Hopper, D. J., Chapman, P. J., and Dagley, S. (1971). The enzymic degradation of alkyl-substituted gentisates, maleates, and malates. *Biochem. J.* **122,** 29.

119. Hopwood, D. A. (1967). Genetic analysis and genome structure in *Streptomyces coelicolor*. *Bacteriol. Rev.* **31,** 373.

120. Hörhold, C., Böhme, K.-H., and Schubert, K. (1969). Über den Abbau von Steroiden durch Mikroorganismen. *Z. Allg. Mikrobiol.* **9,** 225.

121. Horvath, R. S. (1970). Cometabolism of methyl- and chloro-substituted catechols by an *Achromobacter* sp. possessing a new meta-cleaving oxygenase. *Biochem. J.* **119,** 871.

122. Horvath, R. S., and Koft, B. W. (1972). Degradation of alkylbenzene sulfonate by *Pseudomonas* species. *Appl. Microbiol.* **23,** 407.

123. Hosokawa, K., and Stanier, R. Y. (1966). Crystallization and properties of *p*-hydroxybenzoate hydroxylase from *Pseudomonas putida*. *J. Biol. Chem.* **241,** 2453.

124. Hosokawa, K. (1970). Regulation of synthesis of early enzymes of *p*-hydroxybenzoate pathway in *Pseudomonas putida*. *J. Biol. Chem.* **245,** 5304.

125. Ichihara, A., Adachi, K., Hosokawa, K., and Takeda, Y. (1962). The enzymatic hydroxylation of aromatic carboxylic acids; substrate specificities of anthranilate and benzoate oxidases. *J. Biol. Chem.* **237,** 2296.

126. Ihismura, Y., Ulrich, V., and Peterson, J. A. (1971). Oxygenated cytochrome P-450 and its possible role in enzymic hydroxylation. *Biochem. Biophys. Res. Commun.* **42,** 140.

127. Ingledew, W. M., Tresguerres, M. E. F., and Canovas, J. L. (1971). Regulation of the enzymes of the hydroaromatic pathway in *Acinetobacter calcoaceticus*. *J. Gen. Microbiol.* **68,** 273.

128. Jensen, H. L., and Petersen, H. I. (1952). Detoxication of hormone herbicides by soil bacteria. *Nature (London)* **170,** 39.

129. Jerina, D., Daly, J. Witkop, B., Zaltaman-Nirenberg, P., and Udenfriend, S. (1968). Role of the arene oxide–oxepin system in the metabolism of aromatic substrates. 1. *In vitro* conversion of benzene oxide to a premercapturic acid and a dihydrodiol. *Arch. Biochem. Biophys.* **128,** 176.

130. Jerina, D., Daly, J. W., Jeffrey, A. M., and Gibson, D. T. (1971). *Cis*-1,2-dihydroxy-1,2-dihydronaphthalene: A bacterial metabolite from naphthalene. *Arch. Biochem. Biophys.* **142,** 394.

131. Johnson, B. F., and Stanier, R. Y. (1971). Dissimilation of aromatic compounds by *Alcaligenes eutrophus*. *J. Bacteriol.* **107,** 468.

132. Johnson, B. F., and Stanier, R. Y. (1971). Regulation of the β-ketoadipate pathway in *Alcaligenes eutrophus*. *J. Bacteriol.* **107,** 476.

133. Kashiwamata, S., Nakashima, K., and Kotake, Y. (1966). Anthranilic acid hydroxylation by rabbit liver microsomes. *Biochim. Biophys. Acta* **113,** 244.

134. Katagiri, M., Maeno, H., Yamamoto, S., Hayaishi, O., Kito, H., and Oae, T. (1965). Salicylate hydroxylase—A monooxygenase requiring FAD. II. Mechanism of salicylate hydroxylation to catechol. *J. Biol. Chem.* **240,** 3414.

135. Katagiri, M., Takemori, S., Nakazawa, K., Suzuki, H., and Akagi, K. (1967). Benzylalcohol dehydrogenase, a new alcohol dehydrogenase from *Pseudomonas* sp. *Biochim. Biophys. Acta* **139,** 173.

136. Katagiri, M., Ganguli, B. N., and Gunsalus, I. C. (1968). A soluble cytochrome P-450 functional in methylene hydroxylation. *J. Biol. Chem.* **243,** 3542.

137. Katagiri, M., and Wheelis, M. L. (1971). Comparison of the two isofunctional enollactone hydrolases from *Acinetobacter calcoaceticus*. *J. Bacteriol.* **106,** 369.

138. Kearney, P. C., and Kaufman, D. D. (1969). "Degradation of Herbicides." Dekker, New York.

139. Kemp, M. B., and Hegeman, G. D. (1968). Genetic control of the β-ketoadipate pathway in *Pseudomonas aeruginosa*. *J. Bacteriol.* **96,** 1488.

140. Kennedy, S. I. T., and Fewson, C. A. (1968). Metabolism of mandelate and related compounds by *Bacterium* NCIB 8250. *J. Gen. Microbiol.* **53,** 259.

141. Kester, A. S., and Foster, J. W. (1963). Diterminal oxidation of long-chain alkanes by bacteria. *J. Bacteriol.* **85,** 859.

142. Killinger, A. (1970). Der Abbau von Undecan durch ein marines Bakterium. *Arch. Mikrobiol.* **73,** 160.

143. Klug, M. J., and Markovetz, A. J. (1971). Utilization of aliphatic hydrocarbons by microorganisms. *Advan. Microbial Physiol.* **5,** 1.

144. Kojima, Y., Itada, N., and Hayaishi, O. (1961). Metapyrocatechase: A new catechol enzyme. *J. Biol. Chem.* **236,** 2223.

145. Kurz, W. G. W., Dawson, P. S. S., and Blakley, E. R. (1969). A comparative study *in vivo* of enzyme activities in batch, continuous, and phased cultures of a pseudomonad grown on phenylacetic acid. *Can. J. Microbiol.* **15,** 27.

146. Kusunose, M., Kusunose, E., and Coon, M. J. (1964). Enzymatic ω-oxidation of fatty acids. I. Products of octanoate, decanoate, and laurate oxidation. *J. Biol. Chem.* **239,** 1374.

147. Kusunose, M., Kusunose, E., and Coon, M. (1964). Enzymatic ω-oxidation of fatty acids. II. Substrate specificity and other properties of the enzyme system. *J. Biol. Chem.* **239,** 2135.

148. Lack, L. (1959). The enzymic oxidation of gentisic acid. *Biochim. Biophys. Acta* **34,** 117.

149. Lack, L. (1961). Enzymic cis–trans isomerization of maleylpyruvic acid. *J. Biol. Chem.* **236,** 2835.

150. Leadbetter, E. R., and Foster, J. W. (1960). Bacterial oxidation of gaseous alkanes. *Arch. Mikrobiol.* **35,** 92.

151. Lennarz, W. J. (1966). Lipid metabolism in the bacteria. *Advan. Liquid Res.* **4,** 177.

152. Lestrovaja, N. N., and Bukhar, M. I. (1970). Proof for the participation of two enzymes in the process of microbiological 1,2-dehydration and reduction of the Δ′-bond in the A-ring of steroids. *Biokhimiya* **35,** 843.

153. Levy, C. C. (1964). Metabolism of coumarin by a microorganism: *o*-Coumaric acid as an intermediate between coumarin and melilotic acid. *Nature* (*London*) **203,** 1059.

154. Levy, C. C., and Frost, P. (1966). The metabolism of coumarin by a microorganism. V. Melilotate hydroxylase. *J. Biol. Chem.* **241,** 997.

155. Little, C., Olinescu, R., Reid, K. G., and O'Brien, P. J. (1970). Properties and regulation of glutathione peroxidase. *J. Biol. Chem.* **245,** 3632.

156. Lode, E. T., and Coon, M. J. (1971). Enzymatic ω-oxidation. V. Forms of *Pseudomonas oleovorans* rubredoxin containing one or two iron atoms: Structure and function in ω-hydroxylation. *J. Biol. Chem.* **246,** 791.

157. Loos, M. A., Roberts, R. N., and Alexander, M. (1967). Formation of 2,4-dichlorophenol and 2,4-dichloroanisole from 2,4-dichlorophenoxyacetate by *Arthrobacter* sp. *Can. J. Microbiol.* **13,** 691.

158. Lukins, H. B., and Foster, J. W. (1963). Utilization of hydrocarbons and hydrogen by mycobacteria. *Z. Allg. Mikrobiol.* **3,** 251.

159. MacDonald, D. L., Stanier, R. Y., and Ingraham, J. L. (1954). The enzymatic formation of γ-carboxymuconic acid. *J. Biol. Chem.* **210,** 809.

160. McKenna, E. J., and Kallio, R. E. (1965). The biology of hydrocarbons. *Annu. Rev. Microbiol.* **19,** 183.

161. McKenna, E. J., and Coon, M. J. (1970). Enzymatic ω-oxidation. IV. Purification and properties of the ω-hydroxylase of *Pseudomonas oleovorans*. *J. Biol. Chem.* **245,** 3882.

162. MacLennan, D. G., and Pirt, S. J. (1970). The dynamics of decane and glucose utilization by a *Pseudomonas* sp. in batch and chemostat cultures under controlled dissolved oxygen tension. *J. Appl. Bacteriol.* **33,** 390.

163. MacRae, I. C., and Alexander, M. (1963). Metabolism of phenoxyalkyl carboxylic acids by a *Flavobacterium* species. *J. Bacteriol.* **86,** 1231.

164. MacRae, I. C., Alexander, M., and Rovira, A. D. (1963). The decomposition of 4-(2,4-dichlorophenoxy)butyric acid by *Flavobacterium* sp. *J. Gen. Microbiol.* **32,** 69.

165. Mandelstam, J., and Jacoby, G. A. (1965). Induction and multisensitive end-product repression in the enzymic pathway degrading mandelate in *Pseudomonas fluorescens*. *Biochem. J.* **94,** 569.
166. Martonova, M., Škárka, B., and Raděj, Z. (1972). Degradation of naphthalene to salicylic acid by cultures of *Pseudomonas denitrificans* and *Achromobacter* sp. from the effluents of petroleum refinery. *Folia Microbiol.* (*Prague*) **17,** 63.
167. Mason, H. S., Fowlks, W. L., and Peterson, E. (1955). Oxygen transfer and electron transport by the phenolase complex. *J. Am. Chem. Soc.* **77,** 2914 [as cited by Dagley (50)].
167a. Meagler, R. B., and Ornston, L. N. (1973). Relationships among enzymes of the β-ketoadipate pathway. I. Properties of *cis,cis*-muconate-lactonizing enzyme and muconolactone isomerase from *Pseudomonas putida*. *Biochemistry* **12,** 3523.
168. Miles, H. T., Smyrniotis, P. Z., and Stadtman, E. R. (1959). Bacterial degradation of riboflavin. III. Isolation, structure, determination, and biological transformations of 1-ribityl-2,3-diketo-1,2,3,4-tetrahydro-6,7-dimethyl quinoxaline. *J. Amer. Chem. Soc.* **81,** 1964.
169. Milne, G. W. A., Goldman, P., and Holtzman, J. L. (1968). The metabolism of 2-fluorobenzoic acid. II. Studies with $^{18}O_2$. *J. Biol. Chem.* **243,** 5374.
169a. Murray, K., Duggleby, C. J., Sala-Trepat, J. M., and Williams, P. A. (1972). The metabolism of benzoate and methylbenzoates via the meta-cleavage pathway by *Pseudomonas arvilla* mt-2. *Eur. J. Biochem.* **28,** 301–310.
169b. Murray, K., and Williams, P. A. (1974). Role of catechol and the methylcatechols as inducers of aromatic metabolism in *Pseudomonas putida*. *J. Bacteriol.* **117,** 1153.
170. Nakamura, S., Ogura, Y., Yano, K., Hagoshi, N., and Arima, K. (1970). Kinetic studies on the reaction mechanism of *p*-hydroxybenzoate hydroxylase. *Biochemistry* **9,** 3235.
171. Nakazawa, T., Kojima, Y., Fujiwasa, H., Mazaki, M., Hayaishi, O., and Yamano, T. (1965). Studies on the mechanisms of pyrocatechase by ERS spectroscopy. *J. Biol. Chem.* **240,** PC3224.
171a. Nakazara, T., and Yokota, T. (1973). Benzoate metabolism in *Pseudomonas putida* (arvilla) mt-2: Demonstration of two benzoate pathways. *J. Bacteriol.* **115,** 262.
172. Nei, M. (1967). Modification of linkage intensity by natural selection. *Genetics* **57,** 625.
173. Nozaki, M., Kotani, S., Ono, K., and Senoh, S. (1970). Metapyrocatechase. III. Substrate specificity and mode of ring fission. *Biochim. Biophys. Acta* **220,** 213.
174. Nyns, E. J., Zach, D., and Snell, E. E. (1969). The bacterial degradation of vitamin B_6. VIII. Enzymatic breakdown of α-(*N*-acetylaminomethylene)succinic acid. *J. Biol. Chem.* **244,** 2601.
175. Ono, K., Nozaki, M., and Hayaishi, O. (1970). Purification and some properties of protocatechuate, 4,5-dioxygenase. *Biochim. Biophys. Acta* **220,** 224.
176. Ornston, L. N., and Stanier, R. Y. (1966). The conversion of catechol and protocatechuate to β-ketoadipic acid by *Pseudomonas putida*. I. Biochemistry. *J. Biol. Chem.* **241,** 3776.
177. Ornston, L. N. (1966). The conversion of catechol and protocatechuate to β-ketoadipate by *Pseudomonas putida*. III. Enzymes of the catechol pathway. *J. Biol. Chem.* **241,** 3795.
178. Ornston, L. N. (1966). The conversion of catechol and protocatechuate to β-ketoadipate by *Pseudomonas putida*. IV. Regulation. *J. Biol. Chem.* **241,** 3800.
179. Ornston, L. N. (1971). Regulation of catabolic pathways in *Pseudomonas*. *Bacteriol. Rev.* **35,** 87.

179a. Parke, D., Meagher, R. B., and Ornston, L. N. (1973). Relationships among enzymes of the β-ketoadipate pathway. III. Properties of crystalline γ-carboxymuconolactone decarboxylase from *Pseudomonas putida Biochemistry* **12,** 3537.

180. Patel, J. C., and Grant, D. J. W. (1969). The formation of phenol in the degradation of *p*-hydroxybenzoic acid by *Klebsiella aerogenes* (*Aerobacter aerogenes*). *Antonie van Leeuwenhoek; J. Microbiol. Serol.* **35,** 53.

180a. Patel, R. N., Meagher, R. B., and Ornston, L. N. (1973). Relationships among enzymes of the β-ketoadipate pathway. II. Properties of crystalline β-carboxy-*cis,cis*-muconate-lactonizing enzyme from *Pseudomonas putida. Biochemistry* **12,** 3531.

181. Peterson, J. A., Basu, D., and Coon, M. J. (1966). Enzymatic ω-oxidation. I. Electron carriers in fatty acid and hydrocarbon hydroxylation. *J. Biol. Chem.* **241,** 5162.

182. Peterson, J. A., Kusunose, M., Kusunose, E., and Coon, M. J. (1967). Enzymatic ω-oxidation. II. Function of rubredoxin as the electron carrier in ω-hydroxylation. *J. Biol. Chem.* **242,** 4334.

183. Peterson, J. A., and Coon, M. J. (1968). Enzymatic ω-oxidation. III. Purification and properties of rubredoxin, a component of the ω-hydroxylation system of *Pseudomonas oleovorans. J. Biol. Chem.* **243,** 329.

184. Peterson, J. A., McKenna, E. J., Estabrook, R. W., and Coon, M. J. (1969). Enzymic ω-hydroxylation: Stoichiometry of ω-oxidation of fatty acids. *Arch. Biochem. Biophys.* **131,** 245.

185. Peterson, J. A. (1970). Cytochrome content of two pseudomonads containing mixed function oxidase systems. *J. Bacteriol.* **103,** 714.

186. Poglazova, M. N., Fedosseva, G. E., Khesina, A. Y., Meisel, M. N., and Shabad, L. M. (1967). Further investigations of the decomposition of benz(A)pyrene by soil bacteria. *Dokl. Akad. Nauk SSSR* **176,** 1165.

187. Poglazova, M. N., Fedoseeva, G. E., Khesina, A. Y., Meisel, M. N., and Shabad, L. M. (1968). The oxidation of benz(A)pyrene by microorganisms in relation to its concentration in the medium. *Dokl. Akad. Nauk SSSR* **179,** 1460.

188. Proctor, M. H. (1960). A paraffin-oxidizing pseudomonad. *Biochim. Biophys. Acta* **42,** 559.

189. Raab, W., and Windisch, J. (1969). Zur Frage einer Metabolisierung von Cortocosteroiden durch *B. pyocyaneus. Arch. Klin. Exp. Dermatol.* **235,** 234.

190. Rao, D. R., and Rodwell, V. W. (1962). Metabolism of pipecolic acid in a *Pseudomonas* species. I. β-Aminoadipic and glutamic acids. *J. Biol. Chem.* **237,** 2232.

191. Ravdin, R. G., and Crandall, D. I. (1951). The enzymatic conversion of homogentisic acid to 4-fumarylacetoacetic acid. *J. Biol. Chem.* **189,** 137.

192. Reiner, A. M. (1971). Metabolism of benzoic acid by bacteria: 3,5-Cyclohexadiene-1,2-diol-1-carboxylic acid is an intermediate in the formation of catechol. *J. Bacteriol.* **108,** 89.

193. Reitz, M. S., and Rodwell, V. W. (1969). α-Hydroxyglutarate oxidoreductase of *Pseudomonas putida. J. Bacteriol.* **100,** 708.

194. Ribbons, D. W. (1970). Specificity of monohydric phenol oxidation by meta-cleavage pathways in *Pseudomonas aeruginosa* T1. *Arch. Mikrobiol.* **74,** 103.

195. Ripin, M. J., Noon, K. F., and Cook, T. M. (1971). Bacterial metabolism of arylsulfonates. I. Benzene sulfonate as growth substrate for *Pseudomonas testosteroni* A-8. *Appl. Microbiol.* **21,** 495.

196. Robbins, K. C. (1961). Enzymatic omega oxidation of fatty acids. *Fed. Proc., Fed. Amer. Soc. Exp. Biol.* **20,** 273.

197. Robert-Gero, M., Poiret, M., and Stanier, R. Y. (1969). The function of the β-ketoadipate pathway in *Pseudomonas acidovorans*. *J. Gen. Microbiol.* **57,** 207.

198. Rogoff, M. H., and Reid, J. J. (1956). Bacterial decomposition of 2,4-dichlorophenoxyacetic acid. *J. Bacteriol.* **71,** 303.

199. Rogoff, M. H., and Wende, I. (1957). The microbiology of coal. I. Bacterial oxidation of phenanthrene. *J. Bacteriol.* **73,** 264.

200. Rogoff, M. H., and Wende, I. (1957). 3-Hydroxy-2-naphthoic acid as an intermediate in bacterial dissimilation of anthracene. *J. Bacteriol.* **74,** 108.

201. Rosenberg, S. L. (1971). Regulation of the mandelate pathway in *Pseudomonas aeruginosa*. *J. Bacteriol.* **108,** 1257.

202. Rosenberg, S. L., and Hegeman, G. D. (1971). Genetics of the mandelate pathway in *Pseudomonas aeruginosa*. *J. Bacteriol.* **108,** 1270.

203. Saito, J., and Matsura, T. (1967). Chemical studies on riboflavin and related compounds. I. Oxidation of quinoxaline-2,3-diols as a possible model for the biological decomposition of riboflavin. *Biochemistry* **6,** 3602.

204. Sala-Trepat, J. M., and Evans, W. C. (1971). The metabolism of 2-hydroxymuconic semialdehyde by *Azotobacter* species. *Biochem. Biophys. Res. Commun.* **43,** 456.

204a. Sala-Trepat, J. M., and Evans, W. C. (1971). The meta-cleavage of catechol by *Azotobacter* species. 4-Oxalocrotonate pathway. *Eur. J. Biochem.* **20,** 400.

205. Sala-Trepat, J. M. Murray, K., and Williams, P. A. (1971). The physiological significance of the two divergent metabolic steps in the meta-cleavage of catechols by *Pseudomonas putida* NCIB 10105. *Biochem. J.* **124,** 20P.

205a. Sala-Trepat, J. M., Murray, K., and Williams, P. A. (1972). The metabolic divergence in the meta cleavage of catechols by *Pseudomonas putida* NCIB 10015. *Eur. J. Biochem.* **28,** 347.

206. Sanderson, K. E. (1970). Current linkage map of *Salmonella typhimurium*. *Bacteriol. Rev.* **34,** 176.

207. Schubert, A., Kunstmann, F. W., Onken, D., and Zepter, R. (1969). Über einige neue Ergebnisse der enzymatischen Umwandlung von C_{18}-Steroiden durch verschiedene Mikroorganismen. *Acta Biol. Med. Ger.* **22,** 699.

208. Seidman, M. M., Toms, A., and Wood, J. M. (1969). Influence of side-chain substituents on the position of cleavage of the benzene ring by *Pseudomonas fluorescens*. *J. Bacteriol.* **97,** 1192.

209. Sih, C. J., Wang, K. S., and Tai, H. H. (1968). Mechanism of steroid oxidation by microorganisms. XIII. C_{22}-acid intermediates in the degradation of the cholesterol side chain. *Biochemistry* **7,** 796.

210. Sih, C. J., Tai, H. H., Tsong, Y. Y., Lee, S. S., and Coombe, R. G. (1968). Mechanisms of steroid oxidation by microorganisms. XIV. Pathway of cholesterol side chain degradation. *Biochemistry* **7,** 808.

211. Sistrom, W. R., and Stanier, R. Y. (1954). The mechanism of formation of β-ketoadipic acid by bacteria. *J. Biol. Chem.* **210,** 821.

212. Smithe, A., Trauter, E. K., and Cain, R. B. (1968). The utilization of some halogenated aromatic acids by *Nocardia*. Effects on growth and enzyme induction. *Biochem. J.* **106,** 203.

213. Smyrniotis, P. Z., Miles, H. T., and Stadtman, E. R. (1958). Isolation and structure of 3,4-dimethyl-6-carboxy-Δ-pyrone as a bacterial degradation product of riboflavin. *J. Amer. Chem. Soc.* **80,** 2541.

214. Sparrow, L. G., Ho, P. P. K., Sundaram, T. K., Zach, D., Nyns, E. J., and Snell, E. E. (1969). The bacterial oxidation of vitamin B_6. VII. Purification, properties,

and mechanism of action of an oxygenase which cleaves the 3-hydroxypyridine ring. *J. Biol. Chem.* **244,** 2590.

215. Stahl, F. W., and Murray, N. (1966). The evolution of gene clusters and genetic circularity in microorganisms. *Genetics* **53,** 569.
216. Stanier, R. Y. (1947). Simultaneous adaptation: A new technique for the study of metabolic pathways. *J. Bacteriol.* **54,** 339.
216a. Stanier, R. Y., and Ornston, L. N. (1973). The β-ketoadipate pathway. *Advan. Microbial Physiol.* **9,** 89.
217. Steenson, T. I., and Walker, N. (1956). Observations on the bacterial oxidation of chlorophenoxyacetic acids. *Plant Soil* **8,** 17.
218. Steenson, T. I., and Walker, N. (1957). The pathway of breakdown of 2:4-dichloro- and 4-chloro-2-methylphenoxyacetic acid by bacteria. *J. Gen. Microbiol.* **16,** 146.
219. Steenson, T. I., and Walker, N. (1958). Adaptive patterns in the bacterial oxidation of 2:4-dichloro- and 4-chloro-2-methylphenoxyacetic acid. *J. Gen. Microbiol.* **18,** 692.
220. Stevenson, I. L. (1967). Utilization of aromatic hydrocarbons by *Arthrobacter* spp. *Can. J. Microbiol.* **13,** 205.
221. Stevenson, I. L., and Mandelstam, J. (1965). Induction and multisensitive end product repression in two converging pathways degrading aromatic substances in *Pseudomonas fluorescens. Biochem. J.* **96,** 354.
222. Stewart, J. E., Kallio, R. E., Stevenson, D. P., Jones, A. C., and Schissler, D. O. (1959). Bacterial hydrocarbon oxidation I. Oxidation of *n*-hexadecane by a gram-negative coccus. *J. Bacteriol.* **78,** 441.
223. Stewart, J. E., and Kallio, R. E. (1959). Bacterial hydrocarbon oxidation. II. Ester formation from alkanes. *J. Bacteriol.* **78,** 726.
224. Strawinski, R. J., and Stone, R. W. (1943). Conditions governing the oxidation of naphthalene and the chemical analysis of its products. *J. Bacteriol.* **45,** 16.
225. Sundaram, T. K., and Snell, E. E. (1969). The bacterial oxidation of vitamin B_6. V. The enzymatic formation of pyridoxal and isopyridoxal from pyridoxine. *J. Biol. Chem.* **244,** 2577.
226. Takemori, S., Yasuda, H., Mihara, K., Suzuki, K., and Katagiri, M. (1969). Mechanism of the salicylate hydroxylase reaction. II. The enzyme–substrate complex. *Biochim. Biophys. Acta* **191,** 58.
227. Taniuchi, H., Hatanaka, M., Kuno, S., Hayaishi, O., Nakajima, M., and Kurihara, N. (1964). Enzymatic formation of catechol from anthranilic acid. *J. Biol. Chem.* **239,** 2204.
228. Taylor, A. L. (1970). Current linkage map of *Escherichia coli. Bacteriol. Rev.* **34,** 155.
229. Taylor, H. F., and Wain, R. L. (1962). Side chain degradation of certain ω-phenoxyalkane carboxylic acids by *Nocardia coeliaca* and other microorganisms isolated from soil. *Proc. Roy. Soc. Ser. B* **268,** 172.
230. Teng, N., Kotowycz, G., Calvin, M., and Hosokawa, K. (1971). Mechanism of action of *p*-hydroxybenzoate hydroxylase from *Pseudomonas putida.* III. The enzyme–substrate complex. *J. Biol. Chem.* **246,** 5448.
231. Tiedje, J. M., Duxbury, J. M., Alexander, M., and Dawson, J. E. (1969). 2,4-D metabolism: Pathway of degradation of chlorocatechols by *Arthrobacter* sp. *J. Agr. Food Chem.* **17,** 1021.
232. Tiedje, J. M., and Alexander, M. (1969). Enzymatic cleavage of the ether bond of 2,4-dichlorophenoxyacetate. *J. Agr. Food Chem.* **17,** 1080.

233. Tomorkeny, E., Albrecht, K., and Lea, L. (1969). Transformation of 4,5-epoxy-steroids with *Mycobacterium phlei*. I. Transformation under aerobic conditions. *Acta Microbiol.* **16,** 261.

234. Tomorkeny, E., Albrecht, K., and Ila, L. (1970). Transformation of 4,5-epoxy-steroids with *Mycobacterium phlei*. II. Transformation under anaerobic conditions. *Acta Microbiol.* **17,** 199.

235. Treccani, V. (1953). Produzioni di acido salicilico da naftalina per azione di uno schizomicete isolata da terreni petroliferi. *Ann. Microbiol.* **5,** 232.

236. Treccani, V., Walker, N., and Wiltshire, G. H. (1954). The metabolism of naphthalene by soil bacteria. *J. Gen. Microbiol.* **11,** 341.

237. Treccani, V. (1965). Microbial degradation of aliphatic and aromatic hydrocarbons. *Z. Allg. Mikrobiol.* **5,** 332.

238. Treccani, V., Galli, E., Catelani, D., and Sorlini, C. (1968). Induction of 1,2- and 2,3-diphenol oxygenases in *Pseudomonas desmolyticum*. *Z. Allg. Mikrobiol.* **8,** 65.

239. Trippett, S., Dagley, S., and Stopher, D. A. (1960). Bacterial oxidation of protocatechuic acid. *Biochem. J.* **76,** 9P.

240. Trudgill, P. W. (1969). The metabolism of 2-furoic acid by *Pseudomonas* F 2. *Biochem. J.* **113,** 577.

241. Trust, T. J., and Millis, N. F. (1970). The isolation and characterization of alkane-oxidizing organisms and the effect of growth substrate on isocitrate lyase. *J. Gen. Microbiol.* **61,** 245.

242. Tsai, L., Smyrniotis, P. Z., Harkness, D., and Stadtman, E. R. (1963). Bacterial degradation of riboflavin. IV. Oxidation cleavage of 1-ribityl-2,3-diketo-1,2,3,4-tetrahydro-6,7-dimethyl quinoxaline to 6,7-dimethyl quinoxaline-2,3-diol and ribose. *Biochem. Z.* **338,** 561.

242a. Ullrich, V., Ruf, H.-H., and Mimoun, H. (1972). Model systems for monooxygenases. *In* "Biological Hydroxylation Mechanisms" (G. S. Boyd and R. M. S. Smellie, eds.), p. 11. Academic Press Inc., New York.

243. Updegraff, D. M., and Bovey, F. A. (1958). Reduction of organic hydroperoxides by microorganisms and animal tissues. *Nature* (*London*) **181,** 890.

244. Van der Linden, A. C., and Thijsse, G. J. E. (1965). The mechanisms of microbial oxidation of petroleum hydrocarbons. *Advan. Enzymol.* **27,** 469.

245. Van der Linden, A. C. (1967). Dissimilation of 1,6-hexanediol and 1,8-octanediol by a hydrocarbon-oxidizing *Pseudomonas*. *Antonie van Leeuwenhoek; J. Microbiol. Serol.* **33,** 386.

246. Van der Linden, A. C., and Huybregtse, R. (1969). Occurrence of inducible and NAD(P)-independent primary alcohol dehydrogenases in an alkane-oxidizing *Pseudomonas*. *Antonie van Leeuwenhoek; J. Microbiol. Serol.* **35,** 344.

247. Vestal, J. R., and Perry, J. J. (1969). Divergent metabolic pathways for propane and propionate utilization by a soil isolate. *J. Bacteriol.* **99,** 216.

248. Wagner, F., Zahn, W., and Buhring, U. (1967). 1-Hexadecene an intermediate in microbial oxidation of *n*-hexadecane *in vivo* and *in vitro*. *Angew. Chem., Int. Ed. Engl.* **6,** 359.

249. Wakil, S. H., and Barnes, E. M., Jr. (1971). Fatty acid metabolism. *Compr. Biochem.* 18S, 57.

250. Walker, N., and Wiltshire, G. H. (1953). The breakdown of naphthalene by a soil bacterium. *J. Gen. Microbiol.* **8,** 273.

251. Walker, R. L., and Newman, A. S. (1956). Microbial decomposition of 2,4-dichlorophenoxyacetic acid. *Appl. Microbiol.* **4,** 201.

252. Webley, D. M., Duff, R. B., and Farmer, V. C. (1957). Formation of β-hydroxy acid as an intermediate in the microbiological conversion of monochlorophenoxybutyric acids to the corresponding substituted acetic acids. *Nature (London)* **179,** 1130.
253. Wedemeyer, G. (1967). Dechlorination of 1,1,1-Trichloro-2,2-bis(*p*-chlorophenyl)-ethane by *Aerobacter aerogenes*. I. Metabolic products. *Appl. Microbiol.* **15,** 569.
254. Weil-Malherber, H. (1966). Some properties of mandelate racemase from *Pseudomonas fluorescens*. *Biochem. J.* **101,** 169.
255. Wheelis, M. L., Palleroni, N. J., and Stanier, R. Y. (1967). The metabolism of aromatic acids by *Pseudomonas testosteroni* and *Pseudomonas acidovorans*. *Arch. Mikrobiol.* **59,** 302.
256. Wheelis, M. L., and Stanier, R. Y. (1970). The genetic control of dissimilatory pathways in *Pseudomonas putida*. *Genetics* **66,** 245.
257. Wheelis, M. L., and Ornston, L. N. (1972). Genetic control of enzyme induction in the β-ketoadipate pathway of *Pseudomonas putida:* Deleting mapping of cat mutations. *J. Bacteriol.* **109,** 790.
258. Williams, P. A., Murray, K., and Sala-Trepat, J. M. (1971). The coexistence of two metabolic pathways in the meta cleavage of catechol by *Pseudomonas putida* NCIB 10105. *Biochem. J.* **124,** 19P.
259. Wix, G., Albrecht, K., Ambrus, G., and Szabo, A. (1968). Conversion of steroids by *Alcaligenes faecalis*. *Acta Microbiol.* **15,** 239.
260. Wu, C. H., Ornston, M. K., and Ornston, L. N. (1971). The genetic location of a regulatory mutation governing the catechol pathway in *Pseudomonas putida*. *Bacteriol. Proc.*, p. 162.
261. Wu, C. H., and Ornston, L. N. (1972). Genetic control of enzyme induction in the β-ketoadipate pathway of *Pseudomonas putida:* Two point crosses with a regulatory mutant strain. *J. Bacteriol.* **109,** 796.

Questions

1. Outline the differences in the alpha, beta, and omega oxidation of fatty acids.
2. What is the difference between a monoterminal and a diterminal oxidation of alkanes?
3. Describe the action of a "mixed function oxidase" system in comparison to an oxygenase.
4. Name the three major intermediates of aromatic hydrocarbon metabolism that represent the last aromatic hydrocarbons before ring fission occurs.
5. What is meant by ortho- and meta-ring fission of the benzene ring?
6. What is meant by sequential induction of a pathway?
7. Describe the two different types of hydrocarbon utilization that both finally lead to the β-ketoadipate pathway.
8. Describe the reactions involved in ortho and meta cleavage of catechol.
9. Outline the differences between the catechol and gentisate metabolism.
10. Describe the differences between the metabolism of aromatic hydrocarbons and halogenated aromatic hydrocarbons.
11. Discuss the regulatory mechanisms that operate when *p*-hydroxybenzoate is given to *Pseudomonas putida* as growth substrate.

12. *Acinetobacter calcoaceticus*, *Pseudomonas putida*, and *Alcaligenes eutrophus* oxidize *p*-hydroxybenzoate in the same manner but regulate the respective pathways in entirely different fashions. Discuss the main differences in the regulatory mechanisms of all three microorganisms.

13. Describe the metabolism of the hormone herbicide, 2-methyl-4-chlorophenoxyacetate.

14. Describe the metabolism of vitamin B_6.

15. Discuss the differences between the metabolism of halogenated aromatic hydrocarbons and aromatic sulfonates.

16. Discuss the attack of the furan ring by microorganisms.

17. What are the main differences between the β-ketoadipate and the α-keto acid pathways?

9

Fermentation

Introduction

The word "fermentation" has undergone many changes in meaning during the past hundred years. According to the derivation of the term, it signifies merely a gentle bubbling or boiling condition. The term was first applied when the only known reaction of this kind was the production of wine, the bubbling, of course, being caused by the production of carbon dioxide.

It was not until Gay-Lussac studied the chemical aspects of the process that the meaning was changed to signify the breakdown of sugar into ethanol and carbon dioxide (316). It was Pasteur, however, who marked the birth of chemical microbiology with his association of microbes with fermentation in 1857. He used the terms "cell" and "ferment" interchangeably in referring to the microbe. The term "fermentation" thus became associated with the idea of cells, gas production, and the production of organic byproducts.

The evolution of gas and the presence of whole cells were invalidated as criteria for defining fermentation when it was discovered that in some fermentations, such as the production of lactic acid, no gas is liberated. Moreover, other fermentation processes could be obtained with cell-free extracts indicating that the whole cell may not be necessary.

The position was further complicated by the discovery that the ancient process of vinegar production, generally referred to as acetic acid fermentation, which yielded considerable quantities of organic byproducts, was a strictly aerobic process. Fermentation clearly needed to be redefined.

Although carbohydrates are often regarded as essential materials for fermentations, organic acids (including amino acids) and proteins, fats, and other organic compounds are fermentable substrates for selected microorganisms. It was soon realized that these substances play a dual role as a source of food and as a source of energy for the microorganisms (375). The energy produced by total combustion (oxidation) of the substance in a calorimeter is its potential energy. The nearest approach to complete oxidation biologically occurs with acidic oxidations, which, with glucose, yield carbon dioxide and water and result in the liberation of a considerable quantity of energy.

Under anaerobic conditions, only a fraction of the potential energy is liberated because oxidation is incomplete. In order to obtain an amount of energy equivalent to that obtained under aerobic conditions, several times as much glucose must be broken down under anaerobic conditions. There is, in consequence, a high yield of unoxidized organic byproduct.

Fermentation came to be regarded, then, as the anaerobic decomposition of organic compounds to organic products, which could not be further metabolized by the enzyme systems of the cells without the intervention of oxygen. The fermentation products differed with different microorganisms, being governed in the main by the enzyme complex of the cells and the environmental conditions. The economic value of these byproducts led to the development of industrial microbiology.

With the recognition of fermentation as an anaerobic process, parallels were drawn between the biochemistry of microorganisms and that of mammalian tissues. Because the intermediates of the metabolism of glucose were found to be the same, it was postulated that all fermentation processes must follow similar paths. Consequently, the microbial fermentation of carbohydrates was considered to be similar to mammalian glycolysis. This is why many authors use the terms "glycolysis" or "glycolytic pathway" to describe one method of anaerobic breakdown of carbohydrates by microorganisms and why "fermentation" became synonymous with "glycolysis." The two processes differ, however, in two significant ways: (*1*) there is no storage of glycogen in bacteria, and (*2*) lactate is not always an end product or intermediate in the bacterial anaerobic breakdown of carbohydrates. In addition, during the 1950's it was discovered that various bacteria are able to use pathways other than the Embden-Meyerhof-Parnas pathway for anaerobic breakdown of carbohydrates. The application of "fermentation" to all of these processes required some other form of definition.

The intensive research into electron transport systems of microbial metabolism has partly clarified the position, although a number of aspects await attention. From research on the electron donor and acceptor systems, it is now clearly understood that all processes which have as a terminal

TABLE 9.1.
Principal Classes of Carbohydrate Fermentation[a]

Class	End products	Organisms
Alcohol	Ethanol CO_2	Yeasts *Zymomonas mobilis*
Lactic acid (homofermenter)	Lactic acid	*Streptococcus, Lactobacillus*
Mixed lactic acid (heterofermenter)	Lactic acid Ethanol Acetic acid, CO_2	*Leuconostoc*
Mixed acid	Lactic acid Acetic acid $H_2 + CO_2$ Formic acid Trimethylene glycol	*Escherichia, Pseudomonas*
Butylene glycol	As for mixed lactic acid class but also 2,3-butanediol	*Aerobacter, Bacillus polymyxa, Pseudomonas*
Butyric acid, butanol, acetone	Butyric acid Acetic acid H_2, CO_2 Butanol Ethanol Acetone Isopropanol	*Clostridia, Bacillus macerans*
Propionic acid	Propionic acid Acetic acid CO_2	*Propionibacterium*

[a] From Stanier *et al.* (374). Reprinted with permission of Prentice-Hall.

electron acceptor an organic compound are called "fermentations." With this definition, it is possible to state that acetic acid bacteria are not fermentative but respire aerobically. For other bacteria the definition is not restricted to the use of any particular pathway in the fermentative process.

It was also found that fermentative bacteria may dispense with the use of their cytochromes under anaerobic conditions, for their phosphorylation processes are substrate phosphorylations in which the electron donor is an organic substrate that transfers its electrons to an NAD^+ or $NADP^+$ system. The amount of NAD^+ in microorganisms, however, is limited and NAD^+ must therefore be regenerated if metabolism is to continue. Under anaerobic conditions, this regeneration can be accomplished by an oxidation–reduction mechanism involving pyruvate or other compounds derived from pyruvate. These reactions from pyruvate can vary considerably among microorganisms and therefore lead to the formation of characteristic end

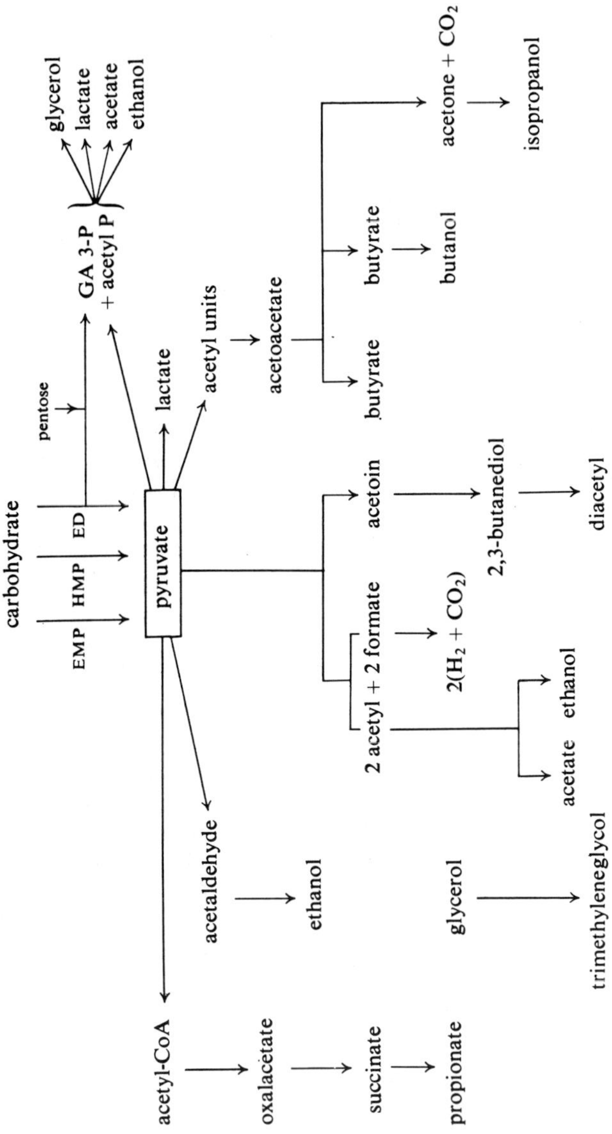

carbohydrate
EMP
HMP
ED
pentose
GA 3-P
+ acetyl P
glycerol
lactate
acetate
ethanol
pyruvate
lactate
acetyl units
acetoacetate
butyrate
butyrate
butanol
acetone + CO2
isopropanol
acetoin
2,3-butanediol
diacetyl
2 acetyl + 2 formate
2(H2 + CO2)
acetate
ethanol
acetaldehyde
ethanol
glycerol
trimethyleneglycol
acetyl-CoA
oxalacetate
succinate
propionate

Fig. 9.1. Fermentation end products of carbohydrate metabolism.

products that are used in bacterial classification. A short summary of the various end products formed from pyruvate is given in Table 9.1 and Fig. 9.1. These end products and their formation are used in classification, and the various bacterial groups will be considered in more detail below.

Carbon, Energy, and Balance

In addition to the identification of initial and final products and the study of carbon and oxidation balances, more concerted effort has been made during the past decade to separate the processes of energy production and storage from those of synthesis. The mass of cells formed is related to the amount of energy-producing substrate used and also to the number of moles of ATP formed during the utilization of that substrate. This applies, however, for conditions in which the substrate is used only for production; an alternate supply of amino acids and other monomers is provided for growth. This relationship can be expressed in two ways:

$$\text{Yield}_{\text{energy substrate (e.g., glucose)}} = Y_G = \frac{\text{dry weight of cells (gm)}}{\text{moles glucose fermented}}$$

$$\text{Yield}_{\text{ATP}} = Y_{\text{ATP}} = \frac{\text{dry weight of cells (gm) per moles ATP formed}}{\text{moles substrate fermented}}$$

As stated above, the number of moles of ATP formed varies with the pathway used. The difficulty of obtaining this information was discussed above, and it is therefore not surprising that only very few microorganisms have been investigated, e.g., *Streptococcus faecalis* ($Y_G = 20$; $Y_{\text{ATP}} = 10$) and *Zymomonas mobilis* ($Y_G = 8.3$; $Y_{\text{ATP}} = 8.3$) (4).

The present methods for end product detection are based on ones that originated from diagnostic methods used in medical microbiology. These have been refined and improved but they still give only a qualitative measurement of the major end product. Our present state of knowledge in carbohydrate fermentation is confined to the main end products formed from hexoses. Some of the organisms responsible for these transformations are listed in Table 9.1 (374).

It must be mentioned, however, that environmental conditions play a most important role in end product formation. Investigations of end products formed during the growth of *Escherichia coli* ML 30 (95) have shown that glucose degradation, which normally leads to the formation of acetate and carbon dioxide, can be altered by an anaerobic shock of the organism. Under the altered conditions, ethanol accumulates in addition to acetate and carbon dioxide. This change can readily be made by appropriate transition from aerobic to anaerobic environmental conditions. This anaerobic mode of growth poses special problems in a great number of heterotrophic organisms because their overall ATP requirement for biosynthesis can be satisfied only by degradation of a relatively large quantity of an organic compound that serves as the energy source. For this reason, various specific control mechanisms are necessary to regulate the electron flow in the metabolism of strict and facultative anaerobes. One of these controls is the ability of many such microorganisms to dispose of "excess" electrons (e^-) in the form of molecular hydrogen (H_2) (149) through the activity of hydrogenases:

$$2e^- + 2\,H^+ \rightarrow H_2$$

In NAD-dependent hydrogenase, the coenzyme NAD^+ is bound to the enzyme and is reduced to $NADH + H^+$ in the presence of hydrogen (3). The velocity of this reaction again is related to the concentration of reduced NAD and the redox ratio, $NADH + H^+/NAD^+$. This allosteric type of inhibition is also a control of the hydrogen transport exhibited by consecutive hydrogen-requiring reactions.

The ability to produce molecular hydrogen is very widespread (149), over entirely different taxonomic and physiological types of microorganisms. These could be catagorized into four groups: (1) photosynthetic microorganisms; (2) heterotrophic anaerobes, the growth of which is inhibited by molecular oxygen; (3) heterotrophic anaerobes that typically contain cytochromes and evolve hydrogen from formate; and (4) heterotrophic anaerobes that contain cytochromes as electron carriers.

Very little is known about the control of anaerobic electron flow, except that there is no hydrogen production on iron-deficient media. In typical saccharolytic clostridia, the absence of such a hydrogen-evolving system leads to a marked shift from the usual metabolic pattern (see "Lactate Formation," p. 587) to lactate production. The loss of the hydrogen production [or "hydrogen valve" (149)] restricts these organisms to more conventional fermentations, which can be observed in organisms that are not able to produce molecular hydrogen because they lack the hydrogenases. This type of depression can come from oxygen in the facultative anaerobes. In general, it could be said that the yield of molecular hydrogen is directly related to the state of reduction of the fermented energy source. Microorganisms must therefore cope with special problems in controlling their energy metabolism. This is probably necessitated by the variation in ATP content in such organisms, which makes it necessary to dispense a variable number of electrons (149).

Fermentation of Propionic Acid Bacteria

Propionic acid is a product of carbohydrate fermentation by species of *Propionibacterium* and some other related species. These bacteria are unicellular, nonphotosynthetic, nonsporing organisms that ferment glucose or lactate to propionic acid under anaerobic conditions. Anaerobically they occur as chains of cocci, but they can also grow quite well under aerobic conditions (93a), in which case they appear as rods. Pleomorphic forms are quite common. Investigations of propionic acid formation with *Propionibacterium pentosaceum* and *Propionibacterium shermanii* 52 W (7) led to the proposal of the pathways shown in Fig. 9.2.

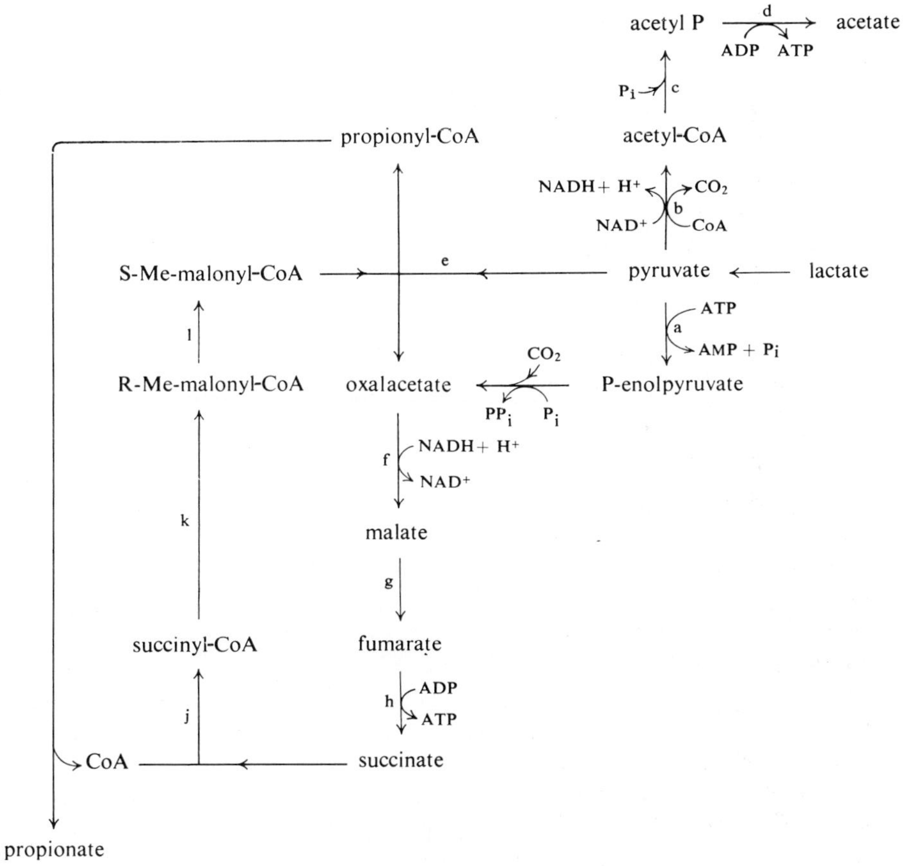
acetyl P
d
acetate
ADP
ATP
Pi
c
propionyl-CoA
acetyl-CoA
NADH+ H+
CO2
b
NAD+
CoA
S-Me-malonyl-CoA
e
pyruvate
lactate
ATP
l
a
AMP + Pi
CO2
R-Me-malonyl-CoA
oxalacetate
P-enolpyruvate
PPi
Pi
NADH+ H+
f
NAD+
k
malate
g
succinyl-CoA
fumarate
ADP
h
j
ATP
CoA
succinate
propionate

Fig. 9.2. The formation of propionate by species of the genus *Propionibacterium*.

The first work of importance on the chemistry of the propionic acid bacteria was that by Fitz (119), who suggested that a dissimilation of lactic acid occurs:

$$3\,CH_3CHOHCOOH \rightarrow CH_3CH_2COOH + CH_3COOH + CO_2 + H_2O$$

The detection of succinic acid (406) and the suggestion by van Niel (404) that pyruvate is an intermediate, and his proposition that one molecule of this C_3 compound was oxidized to acetic acid and carbon dioxide and two others reduced to propionic acid, made it clear that a complex pathway must be involved in the formation of propionic acid.

The greatest problem was the high molecular ratio of propionic to acetic acid:

$$3 \text{ glucose} \rightarrow 4 \text{ propionate} + 2 \text{ acetate} + 2CO_2 + 2H_2O$$

and the role of succinic acid. The propionic–acetic acid ratio of 2:1 had never been observed before. With the discovery of the existence of part of the TCA cycle in these organisms (235, 434), at least the formation of succinic acid was solved. Using the labels ^{11}C and ^{13}C, the pathway for succinic acid (52, 435, 436) was postulated and confirmed to be as follows:

$$\text{oxalacetate} \rightleftharpoons \text{malate} \rightleftharpoons \text{fumarate} \rightleftharpoons \text{succinate}$$

The steps in the conversion of succinic acid to propionic acid were still unknown. The assumption that lactate was an intermediate (225) in the pathway of pyruvate to propionate was opposed by Barker and Lipmann (24). The discovery of the succinate decarboxylation system in *Propionibacterium pentosaceum* (87) supported the evidence that propionate must be formed via succinate. This decarboxylation system was found to be ATP and CoA dependent (417). The dependency suggested that succinyl-CoA might be the activated substrate for the decarboxylase, with the possible involvement of propionyl-CoA:

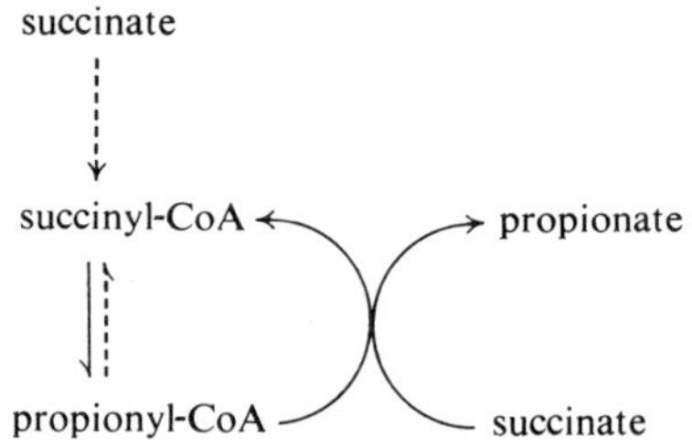
succinate
succinyl-CoA
propionate
propionyl-CoA
succinate

A parallel was drawn at this time to the metabolism of propionic acid in animal tissues (29), and the isomerization of methylmalonyl-CoA drew the particular interest of the investigators. Methylmalonyl-CoA has an asymmetric carbon atom and the corresponding isomerase was found to be stereospecific. The transcarboxylation reaction also was reversible. In other words, the carboxyl group of methylmalonyl- (or succinyl-) CoA would be transferred to an acceptor molecule of propionyl-CoA, which would thereby be converted to succinyl- (or methylmalonyl-) CoA with the release of one molecule of propionyl-CoA (112):

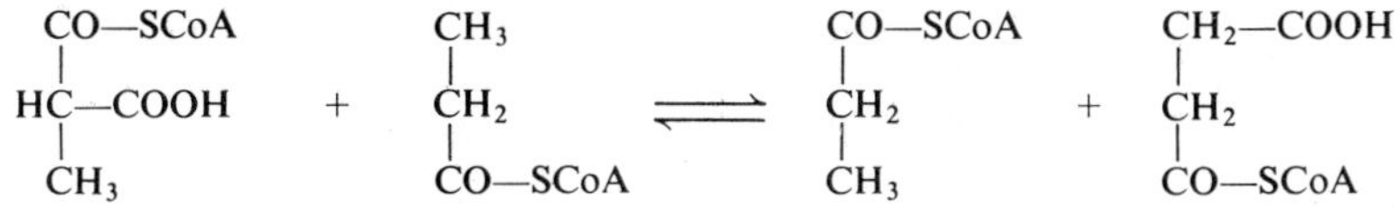

$$\underset{\text{methylmalonyl-CoA}}{\begin{array}{c}\text{CO—SCoA}\\|\\\text{HC—COOH}\\|\\\text{CH}_3\end{array}} + \underset{\text{propionyl-CoA}}{\begin{array}{c}\text{CH}_3\\|\\\text{CH}_2\\|\\\text{CO—SCoA}\end{array}} \rightleftarrows \underset{\text{propionyl-CoA}}{\begin{array}{c}\text{CO—SCoA}\\|\\\text{CH}_2\\|\\\text{CH}_3\end{array}} + \underset{\text{succinyl-CoA}}{\begin{array}{c}\text{CH}_2\text{—COOH}\\|\\\text{CH}_2\\|\\\text{CO—SCoA}\end{array}}$$

This enzymatic carboxylation of propionyl-CoA yields only one of the two isomers of methylmalonyl-CoA. The final solution to the problem came in 1960, when Swick and Wood (386) demonstrated this new type of biochemical reaction in propionic acid fermentation. This reaction permitted a direct carboxylation of pyruvate without the intervention of carbon dioxide:

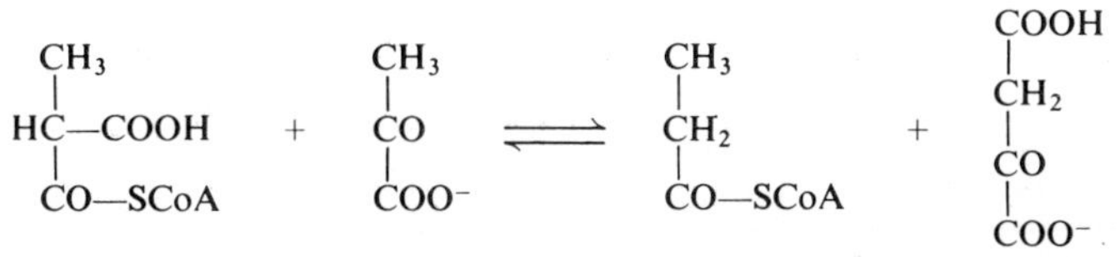

$$\underset{\text{methylmalonyl-CoA}}{\begin{array}{c}\text{CH}_3\\|\\\text{HC—COOH}\\|\\\text{CO—SCoA}\end{array}} + \underset{\text{pyruvate}}{\begin{array}{c}\text{CH}_3\\|\\\text{CO}\\|\\\text{COO}^-\end{array}} \rightleftarrows \underset{\text{propionyl-CoA}}{\begin{array}{c}\text{CH}_3\\|\\\text{CH}_2\\|\\\text{CO—SCoA}\end{array}} + \underset{\text{oxalacetate}}{\begin{array}{c}\text{COOH}\\|\\\text{CH}_2\\|\\\text{CO}\\|\\\text{COO}^-\end{array}}$$

This reaction is catalyzed by a methylmalonyl-CoA carboxyltransferase (EC 2.1.3.1) (437) that appears to have a broad specificity for the CoA esters. This is a very interesting characteristic because it permits the carboxyl group of oxalacetate arising in the TCA cycle or elsewhere to be utilized in fatty acid biosynthesis. From the point of view of economy and control of cellular reactions, it would seem advantageous to transfer carboxyl groups just as ester phosphates are transferred. Oxalacetate is also the only keto acid that has been found to serve as a carboxyl donor. There is no cofactor required, but the enzyme contains biotin (368).

The two methylmalonyl-CoA compounds performing the described reactions, however, were in two different isomeric forms. The enzyme that converted one isomeric form into the other was found to be methylmalonyl-CoA racemase (EC 5.1.99.1) (5). This enzyme did not function by transferring the CoA moiety between carboxyl groups.

Propionic Acid Formation from Glucose

The pathways that can be followed by propionibacteria to produce pyruvate from glucose will be disregarded except to state that phosphoenolpyruvate is formed as an intermediate of the EMP pathway. The first reaction of this intermediate, under normal conditions, is

$$\begin{array}{c} CH_2 \\ | \\ C-OPO_3^- + ADP \\ | \\ COO^- \end{array} \longrightarrow \begin{array}{c} CH_3 \\ | \\ C=O + ATP \\ | \\ COO^- \end{array}$$

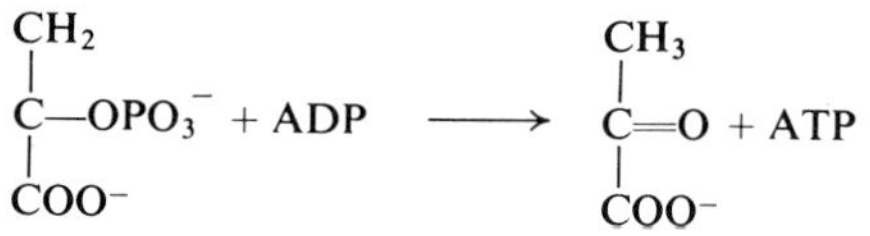

phosphoenolpyruvate → pyruvate

which is catalyzed by pyruvate kinase (ATP:pyruvate phosphotransferase, EC 2.7.1.40). In this reaction, not only ADP but also UDP, GDP, CDP, IDP, and deoxy-ADP can act as acceptor.

Once pyruvate has been formed, the pathway splits in two, one branch terminating in the formation of acetate and carbon dioxide and the second in the formation of propionate. The step toward acetate formation is the conversion of pyruvate to acetyl-CoA catalyzed by a pyruvate dehydro-

$$\begin{array}{c} CH_3 \\ | \\ C=O + NAD^+ + CoA \\ | \\ COO^- \end{array} \longrightarrow \begin{array}{c} CH_3 \\ | \\ C=O + CO_2 + NADH + H^+ \\ | \\ SCoA \end{array}$$

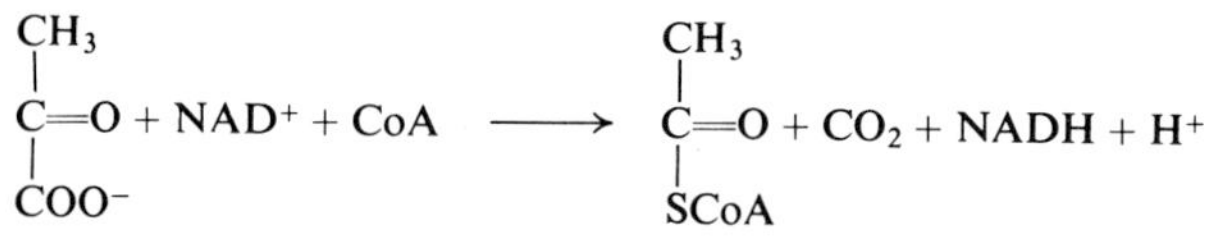

pyruvate → acetyl-CoA

genase, which is very likely pyruvate:ferricytochrome b_1 oxidoreductase (EC 1.2.2.2) as there is no report of the lipoate requirement found in aerobic bacteria (see Chapter 7). This enzyme should not be confused with the other pyruvate dehydrogenase [pyruvate:lipoate oxidoreductase (acceptor-acetylating), EC 1.2.4.1], which also forms acetyl-CoA in its reaction but which involves a system of four other enzymes, as described for the TCA cycle in aerobic respiration.

A phosphate acetyltransferase (acetyl-CoA:orthophosphate acetyltransferase, EC 2.3.1.8) (366) catalyzes the reaction before acetate is formed by transferring the phosphate group to ADP with the help of acetylkinase (ATP:acetate phosphotransferase, EC 2.7.2.1) (333). The other branch,

$$\begin{array}{c} CH_3 \\ | \\ C=O + H_3PO_4 \\ | \\ SCoA \end{array} \longrightarrow \begin{array}{c} CH_3 \\ | \\ C-OPO_3^- + CoA \end{array}$$

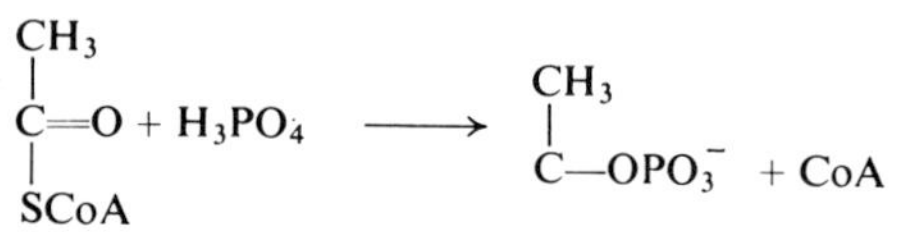

acetyl-CoA → acetyl phosphate

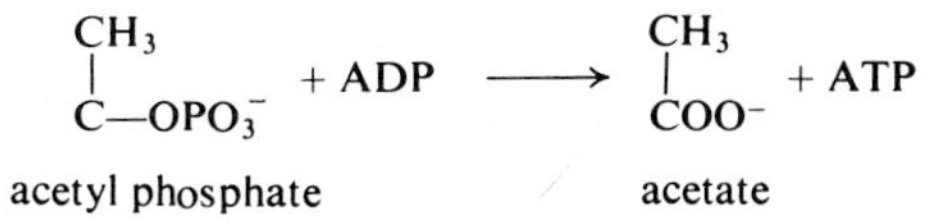
+ ADP
+ ATP
acetyl phosphate
acetate

to propionate, is a double cycling pathway that can start either from pyruvate or from phosphoenolpyruvate. Methylmalonyl-CoA reacts with pyruvate to form propionyl-CoA and oxalacetate. The reaction with pyruvate occurs only with *S*-methylmalonyl-CoA. The *R*-methylmalonyl-CoA (95, 328) is initially formed from succinyl-CoA by the enzyme methylmalonyl-CoA isomerase (86, 176, 292, 368, 379, 386), which is possibly identical with methylmalonyl-CoA:CoA-carbonylmutase (EC 5.4.99.2). It is converted to *S*-methylmalonyl-CoA by a methylmalonyl racemase (5, 292), now called "methylmalonyl-CoA racemase" (EC 5.1.99.1).

Methylmalonyl-CoA carboxyltransferase (386, 437, 438) (methylmalonyl-CoA:pyruvate carboxyltransferase, EC 2.1.3.1), which requires biotin (6), catalyzes the reaction:

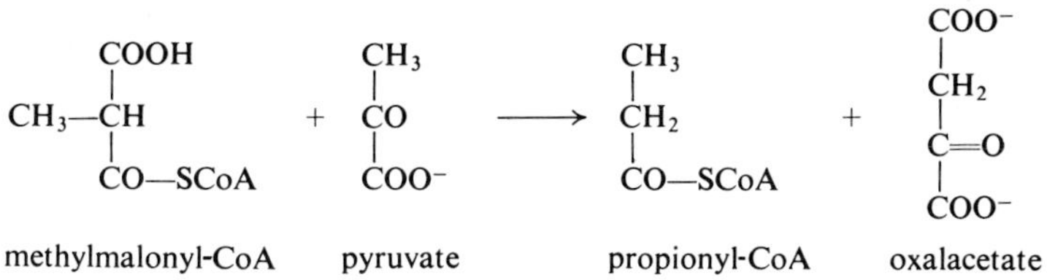
methylmalonyl-CoA
pyruvate
propionyl-CoA
oxalacetate

Malate and succinate are produced finally from the formed oxalacetate via the TCA cycle acting in reverse. The enzymes involved in the first two reactions are L-malate:NAD oxidoreductase (EC 1.1.1.37) and L-malate hydro-lyase (EC 4.2.1.2).

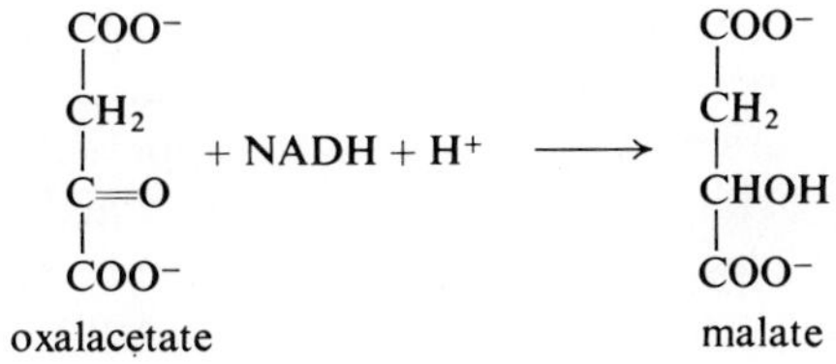
+ NADH + H+
oxalacetate
malate

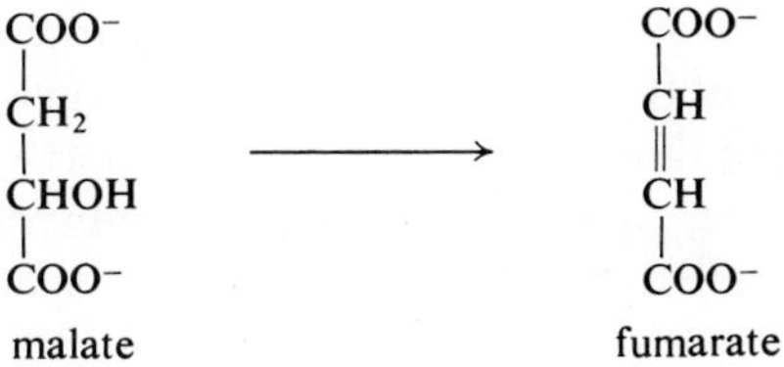
malate
fumarate

A fumarate reductase, which does not occur in the TCA cycle, reduces fumarate to succinate, whereby ATP and FP are formed (936, 364a).

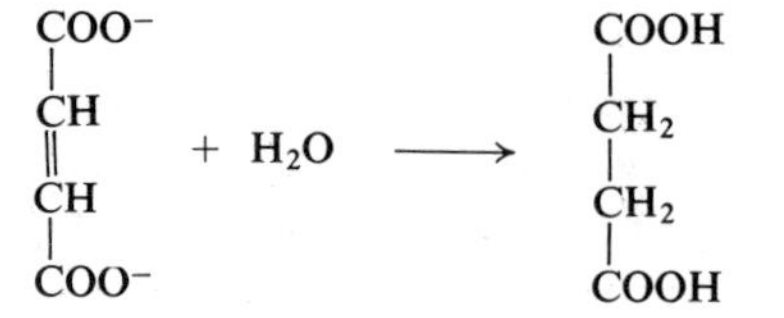

fumarate succinic acid

If the steps from oxalacetate to succinate are considered as a whole, the following energy-yielding process can be said to occur:

$$\text{NADH} + \text{H}^+ + \text{P}_i + \text{ADP} + \text{FP} \rightarrow \text{NAD}^+ + \text{ATP} + \text{FPH} + \text{H}^+$$

Propionyl-CoA, which was formed earlier in the reaction of pyruvate and methylmalonyl-CoA, transfers its coenzyme to succinate with the enzyme succinyl-CoA:3-oxoacid CoA-transferase (EC 2.8.3.5) (367) as catalyst.

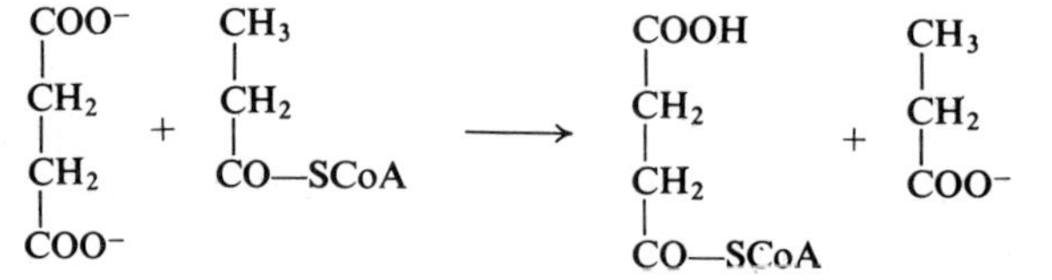

succinate propionyl-CoA succinyl-CoA propionate

With this reaction the actual accumulating end product, propionate, is formed.

The C_1 compound released from *S*-methylmalonyl-CoA and that cleaves with pyruvate to form oxalacetate is suggested almost certainly to be carbon dioxide (6), but this requires further investigation.

An interesting feature of this pathway is the apparent high yield of ATP. On the basis of cell yields it was found that approximately 6 moles of ATP are formed per 1.5 moles of glucose (22). Reactions a and d (see Fig. 9.2) together would yield 4 moles of the 6 moles of ATP, and 2 moles must therefore arise from the reaction yielding propionate. This is possible because of the unique feature of the propionate formation, which involves a complete reduction of fumarate to succinate and the oxidation of pyruvate to acetate and carbon dioxide. An electron transport-coupled phosphorylation may occur during this step through the reduction of a flavoprotein by NADH + H^+ (93b), which occurs also in other anaerobic bacteria (172a, 235a).

Succinate occasionally accumulates in the fermentation of glucose by propionibacteria. In this case, the cycle is broken and the oxalacetate must be generated by fixation of carbon dioxide. This fixation is catalyzed by the enzyme phosphoenolpyruvate carboxykinase (EC 4.1.1.38) (354, 355):

$$\begin{array}{c} CH_2 \\ \| \\ CO-PO_3^- \\ | \\ COO^- \end{array} + CO_2 + P_i \longrightarrow \begin{array}{c} COO^- \\ | \\ CH_2 \\ | \\ C=O \\ | \\ COO^- \end{array} + PP_i$$

phosphoenol-   oxalacetate
pyruvate

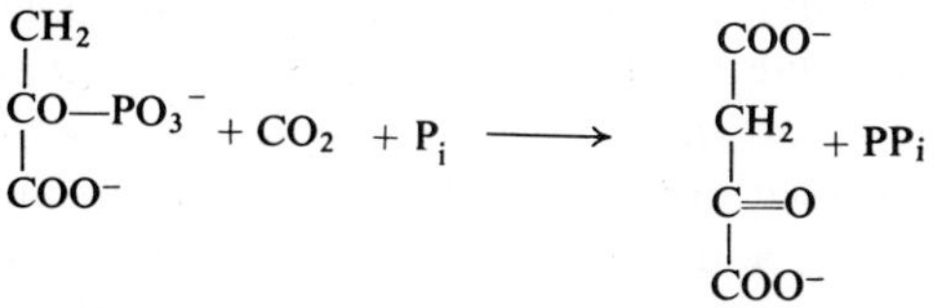

The yield of ATP, of course, would decrease, because phosphoenolpyruvate would only partly be available to form pyruvate.

Under many conditions, succinate has been found as a minor product of the fermentation. This may account for the finding of less than 6 moles of ATP in propionibacteria.

There are other bacteria that produce succinic acid via methylmalonyl-CoA esters (25, 192). It is likely that their intermediary metabolism is closely related to the propionibacteria. These organisms, which occur in the rumen (192), are (a) *Bacteroides* sp. (49, 50a), (b) *Ruminococcus flavefaciens* (193a, 274a, 353); (c) *Succinomonas amylolytica* (49); (d) *Succinovibrio dextrinosolvens* (48); (e) *Borrelia* sp. (47); and (f) *Cytophaga succinicans* (13).

Apart from propionibacteria there are two other bacterial taxa that are able to produce propionic acid from lactate, viz., *Veillonella gazogenes* (197), *V. parvula* (286), and *Clostridium propionicum*. *Veillonella*, *Selenomonas ruminantium* (297), and *Propionibacterium* metabolize very similarly, or even identically, from pyruvate onward. *Veillonella* and *Selenomonas* (93c, 184a), however, do not ferment glucose, but only lactate. Lactate is oxidized to pyruvate by the enzyme lactate–malate transhydrogenase (EC 1.1.99.7), which converts lactate together with oxalacetate to pyruvate and malate (8):

$$\text{Pyruvate} + \text{malate} \rightarrow \text{lactate} + \text{oxalacetate}$$

Kinetic investigations indicate that during the transhydrogenation the first product is liberated before the second substrate binds to the enzyme (106). ATP is generated in the same way described for *Propionibacterium*. Acetate kinase in *V. alcalescens* is active only in the direction of ATP synthesis (299) and could be an important control point in the energy metabolism of this organism.

Veillonella parvula differs from *V. alcalescens* in that its growth is CO_2 independent and in its replacement of methylmalonyl-CoA carboxyl transferase by a propionyl-CoA carboxylase (EC 4.1.1.41) (286). The latter decarboxylates methylmalonyl-CoA to propionyl-CoA and carbon dioxide (134), instead of being coupled to oxalacetate formation. The oxalacetate required for the lactate–malate transhydrogenase reaction is formed from malate or pyruvate with the help of the malic enzyme and pyruvate carboxylase (EC 6.4.1.1) and with the net formation of NADH + H^+. The latter, in turn, would be sufficient to drive the reduction of malate via fumarate to succinate. Carbon balance investigations also revealed that some oxalacetate formed from pyruvate could subsequently be converted to propionate. Indications are that both enzymes, methylmalonyl-CoA carboxyl transferase and propionyl-CoA carboxylase, are probably present. For biosynthetic purposes it is proposed that *Veillonella* is able to obtain its pentose requirement via a nonoxidative limb of the hexose monophosphate pathway, as the enzymes transaldolase, transketolase, ribosephosphate isomerase, and ribulose phosphate 3-epimerase were found to be present (271, 272).

Veillonella alcalescens is also able to degrade 2-oxoglutarate to CO_2, H_2, and propionate (244). The reaction sequence involves an oxidative decarboxylation of 2-oxoglutarate to succinyl-CoA, CO_2, and H_2, which was followed by a further decarboxylation of succinyl-CoA to propionyl-CoA and carbon dioxide.

The report on the presence of pyruvate synthase (EC 1.2.7.1) (113a) in cell free extracts of mixed rumen microorganisms allows these organisms a reductive carboxylation of acetyl phosphate to pyruvate. Since acetate is one of the major products of carbohydrate fermentation in rumen, this acetate utilization could be of significant importance, which would parallel the observation of interspecies transfer of H_2 with *Vibrio succinogenes* and *Ruminococcus albus* (193a).

Propionic Acid Production from Lactate

Clostridium propionicum possesses an entirely different mechanism. It is known that the route from lactate is

$$\text{Lactate} \rightarrow \text{acrylate} \rightarrow \text{propionate}$$

but other intermediates have not been elucidated. It appears that this pathway, which also occurs in *Bacteroides ruminicola* (411), involves acryloyl-CoA as an intermediate.

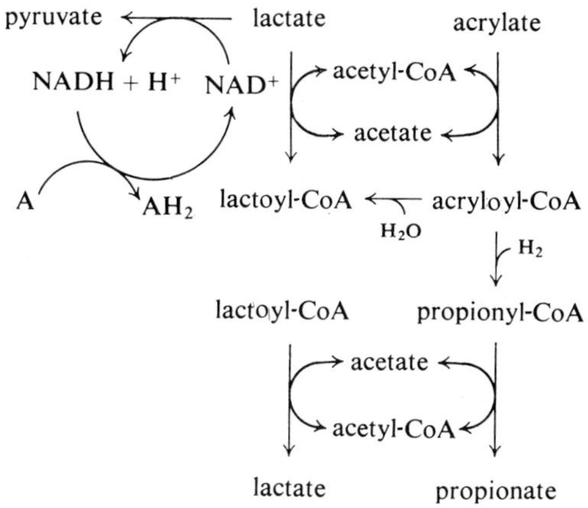
pyruvate
lactate
acrylate
NADH + H+
NAD+
acetyl-CoA
acetate
A
AH2
lactoyl-CoA
acryloyl-CoA
H2O
H2
lactoyl-CoA
propionyl-CoA
acetate
acetyl-CoA
lactate
propionate

Fig. 9.3. Propionic acid formation, with lactate as carbon source, by *Peptostreptococcus elsdenii.*

Although *Peptostreptococcus elsdenii* is not an organism normally classified with the propionic acid bacteria, its lactate metabolism (21) certainly looks very similar to the anticipated pathway in *Clostridium propionicum* and *Bacteroides ruminicola. Peptostreptococcus elsdenii* is strictly anaerobic and metabolizes lactate to propionate according to the pattern given in Fig. 9.3. Lactate can be metabolized to pyruvate, for this organism possesses lactate dehydrogenase (EC 1.1.1.27) and a diaphorase. It is suggested that the conversion of lactate to propionate occurs via the CoA esters of lactate, acrylate, and propionate. Lactate can be converted to lactoyl-CoA under the action of acetate-CoA-transferase (acyl-CoA:acetate-CoA-transferase (EC 2.8.3.8)). Acetyl-CoA required for this step is formed via the phosphoroclastic split (see p. 596) using phosphate acetyltransferase (EC 2.3.1.8) and flavodoxin (269, 270). Lactoyl-CoA dehydratase (EC 4.2.1.54) catalyzes the reaction between lactoyl-CoA and acryloyl-CoA. The conversion of acryloyl-CoA to propionyl-CoA is mediated by acyl-CoA dehydrogenase (EC 1.3.99.3). The final product, propionate, is arrived at in a way similar to that of lactoyl-CoA or acryloyl-CoA, by reaction with acetate and propionate CoA-transferase (acetyl-CoA:propionate CoA-transferase, EC 2.8.3.1). Whether or not this pathway is applicable to *Clostridium propionicum* and *Bacteroides ruminicola* is not known.

Peptostreptococcus elsdenii (364) and *Bacteroides ruminicola* (10) obtain their amino acids from lactate and through the biosynthesis of 2-oxoglutarate via the reductive carboxylation of succinate (10). Such reductive carboxylation reactions are very common among rumen bacteria (9, 332). Both microorganisms also possess aspartate aminotransferase (L-aspartate: 2-oxoglutarate aminotransferase, EC 2.6.1.1) and an $NADP^+$-linked glutamic dehydrogenase [L-glutamate:NADP oxidoreductase (deaminating), EC 1.4.1.4] (207), but as ferredoxin has not as yet been detected, the necessary low-potential electron carrier is not known. It could be flavodoxin (269, 270), or possibly cytochrome b from *Bacteroides ruminicola* (416), however. The latter was reported to form an electron transport system for fumarate reduction to succinate.

Fermentation of Saccharolytic Clostridia

The bacteria that carry out a fermentation with such end products as acetate, butyrate, acetone, and isopropanol belong to the genus *Clostridium* and the genus *Butyribacterium*. The species of the genus *Clostridium* are commonly divided into a number of groups, depending on their carbon-

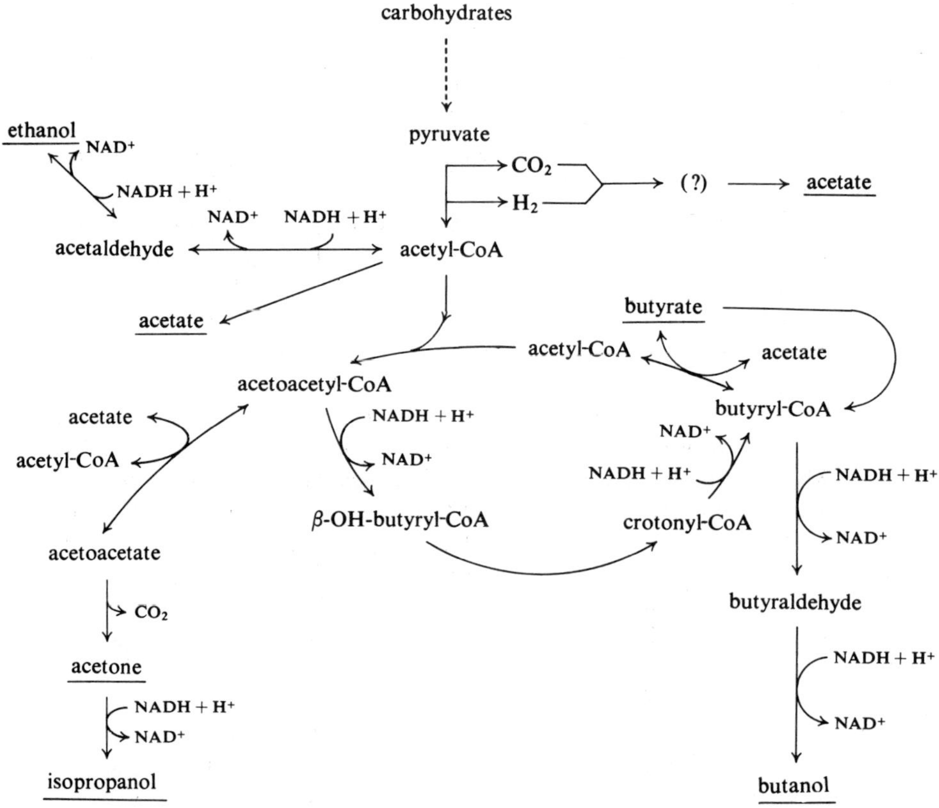
carbohydrates
pyruvate
ethanol
NAD+
NADH + H+
CO_2
H_2
(?)
acetate
NAD+
NADH + H+
acetaldehyde
acetyl-CoA
acetate
butyrate
acetyl-CoA
acetate
acetoacetyl-CoA
butyryl-CoA
acetate
NADH + H+
acetyl-CoA
NAD+
NAD+
NADH + H+
NADH + H+
NAD+
β-OH-butyryl-CoA
crotonyl-CoA
acetoacetate
butyraldehyde
CO_2
acetone
NADH + H+
NADH + H+
NAD+
NAD+
isopropanol
butanol

Fig. 9.4. The formation of acetate, acetone, butanol, and butyric acid by species of the genus *Clostridium*.

TABLE 9.2

Metabolic Activities of Various Species of Clostridia[a]

End products	Moles/100 moles glucose fermented						
	C. butyricum	*C. lactoacidophilum* (*C. tyrobutyricum*)	*C. perfringens*		*C. acetobutylicum*	*C. butylicum*	*B. rettgeri*
			←Fe deficient→				
Butyric acid	76	73	9	34	4	17	29
Acetic acid	42	28	15	60	14	17	88
Lactic acid			160	33			107
CO_2	188	190	24	176	221	203	48
H_2	235	182	21	214	135	77	74
Ethanol			10	26	7		
Butanol					56	58	
Acetone					22	—	
Acetoin					6	—	
Isopropanol						12	
% C recovered	96	91	98	97	99	96	110
O/R Balance	0.97	1.16	0.81	1.05	1.01	1.06	0.74

[a] From W. A. Wood (439).

source metabolism. Those which ferment carbohydrates and have limited proteolytic properties are called "saccharolytic clostridia," and these are the ones dealt with here. The great variety of end products formed is shown in the fermentation balance sheet for clostridia (439) in Table 9.2.

The balance sheet indicates that all strains studied produce butyric acid and acetic acid. *Clostridium butyricum* and *C. lactoacetophilum* produce butyric acid as their major end product, together with carbon dioxide and hydrogen. Other strains produce mostly butanol and acetone (*Clostridium acetobutylicum*) or butanol alone (*Clostridium butylicum*), together with carbon dioxide and hydrogen. There are only a few that can produce appreciable amounts of acetoin (*Clostridium acetobutylicum*) or isopropanol (*Clostridium butylicum*). Studies on *Clostridium perfringens* reveal significant differences in the metabolic activities between iron-deficient strains and the normal strains. The overall pathways in the fermentation of glucose by species of *Clostridium* and *Butyribacterium* are demonstrated in Fig. 9.4.

The Formation of Acetate

The first step in acetate formation is the breakdown of pyruvate to acetyl-CoA. This step has already been described for aerobic conditions. It requires, apart from pyruvate dehydrogenase (EC 1.2.4.1), lipoate and thiamine pyrophosphate (TPP). Under anaerobic conditions there are three systems in operation (439), two of which are similar.

1. Pyruvate is converted into acetyl-CoA or acetyl phosphate, with the formation of hydrogen and carbon dioxide, by a mechanism that does not involve formate as the precursor for hydrogen and carbon dioxide.
2. Pyruvate is converted into acetyl-CoA or acetyl phosphate with the formation of formate, which may be converted to hydrogen and carbon dioxide by the formate hydrogen lyase system.
3. Pyruvate is decarboxylated directly to acetaldehyde and carbon dioxide.

Whereas system 3 occurs mainly in yeast and higher plants, system 1 is characteristic for clostridia and system 2 for the Enterobacteriaceae. Systems 1 and 2 require CoA and TPP. It is well known that in the enterobacteria, the energy available in the thiol ester bond of acetyl-CoA enters the high-energy phosphate pool by means of two reactions. Reaction 1 is catalyzed by phosphate acetyltransferase (acetyl-CoA:orthophosphate acetyltransferase, EC 2.3.1.8) and reaction 2 by acetate kinase (ATP: acetate phosphotransferase, EC 2.7.2.1), which forms the end product acetate:

reaction 1: acetyl CoA + orthophosphate → CoA + acetyl phosphate

reaction 2: acetyl phosphate + ADP → acetate + ATP

For each mole of acetate formed, 1 mole of ATP is generated.

As acetyl phosphate is a product inhibitor of the reaction and as ADP is involved in the acetate kinase reaction, it was proposed that acetyl phosphate acts as a coarse control and ADP as a fine control point (33). This metabolic system is a simple degradation of pyruvate to acetate, with the overall production of 2 moles of acetic acid from 1 mole of glucose. Baker (25), however, found that *Clostridium thermoaceticum* formed not 2 but 3 moles of acetate from 1 mole of glucose, suggesting a more complicated system. The breakdown of pyruvate was therefore reinvestigated, and it was found that carbon dioxide must be reduced in some manner in order that the additional mole of acetate be formed.

To carry out this reduction, a hydrogen acceptor had to be present. It had been known for some time that in clostridia the degradation of pyruvate to acetyl-CoA involved the release of carbon dioxide and hydrogen. Wood, in his review (439), noted that it was assumed from experience in studies on enterobacteria that formate must be an intermediate in clostridia being decomposed in the following reaction with formic hydrogenlyase system as catalyst. This reaction was found to occur in the enterobacteria in two steps:

$$HCOOH + A \rightarrow H_2A + CO_2$$

$$H_2A \rightarrow A + H_2$$

The compound A is known to be ferredoxin and the enzymes involved in the above reactions are formate dehydrogenase (formate:ferricytochrome b_1 oxidoreductase, EC 1.2.2.1) and hydrogenase (hydrogen:ferredoxin oxidoreductase, EC 1.12.7.1). Formate, however, could not be detected in clostridia. Valentine (402) tried to solve this problem in another way, without considering formate as intermediate, and assumed that a ferredoxin-linked cleavage may occur generally in clostridia. It was found in *Clostridium pasteurianum* (281, 401, 414), *C. lactoacetophilum* (402), *C. acidi-urici* (402), *C. thermosaccharolyticum* (*C. nigrificans*) (422, 434), *Micrococcus lactilyticus* (402), *Peptostreptococcus elsdenii* (402), *Butyribacterium rettgeri* (402), *Diplococcus glycinophilus* (402), *Clostridium kluyveri* (130) and *Spirochaeta stenostrepta* (180).

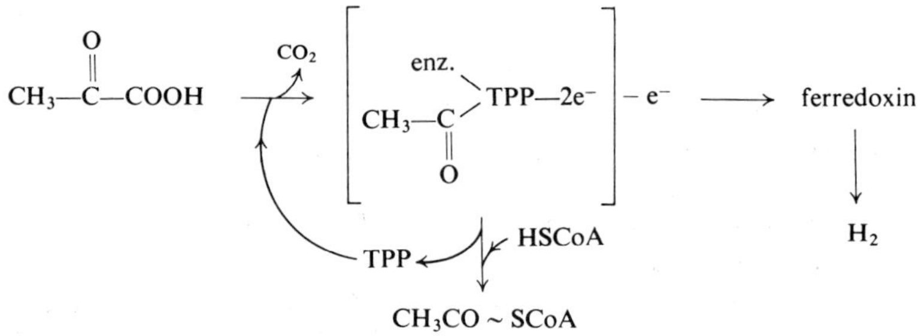

O
CH_3—C—COOH
CO_2
enz.
TPP—2e⁻
CH_3—C
O
– e⁻
ferredoxin
H_2
HSCoA
TPP
$CH_3CO \sim SCoA$

Fig. 9.5. Clostridial type of pyruvate decarboxylation (401) (reprinted with permission of the American Society of Microbiologists).

Pyruvate is first decarboxylated by pyruvate synthase [pyruvate–ferredoxin oxidoreductase (CoA-acetylating), EC 1.2.7.1] (398, 399) or pyrvate dehydrogenase (EC 1.2.2.2) with the formation of a thiamine pyrophosphate enzyme complex (Fig. 9.5), from which point electrons are transferred to ferredoxin. The reduced ferredoxin is then reoxidized by ferredoxin hydrogenase (EC 1.12.7.1) and forms molecular hydrogen (224). The detailed mechanism of this (−)-2α-hydroxyethyl thiamine pyrophosphate formation has been described above (see p. 385; 396). This enzyme–TPP complex is attacked not by lipoic acid, but by a phosphate acetyltransferase (EC 2.3.1.8) and forms, with CoA, acetyl-CoA. This requirement for CoA in the phosphoroclastic reaction is characteristic for the CO_2-pyruvate exchange system in clostridia. It is in contrast to the formate–pyruvate exchange of enterobacteria, where such a CoA requirement does not exist (179). As this reaction is a very rapid exchange reaction, the conversion of the endogenous supply of CoA to the acyl ester results in severe inhibition of the CO_2–pyruvate exchange reaction. It is therefore possible that CoA serves a regulatory function in initiating the degradation of pyruvate. In *C. acidi-urici* this reaction is catalyzed by pyruvate synthase, a reaction that does not require TPP but forms acetyl-CoA, CO_2, and reduced ferredoxin.

Tracer element work (242) showed the existence of formate production and a $NADP^+$-linked formate dehydrogenase in *Clostridium thermoaceticum*, which was supported by the demonstration of a net reduction of CO_2 to formate using cell-free extracts of the same organism (392a). The addition of metals such as molybdate and selenium not only enhanced the cell yield (13a) but was also found to greatly increase the formate dehydrogenase activity (13b), which suggests that this enzyme might be a metalloenzyme with a requirement for molybdate and selenium. Such a requirement is known to exist for the NAD-dependent formate dehydrogenase of enterobacteria (116a, 309a). Whereas *Clostridium thermoaceticum* uses an $NADP^+$-linked formate dehydrogenase for the reduction of CO_2 to formate, *Clostridium acidi-urici* (392b) and *Clostridium pasteurianum* (392c) require reduced ferredoxin and reduced NAD for their formate dehydrogenase activities. These requirements indicate that both types of dehydrogenases are distinctly different from the known formate dehydrogenases, which either are NAD-linked (EC 1.2.1.2) or cytochrome b_1-linked (EC 1.2.2.1).

The reason for the different formate dehydrogenases among the clostridia can be found in the observation that *Clostridium thermoaceticum* does not produce hydrogen. In using pyruvate dehydrogenase (EC 1.2.2.2), CO_2 is produced, which in turn is reduced to formate. In *Clostridium acidi-urici* and *C. pasteurianum*, ferredoxin is necessary for the formation of hydrogen during pyruvate breakdown by pyruvate synthase (EC 1.2.7.1), whereby it is reduced and becomes reoxidized in the CO_2 reduction to formate. A third way of pyruvate metabolism can be visualized for *C. pasteurianum* (414a) in the clostridial type of pyruvate decarboxylation, whereby the reduced ferredoxin again is reoxidized by the formate formation.

Metabolically related to the clostridia appear to be the spirochetes, since all investigated species are exclusively saccharolytic and ferment glucose via the EMP pathway forming CO_2, H_2, acetate, and ethanol as major end products (40a, 180a). Spirochetes also possess the clostridial type of pyruvate utilization (180a), but some species are able to substitute ferredoxin with rubredoxin (202a).

Two mechanisms exist for the conversion of carbon dioxide to the third molecule of acetate:

(*1*) the direct conversion of carbon dioxide to acetate, with the participation of tetrahydrofolate and corrinoids (248, 249, 314, 315), and (*2*) the indirect conversion of carbon dioxide to acetate via α-ketovalerate (238).

THE DIRECT CONVERSION OF CO_2 TO ACETIC ACID. The first step in the formation of the methyl group of acetate from CO_2 (Fig. 9.6) is the formation of formate from carbon dioxide. This reaction could be catalyzed by the nonparticulate $NADP^+$-dependent formate dehydrogenase (242) found in *Clostridium thermoaceticum*. However, the equilibrium of this reduction strongly favors carbon dioxide and $NADPH + H^+$ formation. Ferredoxin therefore appears necessary to pull the reaction together with pyruvate. Pyruvate would therefore serve as an electron donor for the reduction of carbon dioxide.

The demonstration of an NADH-linked ferredoxin reductase (EC 1.6.7.1.?) in *C. pasteurianum*, *C. butyricum* (208), and *C. kluyveri* (209), as well as an ferredoxin NADPH-reductase (EC 1.6.7.1) in *C. kluyveri* (392), certainly indicates the possible involvement of these enzymes in this

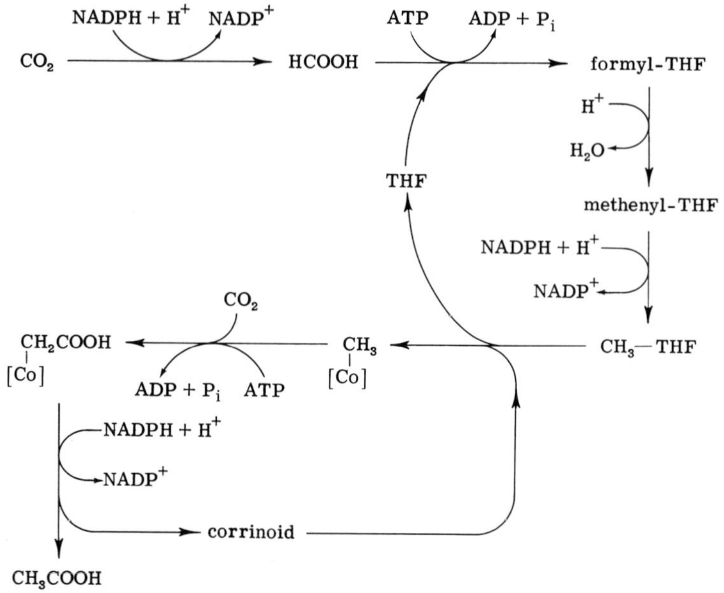
NADPH + H+ NADP+
CO2
HCOOH
ATP ADP + Pi
formyl-THF
H+
H2O
THF
methenyl-THF
NADPH + H+
NADP+
CO2
CH2COOH
[Co]
ADP + Pi ATP
CH3
[Co]
CH3—THF
NADPH + H+
NADP+
corrinoid
CH3COOH

Fig. 9.6. Direct conversion of carbon dioxide to acetate by clostridia (reprinted with permission of the authors and the American Society of Biological Chemists).

$$CH_3COCOOH + Fd_{oxid} + H_2O \longrightarrow CH_3COOH + CO_2 + Fd_{red}$$

$$CO_2 + Fd_{red} \longrightarrow HCOOH + Fd_{oxid}$$

reaction. Both these enzymes appear to be responsible for regulating the phosphoroclastic reaction. If this is the case, the conversion of pyruvate to carbon dioxide and the subsequent reduction of carbon dioxide to formate could involve a complex system consisting of pyruvate synthase (EC 1.2.7.1) ferredoxin NAD(P)H-reductase (EC 1.6.7.1), and a NADP-dependent formate dehydrogenase, with the reductase as a regulator in these reactions. Whereas the ferredoxin NADH-reductase is controlled by the acetyl-CoA/CoA couple (208, 209), the ferredoxin NAD(P)H-reductase is affected by the $NAD^+/NADH + H^+$ couple and the $NADP^+/NADPH + H^+$ couple in the reverse reaction (392).

The reduction of formate to the methyl group of acetate involves the tetrahydrofolate group of compounds. A formyltetrahydrofolate synthetase (183a) [formate:tetrahydrofolate ligase (ADP), EC 6.3.4.3] has been

$$THF + HCOOH + ATP \rightleftharpoons formyl\text{-}THF + ADP + Pi$$

isolated and purified from *C. thermoaceticum* (382). It was shown to catalyze the formation of formyl tetrahydrofolate. The following step involves a reduction to methenyl tetrahydrofolate, which is catalyzed by methenyltetrahydrofolate cyclohydrolase [5,10-methenyltetrahydrofolate 5-hydro-

$$\text{Formyl-THF} + H^+ \rightleftharpoons N^5,N^{10}\text{-methenyl THF}$$

lase (decyclizing), EC 3.5.4.9]. This enzyme has so far only been found in *C. cylindrosporum* and is strongly pH dependent. An acid pH favors the formation of methenyl tetrahydrofolate. A reduced $NADP^+$-linked meth-

$$N^5,N^{10}\text{-Methenyl-THF} + \text{NADPH} + H^+ \rightleftharpoons N^5,N^{10}\text{-methylene-THF} + \text{NADP}^+$$

ylenetetrahydrofolate dehydrogenase (5,10-methylenetetrahydrofolate: NADP oxidoreductase, EC 1.5.1.5) reduces the methenyl group to a methylene group. From here, the pathway depends on whether or not the organism possesses corrinoids. Without corrinoids, the pathway could continue with the incorporation of glycine via serine and pyruvate to acetate (see p. 653).

In *C. thermoaceticum* the methylene group is now reduced to methyl THF by a reduced NADP-linked dehydrogenase.

$$N^5,N^{10}\text{-Methylene-THF} + \text{NADPH} + H^+ \rightleftharpoons N^5\text{-methyl-THF} + \text{NADP}^+$$

This methyl tetrahydrofolate (295) transfers its methyl group to a protein-bound, reduced corrinoid, forming a Co-methyl corrinoid protein complex (247, 249a). This complex is carboxylated by transcarboxylation from

$$N^5\text{-Methyl-THF} + \text{corrinoid-E} \longrightarrow \underset{[\text{Co}]}{\overset{|}{CH_3}} + \text{THF}$$

pyruvate (348c). The final step in acetate formation from carbon dioxide is the NADPH + H^+-dependent cleavage of the Co-carboxymethyl corrinoid, with the regeneration of the reduced protein-bound corrinoid. This final step could also involve a Grignard-type reaction.

THE INDIRECT CONVERSION OF CO_2 TO ACETATE. The indirect conversion of carbon dioxide to acetate originates in the observations (238) that $^{14}CO_2$ was extensively incorporated into a compound that was found to be α-ketoisovalerate. The possibility that this keto acid could be synthesized from pyruvate via acetolactate, acetate, and acetyl-CoA formed the basis of this hypothesis (Fig. 9.7).

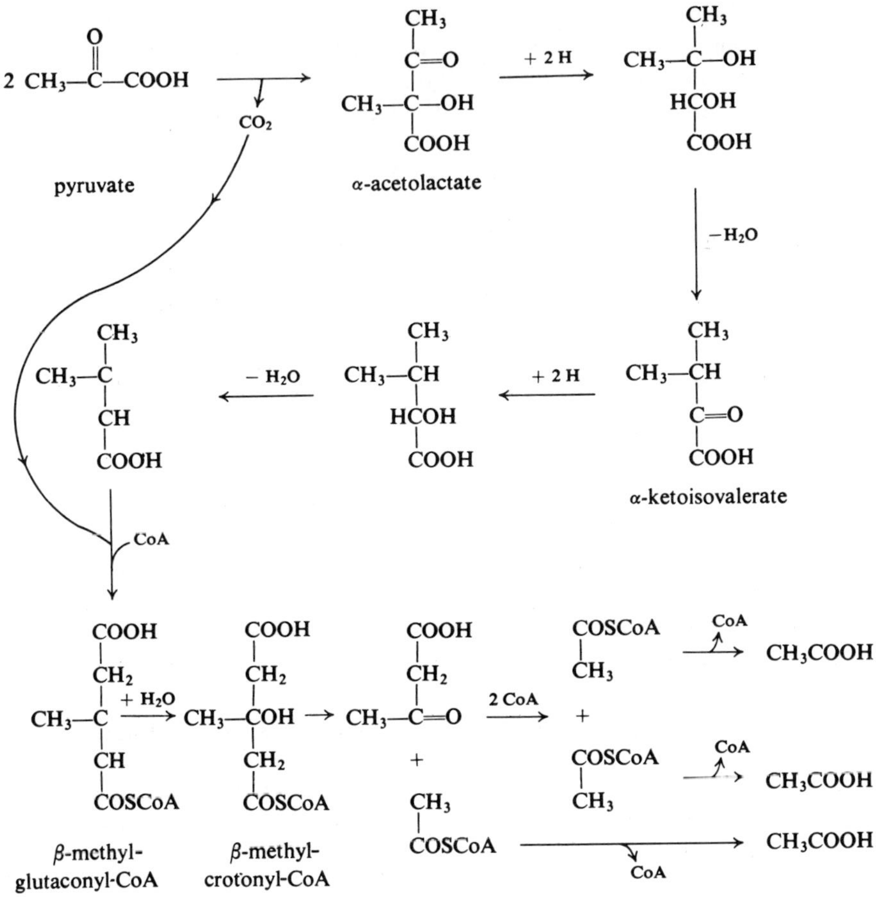
2 CH3—C—COOH
O
CO2
pyruvate
α-acetolactate
+ 2 H
−H2O
α-ketoisovalerate
− H2O
CoA
2 CoA
CH3COOH
β-methyl-glutaconyl-CoA
β-methyl-crotonyl-CoA

Fig. 9.7. The indirect conversion of CO_2 to acetate by clostridia.

Pyruvate is able to undergo condensation and decarboxylation (see the section on the butanediol producers, p. 604) to form α-acetolactate, which in turn is metabolized to α-ketoisovalerate. The fixation of carbon dioxide together with CoASH converts α-ketoisovalerate to α-methylglutaconyl-CoA. The latter is then broken down to acetoacetate, acetyl-CoA, and finally to acetate. The breakdown of acetoacetate may go via acetoacetyl-CoA, as described for *Clostridium kluyveri* (see p. 667). The described mechanism would explain the formation of 3 moles of acetate from 1 mole of glucose and might also explain the unique stoichiometry that is associated with the fermentation of glucose by *C. thermoaceticum* (238). Both theories could apply independently.

Whereas no formate was found in the metabolism of glucose by *Clostridium thermoaceticum*, *C. kluyveri* possesses a formyl-CoA hydrolase (EC 3.1.2.10) (357) apart from the phosphate acetyltransferase. Therefore, *C. kluyveri* appears to be able to form acetate in the following way:

$$\text{acetyl phosphate} + \text{CoASH} \xrightleftharpoons{\text{EC 2.3.1.8}} \text{acetyl-CoA} + P_i$$

$$\text{acetyl-CoA} + \text{HCOOH} \xrightleftharpoons{\text{CoA-transferase}} \text{formyl-CoA} + \text{acetate}$$

$$\text{formyl-CoA} + H_2O \xrightleftharpoons{\text{EC 3.1.2.10}} \text{formate} + \text{CoASH}$$

$$\text{net: acetyl phosphate} + H_2O \rightarrow \text{acetate} + P_i$$

These differences in pathway may be very important in the further classification of clostridia.

The Formation of Butyrate

Saccharolytic clostridia ferment glucose to butyric acid. The intermediate acetyl-CoA will be taken as the starting point. From the point of view of energetics, the production of acetate as sole end product would not be satisfactory because it becomes more and more difficult to reoxidize NADH + H^+ as the pH drops into the acid region (25). It is therefore not surprising to find in the clostridia a cyclic mechanism similar to that in the propionibacteria. This cyclic mechanism brings about the formation of butyric acid, which is much less an acid end product than acetate (Fig. 9.8). Two acetyl-CoA molecules undergo a condensation to form acetoacetyl-CoA, with acetyl-CoA acetyl transferase (acetyl CoA:acetyl-CoA *C*-acetyl-

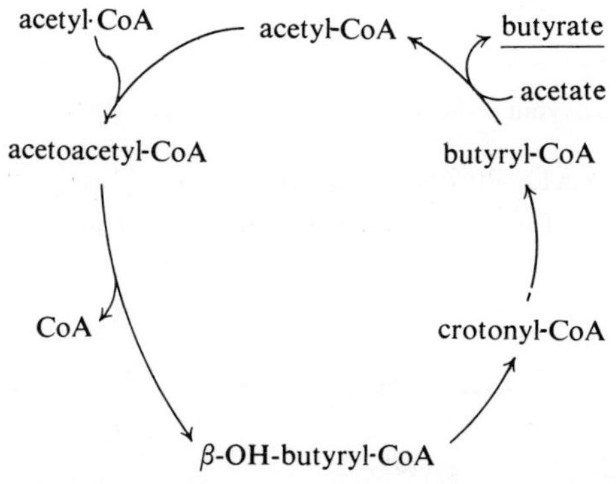
acetyl-CoA
acetyl-CoA
butyrate
acetate
acetoacetyl-CoA
butyryl-CoA
CoA
crotonyl-CoA
β-OH-butyryl-CoA

Fig. 9.8. The formation of butyric acid from acetyl-CoA by clostridia.

transferase, EC 2.3.1.9) as the catalyzing enzyme, liberating 1 mole of CoA:

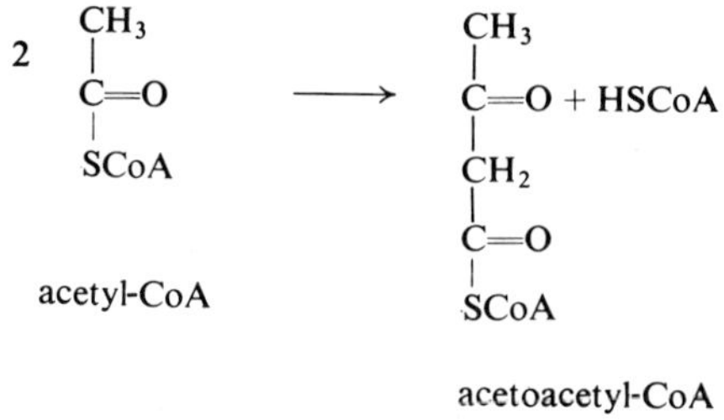
acetyl-CoA
acetoacetyl-CoA

The next step in the cycle is a reductive step, whereby acetoacetyl-CoA is reduced to β-hydroxybutyryl-CoA and reduced NAD is oxidized to NAD^+. This reaction is catalyzed by the action of 3-hydroxybutyrate dehydrogenase (D-3-hydrobutyrate:NAD oxidoreductase EC 1.1.1.30). The

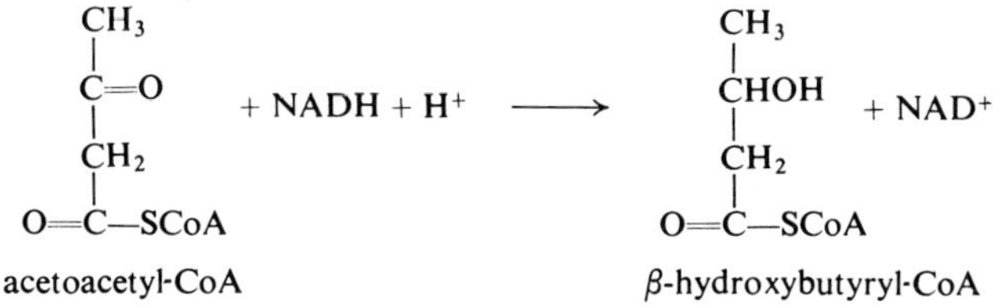
acetoacetyl-CoA
β-hydroxybutyryl-CoA

dehydration of β-hydroxybutyryl-CoA is caused by enoyl-CoA hydratase (L-3-hydroxyacyl CoA-hydro-lyase, EC 4.2.1.17), forming crotonyl-CoA

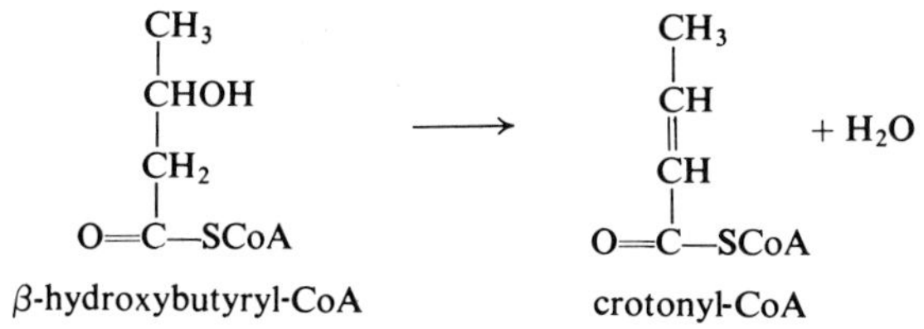
β-hydroxybutyryl-CoA
crotonyl-CoA

and water. An NAD^+-linked dehydrogenase [butyryl-CoA:(acceptor) oxidoreductase, EC 1.3.99.2] is responsible for the further reduction to

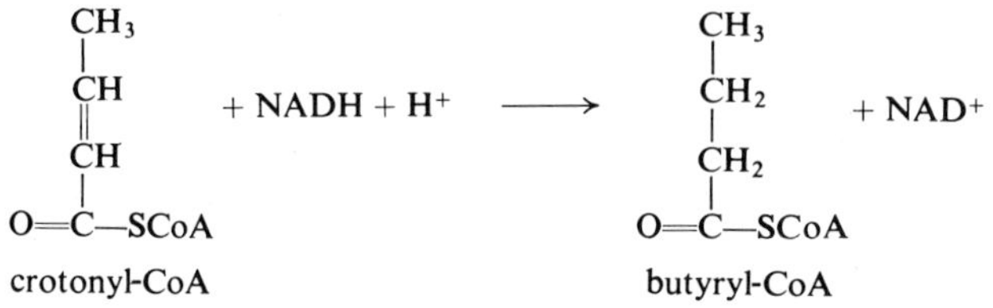
crotonyl-CoA
butyryl-CoA

butyryl-CoA. The last step in the cyclic mechanism is a transfer reaction, whereby acetate and butyryl-CoA act together with a fatty acid CoA-transferase (butyryl-CoA:synthetase, EC 6.2.1.2?) to produce acetyl-CoA and butyrate. Acetyl-CoA is now available to reenter the reaction sequence.

Formation of Acetone, Isopropanol, Butanol, and Ethanol

A number of the saccharolytic clostridia that normally ferment carbohydrates to butyric acid are able to change their system, favoring the production of acetone, and concurrently to convert the butyric acid already produced to butanol. This new system comes into operation as soon as the butyric acid production has lowered the pH of the medium to about 4.0.

ACETONE PRODUCTION. *Clostridium acetobutylicum* possesses a transferase system that diverts acetoacetyl-CoA from the normal cyclic mechanism

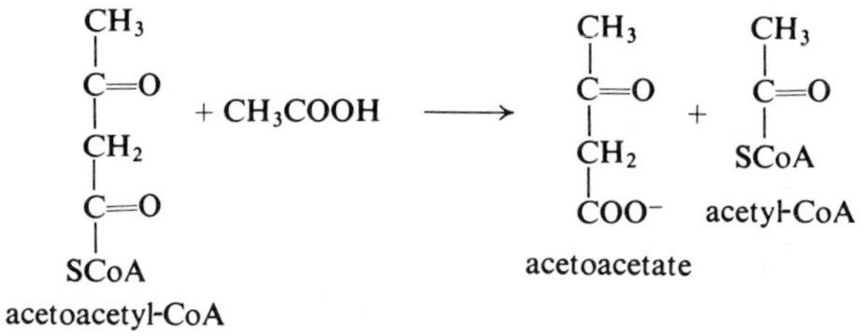
CH3
C=O
CH2
C=O
SCoA
acetoacetyl-CoA
+ CH3COOH
CH3
C=O
CH2
COO−
acetoacetate
+
CH3
C=O
SCoA
acetyl-CoA

to produce acetoacetate. This transferase is different from the fatty acid SCoA-transferase because *Clostridium kluyveri* does not possess the latter enzyme but is able to form acetoacetate from acetoacetyl-CoA. The decarboxylation of acetoacetate to acetone is an irreversible step. The enzyme

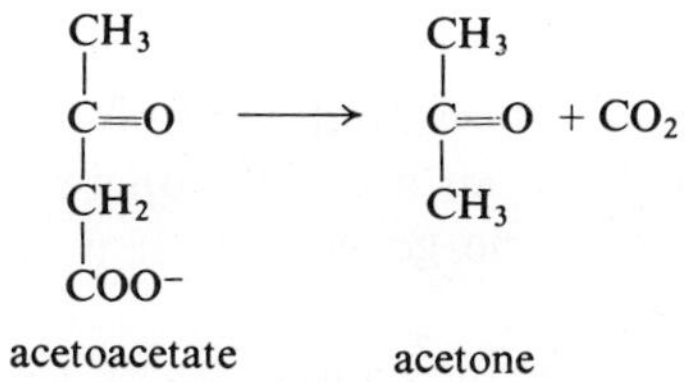
CH3
C=O
CH2
COO−
acetoacetate
CH3
C=O + CO2
CH3
acetone

responsible for this reaction is acetoacetate decarboxylase (acetoacetate carboxy-lyase, EC 4.1.1.4).

FORMATION OF ISOPROPANOL. *Clostridium butylicum* is to our present knowledge the only bacterium that can carry the reaction sequence one step further. It reduces all the acetone produced to isopropanol by means of isopropanol dehydrogenase (EC 1.1.1.80). *Clostridium butylicum*, how-

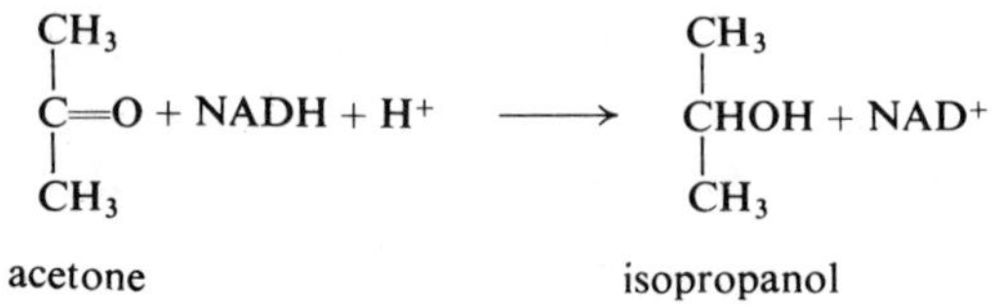
CH3
C=O + NADH + H+
CH3
CH3
CHOH + NAD+
CH3
acetone
isopropanol

ever, does not produce ethanol.

FORMATION OF ETHANOL. It was mentioned above that the production of ethanol occurs together with acetone–butanol formation, in *Clostridium acetobutylicum*. The mode of ethanol production is different from yeast. The chain of production branches off from acetyl-CoA, and an aldehyde dehydrogenase [aldehyde:NAD oxidoreductase (acylating CoA), EC 1.2.1.10] produces acetaldehyde and NAD^+. The NAD-dependent alcohol dehydrogenase (alcohol:NAD oxidoreductase, EC 1.1.1.1) finally produces the end product, ethanol.

FORMATION OF BUTANOL. The diversion of the original cyclic system to form acetone stops further production of butyric acid. In addition to the interruption of the cycle, two steps that generate NAD^+ are also eliminated, which means that some other reduction process must be found. The reduction of butyric acid to butanol is carried out in three consecutive reactions. The first step is virtually a reverse of the last reaction in the

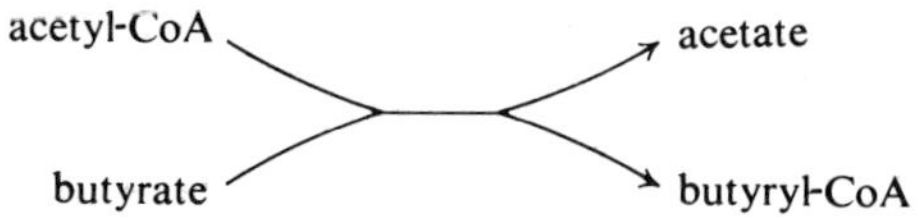
acetyl-CoA
acetate
butyrate
butyryl-CoA

cycle, whereby CoA-transferase (butyryl-CoA synthetase? EC 6.2.1.2) transfers the coenzyme from an acetyl group to the butyryl group. The acetate formed is used for the generation of acetyl-CoA in the acetone production.

Butyryl-CoA may be formed in another way if there is a deficiency in the amount of acetyl-CoA available. Adenosine triphosphate and CoA are required. The reduction of butyryl-CoA to butyraldehyde is NAD^+ linked

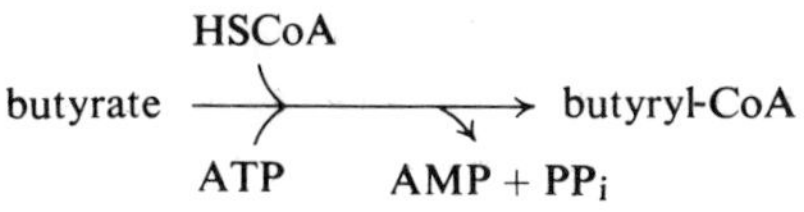
HSCoA
butyrate
butyryl-CoA
ATP
AMP + PPi

and catalyzed by the same aldehyde dehydrogenase (EC 1.2.1.10) that reduces acetyl-CoA to acetaldehyde:

$$\text{butyryl-CoA} + \text{NADH} + \text{H}^+ \rightarrow \text{butyraldehyde} + \text{NAD}^+ + \text{HSCoA}$$

The final reduction to butanol is also carried out with a well-known enzyme, NAD^+-linked alcohol dehydrogenase (EC 1.1.1.1)

$$\text{butyraldehyde} + \text{NADH} + \text{H}^+ \rightarrow n\text{-butanol} + \text{NAD}^+$$

This production of butanol occurs after the change to the production of acetone has taken place.

The metabolic pathway of *Clostridium acetobutylicum* is therefore first from glucose via pyruvate to butyrate until so much butyrate has been accumulated, together with acetate, that the pH of the medium drops to 4.0. At this point, the cyclic mechanism is interrupted and acetone production starts. As a consequence of this change, the accumulated butyrate is further reduced to butanol. The organism uses the same enzyme for the reduction of butyrate to butanol as for ethanol production.

Lactate Formation

Some of the saccharolytic clostridia and other butyrate producers, such as *Butyribacterium* and *Clostridium perfringens*, are able to reduce pyruvate to lactic acid, for they possess a lactate dehydrogenase (EC 1.1.1.27). This reduction, however, occurs only if the bacteria are grown under iron-deficient conditions (439).

Formation of Butanol from Ethanol and Acetate

Clostridium kluyveri can grow on an ethanol–acetate medium, producing butyrate by the mechanism already described. The ethanol is oxidized to acetyl-CoA in two reversible steps, whereby $NADP^+$-linked alcohol dehydrogenase (EC 1.1.1.1) first converts ethanol into acetaldehyde. Subsequently, an NAD^+-linked acetaldehyde dehydrogenase converts acetaldehyde into acetyl-CoA with the help of CoA. If phosphate is added to the medium, a transfer reaction occurs, forming acetyl phosphate and regenerating CoA for further oxidation of aldehyde. In general, this last does not occur and acetyl-CoA joins the cycle to produce butanol, provided acetate predominates in the medium (233).

When ethanol is in excess in the medium, *Clostridium kluyveri* produces not butyrate but caproate as the main product of the fermentation. The

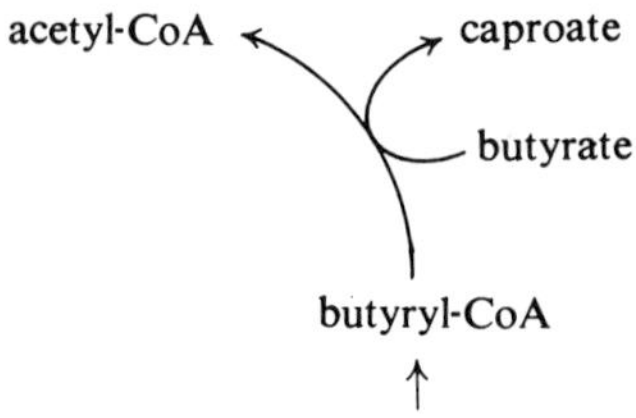
acetyl-CoA
caproate
butyrate
butyryl-CoA

production of caproate is via the same cyclic mechanism as that for butyrate until the last step, where butyrate takes the place of acetate in the transferase reaction, forming caproate and acetyl-CoA.

Succinate Formation

There exists another fermentative pathway for *Clostridium kluyveri*, which produces succinate as the major end product. The appearance of succinate is matched quantitatively with the disappearance of carbon dioxide. The proposed scheme is outlined in Fig. 9.9 (240). The formation of acetyl-CoA from pyruvate was outlined above. Acetyl-CoA carboxylase [acetyl-CoA:carbon-dioxide ligase (ADP), EC 6.4.1.2] incorporated CO_2 and water into acetyl-CoA, forming malonyl-CoA. This reaction is ATP

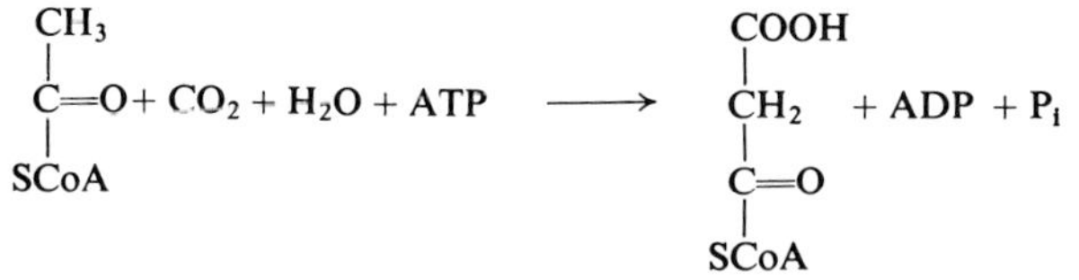
CH3
C=O + CO2 + H2O + ATP
SCoA
COOH
CH2
+ ADP + Pi
C=O
SCoA

acetyl-CoA malonyl-CoA

dependent and requires biotin and Mn^{2+}, as do most of the CO_2-fixation reactions. Malonyl-CoA is then reduced by an NAD-dependent malonate semialdehyde dehydrogenase [malonate semialdehyde:NAD(P) oxidoreductase, EC 1.2.1.15] to malonyl semialdehyde-CoA and NADH + H^+.

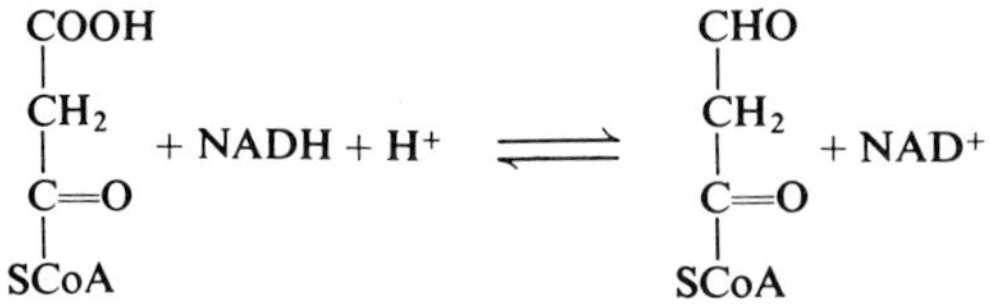
COOH
CH2
+ NADH + H+
C=O
SCoA
CHO
CH2
+ NAD+
C=O
SCoA

malonyl-CoA malonyl semialdehyde-CoA

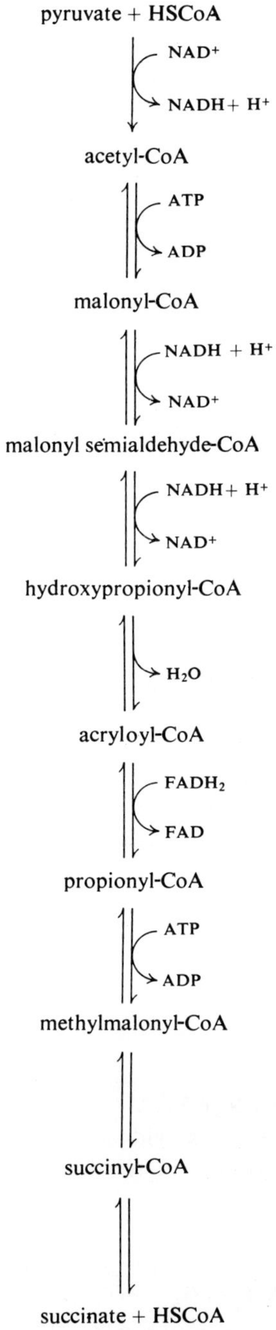
pyruvate + HSCoA
NAD+
NADH+ H+
acetyl-CoA
ATP
ADP
malonyl-CoA
NADH + H+
NAD+
malonyl semialdehyde-CoA
NADH+ H+
NAD+
hydroxypropionyl-CoA
H2O
acryloyl-CoA
FADH2
FAD
propionyl-CoA
ATP
ADP
methylmalonyl-CoA
succinyl-CoA
succinate + HSCoA

Fig. 9.9. The formation of succinate from pyruvate by *Clostridium kluyveri* (240) (reprinted with permission of Williams and Wilkins).

A second reduction step follows immediately thereafter, under the action of 3-hydroxypropionate dehydrogenase (3-hydroxypropionate:NAD oxido-

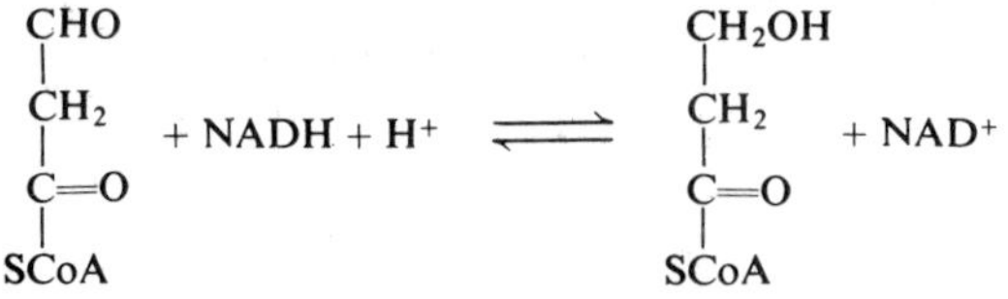

malonyl semialdehyde-CoA hydroxypropionyl-CoA

reductase, EC 1.1.1.59) and forms hydroxypropionyl-CoA and NAD^+. The release of 1 mole of H_2O during the next reaction with enoyl-CoA hydratase

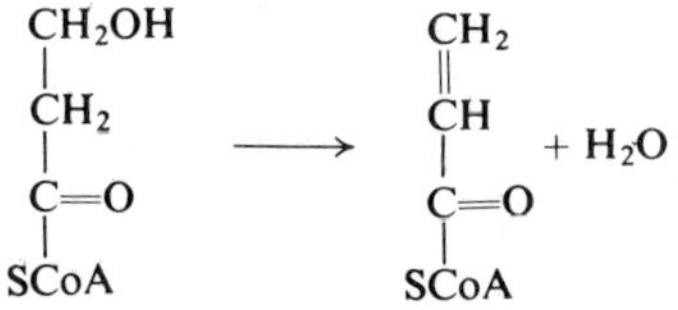

hydroxypropionyl-CoA acryloyl-CoA

(L-3-hydroxyacyl-CoA hydro-lyase, EC 4.2.1.17) leads to the formation of acrylyl-CoA. With $FADH + H^+$ as hydrogen donor, acrylyl-CoA receives one molecule of H_2 and is reduced to propionyl-CoA; $FADH + H^+$

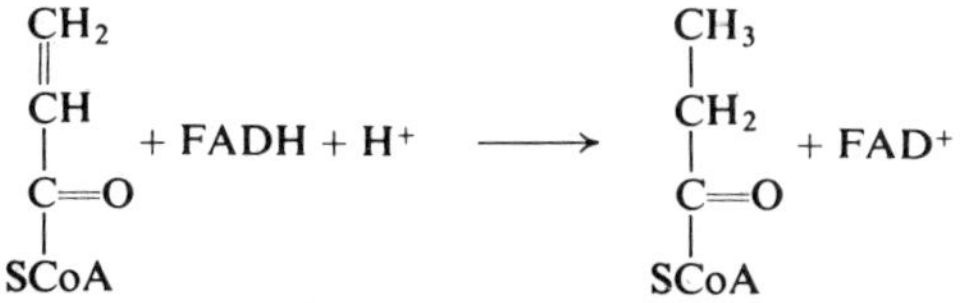

acryloyl-CoA propionyl-CoA

is simultaneously oxidized to FAD^+.

A second molecule of carbon dioxide is now fixed in the presence of biotin, Mn^{2+}, and ATP, which results in the formation of methylmalonyl-CoA.

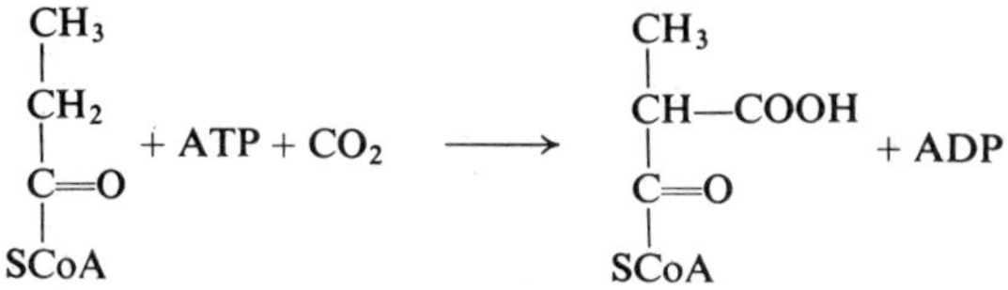

propionyl-CoA methylmalonyl-CoA

The enzyme catalyzing this reaction is propionyl-CoA carboxylase [propionyl-CoA:carbon-dioxide ligase (ADP), EC 6.4.1.3]. The conversion of methylmalonyl-CoA to succinyl-CoA requires a coenzyme of the vitamin B_{12} group and methylmalonyl-CoA mutase (methylmalonyl-CoA:CoA-

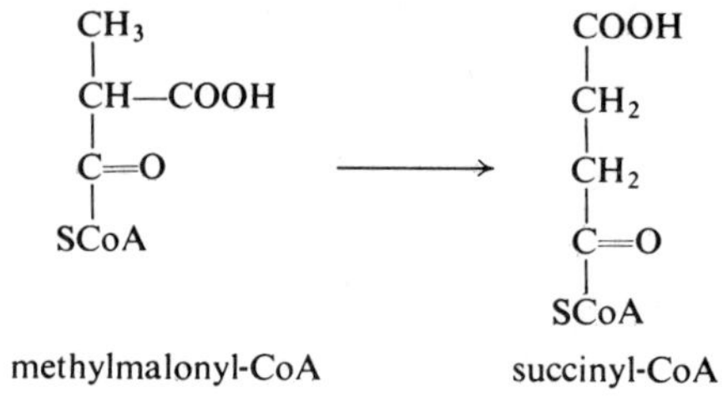
CH3
CH—COOH
C=O
SCoA
COOH
CH2
CH2
C=O
SCoA
methylmalonyl-CoA
succinyl-CoA

carbonylmutase, EC 5.4.99.2). The final product, succinate, is formed with the production of 1 mole of ATP from the interaction of succinyl-CoA and

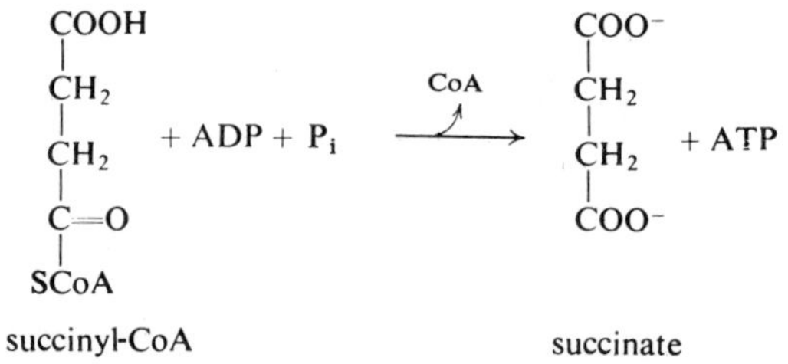
COOH
CH2
CH2
C=O
SCoA
+ ADP + Pi
CoA
COO-
CH2
CH2
COO-
+ ATP
succinyl-CoA
succinate

succinyl-CoA synthetase [succinate:CoA ligase (ADP), EC 6.2.1.5]. Most of the other bacterial species that form succinate from pyruvate do so via the glyoxylate cycle.

Fermentation of Enterobacteriaceae

A great number of strains of the family Enterobacteriaceae have been isolated and their classification is still in a constant state of flux. Metabolically they can be split into three large groups: (1) the mixed acid producer (Methyl Red positive; Voges-Proskauer negative); (2) the butanediol producers (Methyl Red negative; Voges-Proskauer positive); (3) the trimethylene glycol producer. Table 9.3 demonstrates the differences in the ratio and combination of end products formed with the first two groups (439). Although these two groups are usually exclusive, some organisms [e.g., *Klebsiella edwardsii* var. *atlantae* and the oxytoca biotype group (111a, 356)] fail to convert the acids completely and are Methyl Red posi-

TABLE 9.3

Differences in Ratio and Combination of End Products between Mixed Acid and Butanediol Producers[a]

	Mixed acid producer				Butanediol producer							
	Escherichia coli					*Enterobacter indologenes*		*Serratia marcescens*				
End products	pH 6.2	pH 7.8	Resting cells aerobic	*E. aurescens*	*A. aerogenes*	47 hr	209 hr	Anaerobic	Aerobic	*S. plymuthicum*	*S. kielensis*	*E. carotovora*
2,3-Butanediol	0.3	0.3	—	—	19	58	66	64	53	47	—	75
Acetoin	0.06	0.2	—	—	—	2	0	2	8	4	2	—
Glycerol	1.4	0.3	—	—	4	—	0	1	1	2	2	—
Ethanol	50	50	77	40	52	70	69	46	29	51	46	66
Formate	2	86	121	70	68	54	17	48	3	3	2	134
Acetate	36	39	78	40	52	8	0.5	4	9	5	50	64
Lactate	80	70	20	80	10	6	3	10	21	34	104	23
Succinate	11	15	39	10	13	—	—	8	9	7	2	11
CO_2	88	2		2	80	140	172	117	159	145	100	13
H_2	75	26		—		11	35	—	0.2	59	91	—
O/R balance	1.06	0.91	1.04			1.01	0.99	1.02	—	0.98	1.07	0.97

[a] Millimoles per 100 mmoles of glucose fermented.

tive and Voges-Proskauer positive. The catabolism of sugars occurs both by the EMP and by the HMP pathway, at least in all strains that have been investigated so far. Their physiological behavior and the end products of their glucose fermentation, however, suggest a common mechanism. There is no doubt that all the enzymes of the EMP pathway are present in *Escherichia coli* and in species of *Aerobacter*. Several of these enzymes have been detected in *Serratia, Erwinia, Salmonella, Klebsiella,* and *Paracolobacterium*. The enzymes of the HMP pathway were first detected in *Escherichia coli* and later in species of *Aerobacter, Paracolobacterium, Serratia, Klebsiella, Salmonella, Erwinia,* and *Proteus.*

Assuming that the intermediary metabolism of sugars is the same for all the Enterobacteriaceae, the difference in their fermentation patterns is owing to variations in the enzymes concerned with the decomposition of pyruvate. The sugar-fermenting sarcina, *S. ventricula,* employs the same pyruvate utilization mechanisms (51) as do the Enterobacteriaceae, whereas *S. maxima* favors the clostridial type (237).

The Mixed Acid Producers

The most characteristic representative of this group is undoubtedly *Escherichia coli,* which metabolizes pyruvate as shown in Scheme 9.1. Any of the underlined compounds can be major end products. If this system is operating alone, the following ratios should hold

$$\text{Moles (ethanol + acetate)} = \text{moles } (H_2 + \text{formate})$$

and

$$H_2 = (CO_2 + \text{succinate})$$

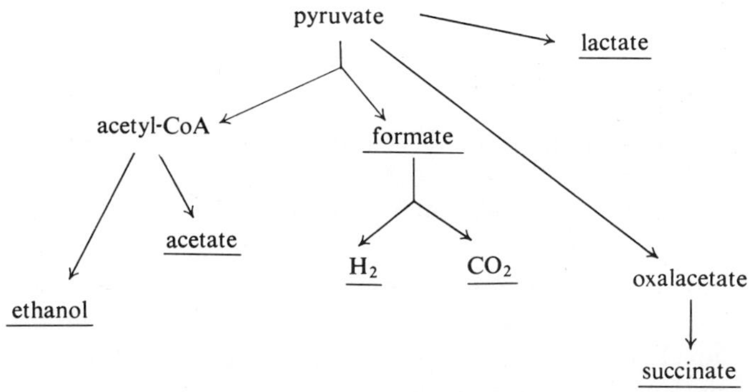
pyruvate
lactate
acetyl-CoA
formate
acetate
H2
CO2
oxalacetate
ethanol
succinate

Scheme 9.1. Pyruvate metabolism of the mixed acid producers of the Enterobacteriaceae.

These ratios hold only for the mixed acid fermenters, for pyruvate is broken down in the phosphoroclastic reaction (see p. 596) to equimolar quantities of acetyl-CoA and formate

$$\text{1 mole of pyruvate} \rightarrow \text{1 mole of acetyl-CoA} + \text{1 mole of formate}$$

Hence

$$\text{Moles pyruvate} = \text{moles acetyl-CoA} = \text{moles formate} \qquad (9\text{-}1)$$

The acetyl-CoA is then converted entirely into ethanol, acetate, or both, but in any case

$$\text{Moles pyruvate} = \text{moles ethanol} + \text{moles acetate} \qquad (9\text{-}2)$$

From Eqs. (9-1) and (9-2) then follows

$$\text{Moles (ethanol + acetate)} \equiv \text{moles formate} \qquad (9\text{-}3)$$

The produced formate may be split into H_2 and CO_2 in equimolar quantities, whereby

$$\text{1 mole } H_2 \text{ or 1 mole } CO_2 \text{ represents 1 mole formate decomposed} \qquad (9\text{-}4)$$

Because it may well be that not all formate is decomposed, the following relation may exist

$$\text{Moles total formate} = \text{moles residual formate} + \text{moles decomposed formate}$$

If Eq. (9-4) is taken into consideration

$$\text{Moles total formate} = \text{moles formate} + \text{moles } H_2 \qquad (9\text{-}5)$$

The final equation is derived by introducing Eq. (9-5) into Eq. (9-3):

$$\text{Moles (ethanol + acetate)} \equiv \text{moles formate} + \text{moles } H_2$$

The end product, hydrogen, was chosen because all H_2 formed comes from formate, whereas this need not necessarily be the case for carbon dioxide. Carbon dioxide is also very likely to be reconsumed. If the requirements of this equation have been filled, a phosphoroclastic split occurs.

If CO_2 is taken into consideration, it is possible that some CO_2 (which has been solely derived from formate) is consumed in the production of

TABLE 9.4
Observations from Fermentation Balances

Organism	millimoles per 100 moles glucose			
	Ethanol + acetate	H_2 + formate	H_2	CO_2 + succinate
E. coli	90	90	0	15
E. coli (pH 6.2)	86	77	75	99
E. coli (pH 7.8)	89	112	26	17
E. aurescens	80	70	0	10
E. carotovora	120	130	0	10
Serratia plymuthicum	56	62	60	156
S. kielensis	100	102	100	102
S. kielensis	96	93	91	102

succinate at the rate of 1 mole of CO_2 per mole succinate

$$\text{Moles } CO_2 \text{ (from formate)} = \text{moles } CO_2 + \text{moles } CO_2 \text{ consumed in succinate production}$$

Because the total moles CO_2 formed from formate equal the total moles H_2 formed from formate

$$\text{Moles } H_2 = \text{moles } CO_2 + \text{moles succinate}$$

This reaction should hold if the CO_2 and succinate are not involved in any further reactions, and if no other CO_2 is produced. This does not apply to the butanediol fermenters, which produce additional CO_2 in the course of butanediol production. The observations given in Table 9.4 can be made from the fermentation balances.

FORMATION OF ACETALDEHYDE. Pyruvic acid can be decarboxylated to acetaldehyde and carbon dioxide with pyruvate decarboxylase (2-oxoacid carboxy-lyase, EC 4.1.1.1). This reaction requires Mn^{2+} and TPP as cofactors. It is assumed that (*a*) an "activated acetaldehyde" in the form of the carbanion CH_3CO^- linked to TPP occurs as an intermediate, and that this hypothetical acetaldehyde-TPP compound reacts with H^+ to give free acetaldehyde and to regenerate TPP (160)

$$CH_3\text{—CO—COOH} \xrightarrow{\text{TPP}} CH_3\text{—CHO} + CO_2$$

or (*b*) a phosphoroclastic split occurs (240), wherein inorganic phosphate, TPP, CoA, lipoate, Mn^{2+}, or Fe^{2+} is required for the reaction. The latter would be a similar process to that in the clostridia, only with the formation of formate instead of carbon dioxide and hydrogen (see p. 578). This phosphoroclastic split is also referred to as the "thioclastic split" (86) because of the involvement of lipoate, in contrast to the yeast system described under (*a*), which also involves a phosphate compound as cofactor. McCormick *et al.* (259), however, distinguished between the phosphoroclastic split and the thioclastic split. They considered the formation of acetyl phosphate, carbon dioxide, and hydrogen as the thioclastic split, whereas the formation of acetyl phosphate and formate is referred to as the "phosphoroclastic split." It is difficult to decide whether or not these two terms should continue to be used to express the formation either of formate or of $CO_2 + H_2$, because acetyl phosphate is obtained in both cases from acetyl-CoA with the aid of the phosphate acetyltransferase. There may be justification for the two terms if acetyl-CoA is the first product in one case (Enterobacteriaceae) and acetyl phosphate in the other (clostridia). Because of the different opinions, "thioclastic" and "phosphoroclastic" will be considered synonymous.

The phosphoroclastic split can form either acetyl-CoA or acetyl phosphate, which are held in equilibrium by a phosphate acetyltransferase

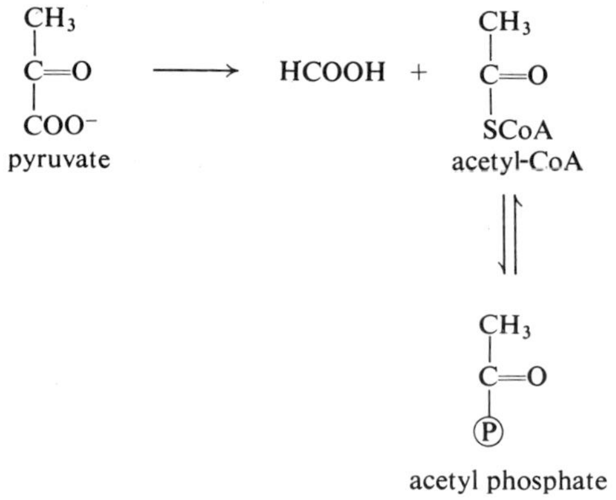
CH3
C=O
COO⁻
pyruvate
HCOOH +
CH3
C=O
SCoA
acetyl-CoA
CH3
C=O
P
acetyl phosphate

(acetyl-CoA:orthophosphate acetyltransferase, EC 2.3.1.8). It is assumed (259, 339) that acetyl-CoA is the first product. This phosphate acetyltransferase has been purified and characterized (351). It was found to be an allosteric enzyme activated by pyruvate and inhibited by $NADH + H^+$ (383, 384) and, most strongly, by ATP. This indicates that phosphate acetyltransferase controls the acetyl-CoA formation and therefore also the use of the tricarboxylic acid cycle, possibly in conjunction with phosphoenolpyruvate carboxylase (EC 4.1.1.31) (384).

The formation of acetaldehyde from acetyl-CoA requires an NAD^+-

linked aldehyde dehydrogenase [aldehyde:NAD oxidoreductase (acylating CoA), EC 1.2.1.10] (336). This is not the only way in which acetalde-

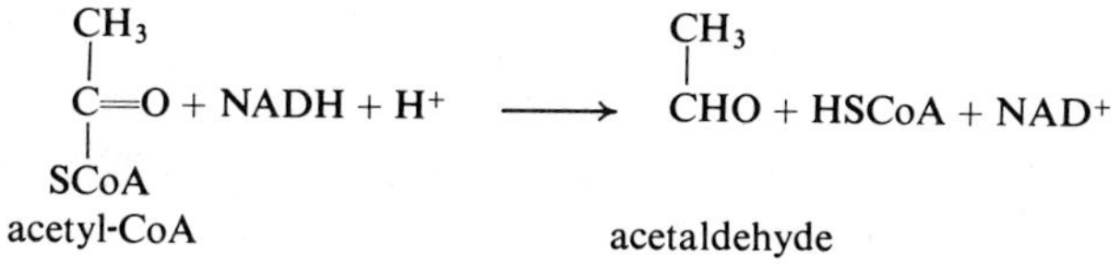

$$\begin{array}{c}CH_3\\|\\C=O\\|\\SCoA\end{array} + NADH + H^+ \longrightarrow \begin{array}{c}CH_3\\|\\CHO\end{array} + HSCoA + NAD^+$$

acetyl-CoA $\qquad\qquad\qquad\qquad$ acetaldehyde

hyde is formed by the Enterobacteriaceae. If the species is producing acetate as well, acetaldehyde can be produced from acetate with a different aldehyde dehydrogenase. The aldehyde dehydrogenase catalyzing this reaction

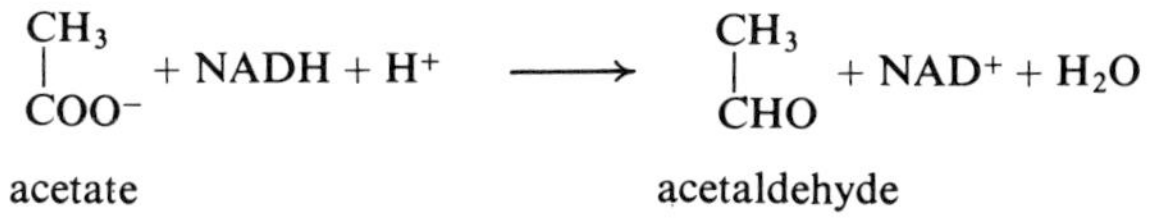

$$\begin{array}{c}CH_3\\|\\COO^-\end{array} + NADH + H^+ \longrightarrow \begin{array}{c}CH_3\\|\\CHO\end{array} + NAD^+ + H_2O$$

acetate $\qquad\qquad\qquad\qquad$ acetaldehyde

is very likely aldehyde:NAD oxidoreductase (EC 1.2.1.3), which has a wide specificity. The second possibility, which has not been reported, would be aldehyde:NAD(P) oxidoreductase (EC 1.2.1.5), which catalyzes the same reaction but can in addition be linked with NADP$^+$.

Special emphasis should be given to the significant difference between *Erwinia amylovora* and the other Enterobacteriaceae, as the former possesses a pyruvate decarboxylase (EC 4.1.1.1) for pyruvate metabolism.

FORMATION OF ETHANOL. The formation of ethanol from acetaldehyde requires alcohol dehydrogenase (alcohol:NAD oxidoreductase EC 1.1.1.1), which acts on primary and secondary alcohols in general, and Zn^{2+}. During

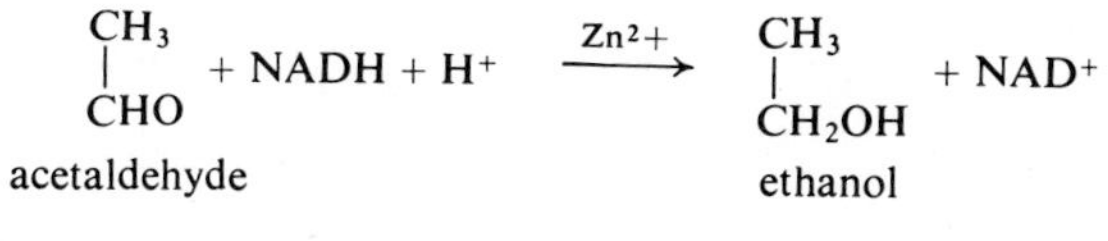

$$\begin{array}{c}CH_3\\|\\CHO\end{array} + NADH + H^+ \xrightarrow{Zn^{2+}} \begin{array}{c}CH_3\\|\\CH_2OH\end{array} + NAD^+$$

acetaldehyde $\qquad\qquad\qquad\qquad$ ethanol

this formation of ethanol from pyruvate, 2 moles of reduced NAD are regenerated per mole of ethanol and the product is neutral.

FORMATION OF ACETATE. Acetate can be formed either directly from pyruvate (73) or from acetyl-CoA via three alternative ways (240):

1. The direct formation of acetate from pyruvate is brought about

by the action of pyruvate oxidase [pyruvate:oxygen oxidoreductase (phosphorylating), EC 1.2.3.3], a soluble flavoprotein that binds both FAD and

$$CH_3COCOOH + E\genfrac{}{}{0pt}{}{FAD}{TPP} \longrightarrow CH_3COOH + CO_2 + E\genfrac{}{}{0pt}{}{FADH_2}{TPP}$$

TPP and that is activated by phospholipids and long-chain fatty acids.

2. An acetyl-CoA hydrolase (EC 3.1.2.1) catalyzes the transforma-

$$\begin{array}{c} CH_3 \\ | \\ C=O + H_2O \\ | \\ SCoA \end{array} \longrightarrow \begin{array}{c} CH_3 \\ | \\ COOH + HSCoA \end{array}$$

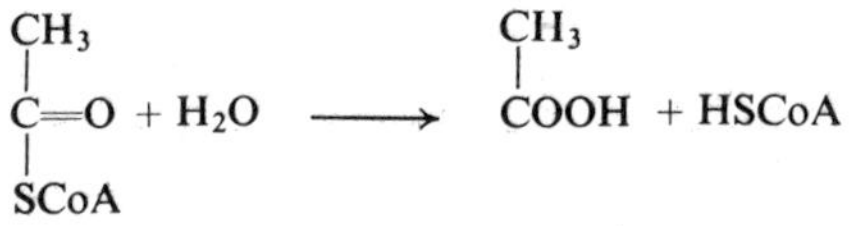

tion of acetyl-CoA to acetate. This is a rather wasteful system, as no energy is gained.

3. Acetyl-CoA synthetase [acetate:CoA ligase (AMP), EC 6.2.1.1] is also present and could convert acetyl-CoA to acetate. This degradation of

$$\begin{array}{c} CH_3 \\ | \\ C=O + AMP + PP_i \\ | \\ SCoA \end{array} \longrightarrow \begin{array}{c} CH_3 \\ | \\ COOH + ATP + HSCoA \end{array}$$

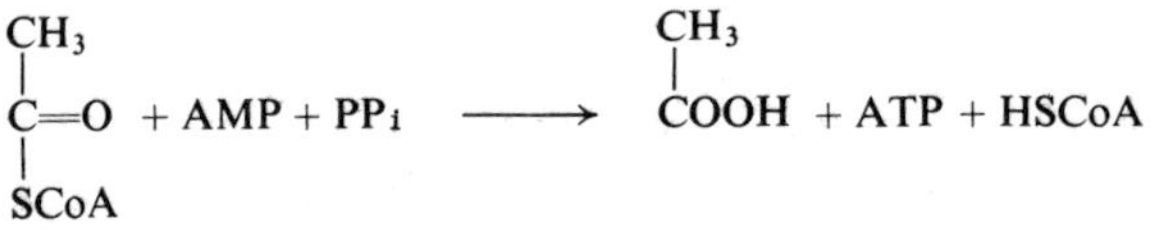

acetyl-CoA produces 1 mole of ATP but requires AMP and pyrophosphate.

4. This last sequence is the most economical and energy-producing utilization step. It is the most widely used one occurring in *Escherichia coli*.

$$\begin{array}{c} CH_3 \\ | \\ C=O + P_i \\ | \\ SCoA \end{array} \longrightarrow \begin{array}{c} CH_3 \\ | \\ C=O \\ | \\ PO_3^- \end{array}$$

$$\downarrow \genfrac{}{}{0pt}{}{-ADP}{\rightarrow ATP}$$

$$CH_3COOH$$

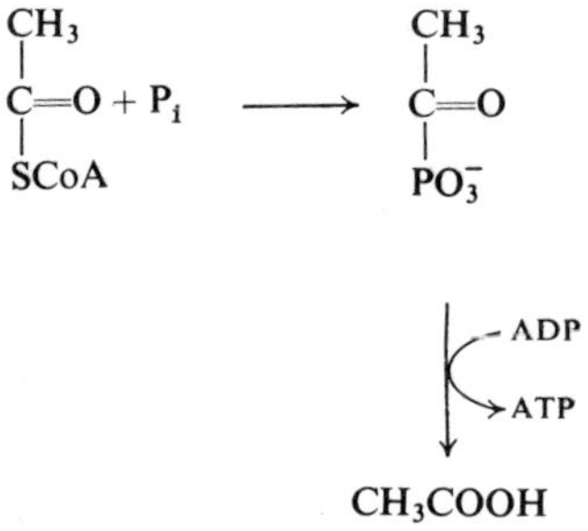

This utilization requires inorganic phosphate and phosphate acetyltransferase (EC 2.3.1.8) as well as acetate kinase (ATP:acetate phosphotransferase, EC 2.7.2.1) to form the end product acetate. The presence of phosphate acetyltransferase in the genus *Escherichia* is indicated by the existence of the equilibrium between acetyl-CoA and acetyl phosphate. As both enzymes play also important roles in acetate utilization (183), kinetic studies (182, 344) give more detailed information into the actual reaction mechanisms.

FORMATION OF CO_2 PLUS H_2 FROM FORMATE. Whether or not the similarity of the clostridial and the coli type phosphoroclastic split is so close is still under dispute. The most important observation is the consistent failure of components isolated from strict anaerobes to interact with those from facultative anaerobes. For example, the formate dehydrogenase of the coliform bacteria will not couple with the hydrogenase of crude clostridial extracts to give the hydrogen lyase system.

The formation of carbon dioxide and hydrogen from formate by the formic hydrogen lyase system is therefore thought (149) to include two enzymatic reactions, a soluble formate dehydrogenase and a membrane-bound hydrogenase, with two unidentified intermediate electron carriers, X_1 and X_2.

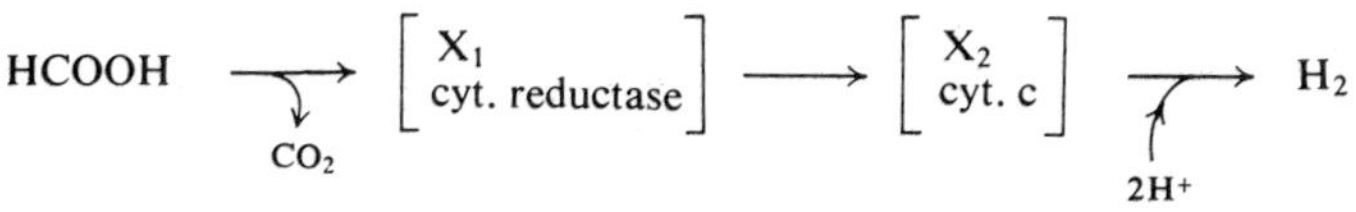
HCOOH
CO2
X1
cyt. reductase
X2
cyt. c
2H+
H2

There exists evidence for the identity of X_2 as a c-type cytochrome of low redox potential ($E_m = -225$ mV). Such low-potential cytochromes are only produced by coliform bacteria during anaerobic growth. Accordingly, X_1 may well have the function of a cytochrome c reductase. Many efforts were undertaken to detect ferredoxin in the facultative anaerobes after this compound had been found in clostridia. It was hoped that the abovementioned ferrodoxin system could also be applied to the hydrogen lyase system, but the results were negative. These results led to the assumption that there were many qualitative differences in the hydrogen-evolving system of strict and facultative anaerobes. Investigations into the pyruvate–formate exchange reaction (60, 61, 226) did not lead to a clarification of this reaction.

Aeromonas hydrophilia, which has a coli-type metabolism, also produces cytochrome c of low potential and can exhibit formate hydrogen lyase activity when grown anaerobically. The possible absence of ferredoxin would also be consonant with studies on *Clostridium thermoaceticum* (242), although this organism was found to possess ferredoxin (see p. 578) and a $NADP^+$-linked formate dehydrogenase. The latter catalyzes an exchange between carbon dioxide and formate. No other electron carrier has, however, been isolated. *Clostridium thermoaceticum* appears to behave differently from all other clostridia, as it also produces formate.

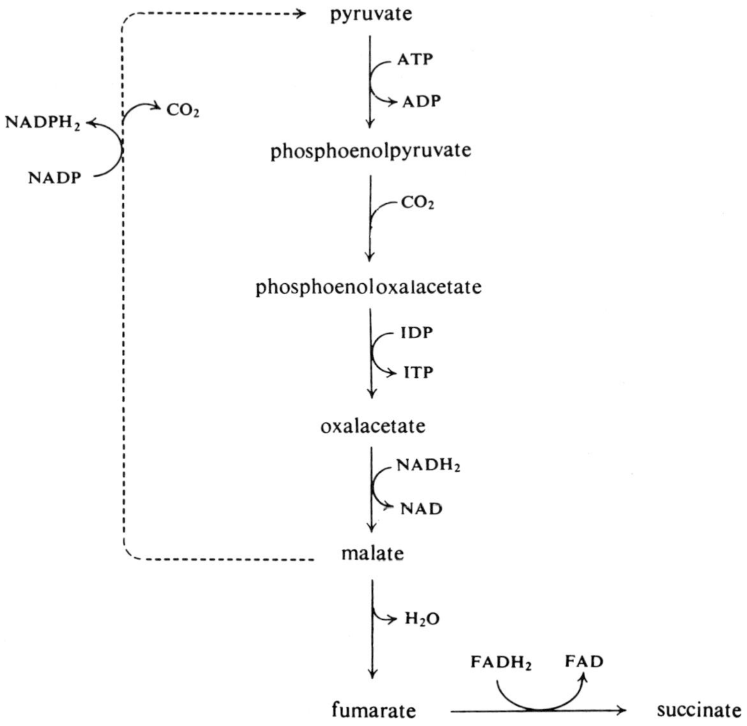
pyruvate
ATP
ADP
NADPH$_2$
NADP
CO_2
phosphoenolpyruvate
CO_2
phosphoenoloxalacetate
IDP
ITP
oxalacetate
NADH$_2$
NAD
malate
H_2O
FADH$_2$
FAD
fumarate
succinate

Fig. 9.10. The formation of succinate from pyruvate by Enterobacteriaceae using the reversed TCA cycle.

The present state of knowledge is therefore that formate is reacted upon by an NAD$^+$-linked formate dehydrogenase (EC 1.2.1.2) that liberates carbon dioxide and reduces NAD$^+$. The latter is reoxidized either by ferredoxin or by a cytochrome c reductase, which in turn is oxidized by a low-potential cytochrome c. This low-potential cytochrome c exchanges its electrons with hydrogenase specific for coli-type bacteria and forms hydrogen. Whether or not this low-potential cytochrome c is identical with ferredoxin has not been elucidated, although the potentials would suggest this possibility.

The importance in this reaction sequence is the breakdown of formate to an equal amount of CO_2 and H_2. There are, however, a number of bacteria that have completely lost the ability to break down formate and therefore accumulate this compound as a major end product, e.g., *Salmonella typhi*, *Shigella*, *Proteus rettgeri*, and *Serratia marcescens*. It may be significant that these organisms normally live in alkaline solutions in the intestine, which could be important for the loss of induction of this enzyme (319).

FORMATION OF SUCCINATE. The aerobic formation of succinate during the TCA cycle was dealt with under aerobic respiration (see p. 382). It was also demonstrated that clostridia are able to form succinate under anaerobic conditions. The Enterobacteriaceae in general use a third pathway (86, 240), which contains a few TCA cycle enzymes (Fig. 9.10). Pyruvate is phosphorylated to phosphoenolpyruvate, requiring Mg^{2+}, K^+, ATP, and

$$\begin{array}{c} CH_3 \\ | \\ C{=}O + ATP \\ | \\ COO^- \end{array} \longrightarrow \begin{array}{c} CH_3 \\ | \\ C{-}O \sim \text{\textcircled{P}} + AMP + P_i \\ | \\ COO^- \end{array}$$

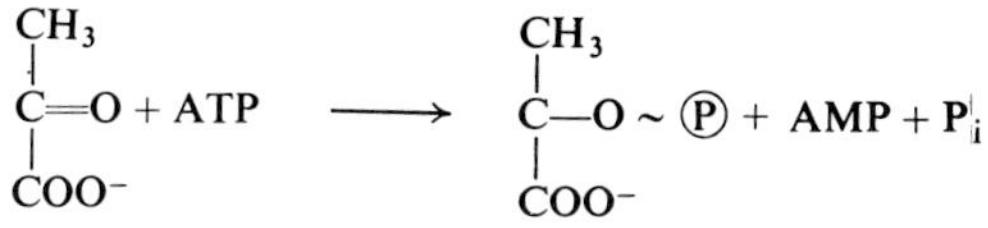

pyruvate   phosphoenolpyruvate

a phosphoenolpyruvate synthase. The incorporation of 1 mole of carbon dioxide and the functioning of a phosphoenolpyruvate carboxylase leads to phosphoenol oxalacetate. The next step in the sequence is an energy-

$$\begin{array}{c} CH_2 \\ | \\ C{-}O{-}PO_3^- + CO_2 \\ | \\ COO^- \end{array} \longrightarrow \begin{array}{c} COOH \\ | \\ CH_2 \\ | \\ C{-}O{-}PO_3^- \\ | \\ COO^- \end{array}$$

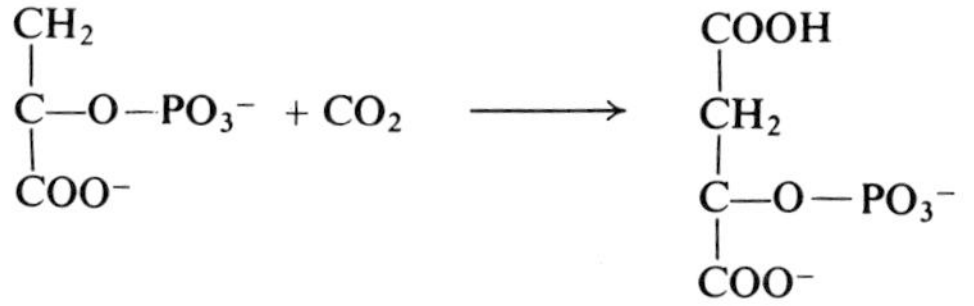

phosphoenolpyruvic   phosphoenoloxalacetic
acid   acid

yielding reaction whereby the phosphate group is transferred to IDP, resulting in the formation of ITP and oxalacetate. No one has reported whether or not an enzyme catalyzes this reaction, but it can be assumed that a phosphate transferase is acting because the reaction requires Mg^{2+}. The further conversion of oxalacetate to succinate via malate and fumarate is identical with the steps in the TCA cycle. The Enterobacteriaceae may also be able to close the cycle by employing a malate dehydrogenase [L-malate:NADP oxidoreductase (decarboxylating), EC 1.1.1.40] for the formation of pyruvate from malate. With this cycle, the organisms can fix the CO_2 evolved during the decarboxylation step from malate to pyruvate in the conversion of phosphoenolpyruvate to phosphoenol oxalacetate, but they are not able to regenerate reduced NADP. It therefore seems more likely that the reaction from malate to pyruvate takes place at an insignificant rate.

Escherichia coli and *Aerobacter* are also able to use the same pathway as described in detail for *Clostridium kluyveri*. Whether this pathway (Fig. 9.11) is significant for taxonomical classification is not clear as yet, and more research must be done. If, however, only a selected group of organisms

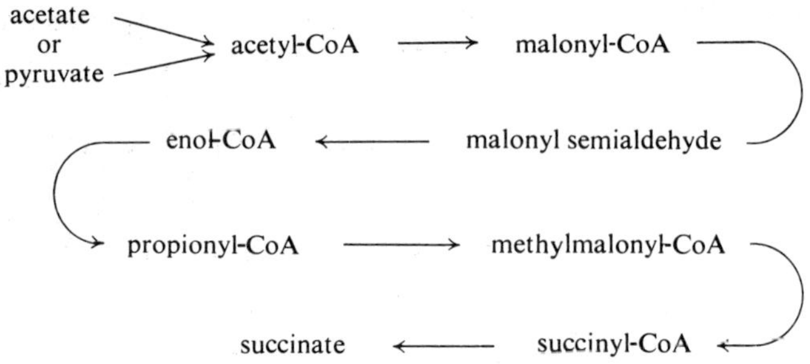
acetate
or
pyruvate
acetyl-CoA
malonyl-CoA
enol-CoA
malonyl semialdehyde
propionyl-CoA
methylmalonyl-CoA
succinate
succinyl-CoA

Fig. 9.11. The formation of succinate from pyruvate, using various CoA derivatives, by *Escherichia coli* and *Aerobacter aerogenes*.

among the Enterobacteriaceae is able to form succinate via the second pathway, this could become a significant characteristic. The overall effect in regard to energy and reduced NAD and FAD regeneration is much the same in both pathways.

FORMATION OF LACTATE. The production of lactate by a reaction of lactate dehydrogenase (L-lactate:NAD oxidoreductase, EC 1.1.1.27, or D-lactate: NAD oxidoreductase, EC 1.1.1.28) on pyruvate results in the regeneration of reduced NAD. It is assumed that both enzymes are present (30, 442),

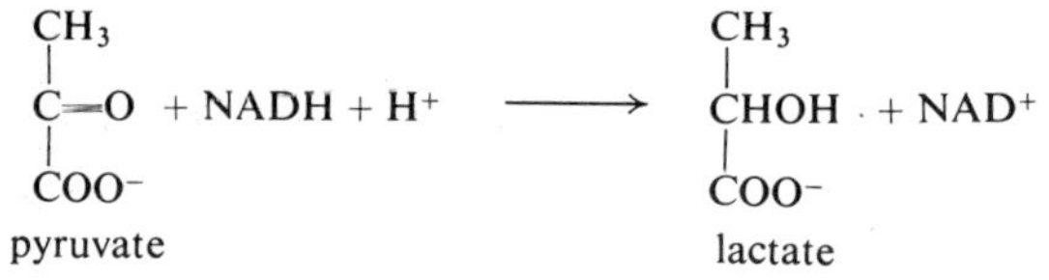
CH3
C=O + NADH + H+
COO−
pyruvate
CH3
CHOH + NAD+
COO−
lactate

although not very much emphasis has been given to which lactate isomer is formed. The amount of lactate varies greatly (439). It may account for as much as half of the pyruvate (*Serratia kielensis*) or may not be produced at all (*Aerobacter aerogenes*). In addition to this variation, lactate production also varies with the conditions under which any one organism is grown (100a, 387, 431).

The Butanediol Producer

The Voges-Proskauer reaction used in studies on the taxonomy of enterobacteria is based on a reaction between diacetyl and the guanidine nucleus found in creatine and in the peptone of the medium (in an alkaline solution). The butanediol producers also produce acetoin, and these two substances, when oxidized by air in alkaline solutions to diacetyl, give the positive Voges-Proskauer reaction.

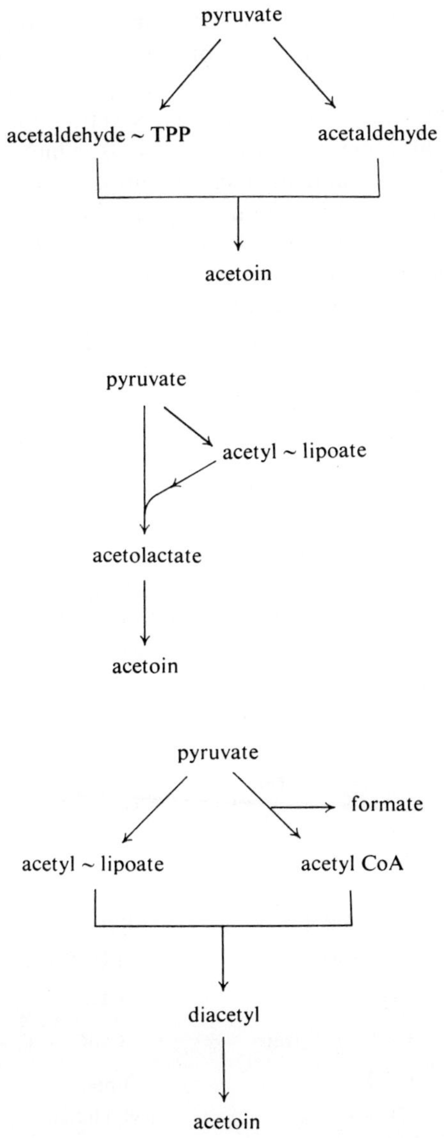
pyruvate
acetaldehyde ~ TPP
acetaldehyde
acetoin
pyruvate
acetyl ~ lipoate
acetolactate
acetoin
pyruvate
formate
acetyl ~ lipoate
acetyl CoA
diacetyl
acetoin

Fig. 9.12. Postulated sequences for the formation of acetoin from pyruvate.

FORMATION OF 2,3-BUTANEDIOL (2,3-BUTYLENE GLYCOL). In addition to the mixed acid products, a number of bacteria are able to produce 2,3-butanediol, which is a neutral compound produced from acetoin (acetylmethyl carbinol) that serves as hydrogen acceptor for reduced NAD regeneration. There are, however, three pathways, which lead from pyruvate to acetoin (240, 439), found in microorganisms. The first postulated reaction sequence has only been found in yeasts and depends on the direct formation of acetaldehyde from pyruvate (Fig. 9.12), with a subsequent carboligase reaction forming acetoin. This sequence is not likely to be present in bacteria, which without exception form acetyl-CoA from pyruvate. This restricts the discussion to the second and third pathways. The significant difference between the two is that the second sequence forms acetoin from acetolactate, whereas in the third pathway diacetyl is the precursor of acetoin.

THE FORMATION OF ACETOIN FROM ACETOLACTATE. *Aerobacter aerogenes, Bacillus subtilis, Clostridium acetobutylicum,* and *Streptococcus faecalis* do not contain a decarboxylase system and are not able to produce appreciable amounts of aldehyde (439). Therefore, a single enzyme is presumably responsible for the pyruvate decarboxylation and aldehyde transfer and the formation of the stable intermediate α-(+)acetolactate and carbon dioxide, although reports on the presence of a catabolic and biosynthetic acetolactate synthase (EC 4.1.3.18) are available (158). The acetyl–lipoate

$$2 \text{ pyruvate} \xrightarrow[\text{Mn}^{2+}]{\text{TPP}} \alpha\text{-}(+)\text{acetolactate} + CO_2$$

complex can be formed and CO_2 released from pyruvate by part of the pyruvate dehydrogenase system [pyruvate:lipoate oxidoreductase (acceptor-acetylating), EC 1.2.4.1, requiring TPP]. This "activated acetate" combines with a second molecule of pyruvate to form acetolactate. Aceto-

$$\begin{array}{c}CH_3\\|\\C=O\\|\\COO^-\end{array} + \text{lipoate} \longrightarrow \begin{array}{c}CH_3\\|\\C=O\\|\\\text{lipoate}\end{array} + CO_2$$

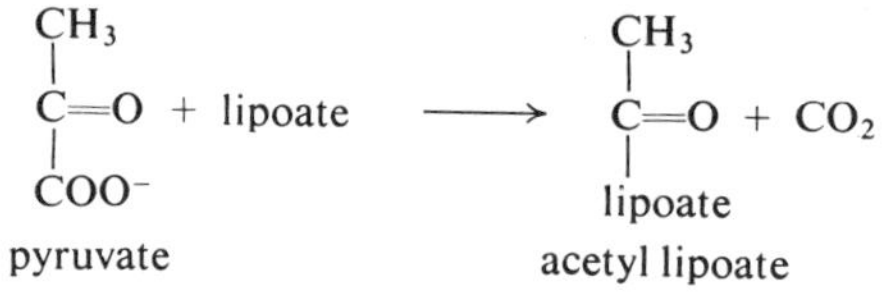

pyruvate   acetyl lipoate

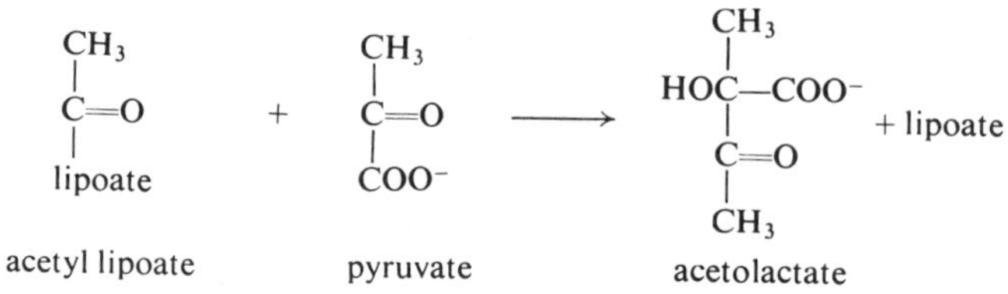

$$\begin{array}{c}CH_3\\|\\C=O\\|\\\text{lipoate}\end{array} + \begin{array}{c}CH_3\\|\\C=O\\|\\COO^-\end{array} \longrightarrow \begin{array}{c}CH_3\\|\\HOC-COO^-\\|\\C=O\\|\\CH_3\end{array} + \text{lipoate}$$

acetyl lipoate   pyruvate   acetolactate

lactate decarboxylase (2-hydroxy-2 methyl-3-oxobutyrate carboxy–lyase, EC 4.1.1.5) decarboxylates acetolactate to acetoin

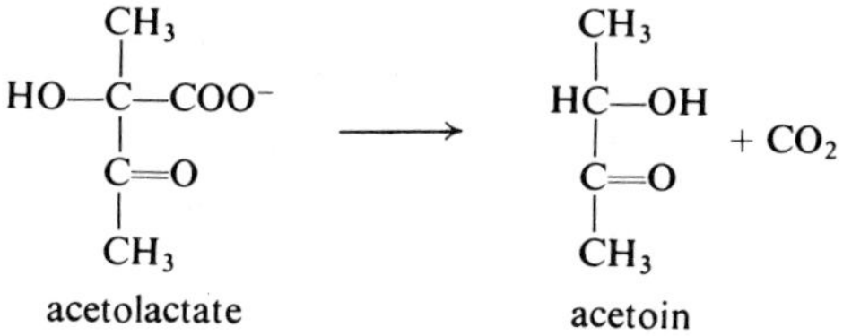
acetolactate
acetoin
+ CO_2

THE FORMATION OF ACETOIN FROM DIACETYL. The third mechanism is a combination of the mixed acid type of degradation of pyruvate and the second mechanism. The organisms are able to form acetyl–lipoate with pyruvate dehydrogenase (EC 1.2.4.1) and at the same time also form acetyl-CoA in the phosphoroclastic (or thioclastic) split. Both "active acetates" now undergo condensation reactions and form diacetyl:

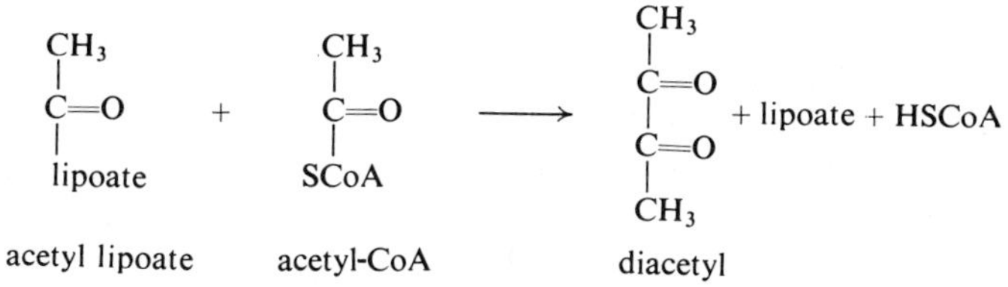
acetyl lipoate
acetyl-CoA
diacetyl
+ lipoate + HSCoA

Acetoin dehydrogenase (acetoin:NAD oxidoreductase, EC 1.1.1.5) reduces diacetyl to acetoin (49a, 181a, 181b, 196a, 353a):

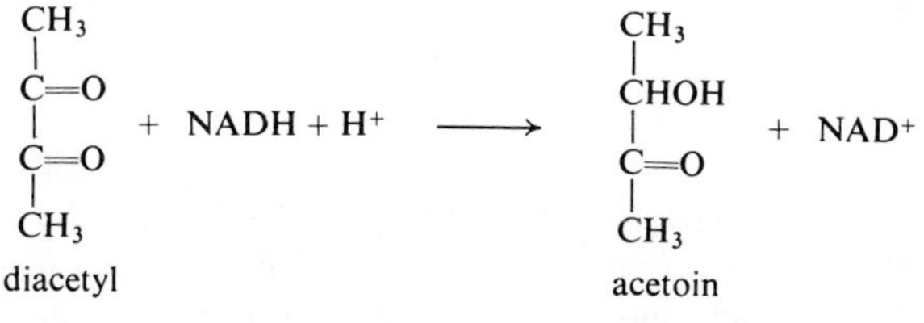
diacetyl
+ NADH + H^+
acetoin
+ NAD^+

Acetoin from the second mechanism can also be oxidized to diacetyl by acetoin dehydrogenase (EC 1.1.1.5). The latter reaction has not been reported in enterobacteria. The overall reactions of both mechanisms are very similar, although the last sequence may be more economical because the organisms can use existing enzymes for the formation of acetyl-CoA. Having the enzymatic complex of the mixed acid fermenters, these organisms produce acetyl-CoA. Because of the presence of an additional enzyme, however, they form butanediol in such a way that virtually no other byproduct can be formed.

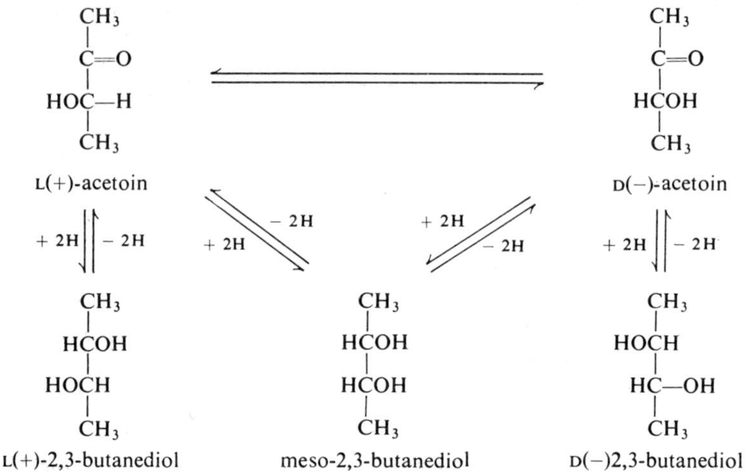

CH3
C=O
HOC—H
CH3
L(+)-acetoin
CH3
C=O
HCOH
CH3
D(−)-acetoin
+ 2H
− 2H
+ 2H
− 2H
+ 2H
− 2H
+ 2H
− 2H
CH3
HCOH
HOCH
CH3
L(+)-2,3-butanediol
CH3
HCOH
HCOH
CH3
meso-2,3-butanediol
CH3
HOCH
HC—OH
CH3
D(−)2,3-butanediol

Fig. 9.13. Postulated mechanism for the reduction of acetoin to 2,3-butanediol (389) (reprinted with the permission of Elsevier, Amsterdam).

It has been found that all bacteria producing 2,3-butanediol as their end product form D(−)-acetoin from pyruvate. The formation of 2,3-butanediol was for a long time thought to be a simple reduction process involving a butanediol dehydrogenase (2,3-butanediol:NAD oxidoreductase, EC 1.1.1.4), with NADH + H^+ being oxidized to NAD^+. Taylor and Juni (389), however, stressed that there exist a number of butanediol dehydrogenases, which are all stereoisomerically specific. *Bacillus polymyxa*, for example, contains D(−)-2,3-butanediol dehydrogenase, *Aerobacter aerogenes* and *Pseudomonas hydrophila* contain the L(+)-, and *Bacillus subtilis* contains both the D(−)- and the L(+)-butanediol dehydrogenase. *Pseudomonas hydrophila* also possesses an acetoin racemase (EC 5.1.2.4). The presence of all these combinations of enzymes can explain the occurrence of various combinations of the three isomers in carbohydrate fermentation. Taylor and Juni (389) therefore postulated the mechanism outlined in Fig. 9.13, which may occur in the reduction of acetoin to butanediol. Almost all bacteria form D(−)-acetoin. The type of butanediol produced depends on whether the organism (*1*) has an acetoin racemase that converts D(−)-acetoin to L(+)-acetoin [D(−)-butanediol dehydrogenase would produce *meso*-2,3-butanediol, whereas a L(+)-butanediol dehydrogenase forms L(+)-2,3-butanediol] or (*2*) has no acetoin racemase, in which case D(−)-butanediol dehydrogenase converts D(−)-acetoin into D(−)-2,3-butanediol and L(+)-butanediol dehydrogenase would produce *meso*-2,3-butanediol.

Because all these reactions are reversible and the dehydrogenases are NAD-linked, an acetoin racemase would undoubtedly lead to the formation of either L(+)-2,3-butanediol or *meso*-2,3-butanediol, with the regeneration of NADH + H^+. If there is no acetoin racemase available, D(−)- and *meso*-2,3-butanediol would be the end products. In each case, the final product depends on the action of one or the other stereospecific dehydrogenase.

Whatever way butanediol is formed, the overall reaction is the same

$$2 \text{ pyruvate} \rightarrow 2CO_2 + 2,3\text{-butanediol}$$

This formation of butanediol seems to be strictly pH dependent (439). If the pH rises above 6.3, acetic acid and formic acid accumulate and the production of CO_2, H_2, acetoin, and 2,3-butanediol is prevented. Below pH 6.3, however, the system switches over to acetoin and 2,3-butanediol production. A number of bacteria are known for their butanediol production: *Aerobacter aerogenes*, *Pseudomonas hydrophila*, *Erwinia carotovora*, *Serratia marcescens*, *Bacillus subtilis*, and others. Some microorganisms, such as *Bacillus polymyxa*, are used for industrial applications and 2,3-butanediol production (1). The production of acetoin is also used for the biochemical diagnosis of *Neisseria*, because *N. intracellularis*, *N. mucosa*, *N. perflava*, *N. sicca*, *N. flava*, *N. subflava*, *N. denitrificans*, *N. flavescens*, *N. ovis*, and *N. cumicola* are positive and *N. animalis*, *N. cinera*, *N. catarrhalis*, and *N. canis* are negative (403).

Whether or not the stereospecific reaction of butanediol formation could play a similar role in classification as does lactic acid for the homofermentative lactobacilli must be further investigated.

Formation of Trimethylene Glycol

Members of the Enterobacteriaceae are able to metabolize glycerol either to trimethylene glycol or to glyceraldehyde 3-phosphate, and subsequently via the EMP pathway.

The formation of trimethylene glycol appears to bear no relationship to any other fermentation or intermediate. *Citrobacter freundii* and *Aerobacter aerogenes* have the mechanisms to reduce glycerol to trimethylene

CH2OH
CHOH
CH2OH
CH2OH
CH2
CH2OH

glycerol trimethylene glycol

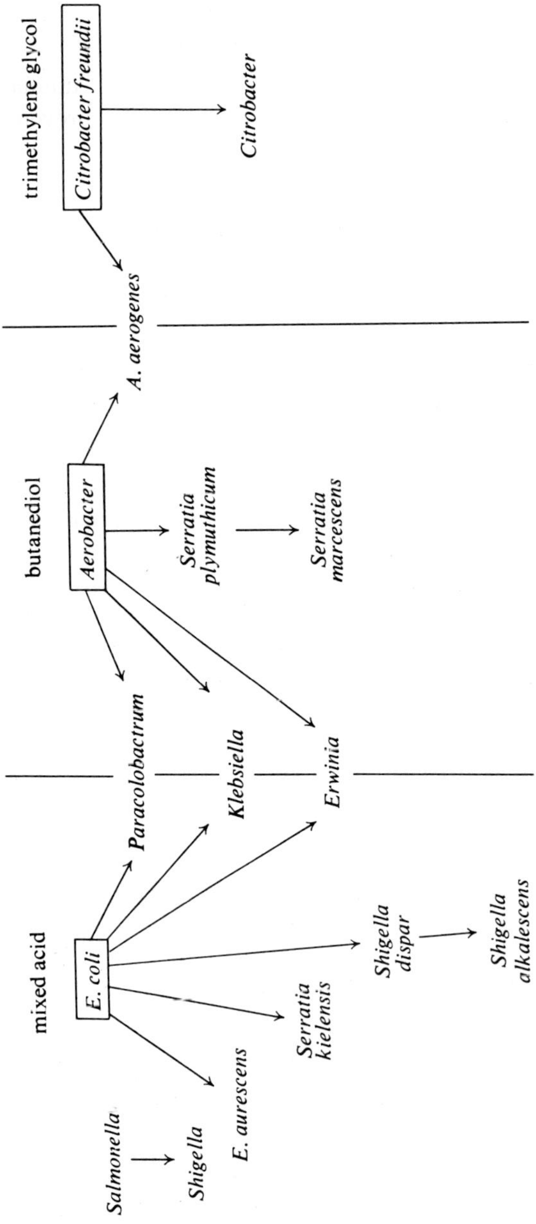
mixed acid
E. coli
butanediol
Aerobacter
trimethylene glycol
Citrobacter freundii
Salmonella
Shigella
E. aurescens
Serratia kielensis
Shigella dispar
Shigella alkalescens
Paracolobactrum
Klebsiella
Erwinia
Serratia plymuthicum
Serratia marcescens
A. aerogenes
Citrobacter

Fig. 9.14. Interrelationships between the three main groups of the Enterobacteriaceae.

glycol. Glycerol seems to compete with ethanol production. In the presence of glycerol, ethanol formation almost ceases and acetate is formed instead. This led to the assumption that the reoxidation of NADH + H^+ must occur in the last step of trimethylene glycol production. With *Citrobacter freundii* it was found that glycerol is converted first to β-hydroxypropionaldehyde, which is then reduced to trimethylene glycol. This process is assumed to be independent from the carbohydrate metabolism, occurring only if glycerol is supplied. There are, however, two processes known among microorganisms whereby glycerol formation can be regarded as a branchoff pathway from dihydroxyacetone phosphate.

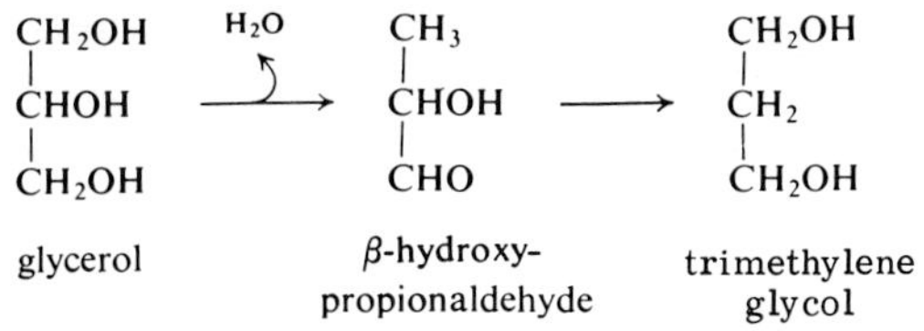
CH2OH
CHOH
CH2OH
glycerol
H2O
CH3
CHOH
CHO
β-hydroxy-
propionaldehyde
CH2OH
CH2
CH2OH
trimethylene
glycol

1. Under alkaline conditions, yeasts are able to produce glycerol from dihydroxyacetone phosphate and acetaldehyde (316). This is known as the "Eoff process" in industrial microbiology. Sodium or potassium carbonate are used for making the reaction alkaline. With this accomplished, the organism is still able to produce acetaldehyde, but the alcohol dehydrogenase becomes inactive.
2. *Acetobacter suboxydans* and *Escherichia coli* are able to oxidize glycerol to dihydroxyacetone phosphate (see Chapter 5). This common metabolic pathway could indicate a close relationship of *Escherichia coli* to the acetic acid bacteria.

The separation of the Enterobacteriaceae into the three groups, mixed acid producer, butanediol producer, and trimethylene glycol producer, is not without overlaps among the species. Figure 9.14 demonstrates the interrelationship of all three groups of Enterobacteriaceae.

Regulation of Carbohydrate Metabolism in Facultative Anaerobic Bacteria (Pasteur and Crabtree Effects)

A large group of bacteria, the so-called facultative anaerobes, are able to carry out either of the two metabolic events, aerobic respiration or fermentation (170a). In simply observing growth and end product formation, it can be seen that the changeover from aerobiosis to anaerobiosis frequently brings about a reduction of the cell yield, the glucose uptake rate, and is reflected in the type of fermentation end product formed. This phenomenon is called the "Pasteur effect" (28).

Under aerobic conditions, the microorganisms metabolize glucose completely to carbon dioxide and water. The means by which this is achieved, as well as the regulatory mechanisms involved, have been dealt with above. The organism overcomes the large redox potential difference from the oxidative couple (approximately −0.3 V) to the H_2O/O_2 couple (+0.8 V) utilizing the electron transport chain, which is connected with energy (ATP) production. The great majority of these respiratory enzymes are membrane bound.

During anaerobic fermentation, organic compounds serve as electron acceptors. They have very often the same redox potential as the oxidative couple, and there is no need for the elaborate electron transport system functioning in aerobic respiration systems. Most of the fermentative electron transport enzymes are soluble enzymes. The organism must therefore have some means by which a change from aerobiosis to anaerobiosis can be carried out and regulated. This phenomenon is one of the oldest, yet one of the most vital problems, of microbiology (74) and has not been fully elucidated. The aim of this review is to give an account of the work being done in this field and finally to suggest a working hypothesis. Because of the enormous literature on hand, the discussion is restricted mainly to the bacterial world but comparisons of these investigations with those carried out on yeast and also on mammalian cells are given.

Oxygen controls the life of microbes by either stimulating or inhibiting their functions. There are indications suggesting that large groups of enzymes, such as those of the TCA cycle and electron transport chain, are subject to induction and repression by oxygen (115). If this is the case, any change in oxygen levels could cause a major upheavel in cell organization. However, the control of anaerobiosis, with its use of the redox potential of the medium, is still an obscure process. A number of questions could be raised in this connection (428):

1. Is oxygen directly involved at the molecular level as a regulator?
2. Is the regulator molecule a single metabolite, controlling the appearance and disappearance of large groups of enzymes, or are there many distinct regulator metabolites, perhaps controlling each enzyme separately?
3. How does oxygen regulate the formation of new structures?

Most of the early work was related to the effect of oxygen on enzyme biosynthesis (148). One of the enzymes in the TCA cycle found to be severely repressed by oxygen (10a) was 2-oxoglutarate dehydrogenase. Without this enzyme (79, 157) *Escherichia coli* uses the reductive pathway to succinate under anaerobic conditions as well as the catabolic pathway leading to 2-oxoglutarate. In order to form succinate, *E. coli* induces the enzyme fumarate reductase (EC 1.3.1.6). It therefore appeared that under aerobic conditions, the TCA cycle served in a cyclic manner, whereas under anaerobic conditions it was modified to provide a branched, noncyclic pathway. The reductive branch, leading to succinate, then serves, either a biosynthetic or an amphibolic role. The enzyme 2-oxoglutarate dehydrogenase is an inducible enzyme in *E. coli* (10) and becomes available again as soon as oxygen is present. *Escherichia coli* grown aerobically corresponded closely to a strict aerobe in that it had an intact TCA cycle and a similar array of terminal respiratory enzymes (150).

As soon as anaerobiosis was obtained, the level of the TCA cycle enzymes was greatly reduced. The membrane-bound cytochromes were maintained at relatively high levels, and a series of enzymes that function under anaerobiosis were derepressed. In contrast to strict or obligate anaerobes, the facultative anaerobes contain many membrane-bound enzymes under anaerobic conditions. During the study of levels of the TCA cycle enzymes in *E. coli* cells, grown either aerobically or anaerobically with varied nutrition, at least three factors were found to influence their biosynthesis: (1) the presence or absence of molecular oxygen; (2) the repressive effect of glucose (catabolite repression); and (3) the balance between catabolic and anabolic demands. The most dramatic nutritional effects were related to glucose metabolism. Glucose markedly repressed the formation of the TCA cycle enzymes. This makes 2-oxoglutarate dehydrogenase an excellent indicator for the presence or absence of glucose catabolite repression under aerobic conditions. This glucose phenomenon is referred to as the "glucose effect" or "Crabtree effect" and should not be confused with the "oxygen" or "Pasteur effect." A separation of these effects is difficult but not impossible, as will be seen below.

Oxygen, of course, can also affect the anaerobic respiration. The more important inorganic compounds that are capable of reduction in place of oxygen are nitrate, nitrite, nitrous oxide, tetrathionate, thiosulfate, sulfate, chlorate, ferricyanide, and hydrogen ions. Many of these reductase reactions are repressed and/or inhibited by oxygen.

Nitrate reduction has been studied thoroughly in terms of oxygen regulation (306). In *Aerobacter aerogenes* less nitrite is produced from nitrate aerobically than anaerobically, as nitrate reductase was both repressed and reversibly inhibited by oxygen. In *Micrococcus denitrificans*, oxygen inhibited the formation of nitrogen from nitrate, nitrite, or nitrous oxide because of the individual reductase repression (307). During these investigations, two types of nitrate reductases could be distinguished (308): type A, which can utilize chlorate in place of nitrate as a substrate; and a type

B, which was inhibited by chlorate. Some organisms possess type A, some type B, and some both.

Oxygen inhibited the *in vivo* activity of all these enzymes, possibly by competition for electrons. This repression of nitrate reductase is widespread (67, 337) and is also reflected in the inhibition of the TCA cycle enzymes (427). Apart from the competition for electrons, there is also some evidence that the repression was related to a low redox potential, soluble c-type cytochrome produced anaerobically by *Escherichia coli* (67, 133). This cytochrome was found to be cytochrome c-552 (147), which is present in all enteric organisms tested and is repressed by oxygen (425). It has been suggested (132, 288) that this cytochrome may well function as a carrier in the formic hydrogen lyase reaction system (see p. 600). Although facultative anaerobic bacteria have long been known to produce hydrogen gas anaerobically but not aerobically, it was shown (304, 306) that both hydrogenase itself and the formate hydrogen lyase complex of which hydrogenase is a part were repressed and inhibited by oxygen. This repression could be correlated with the redox potential of the culture medium (426). Very similar repression was observed with tetrathionate reductase (305, 312) and thiosulfate reductase in *Proteus vulgaris* (306).

A number of dehydrogenases required for anaerobic growth have been shown to be repressed by oxygen. They included the alcohol and glycerol dehydrogenases of *Aerobacter aerogenes* (243, 262) and the acetoin and fumarate reductases of the same organism (306) and of *E. coli* (184, 440). However, formate dehydrogenase could exist in two forms, a membrane-bound form that linked the cytochrome system to oxygen, in which case the aerobic level was higher than the anaerobic one, and a soluble form that functioned in the anaerobic formic hydrogen lyase system.

Although it is possible to extend this catalog of observations, the above should demonstrate that two classes of dehydrogenase enzymes are controlled in opposite ways by oxygen. The aerobic TCA cycle dehydrogenase enzymes and flavoprotein–dehydrogenases linking directly to the aerobic terminal electron transport system are increased by oxygen. In contrast, dehydrogenase enzymes with Michaelis constants favoring substrate reduction operate during anaerobic fermentation reactions and are repressed by oxygen.

All these investigations suggest that a reduction in oxygen supply inhibits first the respiratory enzymes, such as 2-oxoglutarate dehydrogenase, as well as part of the electron transport chain (415). A shortening of the electron transport system automatically means a shorter distance or less difference in the redox potential between the dehydrogenated couple and the final acceptor couple. Under anaerobic fermentation conditions, the hydrogen from the metabolized couple is transferred to another organic compound of the same or similar redox potential, using $NAD^+/NADH + H^+$ as mediator. This means an automatic, severe drop in ATP production (227), which must be followed with a drop in the cell yield. This relationship between energy and growth can also be expressed in terms of values of adenylate energy charge (ATP + $\frac{1}{2}$ ADP)/(ATP + ADP + AMP) (56). It has been found that *E. coli* grows best at an energy charge value of about 0.8. Viability is maintained, however, if this value drops to 0.5, but cells die at values below this. Similar results have been obtained with *Peptococcus prévotii* (279a). Glucose is no longer completely utilized and oxidized to carbon dioxide and water and the first thought, of course, was to investigate the energy regulation. If the TCA cycle and its anaplerotic sequences cease operating, what replaces them?

Kinetic studies into the enzymes of the EMP pathway revealed very quickly that there are two enzymes that exhibit an allosteric reaction toward ATP. These enzymes were phosphofructokinase and glyceraldehyde 3-phosphate dehydrogenase. At present, the most favored mechanism for the oxygen-induced inhibition of glucose uptake rate in facultative anaerobic organisms is still the inhibition by ATP of phosphofructokinase activity. This concept developed from the observation that the influence on the rate of fermentation is related to the changes in the balance of inorganic phosphate (Pi) and ADP plus ATP following a transition from air to nitrogen (28).

There have been many theories proposed to explain the mechanism for the Pasteur effect, although only a few of these theories have been exhaustively tested.

1. Yeasts grown aerobically exhibit full mitochondrial structures, which disintegrate under anaerobic conditions to so-called "promitochondria" (72, 77, 414). When the organism changes back to aerobic conditions these mitochondria reappear (311, 332a). As the mitochondria are responsible for the oxidation of citric acid intermediates and phosphorylation (407, 408), a lack of these structures in yeasts causes inhibition of oxidative phosphorylation and a change from respiration to fermentation (42, 43).

2. There is growing evidence suggesting that phosphofructokinase may be effected under anaerobic conditions (137, 280, 395, 397). This enzyme is activated by inorganic phosphate, AMP, and ADP but is inhibited by relatively low concentrations of ATP. The latter can be bound at an effector site, which leads to conformational changes that modify both the affinity of the enzyme for its substrate and the maximal rate of reaction catalyzed (14).

3. The oxidation of 3-phosphoglyceraldehyde requires inorganic orthophosphate and also a supply of ADP. Orthophosphate and ADP are

also consumed, however, in the oxidation of pyruvate via the tricarboxylic acid cycle, ATP is being generated. There may exist true competition between the reactions. When respiration is low, ample phosphate and ADP are provided for a rapid use of the EMP pathway, whereas in the reverse case they may diminish to the level of nonavailability (317). This leads to the proposal that the control of the rate of anaerobic metabolism is mediated by intracellular orthophosphate or feedback mechanisms involving ATP and ADP (127).

4. A number of the enzymes involved in the EMP pathway contain SH groups, i.e., glyceraldehyde-phosphate dehydrogenase (EC 1.2.1.12) and phosphofructokinase. These may be oxidized, which decreases the activity of the enzymes.

In 1941, Lipmann (246) proposed that oxygen may reduce the rate of operation of the EMP pathway directly by oxidative inactivation of one of the EMP enzymes, such as glyceraldehyde-3-phosphate dehydrogenase (EC 1.2.1.12). Once glucose has been phosphorylated to glucose 6-phosphate, it may serve as substrate for either glucose-6-phosphate isomerase (EC 5.3.1.9), leading to the EMP pathway, or glucose-6-phosphate dehydrogenase (EC 1.1.1.49), leading to the HMP pathway. If the latter reaction were significantly favored at high levels of oxygen, this would reduce the rate of operation of the EMP pathway. This idea is indirectly supported by the fact that the activity of the HMP pathway has been correlated with the demand for biosynthetic precursors for nucleic acids and aromatic amino acids (53, 206). The presence of oxygen leads to a higher growth rate and hence a higher demand for biosynthetic precursors.

The competition theory postulates that in the presence of oxygen certain cofactors required for the EMP pathway are removed by the respiratory systems, causing a decrease in glucose utilization rate and the overall rate of the EMP pathway. Competition between respiration and the EMP pathway for inorganic phosphate (199), or for adenosine diphosphate and inorganic phosphate (320), has been suggested as the mechanism for the Pasteur effect. This theory was further developed (258) with the postulate that the reduced glucose utilization rate of respiring yeast cells is caused by impairment of the hexokinase reaction owing to lack of adenosine triphosphate at the site of glucose phosphorylation. The ATP formed during respiration is accumulated in the mitochondria and, it was suggested, is not readily available for extramitochondrial glucose phosphorylation.

In 1962, however, Passoneau and Lowry (296) proposed a new theory. In measuring the levels of EMP intermediates, they found that the enzyme reaction primarily responsible for the Pasteur effect in yeast (258), ascites tumor cells (250), and diaphragm (285), liver, heart (268), etc., is the phosphofructokinase reaction. They studied the kinetic properties of phosphofructokinase (PFK) and proposed that the control of PFK by its unusual kinetic properties is the mechanism of the Pasteur effect. The levels of intermediates of the EMP pathway, including fructose 6-phosphate and FDP, were assayed in cells cultured with and without oxygen. With cells cultured aerobically, levels of intermediates before the PFK reaction are higher than the levels of intermediates after the PFK reaction. With cells cultured anaerobically, the opposite result is obtained. This indicates that during oxygenation the phosphofructokinase reaction is inhibited, which leads to the decrease in glucose utilization rate associated with the Pasteur effect.

Because of the implied significance of the PFK reaction, the enzyme was purified from rabbit muscle and the kinetic properties of the enzyme were investigated (296). Passoneau and Lowry (296) suggested that their result would be accounted for if there were two ATP sites on the PFK molecule, a primary active site and a second inhibitory one, and that inorganic phosphate, adenylic acid, fructose 6-phosphate, and fructose 1,6-diphosphate act as stimulators and compete with ATP for the second site. These data were subsequently verified (265, 266) and the existence of a separate allosteric site for ATP, in addition to the substrate site, has been demonstrated in PFK from yeast (341) and sheep heart (2) cells. The significance of these results is that a mechanism was suggested by which oxygenation leads to the inhibition of glucose utilization rate. The high efficiency of aerobic energy production leads to a high ATP production rate and low levels of the stimulatory intermediates, so that ATP, although a substrate for PFK, becomes inhibitory. Surveys of the kinetic properties of PFK in cell extracts of *Escherichia coli* (14, 15, 256), *Clostridium perfringens* (256), *Staphylococcus aureus* (256), and *Klebsiella aerogenes* (343) indicated that the Pasteur mechanism as proposed by Passoneau and Lowry may also apply to the facultative anaerobic bacteria.

The rate of the reaction phosphorylating fructose 6-phosphate to FDP will depend not only on the activity of PFK as controlled by ATP, ADP, and F6-P concentrations, but also on the concentration of PFK in the cell. The theory for control of the Pasteur effect by control of PFK activity appears to assume a constant concentration of PFK in the cell and therefore a constant rate of synthesis of PFK. Although the use of rapid sampling techniques (167) confirmed that the control of the PFK reaction is exerted extremely rapidly when oxygenation conditions are altered, this does not preclude the possibility that control of PFK synthesis may exist. This control by repression or induction of enzyme formation is comparatively sluggish, but it would certainly serve as a mechanism enabling the cell to select the most suitable enzymatic composition for growth in a particular environment (294). If oxygenation leads to the inhibition of PFK activity, then a more suitable enzymatic composition would be obtained by reducing the rate of synthesis of PFK so as to remove the necessity for continued inhibition of phosphofructokinase activity.

It can therefore be proposed that the reduction in glucose utilization rate on oxygenation is caused by an immediate inhibition of PFK activity followed by a reduction of the concentration of PFK in the cell in coordination with a removal of the inhibition of activity. The overall result would be as has been deduced from studies of levels of EMP pathway intermediates, namely, a reduction in the rate of the PFK reaction, imposed rapidly and maintained while oxygenation continues.

A control of the EMP pathway by control of the quantities of certain enzymes has been suggested (58) but no conclusive evidence for the control of PFK synthesis by oxygen was obtained. Increased levels of ATP on oxygenation have been observed in *Saccharomyces carlsbergensis* (188) and *S. cerevisiae* (318), but other reports (258, 340) indicate the same level of ATP in *S. cerevisiae* with or without oxygenation. In *Klebsiella aerogenes* (167) the ATP content decreased during anaerobiosis but subsequently increased again. It was suggested that the ATP content is controlled so that any change in its value causes reactions that tend to bring the value back to steady-state conditions, but with sufficient time delays in the system to allow transient overshoots to occur. This also might explain the conflicting results found with yeasts and would be consistent with the proposed hypothesis of immediate control of PFK activity replaced by a control of PFK synthesis.

If this hypothesis is correct, assays of total PFK activity would show higher activity with extracts of cells grown anaerobically than with extracts from cells grown aerobically. Batch culture investigations with *Pasteurella pestis* (116) revealed that although TCA cycle enzymes and cytochromes were greater under aerobic conditions, PFK and glyceraldehyde-3-phosphate dehydrogenase showed greater activities under anaerobic culture conditions. The total PFK activity of anaerobically grown cell extracts was twice that from aerobically grown cells. This is evidence for control of the synthesis of PFK. However, the experiment did not demonstrate conclusively that the effect is caused by oxygen, as the various parameters, e.g., glucose concentration, growth rate, and oxygenation, were not kept constant by the use of a 18–24 hour shaken culture (114). Attention has been drawn to the effect that improper cultivation of yeast can lead to disturbances of the normal Pasteur reaction (190).

In order to demonstrate the presence or absence of oxygen-induced PFK synthesis, it is necessary to (a) define the degree of oxygenation used and (b) prevent the action of other factors that might conceivably alter the rate of synthesis of phosphofructokinase.

Using continuous cultivation (see Chapter 2), a steady state with a constant population size can be maintained in the presence of a constant input concentration of oxygen. This technique has been used by a number of research workers studying the effects of oxygen on microbial metabolic activities (166, 282, 283, 329, 334). A factor potentially capable of causing changes in the rate of PFK synthesis is the difference in specific growth rates between aerobic and anaerobic cultures of facultative anaerobic microorganisms. Aerobic cultures achieve a higher specific growth rate and it has been shown that the oxygen uptake rate per gram dry weight of cells (Q_{O_2}) is directly proportional to the specific growth rate in *Escherichia coli* (352) and *Torula utilis* (390). The specific growth rate also affects the cell wall composition; slow growth leads to smaller cells with a higher cell wall/biomass ratio (113, 391). The chemostate technique can be adopted specifically to prevent effects of changes in specific growth rate (41, 205). Another factor potentially capable of causing changes in the rate of synthesis of PFK is the concentration of glucose used in the growth medium (57, 64, 135). Glucose can prevent the induction by oxygen of 2-oxoglutarate dehydrogenase in *E. coli*, so that glucose is oxidized to CO_2 and acetate (10). Glucose has therefore to be kept at a low concentration.

In summary, it has been shown (327, 393) that the specific activity of PFK in aerobically grown cells is less than that of anaerobically grown cells. The extent of reduction in PFK is comparable to the extent of the reduction in the specific glucose uptake rate of the cells. Because enzyme assays in cell-free extracts from chemostat cultures (172) indicate the total enzyme activity present, this is compatible with a mechanism for the Pasteur effect involving a reduction in the rate of synthesis of PFK in aerobically grown cells rather than a continuing ATP inhibition of the enzyme. When the PFK activity of anaerobically grown cells is compared with that of aerobically grown cells, a ratio of 70:30 exists in *Escherichia coli* K-12 (327). Kinetic investigations of PFK supported these findings (100, 393), as the PFK from aerobically grown cells was ATP insensitive, whereas the same enzyme under anaerobic conditions was ATP sensitive. The existence of these two different types of PFK in *Escherichia coli* were confirmed by kinetic investigations (100b). It was observed that the ATP-insensitive form consists of a dimer (mol. wt. 150,000), whereas the ATP-sensitive form consists of a tetramer (mol. wt. 350,000). Electrophoretic mobility together with molecular weight determinations at 13 different steady states of oxygen tension in a continuous culture (100c) showed that the dimer only was present under aerobic conditions as long as α-ketoglutarate dehydrogenase was active. At the time the T.C.A. cycle was disrupted owing to low oxygen tension, the tetramer appeared with the dimer still present. The dimer disappeared at very low oxygen concentration when a sharp drop on the milli-volt scale occurred, leaving only the tetramer at anaerobic conditions.

The existence of two forms of PFK has so far only been reported from mammalian (267), fluke (268) and *E. coli* (100, 393, 129a) phosphofructokinase. In yeasts there are indications of the existence for an enzyme-catalyzed modification of PFK (405) that causes a desensitization of the enzyme to ATP (1a). This enzyme has been isolated and purified (17) and is tentatively called "phosphofructokinase-desensitizing enzyme." In order to convert the ATP-sensitive PFK to its ATP-insensitive forms with this purified enzyme, a heat-stable factor is required (18, 136), which was replaceable by a mixture of AMP and fructose 6-phosphate. Such a modification would be a further possibility for the regulation of enzymes by the adenylic acid system (187) particularly since the requirement for an enzyme is disputed (1c). If this is the case, the "PFK-desensitizing enzyme" must undergo oxygen induction or activation. Both forms exhibited however, identical molecular weights (1a).

All these investigations indicate that the Pasteur effect is not a single-factor mechanism. PFK is able to react immediately on any change in oxygen levels by the reduction of its activity owing to ATP inhibition. However, prolonged exposure to a certain oxygen level results in a reducing effect on its synthesis together with a change in its protein structure. It appears unlikely, at the present state of our knowledge, that PFK would react as a hysteretic enzyme (131), although further research is required. Nucleic acids could also play significant roles as modulators (185).

It is now necessary to determine how the new approach to the Pasteur effect fits into the overall concept of aerobic–anaerobic transition. Is the Pasteur effect a single function of PFK, which could mean a rapid switch-over, or is the PFK concept only one of a number of factors, which could mean a gradual move from aerobic respiration to fermentation?

This problem has been under investigation for only a few years, as it can only be studied with the help of the continuous cultivation apparatus. Investigations into the relationship between oxygen concentration and respiration in *Candida utilis* at various steady-state levels of oxygen as obtained with continuous cultivation exhibited simulated Michaelis-Menten behavior (200). If no time for adaptation was allowed, the respiration rate of the yeast was then directly proportional to oxygen concentration at low oxygen concentrations and independent at high ones. The transition from one type to the other occurred very rapidly. These findings certainly agree with the gradual disappearance of the mitochondrial structure during deaeration (257) and their reappearance on aeration. Once the mitochondrial structure has been fully restored, the respiration rate would be independent of the oxygen concentration.

As was discussed above, the transition from aerobiosis to anaerobiosis is, in effect, a transition from aerobic respiration to fermentation. As the final electron acceptors change, so does the redox potential. It can therefore be envisaged that under low oxygen concentrations the redox potential may play a significant role (429). The parameter E_h could be of particular use when the dissolved oxygen level in the culture becomes too low to be measured using an oxygen electrode. When E_h was used (430), four separate physiological phases could be distinguished during the transition from anaerobiosis to aerobiosis in both *Escherichia coli* and *Klebsiella aerogenes*:

Phase 1, being when the anaerobiosis is at an E_h value of less than 0 mV
Phase 2, being around +100 mV,
Phase 3, being between +200 and 300 mV
Phase 4, being for values above 300 mV

These different physiological phases, however, do not differ very much from the oxygen states observed with the dissolved oxygen electrode (166, 327, 393). The aim should be to develop the parameter E_h further below 0 mV in order to observe changes in that region, where the oxygen electrode becomes useless.

Recent investigations into the behavior of a number of fermentative, as well as respiratory, enzymes during the transition period showed very convincingly that the transition starts well before the dissolved oxygen becomes zero (393). It could also be demonstrated that *Escherichia coli* utilizes glucose under aerobic conditions primarily via the HMP pathway owing to the low PFK and extremely low FDP aldolase activity levels. As PFK is ATP insensitive, FDP must accumulate (394). It can therefore function as an inhibitor of 6-phosphogluconate dehydrogenase (414b) and, in this respect, would be an important control step in glucose utilization via the HMP pathway. Under these conditions, PFK does not appear to be a limiting factor and it is possible that FDP aldolase and 6-phosphogluconate dehydrogenase are the major enzymes involved. The first enzymes involved in the transition before oxygen reaches zero are 2-oxoglutarate dehydrogenase and isocitrate dehydrogenase, which decrease rapidly. In contrast, FDP aldolase increased, indicating an opening up of the EMP pathway. As soon as the dissolved oxygen level becomes zero, 2-oxoglutarate dehydrogenase becomes zero and PFK synthesis, as well as acid production, commences.

During further decrease in oxygen partial pressure (pO_2), the respiratory enzymes acquire maximal activity (151, 282, 283, 393, 428), followed by a gradual collapse of the respiratory chain. The gradual increase in PFK and FDP aldolase synthesis results in an increased glucose metabolism via the EMP pathway. Because the synthesis of PFK is more rapid than that of FDP aldolase, FDP still accumulates and continues to inhibit 6-phosphogluconate dehydrogenase, which is reflected in the almost constant levels of this enzyme. This greater participation of the EMP pathway and the collapse of the oxidative TCA cycle causes a reduction in CO_2 formation as well as in cell mass. The increase of reduced NAD during glucose utilization via the EMP pathway is facilitated under anaerobic conditions. Reduced NAD is an allosteric effector of glucose-6-phosphate dehydrogenase from *Escherichia coli* (342). This inhibition is absolutely necessary, as an increase in glucose concentration increases FDP aldolase many times, with the PFK activity staying constant. This would release the inhibition of FDP on 6-phosphogluconate dehydrogenase but increases the reduced NAD inhibition on glucose-6-phosphate dehydrogenase. This regulatory mechanism could explain why even under strict anaerobic conditions, *Escherichia coli* utilizes 20–30% of its glucose through the HMP pathway (129, 214). It could well be envisaged that some sort of correlation exists under anaerobic conditions. The ATP-controlled PFK and reduced NAD-controlled glucose-6-phosphate dehydrogenase therefore may or may not be the cause of the oscillations observed with pyridine nucleotides (55, 83, 168), although these have been explained in terms of increased respiration rates at low oxygen tension (83). As far as the enzyme phosphofructokinase is concerned, the above mentioned observations revealed that the mechanism of the Pasteur effect cannot be related to the allostery of this particular enzyme (100c). The reduction of ATP production owing to the cessation of the TCA cycle possibly induces the synthesis of the tetrameric form, which regulates the energy metabolism of the anaerobic cell. This shift of energy regulation from the enzymes of the TCA cycle (e.g., isocitrate dehydrogenase)during aerobiosis to phosphofructokinase is necessary because of the cessation of aerobic respiration and the TCA cycle under anaerobic conditions. The mechanism of induction of the different forms, however, is not known.

A transition from anaerobiosis to aerobiosis is much more difficult to obtain, but the general trend is very similar to that described for the reverse cycle (54, 100c). However, it has been claimed (169, 170) that during the transition a loss of tight coupling between growth and energy-conserving processes occurs.

A very interesting, although completely different, effect of oxygen on glucose metabolism was observed with the strictly aerobic *Azotobacter beijerinckii* (331, 350). This organism uses the ED pathway as its main catabolic route. It also has a functional HMP pathway. On oxygen limitation, the synthesis of the polymer poly-β-hydroxybutyric acid is initiated, which involves coenzyme A esters and the reduction of acetoacetyl-CoA to D(−)-β-hydroxybutyryl-CoA. The latter reaction utilizes both NADH + H^+ and NADPH + H^+. As in the Pseudomonadaceae, it was found that glucose-6-phosphate dehydrogenase is inhibited by ATP, ADP, reduced NAD, and reduced NADP, whereas the NADP-linked 6-phosphogluconate dehydrogenase is strongly inhibited only by reduced NAD and reduced NADP. The ED pathway from 6-phosphogluconate, however, is inhibited by ATP, citrate, and isocitrate. It has been suggested (350) that with oxygen limitation, the TCA cycle will cease first and the accumulation of citrate and reduced NAD and reduced NADP would decrease the glucose oxidation. The formation of poly-β-hydroxybutyric acid would therefore be a regulatory phenomenon to store the surplus of carbon and reduced NAD or reduced NADP being formed. The detailed mechanism, however, has yet to be found.

The glucose or "Crabtree effect" is less well studied, although it is known to be very different from the Pasteur effect. It is known, however, that glucose effects the enzyme 2-oxoglutarate dehydrogenase, as mentioned above, and also the cytochrome system (78, 178). There is evidence accumulating (100a, 189, 393) that phosphofructokinase has nothing to do with this effect, but cyclic AMP formation is inhibited (300). It also was mentioned above that there are indications that FDP aldolase could play an important role. Increasing glucose concentrations stimulate FDP aldolase under anaerobic conditions (100a) and open the EMP pathway. As PFK activity stays constant, FDP aldolase must be regarded as a rate-limiting enzyme. The increase in glucose and FDP aldolase stimulates fermentation and growth. Continuous culture experiments with turbidostat control should make it possible to investigate the glucose effect more closely in the near future.

It has been the aim, here, to give an account and an appreciation of the control and regulation of carbohydrate metabolism of facultative anaerobic bacteria. In most instances, these controls are exerted on enzyme activity or even enzyme synthesis. The main controls appear to be negative feedback controls, if reduced NAD is regarded as an end product. Energy-linked control appears to be predominant when oxygen limitation is involved. However, the control by reduced NAD, the indicator of fermentative pathways, develops as one of the major regulations. Another important regulatory function is the control by duplicate enzymes, as has been mentioned particularly in the case of the EMP pathway and gluconeogenesis. Phosphofructokinase and hexose diphosphatase (EC 3.1.3.11), as well as pyruvate kinase, PEP carboxylase, and PEP synthase, are good examples. It is of great importance to realize that in amphibolic pathways both feedback by energy indicators (for catabolic routes) and negative feedback by end products (for biosynthetic pathways) are present (293). Most of these enzymes are allosteric (278). However, activation also exists, and precursor activation appears to be unique to amphibolic pathways (342). It was shown that FDP appears to play a key role in this respect, as this compound activates pyruvate kinase and PEP carboxylase not only in bacteria, but also in yeast (181) and mammals (20, 388).

The control mechanisms depend, in other words, entirely on the nature of the enzyme. Glucose-6-phosphate dehydrogenase, which is susceptible to ATP and not to reduced NAD in aerobic pseudomonads, reacts exactly the reverse in facultative anaerobic bacteria. Sanwal (342) catagorizes the regulatory enzymes into (a) enzymes with a distinct and relatively specific site that binds the regulatory ligand and (b) enzymes that exhibit sigmoidal (or cooperative) velocity responses to increasing concentrations of substrate. The latter, or allosteric, enzymes are subdivided into those which show modulator-independent cooperativity (sigmoidal rate–concentration plots in the absence of activators and inhibitors), e.g., phosphofructokinase, phosphoenolpyruvate carboxylase, and pyruvate kinase, and those which exhibit modulator-dependent cooperativity (sigmoidal initial velocity plots only in presence of activators and inhibitors), e.g., citrate synthase, glucose-6-phosphate dehydrogenase, malic enzymes. For further reading on allosteric enzymes the reader is referred to the following excellent articles (15, 219, 279, 371, 385).

A new proposal, that amphibolic pathways can be regulated to a certain degree by the energy charge of the adenylate pool (16) by interactions at the catalytic site, must await further confirmation.

Fermentation of Lactic Acid Bacteria

The lactic acid bacteria include species of the Lactobacillaceae, which are well-known spoilage organisms in food. They are morphologically a heterogeneous group and are biochemically characterized by their main end product, lactic acid. These organisms are gram-positive, nonsporulating cocci or rods. Some are catalase positive and some catalase negative. All species of this family are obligate fermenters and are roughly subdivided into five different groups (374), as shown in Table 9.5.

TABLE 9.5
Five Groups of Lactic Acid Species[a]

Cell form	Nature of fermentation	Name of genus
Cocci in chains	Homofermentative	*Streptococcus*
Cocci in chains	Heterofermentative	*Leuconostoc*
Cocci in tetrads	Homofermentative	*Pediococcus*
Rods, usually in chains	Homofermentative	*Lactobacillus*
Rods, usually in chains	Heterofermentative	*Lactobacillus*

[a] From Stanier *et al.* (374). Reprinted with permission of Prentice-Hall.

TABLE 9.6
Configuration of Lactic Acid Produced by Various Microorganisms[a] (439)

Organisms	Configuration
Homofermenter	
Streptococcus sp.	L(+)[b]
Lactobacillus caucascus and *L. lactis*	D(−)[c]
L. leichmanii	D(−)
L. helveticus and *L. bifidus*	DL[d]
L. plantarum	DL
L. thermophilus and *L. delbruckii*	L(+)
L. bulgaricus	DL or D(−)
L. casei	L(+), L(+), and D(−)
Pediococcus sp.	DL
Bacillus sp.	L(+)
Clostridium sp.	DL
Butyribacterium rettgeri	DL
Heterofermenter	
Lactobacillus brevis, *L. buchneri*, *L. pasteurianus*, and *L. fermenti*	DL
Leuconostoc sp.	D(−)
Microbacterium sp.	L(+)
Rhizopus sp.	L(+)
Mixed fermenter	
Serratia sp.	D(−)

[a] From Wood (439).
[b] L(+) −2 μ salt × 2 H_2O D = −8.25° (4%; 25°C); H_2O = 12.89%.
[c] D(−) −2 μ salt × 2 H_2O D = +8.25° (4%; 25°C); H_2O = 12.89%.
[d] DL −2 μ salt × 3 H_2O D = ±0; H_2O = 18.18%.

There is now general acceptance of two broad divisions of the lactobacilli into homofermentative and heterofermentative strains (39), based obviously on the percentage of byproducts other than lactic acid formed during fermentation. If all strains produced either less than 10% or more than 40% byproducts, this division would be easy, but in practice this is not so. Therefore, an indefinite demarcation line exists between homo- and heterofermentative types. Davis and Hayward (80, 82, 173) claim that the main characteristic of taxonomic importance, the proportion of end product formed in fermentation, is reasonably constant, and that borderline cases or apparent changes from the homofermentative character result from inadequate initial purification. There is, however, insufficient evidence presented to support this claim that the byproducts of homofermentation are constant. The heterofermentation end products increase as sugar concentration increases (62). These and other observations (84, 91, 290, 349) support far more the thesis that adverse conditions favor the production of byproducts.

A characteristic feature of the lactic acid produced by the Lactobacillaceae is that the optical activity of the acid depends on the strain of organism (230). Two factors appear to be responsible for this final activity:

1. The stereospecificity of the lactate dehydrogenase involved. Organisms producing DL-lactic acid may contain the enzyme for the D isomer as well as that for the L isomer. These two different enzymes may or may not be NAD linked. The organism also could contain either or both of the isomeric dehydrogenases and form D- or L-lactic acid or DL-lactic acid (racemic).
2. The presence or absence of a lactate racemase. In some cases, strains of the Lactobacillaceae produce racemic mixtures, irrespective of the specificity of the lactate dehydrogenase present. This can only be attributed to a racemase.

In *Lactobacillus plantarum*, for example, both L- and D-stereospecific NAD-linked dehydrogenases operate with equal activity in glucose-metabolizing cells (90, 277). However, Kitahara (220) suggested that the production of DL-lactic acid may result from the joint action of D-lactate dehydrogenase (D-lactate:NAD oxidoreductase, EC 1.1.1.28) and a lactate racemase (EC 5.1.2.1). Table 9.6 gives some idea of the distribution of lactic acid configurations produced by various microorganisms.

Glucose Metabolism and Regulation

All strains of the Lactobacillaceae that produce as much as 1.8 moles of lactic acid per mole of glucose, with minor amounts of acetic acid, ethanol, and carbon dioxide, are grouped as the homofermenters. The genera are *Diplococcus, Streptococcus, Pediococcus, Lactobacillus,* and *Microbacterium,* which range from facultative anaerobic to aerobic microorganisms.

The best evidence for the operation of the EMP pathway comes from experiments using labeled glucose (142). Although it is thought that *Streptococcus faecalis* ferments glucose by a combination of EMP, HMP, and ED pathways, transaldolase and phospho-2-keto-3-deoxy–gluconate aldolase are considered to be inducible enzymes (363). In *Streptococcus faecalis,* the nature of the end products of fermentation very much depends on cultural conditions. At pH 5.0–6.0, at which fermentation occurs, pyruvate is reduced by a stereospecific lactate dehydrogenase to D-, L-, or DL-lactic acid. At neutral or slightly alkaline pH, however, pyruvate metabolism leads to the production of formic and acetic acids and ethanol in the ratio 2:1:1, a typical heterolactic fermentation (155).

As well as glucose, homofermentative microorganisms also ferment fructose, mannose, galactose, and such disaccharides as lactose, maltose, and sucrose as substrates. These sugars are presumably converted into intermediates of the EMP pathway by inducible enzymes. In *Streptococcus pyogenes*, the fermentation of galactose accounts for 50% of the galactose carbon, whereas the reminder is found in acetate, formate, and ethanol (2:1:1). This metabolism is very similar to the gluconate metabolism (362, 363) of *Streptococcus faecalis* at high pH. It is possible that another independent pathway is involved (86).

Organisms that ferment hexoses with the production of less than 1.8 moles of lactic acid per mole of glucose and, in addition, ethanol, acetate, glycerol, mannitol, and carbon dioxide are called "heterofermenters." The main group of these organisms includes strains of *Leuconostoc*, some strains of *Lactobacillus*, the anaerobic peptostreptococci, and the anaerobic species of *Eubacterium*, *Catenabacterium*, *Ramibacterium*, and *Bifidobacterium*. There exist several types of heterofermentative mechanisms (86):

1. *Leuconostoc*-type heterofermentation, which results in the formation of 0.8 mole of lactic acid, 0.1–0.2 mole of acetate, 0.9 mole of CO_2, 0.8 mole of ethanol, and 0.2–0.4 mole of glycerol. Glycerol is not formed in all cases.
2. *Peptostreptococcus*-type heterofermentation, which have lactic and propionic acids as the end products of glucose fermentation.
3. A third type of heterofermentation, characteristic of all those bacteria which produce lactic acid and butyric acid as the end product of glucose metabolism.

Leuconostoc species and *Lactobacillus brevis* possess the so-called "phosphoketolase" pathway (see Chapter 6). All three other pathways, EMP, HMP, and ED, were found to be absent. In the overall reaction, glucose is fermented to carbon dioxide, ethanol, and lactate, with the production of 1 mole of ATP per mole of glucose (325). This is only half the energy produced by the homofermenters.

Investigations into the presence of enzymes in heterofermentative microorganisms have revealed the presence of hexokinase (ATP:D-hexose 6-phosphotransferase EC 2.7.1.1), phosphoglycerate kinase (ATP:3-phospho-D-glycerate 1-phosphotransferase, EC 2.7.2.3), glycerolphosphate dehydrogenase [L-glycerol-3-phosphate:(acceptor) oxidoreductase, EC 1.1.99.5], D-lactate dehydrogenase (D-lactate:NAD oxidoreductase, EC 1.1.1.28), alcohol dehydrogenase (alcohol:NAD oxidoreductase, EC 1.1.1.1) (88), aldehyde dehydrogenase (EC 1.2.1.3), and acetoin dehydrogenase (acetoin: NAD oxidoreductase, EC 1.1.1.5.). Fructosediphosphate aldolase (EC 4.1.2.13) and triosephosphate isomerase (D-glyceraldehyde-3-phosphate ketol-isomerase, EC 5.3.1.1) were absent. The lack of the latter two enzymes makes it impossible for the EMP pathway to function. Enzymes of the HMP pathway present were glucose-6-phosphate dehydrogenase (EC 1.1.1.49) and phosphogluconate dehydrogenase (EC 1.1.1.43) (193, 439). The glucose-6-phosphate dehydrogenase reacted equally well with NAD^+ and $NADP^+$ (291). Its electrophoretic mobility has been found useful in the taxonomy of heterofermentative lactobacilli (424).

It is, of course, possible that two of the above-mentioned patterns may be superimposed to varying degrees. *Lactobacillus brevis*, for example, metabolizes glucose if grown at 24°C heterofermentatively but requires an external hydrogen acceptor if grown at 37°C:

$$\text{Glucose} + H_2O \rightarrow \text{lactic acid} + \text{acetate} + CO_2 + 4\ H^+$$

It has also been observed that fermentation of fructose occurs at 37°C with the formation of mannitol:

$$3\ \text{Fructose} \rightarrow 1\ \text{lactate} + 1\ \text{acetate} + 2\ \text{mannitol} + 1\ CO_2$$

$$1\ \text{Glucose} + 2\ \text{fructose} \rightarrow 1\ \text{lactate} + 1\ \text{acetate} + 2\ \text{mannitol} + 1\ CO_2$$

Under these conditions fructose acts as the hydrogen acceptor. Unlike the clostridia, the heterofermentative lactobacilli are unable to eliminate hydrogen or hydrogen gas. The more common internal electron acceptors of heterofermentation are pyruvate, dihydroxyacetone phosphate, or acetaldehyde. Acetyl-CoA can also serve as electron acceptor, but only in a limited number of genera. Stoichiometry is adhered to in some organisms, such as *Lactobacillus brevis*, but less so in others, and it is never as significant as in the homofermenters.

Streptococcus faecalis is a homofermentative lactic acid bacterium that under slightly acid conditions utilizes glucose stoichiometrically to lactic acid (143, 210). The conversion of 1 mole glucose to 1.8 moles of lactic acid suggests that traces of other end products can occur. Therefore, if homofermenters are only able to utilize the EMP pathway, because of the lack of glucose-6-phosphate dehydrogenase, pyruvate must not only be metabolized to lactic acid, but also to acetic acid and ethanol as well as other end products. It is therefore not surprising to find reports that a phosphoroclastic reaction similar to the *E. coli* type (245), exists in this organism. As there was no evidence that only one component was involved in such a reaction, a protein component or protein complex is thought to be involved. Thiamine pyrophosphate was the only cofactor requirement. Because of the low oxidation–reduction potential requirement, the pyruvate–formate exchange reaction must involve α-lactylthiamine pyrophosphate, which can be then rearranged or an oxidation–reduction may yield acetyl-thiamine enzyme plus formate. Acetyl-CoA or acetyl phosphate could be formed in the presence of acetyltransferase which would transfer the C_2 unit from the pyruvate–formate exchange system to the acceptor. A phosphate acetyltransferase and acetate kinase (333) then convert acetyl-CoA via acetyl phosphate to acetate.

A very similar system works under aerobic conditions (212), although the organism switches from the phosphoroclastic reaction to a lipoic acid-dependent pyruvate oxidation system (289, 326). Although this mechanism has not been elucidated as yet, propionate inhibits and lipoic acid stimulates the metabolism of pyruvate. It is thought that propionate may interfere by competing for the available CoA in forming propionyl-CoA (211).

Another possible diversion from pyruvate is caused by the functioning of pyruvate decarboxylase, which converts pyruvate to oxalacetate with a subsequent transamination to aspartate. This CO_2 fixation was observed in *Streptococcus faecalis* var. *liquefaciens* (171).

The regulatory control mechanism for the different end product formation appears to be centered around FDP (93, 432), which is necessary for the activation of the stereospecific lactate dehydrogenases. Low FDP pools would prevent the lactic acid production in favor of acetate and/or aspartate formation. Fluctuation of cellular pool levels of FDP would thus influence qualitatively and quantitatively the utilization of glucose. This influence of FDP on lactate dehydrogenase appears to be a feature common to all streptococci (433). FDP also regulates negatively the induction of the lactate oxidase system (251) as well as the malate dehydrogenase (decarboxylating) system (252). Therefore, FDP control at the pyruvate level is of importance in connection with the reports that the group D streptococci possess glucose-6-phosphate dehydrogenase activity (423). The presence of this enzyme would suggest that the EMP pathway is not the sole glucose dissimilation pathway in streptococci (153).

The possession of an aerobic metabolism of glucose has been suggested (229, 273, 373) for *Streptococcus faecium*, *Streptococcus agalactiae*, and *Lactobacillus brevis*. Under aerobic conditions *S. agalactiae* produces 52–61% lactic acid, 20–23% acetic acid, 5–6% acetoin, and 14–16% carbon dioxide. It was also suggested that oxygen can be the terminal electron acceptor, as for each mole NADH + H^+ oxidized, the organism consumed 0.5 mole of oxygen. *Streptococcus agalactiae* also possesses a strong NADH dehydrogenase under these conditions. In *S. faecium* an NADH diaphorase was closely associated with the oxygen uptake that resulted in hydrogen peroxide formation. It is quite possible that the streptococci are able to utilize lactate aerobically, because the previously mentioned lactate oxidizing system is induced as soon as aerobic conditions prevail (93). The carbon balance as to the amount of lactic acid produced from glucose could therefore be very much arbitrary (251).

An organism very closely related to the streptococci is *Lactobacillus casei*. It is the only known *Lactobacillus* species that requires FDP for its lactate dehydrogenases (93). This organism also revealed two pathways of pyruvate conversion, a reduction to lactate and a possible functioning of the phosphoroclastic cleavage. Using chemostat cultures, it was shown that high growth rates of the organism predominantly led to lactic acid production, whereas low growth rates increased the volatile acid formation (93), as in *Streptococcus faecalis* (177, 335). Additional reports on anomalous growth yields under anaerobic compared with aerobic cultivation of *L. casei* strain 103 (44) and the possession of acetoin reductase (38) (acetoin: NADH oxidoreductase, EC 1.1.1.5) certainly hinted at an influence of oxygen on the glucose metabolism of lactic acid bacteria. There is, however, no certainty as yet as to what influence oxygen may have on the other lactic acid bacteria. This is a problem not only with the homofermentative streptococci and *L. casei*, but also with almost every homofermentative lactobacillus. Theories on the effect of oxygen on lactic acid bacteria therefore vary from there being an indifference to oxygen (81, 159, 218) to increasing evidence suggesting that a number of lactic acid bacteria can physiologically "use" oxygen (45, 117, 213, 380, 381, 410). Strains of *L. casei* use predominantly the EMP pathway (142) but also possess typical HMP pathway enzymes (50). This may mean that under different environmental conditions changes could occur similar to the adaptive alterations of the metabolic constitution of cells grown on pentoses (284).

Recent investigations into the effect of oxygen on glucose metabolism of *Lactobacillus casei* var. *rhamnosus* ATCC 7469 (264) revealed very slow maximal specific growth rates (0.19 hr^{-1}), exponential glucose utilization rates (0.1 log unit/hour/ml culture), and exponential lactate production rates (0.18 log unit/hour/ml culture) compared to anaerobic cultures. It was found that aeration affects both the rate at which glucose is converted to lactate and the stoichiometry of this conversion. Because there was no difference in molar growth yields between anaerobic and aerobic cultures, it has been suggested that the site of oxygen interference is after that reaction step leading to the second mole of ATP during glycolysis, i.e., at pyruvate or, more specifically, at the lactate dehydrogenase level. Although aeration increased glucose-6-phosphate and 6-phosphogluconate dehydrogenases and reduced FDP aldolase activity (264), a Pasteur effect could be excluded, for phosphofructokinase from both *Lactobacillus casei* var. *rhamnosus* ATCC 7469 and *L. plantarum* ATCC 14917 did not exhibit sigmoidal kinetics (99). Even FDP aldolase could be excluded as a regulatory enzyme (98) because of its extreme stability compared with its counterpart from *Escherichia coli*. Phosphofructokinase, therefore, does not function as an allosteric protein in lactobacilli. It therefore seems possible that the lactobacilli are able to utilize fractions of intracellular glucose 6-phosphate oxidatively, with the production of carbon dioxide. Such a change in metabolism need not be accompanied by an increase in cell yield (45, 410). Such pathway alterations are not unknown in certain lactic acid bacteria growing under aerated conditions (201, 202, 251, 273, 335, 358, 381). Enzyme inhibition as well as repression are possible causes for such alterations. In regard to the end products formed, diacetyl and acetoin production was observed (264) not only in *Lactobacillus casei*, but also in *Leuconostoc citrovorum* and *Leuconostoc dextranicum* (40, 217).

A repression of lactate dehydrogenases by oxygen with Michaelis constants favoring substrate reduction has been suggested (428), and in *Staphylococcus* the aerobic level of lactate dehydrogenase is approximately one-third of that in anaerobically grown cells.

All these investigations indicate very strongly that oxygen influences the glucose metabolism at the lactate dehydrogenase level by an unknown mechanism that causes a homofermentative strain to appear as a heterofermentative strain. This may have repercussions on the taxonomic classification of these microorganisms.

Another very important taxonomic characteristic is the optical activity of the lactic acid produced (230). Bacterial lactate dehydrogenases are of two types, one NAD linked and the other flavine linked. Each of these two types exists in a stereospecific (D and L specific) form. The NAD-linked form has been detected in various bacteria (90, 139–141, 216, 431) and their stereospecificity is shown to correlate with the lactic acid isomers produced (90). Flavine-linked dehydrogenases, however, have been studied only in *Lactobacillus casei* (276), *Lactobacillus plantarum* (360, 361), and *Leuconostoc* (139) and are found to vary in their hydrogen acceptor preference, using either phenazine methosulfate (140) or dichlorophenol indophenol (361). As these flavin-linked lactate dehydrogenases work only unidirectionally from lactate to pyruvate, the following mechanism would exist at the pyruvate level:

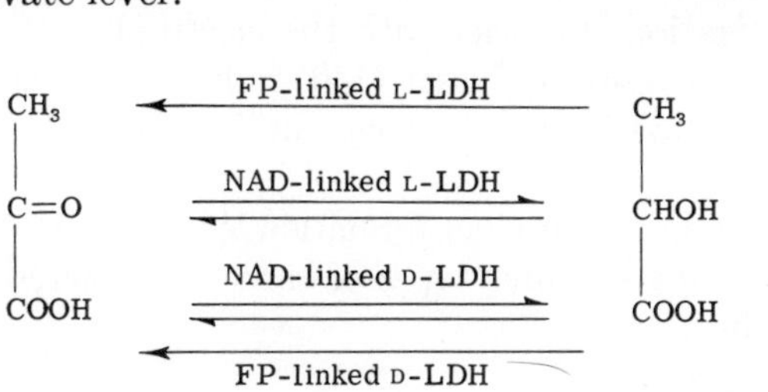
FP-linked L-LDH
CH3
CH3
NAD-linked L-LDH
C=O
CHOH
NAD-linked D-LDH
COOH
COOH
FP-linked D-LDH

pyruvic acid lactic acid

The NAD-linked lactate dehydrogenases of the homofermenters have a pH optimum between 7.8 and 8.8, compared with 9.0–10.0 of the heterofermenter. The pH optima of the FP-linked lactate dehydrogenases lie between 5.8 and 6.6 (96).

The importance of both pairs of lactate dehydrogenases to isomer production under anaerobic condition was revealed by a comparison of the specific NAD-dependent and NAD-independent lactate dehydrogenase activities. This established a direct correlation of the D/L ratios of these activities with the type of lactic acid produced during the growth of the organism. This correlation appears to exist not only with the homofermenters but also with the heterofermenters. *Leuconostoc oenos* (138) did not exhibit any flavin-linked lactate dehydrogenase activity and was therefore different from *Leuconostoc mesenteroides*. The existence of these two pairs of lactate dehydrogenases could therefore provide further evidence that the glucose metabolism of lactobacilli is regulated at the pyruvate–lactate level.

In *Leuconostoc mesenteroides,* the addition of malic acid to a glucose-containing culture causes a switch in the lactic acid isomer production from predominantly D-lactic acid with glucose alone to L-lactic acid (302, 303). Glucose remains the main energy source for this organism (236). During these investigations it was demonstrated that the addition of malic acid to the glucose-containing medium was responsible for a threefold increase in lactic acid production from glucose. This increase resulted in a severe reduction of the D-/L-lactic acid ratio. Malic acid increased the specific activities of NAD-linked L-lactate dehydrogenase 6.5-fold, against a 3.2-fold increase in the case of the corresponding D-lactate dehydrogenase (97). Investigations at the enzyme level revealed that malate has a stimulatory effect on the synthesis of both NAD-linked enzymes, which could result in a rearrangement of the protein structure of the D-lactate dehydrogenase. This rearrangement apparently makes the D-enzyme more susceptible to inhibition of its catalytic activity. The L-lactate dehydrogenase, however, is stimulated not only in its synthesis but also in its activity. Polyacrylamide gel electrophoresis exhibited different mobilities in the presence or absence of malic acid in the culture medium.

These investigations together with the electrophoretic characterization of lactic dehydrogenases by Gasser (140) indicate very strongly that there exists a regulatory mechanism of glucose utilization at the pyruvate–lactate level.

The variability in end product formation led Buyze and co-workers (50) to differentiate between three types of lactic acid bacteria on the basis of enzymatic studies:

1. Obligate homofermenters, containing fructosediphosphate aldolase

(EC 4.1.2.13), but no glucose-6-phosphate dehydrogenase (EC 1.1.1.49) and no phosphogluconate dehydrogenase (EC 1.1.1.43).

2. Obligate heterofermenters, containing the dehydrogenases mentioned under (1), but no fructose diphosphate aldolase.

3. Facultative homofermenters, containing all three enzymes and able to use either of the pathways.

This classification could be adopted under anaerobic growth conditions. It has also been suggested that those homofermenters which fall into the first group be called "homolactic" and that those falling into the last group be called "fermentative homofermenters." This latter terminology could create confusion, however, as all lactic acid bacteria are fermentative under anaerobic growth conditions.

A different approach was taken in using infrared spectra (146) for grouping the different species of lactobacilli. With only a few exceptions, the spectra could be grouped into five distinct types, each of which appeared to correspond to a species of lactobacilli or a group of related species.

Fructose Metabolism

Lactobacillus plantarum and *Lactobacillus pentoaceticum* utilize identical pathways for the fermentation of pentoses and glucose but different pathways if fructose is the main carbon source (191). *Lactobacillus pentoaceticum* produces D-mannitol in addition to lactic acid and acetic acid, whereas *Lactobacillus plantarum* produces lactic and acetic acid (see Scheme 9.2). Because the Lactobacillaceae have a cytochrome-independent electron

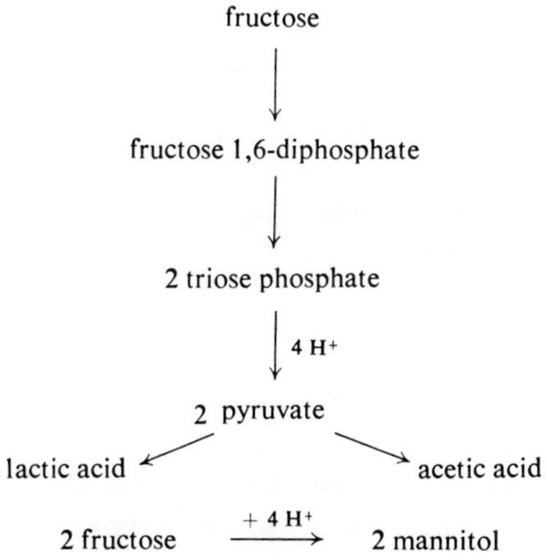
fructose
fructose 1,6-diphosphate
2 triose phosphate
4 H+
2 pyruvate
lactic acid
acetic acid
2 fructose
+ 4 H+
2 mannitol

Scheme 9.2. Metabolism of fructose by lactic acid bacteria.

transport system (see p. 638), the mannitol dehydrogenase is most probably mannitol:NAD oxidoreductase (EC 1.1.1.67). This enzyme must therefore be absent in *Lactobacillus plantarum*. Instead of having 3 moles of fructose as sole carbon source, glucose and fructose could also be present in the ratio of 1:2, in favor of fructose, in order to form mannitol. The reaction is NAD dependent and could therefore be used as a distinct characteristic for the identification and separation of these two species.

Malate Metabolism

The utilization of malic acid to lactic acid is carried out by a large number of lactic acid bacteria (124, 236, 321, 322) and is commercially of extreme importance in the wine industry. This type of fermentation is best known as malolactic fermentation. For further general information the reader is referred to the excellent review by Kunkee (236).

The actual mechanism of malic acid conversion to lactic acid is still unsolved in *Lactobacillus* species. It was thought for a long time that malic acid was decarboxylated to pyruvic acid (215, 231, 232) by an NAD-linked malic enzyme (EC 1.1.1.38), with subsequent reduction of pyruvate to lactic acid by the NAD-dependent lactate dehydrogenases:

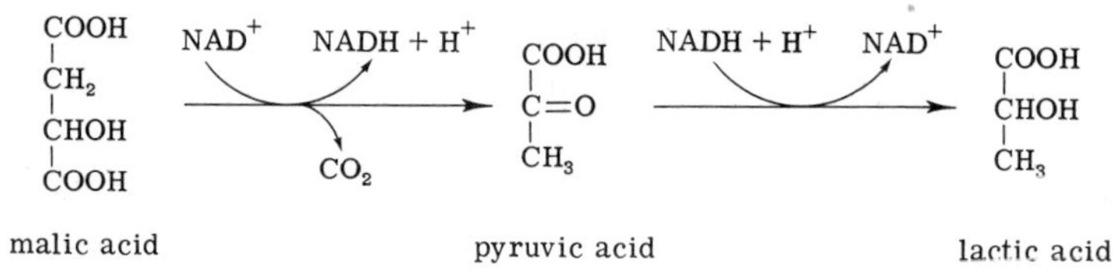
COOH
CH$_2$
CHOH
COOH
NAD$^+$
NADH + H$^+$
CO$_2$
COOH
C=O
CH$_3$
NADH + H$^+$
NAD$^+$
COOH
CHOH
CH$_3$
malic acid
pyruvic acid
lactic acid

A second mechanism was suggested by Charpentie (59). It involves a dehydrogenation step first followed by a decarboxylation; oxalacetate is therefore an additional intermediate between malic and pyruvic acid. Malic acid is first dehydrogenated to oxalacetic acid by a NAD-dependent malate

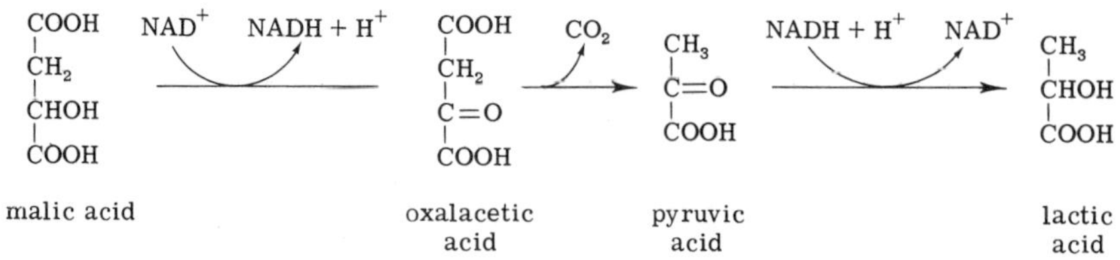
COOH
CH$_2$
CHOH
COOH
NAD$^+$
NADH + H$^+$
COOH
CH$_2$
C=O
COOH
CO$_2$
CH$_3$
C=O
COOH
NADH + H$^+$
NAD$^+$
CH$_3$
CHOH
COOH
malic acid
oxalacetic acid
pyruvic acid
lactic acid

dehydrogenase, which is followed by the action of oxalacetate decarboxylase forming pyruvic acid. This oxalacetate decarboxylase has been isolated from *Lactobacillus plantarum* (120–123, 126).

Apart from the demonstration that both enzyme systems can be used by malic acid-fermenting bacteria (196), a third possibility (303) is that malic acid is decarboxylated directly to lactic acid without the formation of oxalacetic acid or pyruvic acid. The service of lactate dehydrogenase is therefore not required. This suggestion is finding more and more support

$$\underset{\text{malic acid}}{HOOC-CH_2-CHOH-COOH} \xrightarrow[CO_2]{NAD^+} \underset{\text{lactic acid}}{CH_3-CHOH-COOH}$$

because it is known that only L-malic acid is utilized and only L-lactic acid is the end product (348a, 348b). A major step forward was the evidence obtained on purification of the enzyme system (323), which conclusively demonstrated that NAD-dependent lactate dehydrogenase was not involved in the malolactic fermentation of *Lactobacillus plantarum*. This purified enzyme system requires only NAD^+ as cofactor. It was free of any lactate dehydrogenase and converted L-malic acid completely to L-lactic acid with the liberation of carbon dioxide. With pyruvate as substrate, no lactic acid was produced (348a). The heterofermenter *Leuconostoc mesenteroides* 99 (103) did not exhibit any NAD-dependent or NAD-independent L-lactate dehydrogenase and thus converted glucose to D-lactic acid. With the addition of L-malic acid to the culture medium, L-lactate is produced. As malate did not induce any L-lactate dehydrogenase, the L-lactate formed must come totally from L-malic acid. With pyruvate as substrate, D-lactic acid was the only acid produced by the extracts of *Leuconostoc mesenteroides*. The mechanism of this reaction has not been elucidated as yet. It is certain that carbonic acid is not involved in any of the three mechanisms (309). Further work on the mechanisms used by different organisms could be of taxonomic significance. It is almost certain that some organisms have more than one mechanism operative.

The enzymes responsible for the utilization of malate are inducible enzymes that are formed only in the presence of the substrate (125). The induction of an NAD-specific malic enzyme together with a malate permease permits the utilization of malate by *Streptococcus faecalis* (254). The malic enzyme (EC 1.1.1.39) has been purified and shown to be not subject to catabolite repression by intermediates of the EMP pathway (253). The permease, in contrast, is subject to glucose repression (254). Recent investigations, however, indicate that the inducible malate dehydrogenase [L-malate:NAD oxidoreductase (decarboxylating), EC 1.1.1.39] from *Streptococcus faecalis* strain MR and *Lactobacillus casei* strain 64 H could be regulated by two negative effectors of the EMP pathway, namely FDP and 3-phosphoglycerate. At a concentration of 5 m*M* malate, the degree of inhibition followed sigmoidal kinetics (255), indicating an allosteric control.

Tartrate Metabolism

A survey of 78 different strains of the genera *Pediococcus*, *Leuconostoc*, and *Lactobacillus* revealed that only *Lactobacillus plantarum* and *L. brevis* are able to metabolize tartaric acid (324). When resting cells and cell-free extracts were used, 1 mole tartaric acid was metabolized to 1.5 moles CO_2, 0.5 mole acetic acid, and 0.5 mole lactic acid in *L. plantarum* and to 1.33 moles CO_2, 0.67 mole acetic acid, and 0.3 mole succinic acid in *L. brevis*. As the end products indicate, tartrate is utilized by different pathways, depending on the species under investigation. *Lactobacillus plantarum* converts L(+)-tartaric acid to oxalacetic acid first, in a dehydration reaction

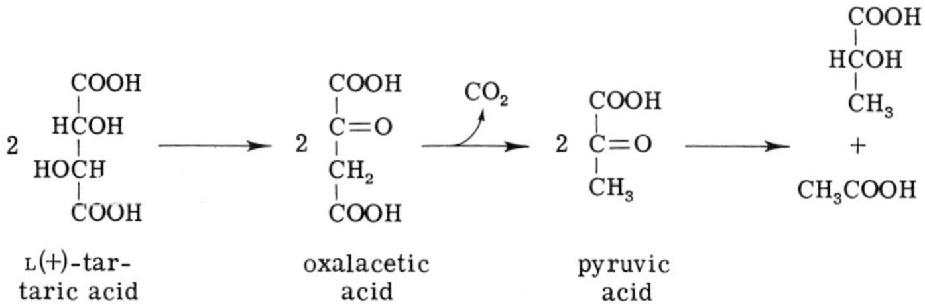
L(+)-tartaric acid
oxalacetic acid
pyruvic acid
CO_2
CH_3COOH

catalyzed by tartrate dehydratase (EC 4.1.1.32). Oxalacetate decarboxylase liberates CO_2 and produces pyruvic acid as the end product of its reaction. The utilization of pyruvate appears to involve two different enzymes. Whereas lactate dehydrogenase is responsible for the formation of lactic acid, a pyruvate dehydrogenase system could be responsible for the formation of carbon dioxide and acetic acid. As no regulatory mechanism has been suggested, the amount of NADH + H^+ available is made responsible for the production of either lactic acid or CO_2 and acetic acid. High amounts of NADH + H^+ tend to increase the formation of lactic acid. As the conversion of pyruvic acid to acetate and carbon dioxide is NAD dependent, it appears that the NAD^+/NADH + H^+ ratio determines the final end product formation.

Lactobacillus brevis has a pathway to pyruvic acid, and finally to acetic acid plus CO_2, similar to the one described above. The lack of lactic acid formation is surprising, however, as this organism is known to possess both stereospecific lactate dehydrogenases. No suggestions have been offered except that one-third of the oxalacetate formed from tartaric acid flows via malate and fumarate to succinate (reversed TCA cycle) and two-thirds of the oxalacetate is decarboxylated to pyruvate and acetate plus carbon dioxide. The individual enzymes have not been elucidated as yet.

The existence of two different pathways in these two species of *Lactobacillus* was supported by the finding that the addition of glucose effects the tartrate metabolism of *L. plantarum* only, and not that of *L. brevis*. In any event, glucose had first to be metabolized before tartrate could be utilized.

Metabolism of Group N Streptococci

This group of bacteria, namely *Streptococcus diacetilactis*, *S. lactis*, and *S. cremoris*, are noted for their ability to ferment the lactose of milk to lactic acid. The lability of β-galactosidase activity led to the suggestion that a phosphoenolpyruvate phosphotransferase-dependent system is responsible for the transport of lactose into the cell (260, 261). It involves factor III and is therefore similar to the staphylococcal system (see p. 241). Once lactose is in the cell, these streptococci utilize it in conjunction with threonine and the formation of glycine and acetaldehyde via the pathway shown in Scheme 9.3.

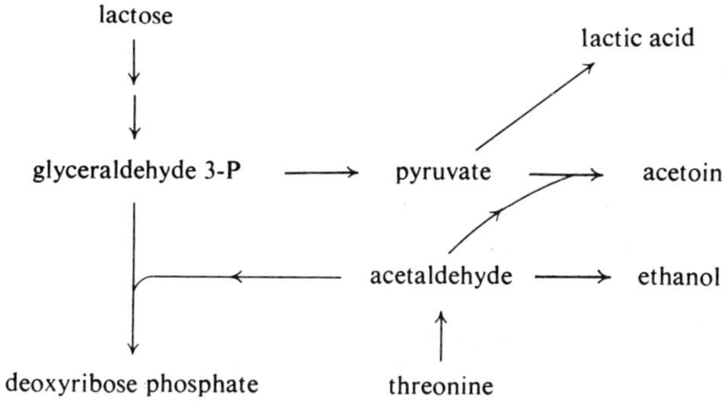
lactose
lactic acid
glyceraldehyde 3-P
pyruvate
acetoin
acetaldehyde
ethanol
deoxyribose phosphate
threonine

Scheme 9.3. Lactose metabolism of Group N streptococci.

The enzyme catalyzing the utilization of threonine is serine hydroxymethyltransferase (EC 2.1.2.1) and pyridoxal phosphate is required as co-

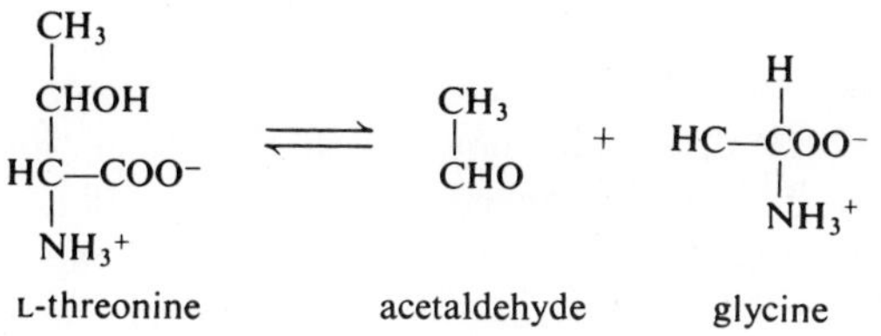
CH3
CHOH
HC—COO−
NH3+
L-threonine
CH3
CHO
acetaldehyde
+
H
HC—COO−
NH3+
glycine

factor. Acetaldehyde itself can be metabolized further via three different paths, depending on the availability of three enzymes as well as on the utilization of lactose via the EMP pathway:

1. Acetaldehyde and glyceraldehyde 3-phosphate are cleaved under the catalytic action of deoxyribose-phosphate aldolase (2-deoxy-D-ribose-5-phosphate acetaldehyde-lyase, EC 4.1.2.4) to form deoxyribose phosphate.

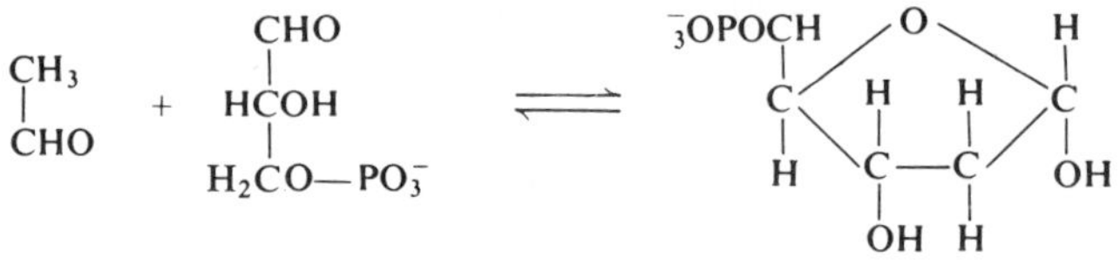

acetaldehyde D-glyceraldehyde 3-phosphate 2-deoxyribose 5-P

2. The group N streptococci also possess an enzyme that cleaves pyruvate with acetaldehyde, with the formation of acctoin. The action of

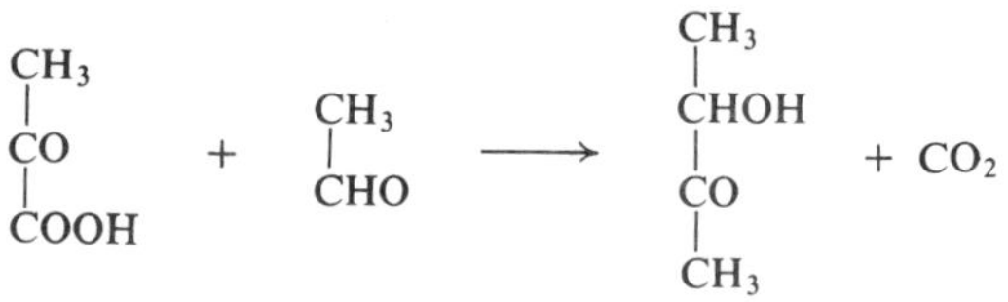

pyruvic acid acetaldehyde acetoin

this enzyme has been regarded as that of a synthetase by Lees and Jago (personal communication), who gave the enzyme the name "acetoin synthetase." *Streptococcus diacetilactis* was found to synthesize acetoin from diacetyl, which was formed by a condensation reaction of acetyl-CoA with hydroxyethylthiamine pyrophosphate (see p. 603) (365a).

3. The third possible acetaldehyde metabolism is directed by an active alcohol dehydrogenase (alcohol:NAD oxidoreductase, EC 1.1.1.1), which converts acetaldehyde to ethanol.

ethanol

There may, of course, be a number of group N streptococci that do not follow this metabolism, as was shown with the glycine-requiring *Streptococcus cremoris* Z. This strain lacked the threonine aldolase.

The main lactose pathway is consistent with the operation of the EMP pathway. Under anaerobic conditions, pyruvate is the major hydrogen acceptor and 96% of the glucose fermented is converted to lactic acid. All group N streptococci possess both NAD-dependent and NAD-independent lactate dehydrogenases (12); the NAD-dependent enzymes require FDP for activation. However, these organisms have other hydrogen acceptors apart from pyruvate, namely acetoin, diacetyl, acetyl-CoA, and acetaldehyde (186). In this respect they differ from the group D streptococci. Acetyl-CoA formation is via a lipoic acid-dependent dehydrogenase system similar to that of *Streptococcus faecalis* (see p. 626). Depending on the availability of acetyl-CoA, acetoin plus diacetyl are formed (68). If cell growth comes to a halt, more acetyl-CoA can be channeled to these neutral end products (63, 365).

Under aerobic conditions no change in pathways occurs (186), but molecular oxygen and hydrogen peroxide are involved. The formation of H_2O_2 by group N streptococci under aerobic conditions is facilitated by the action of NADH dehydrogenase, which catalyzes the oxidation of NADH + H^+ by molecular oxygen. Although some of the H_2O_2 could be removed through the action of an NAD^+ peroxidase (11), H_2O_2 accumulates in sufficient amounts to inhibit growth. The activity of both enzymes appears to vary from strain to strain.

Deoxyribose Metabolism

The metabolism of deoxyriboses does not occur in cells grown on glucose, arabinose, or xylose (191), but it can be induced in *Lactobacillus plantarum* and possibly in other species of the Lactobacillaceae. In the presence of a mixture of glucose and deoxyribose, the metabolic breakdown of the latter begins only after full growth is reached. It is therefore very probable that growth occurs at the expense of glucose and that deoxyribose metabolism has no connection with growth and is also relatively independent of the quantity added or present. It is significant that no growth occurs during deoxyribose utilization. The inducible enzyme for this breakdown of deoxyribose has been shown to be deoxyriboaldolase (2-deoxy-D-ribose-5-phosphate acetaldehyde-lyase, EC 4.1.2.4), which catalyzes the reaction

$$\text{2-deoxy-D-ribose phosphate} \rightarrow \text{D-glyceraldehyde 3-P} + \text{acetaldehyde}$$

D-Glyceraldehyde 3-phosphate, of course, can be broken down to lactic

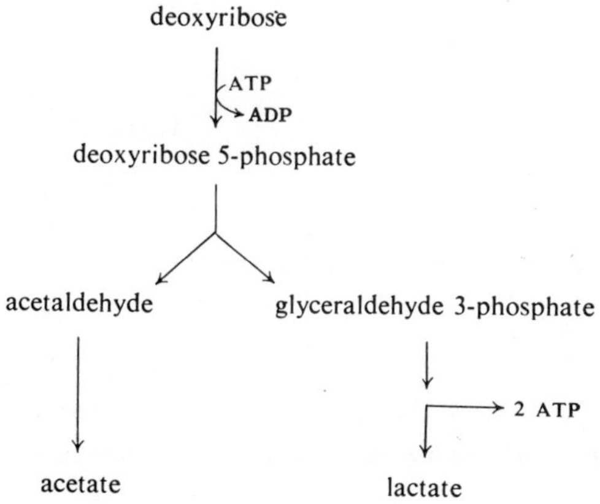
deoxyribose
ATP
ADP
deoxyribose 5-phosphate
acetaldehyde
glyceraldehyde 3-phosphate
2 ATP
acetate
lactate

acid because *L. plantarum* possesses all enzymes of the EMP pathway, and acetaldehyde is converted to acetate. The end products of this metabolism are therefore lactic acid and acetic acid. The net energy yield is 1 mole ATP per mole deoxyribose fermented (107).

Polysaccharide Formation

The production of large mucoid colonies is a well-established feature of *Leuconostoc*, a heterofermentative, chain-forming coccus grown in high-sucrose plant extracts. It is also known that a number of strains of heterofermentative lactobacilli produce dextran following prolonged cultivation in acid media. The production of high molecular weight compounds is only mentioned here; their formation is a biosynthetic feature and not a catabolic one, although it is nonetheless characteristic for some species in the Lactobacillaceae.

Biotin Degradation

Lactobacillus plantarum was found to metabolize D-biotin (34), which was converted to biotin vitamers. The biotin vitamers could be distinguished as being combinable and noncombinable with avidin (35). There is some evidence available (36) suggesting that *L. plantarum* possesses an enzyme system that not only converts biotin into vitamers but also metabolizes oxybiotin and dethiobiotin to products which are not utilizable by *Saccharomyces cerevisiae* and are still unknown. *Lactobacillus plantarum* is not unique in its biotin metabolism; *L. casei* (37) appears to have a similar system.

Phosphorylation and Electron Transport System

The Lactobacillaceae have powerful fermentative properties. They may contain cytochromes but do not use them, as their respiratory system depends entirely on flavoproteins (104):

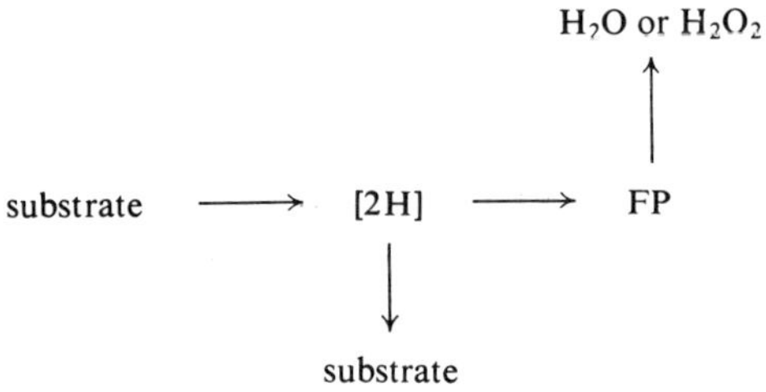
H₂O or H₂O₂
substrate
[2H]
FP
substrate

Under anaerobic conditions a second substrate serves as a hydrogen acceptor (105). Under aerobic conditions, however, the electrons are passed from the substrate to a flavoprotein, which reacts with oxygen to produce either water or hydrogen peroxide:

$$FPH + H^+ + O_2 \rightarrow FP^+ + H_2O_2 \qquad \text{2-electron transfer}$$

$$2(FPH + H^+) + O_2 \rightarrow 2\,FP^+ + 2\,H_2O \qquad \text{4-electron transfer}$$

The two-electron transfer system was mainly observed during the oxidation of glucose by cell suspensions of *Lactobacillus delbruckii* or *L. acidophilus.* A 1:1 equivalence between oxygen uptake and hydrogen peroxide production was observed (31). As this observation was reported also with *L. bulgaricus*, it was postulated that these bacteria contain a flavoprotein which accounts for the entire respiration. There are, however, a number of lactic acid bacteria that exhibit a rapid oxidation without the need for high oxygen concentrations. Moreover, hydrogen peroxide was not always the end product but was broken down by a NAD-peroxidase (reduced-

$$NADH + H^+ + H_2O_2 \rightarrow NAD^+ + 2H_2O$$

NAD:hydrogen-peroxide oxidoreductase, EC 1.11.1.1). Other streptococci and lactobacilli were able to reduce peroxide in the presence of oxidizable substrate with enzymes that did not involve any heavy-metal catalysts and were therefore called "atypical peroxidases."

The combined action of these two enzymes, flavoprotein-linked NAD dehydrogenase (103) and flavoprotein-linked NAD peroxidase, provides the required four-electron step of oxygen reduction to water (101). The

$$\begin{array}{c} NADH + H^+ + O_2 \rightarrow NAD^+ + H_2O_2 \\ NADH + H^+ + H_2O_2 \rightarrow NAD^+ + H_2O \\ \hline 2(NADH + H^+) + O_2 \rightarrow 2\,NAD^+ + 2\,H_2O \end{array}$$

lactic acid bacteria therefore employ diaphorases for their anaerobic and flavoprotein-linked oxidases, alone or together with flavoprotein-linked peroxidases, for their aerobic metabolism.

DIAPHORASES. Diaphorases are enzymes that may couple oxidation of reduced NAD to the reduction of artificial electron acceptors:

$$NAD(P)H + H^+ + A \rightarrow NAD(P)^+ + AH_2$$

Many of these enzymes are flavoproteins (Table 9.7). Because the cell

TABLE 9.7
Purified Diaphorases[a]

Enzymes	Substrate	Oxidant	Prosthetic group	Molecular weight	Turnover no.
Straub's soluble diaphorase (pig heart)	$NADH_2$	Methylene Blue	FAD	67,000	8,500
New yellow enzyme (yeast)	$NADPH_2$	Methylene Blue	FAD	60,000	130
S. faecalis	$NADH_2$	$(FeCN_6)^{2-}$	FMN	53,000	6,000
		2,6-DCIP			29,000
		p-Benzoquinone			88,000
E. coli	$NADH_2$	Menadione	—	—	50
		Quinones	—	—	—
P. fluorescens	$NADH_2$	2,6-DCIP	FMN	—	80
L. mesenteroides	$NADH_2$	Menadione	—	—	80
S. faecalis	$NADH_2$	Menadione	FAD, FMN	—	40–100
L. delbruckii	$NADH_2$	Menadione	—	—	30
		2,6-DCIP			30

[a] From Dolin (104).

apparently does not use them as physiological oxidants, their significance in the cell is questionable.

It is known that the velocity constants for the reaction between reduced diaphorase from *Streptococcus faecalis* and various oxidized quinones are very high. Low concentrations of these quinones can act readily as efficient electron carriers by accepting them from reduced NAD. When the quinones involved are autoxidizable, a direct path from reduced NAD to O_2 would be established. There is some evidence available that supports the view that quinones play an important role in the metabolism of lactic acid bacteria (144, 301).

FLAVOPROTEIN-LINKED OXIDASES. At least four distinct flavoproteins that catalyze the oxidation of reduced NAD have been isolated from *Streptococcus faecalis*: (*1*) flavoprotein-linked oxidase; (*2*) cytochrome c reductase [reduced-NAD: (acceptor) oxidoreductase, EC 1.6.99.3); (*3*) a diaphorase that uses 2,6-dichlorophenol indophenol, ferricyanide, and a series of quinones as oxidants; and (*4*) a menadione reductase [reduced-NAD(P): acceptor) oxidoreductase, EC 1.6.99.2]. These four enzymes can be separated from each other (409), suggesting that all reduced NAD oxidase activity in a crude cell-free extract cannot be attributed to a single "NAD dehydrogenase." These oxidases are very unstable during isolation but can be reactivated with thiol groups (410).

FLAVOPROTEIN-LINKED PEROXIDASE. Flavoprotein-linked peroxidases of lactic acid bacteria are divided into atypical and typical peroxidases.

1. Atypical peroxidases: *Streptococcus mastitidis*, *Streptococcus faecalis* strain B 33 a, *Streptococcus mitis*, *Lactobacillus brevis*, and *Leuconostoc mesenteroides* strain 548 are catalase- and cytochrome-free organisms (108, 152, 203, 349) that can catalyze the reduction of hydrogen peroxide to water with oxidizable substrates as electron donors. The oxygen uptake therefore does not result in a peroxide accumulation. The oxidizable substrate can be alcohol, glucose, glycerol, lactate, or fructose.

2. Typical peroxidase (105): The flavoprotein-linked NAD peroxidase (EC 1.11.1.1) has been isolated from *Streptococcus faecalis* 10C1. The prosthetic group of this enzyme contains FAD but no heme or metal group, and the enzyme is specific only for reduced NAD. Apart from *Leuconostoc mesenteroides* and *L. delbruckii*, this enzyme has been detected in three streptococcal strains, *S. faecalis* 10C1, *S. faecalis* B 33 a, and *S. mastitidis*. The survey of this enzyme is difficult and has not been as extensive as that of atypical peroxidases because catalase and possibly cytochrome c peroxidase offer serious interferences.

DIRECT FLAVOPROTEIN OXIDASES. The absence of oxidases for $NADH_2$ in extracts of *L. delbruckii* suggests that this organism does use direct substrate oxidases in its oxidative metabolism (157).

1. Pyruvate oxidation: the oxidation of pyruvate in *L. delbruckii* follows the pattern

$$\text{pyruvate} + P_i + O_2 \xrightarrow[\text{TPP}]{\text{FAD}} \text{acetyl P} + CO_2 + H_2O_2$$

$$\text{pyruvate} + H_2O_2 \xrightarrow{\text{spontaneous}} \text{acetate} + CO_2 + H_2O$$

$$2\ \text{pyruvate} + P_i + O_2 \rightarrow \text{acetyl P} + \text{acetate} + 2\,CO_2 + H_2O$$

The feature of this reaction is the generation of acetyl phosphate with either oxygen or ferricyanide as electron acceptor. The detailed mechanism of this acetyl phosphate formation is not yet established but is probably different from that of the NAD-linked systems. It may well be that this system could provide a mechanism whereby energy-rich compounds can be generated at the flavoprotein level.

2. Pyruvate dismutation: The dismutation of pyruvate to lactate, acetate, and carbon dioxide also occurs in *L. delbruckii*. Two enzymes are involved in this process, pyruvate oxidase [pyruvate:oxygen oxidoreductase (phosphorylating), EC 1.2.3.3], which is a thiamine pyrophosphate-requiring flavoprotein, and lactate oxidase (L-lactate:oxygen oxidoreductase, EC 1.1.3.2), which catalyzes the following sequence (157):

$$\text{pyruvate} + P_i + \text{FAD} \rightarrow \text{acetyl P} + CO_2 + \text{FADH} + H^+ \quad \text{(a)}$$

$$\text{FADH} + H^+ + \text{riboflavin} \rightarrow \text{FAD}^+ + \text{riboflavin}\cdot H_2 \quad \text{(b)}$$

$$\text{riboflavin}\cdot H_2 + \text{FAD}^+ \rightarrow \text{riboflavin} + \text{FADH} + H^+ \quad \text{(c)}$$

$$\text{FADH} + H^+ + \text{pyruvate} \rightarrow \text{FAD}^+ + \text{lactate} \quad \text{(d)}$$

$$2\ \text{pyruvate} + P_i \rightarrow \text{lactate} + \text{acetyl P} + CO_2$$

In reactions (a) and (b) pyruvate oxidase is linked to FAD^+ or $FADH + H^+$, whereas in reactions (c) and (d) lactate oxidase is linked to the flavoprotein. This reaction sequence, in which high concentrations of riboflavin serve to link two FAD oxidases, may be a model for anaerobic electron transport between flavoproteins.

DEHYDROGENASE ACTIVITY. It was mentioned above that *Streptococcus faecalis* and *Lactobacillus casei* have a lipoic acid-activating pyruvate metabolism (see p. 626) under aerobic conditions. These organisms must therefore use part of the pyruvate system or pyruvate dehydrogenase complex. *Leuco-*

nostoc mesenteroides 39 possesses the lipoamide dehydrogenase (reduced-NAD:lipoamide oxidoreductase, EC 1.6.4.3) that catalyzes the reversible reaction

$$\text{lipoate-SH}_2 + \text{NAD}^+ \rightleftharpoons \text{lipoate} + \text{NADH} + \text{H}^+$$

This enzyme is a flavoprotein. It is also assumed that *Leuconostoc mesenteroides* has a second type of lipoamide dehydrogenase, which must be similar to that of *Escherichia coli*. This enzyme, however, is difficult to demonstrate. *Streptococcus faecalis* also requires lipoic acid for its diacetyl reduction (102), which also involves acetyl-CoA.

FLAVOPROTEIN RESPIRATION. An electron transport system as such need not be coupled with phosphorylation in order to be useful (45). A variety of substrates that would not otherwise be used because of lack of hydrogen acceptors could be made available to fermentative organisms in alternate transport mechanisms to oxygen. In these reactions, energy becomes available through substrate-linked phosphorylation, as has been seen in fermentation processes.

Many lactic acid bacteria, for example, are able to metabolize glycerol only under aerobic conditions (154), whereas under anaerobic conditions this cannot happen without an exogenous hydrogen acceptor. In *Streptococcus faecalis* it is assumed that a direct flavoprotein glycerophosphate dehydrogenase may be used for the transfer of electrons to oxygen. Other strains of a more aerobic character possess a NAD^+ dehydrogenase that allows the coupling between triose phosphate dehydrogenation and oxygen reduction. Even if the reoxidation of reduced NAD is not energetically coupled, the oxidation of glucose with oxygen as electron acceptor furnishes more energy for growth than the lactic acid fermentation. This is because, aerobically, pyruvate is removed from its role as the obligatory hydrogen acceptor and can be oxidized to acetyl-CoA, yielding one more energy-rich bond per triose.

Such oxidative phosphorylations have certainly been substantiated in *Streptococcus agalactiae* (274) and *Streptococcus faecalis* (101, 358), which do not involve any cytochromes. ATP formation appears to occur predominantly nonoxidatively but 20–40% of this ATP formation is thought to come from such oxidative phosphorylations. In addition to these systems, fumarate was found as an electron acceptor to the oxidation of reduced NAD in the cytochrome-free *Streptococcus faecalis* 10 C 1 (118). The investigations into oxidative phosphorylation of lactobacilli by using uncouplers can be hampered by the fact that these uncouplers also interfere with the utilization of metabolic energy for membrane transport (164).

N,N'-Dicyclohexylcarbodiimide, for example, is a potent inhibitor of the membrane-bound ATPase of *Streptococcus faecalis,* which is responsible for the transport of cations and other metabolites (165, 348).

Such oxidative reactions become useful if an enzyme is present to dispose of hydrogen peroxide. The accumulation of hydrogen peroxide eventually would inhibit the oxidation and growth of the organisms. As bifidobacteria are not able to dispose of hydrogen peroxide, it is assumed that these bacteria require a low oxidation–reduction potential for growth and fermentation, which also determines its degree of anaerobiosis (92). However, organisms, such as *Streptococcus faecalis,* that do not possess a peroxidase appear to use this pathway and seem to be more resistant toward hydrogen peroxide. In species of *Pediococcus* the use of the above-mentioned system of glycerol oxidation is dependent on the production of catalase. There appears to exist a relationship between catalase production and glycerol utilization; the higher the catalase content of cells, the better the growth on glycerol.

At least two types of catalase are observed in lactic acid bacteria (421):

1. The classical catalase (hydrogen-peroxide:hydrogen-peroxide oxidoreductase, EC 1.11.1.6) (94), which vigorously destroys hydrogen peroxide, is azide sensitive but not sensitive to acid pH. This catalase was observed in certain strains of homo- and heterofermentative lactobacilli, *Pediococcus,* and *Leuconostoc* only when they were grown on heme-containing medium. Unlike most catalase-producing bacteria, these organisms cannot synthesize the heme component of the catalase.
2. A pseudocatalase, which has no heme prosthetic group and is often acid sensitive, can be found in strains of *Leuconostoc, Pediococcus,* and *L. plantarum* (one strain only).

Whittenbury (421) has suggested that lactic acid bacteria have a rudimentary respiratory system and that only some of them have the ability to form hemeproteins if provided with a source of iron porphyrins (402a). Both types of catalases have been found in the same organism (204), which would support Whittenbury's suggestion. On heme-supplemented media, *Streptococcus faecalis* was able to form cytochromes (b_2 type), and a complete oxidative phosphorylation system was demonstrated (46). The phosphorylation was sensitive to cyanide, azide, and antimycin A.

The addition of hemin to anaerobic cultures of *Staphylococcus* (174) caused an increase in nitrate reductase activity (see Chapter 7) and an increase in catalase to the level found in cultures grown under oxygen. This suggests that the primary part which oxygen plays in the formation of staphylococcal catalase could be associated with heme biosynthesis. This theory is supported by earlier findings, with *Rhodopseudomonas spheroides*

(65) and *Staphylococcus aureus* (194, 234), that oxygen induces catalase activity. Large parts of the protein moiety of catalase are formed in catalase-negative staphylococci (195). The proof for the suggestion that the heme biosynthesis is interrupted by lack of oxygen was provided by quantitative catalase tests carried out on 14 strains of *Haemophilus* species cultured with graded amounts of hemin. These revealed that the catalase activity increased in proportion to the hemin concentration in the medium (32).

This behavior of the catalase is reflected in the division of the heterofermentative bacteria into two groups (421). The first group is comprised of those heterofermenters which show a positive catalase test on normal agar and on blood agar, and the second group contains those which show a positive catalase test only on blood agar. In the Lactobacillaceae a parallel could be drawn to the cytochrome system that leads to nitrate reductase.

Apart from the catalase, which depends on heme biosynthesis and therefore oxygen in the medium and is possibly affected also by glucose concentrations (143), there exists a second hydrogen peroxide enzyme, which is provisionally named "pseudocatalase." The latter appears to be an acid-sensitive, nonheme-containing enzyme, which is mostly found in *Leuconostoc* and *Pediococcus* (310, 324). When these organisms are grown under

TABLE 9.8
Cytochrome Formation in Lactic Acid Bacteria[a]

H_2O_2 splitting characteristics and organisms	Medium used[b]	Cytochromes detected after $Na_2S_2O_4$ addition
Catalase positive; pseudocatalase negative		
L. plantarum NCIB 5914	A	None
	B	a_2; b_1
L. brevis NCIB 947	A	None
	B	b_1
Leuconostoc mesenteroides NCIB 8018	A	None
	B	None
S. faecalis H 69 D5	A	None
	B	a_1; a_2; b_1
Catalase negative; pseudocatalase negative		
L. plantarum NCIB 6105	A	b_1 (very weak)
	B	b_1 (very weak)
Catalase negative; pseudocatalase positive		
Leuconostoc citrovorum NCIB 7837	C	None
Leuconostoc mesenteroides RW 66	C	None

[a] From Whittenbury (421). Reprinted with permission of Cambridge University Press.
[b] Here, A = 1.0% (w/v) glucose agar; B = 1.0% glucose agar plus heated blood; and C = 0.05% glucose.

anaerobic conditions and fail to split H_2O_2 in a "normal" catalase test, it is absolutely necessary to add heme before it can be concluded whether the organism has the ability to produce catalase or not. Catalase-negative organisms can therefore become catalase positive with changes in nutrition. This change would not result from the formation of an adaptive or constitutive enzyme but from the synthesis of the prosthetic group of catalase, heme.

Four organisms that formed catalase also possessed cytochromes when grown in the presence of heated blood (Table 9.8) but not in its absence. Heavier suspensions of these strains may probably have given the same result as *Streptococcus faecalis* H 69 D5. Whether these cytochromes are functional and why they are formed remains to be determined. Anticipating the difficulties of measuring and finding cytochromes in anaerobic organisms, the question remains whether there are more cytochromes present that are still undetectable and whether they play a function similar to that in the anaerobic microorganisms mentioned in Chapter 6.

Fermentation of Proteolytic Clostridia

The ability to break down proteins to peptones, polypeptides, and amino acids is not shared equally by all groups of bacteria. Because the majority of these organic nitrogenous compounds are at an oxidation level between carbohydrates and fats, however, they are potentially useful as a source of carbon, nitrogen, and energy for both aerobic and anaerobic microorganisms.

There are two possibilities open for the cell that metabolizes these large molecules, which are normally outside the cell. There is plenty of evidence (313) available to show that the passage of molecules across the cell membrane is not determined solely by size, although size is certainly an important factor. Mechanisms for the passage of large molecules seem to be not as frequent as for small molecules (permeases). This is particularly so for bacteria, for which nearly all the enzymes liberated outside the cell seem to be concerned with the metabolism of large molecules. These enzymes are called "exoenzymes," in contrast to the endoenzymes.

Very little work has so far been directed to the problem of exoenzymatic liberation in bacteria. Pollock (313) states the reason for this as follows: "The only method of formally proving an enzyme to be extracellular according to the definition would be to demonstrate its liberation from a series of individual cells subsequently shown to be capable of normal growth and metabolism. So far, this has not been attempted . . . ." Pollock continues: "In most reported work on exoenzymes little effort is made to control or estimate the extent of cell autolysis. There are, however, relatively simple, although indirect, ways of measuring the degree of lysis in a bacterial culture, and it should not be difficult to investigate the liberation of enzymes under conditions where the possibility of autolysis is minimized. Too often, exoenzymatic studies are done with old cultures, whereas young cultures in the logarithmic phase of growth should be used."

The bacteria involved in protein metabolism therefore must have proteolytic enzymes to break these polymeric compounds into their monomeric components before they can enter the cell and be used either as building blocks or as fermentative substrates. The present discussion will be limited to the decomposition of nitrogenous compounds by bacteria under anaerobic conditions without the intervention of inorganic oxidizing agents, such as sulfate, nitrate, or carbonate. It also will exclude such processes as denitrification.

Proteolytic enzymes fall into two classes, endopeptidases and exopeptidases, according to their mode of action on a polypeptide chain. Such a chain can be attacked in two ways—at either end or at random points along its length:

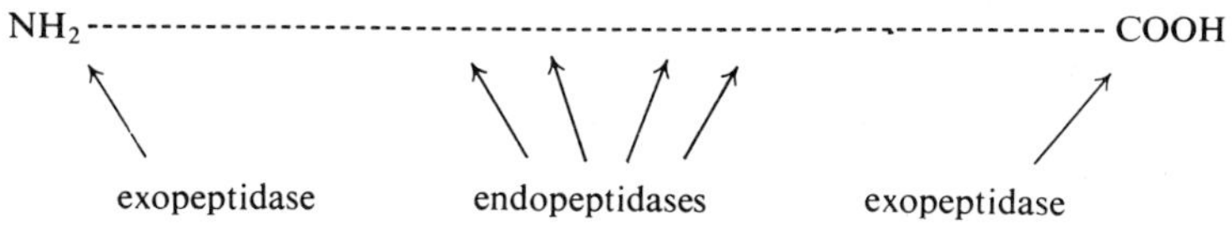
NH2
COOH
exopeptidase
endopeptidases
exopeptidase

The enzyme pepsin (EC 3.4.23.1), for example, is a typical endopeptidase, attacking polypeptide chains wherever aromatic R groups (L-phenylalanine or L-tyrosine) occur.

A further basis for their differentiation lies in their activation by metals, such as Fe, Mn, and Mg, and by reducing agents, such as —CN or —SH. Exopeptidases are subdivided as follows: (*a*) aminopeptidases, which require a free terminal —NH_2 group and are dependent on metal ions for their activity, and (*b*) carboxypeptidases, which hydrolyze peptides with a free terminal —COOH group. Both types of enzymes can break down the chain, liberating amino acids until the di- or tripeptidases complete the task of hydrolysis by forming free amino acids. These monomers are now able to diffuse into the cell and be utilized. For further biochemical details the reader is referred to appropriate textbooks.

The catabolism of amino acids is carried out by a great number of anaerobic or facultative anaerobic bacteria. This is done either with single amino acids or pairs of amino acids or with one amino acid in combination with one nonnitrogenous compound.

Single Amino Acid Metabolism

All processes that liberate ammonia from amino acids are called "deaminations." Bacterial deamination proceeds in several ways, which differ according to the enzymatic constitution of the organism and the conditions prevailing in the medium.

Oxidative deamination is a reaction wherein the products of the process are a keto acid and ammonia:

$$R\text{—}\overset{\displaystyle H}{\underset{\displaystyle NH_2}{C}}\text{—}COOH + \tfrac{1}{2}\,O_2 \rightleftharpoons R\text{—}\underset{\displaystyle NH_2}{C}\text{—}COOH$$

$$\Big\updownarrow$$

$$R\text{—}\overset{\displaystyle O}{\overset{\|}{C}}\text{—}COOH + NH_3$$

An oxidative deamination of DL-alanine, for example, results in the accumulation of pyruvate.

In a desaturation deamination, the end product is an unsaturated fatty acid:

$$R\text{—}\underset{\displaystyle NH_2}{CH_2}\text{—}COOH \rightleftharpoons R\text{—}CH{=}CH\text{—}COOH + NH_3$$

In general, however, the compound resulting from deamination by a strict anaerobe or by a facultative anaerobe acting anaerobically is the corresponding saturated fatty acid, which is obtained by a reductive deamination:

$$R\text{—}CH_2\text{—}\underset{\displaystyle NH_2}{CH}\text{—}COOH + 2\,H^+ \longrightarrow R\text{—}CH_2\text{—}CH_2\text{—}COOH + NH_3$$

This reductive deamination can be clearly demonstrated when the organism contains dehydrogenase (EC 1.12.7.1) and hydrogen acts as the hydrogen donor.

Clostridium tetanomorphum exhibits an anaerobic oxidative deamination that is accompanied by the evolution of hydrogen. This type of reaction may perhaps be regarded as an anaerobic device for obtaining energy from amino acids without the use of a hydrogen acceptor. Stickland (376–378) established the following reactions and end products:

alanine → acetic acid + NH_3 + CO_2

glycine → acetic acid + NH_3

proline → α-aminovaleric acid

Cohen-Bazire *et al.* (66) added to these the end products derived from suspensions of *Clostridium caproicum* and/or *Clostridium valerianum*

valine → isobutyric acid + CO_2

leucine → isovaleric acid + CO_2

isoleucine → valeric acid + CO_2

ARGININE METABOLISM. Arginine metabolism is a significant metabolic step for species of *Mycoplasma* that have been isolated from cell cultures. The conversion of arginine is rapid and appears to be the major source of ATP (23, 346, 347). L-Arginine is deaminated first by an arginine deiminase (L-arginine iminohydrolase, EC 3.5.3.6), which step is followed by a reversible transferase reaction, converting L-citrulline to ornithine. The en-

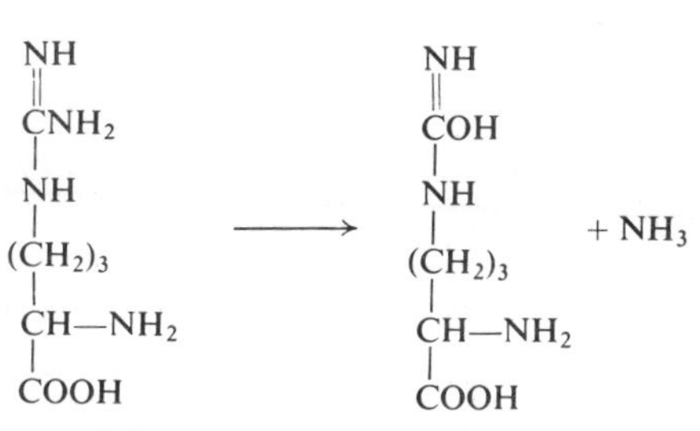

L-arginine L-citrulline

zyme responsible for this reaction is ornithine carbamoyltransferase (carba-

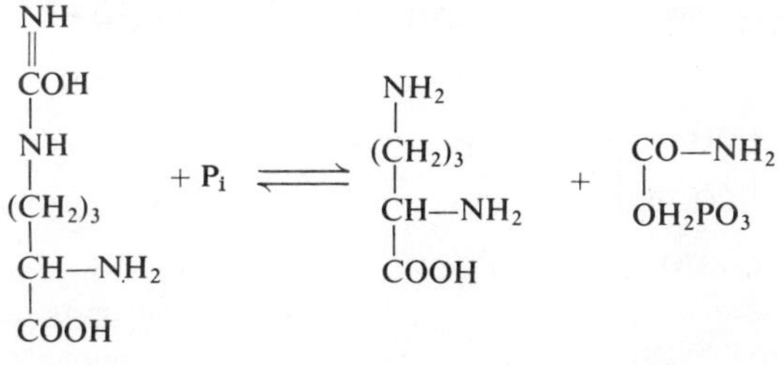

L-citrulline L-ornithine carbamoyl phosphate

moylphosphate: L-ornithine carbamoyltransferase, EC 2.1.3.3) and requires inorganic phosphate. This is followed by the vital energy-yielding reaction, the degradation of carbamoylphosphate to ammonia and carbon dioxide with carbamate kinase (ATP: carbamate phosphotransferase, EC 2.7.2.2), which is an ATP-dependent reaction.

$$\mathrm{ADP} + \overset{\mathrm{CO{-}NH_2}}{\overset{|}{\mathrm{OH_2PO_3}}} \rightleftharpoons \mathrm{NH_3} + \mathrm{CO_2} + \mathrm{ATP}$$

For earlier work on the energy metabolism of PPLO and L forms of bacteria, the reader is referred to the review by Smith (359). The reaction sequence of this arginine metabolism was also found in *Halobacterium salinarum* (109) and *Clostridium botulinum* (275) up to ornithine. *Clostridium botulinum* degrades ornithine further to volatile acids. All evidence suggests that L-ornithine is first reductively deaminated to δ-aminovalerate and from here into the various end products acetate, propionate, valerate, and butyrate. The proper sequence, however, is not known as yet.

HISTIDINE AND GLUTAMIC ACID METABOLISM. *Clostridium tetanomorphum* converts histidine to butyrate, acetate, CO_2, ammonia, and formamide (25) (Fig. 9.15). The first reaction is a desaturation deamination catalyzed by L-histidine ammonia-lyase (EC 4.3.1.3), which is followed by an uro-

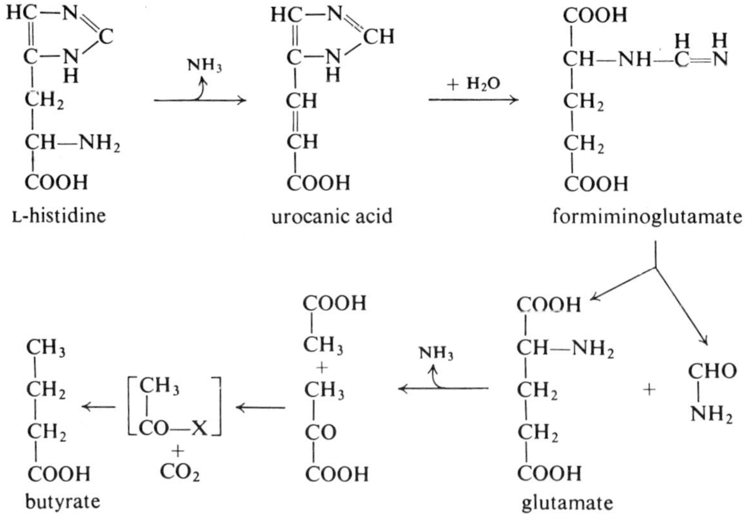
L-histidine
NH3
urocanic acid
+ H2O
formiminoglutamate
glutamate
NH3
butyrate

Fig. 9.15. Metabolism of histidine by *Clostridium tetanomorphum* (reprinted with permission of John Wiley and Sons).

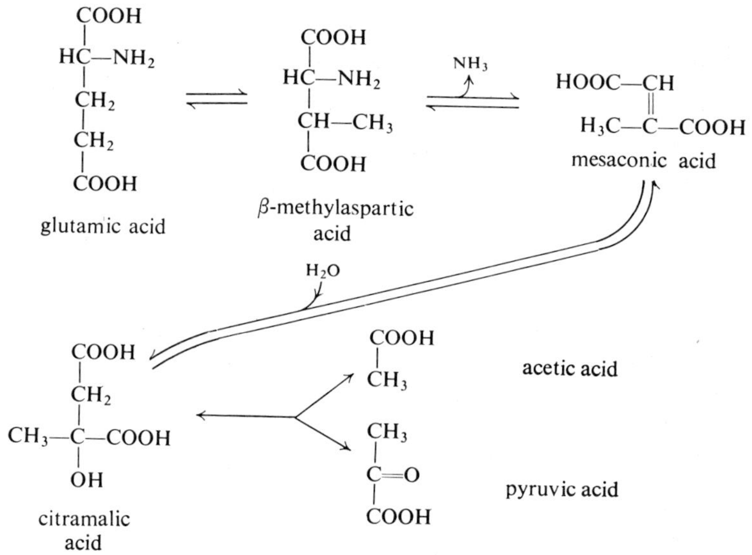
COOH
HC—NH2
CH2
CH2
COOH
glutamic acid
COOH
HC—NH2
CH—CH3
COOH
β-methylaspartic acid
NH3
HOOC—CH
H3C—C—COOH
mesaconic acid
H2O
COOH
CH2
CH3—C—COOH
OH
citramalic acid
COOH
CH3
acetic acid
CH3
C=O
COOH
pyruvic acid

Fig. 9.16. Metabolism of glutamic acid by *Clostridium tetanomorphum.*

canase (urocanate hydratase, EC 4.2.1.49 ?) system. The step to glutamate is still obscure, but it is assumed that a tetrahydrofolate-dependent reaction is involved. The breakdown of glutamate, however, is a reductive deamination.

A different route of L-histidine degradation has been reported (175) from studies with *Escherichia coli* B. It seems from these results that L-histidine is oxidatively deaminated to imidazolepyruvic acid, which is followed by a reduction to imidazole, lactic acid, and then further to acyclic compounds. The reduction that immediately follows the oxidative deamination indicates a cofactor requirement, which may possibly be reduced NAD or NADP.

Clostridium tetanomorphum grows very well on glutamate (49b) and rapidly utilizes this amino acid via mesaconate (Fig. 9.16). The deamination reaction is a desaturation reaction and is catalyzed by L-*threo*-3-methylaspartate ammonia-lyase (EC 4.3.1.2). Mesaconate is reversibly converted to citramalate (26) by citramalate-lyase (EC 4.1.3.22) (mesaconase). This enzyme has been purified (412) and found to require the presence of two different proteins (413), neither of which exhibits significant, catalytic activity alone. Citramalate lyase, a very active but labile enzyme (27), converts L(+)-citramalate in an aldolase-type reaction into acetate and pyruvate.

Neisseria meningitis (263) appears to have a different type of glutamate metabolism. It has not been fully elucidated but seems to have succinate as an intermediate.

The purification of glutamate dehydrogenase and 2-hydroxyglutarate dehydrogenase (EC 1.1.99.2), suggests a different pathway for *Micrococcus aerogenes* (49a, 241). Although the end products of this glutamate degradation are like the ones of *Clostridium tetanomorphum* with butyrate, acetate, CO_2, and NH_3, glutamate is first converted to α-ketoglutarate and the dehydrogenase forms the α-hydroxyglutarate. The latter intermediate must

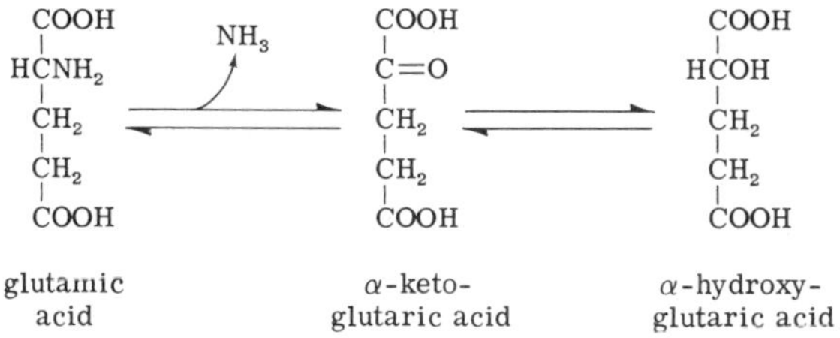
COOH
HCNH2
CH2
CH2
COOH
NH3
COOH
C=O
CH2
CH2
COOH
COOH
HCOH
CH2
CH2
COOH
glutamic acid
α-keto-glutaric acid
α-hydroxy-glutaric acid

then undergo conversion to butyrate, acetate, and carbon dioxide, in a mechanism possibly similar to the one in *C. tetanomorphum.*

GLYCINE METABOLISM. The only anaerobic bacteria known to carry out a true glycine metabolism are *Diplococcus glycinophilus, Micrococcus anaerobius*, and *Micrococcus variabilis*:

$$4\ \text{glycine} + 2H_2O \rightarrow 4NH_3 + 2CO_2 + 3CH_3COOH$$

In addition to this reaction, the two *Micrococcus* species also evolve hydrogen (26). The pathway of glycine metabolism in *Diplococcus glycinophilus* was found (221, 338) to involve tetrahydrofolate, pyridixal phosphate, and NAD^+, as well as an acetate-generating system (Fig. 9.17). This oxidative decarboxylation is catalyzed by a system containing four proteins (19): pyridoxal phosphate-containing protein, heat-stable protcin, flavoprotein, and a still uncharacterized protein. The first reaction is an oxidation–reduction step, whereby CO_2, NH_3, and hydrogen are formed, with

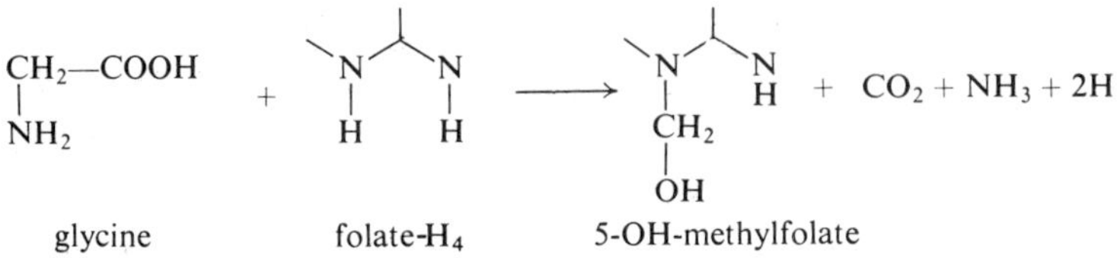
CH2—COOH
NH2
N
H
N
H
N
CH2
OH
N
H
+ CO2 + NH3 + 2H
glycine
folate-H4
5-OH-methylfolate

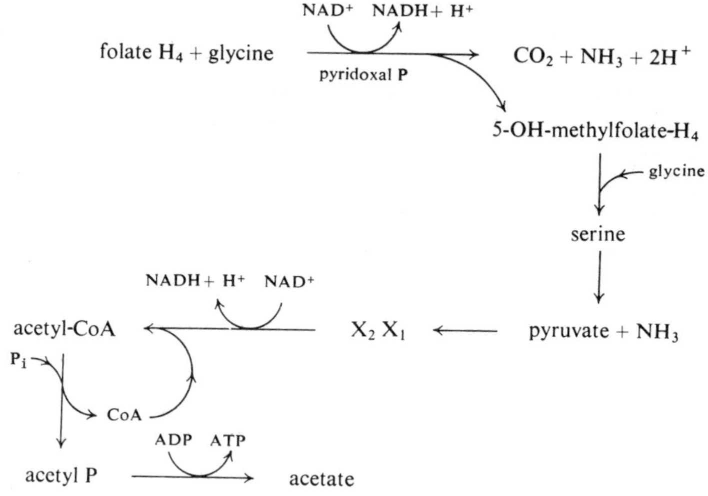
NAD+ NADH+ H+
folate H4 + glycine
pyridoxal P
CO2 + NH3 + 2H+
5-OH-methylfolate-H4
glycine
serine
NADH+ H+ NAD+
acetyl-CoA
X2 X1
pyruvate + NH3
Pi
CoA
ADP ATP
acetyl P
acetate

Fig. 9.17. Metabolism of glycine by anaerobic bacteria.

pyridoxal phosphate and NAD^+ as cofactors (222, 223). A second molecule of glycine is now necessary to react with 5-hydroxymethylfolate-H_4 to form

$$\text{folate-}H_4\text{—}CH_2OH + \underset{\displaystyle NH_2}{CH_2COOH} \longrightarrow \underset{\text{serine}}{\underset{\displaystyle OH\ \ NH_2}{CH_2CHCOOH}} + \text{folate-}H_4$$

serine and reoxidize 5-hydroxymethylfolate-H_4. Methylfolate-H_4 is used by this organism simply for the transfer of a methyl group from glycine to a second molecule of glycine, producing serine. A deamination step

$$\underset{\text{serine}}{\underset{\displaystyle OH\qquad NH_2}{CH_2\text{—}CH\text{—}COOH}} \longrightarrow \underset{\text{pyruvate}}{\underset{\displaystyle \| \atop O}{CH_3\text{—}C\text{—}COOH}} + NH_3$$

follows, with the production of pyruvate. The reduction of pyruvate to acetate may well involve a phosphoroclastic split, with the formation of

$$\underset{\displaystyle \| \atop O}{CH_3\text{—}C\text{—}COOH} \xrightarrow[CO_2]{} \underset{\displaystyle \| \atop O}{CH_3\text{—}C\text{—}SCoA} \longrightarrow$$

$$\underset{\displaystyle \| \atop O}{CH_3\text{—}C\text{—}OPO_3H_2} \longrightarrow CH_3COOH$$

acetyl-CoA, followed by a phosphate acetyltransferase (EC 2.3.1.8) and acetate kinase (EC 2.7.2.1).

SERINE METABOLISM. The metabolism of serine to pyruvate is frequently found in anaerobic bacteria (128) and is catalyzed by serine dehydratase [L-serine hydro-lyase (deaminating), EC 4.2.1.13]. It is believed that serine dehydratase is sometimes specific for L-serine and can

$$\underset{\displaystyle NH_2}{CH_2OH—\underset{|}{CH}—COOH} \longrightarrow CH_3—CO—COOH + NH_3$$

also act on L-threonine. When these cases occur, a second enzyme would be necessary for the degradation of D-serine and D-threonine (239).

THREONINE METABOLISM. Such organisms as *Clostridium propionicum*, *Micrococcus aerogenes* (418), *M. lactilyticus* (419), and *Clostridium tetanomorphum* (420) require a threonine dehydratase [L-threonine hydro-lyase (deaminating), EC 4.2.1.16] for the utilization of threonine and the pro-

$$CH_3—CHOH—\underset{\displaystyle NH_2}{\underset{|}{CH}}—COOH \longrightarrow CH_3—CH_2—CO—COOH + NH_3$$

duction of α-oxobutyrate. *Micrococcus aerogenes* carries this step further and forms butyric acid with the evolution of hydrogen. *Clostridium pasteurianum*, however,

$$CH_3—CH_2—CO—COOH \rightarrow CH_3—CH_2—COOH + CO_2 + H_2$$

utilizes serine hydroxymethyltransferase (5,10-methylenetetrahydrofolate: glycine hydroxymethyltransferase EC 2.1.2.1) to cleave threonine to glycine

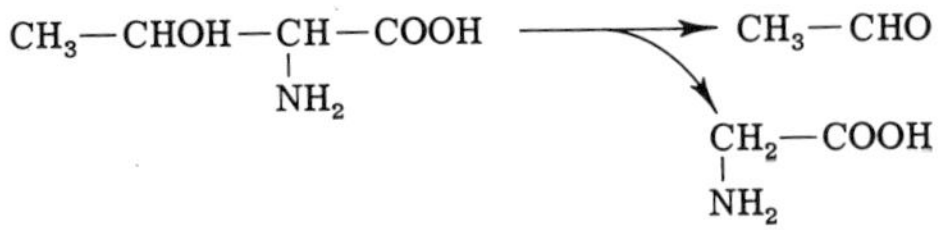

CH₃—CHOH—CH—COOH
NH₂
CH₃—CHO
CH₂—COOH
NH₂

and acetaldehyde (75, 76). Acetaldehyde is then converted to alcohol, with the oxidation of reduced NAD. If all these reactions are taken together in a general scheme, the mechanism of threonine metabolism shown in Scheme 9.4 emerges. With these different products from threonine, a metabolic differentiation of the organism can be obtained.

TRYPTOPHAN METABOLISM. Tryptophan is only slowly fermented by *Clostridium sporogenes* and a number of diverse bacterial species (89), with

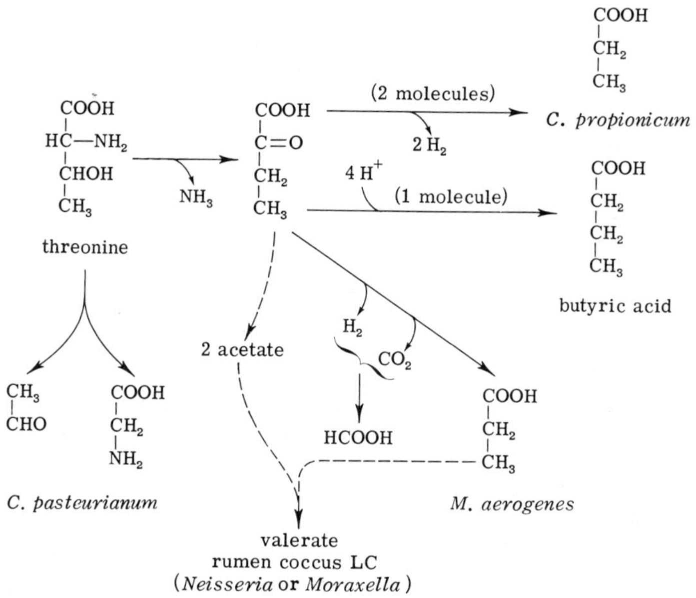
threonine
NH3
(2 molecules)
2 H2
4 H+
(1 molecule)
C. propionicum
butyric acid
2 acetate
H2
CO2
HCOOH
C. pasteurianum
M. aerogenes
valerate
rumen coccus LC
(Neisseria or Moraxella)

Scheme 9.4. Threonine metabolism by proteolytic clostridia and other bacterial genera.

the formation of indole propionic acid. The degradation of tryptophan,

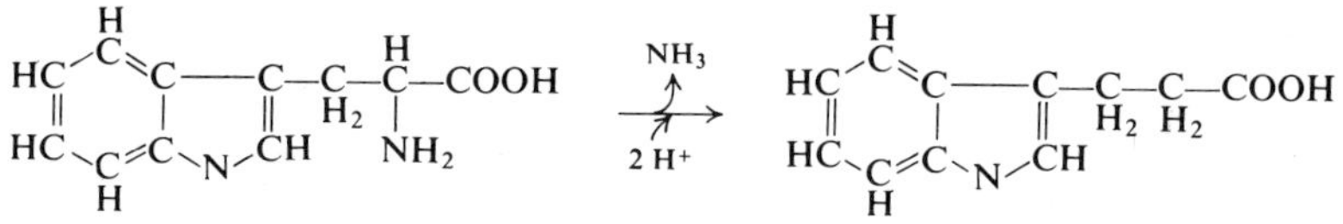

however, plays an important role in bacterial classification. A number of bacteria exhibit tryptophanase [L-tryptophan indole-lyase (deaminating), EC 4.1.99.1] (89, 210), which forms pyruvate and indole. The formation of

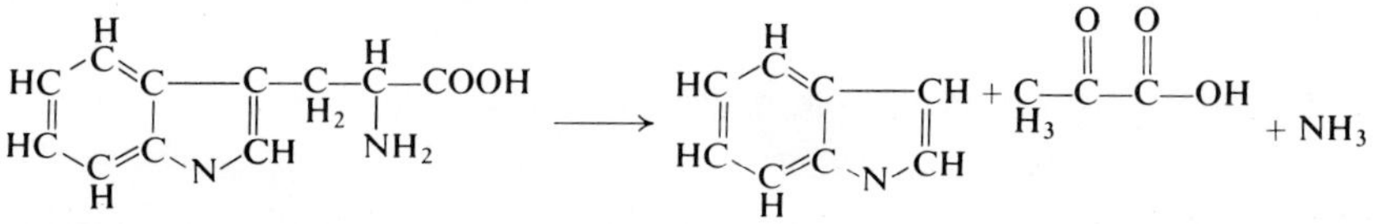

indole is used as an index for tryptophan utilization.

LYSINE METABOLISM. *Clostridium sticklandii* and *Clostridium* SB 4 are able to utilize lysine as their sole carbon and energy source (369, 370, 372)

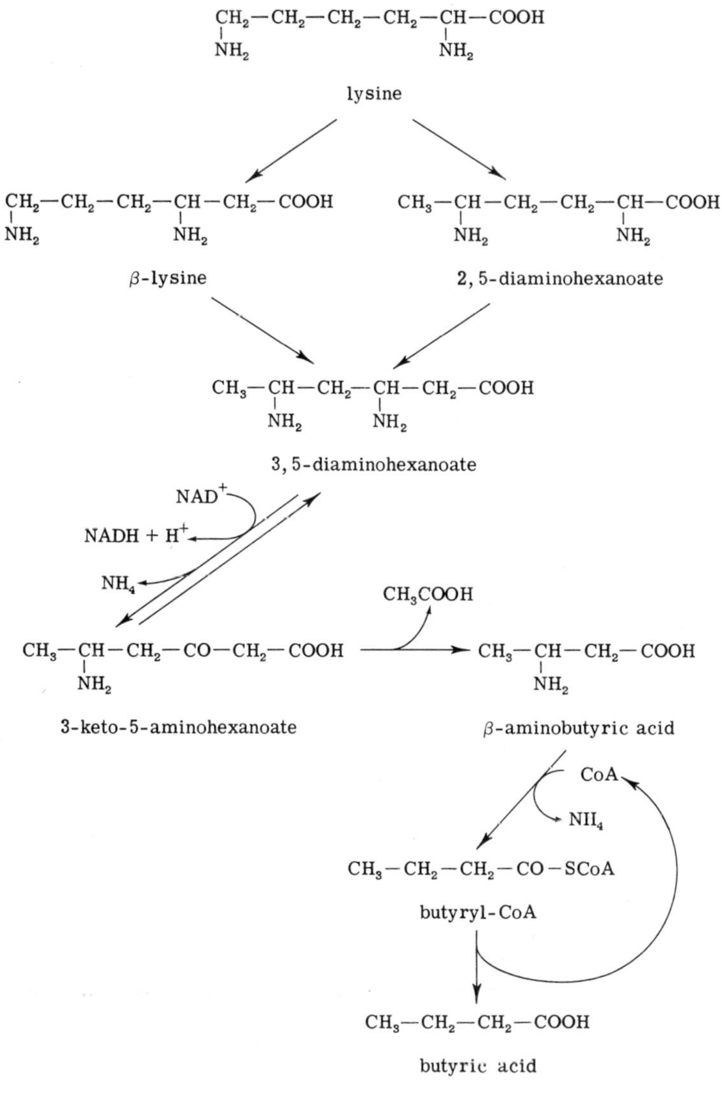
lysine
β-lysine
2, 5-diaminohexanoate
3, 5-diaminohexanoate
NAD+
NADH + H+
NH4
CH3COOH
3-keto-5-aminohexanoate
β-aminobutyric acid
CoA
NH4
butyryl-CoA
butyric acid

Fig. 9.18. Metabolism of lysine by *Clostridium sticklandii* and *Clostridium* SB 4.

(Fig. 9.18). The first attack on lysine is a cobamide-dependent reaction, which shifts the amino group either from carbon 2 to 3, giving 2,5-diaminohexanoate (372), or from position 6 to 5, resulting in the formation of β-lysine (L-β-ε-diaminohexanoate) (69). In this reaction, α-ketoglutarate serves as cofactor, but the mechanism of this requirement is not known. Both intermediates are converted to 3,5-diaminohexanoate (85) and an

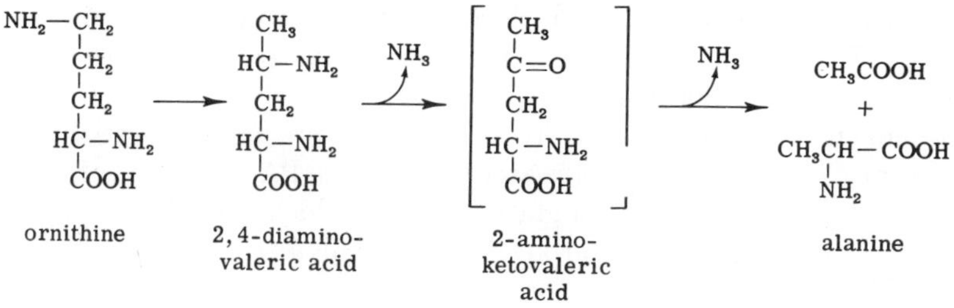
NH₂—CH₂
CH₂
CH₂
HC—NH₂
COOH
ornithine
CH₃
HC—NH₂
CH₂
HC—NH₂
COOH
2,4-diamino-valeric acid
NH₃
CH₃
C=O
CH₂
HC—NH₂
COOH
2-amino-ketovaleric acid
NH₃
CH₃COOH
+
CH₃CH—COOH
NH₂
alanine

Scheme 9.5. Ornithine metabolism of *Clostridium sticklandii.*

oxidative deamination leads to the formation of 3-keto-5-aminohexanoic acid (330). This reaction requires NAD^+. The fact that coenzyme A prevents the accumulation of the keto amino acid suggests that the keto amino acid may undergo a thiolytic cleavage to form acetate and the CoA ester of, possibly, β-aminobutyrate. A deamination would produce butyryl-CoA and a deacylation of butyryl-CoA produces not only the end product butyric acid, but also 1 mole of ATP (330).

ORNITHINE METABOLISM. The utilization of ornithine is similar to the metabolism of lysine, for the first reaction is a cobamide-dependent cleavage between carbons 3 and 4 in *Clostridium sticklandii* (110). All indications are that 2,4-diaminovalerate is the first intermediate in this pathway (111). The pathway shown in Scheme 9.5 could therefore be envisaged, although details are not known.

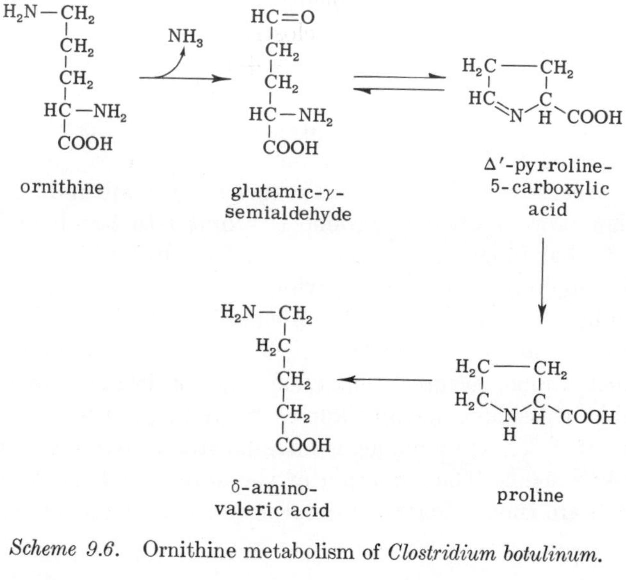
H₂N—CH₂
CH₂
CH₂
HC—NH₂
COOH
ornithine
NH₃
HC=O
CH₂
CH₂
HC—NH₂
COOH
glutamic-γ-semialdehyde
H₂C—CH₂
HC
N
C
H
COOH
Δ′-pyrroline-5-carboxylic acid
H₂N—CH₂
H₂C
CH₂
CH₂
COOH
δ-amino-valeric acid
H₂C—CH₂
H₂C
N
H
C
H
COOH
proline

Scheme 9.6. Ornithine metabolism of *Clostridium botulinum.*

In *Clostridium botulinum*, the degradative pathway appears to be entirely different, as proline and α-aminovalerate are prospective intermediates (70, 71) (Scheme 9.6).

An enzyme system that converts ornithine to proline was partially purified from *Clostridium botulinum* and *Clostridium* PA 3670. Ornithine was found to be converted first to glutamic-γ-semialdehyde, which is in equilibrium with Δ′-pyrroline-5-carboxylic acid. This reaction is catalyzed by an ornithine-γ-transaminase (ornithine-oxo acid aminotransferase, EC 2.6.1.13?) (345). The latter compound is reduced to proline, with reduced NAD as electron donor (298). A split in the proline ring would then lead to the δ-aminovaleric acid. Whether or not this system also functions in *Clostridium sticklandii* has not been fully elucidated.

Metabolism of Pairs of Amino Acids

Many clostridia, growing on protein hydrolyzates or amino acid mixtures, appear to obtain most of their energy by a coupled oxidation–reduction reaction between suitable amino acids, or amino acids and nonnitrogenous compounds. The coupled decomposition of amino acids is commonly referred to as the "Stickland reaction." The characteristic feature of this Stickland reaction is that single amino acids are not decomposed appreciably, but appropriate pairs are decomposed very rapidly. One member of the pair is oxidized while the other is reduced.

Evidence for this formulation was initially obtained by studying the interaction between amino acids and suitable redox dyes in the presence of cell suspensions of *Clostridium sporogenes*. Bacteria using this Stickland reaction are mainly the proteolytic clostridia: *C. acetobutylicum*, *C. aerofoetidum*, *C. bifermentans*, *C. botulinum* (A + B), *C. butyricum*, *C. caproicum*, *C. histolyticum*, *C. sporogenes*, *C. sticklandii*, etc.

The mechanism of individual oxidations and reductions varies according to the individual acid involved. Alanine, for example, is oxidized in a deamination process whereby glycine is reduced to acetate and also in a non-deamination process whereby proline is reduced to δ-aminovalerate. All amino acids that function as reductants can be divided into three groups (287): (1) aliphatic amino acids, which are more reduced than α-keto acids (alanine, leucine, isoleucine, norleucine, and valine); (2) aliphatic amino acids in the same oxidation state as α-keto acids (serine, threonine, cystine, methionine, arginine, citrulline, and ornithine); and (3) other amino acids (histidine, phenylalanine, tryptophan, tyrosine, aspartate, and glutamate). These amino acids have mostly an oxidation state below that of α-keto acids. The oxidation of the amino acids is via the α-keto acids, which are then oxidatively decarboxylated. The following sequence is well established:

$$\underset{\substack{|\\ NH_2}}{R\text{—}CH\text{—}COOH} + H_2O \longrightarrow NH_3 + \underset{\substack{\|\\ O}}{R\text{—}C\text{—}COOH} \longrightarrow R\text{—}COOH + CO_2$$

The most difficult reaction mechanisms, which are not yet completely resolved, are involved in the first group of amino acids. It has been suggested that they undergo either a direct oxidative deamination, which would be similar to glutamate dehydrogenase reactions, or a transamination followed by an oxidative deamination of the new amino acid (Fig. 9.19). This reaction cycle would involve a transaminase and glutamate dehydrogenase (EC 1.4.1.2). In *Clostridium saccharobutyricum*, a species that does not catalyze Stickland reactions, the oxidation of alanine, valine, and leucine requires the presence of α-oxoglutarate.

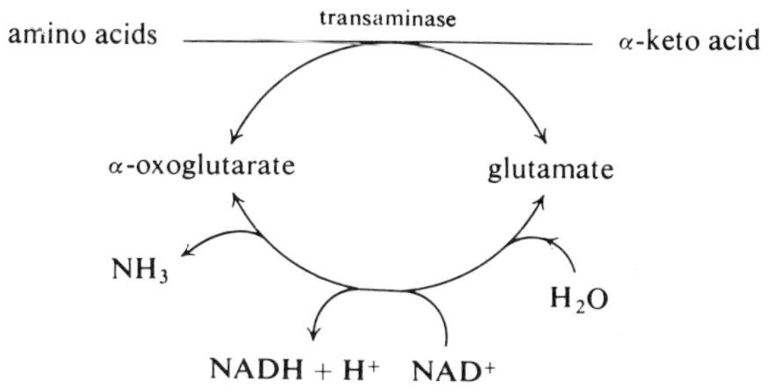
amino acids
transaminase
α-keto acid
α-oxoglutarate
glutamate
NH_3
H_2O
$NADH + H^+$
NAD^+

Fig. 9.19. Transamination reactions in microorganisms.

Working with extracts of *Clostridum sporogenes*, Stickland himself (366a) found the coupled reaction of glycine–alanine and proline–alanine and established the overall mechanism as follows:

$$\underset{\substack{|\\ NH_3^+}}{R\text{—}CH\text{—}COO^-} + 2\ \underset{\substack{|\\ NH_3^+}}{R'\text{—}CH\text{—}COO^-} + H_2O \longrightarrow$$

$$R\text{—}COO^- + CO_2 + 2\ R'CH_2COOH + 3\ NH_4$$

The two acids formed may then be metabolized further (240). Kocholaty and Hoogerheide (228) demonstrated the reduction of proline and glycine by hydrogen in cell suspensions.

The present evidence suggests that the Stickland reaction is composed of a number of steps. The first step involves NAD^+ as primary hydrogen acceptor and an amino-acid dehydrogenase system. The second step involves the reoxidation of $NADH_2$ by the acceptor amino acid together with an amino-acid reductase system:

$$\underset{\displaystyle NH_3^+}{R\text{—}CH\text{—}COOH} + NAD^+ + H_2O \longrightarrow \underset{\displaystyle O}{\overset{}{R\text{—}\underset{\|}{C}\text{—}COO^-}} + NH_4^+ + NADH + H^+$$

$$\underset{\displaystyle NH_3^+}{R'\text{—}CH\text{—}COO^-} + NADH + H^+ \longrightarrow NAD^+ + R'CH_2\text{—}COO^- + NH_4^-$$

This latter reaction could be quite independent of the oxidation step in the first reaction. The oxidation step, in contrast, could be a multienzymatic step, as outlined above, also involving a transaminase system. The overall mechanism for the Stickland reaction may therefore be as shown in Fig. 9.20. The numbers in parentheses represent the action of the following enzymes or enzyme systems: (1) amino-acid dehydrogenase system; (2) 2-oxodehydrogenase with lipoamide, TPP, and HSCoA requirements; and (3) phosphate acetyltransferase.

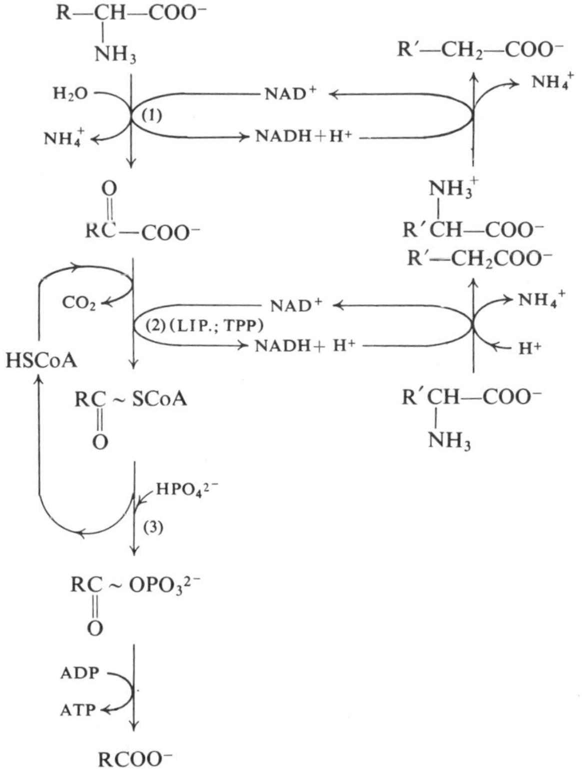
R—CH—COO⁻
NH₃
R′—CH₂—COO⁻
NH₄⁺
H₂O
NAD⁺
(1)
NH₄⁺
NADH+H⁺
O
RC—COO⁻
NH₃⁺
R′CH—COO⁻
R′—CH₂COO⁻
CO₂
NAD⁺
NH₄⁺
(2)(LIP.; TPP)
NADH+ H⁺
H⁺
HSCoA
RC ~ SCoA
O
R′CH—COO⁻
NH₃
HPO₄²⁻
(3)
RC ~ OPO₃²⁻
O
ADP
ATP
RCOO⁻

Fig. 9.20. Stickland reaction. See text for explanation.

During the Stickland reaction three main enzyme systems are involved: (1) an L-amino-acid dehydrogenase system, (2) an α-ketoacid dehydrogenase system, and (3) an amino-acid reductase system of a hydrogenase type. The hydrogen transfer proceeds anaerobically to an enzyme system whereby the acceptor amino acid is reduced.

Metabolism of a Single Amino Acid Together with a Keto Acid

ALANINE METABOLISM. *Clostridium propionicum* (145) uses the Stickland reaction to metabolize β-alanine to propionic acid, with pyruvate playing a key role in the catalytic functions of the two cycles (Fig. 9.21). The amino

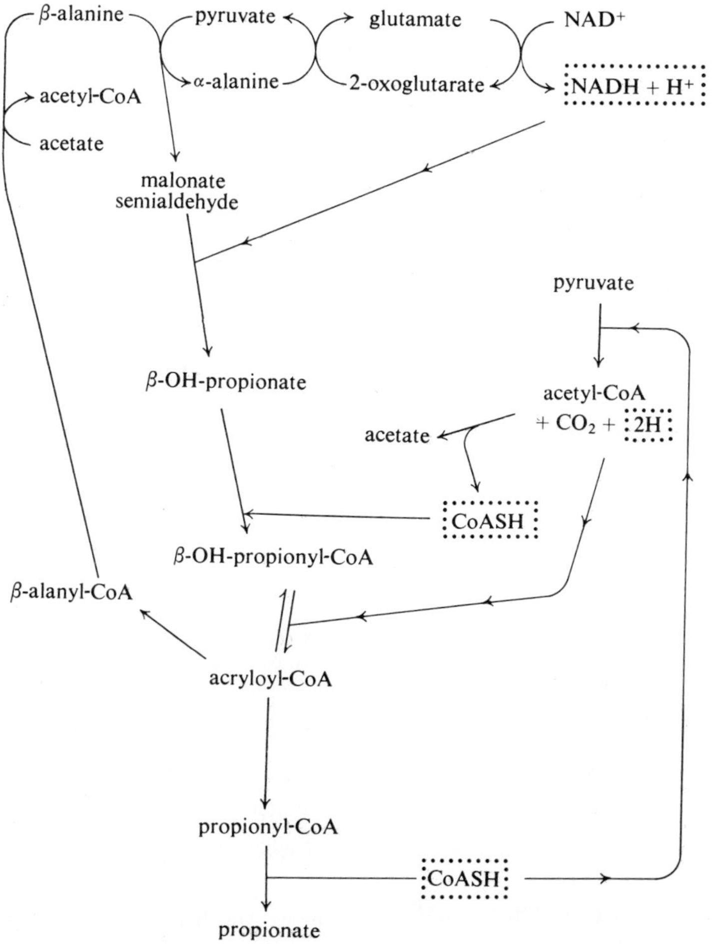
β-alanine
pyruvate
glutamate
NAD+
acetyl-CoA
α-alanine
2-oxoglutarate
NADH + H+
acetate
malonate
semialdehyde
pyruvate
β-OH-propionate
acetyl-CoA
acetate
+ CO2 + 2H
CoASH
β-OH-propionyl-CoA
β-alanyl-CoA
acryloyl-CoA
propionyl-CoA
CoASH
propionate

Fig. 9.21. Alanine metabolism by *Clostridium propionicum.*

group of β-alanine is first transferred to pyruvate for the formation of α-alanine and malonate semialdehyde. This transamination reaction is the

$$\begin{array}{c} H_2N-CH_2 \\ | \\ CH_2 \\ | \\ COOH \end{array} + \begin{array}{c} CH_3 \\ | \\ C=O \\ | \\ COOH \end{array} \longrightarrow \begin{array}{c} CH_3 \\ | \\ H_2N-CH \\ | \\ COOH \end{array} + \begin{array}{c} OCH \\ | \\ CH_2 \\ | \\ COOH \end{array}$$

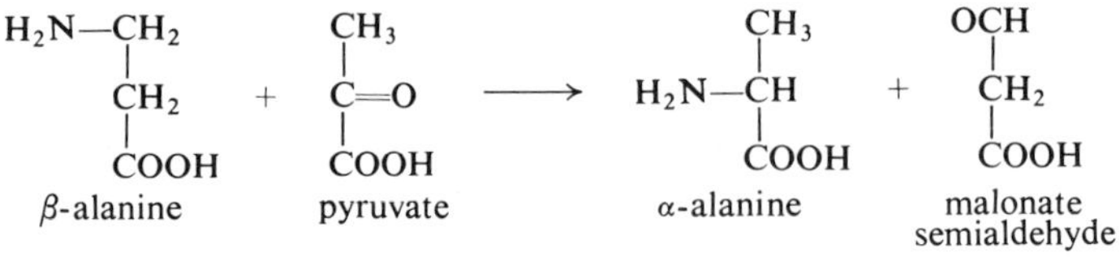

β-alanine   pyruvate   α-alanine   malonate semialdehyde

function of β-alanine-pyruvate aminotransferase (L-alanine:malonate-semialdehyde aminotransferase, EC 2.6.1.18). Both starter substances can be regenerated by two different systems.

1. *Regeneration of pyruvate.* The first transamination is followed by a second one, whereby the amino group from α-alanine is transferred to 2-oxoglutarate for the formation of glutamate and pyruvate. This trans-

$$\begin{array}{c} CH_3 \\ | \\ H_2N-CH \\ | \\ COOH \end{array} + \begin{array}{c} COOH \\ | \\ C=O \\ | \\ CH_2 \\ | \\ CH_2 \\ | \\ COOH \end{array} \longrightarrow \begin{array}{c} COOH \\ | \\ H_2N-CH \\ | \\ CH_2 \\ | \\ CH_2 \\ | \\ COOH \end{array} + \begin{array}{c} CH_3 \\ | \\ C=O \\ | \\ COOH \end{array}$$

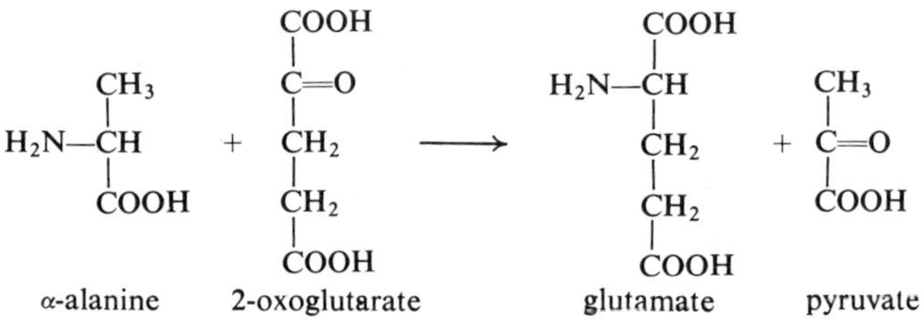

α-alanine   2-oxoglutarate   glutamate   pyruvate

amination requires an α-alanine aminotransferase (L-alanine:2-oxoglutarate aminotransferase, EC 2.6.1.2). The final step of this first cycle is brought about by glutamate dehydrogenase [L-glutamate:NAD oxidoreductase (deaminating), EC 1.4.1.2], which acts on glutamate in the

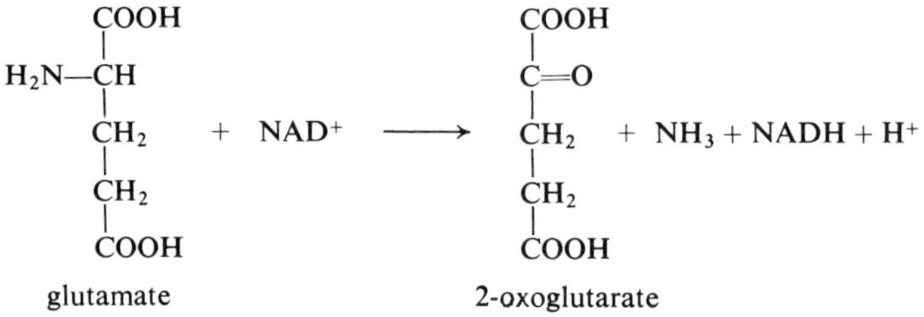

$$\begin{array}{c} COOH \\ | \\ H_2N-CH \\ | \\ CH_2 \\ | \\ CH_2 \\ | \\ COOH \end{array} + NAD^+ \longrightarrow \begin{array}{c} COOH \\ | \\ C=O \\ | \\ CH_2 \\ | \\ CH_2 \\ | \\ COOH \end{array} + NH_3 + NADH + H^+$$

glutamate   2-oxoglutarate

presence of NAD^+. This reaction completes the first of the two cycles in β-alanine metabolism.

2. Regeneration of β-alanine and the formation of propionic acid. The reduced NAD produced in the last reaction of the first cycle is used to reduce, in a NAD-linked process, malonate semialdehyde to β-hydroxypro-

$$\begin{array}{c} \text{OCH} \\ | \\ \text{CH}_2 \\ | \\ \text{COOH} \end{array} + \text{NADH} + \text{H}^+ \longrightarrow \begin{array}{c} \text{CH}_2\text{OH} \\ | \\ \text{CH}_2 \\ | \\ \text{COOH} \end{array} + \text{NAD}^+$$

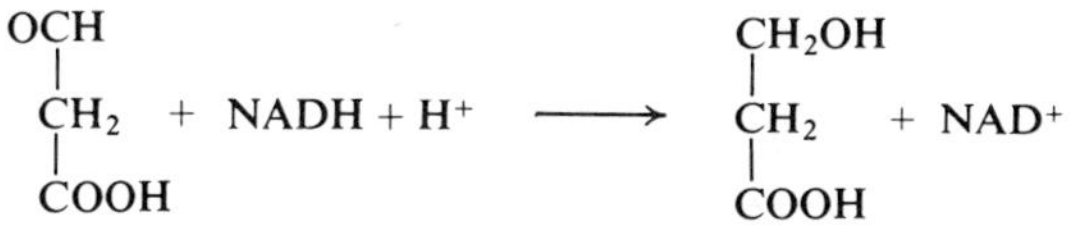

malonate semialdehyde   β-hydroxypropionate

pionate. The enzyme responsible for this reaction may be 3-hydroxypropionate dehydrogenase (3-hydroxypropionate:NAD oxidoreductase, EC 1.1.1.59), which reoxidizes the reduced NAD. β-Hydroxypropionate now takes up CoA and forms its CoA ester. The required coenzyme for this re-

$$\begin{array}{c} \text{CH}_2\text{OH} \\ | \\ \text{CH}_2 \\ | \\ \text{COOH} \end{array} + \text{HSCoA} \longrightarrow \begin{array}{c} \text{CH}_2\text{OH} \\ | \\ \text{CH}_2 \\ | \\ \text{C}=\text{O} \\ | \\ \text{SCoA} \end{array}$$

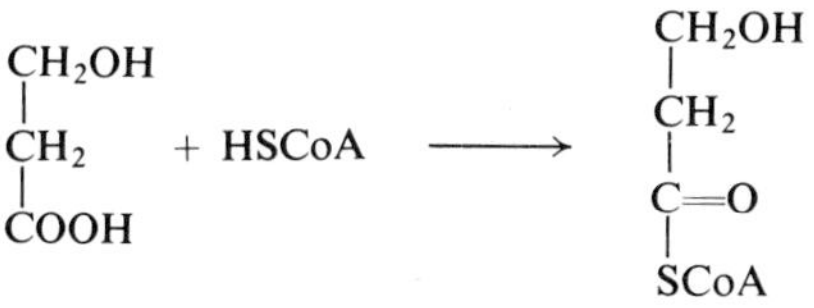

β-hydroxypropionate   β-hydroxypropionyl-CoA

action comes from the pyruvate degradation to acetate by this organism:

$$\text{CH}_3\text{COCOOH} \rightarrow \text{CH}_3\text{COSCoA} + \text{CO}_2 + 2\text{H}^+ \rightarrow \text{CH}_3\text{COOH}$$

The liberated hydrogen in this side reaction is used in the next reduction step, whereby β-hydroxypropionyl-CoA is reduced to acryloyl-CoA. Acryloyl-CoA itself can go two different ways. It can form β-alanyl-CoA

$$\begin{array}{c} \text{CH}_2\text{OH} \\ | \\ \text{CH}_2 \\ | \\ \text{C}=\text{O} \\ | \\ \text{SCoA} \end{array} + 2\text{H}^+ \rightleftharpoons \begin{array}{c} \text{CH}_2 \\ \| \\ \text{CH} \\ | \\ \text{C}=\text{O} \\ | \\ \text{SCoA} \end{array}$$

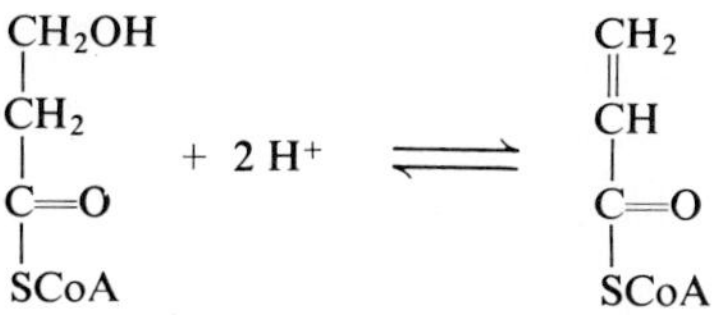

with the help of an aminase (possibly β-alanyl-CoA ammonia-lyase, EC 4.3.1.6) and regenerate β-alanine with an acid and a CoA-transferase, or it can be converted into propionic acid via propionyl-CoA. The conversion of acryloyl-CoA to propionate is suggested to be similar to that described for the propionic acid bacteria. Pyruvate is necessary in this reaction sequence in two steps. It has a catalytic function in the deamination of β-alanine and later in the dissimilation of β-hydroxybutyrate.

In *Clostridium kluyveri* (400), it appears that β-hydroxypropionate is not formed. Malonyl-CoA is, however, and is converted via malonyl semialdehyde CoA to β-hydroxypropionyl-CoA. There seem to be two differences compared with *C. propionicum*: (*1*) the level at which CoA enters the cycle, and (*2*) the dependence on $NADP^+$ instead of NAD^+.

γ-AMINOBUTYRIC ACID METABOLISM. A degradative system very similar to that of alanine exists in the metabolism of ω-amino acids, such as γ-aminobutyrate, by *Clostridium aminobutyricum* (Fig. 9.22). The initial steps

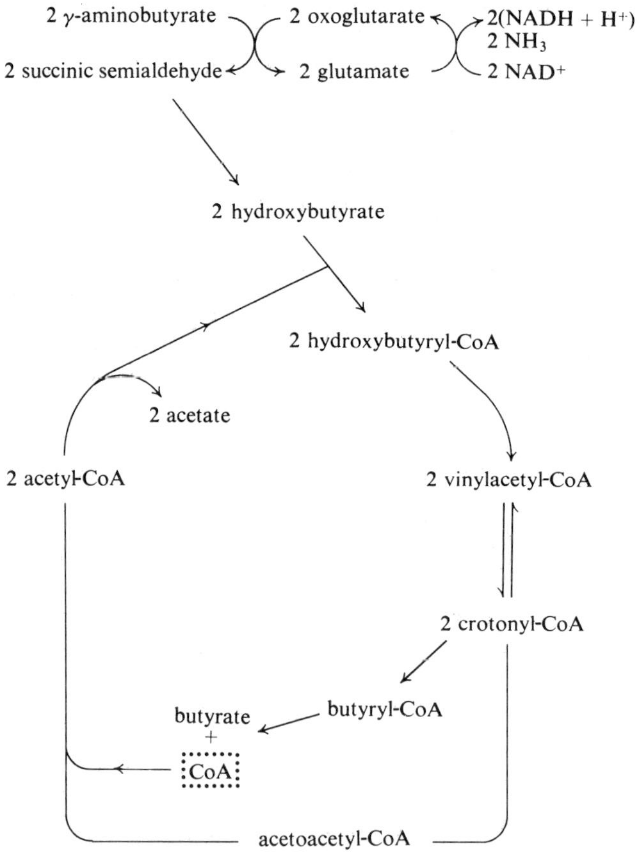
2 γ-aminobutyrate
2 oxoglutarate
2(NADH + H+)
2 NH3
2 succinic semialdehyde
2 glutamate
2 NAD+
2 hydroxybutyrate
2 hydroxybutyryl-CoA
2 acetate
2 acetyl-CoA
2 vinylacetyl-CoA
2 crotonyl-CoA
butyrate
+
CoA
butyryl-CoA
acetoacetyl-CoA

Fig. 9.22. Aminobutyric acid metabolism by *Clostridium aminobutyricum.*

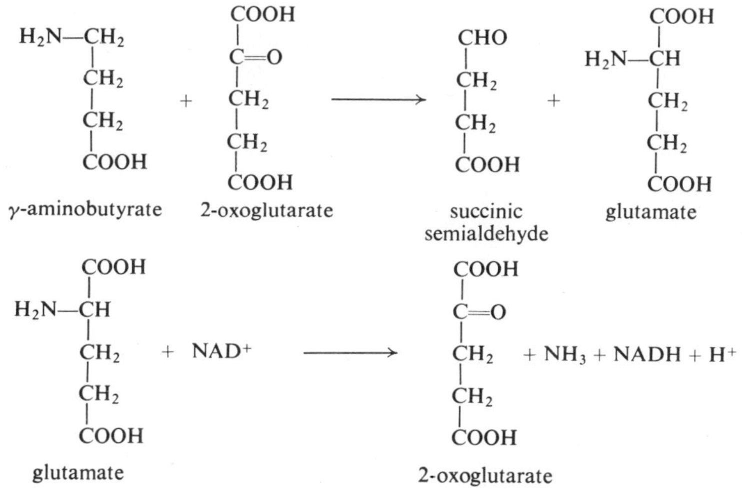
γ-aminobutyrate
2-oxoglutarate
succinic semialdehyde
glutamate
glutamate
2-oxoglutarate

in the catabolism of γ-aminobutyrate by *Clostridium aminobutyricum* are balanced oxidation–reduction reactions whereby γ-aminobutyrate is converted to succinic semialdehyde. These reactions are catalyzed by an γ-aminobutyrate aminotransferase (EC 2.6.1.19) (161) and glutamic dehydrogenase (EC 1.4.1.2). The formed succinic semialdehyde is then reduced to γ-hydroxybutyrate by γ-hydroxybutyrate dehydrogenase and NADH + H^+. In this process NADH + H^+ is reoxidized to NAD^+. The

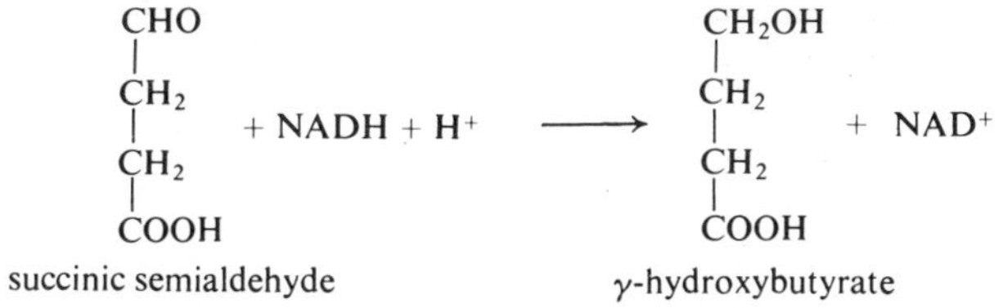
succinic semialdehyde
γ-hydroxybutyrate

second part of the pathway starts with a CoA-transferase reaction, which converts γ-hydroxybutyrate to γ-hydroxybutyryl-CoA and acetyl-CoA

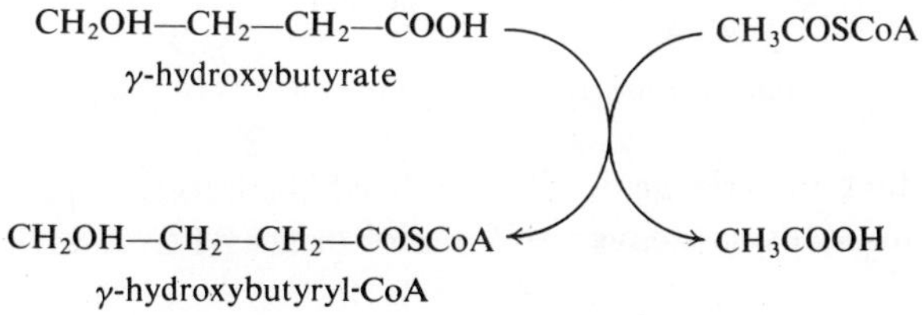
γ-hydroxybutyrate
γ-hydroxybutyryl-CoA

to acetate. The CoA-transferase has been given the name "acetyl CoA: γ-hydroxybutyrate CoA-transferase" (162). The unsaturated compound vinylacetyl-CoA is formed under the action of a dehydrase, and vinylacetyl-

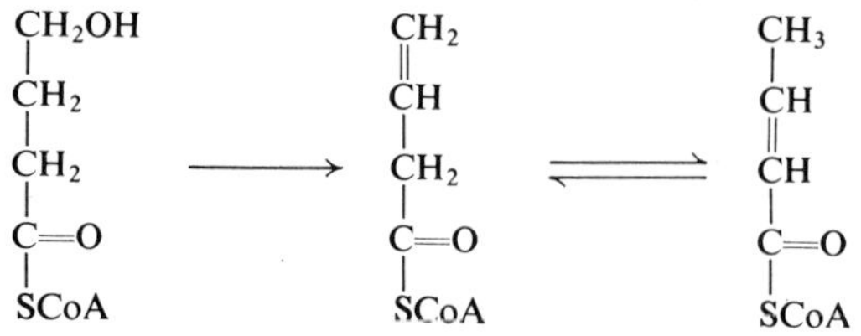

γ-hydroxybutyryl-CoA vinylacetyl-CoA crotonyl-CoA

CoA is held in equilibrium with crotonyl-CoA by an isomerase. The intermediate crotonyl-CoA, in analogy with fatty acid synthesis, undergoes a dismutation whereby 1 mole is reduced to butyryl-CoA and the second

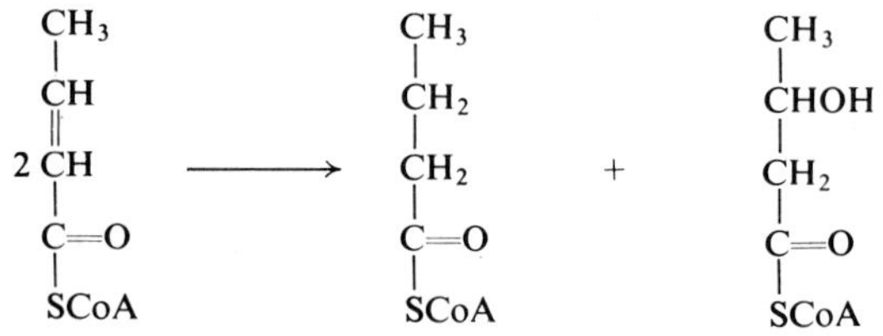

crotonyl-CoA butyryl-CoA acetoacetyl-CoA

mole is oxidized to acetoacetyl-CoA. A NAD^+-linked butyryl-CoA dehydrogenase [butyryl-CoA (acceptor) oxidoreductase, EC 1.3.99.2] is responsible for the butyryl-CoA formation, which in turn is converted to

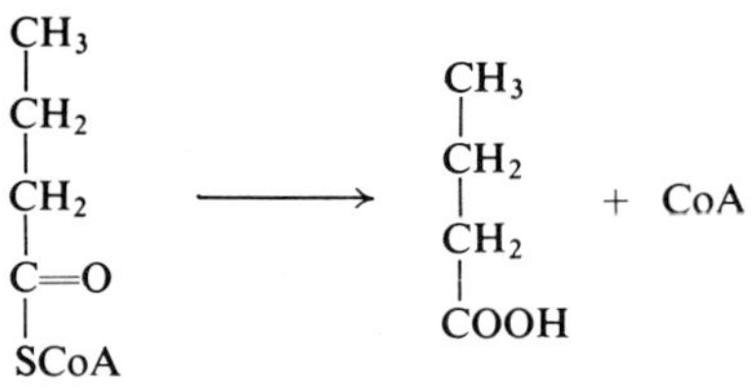

butyryl-CoA butyrate

the end product butyric acid. The oxidized dismutation product acetoacetyl-CoA, however, is cleaved to 2 moles of acetyl-CoA with an acetyl-

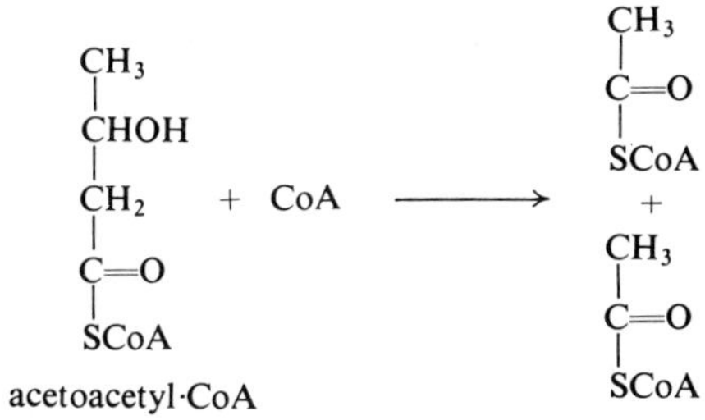

acetoacetyl·CoA

CoA acetyltransferase (acetyl-CoA:acetyl-CoA *C*-acetyltransferase, EC 2.3.1.9). Acetyl-CoA reenters the cycle in the first CoA-transferase step of this system. The net result of energy is 1 mole of ATP per 2 moles of amino acids fermented (163). An additional mole comes from the reduction of crotonyl-CoA by NADH + H^+ to butyryl-CoA, which is sufficiently exergonic to be coupled to ATP synthesis.

References

1. Adams, G. A., and Stanier, R. Y. (1945). Production and properties of 2,3-butanediol. III. Studies on the biochemistry of carbohydrate fermentation by *Aerobacillus polymyxa*. *Can. J. Res., Sect. B* **23,** 1.

1a. Afting, E. G., and Ruppert, D. (1973). Yeast phosphofructokinase. III. Comparison of the allosteric properties of PFKs and PFKd. Archs. Biochem. Biophys. **156,** 720.

2. Ahlfons, C. E., and Mansour, T. E. (1969). Studies on heart phosphofructokinase. Desensitization of the enzyme to adenosine triphosphate inhibition. *J. Biol. Chem.* **244,** 1247.

3. Ahrens, J., and Schlegel, H. G. (1966). Zur Regulation der NAD-abhängigen Hydrogenase-Aktivität. *Arch. Mikrobiol.* **55,** 257.

4. Aiba, S., Humphrey, A. E., and Millis, N. F. (1965). "Biochemical Engineering." Univ. of Tokyo Press, Tokyo.

5. Allen, S. H. G., Kellermeyer, R. W., Stjernholm, R. C., Jacobson, B. E., and Wood, H. G. (1963). The isolation, purification, and properties of methylmalonyl racemase. *J. Biol. Chem.* **238,** *1637.*

6. Allen, S. H. G., Jacobson, B. E., and Stjernholm, R. (1964). Biocytin as a constituent of methylmalonyl-oxaloacetate transcarboxylase and propionyl-CoA carboxylase of bacterial origin. *Arch. Biochem. Biophys.* **105,** 494.

7. Allen, S. H. G., Kellermeyer, R. W., Stjernholm, R., and Wood, H. G. (1964). Purification and properties of enzymes involved in the propionic acid fermentation. *J. Bacteriol.* **87,** 171.

8. Allen, S. H. G. (1966). The isolation and characterization of malate-lactate transhydrogenase from *Micrococcus lactilyticus*. *J. Biol. Chem.* **241,** 5266.

9. Allison, M. J. (1969). Biosynthesis of amino acids by ruminal microorganisms. *J. Anim. Sci.* **29,** 797.

10. Allison, M. J., and Robinson, I. M. (1970). Biosynthesis of α-ketoglutarate by

the reductive carboxylation of succinate in *Bacteroides ruminicola*. *J. Bacteriol.* **104,** 50.

10a. Amarasingham, C. R., and Davis, B. D. (1965). Regulation of α-ketoglutarate dehydrogenase formation in *Escherichia coli*. *J. Biol. Chem.* **240,** 3664.

11. Anders, R. F., Hogg, D. M., and Jago, G. R. (1970). Formation of hydrogen peroxide by group N streptococci and its effect on their growth and metabolism. *Appl. Microbiol.* **19,** 608.

12. Anders, R. F., Jonas, H. A., and Jago, G. R. (1970). A survey of the lactate dehydrogenase activities in group N streptococci. *Aust. J. Dairy Technol.* **25,** 73.

13. Anderson, R. L., and Ordal, E. J. (1961). *Cytophaga succinicans* sp. n., a facultatively anaerobic, aquatic myxobacterium. *J. Bacteriol.* **81,** 130.

13a. Andreesen, J. R., Schaupp, A., Neurauter, C., Brown, A., and Ljungdahl, L. G. (1973). Fermentation of glucose, fructose and xylose by *Clostridium thermoaceticum*. Effects of metals on growth yield, enzymes, and the synthesis of acetate from CO_2. *J. Bacteriol.* **114,** 743.

13b. Andreesen, J. R., and Ljungdahl, L. G. (1973). Formate dehydrogenase of *Clostridium thermoaceticum:* Incorporation of Selenium-75, and the effects of selenite, molybdate, and tungstate on the enzyme. *J. Bacteriol.* **116,** 867.

14. Atkinson, D. E., and Walton, G. M. (1965). Kinetics of regulatory enzymes. *Escherichia coli* phosphofructokinase. *J. Biol. Chem.* **240,** 757.

15. Atkinson, D. E. (1966). Regulation of enzyme activity. *Annu. Rev. Biochem.* **35,** 85.

16. Atkinson, D. E. (1968). The energy change of the adenylate pool as a regulatory parameter. Interaction with feedback modifiers. *Biochemistry* **7,** 4030.

17. Atzpodien, W., and Bode, H. (1970). Purification and regulatory properties of ATP-sensitive phosphofructokinase from yeast. *Eur. J. Biochem.* **12,** 6. [as cited by Holzer (187)].

18. Atzpodien, W., Gancedo, J. M., Hagmaier, K., and Holzer, H. (1970). Desensitization of yeast phosphofructokinase to ATP inhibition by a protein fraction from yeast. *Eur. J. Biochem.* **12,** 126. [as cited by Holzer (187)].

19. Baginski, M. L., and Huennekens, F. M. (1967). Further studies on the electron proteins involved in the oxidative decarboxylation of glycine. *Arch. Biochem. Biophys.* **120,** 703.

20. Bailey, E., Stirpe, F., and Taylor, C. B. (1968). Regulation of liver pyruvate kinase. *Biochem. J.* **108,** 427.

21 Baldwin, R. L., Wood, W. A., and Emergy, R. S. (1965). Lactate metabolism by *Peptostreptococcus elsdenii*. Evidence for lactyl-coenzyme A dehydrase. *Biochim. Biophys. Acta* **97,** 202.

22. Banchop, T., and Elsden, S. R. (1960). The growth of microorganisms in relation to their energy supply. *J. Gen. Microbiol.* **23,** 457.

23. Barile, M. F., Schimke, R. T., and Riggs, D. B. (1966). Presence of the arginine dihydrolase pathway in *Mycoplasma*. *J. Bacteriol.* **91,** 189.

24. Barker, H. A., and Lipmann, F. (1944). On lactic acid metabolism in propionic acid bacteria and the problem of oxidoreduction in the system fatty hydroxyketo acid. *Arch. Biochem.* **4,** 361 [as cited by Johns (198)].

25. Barker, H. A. (1956). "Bacterial Fermentations," CIBA Lectures in Microbial Chemistry. Wiley, New York.

26. Barker, H. A. (1961). Fermentation of nitrogenous organic compounds. *In* "The Bacteria" (I. C. Gunsalus and R. Y. Stanier, eds.), Vol. 2, p. 151. Academic Press, New York.

27. Barker, H. A. (1967). Citramalate lyase of *Clostridium tetanomorphum. Arch. Mikrobiol.* **59,** 4.

28. Barker, J., Khan, M. A. A., and Solomos, T. (1964). Mechanism of the Pasteur effect. *Nature* (*London*) **201,** 1126.

29. Beck, W. S., and Ochoa, S. (1958). Metabolism of propionic acid in animal tissues. *J. Biol. Chem.* **232,** 931.

30. Bennett, R., Taylor, D. R., and Hurst, A. (1966). D- and L-lactate dehydrogenases in *Escherichia coli. Biochim. Biophys. Acta* **118,** 512.

31. Bertho, A., and Gluck, H. (1931). *Naturwissenschaften* **19,** 88 [as cited by Dolin (105)].

32. Bieberstein, E. L., and Gills, M. (1961). Catalase activity of *Haemophilus* species grown with graded amounts of hemin. *J. Bacteriol.* **81,** 380.

33. Biggins, D. R., and Dilworth, M. J. (1968). Control of pyruvate phosphoroclastic activity in extracts of *Clostridium pasteurianum* by ADP and acetyl phosphate. *Biochim. Biophys. Acta* **156,** 285.

34. Birnbaum, J., and Lichstein, H. C. (1965). Conversion of D-biotin vitamers by *Lactobacillus arabinosus. J. Bacteriol.* **89,** 1035.

35. Birnbaum, J., and Lichstein, H. C. (1966). Metabolism of biotin and analogs of biotin by microorganisms. II. Further studies on the conversion of D-biotin to biotin vitamers by *Lactobacillus plantarum. J. Bacteriol.* **92,** 913.

36. Birnbaum, J., and Lichstein, H. C. (1966). Metabolism of biotin and analogs of biotin by microorganisms. III. Degradation of oxybiotin and desthiobiotin by *Lactobacillus plantarum. J. Bacteriol.* **92,** 920.

37. Birnbaum, J., and Lichstein, H. C. (1966). Metabolism of biotin and analogs of biotin by microorganisms. IV. Degradation of biotin, oxybiotin, and desthiobiotin by *Lactobacillus casei. J. Bacteriol.* **92,** 925.

38. Branen, A. L., and Keenan, T. W. (1970). Diacetyl reductase of *Lactobacillus casei. Can. J. Microbiol.* **16,** 947.

39. Breed, R. S., Murray, E. G. D., and Smith, N. R. (1957). "Bergey's Manual of Determinative Bacteriology," 7th ed. Williams & Wilkins, Baltimore, Maryland.

40. Brewer, C. R., Werkman, C. H., Michaelian, M. B., and Hammer, B. W. (1968). Effect of aeration under pressure on diacetyl production in butter cultures. *Iowa, Agr. Exp. Sta., Res. Bull.* **233,** 42.

40a. Breznak, J. A., and Canale-Parola, E. (1972). Metabolism of *Spirochaeta aurantia.* I. Anaerobic energy-yielding pathways. *Arch. Mikrobiol.* **83,** 261.

41. Brown, C. M., and Rose, A. H. (1969). Fatty acid composition of *Candida utilis* as affected by growth temperature and dissolved oxygen tension. *J. Bacteriol.* **99,** 371.

42. Brown, C. M., and Johnson, B. (1970). Influence of the concentration of glucose and galactose on the physiology of *Saccharomyces cerevisiae* in continuous culture. *J. Gen. Microbiol.* **64,** 279.

43. Brown, C. M., and Johnson, B. (1971). Influence of oxygen tension on the physiology of *Saccharomyces cerevisiae* in continuous culture. *Antonie van Leeuwenhoek; J. Microbiol Serol.* **37,** 477.

44. Brown, J. P. (1968). Anomalous anaerobic growth yields of *Lactobacillus casei* strain 103 in complex medium. *Appl. Microbiol.* **16,** 805.

45. Brown, J. P., and Vandemark, P. J. (1968). Respiration of *Lactobacillus casei. Can. J. Microbiol.* **14,** 829.

46. Bryan-Jones, D. G., and Whittenbury, R. (1969). Hematin-dependent oxidative phosphorylation in *Streptococcus faecalis. J. Gen. Microbiol.* **58,** 247.

47. Bryant, M. P. (1952). The isolation and characteristics of a spirochaete from the bovine rumen. *J. Bacteriol.* **64,** 325.
48. Bryant, M. P., and Small, N. (1956). Characteristics of two new genera of anaerobic curved rods isolated from the rumen of the cattle. *J. Bacteriol.* **72,** 22.
49. Bryant, M. P., Small, N., Bouma, C., and Chum, H. (1958). *Bacteroides ruminicola* n.sp. and *Succinomonas amylolytica* the new genus and species. *J. Bacteriol.* **76,** 15.
49a. Bryn, K., Hetland, O., and Stormer, F. C. (1971). The reduction of diacetyl and acetoin in *Aerobacter aerogenes*. Evidence for one enzyme catalyzing both reactions. *Eur. J. Biochem.* **18,** 116.
49b. Buckel, W., and Barker, H. A. (1974). Two pathways of glutamate fermentation by anaerobic bacteria. *J. Bacteriol.* **117,** 1248.
50. Buyze, G., van der Hamer, C. J. A., and de Haan, P. G. (1957). Correlation between hexose monophosphate shunt, glycolytic system, and fermentation type in lactobacilli. *Antonie van Leeuwenhoek; J. Microbiol. Serol.* **23,** 345.
50a. Caldwell, D. R., Keeney, M., Barton, J. S., and Kelley, J. F. (1973). Sodium and other inorganic growth requirements of *Bacteroides amylophilus*. *J. Bacteriol.* **114,** 782.
51. Canale-Parola, E. (1970). Biology of the sugar-fermenting sarcinae. *Bacteriol. Rev.* **34,** 82.
52. Carson, S. F., and Ruben, S. (1940). CO_2 assimilation by propionic acid bacteria studied by the use of radioactive carbon. *Proc. Nat. Acad. Sci. U.S.* **26,** 418 [as cited by Johns (198)].
53. Carter, B. L. A., and Bull, A. T. (1969). Studies on fungal growth and intermediary carbon metabolism under steady- and nonsteady-state conditions. *Biotechnol. Bioeng.* **11,** 785.
54. Cavari, B. Z., Avi-Dor, Y., and Grossowicz, N. (1968). Induction by oxygen of respiration and phosphorylation of anaerobically grown *Escherichia coli*. *J. Bacteriol.* **96,** 751.
55. Chance, B., Ghosh, A., Higgins, J. J., and Maitra, P. K. (1964). Cyclic and oscillatory response of metabolic pathways involving chemical feedback and their computer representations. *Ann. N.Y. Acad. Sci.* **115,** 1010.
56. Chapman, A. G., Fall, L., and Atkinson, D. E. (1971). Adenylate energy charge in *Escherichia coli* during growth and starvation. *J. Bacteriol.* **108,** 1072.
57. Chapman, C., and Bartley, W. (1968). The kinetics of enzyme changes in yeast under conditions that cause loss of mitochondria. *Biochem. J.* **107,** 455.
58. Chapman, C., and Bartley, W. (1969). Adenosine phosphates and the control of glycolysis and gluconeogenesis in yeast. *Biochem. J.* **111,** 609.
59. Charpentie, Y. (1954). Contribution à l'étude biochimique des facteurs de l'acidité des vins. Thèse Docteur-Ingeneur, Bordeaux [as cited by Peynaud *et al.* (302)].
60. Chase, T., Jr., and Rabinowitz, J. C. (1968). Role of pyruvate and *S*-adenosyl methionine in activating the pyruvate-formate lyase of *Escherichia coli*. *J. Bacteriol.* **96,** 1065.
61. Chautrenne, H., and Lipmann, F. (1950). Coenzyme A dependence and acetyl donor function of the pyruvate–formate exchange system. *J. Biol. Chem.* **187,** 757.
62. Christensen, M. D., Albury, M. N., and Peterson, C. S. (1958). Variation in the acetic acid–lactic acid ratio among the lactic acid bacteria. *Appl. Microbiol.* **6,** 316.
63. Chuang, L. F., and Collins, E. B. (1968). Biosynthesis of diacetyl in bacteria and yeast. *J. Bacteriol.* **95,** 2083.
64. Clarke, P. H., Holdsworth, M. A., and Lilly, M. D. (1968). Catabolite repression and the induction of amidase synthesis by *Pseudomonas aeruginosa* 8602 in continuous culture. *J. Gen. Microbiol.* **51,** 225.

65. Clayton, R. K. (1960). Protein synthesis in the induced formation of catalase in *Rhodopseudomonas spheroides*. *J. Biol. Chem.* **235,** 405.
66. Cohen-Bazire, G., Cohen, G.-N., and Prevot, A.-R. (1948). Nature et mode d'formation des acides volatiles dans les cultures de quelques bactéries anaérobies protéolytique du groupe de *Clostridium sporogenes*. Formation par réaction de Stickland des acides isobutyrique, isovalérianique, et valérianique optiquement actif. *Ann. Inst. Pasteur, Paris* **75,** 291.
67. Cole, M. A. (1967). The function and control of anaerobic respiratory pigments in facultative bacteria. Ph.D. Thesis, University of Oxford.
68. Collins, E. B., and Bruhn, J. C. (1970). Roles of acetate and pyruvate in the metabolism of *Streptococcus diacetilactis*. *J. Bacteriol.* **103,** 541.
69. Costilow, R. N., Rochovanski, O. M., and Barker, H. A. (1966). Isolation and identification of β-lysine as an intermediate in lysine fermentation. *J. Biol. Chem.* **241,** 1573.
70. Costilow, R. N., and Laycock, L. (1968). Proline as an intermediate in the reductive deamination of ornithine to δ-aminovaleric acid. *J. Bacteriol.* **96,** 1011.
71. Costilow, R. N., and Laycock, L. (1969). Reactions involved in the conversion of ornithine to proline in clostridia. *J. Bacteriol.* **100,** 662.
72. Criddle, R. S., and Schatz, G. (1969). Promitochondria of anaerobically grown yeast. 1. Isolation and biochemical properties. *Biochem. J.* **8,** 322.
73. Cunningham, C. C., and Hager, L. P. (1971). Crystalline pyruvate oxidase from *Escherichia coli*. III. Phospholipid as an allosteric effector for the enzyme. *J. Biol. Chem.* **246,** 1583.
74. Dagley, S., Dawes, E. A., and Morrison, G. A. (1951). The effect of aeration on the growth of *Aerobacter aerogenes* and *Escherichia coli* with reference to the Pasteur mechanism. *J. Bacteriol.* **61,** 433.
75. Dainty, R. H., and Peel, J. L. (1970). Biosynthesis of amino acids in *Clostridium pasteurianum*. *Biochem. J.* **117,** 573.
76. Dainty, R. H. (1970). Purification and properties of threonine aldolase from *Clostridium pasteurianum*. *Biochem. J.* **117,** 585.
77. Damsky, C. H., Nelson, W. M., and Claude, A. (1969). Mitochondria in anaerobically grown, lipid-limited brewer's yeast. *J. Cell. Biol.* **43,** 174.
78. Daniel, R. M., and Erickson, S. K. (1969). The effect of glucose on the electron transport systems of the aerobic bacteria *Azotobacter vinelandii* and *Acetobacter suboxydans*. *Biochim. Biophys. Acta* **180,** 63.
79. Davis, B. D., Kornberg, H. L., Nagle, A., Miller, P., and Mingiolo, E. S. (1959). Formation and functions of succinate in *Escherichia coli*. *Fed. Proc., Fed. Amer. Soc. Exp. Biol.* **18,** 211.
80. Davis, G. H. G., and Hayward, A. C. (1955). The stability of *Lactobacillus* strains. *J. Gen. Microbiol.* **13,** 533.
81. Davis, J. G. (1933). Über Atmung und Gärung von Milchsäurebakterien. *Biochem. Z.* **267,** 357.
82. Davis, J. G. (1960). The Lactobacilli. *Progr. Ind. Microbiol.* **2,** 3.
83. Degn, H., and Harrison, D. E. F. (1968). Theory of oscillations of respiration rate in continuous culture of *Klebsiella aerogenes*. *J. Theor. Biol.* **22,** 238.
84. Deibel, R. H., and Kvetkas, M. J. (1964). Fumarate reduction and its role in the diversion of glucose fermentation by *Streptococcus faecalis*. *J. Bacteriol.* **88,** 858.
85. Dekker, E. E., and Barker, H. A. (1968). Identification and cobamide-dependent formation of 3,5-diaminohexanoic acid, an intermediate in lysine fermentation. *J. Biol. Chem.* **243,** 3232.

86. DeLey, J. (1962). Comparative biochemistry and enzymology in bacterial classification. *Symp. Soc. Gen. Microbiol.* **12,** 190.

87. Delwiche, E. A., Phares, E. F., and Carson, S. F. (1956). Succinic acid decarboxylation system in *Propionibacterium pentosaceum* and *Veillonella gazogenes*. I. Activation decarboxylation and related reactions. *J. Bacteriol.* **71,** 589.

88. DeMoss, R. D. (1954). A triphosphopyridine nucleotide-dependent alcohol dehydrogenase from *Leuconostoc mesenteroides J. Bacteriol.* **68,** 252.

89. DeMoss, R. D., and Moser, K. (1969). Tryptophanase in diverse bacterial species. *J. Bacteriol.* **89,** 167.

90. Dennis, D., and Kaplan, N. O. (1960). D- and L-lactic acid dehydrogenases in *Lactobacillus plantarum*. *J. Biol. Chem.* **235,** 810.

91. DeVleeschanwer, A., De Grotte, J., and Wilsens, A. (1961). Fermentation of glucose-^{14}C by *Lactobacillus brevis*. *Meded. Landbouwhogesch. Opzoekingssta. Staat Gent* **26,** 165 [as cited by *Chem. Abstr.* **60,** 4487 (1964)].

92. DeVries, W., and Stouthamer, A. H. (1969). Factors determining the degree of anaerobiosis of *Bifidobacterium* strains. *Arch. Mikrobiol.* **65,** 275.

93. DeVries, W., Kapteijn, W. M. G., van der Beek, E. G., and Stouthamer, A. H. (1970). Molar growth yields and fermentation balances of *Lactobacillus casei* L 3 in batch cultures and in continuous cultures. *J. Gen. Microbiol.* **63,** 333.

93a. DeVries, W., van Wyck-Kapteyn, W. M. C., and Stouthamer, A. H. (1972). Influence of oxygen on growth, cytochrome synthesis and fermentation pattern in propionic acid bacteria. *J. Gen. Microbiol.* **71,** 515.

93b. DeVries, W., van Wyck-Kapteyn, W. M. C., and Stouthamer, A. H. (1973). Generation of ATP during cytochrome-linked anaerobic electron transport in propionic acid bacteria. *J. Gen. Microbiol.* **76,** 31.

93c. DeVries, W., van Wyck-Kapteyn, W. M. C., and Oosterhuis, S. K. H. (1974). The presence and function of cytochromes in *Selenomonas ruminantium, Anaerovibrio lipolytica* and *Veillonella alcalescens*. *J. Gen. Microbiol.* **81,** 69.

94. Dixon, M., and Webb, E. C. (1964). "Enzymes," 2nd ed. Longmans, Green, New York.

95. Dobrogosz, W. J. (1966). Altered end product patterns and catabolic repression in *Escherichia coli*. *J. Bacteriol.* **91,** 2263.

96. Doelle, H. W. (1971). Nicotinamide adenine dinucleotide-dependent and nicotinamide adenine dinucleotide-independent lactate dehydrogenases in homofermentative and heterofermentative lactic acid bacteria. *J. Bacteriol.* **108,** 1284.

97. Doelle, H. W. (1971). Influence of carboxylic acids on the stereospecific nicotinamide adenine dinucleotide-dependent and nicotinamide adenine dinucleotide-independent lactate dehydrogenases of *Leuconostoc mesenteroides*. *J. Bacteriol.* **108,** 1290.

98. Doelle, H. W., and Manderson, G. J. (1971). Comparative studies of fructose-1,6-diphosphate aldolase from *Escherichia coli* 518 and *Lactobacillus casei* var. *rhamnosus* ATCC 7469. *Antonie van Leeuwenhoek; J. Microbiol. Serol.* **37,** 21.

99. Doelle, H. W. (1972). Kinetic characteristics of phosphofructokinase from *Lactobacillus casei* var *rhamnosus* ATCC 7469 and *Lactobacillus plantarum* ATCC 14917. *Biochim. Biophys. Acta* **258,** 404.

100. Doelle, H. W. (1972). Aerobic and anaerobic phosphofructokinase from *Escherichia coli* K-12. *Proc. Aust. Biochem. Soc.* **5,** 30.

100a. Doelle, H. W., Hollywood, N., and Westwood, A. W. (1974). Effect of glucose concentration on a number of enzymes involved in the aerobic and anaerobic utilization of glucose in turbidostat-cultures of *Escherichia coli*. *Microbios* **9,** 221.

100b. Doelle, H. W. (1975). ATP-sensitive and ATP-insensitive phosphofructokinase in *Escherichia coli* K-12. *Eur. J. Biochem.* **50,** 335.

100c. Doelle, H. W. (1974). Dimeric and tetrameric phosphofructokinase and the Pasteur effect in *Escherichia coli* K-12. *FEBS Lett.* **49,** 220.

101. Dolin, M. I. (1955). The DPNH-oxidizing enzymes of *Streptococcus faecalis.* II. The enzymes utilizing oxygen, cytochrome c, peroxide, and 2,6-dichlorophenol-indophenol or ferricyanide as oxidants. *Arch. Biochem. Biophys.* **55,** 415.

102. Dolin, M. I. (1955). Diacetyl oxidation by *Streptococcus faecalis,* a lipoic acid-dependent reaction. *J. Bacteriol.* **69,** 51.

103. Dolin, M. I. (1957). The *Streptococcus faecalis* oxidase for reduced pyridine nucleotide. III. Isolation and properties of a flavin peroxidase for reduced diphosphopyridine nucleotide. *J. Biol. Chem.* **225,** 557.

104. Dolin, M. I. (1961). Survey of microbial electron transport mechanisms. *In* "The Bacteria" (I. C. Gunsalus and R. Y. Stanier, eds.), Vol. 2, p. 319. Academic Press, New York.

105. Dolin, M. I. (1961). Cytochrome-independent electron transport enzymes of bacteria. *In* "The Bacteria" (I. C. Gunsalus and R. Y. Stanier, eds.), Vol. 2, p. 425. Academic Press, New York.

106. Dolin, M. I. (1969). Kinetics of malic-lactic transhydrogenase. Abortive complex formation with substrates and products. *J. Biol. Chem.* **244,** 5273.

107. Domagk, G. F., and Horecker, B. L. (1958). Pentose fermentation by *Lactobacillus plantarum. J. Biol. Chem.* **233,** 283.

108. Douglas, H. C. (1947). Hydrogen peroxide and the metabolism of *Lactobacillus brevis. J. Bacteriol.* **54,** 272.

109. Dundas, I. E. D., and Halvorson, H. W. (1966). Arginine metabolism in *Halobacterium salinarum,* an obligately halophilic bacterium. *J. Bacteriol.* **91,** 113.

110. Dyer, J. K., and Costilow, R. N. (1968). Fermentation of ornithine by *Clostridium sticklandii. J. Bacteriol.* **96,** 1617.

111. Dyer, J. K., and Costilow, R. N. (1970). 2,4-Diaminovaleric acid an intermediate in the anaerobic oxidation of ornithine by *Clostridium sticklandii. J. Bacteriol.* **101,** 77.

111a. Edwards, P. R., and Ewing, W. H. (1972). "Identification of Enterobacteriaceae." Burgess, Minneapolis, Minnesota.

112. Eggerer, H., Stadtman, E. R., Overath, P., and Lynen, F. (1960). Zum Mechanismus der durch Cobalamin-Coenzym katalysierten Umlagerung von Methylmalonyl-CoA in Succinyl-CoA. *Biochem. Z.* **333,** 1.

113. Ellwood, D. C., and Tempest, D. W. (1967). Influence of growth conditions on the cell wall content and wall composition of *Aerobacter aerogenes. Biochem. J.* **105,** 9P.

113a. Emmanuel, B., and Milligan, L. P. (1973). Reductive carboxylation of acetyl phosphate by cell-free extracts of mixed rumen microorganisms. *J. Gen. Microbiol.* **77,** 537.

114. Englesberg, E., Gibor, A., and Levy, J. B. (1954). Adaptive control of terminal respiration in *Pasteurella pestis. J. Bacteriol.* **68,** 146.

115. Englesberg, E., Levy, J. B., and Gibor, A. (1954). Some enzymatic changes accompanying the shift from anaerobiosis to aerobiosis in *Pasteurella pestis. J. Bacteriol.* **68,** 178.

116. Englesberg, E., and Levy, J. B. (1955). Induced synthesis of tricarboxylic acid cycle enzymes as correlated with the oxidation of acetate and glucose by *Pasteurella pestis. J. Bacteriol.* **69,** 418.

116a. Enoch, H. G., and Lester, R. L. (1972). Effects of molybdate, tungstate, and selenium compounds on formate dehydrogenase and other enzyme systems in *Escherichia coli*. *J. Bacteriol.* **110,** 1032.
117. Erkama, J., Kauppinen, V., and Heino, K. (1968). Oxygen and the microaerophilic bacteria: Growth and peroxidase activity of *Lactobacillus casei*. *Acta Chem. Scand.* **22,** 2166.
118. Faust, P. J., and Vandemark, P. J. (1970). Phosphorylation coupled to NADH oxidation with fumarate in *Streptococcus faecalis* 10 C1. *Arch. Biochem. Biophys.* **137,** 392.
119. Fitz, R. (1878). *Ber. Deut. Chem. Ges.* **11,** 1896 [as cited by Wood and Werkman (434)].
120. Flesch, P., and Holbach, B. (1965). Zum Abbau der L-Äpfelsäure durch Milchsäurebakterien. 1. Über die Malat-abbauenden Enzyme des Bakterium L unter besonderer Berücksichtigung der Oxalessigsäure-Decarboxylase. *Arch. Mikrobiol.* **51,** 401.
121. Flesch, P., and Holbach, B. (1965). Zum Abbau der L-Äpfelsäure durch Milchsäurebakterien. II. Vergleichende Untersuchungen der Malat-abbauenden Enzyme bei verschiedenen Arten von *Lactobacillus plantarum*. *Arch. Mikrobiol.* **52,** 147.
122. Flesch, P., and Holbach, B. (1965). Zum Abbau der L-Äpfelsäure durch Milchsäurebakterien. III. Über die Trennung der Malic Enzym Aktivität von der Oxalessigsäure-Decarboxylase Aktivität bei *Lactobacillus plantarum*. *Arch. Mikrobiol.* **52,** 297.
123. Flesch, P., and Holbach, B. (1967). Zum Abbau der L-Äpfelsäure durch Milchsäurebakterien. IV. Die Aktivität intakter *Lactobacillus plantarum*-Zellen unter besonderer Berücksichtigung der Brenztraubensäure-Decarboxylierung. *Arch. Mikrobiol.* **58,** 63.
124. Flesch, P. (1968). Morphologie, Stoffwechselphysiologie und Characterisierung der Malic-Enzym-Aktivität L-Äpfelsäure-abbauender Bakterien. *Arch. Mikrobiol.* **60,** 285.
125. Flesch, P. (1968). Malic-Enzym Synthese und pH-Adaptation bei L-Äpfelsäure abbauenden Bakterien. *Arch. Mikrobiol.* **64,** 9.
126. Flesch, P. (1969). Über die Malat-Dehydrogenase- und Lactat-Dehydrogenase-Aktivität L-Äpfelsäure-abbauender Bakterien. *Arch. Mikrobiol.* **68,** 259.
127. Forrest, W. W., and Walker, D. J. (1965). Control of glycolysis in washed suspensions of *Streptococcus faecalis*. *Nature (London)* **207,** 46.
128. Foulbert, E. L., and Douglas, H. C. (1948). Studies on the anaerobic Micrococci. II. The fermentation of lactate by *Micrococcus lactilyticus*. *J. Bacteriol.* **56,** 35.
129. Fraenkel, D. G., and Horecker, B. L. (1965). Fructose-1,6-diphosphatase and acid hexose phosphatase of *Escherichia coli*. *J. Bacteriol.* **90,** 837.
129a. Fraenkel, D. G., Kotlarz, D., and Buc, H. (1973). Two fructose-6-phosphate kinase activities in *Escherichia coli*. *J. Biol. Chem.* **248,** 4865.
130. Fredericks, W. W., and Stadtman, E. R. (1965). The role of ferredoxin in the hydrogenase system from *Clostridium kluyveri*. *J. Biol. Chem.* **240,** 4065.
131. Frieden, C. (1970). Kinetic aspects of regulation of metabolic processes. The hysteretic concept. *J. Biol. Chem.* **245,** 5788.
132. Fujito, T. (1966). Studies on soluble cytochromes in Enterobacteriaceae. I. Detection, purification, and properties of cytochrome c-552 in anaerobically grown cells. *J. Biochem. (Tokyo)* **60,** 204.

133. Fujito, T., and Sato, R. (1967). Studies on soluble cytochromes in Enterobacteriaceae. V. Nitrite dependent gas evolution in cells containing cytochrome c-552. *J. Biochem.* (*Tokyo*) **62,** 230.
134. Galivan, J. H., and Allen, S. H. G. (1968). Methylmalonyl coenzyme A decarboxylase. Its role in succinate decarboxylation by *Micrococcus lactilyticus*. *J. Biol. Chem.* **243,** 1253.
135. Gancedo, C., Salas, M. L., Giner, A., and Sols, A. (1965). Reciprocal effect of carbon sources on the levels of an AMP-sensitive fructose-1,6-diphosphatase and phosphofructokinase in yeast. *Biochem. Biophys. Res. Commun.* **20,** 15.
136. Gancedo, J. M., Atzpodien, W., and Holzer, H. (1969). Stimulation of the protein-dependent interconversion of two forms of yeast phosphofructokinase by a heat-stable fraction from yeast. *FEBS* (*Fed. Eur. Biochem. Soc.*) *Lett.* **5,** 199. [as cited by Holzer (187)].
137. Garfinkel, D. (1965). A computer simulation study of the metabolic control behavior of phosphofructokinase. *Contr. Energy Metab., Colloq. Metab. Contr., Symp. Contr. Energy Metab., 1965* p. 101; *Chem. Abstr.* **65,** 12489 (1966).
138. Garvie, E. I. (1967). *Leuconostoc oenos* sp. nov. *J. Gen. Microbiol.* **48,** 431.
139. Garvie, E. I. (1969). Lactic dehydrogenases of strains of the genus *Leuconostoc*. *J. Gen. Microbiol.* **58,** 85.
140. Gasser, F. (1970). Electrophoretic characterization of lactic dehydrogenases in the genus *Lactobacillus*. *J. Gen. Microbiol.* **62,** 223.
141. Gasser, F., Doudoroff, M., and Contopoulos, R. (1970). Purification and properties of NAD-dependent lactic dehydrogenases of different species of *Lactobacillus*. *J. Gen. Microbiol.* **62,** 241.
142. Gibbs, M., Dumrose, R., Bennett, F. A., and Bubeck, M. R. (1950). On the mechanism of bacterial fermentation to lactic acid studied with ^{14}C-glucose. *J. Biol. Chem.* **184,** 545.
143. Gibbs, M., Sokatch, J. T., and Gunsalus, I. C. (1955). Product labeling of glucose-1-^{14}C fermentation of homofermentative and heterofermentative lactic acid bacteria. *J. Bacteriol.* **70,** 572.
144. Glick, M. C., Zillikar, F., and Gyorgy, P. (1959). Supplementary growth promoting effect of 2-methyl-1,4-naphthoquinone on *Lactobacillus bifidus* var. *pennsylvanicus* *J. Bacteriol.* **77,** 230.
145. Goldfine, H., and Stadtman, E. R. (1960). Propionic acid metabolism. V. The conversion of β-alanine to propionic acid by cell-free extracts of *Clostridium propionicum*. *J. Biol. Chem.* **235,** 2238.
146. Goulden, J. D. S., and Sharpe, M. E. (1958). The infrared absorption spectra of lactobacilli. *J. Gen. Microbiol.* **19,** 76.
147. Gray, C. T., Wimpenny, C. W. T., Hughes, D. E., and Ranlett, M. (1963). A soluble c-type cytochrome from anaerobically grown *Escherichia coli* and various Enterobacteriaceae. *Biochim. Biophys. Acta* **67,** 157.
148. Gray, C. T. (1964). Atmospheric control of enzyme biosynthesis in facultative anaerobes. *Biochem. J.* **90,** 23P.
149. Gray, C. T., and Gest, H. (1965). Biological formation of molecular hydrogen. A "hydrogen valve" facilitates regulation of anaerobic energy metabolism in many microorganisms. *Science* **148,** 186.
150. Gray, C. T., Wimpenny, J. W. T., and Mossman, M. R. (1966). Regulation of metabolism in facultative bacteria. T. Structural and functional changes in *Escherichia coli* associated with shifts between aerobic and anaerobic states. *Biochim. Biophys. Acta* **117,** 22.

151. Gray, C. T., Wimpenny, J. W. T., and Mossman, M. R. (1966). Regulation of metabolism in facultative bacteria. II. Effects of aerobiosis and anaerobiosis and nutrition on the formation of Krebs cycle enzymes in *Escherichia coli*. *Biochim. Biophys. Acta* **117,** 33.

152. Greisen, E. C., and Gunsalus, I. C. (1943). Hydrogen peroxide destruction by streptococci. *J. Bacteriol.* **45,** 16.

153. Gunsalus, I. C., and Niven, C. F. (1942). The effect of pH on the lactic acid fermentation. *J. Biol. Chem.* **145,** 131.

154. Gunsalus, I. C., and Sherman, G. M. (1943). The fermentation of glycerol by streptococci. *J. Bacteriol.* **45,** 155.

155. Gunsalus, I. C., Horecker, B. L., and Wood, W. A. (1955). Pathways of carbohydrate metabolism in microorganisms. *Bacteriol. Rev.* **19,** 79.

156. Hager, L. P., Geller, D. M., and Lipmann, F. (1954). Flavoprotein-catalyzed pyruvate oxidation in *Lactobacillus delbruckii*. *Fed. Proc., Fed. Amer. Soc. Exp. Biol.* **13,** 734.

157. Hager, L. P., and Kornberg, H. L. (1961). On the mechanism of 2-oxoglutarate oxidation in *Escherichia coli*. *Biochem. J.* **78,** 194.

158. Halpern, Y. S., and Even-Shoshan, A. (1967). Further evidence for two distinct acetolactate synthetases in *Aerobacter aerogenes*. *Biochim. Biophys. Acta* **139,** 502.

159. Hansen, P. A. (1938). The respiration of the rod-shaped lactic acid bacteria. *Zentralbl. Bakteriol., Parasitenk. Infektionskr., Abt. 1: Orig.* **98,** 289.

160. Haq, A., and Dawes, E. A. (1971). Pyruvic acid metabolism and ethanol formation in *Erwinia amylovora*. *J. Gen. Microbiol.* **68,** 295.

161. Hardman, J. K., and Stadtman, T. C. (1963). Metabolism of ω-amino acids. III. Mechanism of conversion of γ-aminobutyrate to γ-hydroxybutyrate by *Clostridium aminobutyricum*. *J. Biol. Chem.* **238,** 2081.

162. Hardman, J. K., and Stadtman, T. C. (1963). Metabolism of ω-amino acids. IV. γ-Aminobutyrate fermentation by cell-free extracts of *Clostridium aminobutyricum*. *J. Biol. Chem.* **238,** 2088.

163. Hardman, J. K., and Stadtman, T. C. (1963). Metabolism of ω-amino acids. V. Energetics of the γ-aminobutyrate fermentation by *Clostridium aminobutyricum*. *J. Bacteriol.* **85,** 1326.

164. Harold, F. M., and Baarda, J. R. (1968). Inhibition of membrane transport in *Streptococcus faecalis* by uncouplers of oxidative phosphorylation and its relationship to proton conduction. *J. Bacteriol.* **96,** 2025.

165. Harold, F. M., Baarda, J. R., Baron, C., and Abrams, A. (1969). Inhibition of membrane-bound adenosine triphosphatase and of cation transport in *Streptococcus faecalis* by *N,N'*-dicyclohexylcarbodiimide. *J. Biol. Chem.* **244,** 2261.

166. Harrison, D. E. F., and Pirt, S. J. (1967). The influence of dissolved oxygen concentration on the respiration and glucose metabolism of *Klebsiella aerogenes* during growth. *J. Gen. Microbiol.* **46,** 193.

167. Harrison, D. E. F., and Maitra, P. K. (1969). Control of respiration and metabolism in growing *Klebsiella aerogenes*. The role of adenine nucleotides. *Biochem. J.* **112,** 647.

168. Harrison, D. E. F. (1970). Undamped oscillations of pyridine nucleotide and oxygen tension in chemostat cultures of *Klebsiella aerogenes*. *J. Cell Biol.* **45,** 514.

169. Harrison, D. E. F., and Loveless, J. E. (1971). The effect of growth conditions in respiratory activity and growth efficiency in facultative anaerobes grown in chemostat culture. *J. Gen. Microbiol.* **68,** 35.

170. Harrison, D. E. F., and Loveless, J. E. (1971). Transient response of facultatively

anaerobic bacteria growing in chemostat culture to a change from anaerobic to aerobic conditions. *J. Gen. Microbiol.* **68,** 45.

170a. Harrison, D. E. F. (1973). Growth, oxygen and respiration. *CRC Critical Rev. Microbiol.* **2,** 185.

171. Hartman, R. E. (1970). Carbon dioxide fixation by extracts of *Streptococcus faecalis* var. *liquefaciens*. *J. Bacteriol.* **102,** 341.

172. Harvey, R. J. (1970). Metabolic regulation in glucose-limited chemostat cultures of *Escherichia coli*. *J. Bacteriol.* **104,** 698.

172a. Hatchikian, E. C., and LeGall, J. (1972). Evidence for the presence of a b-type cytochrome in the sulfate-reducing bacterium *Desulfovibrio gigas*, and its role in the reduction of fumarate by molecular hydrogen. *Biochim. Biophys. Acta* **267,** 479.

173. Hayward, A. C. (1957). Detection of gas production from glucose by heterofermentative lactic acid bacteria. *J. Gen. Microbiol.* **16,** 9.

174. Heady, R. E., Jacobs, N. J., and Deibel, R. H. (1964). Effect of haemin supplementation on porphyrin accumulation and catalase synthesis during anaerobic growth of *Staphylococcus*. *Nature (London)* **203,** 1285.

175. Hedegaard, J., Brevet, J., and Roche, J. (1966). Imidazole lactic acid: An intermediate in L-histidine degradation in *Escherichia coli* B. *Biochem. Biophys. Res. Commun.* **25,** 335.

176. Hegre, C. S., Miller, S. J., and Lane, M. D. (1962). Studies on methylmalonyl isomerase. *Biochim. Biophys. Acta* **56,** 538.

177. Hempfling, W. P., Mainzer, S. E., and Vandemark, P. J. (1969). Invariance of Y (adenosine triphosphate) of *Streptococcus faecalis* 10 C1 during anaerobic continuous culture. *Bacteriol. Proc.* p. 172.

178. Hempfling, W. P. (1970). Repression of oxidative phosphorylation in *Escherichia coli* B by growth in glucose and other carbohydrates. *Biochem. Biophys. Res. Commun.* **41,** 9.

179. Hespell, R. B., Joseph, R., and Mortlock, R. P. (1969). Requirement for coenzyme A in the phosphoroclastic reaction of anaerobic bacteria. *J. Bacteriol.* **100,** 1328.

180. Hespell, R. B., and Canale-Parola, E. (1970). Carbohydrate metabolism in *Spirochaeta stenostrepta*. *J. Bacteriol.* **103,** 216.

180a. Hespell, R. B., and Canale-Parola, E. (1973). Glucose and pyruvate metabolism of *Spirochaeta litoralis*, an anaerobic marine spirochete. *J. Bacteriol.* **116,** 931.

181. Hess, B., Haeckel, R., and Brand, K. (1966). FDP-activation of yeast pyruvate kinase. *Biochem. Biophys. Res. Commun.* **24,** 824.

181a. Hetland, O., Bryn, K., and Stormer, F. C. (1971). Diacetyl (acetoin) reductase from *Aerobacter aerogenes*. Evidence for multiple forms of the enzyme. *Eur. J. Biochem.* **20,** 206.

181b. Hetland, O., Olsen, B. P., Christensen, T. B., and Stormer, F. C. (1971). Diacetyl-(acetoin) reductase from *Aerobacter aerogenes*. Structural properties. *Eur. J. Biochem.* **20,** 200.

182. Hibbert, F., Kyrtopoulos, S. A., and Satchell, D. P. N. (1971). Kinetic studies with phosphotransacetylase. *Biochim. Biophys. Acta* **242,** 39.

183. Higgins, T. E., and Johnson, M. J. (1970). Pathways of anaerobic acetate utilization in *Escherichia coli* and *Aerobacter cloaceae*. *J. Bacteriol.* **101,** 885.

183a. Himes, R. H., and Harmony, J. A. K. (1973). Formyltetrahydrofolate synthetase. *CRC Critical Rev. Biochem.* **1,** 501.

184. Hirsch, C. A., Rasminsky, M., Davis, B. D., and Lin, E. C. C. (1963). A fumarate

reductase in *Escherichia coli* distinct from succinate dehydrogenase. *J. Biol. Chem.* **238,** 3770.

184a. Hobson, P. N., and Summers, R. (1972). ATP pool and growth yield in *Selenomonas ruminantium. J. Gen. Microbiol.* **70,** 351.

185. Hofer, H. W., and Pette, D. (1966). The role of the nucleic acid in phosphofructokinase as an enzyme modulator. *Life Sci.* **5,** 199.

186. Hogg, D. McG., and Jago, G. R. (1970). The influence of aerobic conditions in some aspects of the growth and metabolism of group N streptococci. *Aust. J. Dairy Technol.* **25,** 17.

187. Holzer, H. (1970). Some aspects of regulation of metabolism by ATP. *Advan. Enzyme Regul.* **8,** 85.

188. Hommes, F. A., and Stekhoven, F. (1964). Aperiodic changes of reduced NAD during anaerobic glycolysis in brewer's yeast. *Biochim. Biophys. Acta* **86,** 427.

189. Hommes, F. A. (1966). Effect of glucose on the level of glycolytic enzymes in yeast. *Arch. Biochem. Biophys.* **114,** 231.

190. Hoogerheide, J. C. (1971). On a disturbance of the normal Pasteur reaction in baker's yeast. *Antonie van Leeuwenhoek; J. Microbiol. Serol.* **37,** 435.

191. Horecker, B. L. (1962). "Pentose Metabolism in Bacteria," CIBA Lectures in Microbial Chemistry. Wiley, New York.

192. Hungate, R. E. (1966). "The Rumen and its Microbes." Academic Press, New York.

193. Hurnitz, J. (1958). Pentose phosphate cleavage by *Leuconostoc mesenteroides. Biochim. Biophys. Acta* **28,** 599.

193a. Iannotti, E. L., Kafkewitz, D., Wolin, M. J., and Bryant, M. P. (1973). Glucose fermentation products of *Ruminococcus albus* grown in continuous culture with *Vibrio succinogenes:* Changes caused by interspecies transfer of H_2. *J. Bacteriol.* **114,** 1231.

194. Jensen, J. (1957). Biosynthesis of hematin compounds in a hemin-requiring strain of *Micrococcus pyogenes* var. *aureus.* I. The significance of coenzyme A for the terminal synthesis of catalase. *J. Bacteriol.* **73,** 324.

195. Jensen, J., and Hyde, M. O. (1963). "Apocatalase" of catalase-negative staphylococci. *Science* **141,** 45.

196. Jerchel, D., Flesch, P., and Bauer, E. (1956). Untersuchungen zum Abbau der L-Äpfelsäure durch *Bacterium gracile. Justus Liebigs Ann. Chem.* **601,** 40.

196a. Johansen, L., Larsen, S. H., and Stormer, F. C. (1973). Diacetyl(acetoin) reductase from *Aerobacter aerogenes.* Kinetic studies of the reduction of diacetyl to acetoin. *Eur. J. Biochem.* **34,** 97.

197. Johns, A. T. (1951). The mechanism of propionic acid formation by *Veillonella gazogenes. J. Gen. Microbiol.* **5,** 326.

198. Johns, A. T. (1951). The mechanism of propionic acid formation by propionibacteria. *J. Gen. Microbiol.* **5,** 337.

199. Johnson, M. J. (1941). The role of aerobic phosphorylation in the Pasteur effect. *Science* **94,** 200.

200. Johnson, M. J. (1967). Aerobic microbial growth at low oxygen concentrations. *J. Bacteriol.* **94,** 101.

201. Johnson, M. K., and McClesky, C. S. (1957). Studies on the aerobic carbohydrate metabolism of *Leuconostoc mesenteroides. J. Bacteriol.* **74,** 22.

202. Johnson, M. K., and McClesky, C. S. (1958). Further studies on the aerobic metabolism of *Leuconostoc mesenteroides. J. Bacteriol.* **75,** 98.

202a. Johnson, P. W., and Canale-Parola, E. (1973). Properties of rubredoxin and ferredoxin isolated from spirochetes. *Arch. Mikrobiol.* **89,** 341.

203. Johnston, M. A., and McClesky, C. S. (1958). Further studies on the aerobic metabolism of *Leuconostoc mesenteroides*. *J. Bacteriol.* **75,** 98.

204. Johnston, M. A., and Delwiche, E. A. (1965). Distribution and characteristics of the catalases of Lactobacillaceae. *J. Bacteriol.* **90,** 347.

205. Jones, R. C., and Hough, J. S. (1970). The effect of temperature on the metabolism of baker's yeast growing on continuous culture. *J. Gen. Microbiol.* **60,** 107.

206. Josephson, S. L., and Fraenkel, D. G. (1969). Transketolase mutants of *Escherichia coli*. *J. Bacteriol.* **100,** 1289.

207. Joyner, A. E., Jr., and Baldwin, R. L. (1966). Enzymatic studies of pure cultures of rumen microorganisms. *J. Bacteriol.* **92,** 1321.

208. Jungermann, K., Leimenstoll, G., Rupprecht, E., and Thauer, R. K. (1971). Demonstration of NADH-ferredoxin reductase in two saccharolytic clostridia. *Arch. Mikrobiol.* **80,** 370.

209. Jungermann, K., Rupprecht, E., Ohrloff, C., Thauer, R., and Decker, K. (1971). Regulation of the reduced nicotinamide adenine dinucleotide-ferredoxin reductase system in *Clostridium kluyveri*. *J. Biol. Chem.* **246,** 960.

210. Kagamiyama, H., Morino, Y., and Snell, E. E. (1970). The chemical structure of tryptophanase from *Escherichia coli*. *J. Biol. Chem.* **245,** 2819.

211. Kamihara, T., Yabushita, H., and Fukui, S. (1968). Studies on pyruvate oxidation and related metabolism in *Streptococcus faecalis*. I. On the mechanism of growth inhibition by propionate. *J. Agr. Chem. Soc. Jap.* **42,** 146.

212. Kamihara, T. (1969). Role of pyruvate metabolism in the growth of *Streptococcus faecalis* in the presence of propionate. *J. Bacteriol.* **97,** 151.

213. Kats, L. N., and Kharat'yan, E. F. (1969). Membraneous structures of lactic acid bacteria and their variation in different cultivation condition. *Mikrobiologiya* **38,** 229.

214. Katz, J., and Wood, H. G. (1960). The use of glucose-^{14}C for the evaluation of the pathways of glucose metabolism. *J. Biol. Chem.* **235,** 2165.

215. Kaufman, S., Korkes, S., and Del Campillo, A. (1951). Biosynthesis of dicarboxylic acids by carbon dioxide fixation. V. Further study of the "malic" enzyme of *Lactobacillus arabinosus*. *J. Biol. Chem.* **192,** 301.

216. Kaufmann, E., and Dikstein, S. (1961). A D-lactic dehydrogenase from *Leuconostoc mesenteroides*. *Nature (London)* **190,** 346.

217. Keenan, T. W. (1968). Production of acetic acid and other volatile compounds by *Leuconostoc citrovorum* and *Leuconostoc dextranicum*. *Appl. Microbiol.* **16,** 1881.

218. Kemp, L. L., and West, R. G. (1959). The effect of agitation on the rate of acid formation by *Lactobacillus delbruckii*. *J. Biochem. Microbiol. Technol. Eng.* **1,** 335.

219. Kirtley, M. E., and Koshland, D. E., Jr. (1967). Models of cooperative effects in proteins containing subunits. Effects of two interacting ligands. *J. Biol. Chem.* **242,** 4192.

220. Kitahara, K. (1952). *J. Agr. Chem. Soc. Jap.* **26,** 162 [as cited by Mizushima *et al.* (277)].

221. Klein, S. M., and Sagers, R. D. (1962). Intermediary metabolism of *Diplococcus glycinophilus*. II. Enzymes of the acetate-generating system. *J. Bacteriol.* **83,** 121.

222. Klein, S. M., and Sagers, R. D. (1966). Glycine metabolism. I. Properties of the system catalyzing the exchange of bicarbonate with the carboxyl group of glycine in *Peptococcus glycinophilus*. *J. Biol. Chem.* **241,** 197.

223. Klein, S. M., and Sagers, R. D. (1966). Glycine metabolism. II. Kinetic and optical studies on the glycine decarboxylation system from *Peptococcus glycinophilus*. *J. Biol. Chem.* **241,** 206.

224. Kleiner, D., and Burris, R. H. (1970). The hydrogenase of *Clostridium pasteurianum*. Kinetic studies and the role of molybdenum. *Biochim. Biophys. Acta* **212,** 417.

225. Kluyver, A. J. (1931). "The Chemical Activities of Microorganisms." Oxford Univ. Press, London and New York.

226. Knappe, J., Schacht, J., Mockel, W., Hopner, T., Vetter, H., Jr., and Edenharder, R. (1969). Pyruvate-formate lyase reaction in *Escherichia coli*. The enzymatic system converting an inactive form of the lyase into the catalytically active enzyme. *Eur. J. Biochem.* **11,** 316.

227. Knowles, C. J., and Smith, L. (1970). Measurement of intact *Azotobacter vinelandii* under different conditions. *Biochim. Biophys. Acta* **197,** 152.

228. Kocholaty, W., and Hoogerheide, J. C. (1938). Dehydrogenation reactions by suspensions of *C. sporogenes*. *Biochem. J.* **32,** 437.

229. Koditschek, L. K., and Umbreit, W. W. (1969). α-Glycerophosphate oxidase in *Streptococcus faecium* F 24. *J. Bacteriol.* **98,** 1063.

230. Kopeloff, L. M., Kopeloff, N., Etchells, J. L., and Posselt, E. (1937). Optical activity of lactic acid produced by *Lactobacillus acidophilus* and *Lactobacillus bulgaricus*. *J. Bacteriol.* **33,** 89.

231. Korkes, S., and Ochoa, S. (1948). Adaptive conversion of malate to lactate and carbon dioxide by *Lactobacillus arabinosus*. *J. Biol. Chem.* **176,** 463.

232. Korkes, S., DelCampillo, A., and Ochoa, S. (1950). Biosynthesis of dicarboxylic acids by carbon dioxide fixation. IV. Isolation and properties of an adaptive "malic" enzyme from *Lactobacillus arabinosus*. *J. Biol. Chem.* **187,** 891.

233. Kornberg, H. L., and Elsden, S. R. (1961). The metabolism of 2-carbon compounds by microorganisms. *Advan. Enzymol.* **23,** 401.

234. Kovacs, E., Mazarean, H. H., and Jaki, A. (1965). Isolation between the synthesis of catalase and the intensity of respiration in cultures of *Staphylococcus aureus*. *Enzymologia* **28,** 316.

235. Krebs, H. A., and Eggleston, L. V. (1941). Biological synthesis of oxaloacetate from pyruvic acid and carbon dioxide. 2. The mechanism of CO_2 fixation in propionic acid bacteria. *Biochem. J.* **35,** 676.

235a. Kroger, A. (1974). Electron transport phosphorylation coupled to fumarate reduction in anaerobically grown *Proteus rettgeri*. *Biochim. Biophys. Acta* **347,** 273.

236. Kunkee, R. E. (1967). Malolactic fermentation. *Advan. Appl. Microbiol.* **9,** 235.

237. Kupfer, D. G., and Canale-Parola, E. (1967). Pyruvate metabolism in *Sarcina maxima*. *J. Bacteriol.* **94,** 984.

238. Kuratomi, K., and Stadtman, E. R. (1966). The conversion of carbon dioxide to acetate. II. The role of α-ketoisovalerate in the synthesis of acetate by *Clostridium thermoaceticum*. *J. Biol. Chem.* **241,** 4217.

239. Labow, R., and Robinson, W. G. (1966). Crystalline D-serine dehydrase. *J. Biol. Chem.* **241,** 1239.

240. Lamanna, C., and Mallett, M. F. (1965). "Basic Bacteriology—Its Biological and Chemical Background." Williams & Wilkins, Baltimore, Maryland.

241. Lerud, R. F., and Whiteley, H. R. (1971). Purification and properties of α-ketoglutarate reductase from *Micrococcus aerogenes*. *J. Bacteriol.* **106,** 571.

242. Li, L. F., Ljungdahl, L. G., and Wood, H. G. (1966). Properties of NAD-dependent formate dehydrogenase from *Clostridium thermoaceticum*. *J. Bacteriol.* **92,** 405.

243. Lin, E. C. C., Levin, A. P., and Magasanik, B. (1960). The effect of aerobic metabolism on the inducible glycerol dehydrogenase of *Aerobacter aerogenes*. *J. Biol. Chem.* **235,** 1824.

244. Lindley, R. W., and Delwiche, E. A. (1969). Degradation of α-ketoglutarate by *Veillonella alcalescens*. *J. Bacteriol.* **98,** 315.
245. Lindmark, D. G., Paoblea, P., and Wood, H. G. (1969). The pyruvate-formate lyase system of *Streptococcus faecalis*. I. Purification and properties of the formate–pyruvate exchange. *J. Biol. Chem.* **244,** 3605.
246. Lipmann, F. (1941). Metabolic generation and utilization of phosphate bond energy. *Advan. Enzymol.* **1,** 99.
247. Ljungdahl, L. G., Irion, E., and Wood, H. G. (1965). Total synthesis of acetate from CO_2. I. Comethylcobric acid and co(methyl)-5-methoxy-benzimidazolyl-cobamide as intermediates with *Clostridium thermoaceticum*. *Biochemistry* **4,** 2771.
248. Ljungdahl, L. G., Irion, E., and Wood, R. G. (1966). Role of corrinoids in the total synthesis of acetate from CO_2 by *Clostridium thermoaceticum*. *Fed. Proc., Fed. Amer. Soc. Exp. Biol.* **25,** 1643.
249. Ljungdahl, L. G., and Wood, H. G. (1969). Total synthesis of acetate from CO_2 by heterotrophic bacteria. *Annu. Rev. Microbiol.* **23,** 515.
249a. Ljungdahl, L. G., LeGall, J., and Lee, J. P. (1973). Isolation of a protein tightly bound 5-methoxybenzimidazolyl-cobamide (Factor 3M) from *Clostridium thermoaceticum*. *Biochemistry* **12,** 1802.
250. Lonberg-Holm, K. K. (1959). A direct study of intracellular glycolysis in Ehrlich's ascites tumor. *Biochim. Biophys. Acta* **35,** 464.
251. London, J. (1968). Regulation and function of lactate oxidation in *Streptococcus faecium*. *J. Bacteriol.* **95,** 1380.
252. London, J., and Meyer, E. Y. (1969). Malate utilization by group D streptococci. II. Evidence for allosteric inhibition of an inducible malate dehydrogenase (decarboxylating) by ATP and glycolytic intermediate products. *Biochim. Biophys. Acta* **178,** 205.
253. London, J., and Meyer, E. Y. (1969). Malate utilization by group D streptococci: Physiological properties and purification of an inducible malic enzyme. *J. Bacteriol.* **98,** 705.
254. London, J., and Meyer, E. Y. (1970). Malate utilization by group D streptococci. Regulation of malic enzyme synthesis by an inducible malate permease. *J. Bacteriol.* **102,** 130.
255. London, J., Meyer, E. T., and Kubczyk, S. (1970). Allosteric control of a *Lactobacillus* malate dehydrogenase (decarboxylating) by two glycolytic intermediate products. *Biochim. Biophys. Acta* **212,** 512.
256. Lowry, O. H., and Passoneau, J. V. (1964). A comparison of the kinetic properties of phosphofructokinase from bacterial plant and animal sources. *Naunyn Schmiedebergs Arch. Exp. Pathol. Pharmakol.* **248,** 185.
257. Luzikov, V. N., Zubatov, A. S., Rainina, E. I., and Bakeyera, L. E. (1971). Degradation and restoration of mitochondria upon deaeration and subsequent aeration of aerobically grown *Saccharomyces cerevisiae* cells. *Biochim. Biophys. Acta.* **245,** 321.
258. Lynen, F., Hartmann, C., Netter, K. F., and Schnegraf, A. (1959). Phosphate turnover and the Pasteur effect. *Regul. Cell Metab., Ciba Found. Symp., 1958,* p. 256.
259. McCormick, N. G., Ordal, E. J., and Whiteley, H. R. (1962). Degradation of pyruvate by *Micrococcus lactilyticus*. *J. Bacteriol.* **83,** 887.
260. McKay, L. L., Walker, L. A., Sandine, W. E., and Elliker, P. R. (1969). Involvement of phosphoenolpyruvate in lactose utilization by group N streptococci. *J. Bacteriol.* **99,** 603.

261. McKay, L. L., Miller, A., III, Sandine, W. E., and Elliker, P. R. (1970). Mechanism of lactose utilization by lactic acid streptococci: Enzymatic and genetic analysis. *J. Bacteriol.* **102,** 804.
262. McPhedran, P., Sommer, B., and Lin, E. C. C: (1961). Control of ethanol dehydrogenase levels in *Aerobacter aerogenes. J. Bacteriol.* **81,** 852.
263. Mallavia, L. P., and Weiss, E. (1970). Catabolic activities of *Neisseria meningitidis:* Utilization of glutamate. *J. Bacteriol.* **101,** 127.
264. Manderson, G. J., and Doelle, H. W. (1972). The effect of oxygen and pH on the glucose metabolism of *Lactobacillus casei* var. *rhamnosus* ATCC 7469. *Antonie van Leeuwenhoek; J. Microbiol. Serol.* **38,** 223.
265. Mansour, T. E., and Mansour, J. M. (1962). Effects of serotonin (5-hydroxytryptamine) and adenosine 3′,5′-phosphate on phosphofructokinase from the liver fluke *Fasciola hepatica. J. Biol. Chem.* **237,** 629.
266. Mansour, T. E. (1963). Studies on heart phosphofructokinase: Purification, inhibition, and activation. *J. Biol. Chem.* **238,** 2285.
267. Mansour, T. E. (1965). Studies on heart phosphofructokinase: Active and inactive forms of the enzyme. *J. Biol. Chem.* **240,** 2165.
268. Mansour, T. E. (1970). Kinetic and physical properties of phosphofructokinase. *Advan. Enzyme Regul.* **8,** 37.
269. Mayhew, S. G., and Massey, V. (1969). Purification and characterization of flavodoxin from *Peptostreptococcus elsdenii. J. Biol. Chem.* **244,** 794.
270. Mayhew, S. S., Foust, G. P., and Massey, V. (1969). Oxidation–reduction properties of flavodoxin from *Peptostreptococcus elsdenii. J. Biol. Chem.* **244,** 803.
271. Michaud, R. N., and Delwiche, E. A. (1970). Multiple impairment of glycolysis in *Veillonella alcalescens. J. Bacteriol.* **101,** 138.
272. Michaud, R. N., Carrow, J. A., and Delwiche, E. A. (1970). Nonoxidative pentose phosphate pathway in *Veillonella alcalescens. J. Bacteriol.* **101,** 141.
273. Mickelson, M. N. (1967). Aerobic metabolism of *Streptococcus agalactiae. J. Bacteriol.* **94,** 184.
274. Mickelson, M. N. (1969). Phosphorylation and the reduced nicotinamide adenine dinucleotide oxidase reaction in *Streptococcus agalactiae. J. Bacteriol.* **100,** 895.
274a. Miller, T. L., and Wolin, M. J. (1973). Formation of hydrogen and formate by *Ruminococcus albus. J. Bacteriol.* **116,** 836.
275. Mitruka, B. M., and Costilow, R. N. (1967). Arginine and ornithine catabolism by *Clostridium botulinum. J. Bacteriol.* **93,** 295.
276. Mizushima, S., and Kitahara, K. (1962). Purification and properties of lactic dehydrogenase of *Lactobacillus casei. J. Gen. Appl. Microbiol.* **8,** 130.
277. Mizushima, S., Machida, Y., and Kitahara, K. (1963). Quantitative studies on glycolytic enzymes in *Lactobacillus plantarum. J. Bacteriol.* **86,** 1295.
278. Monod, J., Changeux, J. P., and Jacob, F. (1963). Allosteric proteins and cellular control systems. *J. Mol. Biol.* **6,** 306.
279. Monod, J., Wyman, J., and Changeux, J. P. (1965). On the nature of allosteric transitions: A plausible model. *J. Mol. Biol.* **12,** 88.
279a. Montague, M. D., and Dawes, E. A. (1974). The survival of *Peptococcus prevotii* in relation to the adenylate energy change. *J. Gen. Microbiol.* **80,** 291.
280. Moore, C. C. (1965). Control properties of phosphofructokinase from *Saccharomyces cerevisiae. Contr. Energy Metab., Colloq., Metab. Contr., Symp. Centr. Energy Metab., 1965* p. 97. *Chem. Abstr.* **65,** 12490 (1966).
281. Mortensen, L. E., Valentine, R. C., and Carnahan, J. E. (1963). Ferredoxin in

the phosphoroclastic reaction of pyruvic acid and its relation to nitrogen fixation in *Clostridium pasteurianum*. *J. Biol. Chem.* **238,** 794.
282. Moss, F. (1952). The influence of oxygen tension on respiration and cytochrome a_2 formation of *Escherichia coli*. *Aust. J. Exp. Biol. Med. Sci.* **30,** 531.
283. Moss, F. (1956). Adaptation of the cytochromes of *Aerobacter aerogenes* in response to environmental oxygen tension. *Aust. J. Exp. Biol. Med. Sci.* **34,** 395.
284. Naik, V. R., and Nadkarni, G. B. (1968). Adaptive alterations in the fermentative sequence of *Lactobacillus casei*. *Arch. Biochem. Biophys.* **123,** 431.
285. Newsholme, E. A., and Randle, P. J. (1961). Regulation of glucose uptake by muscle. *Biochem. J.* **80,** 655.
286. Ng, S. K. C., and Hamilton, I. R. (1971). Lactate metabolism by *Veillonella parvula*. *J. Bacteriol.* **105,** 999.
287. Nisman, B. (1954). The Stickland reaction. *Bacteriol. Rev.* **18,** 16.
288. O'Hara, J., Gray, C. T., Puig, J., and Pichinoty, F. (1967). Defects in formate hydrogen lyase in nitrate-negative mutants of *Escherichia coli*. *Biochem. Biophys. Res. Commun.* **28,** 951.
289. O'Kane, D. J., and Gunsalus, I. C. (1948). Pyruvic acid metabolism. A factor required for oxidation by *Streptococcus faecalis*. *J. Bacteriol.* **56,** 499.
290. O'Kane, D. J. (1950). Influence of the pyruvate oxidation factor on the oxidative metabolism of glucose by *Streptococcus faecalis*. *J. Bacteriol.* **60,** 449.
291. Olive, C., Gerock, M. E., and Levy, H. R. (1971). Glucose-6-phosphate dehydrogenase from *Leuconostoc mesenteroides* *J. Biol. Chem.* **246,** 2047.
292. Overath, P. E., Stadtman, E. R., Kellerman, G. M., and Lynen, F. (1962). Zum Mechanismus der Umlagerung von Methylmalonyl-CoA in Succinyl-CoA. III. Reinigung und Eigenschaften der Methylmalonyl-CoA Isomerase. *Biochem. Z.* **336,** 77.
293. Paigen, K., and Williams, P. (1970). Catabolite repression and other control mechanisms in carbohydrate utilization. *Advan. Microbial Physiol.* **4,** 251.
294. Pardee, A. B. (1959). Mechanisms for control of enzyme synthesis and enzyme activity in bacteria. *Regul. Cell Metab., Ciba Found. Symp., 1958* p. 295.
295. Parker, D. J., Wu, T. F., and Wood, H. G. (1971). Total synthesis of acetate from CO_2: Methyltetrahydrofolate, an intermediate and a procedure for separation of the folates. *J. Bacteriol.* **108,** 770.
296. Passoneau, J. V., and Lowry, O. (1962). Phosphofructokinase and the Pasteur effect. *Biochem. Biophys. Res. Commun.* **7,** 10.
297. Paynter, N. J. B., and Elsden, S. R. (1970). Mechanism of propionate formation by *Selenomonas ruminantium*, a rumen microorganism. *J. Gen. Microbiol.* **61,** 1.
298. Peisack, J., and Strecker, H. J. (1962). The interconversion of glutamic acid and proline. V. The reduction of Δ'-pyrroline-5-carboxylic acid to proline. *J. Biol. Chem.* **237,** 2255.
299. Pelroy, R. A., and Whiteley, H. R. (1971). Regulatory properties of acetokinase from *Veillonella alcalescens*. *J. Bacteriol.* **105,** 259.
300. Peterkofsky, A., and Gazdar, C. (1971). Glucose and the metabolism of adenosine 3',5'-cyclic monophosphate in *Escherichia coli*. *Proc. Nat. Acad. Sci. U.S.* **68,** 2794.
301. Peterson, W. H., and Peterson, M. S. (1945). Relation of bacteria to vitamin and other growth factors. *Bacteriol. Rev.* **9,** 49.
302. Peynaud, E., Lafon-Lafourcade, S., and Guimberteau, G. (1966). L(+)-Lactic acid and D(−)-lactic acid in wines. *Amer. J. Enol. Viticult.* **17,** 302.
303. Peynaud, E., Lafon-Lafourcade, S., and Guimberteau, G. (1968). Über den Mecha-

nismus der Äpfelsäure-Milchsäure-Gärung. *Mitt. Rebe Wein, Obstbau Fruechteverwertung* **17,** 343.

304. Pichinoty, F. (1962). Inhibition by oxygen of the biosynthesis and activity of hydrogenase and hydrogen lyase in some anaerobic bacteria. *Biochim. Biophys. Acta* **64,** 111.

305. Pichinoty, F., and Bigliardi-Rouvier, J. (1962). Etude et mise au point d'une méthode permettent de medurer l'activité des tetrathionate reductases d'origine bactérienne. Inhibition par l'oxygene du la biosynthèse et de l'activité de l'enzyme à intermedia. *Antonie van Leeuwenhoek; J. Microbiol. Serol.* **28,** 134.

306. Pichinoty, F. (1965). "L'effect oxygène et la biosynthèse des enzymes d'oxydoreduction bactériens," No. 24. CNRS, Paris.

307. Pichinoty, F. (1965). L'inhibition par l'oxygène de la dénitrification bactérienne. *Ann. Inst. Pasteur, Paris* **109,** 248.

308. Pichinoty, F. (1966). Properties, regulation, et functions physiologique des nitrate-reductases bactériennes. A et B. *Bull. Soc. Fr. Physiol. Veg.* **12,** 97.

309. Pilone, G. J., and Kunkee, R. E. (1970). Carbonic acid from decarboxylation by "malic" enzyme in lactic acid bacteria. *J. Bacteriol.* **103,** 404.

309a. Pinsent, J. (1954). The need of selenite and molybdate in the formation of formic dehydrogenase by members of the coli-aerogenes group of bacteria. *Biochem. J.* **57,** 10.

310. Platt, T. B., and Foster, E. M. (1957). Products of glucose metabolism by homofermentative streptococci under anaerobic conditions. *J. Bacteriol.* **75,** 453.

311. Polakis, E. S., Bartley, W., and Meek, G. A. (1964). Changes in the structure and enzyme activity of *Saccharomyces cerevisiae* in response to changes in the environment. *Biochem. J.* **90,** 369.

312. Pollock, M. R., and Knox, R. (1943). Bacterial reduction of tetrathionate. *Biochem. J.* **90,** 369.

313. Pollock, M. R. (1962). Exoenzymes. *In* "The Bacteria" (I. C. Gunsalus and R. Y. Stanier, eds.), Vol. 4, p. 121. Academic Press, New York.

314. Poston, J. M., and Kuratomi, K. (1965). Roles of methylcobalamin and ferredoxin in the conversion of CO_2 to acetate by extracts of *Clostridium thermoaceticum. Fed. Proc., Fed. Amer. Soc. Exp. Biol.* **24,** 421 [as cited in Poston *et al.* (315)].

315. Poston, J. M., Kuratomi, K., and Stadtman, E. R. (1966). The conversion of carbon dioxide to acetate. I. The use of cobalt–methylcobalamin as a source of methyl groups for the synthesis of acetate by cell-free extracts of *Clostridium thermoaceticum. J. Biol. Chem.* **241,** 4209.

316. Prescott, S. C., and Dunn, C. G. (1959). "Industrial Microbiology," 3rd ed. McGraw-Hill, New York.

317. Pye, E. K. (1965). The control of glycolysis in yeast. *Contr. Energy Metab., Colloq., Metab. Contr., Symp. Contr. Energy Metab., 1965* p. 193; *Chem. Abstr.* **65,** 11002 (1966).

318. Pye, E. K., and Eddy, A. A. (1965). The regulation of glycolysis in yeast. *Fed. Proc., Fed. Amer. Soc. Exp. Biol.* **24,** 537.

319. Quiste, R. G., and Stokes, J. L. (1969). Temperature range for formic hydrogen lyase induction and activity in psychrophilic and mesophilic bacteria. *Antonie van Leeuwenhoek; J. Microbiol. Serol.* **35,** 1.

320. Racker, E. (1954). Alternative pathways of glucose and fructose metabolism. *Advan. Enzymol.* **15,** 141.

321. Radler, F. (1962). Über die Milchsäurebakterien des Weines und den biologischen Säureabbau. Übersicht. I. Systematik und chemische Grundlagen. *Vitis* **3,** 144.

322. Radler, F. (1962). Über die Milchsäurebakterien des Weines und den biologischen Säureabbau. Übersicht. II. Physiologie und Ökologie der Bakterien. *Vitis* **3,** 207.
323. Radler, F., Schütz, M., and Doelle, H. W. (1970). Die beim Abbau von L-Äpfelsäure durch Milchsäurebakterien entstehenden Isomeren der Milchsäure. *Naturwissenschaften* **57,** 672.
324. Radler, F., and Yanissis, C. (1972). Weinsäureabbau bei Milchsäurebakterien. *Arch. Mikrobiol.* **82,** 219.
325. Rainbow, C., and Rose, A. H., eds. (1963). "Biochemistry of Industrial Microorganisms." Academic Press, New York.
326. Reed, L. J., Leach, F. R., and Kolke, M. (1958). Studies on a lipoic acid-activating system. *J. Biol. Chem.* **232,** 123.
327. Reichelt, J. L., and Doelle, H. W. (1971). The influence of dissolved oxygen concentration on phosphofructokinase and the glucose metabolism of *Escherichia coli* K-12. *Antonie van Leeuwenhoek; J. Microbiol. Serol.* **37,** 497.
328. Rety, J., and Lynen, F. (1964). The absolute configuration of methylmalonyl-CoA. *Biochem. Biophys. Res. Commun.* **16,** 358.
329. Rickard, P. A. D., Moss, F., and Sam, C. T. (1969). Variation in the ability of yeasts cultured in oxygen-controlled environments to oxidize NADH. *Aust. J. Exp. Biol. Med. Sci.* **47,** 521.
330. Rimerman, E. A., and Barker, H. A. (1968). Formation and identification of 3-keto-5-aminohexanoic acid, a probable intermediate in lysine fermentation. *J. Biol. Chem.* **243,** 6151.
331. Ritchie, G. A. F., Senior, P. J., and Dawes, E. A. (1971). The purification and characterization of acetoacetyl-coenzyme A reductase from *Azotobacter beijerinckii. Biochem. J.* **121,** 309.
332. Robinson, I. M., and Allison, M. J. (1969). Isoleucine biosynthesis from 2-methylbutyric acid by anaerobic bacteria from the rumen. *J. Bacteriol.* **97,** 1220.
332a. Rogers, P. J., and Stewart, P. R. (1973). Respiratory development in *Saccharomyces cerevisiae* grown at controlled oxygen tension. *J. Bacteriol.* **115,** 88.
333. Rose, I. A., Grunberg-Manago, M., Korey, S. R., and Ochoa, S. (1954). Enzymatic phosphorylation of acetate. *J. Biol. Chem.* **211,** 737.
334. Rosenberger, R. F., and Kogut, M. (1958). The influence of growth rate and aeration on the respiratory and cytochrome system of a fluorescent pseudomonad grown in continuous culture. *J. Gen. Microbiol.* **19,** 228.
335. Rosenberger, R. F., and Elsden, S. R. (1960). The yields of *Streptococcus faecalis* grown in continuous culture. *J. Gen. Microbiol.* **22,** 726.
336. Rudolph, F. B., Purich, D. L., and Fromm, H. J. (1968). Coenzyme A-linked aldehyde dehydrogenase from *Escherichia coli* I. Partial purification, properties, and kinetic studies of the enzyme. *J. Biol. Chem.* **243,** 5539.
337. Sachs, L. E., and Barker, H. A. (1949). The influence of oxygen on nitrate and nitrite reduction. *J. Bacteriol.* **58,** 11.
338. Sagers, R. D., and Gunsalus, I. C. (1961). Intermediary metabolism of *Diplococcus glycinophilus.* I. Glycine cleavage and one-carbon interconversion. *J. Bacteriol.* **81,** 541.
339. Sagers, R. D., Benziman, M., and Klein, S. M. (1963). Failure of arsenate to uncouple the phosphotransacetylase system in *Clostridium acidi-urici. J. Bacteriol.* **86,** 978.
340. Salas, M. L., Vinuela, E., Salas, M., and Sols, A. (1965). Citrate inhibition of PFK and the Pasteur effect. *Biochem. Biophys. Res. Commun.* **19,** 371.
341. Salas, M. L., Salas, J., and Sols, A. (1968). Desensitization of yeast phospho-

fructokinase to ATP inhibition by treatment with trypsin. *Biochem. Biophys. Res. Commun.* **31,** 461.

342. Sanwal, B. D. (1970). Allosteric controls of amphibolic pathways in bacteria. *Bacteriol. Rev.* **34,** 20.

343. Sapico, V., and Anderson, R. L. (1969). D-Fructose-1-phosphate kinase and D-fructose-6-phosphate kinase from *Aerobacter aerogenes.* A comparative study of regulatory properties. *J. Biol. Chem.* **244,** 6280.

344. Satchell, D. P. N., and White, G. F. (1970). Kinetic studies with acetate kinase. *Biochim. Biophys. Acta* **212,** 248.

345. Scher, W. I., and Vogel, H. J. (1957). Occurrence of ornithine γ-transaminase: A dichotomy. *Proc. Nat. Acad. Sci. U.S.* **43,** 796.

346. Schimke, R. D., and Baril, M. F. (1963). Arginine metabolism in *Pleuropneumonia*-like organisms isolated from mammalian cell culture. *J. Bacteriol.* **86,** 195.

347. Schimke, R. D., Berlin, C. M., Sweeney, E. W., and Carroll, W. R. (1966). The generation of energy by the arginine dihydrolase pathway in *Mycoplasma hominis* 07. *J. Biol. Chem.* **241,** 2228.

348. Schnebli, H. P., and Abrams, A. (1970). Membrane adenosine triphosphatase from *Streptococcus faecalis.* Preparation and homogeneity. *J. Biol. Chem.* **245,** 1115.

348a. Schütz, M., and Radler, F. (1973). Das "Malatenzym" von *Lactobacillus plantarum* und *Leuconostoc mesenteroides. Arch. Mikrobiol.* **91,** 183.

348b. Schütz, M., and Radler, F. (1974). Das Vorkommen von Malat-enzym und Malo-Lactat-Enzym bei verschiedenen Milchsäurebakterien. *Arch. Mikrobiol.* **96,** 329.

348c. Schulman, M., Ghambeer, R. K., Ljungdahl, L. G., and Wood, H. G. (1973). Total synthesis of acetate from CO_2. VII. Evidence with *Clostridium thermoaceticum* that the carboxyl of acetate is derived from the carboxyl of pyruvate by transcarboxylation and not by fixation of CO_2. *J. Biol. Chem.* **248,** 6255.

349. Seeley, H. W., and Vandemark, P. J. (1951). An adaptive peroxidation of *Streptococcus faecalis. J. Bacteriol.* **61,** 27.

350. Senior, P. J., and Dawes, E. A. (1971). Poly-β-hydroxybutyrate biosynthesis and the regulation of glucose metabolism in *Azotobacter beijerinckii. Biochem. J.* **125,** 55.

351. Shimizu, M., Suzuki, T., Kameda, K. Y., and Abiko, Y. (1969). Phosphotransacetylase of *Escherichia coli* B, purification and properties. *Biochim. Biophys. Acta* **191,** 550.

352. Shulze, K. L., and Lipe, R. S. (1964). Relationship between substrate concentration, growth rate, and respiration rate of *Escherichia coli* in continuous culture. *Arch. Mikrobiol.* **48,** 1.

353. Sijpesteyn, A. (1951). On *Ruminococcus flavefaciens,* a cellulose-decomposing bacterium from the rumen of the sheep and cattle. *J. Gen. Microbiol.* **5,** 869.

353a. Silber, P., Chung, H., Gargiulo, P., and Schulz, H. (1974). Purification and properties of a diacetyl reductase from *Escherichia coli. J. Bacteriol.* **118,** 919.

354. Sin, P. M. L., Wood, H. G., and Stjernholm, R. L. (1961). Fixation of CO_2 by phosphoenolpyruvic carboxy transphosphorylase. *J. Biol. Chem.* **236,** PC21.

355. Sin, P. M. L., and Wood, H. G. (1962). Phosphoenolpyruvic carboxytransphosphorylase, a CO_2 fixation enzyme from propionic acid bacteria. *J. Biol. Chem.* **237,** 3044.

356. Skerman, V. B. D. (1967). "A Guide to the Identification of the Genera of Bacteria," 2nd ed. Williams & Wilkins, Baltimore, Maryland.

357. Sly, W. S., and Stadtman, E. R. (1963). Formate metabolism. I. Formyl-coenzyme

A, an intermediate in the formate-dependent decomposition of acetyl phosphate in *Clostridium kluyveri*. *J. Biol. Chem.* **238,** 2632.

358. Smalley, A. J., Jahrling, P., and Vandemark, P. J. (1968). Molar growth yields as evidence for oxidative phosphorylation in *Streptococcus faecalis* strain 10 C1. *J. Bacteriol.* **96,** 1595.

359. Smith, P. F. (1964). Comparative physiology of *Pleuropneumonia*-like and L-type organisms. *Bacteriol. Rev.* **28,** 97.

360. Snoswell, A. M. (1963). Oxidized nicotinamide adenine dinucleotide-independent lactate dehydrogenases of *Lactobacillus arabinosus* 17.5. *Biochim. Biophys. Acta* **77,** 7.

361. Snoswell, A. M. (1966). NAD-independent DL-lactate dehydrogenases from *Lactobacillus arabinosus*. *In* "Methods in Enzymology" (W. A. Wood, ed.), Vol. 9, p. 321. Academic Press, New York.

362. Sokatch, J. T., and Gunsalus, I. C. (1954). The enzymes of an adaptive gluconate fermentation pathway in *Streptococcus faecalis*. *Bacteriol. Proc.* p. 109.

363. Sokatch, J. T., and Gunsalus, I. C. (1957). Aldonic acid metabolism. I. Pathway of carbon in an inducible gluconate fermentation by *Streptococcus faecalis*. *J. Bacteriol.* **73,** 452.

364. Somerville, H. J. (1968). Enzymatic studies on the biosynthesis of amino acids from lactate by *Peptostreptococcus elsdenii*. *Biochem. J.* **108,** 107.

364a. Sone, N. (1972). The redox reactions in propionic acid fermentation. I. Occurrence and nature of an electron transfer system in *Propionibacterium arabinosum*. *J. Biochem.* **71,** 931.

365. Speckman, R. A., and Collins, E. B. (1968). Diacetyl biosynthesis in *Streptococcus lactis* and *Leuconostoc citrovorum*. *J. Bacteriol.* **95,** 174.

365a. Speckman, R. A., and Collins, E. B. (1973). Incorporation of radioactive acetate into diacetyl by *Streptococcus diacetilactis*. *Appl. Microbiol.* **26,** 744.

366. Stadtman, E. R. (1952). The purification and properties of phosphotransacetylase. *J. Biol. Chem.* **196,** 527.

367. Stadtman, E. R. (1953). The CoA-transphorase system in *Clostridium kluyveri*. *J. Biol. Chem.* **203,** 501.

368. Stadtman, E. R., Overath, P., Eggerer, H., and Lynen, F. (1960). The role of biotin and vitamin B_{12} coenzyme in propionate metabolism. *Biochem. Biophys. Res Commun.* **2,** 1.

369. Stadtman, T. C. (1962). Lysine fermentation to fatty acids and ammonia: A cobamide coenzyme dependent process. *J. Biol. Chem.* **237,** PC2409.

370. Stadtman, T. C. (1963). Anaerobic degradation of lysine. II. Cofactor requirements and properties of the soluble enzyme system. *J. Biol. Chem.* **23,** 2766.

371. Stadtman, E. R. (1966). Allosteric regulation of enzyme activity. *Advan. Enzymol.* **28,** 41.

372. Stadtman, T. C., and Tsai, L. (1967). A cobamide-dependent migration of the ε-amino group of D-lysine. *Biochem. Biophys. Res. Commun.* **28,** 920.

373. Stamer, J. R., and Stoyla, B. O. (1967). Growth response of *Lactobacillus brevis* to aeration and organic catalysts. *Appl. Microbiol.* **15,** 1025.

374. Stanier, R. Y., Doudoroff, M., and Adelberg, E. A. (1969). "The Microbial World," 3rd ed. Prentice-Hall, Englewood Cliffs, New Jersey.

375. Stephenson, M. (1949). "Bacterial Metabolism," 3rd ed. Longmans, Green, New York.

376. Stickland, L. H. (1934). The chemical reaction by which *Clostridium sporogenes* obtains its energy. *Biochem. J.* **28,** 1746.

377. Stickland, L. H. (1935). The oxidation of alanine by *Clostridium sporogenes*. *Biochem. J.* **29,** 288.

378. Stickland, L. H. (1935). The reduction of glycine by *Clostridium sporogenes*. *Biochem. J.* **29,** 896.

379. Stjernholme, R., and Wood, H. G. (1961). Methylmalonyl isomerase. II. Purification and properties of the enzyme from propionibacteria. *Proc. Nat. Acad. Sci. U.S.* **47,** 303.

380. Strittmatter, C. F. (1959). Electron transport to oxygen in lactobacilli. *J. Biol. Chem.* **234,** 2789.

381. Strittmatter, C. F. (1959). Flavin-linked oxidative enzymes of *Lactobacillus casei*. *J. Biol. Chem.* **234,** 2794.

382. Sun, A. Y., Ljungdahl, L. G., and Wood, H. G. (1969). Total synthesis of acetate from CO_2. II. Purification and properties of formyltetrahydrofolate synthetase from *Clostridium thermoaceticum*. *J. Bacteriol.* **98,** 842.

383. Suzuki, T. (1969). Phosphotransacetylase of *Escherichia coli* B, activation by pyruvate and inhibition by NADH and certain nucleotides. *Biochim. Biophys. Acta* **191,** 559.

384. Suzuki, T., Abiko, Y., and Shimizu, M. (1969). Activation and inhibition of purified phosphotransacetylase of *Escherichia coli* B by pyruvate and by $NADH_2$ and certain nucleotides. *Biochem. Biophys. Res. Commun.* **35,** 102.

385. Sweeney, J. R., and Fisher, J. R. (1968). An alternative to allosterism and cooperativity in the interpretation of enzyme kinetic data. *Biochemistry* **7,** 561.

386. Swick, R. W., and Wood, H. G. (1960). The role of transcarboxylation in propionic acid fermentation. *Proc. Nat. Acad. Sci. U.S.* **46,** 28.

387. Tarmy, E. M., and Kaplan, N. O. (1968). Kinetics of *Escherichia coli* B D-lactate dehydrogenase and evidence for pyruvate-controlled change in conformation. *J. Biol. Chem.* **243,** 2587.

388. Taylor, C. B., and Bailey, E. (1967). Activation of liver pyruvate kinase by fructose 1,6-diphosphate. *Biochem. J.* **102,** 326.

389. Taylor, M. B., and Juni, E. (1960). Stereoisomeric specificities of 2,3-butanediol dehydrogenases. *Biochim. Biophys. Acta* **39,** 448.

390. Tempest, D. W., and Herbert, D. (1965). Effect of dilution rate and growth-limiting substrate on the metabolic activity of *Torula utilis* cultures. *J. Gen. Microbiol.* **41,** 143.

391. Tempest, D. W., and Ellwood, D. C. (1969). The influence of growth composition of some cell wall components of *Aerobacter aerogenes*. *Biotechnol. Bioeng.* **11,** 775.

392. Thauer, R. K., Rupprecht, E., Ohrloff, C., Jungermann, K., and Decker, K. (1971). Regulation of the reduced nicotinamide adenine dinucleotide phosphate-ferredoxin reductase system in *Clostridium kluyveri*. *J. Biol. Chem.* **246,** 954.

392a. Thauer, R. K. (1972). CO_2-reduction to formate by NADPH. The initial step in the total synthesis of acetate from CO_2 in *Clostridium thermoaceticum*. *FEBS Lett.* **27,** 111.

392b. Thauer, R. K. (1973). CO_2-reduction to formate in *Clostridium acidi-urici*. *J. Bacteriol.* **114,** 443.

392c. Thauer, R. K., Fuchs, G., and Jungermann, K. (1974). Reduced ferredoxin:CO_2 oxidoreductase from *Clostridium pasteurianum:* Its role in formate metabolism. *J. Bacteriol.* **118,** 758.

393. Thomas, A. D., Doelle, H. W., Westwood, A. W., and Gordon, G. L. (1972). The effect of oxygen on a number of enzymes involved in the aerobic and anaerobic utilization of glucose in *Escherichia coli* K-12. *J. Bacteriol.* **112,** 1099.
394. Tiedemann, H., and Born, J. (1959). Versuche zum Mechanismus der Pasteur-Reaktion. Der Einfluss von Phosphationen auf die Aktivität der strukturgebundenen Hexokinase. *Z. Naturforsch. B* **14,** 477.
395. Ui, M. (1966). A role of phosphofructokinase in pH-dependent regulation of glycolysis. *Biochim. Biophys. Acta* **124,** 310.
396. Ullrich, J., and Mannschreck, A. (1967). Studies on the properties of (−)-2-α-hydroxyethylthiamine pyrophosphate ("active acetaldehyde"). *Eur. J. Biochem.* **1,** 110.
397. Uyeda, K., and Racker, E. (1965). Coordinated stimulation of hexokinase and phosphofructokinase by phosphate in a reconstituted system of glycolysis. *Contr. Energy Metab., Colloq., Metab. Contr., Symp. Contr. Energy Metab., 1965* p. 127; *Chem. Abstr.* **65,** 12489 (1966).
398. Uyeda, K., and Rabinowitz, J. C. (1971). Pyruvate-ferredoxin oxidoreductase. III. Purification and properties of the enzyme. *J. Biol. Chem.* **246,** 3111.
399. Uyeda, K., and Rabinowitz, J. C. (1971). Pyruvate-ferredoxin oxidoreductase. IV. Studies on the reaction mechanism. *J. Biol. Chem.* **246,** 3120.
400. Vagelos, P. R., and Earl, J. M. (1959). Propionic acid metabolism. III. β-Hydroxypropionyl-coenzyme A and malonyl semialdehyde coenzyme A, intermediates in propionate oxidation by *Clostridium kluyveri*. *J. Biol. Chem.* **234,** 2272.
401. Valentine, R. C., Mortensen, L. E., and Carnahan, J. E. (1963). The hydrogenase system of *Clostridium pasteurianum*. *J. Biol. Chem.* **238,** 1141.
402. Valentine, R. C. (1964). Bacterial ferredoxin. *Bacteriol. Rev.* **28,** 497.
402a. Van der Wiel-Korstanje, J. A. A., and DeVries, W. (1973). Cytochrome synthesis by Bifidobacterium during growth in media supplemented with blood. *J. Gen. Microbiol.* **75,** 417.
403. Vandekarkove, M., Fancon, R., Andiffren, P., and Oddon, A. (1965). Metabolism of carbohydrates by *Neisseria intracellularis*. V. Evidence of acetylmethylcarbinol produced from glucose. *Med. Trop. Chem.* **25,** 457; *Chem. Abstr.* **64,** 5483 (1966).
404. Van Niel, C. B. (1928). Dissertation, Delft [as cited by Wood and Werkman (434)].
405. Vinuela, E., Salas, M. L., Salas, M., and Sols, A. (1964). Two interconvertible forms of yeast phosphofructokinase with different sensitivity to end product inhibition. *Biochem. Biophys. Res. Commun.* **15,** 243.
406. Virtanen, A. I. (1923). *Comment. Phys.-Math.* **1,** 36 [as cited by Wood and Werkman (434)].
407. Vitols, E., North, R. F., and Linnane, A. W. (1961). Studies on the oxidative metabolism of *Saccharomyces cerevisiae*. I. Observations on the fine structure of the yeast cell. *J. Biophys. Biochem. Cytol.* **9,** 689.
408. Vitols, E., and Linnane, A. W. (1961). Studies on the oxidative metabolism of *Saccharomyces cerevisiae*. II. Morphology and oxidative phosphorylation capacity of mitochondria and derives particles from Baker's yeast. *J. Biophys. Biochem. Cytol.* **9,** 701.
409. Walker, G. A., and Kilgour, G. L. (1965). Pyridine nucleotide oxidizing enzymes of *Lactobacillus casei*. I. Diaphorase. *Arch. Biochem. Biophys.* **111,** 529.
410. Walker, G. A., and Kilgour, G. L. (1965). Pyridine nucleotide oxidizing enzymes of *Lactobacillus casei*. II. Oxidase and peroxidase. *Arch. Biochem. Biophys.* **111,** 534.

411. Wallhofer, P., and Baldwin, R. L. (1967). Pathway of propionate formation in *Bacteroides ruminicola. J. Bacteriol.* **93,** 504.
412. Wang, C. C., and Barker, H. A. (1969). Purification and properties of L-citramalate hydrolyase. *J. Biol. Chem.* **244,** 2516.
413. Wang, C. C., and Barker, H. A. (1969). Activation of L-citramalate hydrolyase from *Clostridium tetanomorphum. J. Biol. Chem.* **244,** 2527.
414. Watson, K., Haslam, J. M., and Linnane, A. W. (1970). Biogenesis of mitochondria. XIII. The isolation of mitochondrial structure from anaerobically grown *Saccharomyces cerevisiae. J. Cell Biol.* **46,** 88.
414a. Westlake, D. W. S., Shug, A. L., and Wilson, P. W. (1961). The pyruvic dehydrogenase system of *Clostridium pasteurianum. Can. J. Microbiol.* **7,** 515.
414b. Westwood, A. W., and Doelle, H. W. (1974). Glucose-6-phosphate and 6-phosphogluconate dehydrogenases and their control mechanisms in *Escherichia coli* K-12. *Microbios* **9,** 143.
415. Whistance, G. R., and Threlfall (1968). Effect of anaerobiosis on the concentrations of demethylmenaquinone, menaquinone, and ubiquinone in *Escherichia coli, Proteus mirabilis,* and *Aeromonas punctata. Biochem. J.* **108,** 505.
416. White, D. C., Bryant, M. P., and Caldwell, D. R. (1962). Cytochrome-linked fermentation in *Bacteroides ruminicola. J. Bacteriol.* **84,** 822.
417. Whiteley, H. R. (1953). Mechanism of propionic acid fermentation by succinate decarboxylation. *Proc. Nat. Acad. Sci. U.S.* **39,** 772.
418. Whiteley, H. R. (1957). Fermentation of amino acids by *Micrococcus aerogenes. J. Bacteriol.* **74,** 324.
419. Whiteley, H. R., and Ordal, E. J. (1957). Fermentation of alpha ketoacids by *Micrococcus aerogenes* and *Micrococcus lactilyticus. J. Bacteriol.* **74,** 331.
420. Whiteley, H. R., and Tahara, M. (1966). Threonine deaminase in *Clostridium tetanomorphum.* I. Purification and properties. *J. Biol. Chem.* **241,** 4881.
421. Whittenbury, R. (1960). Two types of catalase-like activity. *Nature (London)* **187,** 433.
422. Wilder, M., Valentine, R. C., and Akagi, J. M. (1963). Ferredoxin of *Clostridium thermosaccharolyticum. J. Bacteriol.* **86,** 861.
423. Williams, R. A. D., and Bowden, E. (1968). The starch–gel electrophoresis of glucose-6-phosphate dehydrogenase and glyceraldehyde-3-phosphate dehydrogenase of *Streptococcus faecalis,* Str. *faecium* and Str. *durans. J. Gen. Microbiol.* **50,** 329.
424. Williams, R. A. D., and Sadler, S. A. (1971). Electrophoresis of glucose-6-phosphate dehydrogenase, cell wall composition, and the taxonomy of heterofermentative lactobacilli. *J. Gen. Microbiol.* **65,** 351.
425. Wimpenny, J. W. T., Ranlett, M. R., and Gray, C. T. (1963). Repression and derepression of cytochrome c biosynthesis in *Escherichia coli. Biochim. Biophys. Acta* **73,** 170.
426. Wimpenny, J. W. T. (1967). ATP pools and turnover in microorganisms. *Biochem. J.* **102,** 34P.
427. Wimpenny, J. W. T., and Warmsley, A. M. H. (1968). The effect of nitrate on Krebs cycle enzymes in various bacteria. *Biochim. Biophys. Acta* **156,** 297.
428. Wimpenny, J. W. T. (1969). Oxygen and carbon dioxide as regulators of microbial metabolism. *Symp. Soc. Gen. Microbiol.* **19,** 161.
429. Wimpenny, J. W. T. (1969). The effect of E_h on regulatory processes in facultative anaerobes. *Biotechnol. Bioeng.* **11,** 623.

430. Wimpenny, J. W. T., and Necklen, D. K. (1971). The redox environment and microbial physiology. 1. The transition from anaerobiosis to aerobiosis in continuous cultures of facultative anaerobes. *Biochim. Biophys. Acta* **253,** 352.
431. Wittenberger, C. L. (1968). Kinetic studies on the inhibition of a D(−)-specific lactate dehydrogenase by adenosine triphosphate. *J. Biol. Chem.* **243,** 3067.
432. Wittenberger, C. L., and Angelo, N. (1970). Purification and properties of a fructose 1,6-diphosphate-activated lactate dehydrogenase from *Streptococcus faecalis*. *J. Bacteriol.* **101,** 717.
433. Wolin, M. J. (1964). Fructose 1,6-diphosphate requirement of streptococcal lactate dehydrogenase. *Science* **146,** 775.
434. Wood, H. G., and Werkman, C. H. (1936). Mechanisms of glucose dissimilation by the propionic acid bacteria. *Biochem. J.* **30,** 618.
435. Wood, H. G., Werkman, C. H., Hemingway, A., and Nier, A. O. (1940). Heavy carbon as a tracer in bacterial fixation of carbon dioxide. *J. Biol. Chem.* **135,** 789.
436. Wood, H. G., Werkman, C. H., Hemingway, A., and Nier, A. O. (1940). Heavy carbon as a tracer in heterotrophic carbon dioxide assimilation. *J. Biol. Chem.* **139,** 365.
437. Wood, H. G., and Stjernholm, R. (1961). Transcarboxylase. II. Purification and properties of methylmalonyloxaloacetate transcarboxylase. *Proc. Nat. Acad. Sci. U.S.* **47,** 289.
438. Wood, H. G., Allen, S. H. G., Stjernholm, R., and Jacobson, B. (1963). Transcarboxylase. III. Purification and properties of methylmalonyloxaloacetic transcarboxylase containing tritiated biotin. *J. Biol. Chem.* **238,** 547.
439. Wood, W. A. (1961). Fermentation of carbohydrates and related compounds. *In* "The Bacteria" (I. C. Gunsalus and R. Y. Stanier, eds.), Vol. 2, p. 59. Academic Press, New York.
440. Wyer-Jones, R. G., and Lascelles, J. (1967). The relationship of 4-hydroxybenzoic acid to lysine and methionine formation in *Escherichia coli*. *Biochem. J.* **103,** 709.
441. Yoshida, A., and Freese, E. (1965). Purification and chemical characterization of lactate dehydrogenase of *Bacillus subtilis*. *Biochim. Biophys. Acta* **99,** 56.

Questions

1. What does the term "fermentation" mean?
2. What is the yield coefficient of a microorganism?
3. Describe the importance of carbon balance sheets in fermentation.
4. Outline the formation of propionic acid by *Propionibacterium shermanii.*
5. What is the importance of the transcarboxylation reaction discovered by Swick and Wood in 1960?
6. Briefly describe the differences in the anaerobic formation of propionic acid by *Propionibacterium* and *Peptostreptococcus elsdenii.*
7. Describe the different modes of pyruvate degradation under fermentative conditions.
8. Describe the direct and indirect conversion of carbon dioxide to acetate by clostridia.
9. Outline the formation of butyric acid and butanol by clostridia.
10. Discuss the reasons for the metabolic classification of the Enterobacteriaceae into three main groups.

11. Describe the phosphoroclastic split of pyruvate.
12. Describe the mode of action of the formic hydrogen lyase system and its importance in bacterial taxonomy.
13. Outline the differences in succinate formation under anaerobic conditions by *Clostridium kluyveri* and enterobacteria.
14. What is the Voges-Proskauer reaction?
15. Describe the different pathways leading from pyruvate to acetoin.
16. What is meant by the "Pasteur effect" and the "Crabtree effect"?
17. Describe the basic principles of the Pasteur effect.
18. Describe the basic principles of the Crabtree effect.
19. Discuss the basic principles of homo- and heterofermentation of lactic acid bacteria.
20. Describe the regulatory mechanism of glucose utilization by lactic acid bacteria.
21. Discuss the effect of oxygen on the glucose metabolism of lactic acid bacteria.
22. Describe the malolactic fermentation.
23. Discuss the main differences between the electron transport system of lactic acid bacteria and *Escherichia coli.*
24. Outline the main differences between the aerobic and fermentative metabolism of amino acids.
25. What is the Stickland reaction?

Subject Index

B

C

D

E

F

I

K

L

M

P

Q

R

T

Y

Z

Microorganism Index

Numbers in parentheses are reference numbers and indicate that studies have been carried out with the particular microorganism, although the actual name is not cited in the text.

R

S

T

V

Z

B
C 7
D 8
E 9
F 0
G 1
H 2
I 3
J 4